Springer Biographies

More information about this series at http://www.springer.com/series/13617

Robert Smail Jack · Fritz Scholz
Editors

Wilhelm Ostwald

The Autobiography

Translated by Robert Jack, Edited by Fritz Scholz
and Robert Jack

Springer

Editors
Robert Smail Jack
Institute for Immunology and Transfusion Medicine
University of Greifswald
Greifswald
Germany

Fritz Scholz
Institute of Biochemistry
University of Greifswald
Greifswald
Germany

Springer Biographies

ISBN 978-3-319-46953-9 ISBN 978-3-319-46955-3 (eBook)
DOI 10.1007/978-3-319-46955-3

Library of Congress Control Number: 2016954018

Translation from Wilhelm Ostwald: Lebenslinien I-III, Klasing & Co., GmbH/Berlin/1926

Cover photo source: Leipziger Professorenporträts, Verlag Louis Pernitzsch, original postcard in possession of Fritz Scholz

Printed on acid-free paper

This Springer imprint is published by Springer Nature
The registered company is Springer International Publishing AG
The registered company address is: Gewerbestrasse 11, 6330 Cham, Switzerland

Dedicated to German youth[1]

[1]This is Ostwald's dedication of volume I of his autobiography.

Translator's Note

Why do we need a translation of Wilhelm Ostwald's autobiography? George Orwell once remarked that autobiographies are almost never worth reading: *A man who gives a good account of himself is probably lying, since any life viewed from the inside is simply a series of defeats.* Ostwald's autobiography is different: Its strength lies in the fact that it is both an idiosyncratic account of the life of a gifted and important scientist and at the same time it provides a glimpse into the soul of a brilliant, arrogant, sometimes foolish, often self-satisfied and yet deeply disappointed and frustrated individual. And so when Fritz Scholz invited me to join him in preparing this annotated translation I jumped at the chance.

Ostwald's tale as he lets it unfold, is rather like Dick Whittington without the cat. A poor but gifted boy from the impoverished Latvian province propels himself by sheer hard work, discipline and will power into the big bright world of German science. Once there he eagerly seized the opportunity to propagate the important discoveries made by van't Hoff and Arrhenius and, by playing Huxley to van't Hoff's Darwin, he contributed significantly to the establishment of physical chemistry. However, he also makes clear in his autobiography that this important role was, for him, not enough. He wanted to produce a truly revolutionary idea of his own and the first attempt was what he called "Energetics". His contemporaries, however, did not take Energetics seriously. Ostwald endured a humiliating and very public defeat at the Natural Scientist's meeting in Lübeck in the autumn of 1895 and, in the aftermath, he suffered a severe nervous breakdown. He recovered slowly while at the same time gradually losing interest in chemistry—and during this process Dick Whittington turned into Don Quixote.

As related in the autobiography, Ostwald had largely lost interest in chemistry by the time he was 55 and turned instead to making his mark in a wide range of other issues, in many of which he was quickly out of his depth. In some of these matters his opinions were far sighted, in others they were merely commonplace and in yet others, such as his ideas on how to write a splendid poem or his Doctrine of Happiness, his views were simply ridiculous.

The autobiography was written in the period 1925–1927 and may owe some of its quirks to the manner of its birth. At the beginning of his career Ostwald wrote all his drafts in longhand, though later he speeded things up by learning to use a typewriter. It was only after his retirement that, in 1910, he switched to the use of a dictaphone and this is what he was using for his autobiography. He was well aware that the dictaphone brings with it the danger of repetition, for instead of working hard to get the meaning across accurately the first time—as he did when writing in longhand—there was a tendency to just repeat an argument in several imperfect versions. He promised to try to avoid this pitfall, but he was not always successful. Perhaps in his old age Ostwald simply lacked the energy to proofread the dictaphone transcripts, for such repetitions abound. German sentence structure is more forgiving of this sin than is English and so, in the interests of clarity; many of the unhelpful repetitions in the German text have been removed.

Lack of adequate proofreading might also explain some aspects of the organisation of the text. Ostwald frequently seemed to dictate in "lung-fulls" of words and the inevitable pauses, which then entailed, may have been transcribed as the start of new paragraphs. Certainly, the text does tend to jump about in a plethora of one sentence and two sentence paragraphs. Though this layout is at times challenging, it has been left intact in the translation.

Yet despite all that, the text is eminently readable for Wilhelm Ostwald—winner of the Nobel Prize in Chemistry in 1909—was a prolific and accomplished writer both of scientific and of popular texts. He was someone who could and did write clearly and effectively. Here in his autobiography, which he kept largely free of any hard to understand technical matters, his style is dictated by his view that, *every time a writer treats a subject in exalted, solemn, touching or uplifting terms then he has probably abandoned logic and clarity.* True to this motto, he left the story of his life free to emerge in plain language and in a largely conversational style.

Berlin

Robert Smail Jack

Preface

Wilhelm Ostwald was a genius who was equipped with a knack for unconventional thinking in science and philosophy and with skills in both painting and music. On the other hand he was, surprisingly, a much more conventional and average personality in everyday life. Whereas his scientific achievements are part of the fundaments of modern chemistry, and his philosophic views have been extensively and critically reviewed,[2] the world of his personal thinking and feeling, laid out in his autobiography, has so far been rather closed for non-German readers. The present translation will hopefully change that. The translation of his autobiography is certainly of special interest for readers in the US, the UK and France, as it shows what impressions Ostwald took from these countries during the course of his extensive travels. They will, of course, not agree with all of his views.

Ostwald published the first volume of his autobiography in 1926, followed by the second and third volumes in 1927.[3] An abridged and commented (German) version was published in 2003.[4] That edition contains a number of regrettable cuts and also errors in some comments. Beside his autobiography, two biographies have been published in German; one by his daughter Grete Ostwald[5] in 1953 and one as early as 1904 by his former student Paul Walden.[6] There are also other biographical

[2]Görs B, Psarros N, Ziche P (eds) (2005) Wilhelm Ostwald at the Crossroads between Chemistry, Philosophy and Media Culture. Universitätsverlag, Leipzig.

[3]Ostwald W (1926/1927) Lebenslinien. Eine Selbstbiographie. Klasing & Co GmbH, Berlin.

[4]Ostwald W (2003) Lebenslinien. Eine Selbstbiographie. Nach der Ausgabe von 1927/27 überarbeitet und kommentiert von Karl Hansel. Verlag der Sächsischen Akademie der Wissenschaften, Leipzig (in Kommission bei S. Hirzel, Stuttgart, Leipzig)

[5]Ostwald G (1953) Mein Vater. Berliner Union, Stuttgart

[6]Walden P (1904) Wilhelm Ostwald. Engelmann, Leipzig

publications about Ostwald in German.[7,8,9] Although the corner stones of Ostwald's life are known in the English speaking world,[10,11] the subtle details and especially his personal attitude towards his colleagues in Germany and abroad are still hidden in his German writings. The fact that the autobiography has never been translated to English may be surprising, especially because Ostwald and his students played an outstanding role in the establishment of physical chemistry in the USA[12,13,14] and his contributions to chemistry are basic to this science and part of the chemistry curricula world wide. The lack of a translation may be due at least in part to the anti-German feelings that prevailed after World War I, during the Nazi period (1933–1945), and also after World War II. These times were not conducive to the popularisation of the biography of a German scientist, especially since Ostwald, as a child of his times, was a strong believer in a nationalistically biased world view. There is however an interesting dichotomy here for, as the autobiography shows, Ostwald was at the same time a tireless propagator in many diverse areas of, international cooperation. Perhaps now that the dust of these historical struggles has settled back down, non-German readers may be better prepared to understand that some of Ostwald's seemingly outrageous remarks merely reflect the spirit of the times in which he lived. The present publication, together with a recently published book on the development of electrochemistry in Eastern Europe[15] is aimed at opening up the history of science in Europe and making it accessible to English speakers.

This translation is based on the German original 1926/27 edition. In addition to the text and photos, it contains a Name Index (see end) and a pedigree of the Ostwald family. I am very happy that my colleague at the University of Greifswald, Professor Robert Jack, carried out the translation, because this was a task which needed someone who is both a "native speaker" and a scientist. We have

---

[7]Ertl G (2009) Chemie in unserer Zeit 121:6724-6730.

[8]Bartel HG (1999) Ostwald, Friedrich Wilhelm. In: Neue Deutsche Biographie 19:630-631 (online: http://www.deutsche-biographie.de/pnd11859057X.html)

[9]Messow U, Krause K (1998) Physikalische Chemie in Leipzig. Festschrift zum 100. Jahrestag der Einweihung des Physikalisch-chemischen Instituts an der Universität Leipzig. Universitätsverlag, Leipzig.

[10]Donnan FG (1933) Ostwald Memorial Lecture. J Chem Soc 316-332

[11]Sutton M (2003) The Father of Physical Chemistry. Chem Brit 39(5):32-34

[12]Stock JT (2003) Ostwald's American Students. Apparatus, Techniques and Careers. Plaidswede Publishing, Concord

[13]Servos JW (1996) Physical Chemistry from Ostwald to Pauling: The Making of a Science in America. University Press, Princeton

[14]Coffey P (2008) Cathedrals of science. The Personalities and Rivalries that Made Modern Chemistry. University Press, Oxford

[15]Scholz F (ed.) (2015) Electrochemistry in a divided world. Innovations in Eastern Europe in the 20th Century. Springer, Berlin

intensively—sometimes heatedly—discussed the details of the translation, because it is no mean task to make Ostwald's text understandable to contemporary English readers. Robert gives some explanations of his translation in the Translator's note.

Greifswald, Germany Fritz Scholz

Contents

Part I
Riga—Dorpat—Riga

Chapter 1
My Parental Home and Childhood

My hometown Riga. I was born in the autumn of 1853 in Riga, Livonia.[1] At that time Riga belonged to the Russian empire in which the old Julian calendar was still used and, despite its heathen origin, the Greek Catholic Church preferred it to the newer Gregorian calendar which, in their view, was the work of "schismatics". By the old calendar my birthday fell on the 21st of August, by the new one on the 2nd of September. I was my parents' second son; my brother Eugen was born 2 years before and my younger brother Gottfried 2 years after me. I had no sister.

At that time Riga was essentially a German city whose architecture and constitution resembled that of Lübeck from where Riga's founders had come some 1000 years previously. The entire upper and middle social strata, aristocrats, owners of large estates, academics (they were referred to as "men of letters"), tradesmen and artisans spoke German as their mother tongue and their mental world was rooted in German culture. Indeed the first edition of Kant's "Critique of Pure Reason" was published by J.F. Hartknoch in Riga. The rural population was made up of native Latvians who had been freed by the estate owners at the beginning of the nineteenth century—long before this happened in Russia—and who had been made owners of tolerable amounts of land. They lived on small scattered farms rather than in villages. From these farms a steady stream of excess workers moved into the cities, particularly into Riga, where they easily found jobs as servants, apprentices and so on and, in the second generation, by picking up the German language and German habits, they formed a new social tier in the city. They were, however, considered lower class.

The Russians were represented by a number of government administrators and army personnel; a few were also traders, artisans or gardeners. They were considered to be culturally and socially inferior, but were tolerated as an unavoidable evil and were treated with cool politeness. A small minority of them, those who had

[1]Governorate of Livonia (in Russian Liflyandskaya Guberniya) was part of the Russian empire. Before 1796 it was called the Riga Governorate.

R.S. Jack and F. Scholz (eds.), *Wilhelm Ostwald*, Springer Biographies,
DOI 10.1007/978-3-319-46955-3_1

benefited from a European education, might on the basis of their personal merit be accepted into society, though they retained the stigma of being basically foreign.

Russia. This Baltic German attitude towards Russia was not unfounded. Tsar Peter the Great's attempt to bring Russia out of barbarism and lead it towards European culture had been at best a superficial success, because Peter himself remained a barbarian to the end of his life. The real civilising work was begun by Alexander II who realised that the standards of the broad mass of the people must be raised, and he implemented the first measures to this end. However this sort of work is like planting a wood; in the best of circumstances the gardener only lives to see the first beginnings and he must place his trust in an uncertain future, for much time and favourable weather will be needed to allow the trees to flourish. Because Peter, driven by his coarse nature, expected to reap the fruits of his work almost before the seeds had germinated, all he managed to achieve was a mere illusion of culture that one patriotic Russian described as being "rotten before it was ripe". The enormous upheavals which have shaken Russia for the last two decades and which are even today not completed, have to be viewed as a movement by which the Russian people rid themselves of Peter the Great's half digested "reforms". Only when this process is complete will an organic development of culture be possible.

My native country. In contrast to the volatility which for centuries had held things up in Russia, cultural development in the German population in the Baltic provinces had proceeded steadily, thanks to the unbroken connections to the mother land. People from these provinces tried, whenever possible, to study in Germany or at least to finish their education there. At each of the larger German universities, particularly in Jena and Göttingen youths from Courland and Livonia made up a significant and easily recognisable part of the student body.

The conditions of life in the Baltic States were heavily influenced by German culture and, as mentioned above, the freeing of the serfs had been achieved here half a century earlier than in Russia. Schools had been built all over the country and they were run in close cooperation with the church. The language of the Letts was studied by the priests because it provided the only route to the souls of the rural population. Because of all this, the cultural border between Europe and Asia ran north and east across the Baltic provinces rather than along the political border of the Russian Empire.

My grandparents. My paternal grandfather—the family memories go no further back—was a master cooper in Riga. He had emigrated from Berlin to Riga where he was not particularly successful in business but did manage to give his five children—four sons and a daughter—sufficient education that they could provide for themselves. My uncles all stayed in the same trade but did manage to give their sons access to higher goals. I remember my grandfather Ostwald only as a decrepit old man who day in and day out sat almost motionless in his armchair smoking a long pipe. His wife, my grandmother, was by comparison spry. She kept the house and the cooperage going till her sons, who all worked as coopers, had settled down. She died long after, in her late eighties when I, her favourite grandson, was about to get married. My bride-to-be was a good listener and grandmother was happy when she had the chance to tell her all sorts of stories from my childhood. When illness finally chained her to her bed her carefree spirit remained unbroken. When she realised that death was near she put her affairs in order and made all the arrangements for her own funeral. The priest responsible was a bigot who wished "to prepare" her for death by graphic descriptions of hell and its punishments. She listened to him patiently rather than devoutly. When he finally left she pulled at one of the ribbons holding her nightcap and with a gesture of indescribable irony threw the bonnet at the door through which he'd left. She'd been a practicing Christian all her life and an admirer of Dietrich, the old village pastor, but she could not find Christianity in the words of this foolish zealot. Soon after, she closed her lively eyes and passed into eternal peace.

My father. At the time of my birth my father was a very poor but adept and enterprising artisan who, long after his journeyman years as a cooper in Russia, had settled down in his home town and married an equally poor baker's daughter. With the first money that he earned he bought his own little house. The money was just enough for a small house in the cheapest neighbourhood which was on the outskirts of the so-called Moscow suburb in the "Sand Hills", a long barren ridge of sand dunes on the eastern periphery of the city. Soon after my birth this house disappeared under an embankment of the Riga–Düneburger railway, so that I never knowingly saw it.

My father was the most talented of the brothers according to his teacher, H. Fromm, from whom I, 30 years later, took my first lessons. Since Fromm lived to be nearly one hundred, he could also have taught my eldest son had I not just at that time left Riga. My father had shown a special skill in drawing. In fact he was so good that it almost cost me my existence. This happened as follows: At that time Russia was ruled by the strong and ruthless Tsar Nicholas I who, in order to ensure a good supply of students to the imperial Academy of Fine Arts in Petersburg, ordered the administrators of all the schools in the enormous empire to select and forward the names of the most gifted schoolboys in their area. A certain number of them—I believe around a hundred—were then to be given an imperial stipend to study at the Petersburg academy. My father was one of those to be nominated from Riga. However my grandfather viewed the whole matter sceptically and exercised his parental rights to block the nomination. Had this not happened my father would have gone to Petersburg, would have married a different woman and the special mix of paternal and maternal genetic material which is responsible for the special properties that make me would never have come into being.

My grandfather's decision ended any possible artistic career for my father and instead he started an apprenticeship. As a memento of these times, there hung in my parents house two of my father's sketches, very carefully executed with special sharpened sticks of chalk and then framed under glass. These were objects of admiration to me throughout my childhood. One was a copy of Raphael's angel from the Madonna in Dresden; the other was a woman's head in oriental costume the original of which is unknown to me. Later my father gave up drawing; he said that the work in the cooperage had made his hand too heavy for it.

Once his apprenticeship was over my father became a travelling journeyman in inner Russia and there he experienced many different sorts of adventure. Apart from the repeated danger of dying in winter from snow and loneliness, he also luckily survived other dangers which threatened to throw the young, fresh and energetic youth off the track. For example he was for a while a teacher in the household of a Count Tolstoy. I have been unable to find out whether the famous Count Leo Tolstoy was one of his pupils; however it is certainly possible because Leo Tolstoy was born in 1828 and my father's years as a journeyman were in the early forties. However, the Tolstoy family is many branched and direct evidence to support the notion is missing.

I asked my father, how he, as a journeyman-cooper, got a job as a teacher. I don't remember the details but it seems that his determined intervention during a chance encounter had rescued the Count from a difficult situation and this had so impressed the Count that he wanted to retain his helper permanently. At that time in Russia almost any German was considered erudite and in any case teaching standards were not very high. By staying up half the night to learn whatever he had to teach the next day he managed to carry out his duties satisfactorily and so keep himself in his job. Out on a walk with his handsome pupils he was once even spoken to by Tsar Nickolas himself; he liked to tell us children of the great impression the sharp look of the emperor's steel blue eyes had made on him.

This situation came to a sudden end when a beautiful grown up sister of his pupils returned home. She soon became the object of the teacher's unspoken desire and she seemed to be not averse to this homage. The family felt it advisable to avoid any future difficulties by discharging the teacher, and a tortoiseshell needle box, daintily inlaid with mother of pearl, which my father long kept as one of his treasures, leads me to suppose that the parting was accompanied by tears from pretty eyes and this gift as a pledge to memory.

However this episode only temporarily separated the young cooper from his trade. He soon found work in the barrel factory of a large brewery in Petersburg, where he quickly rose to a leading position. But here also he did not stay long, because the owner—a rich widow of a certain age—gave him to understand that she would happily promote him to an even higher position. He said goodbye and returned to Riga to settle down and take the exam for his master craftsman's certificate.

My grandparents on my mother's side. My mother was also a pure bred German. She was born in Moscow where her father, who had emigrated there from Hesse, ran a baker's business. I have no idea what induced him to leave Moscow and settle in Riga for such a journey was at that time quite a difficult venture. One made a contract with a wagoner for the transport, and the journey took 30 to 40 days. It could only be undertaken in winter when frost and snow made the roads passable; in summer this was not possible. Food and shelter for the night had to be carried along because accommodation along the way was rarely to be found. In this way my mother came as an adolescent to the town in which she would spend the rest of her life. Her two sisters and two brothers completed the family.

I can still well recall my mother's father Heinrich Leukel. In his later years he gave up the bakery trade and became toll collector on the long bridge which at that time was the sole connection between the two parts of the town on opposite banks of the river Duena. The bridge and the adjacent rampart of the river bank were the landing places for the numerous ships which visited the port. Because of this, certain nautical traits became evident in the former baker's dress and language, all of which left a strong impression on me and this was increased by the occasional gift of ships biscuits and other naval victuals. Altogether grandfather was a good hearted man who made no secret of his love for his grandchildren. He died when I was still just a boy.

Grandmother was a buxom, busy woman who was confined to the house by an ailment. I remember her best for the dry humour with which she countered life's ups and downs. She loved to emulate her husband's trade at the household level and delighted us children at every celebration—and she was always inventing new reasons for celebrations—with homemade "Saftpiroggen", a pastry in the shape of a shoe sole—but much bigger and filled with preserved cranberries. She died soon after my grandfather.

As one can see, I grew up in a lower middle class milieu. Nobody was destitute, but everybody had to be careful to make ends meet and there was nothing left over for luxuries. My mother often told us how, in the first years of their marriage, she'd go every evening to the workshop to collect the left over wood shavings with which to cook the evening meal.

But things soon started to look up. Just as I was born, my father had his first real earnings in his job and from then on he became ever more successful. The hut in the "sand dunes" was changed for a better situated house whose large inner court was conveniently used for a necessary extension of the cooperage. The inner court bordered a stream called the "Speckgraben" along which lay many small businesses, particularly tanneries. They had driven away almost all the fish and, apart from water beetles, I can only remember the large numbers of leeches which filled its turbid waters.

My first memories are connected with this place. I can see myself trying to catch leeches with a rod held in the water. One of my father's cooper apprentices tried to frighten me by telling me that the leech would pull me and the rod into the water and then eat me; I thought however that if the worst came to the worst I could simply let the rod go. Then again I see myself and my brothers together with the cooperage apprentices in the first snow shouting with joy as we used round pieces of wood to toboggan off the low barn roof.

In summer we used the stream for boat trips, which were strictly forbidden, and for which we were severely punished when found out; in winter it served as an ice rink. There remains fixed in my memory a picture of a boy in black velvet knickerbockers, red jacket and yellow cap who elegantly glided on a sunny winter morning along the stream's further bank while on this side I struggled with the rudiments of the sport.

My brothers had inherited my father's robust build and in comparison to them I appeared to be a bit of a weakling for I had more the build of my mother and I often preferred to play quietly alone rather than join in their loud games.

Life at home. At this time the most important person in my life was my mother whose favourite I was. My father worked from dawn till dusk and often did not make it home for the family meals. He was, as I said, tall and strong and later when I met the great chemist R. Bunsen his face reminded me of my father. My father was impetuous and easily roused to anger so that we children were rather shy of him especially as we didn't see him very often. When he wasn't working he devoted his spare time to his one great passion which was hunting and this also kept him from home. Regularly on Saturday afternoons during the summer he'd set out with his little cart to Babitsee, a small reed covered stretch of water around 15 km away, where countless wild ducks lived. He hunted these from a small boat that a local farmer pushed through the reeds with a pole. Late in the evening he'd return home with his take which was often enough 30–40 ducks, coots and other wildfowl. On Monday morning we boys had to bring most of them to the friends and relatives who were regular recipients of these gifts. The rest would be consumed at home and my mother soon developed an extraordinary skill in their tasty preparation.

In winter father hunted hares, deer and foxes; in the spring there was the snipe and blackcock mating season, so that there were only brief intervals when father stayed at home and spent time with the family.

I inherited nothing of my father's love of hunting. In fact I soon found myself in opposition to it. This started as I began to find the distribution of the huntsman's bag to the friends and relatives increasingly intolerable. On top of that there was a certain animal loving influence from popular science tracts that I'd come across by chance. Moreover I was bored by the endless hunting stories that I had to listen to at the family table, for my elder brother shared father's love of hunting and had accompanied him from an early age. When we schoolboys went for a walk on a free afternoon, many armed themselves with reed blowpipes to hunt birds. Although the quarry was rarely hit, I would go ahead and try to chase off the potential victims of this youthful bloodlust before the hunters could get in range. Since the others were not happy about that, I often enough suffered for my love of animals which, of course, only strengthened me in my opinion and in my opposition to it all.

My mother was both by nature and by education an excellent housewife. The rapid growth of the cooperage considerably added to her duties, because at that time it was the custom that the wife of the master craftsman cooked for the journeymen and apprentices. And so my mother had to daily defend her reputation as a housewife by feeding, in addition to the family, 6–12 boarders whose hunger had been sharpened by hard work. The workers were fed first and only after they had finished was the table set for the family, for the parlour wasn't big enough to hold all at once.

I'm still amazed how my mother, with very little help, managed to keep this large household in perfect order and at the same time found the time to read books and newspapers and so to keep alive a by no means narrow field of mental activities. Early on, as soon as the financial situation had permitted, my father had organised for her a regular subscription to the town theatre which she eagerly visited. There numerous excellent artists were to be seen and heard for they were glad of the opportunity to break their long and tiring journeys to guest performances in Petersburg by resting in our appreciative and friendly city on the Duena.

My mother's lively interest in the arts had a great influence on my personal development. Through her I was early led to an appreciation of the enrichment of the inner life which can flow from such sources, and there arose an area of common interest between us that set us apart from the other family members who did not share this disposition. As is usual in such cases I soon grew beyond the spheres to which my mother was restricted by the burden of daily work in the house and with the children. I became on omnivorous reader of everything I could lay my hands on and went my own way—a development which she watched fondly though sometimes disapprovingly.

Support from my father. From his journeyman years my father had brought home with him the strong belief that education was of irreplaceable value. Because of this he had made it an unbreakable rule to open up for his sons every possibility for advancement.

I don't know if he realised from the start the sacrifices he would have to make; sacrifices which went beyond just the financial burden. I mean by this that he had to accept that his children sooner or later would grow beyond his mental horizons and that their rise into the higher social classes could bring him into conflicts which, in the sharp caste system operating in the Riga population, were bound to hit him hard. Whatever the answer may be, he accepted all this without hesitation and never showed the slightest wish to limit the opportunities which he had opened for us.

My first school. As a result we children were sent to the elementary school run by the teacher Mr. Fromm where I started before I was 6 years old. I was a precocious child and had learned to read from what I had, without paying too much attention, overheard from my older brother's lessons. In the school I was diligent without coercion and made my teacher happy, particularly in maths. He was less impressed with my achievement in writing. One says that the sins of youth are paid for later. That seemed to apply in my case for I am certain that of all of my brothers and classmates none would later be faced more than I with the need to write. I did however always like to do it.

Shadow of the future. Fromm's school was a so-called crown school, which meant that it was financed and run by the government. Parallel to these schools there were the city schools which were financed by the local authority and which were considered better both in their teaching and in their social status.

Crown schools were largely for the children of the lower social classes and in particular for the sons of ambitious Latvian and Russian families. Because of this we always had a small number of misfits in the school—Letts who mostly spoke only broken German, whose cleanliness and behaviour left much to be desired and who had little in common with the other pupils. Those of them who were talented quickly learned to adopt our ways of thinking and behaving, but there were also a few who brought with them a determination to move up at all costs though they lacked any real ability. They adopted a hostile attitude to the rest of the pupils and to the teachers and blamed their lack of progress on the malevolence of the school. There were often differences of opinion which, as is the way of boys, would be settled by fisticuffs. I can see today the final scene of such an incident. The Latvian boy, who'd fought with nails and teeth, had been subdued and now sat as a loser with dishevelled hair and clothes, bloody foam on his lips and hate in his face. With a sunken head he squinted maliciously at his opponent. Latvians are, at best, not handsome and this youth was quite some degrees uglier than most. Although boys try not to show these things, I was shocked at the sight of bestial anger in this scarcely human guise.

The impression faded as the boy soon left the school, though it was strong enough to be stamped on my memory. This memory returned to me when, 60 years later, close to the end of the Great War, the Latvian nationalists gained the upper hand in Livonia and Courland and inflicted horrible atrocities on their German fellow citizens.

Chapter 2
Youth

My secondary school. The question of our further education was painstakingly considered by my father. From the time of Tsar Alexander 1st, who had set out with such zeal and success to improve the cultural development of his empire's Baltic provinces, there existed in Riga a Latin school administered from the government in St. Petersburg. This secondary school laid emphasis on Greek and Latin and was organised along the lines of the schools which had been established in Germany since the start of the nineteenth century under the pernicious influence of Wilhelm von Humboldt. It was the "obvious" preparation for an academic career and, except for a short period in the eighteen seventies, the University in Dorpat, just like the universities in Germany, only accepted as students those who had passed the school leaving certificate exams of a secondary school specialising in the classics. As everybody knows there exists to this day in Germany the absurd situation that a university does not have the right to decide who may enter it, but must leave this central decision in the hands of the school examination committees.

The second possible school was the old Cathedral School that had been founded as a priests' seminary before the reformation. It was now run by the town administration which had organised and expanded it as the town's secondary school. The competence of the town council and guild authorities in educational matters can be judged by the fact that they appointed the brilliant young Johann Gottfried Herder as Director of the Cathedral School and gave the 20 year old a free hand to reform and organise it. His success there was outstanding. Herder probably had his happiest years in Riga where he developed the main ideas of his influential work. He left after 5 years in 1769 to develop his work in a broader arena and in the following year he met Goethe in Strasbourg.

In the meantime the progressive attitude of the school authorities in Riga had led to the founding of a Polytechnic which, it was hoped, would stimulate the development of productive industry which the country lacked. In order to attract suitable pupils, the Cathedral School was converted into a technically oriented secondary school and this was pushed through in the face of opposition from the monopoly of educational classicists, although some considerable concessions had to be made to

R.S. Jack and F. Scholz (eds.), *Wilhelm Ostwald*, Springer Biographies,
DOI 10.1007/978-3-319-46955-3_2

them. In particular the Director of the new school was a dyed in the wool classicist called Haffner of whom I will have more to tell later on.

My father was luckily so well advised that he agreed to entrust his sons to this new school. This was a most important decision for my whole future development, because there is no doubt at all that my scientific and organization potential would have been stunted or even smothered at the secondary school specialising in classics. The technical secondary school in Riga was a thoughtfully constructed institution. The entrance exam showed that attendance at the elementary school had been sufficient to satisfy its entrance criteria. The school education was spread over 5 years and hence the institute was divided into five classes. The curriculum was such that only highly gifted and diligent pupils could complete the course in this time so that even a "good" pupil was forced to repeat one or other of the classes. As to foreign languages; there was only French in the first class together with the obligatory Russian, while Latin started in the second class and ended in the fourth where it was replaced by English. Physics was taught from the second class on and Chemistry in the fifth. Maths was taught up to the stage of analytical geometry but did not include differential calculus. At the end of the fifth year there was a final exam which initially only qualified one for entrance to the Polytechnic but not to the University. Shortly before I left the rules were changed so that this school leaving certificate enabled one to study mathematics and natural sciences at the University. This was the result of a long fought battle and my father's decision had been based on the hope that it would turn out this way. Not long afterwards the classicists saw to it that this decision was rescinded.

My best teacher. I look back happily on the 7 years in this secondary school. Not, I must say, that the school itself gave me much, but it introduced me to the comradeship of the large and varied pupil population and this opened many new vistas for me beyond what was available in the narrower family circle. There was, however, at least one of the teachers to whom I am indebted for his outstanding support. This was Gottfried Schweder who had studied astronomy and now taught maths, physics, and biology at the school.

Schweder was a tall man with a broad chest and powerful shoulders, who had the reputation of having been one of the best fencers at the university. His short curly beard, full blond hair and blue eyes made him look like an archetypal German. He had a cheerful and friendly nature so that it was no wonder that we boys, from the youngest to the oldest, were ready to follow his lead and would have gone through fire for him. Afterwards he was for many years director and this was a blessing for the school.

The beginnings of independent thought. In the first years I was the same willing and diligent pupil I'd been in the preparatory school. Then, however, a new vista opened up which was to have the greatest effect on my development. This was the opportunity of getting hold of all sorts of books from my fellow pupils.

Books in my circle were quite rare and costly. They were largely restricted to what one could get from the lending libraries which stocked mainly contemporary novels. At home we also had the weekly magazine "Gartenlaube"[1] whose well bound volumes constituted the main part of our house library and for many years they were my main source of intellectual nourishment. Looking back it's fair to say that this nourishment was both ample and healthy. At this time—the start of the 1860s—there was the great outgrowth of science and the start of modern technology and industry in Germany. The editor of the "Gartenlaube" had responded to this by publishing many articles on these subjects which both stimulated and informed me. In addition, the strong patriotism which was constantly expressed in the articles had the effect of making our family more conscious that we were unquestionably German.

Our German heritage. The political situation in my home country was that the strident pestilence of Slavic nationalism had begun to threaten our special status, within which a rich and fruitful culture had flourished. We felt—without any special pride but also with no particular wish for change—that we were politically part of the Russian empire but intellectually part of German culture. No political alignment with Germany was possible at that time because the German empire (Reich) had not yet been formed.[2] The Russian rulers had always acknowledged and fostered the special status of the Baltic provinces from which they drew many of the leaders of the military, the administration and the professions—people who could not be so easily recruited from Russian sources. Because of this there was a considerable degree of gratitude and affection felt in the Baltic provinces not to the empire or to the dynasty but rather to the individual Tsar.

Russification. The special status of the Baltic provinces was anathema to the developing pan-Slavic movement. Already during my childhood this movement had achieved sufficient influence to force through an increasing use of the Russian language in schools where up till then the lessons had been entirely in German. We pupils despised everything Russian as belonging to a lower class and met the increasing load of Russian lessons with passive resistance. As a means of justifying and anchoring Russian ways of thinking, the theorists of the pan-Slavic movement demanded and got an increasing weight given to Russian history in the schools. At least for the mentally alert pupils this had the opposite effect than that intended, because Russian history is a good deal bloodier than that of Western nations and thus gave them a historical basis for their anti Russian attitude.

---

[1]Founded in 1853, this was the first mass-circulation German newspaper and a forerunner of modern magazines.

[2]Ostwald means here the *German Reich*, which was proclaimed in 1871 in Versailles and ended in 1918 with the abdication of the German emperor following World War I. The First *Reich*, better known as the Holy Roman Empire, existed from 962 to 1806. The Third *Reich* was the Nazi state existing from 1933 to 1945 (1943–45 called Greater German Reich) and ending with World War II.

Russian lessons were in the hands of a German renegade called Haller whose political and ethical decay was instinctively detected by us children and so we made his life rather hard. The uncontrolled fury of the harsh punishments with which he'd respond from time to time, only made matters worse, for we regarded it as more or less a matter of honour to do as little as possible in his lessons. On top of that Russian is a very primitive language full of forms and endings but lacking rules, and therefore hard to learn. Even the attempts of Russian linguists to remedy these faults with appropriate rules did not improve the situation much. Meanwhile the government tried by imposing strict regulations to force us to learn Russian. This led me often enough into unnecessary conflicts.

Order and Beauty. To begin with, all this didn't bother me. The stimulating lessons of Schweder released a latent love of nature in me which to begin with expressed itself in the usual collection of plants, butterflies and beetles. We were expected to prepare for our teacher a certain number of pages on which herbs had been properly dried and pressed. I remember how I took up the challenge of arranging stems, leaves and flowers all to the best advantage and how this somehow resulted in beautiful sheets for which I got special praise. This was the first flowering of a perception which only ripened in my old age.

Another experience, whose relationship to that one I only later discerned, took place about the same time in the drawing class. We were supposed to draw parallel lines by moving a triangle along a ruler and had been told to do this for the different sorts of lines—bold, soft, close together, far apart, continuous, punctate, dashed, and so on. I wanted not just to draw the lines but to give them some sort of context. To do this, I used an ornamental border to define the top of the wall of a room and accentuated this impression by the choice and distribution of the lines. This drawing earned me special praise from the teacher who was otherwise not particularly pleased with my work. I wondered how the effect had been achieved, but I didn't get very far with this line of thought.

The beginnings of chemistry. In my first year in the secondary school I was a model pupil who quickly absorbed and mastered whatever was being taught and so I had no trouble being moved to the next higher class at the end of the year. After that, however, my personal development began. Because of the lively exchange of books between the pupils I managed to get hold of one about making fireworks—something that had interested me for some time. The author was someone called Websky[3] and he had treated the methodological aspects thoroughly. He began appropriately enough with a description of the materials required and used here, in addition to their trivial names, also the chemical formulae. To begin with they didn't mean anything to me because I was just fascinated by the instructions for

[3]Probably, Ostwald has had either the book (i) Websky M (1850) Schule der Lustfeuerwerkerei. Hirt, Breslau, or (ii) Websky M (1842) Lustfeuerwerkkunst, oder leicht fassliche und bewährte Anweisung zur Verfertigung von Lustfeuerwerken. Hirt, Breslau.

producing the various different fireworks. For the future however they were going to play an important role.

At the beginning it wasn't so easy for an 11 year old boy to follow the instructions because there was nobody to offer any help and, in addition, it was not easy to get hold of the necessary materials and equipment. Finally I found a friendly apothecary from whom I bought small amounts of saltpetre, sulphur, antimony and so on, and he gave me some help. However, for the most part I had to rely on the printed instructions and it was decisive for my future life that I discovered at this early age that all the art and science of mankind is stored in the form of printed words and that it can all be recovered and brought back to life at any time by an eager and dedicated reader. I also learned, even if I could not have consciously expressed it so, that the written words alone are seldom enough, and that the more experience the reader brings to the matter the more he will be able to extract from the text. That was a problem because none of my school friends could help me, though they were, to be sure, happy to join in when the fireworks were set off. So playing with fire opened for me a door to the world with all its wonders, and by wonders I mean the opportunity to experience myself all sorts of strange things that grabbed my youthful imagination.

I am thankful to my parents who tolerated all this, despite the fact that my new hobby was potentially dangerous given that my father's business meant that there was always a large stock of wood and readily flammable shavings lying around. My mother helped out by letting me have all sorts of kitchen utensils like a mortar, sieves, bowls and so on. Even when once a whole batch of flares which I was drying in the oven went off and terrified the kitchen girl they did not forbid my activities but rather gave me a little room in the attic for my magic games. I didn't disappoint their trust for I never caused a fire. My experience with fire did, however, make it possible for me occasionally to help put out a fire in the workshop when the guilty worker who'd accidently started it had rather lost his head.

The fireworks gave me for the first time that happiness which comes from putting into practice things which had been up till then mere thoughts and ideas and this is exactly what spurs a researcher or explorer to his efforts. At the beginning, of course, youth is satisfied with much less. Looking back subjectively, I see that this sort of experience was a constant and inexhaustible source of joy throughout my whole life. Even today when I am much less active I experience that same anticipatory excitement before a decisive experiment and the almost painful happiness when it succeeds that I did when I was a boy. In this case the strength and duration of the emotion are not as in most situations inversely related; on the contrary the joy of the researcher is both strong and lasting.

Work style. Although the family became more affluent around this time, life remained frugal. I had only very meagre resources for my experiments; my small amount of pocket money would sometimes be supplemented by a little additional allowance from my mother—but only in the most urgent cases. Because of this I got used to making for myself almost everything that a boy my age could, and I enjoyed doing that. This habit of getting by with few resources and with simple equipment and, as far as possible, putting together whatever was needed myself, is a work style that remained with me for my whole life.

Even in times when money was not limiting I kept to this way of doing things and I usually spent as much time and effort on simplifying the technical aspects of my experiments as on their conception. This style later made it possible for me at a relatively young age to give up my teaching position, together with the large laboratory and all the equipment and material that went with it, because I was sure that I'd be able to finance all the research I wanted to do out of my own pocket. And that is exactly how it turned out.

Another point that characterised my work style back then was that I was not averse to rather monotonous repetitive work. For my fireworks I sometimes had to glue and fill hundreds of paper casings or do similar repetitive chores. I never found this horribly boring, but instead tried to find ways to get it done in a more efficient way, and the search for such a solution gave the work a new attraction. Moreover I gained the satisfaction that comes when repetition leads to ever increasing dexterity and efficiency in the work. Finally, I have to admit that I got real pleasure from seeing the products pile up, perhaps rather in the way that the miser gets the joy of his life from the accumulation of money. There is no doubt that my parents' genetic constitution and habits played a role in all this. My mother was endlessly involved in housework and my father, even after he was well off, spent his mornings in business meetings and then, after lunch, put on his working clothes and went to the workshop to do himself all the things which needed particular care and skill. Nevertheless, there was some personal element at work here for this character trait was stronger in me than in my brothers.

In this way I found myself involved in an ever increasing round of work and interests. I'd have liked to have a lathe to make the rods round which the paper casings were formed. Sadly there was no lathe in my father's workshop, and so I decided to make one myself and actually managed to build one out of simple material. Once it was finished—which took quite a while—my interests had turned elsewhere and in the end I scarcely used it. I have to admit that this also is a characteristic that was typical of my later life.

Painting. Apart from fireworks my main interests then were collecting butterflies and beetles as well as sketching and painting. A neighbouring family called Schwendowski was a major influence on me in the matter of painting. The father was a minor official in Riga's town administration where his drawing abilities were put to use in preparing Diplomas and other calligraphic works of art. Most of his many children were considerably older than me, and one of them was a professional painter. It was always a great joy to me to be allowed to look through his sketch book with its lively water colour pictures whose colours were astonishingly appealing. Some of these I can still see in my mind's eye to this day.

I very much longed to be able to produce similar things myself. I tried, but given the small amounts of unsuitable material available to me it was a vain endeavour. There was no other way out—I'd have to manufacture the paints myself! By busily questioning my school friends I managed to get hold of a book in which I found the description of a milling stone and instructions for the proper amount of gum Arabic to use as a binding agent. I bought these things from the friendly apothecary and set to work. It turned out, however, that the colours alone don't make an artist, for my new paintings were no better than the old ones. I couldn't find anybody to teach me how to paint. The artist Schwendowski lived elsewhere and came home only now and then for a few days, and apart from him no one else in the neighbourhood knew anything about painting. Once he gave me some left over oil paint. It didn't go very far, but it was enough to show me that I'd get a lot further with it than with the more difficult water colours. And so I began to make oil paint which, in the old fashioned way, I stored in pigs' bladders because zinc tube had not yet crossed my Riga horizon.

In a paper shop there was a large stock of printed lithographs which I'd look through for hours on end, and then if my pocket money was enough I might buy one or two which I'd colour in using the water colours. To be honest the results were disappointing even for me. There were no good examples from which I could learn. The only paintings I ever got to see were in a gilder's shop window where occasionally an oil painting he'd been given to frame would be displayed for a few days. There was at that time not a trace of the flood of pictures that threatens to drown humanity today. The reproduction of works of art was restricted to lithographs and to the woodcuts which were coming back into fashion. Coloured reproductions were rare and costly and the first attempts at photomechanical techniques were just being carried out in their inventors' workshops. Because of this I was restricted to the odd example that passed by chance before my hungry eyes and I tried with my scanty resources to reproduce the strong effect it had made on me. Right from the beginning I had more luck with colours than with form: I could easily envision the impact of the colours I was applying while, on the other hand, I never dared compose a picture of my own in my younger years.

The relationship to later work. I mention these childish things because they later had consequences for me. I want to stress here that I consider the establishment of the quantitative theory of colours my most important work. I have not the slightest doubt that I would never have solved this problem, on which the best and brightest from Goethe to Hering racked their brains in vain, had I not from my youth been involved in producing dyes and, by doing so, been made constantly aware of the most important problems in colour theory which always have to do with non luminous colour (Körperfarbe). In particular one sees how the commonly available physical techniques using lenses and prisms available to Helmholtz led him to conclusions which have nothing to do with non luminous colours.

All in all, when I look back on my working life it becomes clear that all of the many interests and hobbies of my youth, no matter how useless my parents and teachers viewed them, all turned out later to be worthwhile and some even essential for those of my efforts which my peers hold to be not just useful but valuable. By collecting facts and the relationships between them in the boundlessly compliant memory of youth, a store of data is compiled from which the building blocks for the mental efforts of the future researcher can be drawn, for he has to form things which do not currently exist—and yet they must not be mere castles in the sky. For this he has no other guide than his memories of earlier experiences and so the mental constructs he produces take on a form dictated by these memories. We are therefore not dealing here with the operation of some mystical power that had given me in my childhood just those things which I would later need. Rather the style, and to a large extent the focus, of my later work was largely determined by what fell into my hands and senses in my youth.

What I am describing here from my personal experience is undoubtedly generally applicable. Goethe for example emphasised that the concept of the main figures and events of his literary works were established in his youth and that his whole long life was devoted to developing these early structures. In this case one can see even more clearly how the work of the man was determined by the material which he accumulated at a young age.

Outlook. If one follows this line of thought then one can come a long way. Goethe's instinctive rejection of Newton's theory of colour had its basis (Goethe was not conscious of this) in the fact that though it provided information about the "physical" nature of colour, that is to say information derived from the colours of refracted light, it told nothing either about the perception of the colour of light reflected from a surface nor about the properties of dyes. Newton's explanation of non luminous colour as reflecting the colour of thin sections was apparently not known to Goethe. He certainly never referred to it and in any case it would not have led to any scientific explanation because it is quite simply wrong. Goethe, for his part, was acquainted with dyes and non luminous colours because of his personal involvement in drawing and painting and he had the strong feeling that these colours are somehow completely different from the spectral colours. His warning, "My friends, abandon the darkroom", was surely largely due to the fact that he felt unsure of himself there and therefore made a virtue from necessity—but all the same his instinct was correct. He felt sure that there was no justification for the simple transfer of the spectral colour theory to the perception of colour in the environment, though he was unable to justify this view in physical terms. Physicists on the other hand viewed the whole matter as solved and therefore not worth further discussion.

If we look at the work on this theme of the great physicist Helmholtz, then we see how his lack of knowledge of painting techniques and of paints rendered him unable to grasp these problems. In his three volume book on the physiology of optical perception the term "non luminous colours" does not even appear in the index and on the page where colours and dyes are supposed to be dealt with one looks in vain for something on this topic. On the other hand, in his insightful speech on his seventieth birthday he told us that already as a boy he'd spent the Latin lessons, which bored him stiff, working out the light path in optical instruments. He said nothing about experiments at sketching or painting. The structure of his mental world was set in his youth and did not change at all throughout his life.

Technical and commercial experiments. No matter how elementary my drawing and painting skills were, they nevertheless turned out to be a great help in developing my many interests all of which suffered grievously from the limitations imposed by my meagre pocket money. My parents kept me on a short rein partly from thriftiness and partly because they rightly feared that my skylarking with this and that would interfere with my school work. I therefore had to find some other way to get hold of the necessary "kopecks" as the smallest coins were called.

Just at that time transfers had become all the rage. They would be moistened and then pressed onto any surface to which they'd stick after which the paper support could be peeled off. Often the outer side would be varnished bronze so that the picture only became visible after transfer. All of this was a great attraction for us boys. In line with my passion for doing everything myself, I invested a lot of thought and experiments in making these transfers myself. To begin with I didn't have any way of making the delicate transfer film on which the picture is painted. However, I did know that soaking silk paper with oil or better with turpentine produced a glassy translucent sheet. I therefore painted pictures on silk paper, made them translucent, cut them out, stuck them onto ordinary writing paper and covered the front side with a good thick layer of Arabic gum. In this way I produced pictures which, just like the bought ones, could be transferred—but they had the advantage that they showed scenes which were of particular interest to us. My school friends were more than ready to give me a few kopecks for them and so I could now buy mortars, glass tubes, and other necessary treasures which I'd wanted for a long time. The joy, however, did not last long because in some way a teacher came to hear of this and he strictly forbade the trade. The usual exchanges between boys were tolerated, for there was in any case no way to stop them, but he believed that as soon as money, no matter how little, was involved it became a serious and dishonourable offence. How much I would have liked to get hold of the things I needed by way of exchange rather than having to pay money to buy them, but there was nobody with whom I might have exchanged these things.

Music. The memories of one's own childhood are like the experience of entering the Finnish or Swedish Archipelago (Schären) To begin with only the occasional bare islands of consciousness give form to the featureless sea. They are at first very small and have little content. Then the islands become more numerous, larger and contain more diversity until they fuse into the solid land of permanent memory. Even here some incidents stand out like mountains in the plain of experience but they are now embedded in a continuum instead of being just perceived as islands of memory. Exactly this sort of island contains the memory of the first time I experienced great music. My parents had decided that the best way to introduce me to music was to take me to a church where a Christmas or Easter oratorio was to be performed. This indeed fitted to the "biogenetic law" of Haeckel, which at that time had not yet been formulated, that every organism during its development runs through a synopsis of the developmental history of the species. In this sense music had been developed first in the service of the church and all of us children knew at least some of the first steps along the path of its development in the form of songs. I no longer recall what it was I heard. However the ethereal sound of the violins soaring above the choir in a peaceful section, to be followed by the power of the combined organ, orchestra and choir, was an unforgettable experience. By chance, from where I sat I was be able to see the conductor, and the power which he exercised over all the musicians with his short thin baton seemed to me to be the most wonderful height of human striving that I'd ever seen or could imagine. From all the many ideals that such experiences awoke in me and whose achievement in later life gave me pleasure, this was one of the few that I never managed to achieve. Of course, realising my limitations, I never really tried to reach it.

At an appropriate time my parents also took us to the theatre. "We" in this case means my elder brother and me. I was a precocious child, while my two brothers developed somewhat more slowly, and so it was only natural that I was taken along for company to all of the events to which my elder brother, on account of his age, was initiated. That saved me a lot of time.

The first visit to the theatre was to an opera—The Magic Flute. The fact that we children didn't understand a word of the text didn't in the least detract from our enjoyment.

The serpent at the beginning and after that the comical Papageno and Papagena, and the evil Moor went straight to our hearts. Tamino and Pamina were less interesting until the magic of the theatre let them stroll through fire and water. In line with my natural predispositions the optical memories are much stronger than the acoustic ones. And when I come to think of it, the memories of the oratorio seem to be a mix of the acoustic memory with the sight of the illuminated choir against the dark background of the church nave.

My parents regarded music as an important part of general education. My elder brother was given piano lessons from an early age and became quite adept. His teacher was called Askenfeld. He was a peculiar little man in an old faded coat and with a white sailor's beard shaved around his pock marked, brown-red cheeks. But he was a good musician. It was his habit to play the first fuge in C major from "The Well-Tempered Clavier" ("Das Wohltemperierte Klavier") to recover from the drudgery of the lesson. For a long time I had no idea what this strange piece of music was until, hearing it much later, I recognised again the sounds from my youth. The violin was chosen for me, and I was to take lessons from a member of the theatre orchestra called Scholz. He was no use as a teacher and some years later he lost his job through drunkenness. But though I never managed to learn the violin properly, my incompetent teacher can be blamed only to a small extent for I simply lacked the acuteness of hearing on which the surety of touch on the strings and the melodiousness of the bowing depend. Later I transferred my meagre skills to the less demanding viola and in this way gained access to the inexhaustible treasures of our chamber music. Domestic string quartets were a source of happiness to me from my last years in school and throughout my years as a student and professor. I got from them not only the direct joy of music but also insights into the Aladdin's cave of thematic work which was of enormous help to me in developing my own theories of art.

Since I was making no progress with the violin I asked instead for piano and harmony theory lessons and for years I practiced Richter's strict harmony with Askenfeld. This early hankering after the scientific side of music, and the fact that it was supported by my parents, was an important part of my inner development, even though it only bore practical and scientific fruit in my old age.

By way of the book exchanges with my school friends, which I mentioned above, I came across some fragments of A. Hoffmann's "Kater Murr" and they made such an impression on me that I didn't rest till I'd managed to get hold of other texts from him. I learned from Hoffmann to understand the emotional side of music. It was a truly decisive experience for me to listen to a performance of "Don Juan" after reading his masterly analysis of this profound work. The deep veneration for Mozart which I developed at that time has stayed with me in undiluted form ever since.

Literature. Since my mother's thirst for reading was not quenched by the two family magazines which came every week, she also always had books from the local lending library. Early on I asked for and was given permission to read them as well. It was soon my job to go to exchange the books and I had a say in which books would be borrowed. Since I was always finished with them more quickly than my mother, who could only read in the odd quarter of an hour that the house work left her, she generously let me exchange most of the books as I wanted.

And so I got to learn the ways of the world beyond the narrow circle of family and school from the inadequate and one sided descriptions in the romantic novels of the 1850s and 1860s. The prevailing view was that of, for example F. Spielhagen, whose novel "Problematic Characters" ("Problematische Naturen") absorbed me in my last school years. The naturalistic movement was just beginning to appear on the horizon and had not yet challenged the domination of the newer romantics. Soon after this, science totally engaged me and I rather lost interest in questions of human relationships. The consequence of this was that I frequently made the gravest mistakes in judging the thoughts and actions of those I came into contact with later in life. Science never lost its hold on me and so to this day I devote less time than I should to the proper assessment of my personal relationships. I'd probably have managed to push through some of the scientific advances which I felt it my duty to support, quicker and more easily had I had more opportunity and interest in the ways of the world in my youth.

The art of human interactions. In particular I have always found it—I don't quite know where this came from—ignoble and even rude to try to influence others to act in a way that seemed preferable to me. I'm not talking here about personal relationships but rather about my scientific efforts in which it never became clear to me that success was almost always dependent on the goodwill of people who, at least initially, felt wounded by my impetuous reform zeal. Every advance faces all the leaders and authorities in the field with the question why they themselves hadn't seen it. In this sense each advance brings with it an implied criticism which thus tends to result in its rejection. I had always just assumed that others would share the passionate joy I felt at each advance and, though sometimes they did, in the majority of cases they did not. I failed to learn from my experiences.

I recently read H. Zschokke's book "Selbstschau" in which he describes his efforts to increase the level of education in his adopted homeland Switzerland. He first posed and then answered the question of how he should present his thoughts and evidence without drawing general antagonism, because, “no one is surer of his knowledge than he who knows nothing and no one believes he understands things better than he who grasps nothing”. Zschokke knew this at the age of twenty eight. I still hadn’t learned it at the age of seventy.

The embryonic author. As to my own writing ability, I soon got praise from my teacher for my German essays. However, from an early age the wish to express my thoughts and feelings in written form went beyond these school assignments. Of all printed matter, novels were most easily accessible to me and so it followed that my wish to do everything myself manifested itself here too.

True, the novel I tried to write got stuck after the first page or two, because the breadth of personal experience which I could call on was just too narrow. However, things went somewhat better with a newspaper which I brought out with the title “Humour”. It was produced by hand and given to my friends to read. I suppose I managed to produce half a dozen issues. When, however, I thought about writing criticism of the performances in the reading of assigned roles in classical drama, which we were doing at the time, I was met with a storm of opposition, because I had decorated the first letter with a huge scissor that was cutting a pitiable victim through the middle. I had to abandon any further issues. Even before that I’d earned the serious disapproval of my readers when I’d described, in the style of Amadeus Hoffmann, incidents which all my school friends could recognise about ice skating with girls of our age. The stories were written as tragic and passionate tales which, at the moment of greatest tension, turned into farce. My readers were perfectly happy to be overcome with emotion—but they didn’t want it to be interrupted.

Difficulties at school. It will be easy to understand that these diverse interests soon enough came into conflict with the demands of schoolwork. Already in the second class my results at the end of the year were insufficient to allow me to be transferred into the next class and I had to repeat half a year. In the third year class where, as I’ll shortly relate, I fell for chemistry, I had to repeat the whole year. But then things got better: I completed the fourth year in three terms and the fifth year in two. Not that I’d developed any greater diligence for my school work in the upper classes—quite the opposite—but rather because I’d gained the goodwill of some of my teachers who were prepared to accept my general intellectual progress as being, at least in part, equivalent in value to the school curriculum, and in consequence they allowed me a lot more freedom.

In particular I’m indebted to that splendid teacher Schweder, whom I already mentioned, because he put in a good word for me at the teachers’ conferences and in this way spared me the enormous waste of energy which I would otherwise have had to make to fulfil the strict letter of the regulations.

My marks in maths were always good; even better were those in physics which began in the third year (chemistry didn’t start until the final year). Schweder’s excellent and stimulating lessons caused me once again to want to repeat for myself the beautiful experiments he’d demonstrated. He lent me the practical textbook from Frick the early editions of which were splendidly adapted to the making of useful instruments from simple components. Here I remember how excited I was to discover that for a collecting lens there are two positions where a sharp image is formed and that for this the distances between object and image remain the same and are merely exchanged. Afterwards I found out that this had been known for ages, but I'd tasted the delight that follows an independent discovery and that left me with a craving for more.

I was soon allowed to help the teacher I venerated with the experiments and could suggest and carry out some technical improvements. For example I cut a radial slit in the disks of paper for the colour top so that one could put two or more disks with different sized coloured sectors on top of each other and thus mix these colours. Here as well someone else got there first; in this case, as I discovered years later, it was the great physicist J.C. Maxwell.

On one occasion I actually managed to catch my honoured teacher out in an error in physics. It was typical of him that from that point on he was even friendlier and supported me even more strongly in the teachers' conferences.

Chemistry. From physics the dominant passion of my youth switched to chemistry. It had all started with the fireworks. As I mentioned before, in Websky's book chemical formulae—which meant nothing to me—were written next to the trivial names of the chemicals. Slowly I worked out that wherever the word sulphur appeared in the name of a chemical the letter S was present in the formula. But the other letters remained un-interpretable. A question to the teacher elicited the short answer that these were chemical formulae and we'd do them in the last year. I, however, didn't want to wait till the last year, especially since at that time it looked as if it would take an age for me to reach it.

So once again I turned to that small but many sided source of knowledge: the books available to me through my school friends. This time luck was really on my side because I got hold of a copy of the "School of Chemistry" ("Schule der Chemie"[4]) from the excellent agricultural chemist Stöckhardt. It was well worn and, having more or less fallen apart, it consisted mostly of loose pages. However, I soon learnt to treat it as the greatest treasure that had so far fallen into my hands.

This book turned out to be a pedagogic masterpiece. Of course I could only judge the book subjectively, but far more than Frick's experimental physics—"Practical Physics" (Praktische Physik)—Stöckhardt's book satisfied my desire to do for myself all the wonderful things I read in it. At the beginning he set only very low hurdles in terms of the materials needed and the skill of the pupil and then, in carefully considered steps, he moved on to more difficult things. Because of this, these chemical experiments were for me much more accessible than the physics ones had been and I drank deeply from this spring.

Of course here as well I was short of money. My father was less ready than ever to support my dawdling with my schoolwork, and my mother was placed in a

[4]This book was published in 22 editions from 1846 to 1920. Several English editions were published under the title "The principles of chemistry: illustrated by simple experiments". In its time, it was one of the most important books popularising chemistry.

difficult position between her kind-heartedness towards her favourite son and her duty to her husband. What little she gave me was nothing like enough and so I looked around for some means to earn the desperately needed money. To buy the retort which I needed to make concentrated nitric acid for gun cotton I once cleared the ice from the whole yard and this cost me two days hard work in the Easter holidays.

The goal of my chemical experiments, in so far as I could carry them out according to Stöckhardt's directions, was to try to understand the underlying phenomena. Only those who have seen a child becoming totally absorbed with something new, taking it in through every pore and anchoring it in his memory, can imagine how I abandoned myself to this new world and with what eagerness I sought for ways to get round the problems that lack of materials posed. Of course I showed my school friends as much as they wanted to see and so my teacher also got to hear of it. In a friendly exchange Schweder made me promise to work at the other school subjects—which I could do quite easily. At the same time he examined my knowledge of chemistry and then lent me other text books the most important of which was Strecker's German edition of Regnault's textbook of chemistry. By this I was introduced to a more purely scientific view of my favourite subject and was so well prepared that my wasted first year as a student at university didn't hold me up.

Photography. All these activities brought me a special status amongst the other pupils and my head became quite swollen, so that I was not shy to boast about the many important things that I was going to do. This elicited of course protest and derision; one thing led to another and then I declared that I'd provide proof of my abilities by making a photograph of one of those present using only self made materials. A date for this was set. I didn't have a camera or any other equipment and what I knew about photography was restricted to the short description of its chemical basis given in the textbooks. None of my acquaintances was a photographer and all I knew about the technical aspects of taking pictures was what the "object" could glean when being photographed. At that time—in the 1860s—a wet collodion plate served as the light sensitive surface. Ammonium iodide and cadmium iodide were dissolved in a cellulose nitrate solution which was then poured over a carefully cleaned glass plate. This had to "ripen" for the proper period of time before it could be used. At the right moment, when the coating had gelled but was still damp, the plate was immersed in a tray of silver solution that also had to have just the right constitution (weakly acidic and saturated with silver iodide). Once silver iodide crystals had been formed, the plate had to be drained and placed wet in the cassette. From this point on it could be used for 5–10 min so that photography was only possible close to the darkroom. I gleaned these details from a book on photography by Monkhoven which I'd managed to get hold of.

I don't want to go in detail into all the difficulties I faced, but I built the camera out of an empty cigar box of my father. My mother's opera glass provided the lens. Thus far it was fairly easy. To make the silver solution I scrounged a broken tea spoon which unfortunately turned out to be rather rich in copper and so I had a lot of work before I could make pure silver nitrate from it. The trays were made of varnished paper while bits of broken window were cut to make the plates. Finally everything was ready and with breathless excitement I watched the development of the negative of my first picture—the view from my window. The feeling of happiness was no less than that I'd experienced when I'd set off my first rocket. A few further experiments gave me the necessary experience and I did indeed manage to produce a photograph of my school friend and print it on albumin paper by the appointed date. He'd managed to freeze the scorn he'd felt for what he thought was a hopeless venture on his very expressive face and kept it there for the thirty second exposure. In this way an extraordinarily lively portrait emerged and I am only sorry that I no longer have it.

I learnt a very great deal from this bet. The necessary chemicals could not be bought in Riga and so I had to prepare them from available substances. Even the gun cotton needed to make the collodion was on this list and the retort for which I'd cleared ice from the yard was part of the story, because I'd needed it to make concentrated nitric acid. The ammonium iodide and cadmium iodide both had to be prepared and the latter pleased me particularly by the sheen of its crystals. I also made my own ether. In addition to all this there were quite a few mechanical and physics problems to be solved such as getting the exact fit of the cassette and the best position for the lens etc. From making fireworks I'd learnt that enough information is available in books to let you do almost anything you want—and that was confirmed here. This all had a decisive influence on my future because I knew that my progress was not dependent on personal instruction. I was, of course, able to benefit from personal instruction, and I'll relate with thanks incidents of this as we go along, but it was not an indispensible necessity for me and in fact I owe a lot more to my books than to my teachers.

The moral flywheel. It was here that I experienced for the first time the operation of the "moral flywheel". As everybody knows a flywheel serves to store a certain amount of energy which can be used to iron out the peaks and troughs of the demands on a machine. This enables a more or less regular speed of operation, despite fluctuating use and load. If a machine is overloaded for a while then it would simply grind to a halt were it not for the flywheel which can cover the excess load with its stored energy. In a similar way I—unconsciously at the time but later quite deliberately—forced myself to an above average effort. In this case I used my ambition as a flywheel to draw the extra energy needed to conquer the field of photography, which combined my twin interests of pictures and chemistry, and thus to win the apparently impossible bet. The success of this first attempt and of a second which I'll tell when we come to the end of my studies, encouraged me to use this procedure often, perhaps oftener than was altogether wise. Nevertheless, although there were failures along the way, it worked pretty well in most cases.

Other activities. These diverse indoor activities didn't prevent me from spending a large part of my time outdoors. Because my father was a cooper there was always a large court yard to store the wood adjacent to our house. After we moved from the house on the Speckgraben, which was soon buried under the Riga-Dünaburger railway, we lived in the Romanowka Street which was also in the "Moscow" suburb. Outside Riga's old city wall, which I saw in my early childhood and which was later replaced with pretty gardens, there were three suburbs—"Petersburg", "Moscow" and "Mitau", whereby the first was the most genteel one. Mostly non-Germans, Russians and Letts lived in the "Moscow" suburb while the "Mitau" suburb lay on the other side of the three quarter of a kilometre wide river Düna, and it was a world on its own. In spring and autumn, when the river was full of drifting ice, that suburb was cut off from the town for weeks on end. In summer traffic passed on a pontoon bridge and on some small steamers, while in winter the river was passable on the ice. Because of this the people from "over the river" were regarded as a bit foreign and they were thought capable of getting up to things that a proper citizen of Riga would never do.

The whole family saw it as an important step in our social rise when around 1860 my father managed to buy a house on Alexander Street, the broad main street of the "Petersburg" suburb. That our house was number 100 increased my respect for it considerably. It lay a bit at the edge of the suburb but still inside the built up area which came to an end a few hundred paces further on at the "Great Pump"—a public fountain in a square surrounded by poplars. Beyond this point there were only some scattered summer houses, farms, windmills and factories before the start of the endless pine forest, extensive high moors and beautiful lakes which make up Riga's surroundings. The "Mitau" suburb was also surrounded by woods and moors while on the edge of the "Moscow" suburb lay the sand dunes, which I mentioned before, and these one had to cross to get into the woods.

This was the stomping ground of the explorations which my elder brother and I and two or three school friends made in ever wider circles. Along the way we'd collect beetles and butterflies while a pond with wonderful crested newts gave us the chance to take a few especially beautiful specimens home where we fed them on earthworms. For us it was enormous fun when two of the newts grabbed different ends of the same worm and then, swallowing slowly, met in the middle. Then there started a sort of wrestling match in which the stronger one pulled the already swallowed part of the worm back out of the loser. And then one time I suffered pangs of conscience when I stepped by accident on one of the newts that had escaped from its bowl. The half squashed animal did not recover from its injuries and I had to witness its slow death. After that I carried the other newts back to their pond.

The director. All in all I had a very happy youth. The schoolwork wasn't onerous and I didn't take it all too seriously. For the 7 years I was at the secondary school Mr. Haffner was the director. He was an old ultra-conservative old-fashioned classical philologist who had previously served as the government appointed rector of Dorpat University.[5] There many stories circulated that centred on his reputation as a petty fogging bureaucrat. Once, when his immediate superior, the chairman of the trustees, was out of town Haffner deputised for him and in this capacity he discovered some minor breach of the regulations which he as rector had committed.

[5]Since 1919 Tartu University, Estonia.

As acting chairman of the trustees, he sent a sharply worded reprimand to himself as rector, and was seen the next day in tears as he read this letter to himself.

However, since he was in essence a benevolent man and since he did not consider himself responsible for the mathematics and science subjects we didn't suffer too much from his classicism. At the start of each semester in the second class he organised a big celebration in which he'd spend one hour telling us how a good pupil would prepare himself for the upcoming Latin lessons. He had a large closely shaved face in which he manage to use the unusually large realm between his nose and his mouth to so dramatically express the various stages of cogitation between the start and the triumphal end of a translation that we boys could scarcely suppress our laughter.

He was, one must add, altogether a conscientious director who every Saturday made the round of all the classes to look in the register books in which each teacher had to confirm that the lessons had taken place and in which the more serious disciplinary offences together with the perpetrator's name were recorded. These culprits then had to suffer a stern philippic and I can assure you that he took care of every single case. He even managed to enliven the inevitable monotony of his punishment sermons by bringing some personal twist to each case.

The teacher. Each pupil at the start of his school career had to nominate a teacher to be his mentor who would keep an eye on him and advise him if he ran into difficulties. My heart had been set on Schweder, but when I asked him it turned out that he already had the maximal permitted number of pupils and so I looked for someone else. I took the teacher my elder brother had chosen. He was called Dr. Groß: his subject was German. He was a short plump man with a round white face, a bald head and a black beard. He was taciturn and had a permanent air of dissatisfaction. I never saw him smile. He was one of those teachers whom a university degree had completely alienated from teaching so that like "Pegasus chained"[6] he looked down on his job with anger and contempt. He probably spent his free time writing tragedies, though none have seen the light of day. He could have had a great influence on me since he was responsible for setting and marking the German essays and this was an area in which I wanted to make my mark. However, he completely refused to cater to my aspirations and stayed instead with the usual stupid topics: "On the life of man", "Seen from a tower", "Wallenstein's guilt", "Tell's monologue". In other words the sort of thing which then—and even long after—was used to train pupils to spew out empty phrases. Once, when the topic was "Where would we be without hope?", I was so annoyed about the silliness of the subject that to taunt him I wrote the superficial banalities which he expected from us in doggerel modelled on Kortum's "Jobsiade"[7] which I'd just read with

---

[6]A poem by Schiller in which the winged horse of Greek mythology was bought by a farmer who used it as a plough horse.

[7]Carl Arnold Kortum was a German physician and writer. The "Jobsiade" was a comical heroic poem about the life of the down at heel theology student Hieronymus Jobs. It is a satire on student life and German philistinism.

delight. He reacted to this by underlining in red ink every error I'd committed in grammar, poetic meter and rhymes and gave the now blood red essay an appropriately terrible mark. But the fact that my cheekiness had no further consequences showed me that he hadn't read it without some amusement and that he had used the formalism of the red ink to spare me any worse punishment. Many years later when he was long dead I heard to my great astonishment from one of his friends that I had been one of his favourite pupils and that he had always put in a good word for me in the teachers' conferences. He'd never said a word of this to me.

The other language teachers also had little influence on me. We learnt French to begin with from a very old, very small gentleman with a leathery face that was always shaved smooth. His clothes were always neat and clean, he used silk handkerchiefs and scent, carried a flask of smelling salts in his waistcoat pocket in a ruby-red glass flask decorated with golden arabesques and was called (or called himself) Sire. Soon after I started he died and was replaced by a jaunty young man called Dubois from the west of Switzerland who, apart from his French was so uneducated that it was obvious even to us third year pupils. He was always friendly to me but didn't hide his annoyance that I didn't really apply myself to his subject, for he considered French to be extremely important.

We had English from a long thin middle-aged teacher called Riecke—a talented man, educated in many areas. He was a good mathematician and had a lively feel for the beauty of poetry which he tried hard to awaken in the older pupils. Sadly he was one of those unfortunates who are completely incapable of keeping a horde of active boys under control. He had to instruct the youngest classes in handwriting and the noise we made in that class and the tricks we got up to are simply indescribable. When we got to the last 2 years and discovered what an injustice we had inflicted on this benevolent and sensitive man it was just no longer possible to develop a suitably respectful attitude to the one-time victim of our impudence. Though he would have liked nothing better, an open and friendly relationship with us never developed. Once much later, when I already held office, I met him by chance and was happy to learn that he had retired long ago and had a comfortable house in South Germany and a circle of attentive friends.

I've already described the Russian teacher Haller. If he had any influence on me then it was only negative. Since my progress in Russian was inadequate I took private lessons with him in the fourth class and this at least had the hoped for result that he gave up his opposition to my transfer into the final year. I learnt nothing from him.

As is so often the case, our history teacher was the reason that we regarded the subject as useless drudgery. In the lower classes the lessons consisted of learning by heart lists of names and dates and this was only rarely leavened by an anecdote. Our teacher was called Matschewski. He was an elderly man with a bald head and just a rim of hair at the nape of his neck, the slow growth of which we'd follow with a naturalist's interest until someone could proclaim, "Matschewski has had his hair cut!" As his name suggests he was of Polish descent. Though he was originally well educated he'd fallen on hard times and his family lived in such a shambles that the squalidness of his clothes was obvious even to us boys. He was unable to keep order and forgave us every irregularity so long as we didn't make too much noise in class. In his younger years he'd studied the French revolution and he had a collection of books about it which, over the years, had fallen apart and were now extraordinarily dirty. In the fourth year he lectured us from these in enormous detail. His strong bias for the French way of life didn't suit us at all because we saw ourselves as German. This became very obvious in 1870 when the French–German war broke out. At the beginning he delivered hate filled prophecies of a sound German defeat and we in turn loved to tell him every day the news of the German victories. These he angrily denounced as plain lies and, when this was no longer possible, retreated behind bitter threats of the retribution which would soon come.

Faith. Religion for all classes was taught by John Helmsing. He was a quiet man of average height with greying blond hair. His long face bore side whiskers and a melancholy expression. In line with his first name, which he always wrote out in full, he tried to dress and act like an Englishman and even lisped a little bit like one. Presumably there was some English blood in his family. In his class he was able to keep order without any great effort. This was made easy for him because the clergy had high standing in Riga society and our respect for his religious office helped him as a teacher.

Helmsing, however, had no great influence on me or on my school friends. My parents stuck to the religion they'd inherited without it impinging much on life at home. My father scarcely went to church, but my mother did to begin with, though later the increasing burden of household duties kept her at home. We children were sent to church—regularly at the beginning and then less so as the spiritual influence on us receded. For a long time I tried hard to sustain the belief in Christianity which I'd picked up as a child. My first teacher (his name was Mr. Fromm which means "pious"), tried to support me in this, but my faith wouldn't last.

Once, worried about some "sin", which I no longer recall, I went into a dark corner and kneeling fervently, prayed to God that he forgive me and free me of the sin. I received no answer, neither from outside nor from within. Everything stayed silent and empty. This was a shattering experience for me and robbed me of my earlier absolute trust. However I didn't fall into a state of religious melancholy as occasionally others of my age faced with a similar problem did. Instead I realised I just had to solve the problem on my own and, given the happy and active nature I'd inherited from my parents, that wasn't so difficult.

I brought this experience with me to the secondary school where my teacher could not bring me back to the path that I had left with so much inner struggle. Most of my schoolmates treated religion with indifference and contempt and the few pious ones were such uninteresting people that I had no connection to them. To begin with I defended my old position against the scoffers but as time went by I gave that up. Confirmation was in the hands of a mild old pastor who had to expend what little energy he had on the uncouth boys and girls of the lower orders and so had little left over for us secondary school pupils. To be honest I was still a little worried about the sinister threat that church doctrine held out for those who dared to take communion while not in a state of grace. But this threat was so inconsistent with all I had by this time learned from science and art that I decided to take a chance. And when nothing happened I felt that I had separated myself without pain from the religion I'd grown up with.

The others felt much the same so that religious classes were treated as another boring school subject that one had to put up with and for which one should do as little as possible. This terrible inconclusiveness of all his work seemed to bear heavily on Helmsing. He became ever more shy and quiet and after I'd left the school I heard that he had to be treated in a psychiatric hospital for religious melancholia.

Chapter 3
The Growing Boy

Dance lessons. The time came when the growing boy carefully examined his chin and lip for sprouting hairs and began to feel a mix of dread and delight when confronted with girls. In my case the dread was uppermost at first because I had no sister to introduce me to this other form of humanity.

My father had a high opinion of the value of good manners for social advancement and he thought hard about how he could help us to rise further than had been possible for him. Since he'd advanced beyond his station there was no one amongst his acquaintances or relatives who could help in the matter, and so there was nothing left for it than a course of dance lessons.

The teacher's name was Krickmeyer. He was a failed actor, tall, thin and supple, with a reputation of being able to lick young bears from my social milieu and age into shape. My elder brother and I, together with around a dozen others, were sent to him to learn the elements of the art of dance. For me these lessons, which I'd initially joined only under protest, became more interesting because the teacher treated his subject scientifically, drawing out the various positions of the limbs and the development of the dances on large posters. In addition none of the feared girls were to be seen because, like us boys, they too were first instructed in the basics on their own before we were let loose on one another.

Unfortunately I believed that once I'd grasped the diagrams I'd pretty much understood it all and so I didn't spend much time on practising the movements needed to actually perform the dance. Because of this I have only seldom experienced the joy of being carried away by the rhythm of the dance and even when it did happen the feeling was not strong enough to make me try hard to recapture it.

At some appropriate time the girls joined the class. They were from the same social circle as us and, as the rigid rules of demeanour demanded, they behaved demurely and with almost excessive good manners. Since Mr. Krickmeyer had also instructed us scientifically in how we should treat these strange creatures I found this new experiment much less difficult than I'd feared it would be. As soon as there was a pause, the two groups immediately separated. While we boys exchanged short meaningful comments on our impressions, we could hear an excited many-voiced twittering from the girls' room.

R.S. Jack and F. Scholz (eds.), *Wilhelm Ostwald*, Springer Biographies,
DOI 10.1007/978-3-319-46955-3_3

Of course each of us soon had his flame whom he shyly worshipped. Mine was called Eveline. She had long light blond curls, a bent nose and a self confident air about her. She didn't have much to say. I tried to get into her good books by telling her that she was very like one of my cousins whom I liked a lot and was astonished that she reacted coldly to this information. Only later did I learn that there is no more certain way of annoying a woman than to compare her to another—unless of course one immediately adds that the other cannot hold a candle to her.

At the end of the winter the dance lessons came to an end. In the following spring I went with some friends on a hiking tour for several days and one evening, sitting by the camp fire, I took out the little photograph of Eveline I'd kept and wanted to ritually burn it. However, the proper emotions just wouldn't gel and in the middle of the ceremony I burst out laughing.

My friend. Since the worthy Mr. Kirkmeyer had not managed to put much in the way of social polish on me, my parents were glad when a new opportunity arose from which they hoped for more substantial results. This was a private dance course which the mother of my friend Fritz Seeck organised for her son, her two daughters and two nieces. My friend gave me an invitation to it, despite his mother's reservations, for from my occasional visits to their house she knew me as something of a tearaway.

Fritz Seeck was my closest friend at school and like me was the son of a self made man. His father was a locksmith or mechanical engineer who had invented a machine which made it easier to pack flax and, since flax was a major part of Riga's growing business, ever growing numbers of the Seeck's packer were required. The inventor and producer—who was called Packer-Seeck to distinguish him from his brothers who were in different lines of business—was soon a rich man. He'd married a young teacher by whom he had four children, two boys and two girls. I never knew him because he died young. He left his widow a considerable fortune and the task of bringing up the highly gifted children—a task to which she, as a teacher, was well suited.

Fritz was the second son. The first was called Otto and was 4 or 5 years older than Fritz. He studied history in Dorpat, became Mommsens's[1] favourite pupil in Berlin and from there went on in Münster and Greifswald to gain a reputation that went far beyond just the circle of academic historians. Back then I only saw him occasionally when he came home for the holidays, but later, when my career led me as well to Germany I got to know him better and had a great regard for him. He died a few years ago.

Fritz was even more talented than his brother. He was my age and our shared interest in science soon brought us together. His mother, Mrs Seeck, was an intellectual. She had a library in her house and made sure that her children were

[1]Theodor Mommsen was an influential German historian. Ostwald however overestimates Mommsen's opinion about Seek: Mommsen supported Seek's appointment as Professor at the University of Greifswald, merely because he considered him to be a lesser evil than the other candidates. See: *Mommsen und Wilamowitz. Briefwechsel 1872–1903*, edited by Friedrich and Dorothea Hiller von Gaertringen, Berlin 1935, 105f. (Nr. 90).

well acquainted with the treasures of literature. Fritz passed this motivation and encouragement along with the books to me, and so I have to thank the kind hearted but strict Mrs Seeck—her bearing reminded one of her previous profession—for a solid introduction to literature in the widest sense. In a similar way I was led to appreciate music. Often enough she had occasion to criticise my heedlessness and lack of polish, but she did it in such a way that I lost neither my respect nor my trust in her. When I was at my friend's house I could sit with him and his sisters at their coffee table and thus had a chance to lose some of my shyness and awkwardness which made interaction with girls my age so difficult.

Of the two girls Helene, the elder, was closer to her brother in age and interests so that I already had things in common with her. She was well educated and seemed to my youthful eye to be a perfect beauty. I soon felt a deep affection for her which I naturally sheepishly tried to hide, though I don't believe that any of the others was fooled for a moment. My crush on her was treated with humorous friendliness and Frau Seeck sometimes exploited it in her attempts to civilise me, for I had no other wish than to be allowed to worship my ideal from afar. The object of my worship was satisfied with that and would sometimes make me happy by sharing an interest in some hobby of mine and this was easy since our interests were in many areas parallel. She read a lot—and widely. Her elder brother Otto also helped with her education by giving her history books to read and some of them were pretty advanced—even academic. She showed little interest in the activities of her lively peers. This was something that interested her younger sister much more. She in turn found my exclusive interest in her elder sister stupid and she ignored me completely—which was a great relief.

From all of this you can imagine what a great joy it was to me when one day in the school Fritz asked me if I wanted to join their dance evenings. I accepted eagerly and even managed to persuade my parents to buy me a suitable suit. They were more than a little astonished by this because up till then I'd shown little interest in my personal appearance.

After surviving the first difficult evenings among the numerous unknown ladies and daughters of well-off families, all of whom outshone me in matters of social polish, I began to enjoy myself. The liveliness and quaintness of my conversation brought me mostly mockery but also some respect. I however didn't care about mockery or approval, but swam in happiness whenever I could be "her" partner at table or dance. Of course other couples had established tender relationships as well and such a sense of harmony now encompassed the whole circle that none of us was looking forward to the day when it would all come to an end.

Our schooling had been entrusted to the best dance teacher in Riga. She was called Mrs. Weller and was the wife of a respected first violinist in the theatre orchestra. She was thin and supple and had a loud voice. She knew exactly how to control her little lambs and bucks and how to keep clumsy slowcoaches up to the mark.

She was constantly dissatisfied with me. For a start I had to do everything different from how I'd been taught in the first course and then my head and heart were increasingly elsewhere so that my legs didn't know what to do. Mrs. Weller summed up her view of me near the end of the course in words that echoed down the hall, "Mr. Ostwald, you'll never be any good". This was a self-fulfilling prophecy, because it destroyed the last of my courage and I gave up dancing after that.

The poet of the dance evening. As the last dance evening approached we wondered what we could do to bring it all to a worthy conclusion. Since from my school newspaper days I had a certain reputation for coming up with amusing ideas I was given the task of arranging a dramatic finale. I looked for a way to humorously portray various little incidents that had occurred during the dance course in such a way that each of the participants would be easily recognisable. This way of doing things provided the content and freed me from the ordeal of coming up with lots of platitudes. Two of the brightest participants who had good strong voices were to dress up as an old married couple and discuss before us whether or not they should permit their children to take part in a dance course. Both of them would have the strongest reservations and would justify them by providing all sorts of examples of the sort of terrible things that could happen on such occasions. "Papa" told the stories which related to the boys, "mama" those of the girls and I took care that each one of us was readily recognisable. It was particularly funny when the two of them had to recite the verses of the sketch about themselves. The biggest problem for me was to make sure that even the object of my veneration got her fair share of mockery; I did this by hinting at her well known choice of reading material which was unusual for a young lady:

> "Her wisdom is also vast. It extends even to Plutarch".

As one can easily see the doggerel of the immortal "Jobsiade" had been my model for the poetic form. The performance was set up so that the two speakers could discreetly leave during dinner and then return in their costumes to play out their roles across the table. Every new innuendo was greeted with loud applause: those who had not yet been "done" waited in suspense while the others relished what was to come with glee and the whole thing ended in laughter.

End of school. I was now nearly eighteen and it was time to prepare myself for the school leaving exams. Maths, the science subjects, German language and literature didn't give me the least worries. I thought I'd also get through all right in English and French—but history and Russian were a different matter. Since my memory was good, I got hold of several different books on world history and compared what was written about the various periods in them and this made it easy to remember the basic facts. In this way I was relatively sure I'd pass this subject too. The big problem was Russian. My class mates hinted that there were ways to get round this, but I was too proud to take them up on it. The result was that though I passed all the other subjects, I failed the Russian exam. However, because Petersburg had decreed that a first class performance in Russian in the school leaving exam was essential, this meant that I would not get my school leaving certificate. Because of this I had to put off my dream of studying chemistry at Dorpat University for half a year.

The Russian exam. I turned down the well meant suggestion of the school director Haffner that I re-sit all the subjects and get even higher marks the next time round. Instead I exercised my right to re-sit only the Russian exam. Because of my many interests the free time before me was more than welcome and I used it to the full. Having learnt the hard way I now resolved to make use of those warped and twisted means that would lead unerringly to my goal of passing the Russian exam. It was a typically Russian ploy.

A Russian priest called Sokolow was in charge of the religious education of the handful of Greek Catholics in the school. He was also a member of the exam committee. Two of the final year boys facing the exam would take personal tuition in Russian from him, and the considerable sum of money required for this was scraped together from all those who took part in the scheme. In this case I and another final year boy were given the job. We went only now and then to the lessons, though the priest would have got annoyed if we'd never turned up. He gave us tea and biscuits and told us all sorts of things in his own language which sounded amusing, for he was a jovial man with long curly hair and a beautiful beard and he didn't play the holy man. Now and then he'd launch into an account of Russian history as he understood it. Once he related the tale of the conversion of the Russians to Christianity. Tsar Wladimir, later known as the sainted Wladimir, had called together priests of all the known religions to find out which of them he should prescribe for his subjects. The Jews were struck off the list right away because their God was angry with them and since they lived in permanent exile they'd obviously still not managed to placate him. The catholic priest didn't stand a chance either because his theology was so complicated. Wladimir liked Islam the best because paradise sounded so nice and God granted the believers as many women as they wanted. However there was no way round their proscription of alcohol. "After all", added Sokolow thoughtfully, "can you imagine a Russian without vodka?" The Greek Catholic priest had brought along a large painting of hell in which the tortures of unbelievers were shown in all their gory details. The picture made such a great impression on Wladimir that he chose this religion for fear of having to suffer these tortures for all eternity. The technical problem of immediately converting all his people to Christianity was solved in an elegant way: the people were driven through a shallow stream after which they were judged to have been christened and thus converted.

This sort of entertaining story, however, was not actually the object of the exercise. As a member of the exam committee Sokolow had prior knowledge of all the exam questions and that meant he also knew which text we would have to translate from German into Russian. This was the centre piece of the Russian exam. A few days before the exam a piece of paper fell by chance out of his pocket as he left the private tuition class for a few moments. Quite by chance we picked it up. When he returned he glanced briefly at the spot where the paper had fallen and nodded in satisfaction. We took our leave with thanks and found on the piece of paper the reference to the exam text—it was something from Schiller. This turned out indeed to be what we had to translate. Each of us was able to turn up for the exam with a good translation all ready to be copied out.

This was all done unselfconsciously, without any of us feeling pangs of guilt. We all knew that in dealing with Russian officials anything could be achieved so long as an adequate tip was paid. The fact that even the German teachers in the school accepted this as one of the facts of life was very clearly shown in this case. The piece of paper we'd picked up identified the start of the text but not the end, so that we'd all brought along a translation which was much longer than required. We then had to copy from this as much as was demanded in the exam. One of us, with quite unparalleled impudence, found all this too laborious. He simply crossed out the excess text on the translation he'd brought with him and handed it in. Although this was a clear proof of the swindle, none of the teachers mentioned it because without this little short cut the school wouldn't have got anybody through the final exams for none of the pupils was prepared to learn as much Russian as the government demanded.

Introduction to teaching. So as to have some sort of regular job for the extra 6 months that I had to stay in Riga, I undertook to teach some children who were being readied for starting school. This was my first attempt to teach. The work was fun, and I think my young boys and girls enjoyed seeing me. They belonged to a rich titled German–Russian family from the empire with rather russophillic views and this was interesting for me because I knew so little about such peoples' way of life. The most astonishing thing for me was that they were willing to pay me for the pleasure of teaching their children. Probably it was, objectively seen, little enough, but from my point of view it was substantial because I'd never till then possessed so much ready cash. I used some of it to buy my mother a sewing machine. Talk of this new contraption had only just reached Riga and the joy it brought my mother moves me even today.

I retained this astonishment at being well paid for doing something I enjoyed for my whole life. Research, teaching and writing have remained to this day the richest and purest sources of joy in my life and they also easily supplied the means of support for me and my family.

Shaping the future. I never had the slightest doubt about what I wanted to study. My father wanted me to become an engineer and therefore tried to persuade me to go to the Riga technical college, but I wasn't going to do that. Not that I had anything against such a career—after all, ever since childhood, I'd had a strong technical bent and would doubtless have turned out to be an above average engineer. It was just that the idea of free basic research in the uncharted seas of the as yet unknown seemed to me to be so magical that I didn't worry at all that the lack of a chemical industry in my homeland meant that the economic future of an academic chemist would be difficult at best. I didn't worry about money. My highest dream for the future was a job as the scientific assistant to a chemistry Professor in Dorpat and I knew from my own home that it was easily possible to lead a happy and fulfilled life with little money. Even though I'd now seen and appreciated the differences between my parents' modest lifestyle and life in the more opulent homes of some of my school friends, I had no particular wish for a fancy flat, fine clothes, gourmet food and all the rest.

Just as there was no question as to which subject I'd study, so there was also no question as to which student society I'd join.

About half of the students at Dorpat belonged to large student organisations each of around a hundred members. Each society was distinguished by a uniform denoting its home region and all of them had constitutions based on that of the nationalistic fraternity of Jena which had been founded at the start of the nineteenth century. Three of these student societies were named after the Baltic provinces Courland, Livonia and Estonia while my home town, as the largest city in the whole land had its own society—the "Fraternitas Rigensis". These societies had all been founded in the first decades of the century so that in my student years they would one after another celebrate their fiftieth year. Their constitutions had remained essentially unchanged. They exercised a large degree of self management and they also had a large influence on the whole student body. Even the students who had not joined a society were subject to the Societies' Court which decided on matters of honour and student behaviour. This was because its severest punishment—known as "Disrepute"—would make the life of an individual at the university almost impossible, and so no one dared to ignore the decisions of the court. The court proceedings were public and the first year members of the societies were required to attend them in order to gain from these real incidents in student life the wisdom with which to organise their own future conduct. The judges of the court were elected from amongst the society members and I can assure you that they did their very best to reach fair and just decisions. All the students normally addressed each other using the familiar "Du" form.[2]

Since I came from Riga my choice of a society was a foregone conclusion, but even if it hadn't been, there was another factor which would have made me reach the same decision. My closest friend Fritz Seeck had sat the school leaving exam at the same time as me, but had managed to avoid the Russian pitfall. He'd entered the university one semester ahead of me and had joined the "Rigenser". Already in the first term he'd been elected to the inner circle which gave him the right to wear the society's cap and coloured band. For both of us it went without saying that I'd also join the "Rigenser".

---

[2]In German close friends and family members are addressed as "Du". All others are addressed as "Sie".

Chapter 4
Student Years

As Freshman on the way to Dorpat. And so in January 1872, at the age of eighteen and a half, I set out with other school leavers from Riga for Dorpat. We travelled in the care of an "olderman" and, since there was no rail connection, we rode in very simple horse drawn open sledges, protected from the cold by bundles of straw and large amounts of alcohol. Every 20 km or so there was a post station where the sledge, the horses and the driver would be changed. Under good conditions the journey took around 30 h; under bad it could be up to 3 days.

This journey gave us a clear taste of what student life would be like. The most important element was clearly alcohol, and getting drunk was not in any way a disgrace but rather something quite normal—in fact almost a duty. The "olderman" set a good example and his herd of "mules", as youths between school and matriculation were known, tried their best to emulate him. For hours on end one or the other lay unconscious amongst the baggage in the sledge and it was astonishing that there were no serious accidents. I was protected from the poison by my sturdy build and so managed to keep my senses intact.

My friend had warned me of this. I'd got to know beer and stronger drink in my last school years and so I saw it as a natural challenge to get used to this aspect of student life. Once in Dorpat I moved into a student flat with a medical student called Hermann Meyer. He was a nice, though not very talented boy, and neither of us minded that the furnishings were pretty primitive. Since the strict local regulations forced the pubs to close at 10 pm, student society members would continue an evening's binge in the flat of one of the freshmen who was required to make tobacco and beer available. On the first evening they came to Meyer and me. There were many of them and they were already rather drunk because they'd been celebrating the valediction of a well liked society member who, having passed his exams, was now leaving. I watched the revelry in astonishment—stunned by the enormous quantities of beer that were consumed. Around 3 am the guest of honour

R.S. Jack and F. Scholz (eds.), *Wilhelm Ostwald*, Springer Biographies,
DOI 10.1007/978-3-319-46955-3_4

passed out and was put into my bed. The others left. My unexpected guest relieved his over loaded stomach by puking up rivers of beer.

Since it was −20 °C outside I could only briefly let in some fresh air. I settled down on the little sofa, which was standard equipment for every student digs, to try to get some sleep. At break of dawn I woke at the same time as my guest, whom I hadn't met till the night before. I watched as he shakily got to his feet. He needed quite some time to work out who and where he was. Then he spat in anger into the mess he'd made in the bed, grabbed his coat and hat and left without saying a word. I met him later and we became good friends.

I fell asleep again and when I woke up I rather hesitatingly opened my eyes because I wasn't looking forward to seeing the mess. Not trace of it was left, for in the meantime the beefy elderly Estonian woman who came with the flat had been through the place. She'd cleaned up the mess, made the bed afresh, cleaned the room and started the tin coffee pot that was to be found in every student flat. The smell of fresh coffee nicely covered the stink of old tobacco smoke which clung to everything. So then I put myself back together again and refreshed my burning eyes in the ice cold water in the washing bowl. In the meantime Meyer was also awake and soon we were sitting at the breakfast table and excitedly discussing the previous evening's events.

I must admit that I'd been rather shocked at this first sample of my future life and as soon as possible went to talk it over with my friend Seeck. He calmed me down by telling me that this had been a special occasion, and that it was not every day that a senior member of the society left. At midday there was the ritual leave-taking at which my guest of the night before was surrounded by the officers of the society with their colour bands and swords. Off we went with the flag in front and all the members of the society following and singing together the traditional song:

As an old student I now leave,
God be with you, philistine house.
I'm going back to the old country
Have to become a philistine again
—and so on

The procession marched through the city to the post station where with final speeches and responses leave was taken of the man's student years. I was carried away by the poetry of the ritual and that made me ready to accept the events of the previous evening as an unpleasant but necessary part of the multifaceted life in a student society.

The matriculation of the new students took place on one of the following days after which we were really members of the student body. The rector at this time was the ophthalmologist Georg von Öttingen one of the three Öttingen brothers (the second was a theologian and the third a physicist) who pretty much ran the university. Previously the rector had been appointed by the government and the last "crown rector" had been my school director Haffner whose title "Excellency" derived from that appointment. However, Tsar Alexander II had given the right to choose the rector to the university, and this had immediately led to a remarkable scientific flowering for new appointment were now in the hands of the professors themselves and the government had only retained the right to approve them. That right was seldom used to overturn the professors' decision. Georg von Öttingen was one of the first rectors chosen by the university. He ran a tight ship and kept provocative dissension of youthful exuberance under control by the use of detention sentences in the university detention room (Karzer). One of his victims revengefully wrote on the wall of his cell a citation about Göttingen from Heinrich Heine's "Journey through the Harz": "One faces G. Öttingen best with one's back". The rector thought the joke was rather good and let this piece of graffiti stand. For many years it was pointed out to willing and unwilling visitors to the karzer.

Student society life. In the meantime we'd been introduced by the olderman to life in the university and in the student society. He read out to us the rules of the society pertaining to the daily life of us black capped freshmen. We were required to report to the pub at 10 am and had to remain there till nearly 1 pm. The afternoon was free unless there was some good reason for drinking sweet wine in one of the freshmen's lodgings. At 7 pm there was "freshman's tea" which was a particularly fine feature of student society life. Even the older students shared a flat with one or two friends and often several such flats would be in one building. In such cases a larger group would join together for the evening meal. The rule was that depending on the size of the group one or more places at the table were reserved for freshmen. The only exceptions were those older students who soon faced exams and therefore were short of time.

We freshmen had the right to go shortly before 7 pm to the "castles" where Freshman's tea was served and ask if there was a free place at the table. If there was, then one was their guest. If not then one went on the next "castle". Since the number of such free places was much greater than the number of freshmen, one usually found a place somewhere.

At these occasions there was, for once, no beer but rather tea, and so they had a very different character from the other meetings. Here there was the possibility to talk about general interests, poetry, music, art, politics, philosophy and sometimes even science. The older students had a chance to get to know the freshmen and could influence their development. These chance daily meetings meant that in the course of the semester everybody met everybody else several times and this made possible one's successful integration into the student society—without the fog of alcohol poisoning. In fact these freshmen's teas provided me with the most enjoyable part of my first year in Dorpat and there I collected that small degree of understanding of human nature which I've had to make do with ever since.

From the freshmen's tea one went straight to the pub and at 10 pm from there to one of the freshmen's rooms. Here also one took turn about so that the burden was quite evenly spread. As you can see, the organisation of our student society was quite communistic as far as the sharing of possessions, which were likely to be eaten by moths and rust, but there was in addition a strict hierarchy. The freshmen's duty of supplying beer and tobacco for the nightly binges was more than could be covered by the money available to most of the students. However, the merchants in the town were happy to give credit so that already in the first semester the society student built up a debt which he'd spend a long time thereafter paying off.

One's level in the hierarchy in the society was determined in part by how long one had been at the university, and in part by whether one belonged to the general membership or to the inner circle. By our statutes the only requirements for membership were that one be "Christian and matriculated". There were hardly any Jews in academic circles in Riga, and in any case the first point was not strictly observed and there was not yet systematic anti-Semitism at this time. The new student member was called a "Fechtbodist" and took part in all the social activities of the society. He was however not part of the governing inner circle and took no part in their discussions. He wore no badge of membership on his sash. After a probationary period of at least one term he could be proposed for admittance to the inner circle of the society. For this a two thirds majority of the inner circle members had to approve. Normally a senior member who knew the candidate well would propose him. A member of the inner circle—a so-called "Landsmann"—had the right to wear the coloured cap and sash, had the right to join the meetings of the inner circle and could be elected to society posts if he'd been a member for a certain minimal period.

The length of membership dictated all of our social intercourse. Since it took two to drink beer, the elder member filled the glass, which only held about a fifth of a litre, and said to the chosen younger one, "I'll mount before you", before drinking half the glass. Then he handed the rest to the younger one who said "Cheers" and emptied it in one pull. Only between good friends was it possible for the order to be reversed. If someone had said or done something which was disapproved of, then one of the older students would sentence him to empty a full glass. This quickly developed into a social occasion because the sentencing student would sing in a loud voice the appropriate sentencing song, the words of which were determined by the level in the hierarchy of the sentenced student. For freshmen the song went, "The freshman is in the dog's house, and so we laugh at him". The higher levels were known as "Young House", "Old House", and "Moss Covered House"; and the song, for example for an "Old House" would be "Our Old House has had bad luck, pull grey horse, pull". The younger students had no right to punish older ones but if necessary a younger student could protest loudly and if his case was justified then most likely an even older society member would sentence the one who had unfairly punished him to "Bad Luck". In this way the aggressiveness of young, cocky and frequently inebriated youths was kept in hand.

There was no requirement for sword duels though one was expected to regularly visit the fencing ring where one fought with blunt rapiers. Fighting duels with sharpened swords was only permitted in the most serious cases. These were seldom and permission for a duel would only be approved after an exhaustive hearing before an honour court on which the most respected members of the society sat.

The declaration by either of the parties that he disapproved of duelling in principle was sufficient to ensure that the court demand that the matter be settled by the offending party giving an oral apology. Duels with pistols scarcely happened and when they did they were usually held outside of Dorpat.

Science. As far as studying was concerned we freshmen were repeatedly told that learning to live in the spirit of the student society was the most important thing of all. One had to concentrate on this and therefore going to lectures was a corruptive distraction which was to be avoided. We could study later once we'd acquired in our "golden student society time" all the experiences which would serve us well in our later life as "philistines". I was not completely convinced of this, because I thought that all of life must have its joys for which pleasant memories were but poor surrogates, and in any case I wanted now to experience science at its source. However, to begin with, I let myself be led by the unanimous opinion of my peers. After all, coming from the narrow circle at home where I'd been practically cut off from wider conviviality, I was mightily impressed by the large and self-assured circle in which I now found myself.

The sons of the first families of the city—of the reigning mayor, of the councillors and of the church dignitaries—shared the same beer table with the children of merchants, artisans and other nameless people. An auspicious family background brought no particular advantage. Here the bearer of one of the great names of the city could be mercilessly teased for having had to sit down after losing the thread during a speech at a society meeting. Here the son of a minor liquor merchant could dominate his peers with his wit and ability to assert himself. Though the ruling upper crust of my home town kept itself socially strictly separate from the lower orders, a student, once he'd been accepted into the inner circle of the student society, was treated as a social equal. And if there was any special talent at the University then the student societies competed for them so as to always have the best brains with which to rejuvenate themselves.

My personal view. Although the secondary school I'd been at was looked down upon by the vast majority of the students, who'd gone to the government school specialising in classical languages, most of them listened without demur to my fierce defence of it. They must have known all about my diverse interests and activities because the older members of the societies treated the screening of prospective entrants as a serious business. For my part I let myself go and regarded democratic egalitarianism as a serious business. I found amused and sometimes sympathetic listeners for the many, often completely absurd bubbles of my imagination, but I soon found that I couldn't count on such a friendly reaction to my airing scientific topics. Among the hundred odd members of the society there were at that time only two chemistry students. One was just about finished, so that I scarcely got to know him, and the other, who was only slightly older than me, regarded chemistry as a bread and butter matter. Only later did I come across the only society member with an enthusiasm for science which matched mine. He was a mineralogist called Lagorio and I've remained on friendly terms with him ever since. Probably these circumstances contributed to the fact that with so many new experiences my old love of chemistry got put on the back burner for a while. I actually cannot remember whether I heard even a single lecture in my first semester. Probably I did once or twice but the experience certainly left no lasting impression.

My admission. And so my first term neared its end. Dorpat shared with the Scandinavian universities the division of the semesters such that they fell naturally into the rhythm of the year with the holidays being at Christmas and in the middle of summer. That's much more sensible than the German system in which the Easter holidays are too early in the year and, as a moveable feast, the date of Easter varies from year to year. The autumn holidays are far too long and are at a time of the year when one would rather be sitting inside. The consequence is the well known difference in the length and usefulness of the winter and summer semesters, the first of which is further broken up by the pause between Christmas and New Year. After the world war there was a good chance to correct this old mistake, but nobody thought to use it.

In the budding spring of 1872 there started the agitated times when the older society members told us that the deliberations about the admission of freshmen to the inner circle had begun. These discussions were taken very seriously and required several extended meetings. It was considered most tactless to ask where one stood and doing this could destroy a candidate's chances—and so everybody tried their best to avoid the subject. Once the decisions had been made (we weren't told when this had happened) the result was never directly given to the lucky winners and instead there now ensued a game of innuendo, involving trick questions and partial explanations all of which were a torture to the candidates. Since I now experienced all this for the first time, I was of course less worried than those older students who'd already experienced one or more disappointments. On top of that an older student who had nearly finished his studies but who'd only appeared in Dorpat near the end of the semester tried to become friendly towards to me. I'd avoided him because he seemed to me to be a pretty rude person with no higher interests. I treated his insinuating jokes coldly, thinking them merely an expression of his crude nature and had no intention of becoming his friend. Because of this I was completely astonished when he finally took my black cap from my head and replaced it with his coloured one to symbolise my admission to the inner circle. Only when my friend Seeck flung his arms around me with laughter and congratulations did I begin to understand what had happened. It turned out that I was the only one of the 14 freshmen who started with me that had been taken up. No one else got the honour, though almost all made it later. At the celebration which followed all my new brothers drank with me so that I collapsed on my bed filled with more beer than ever before in my life.

The young member. This was my first public success and for the first time I experienced the joy of being the only one of many to be honoured. In the course of the next few days there followed the ceremonial admission in front of the entire inner circle and my oath to always preserve the honour and the welfare of the society. Soon after that it was time to go home for the holidays. It was more than

5 months since we'd driven in the opposite direction through the bleak landscape with its endless snow fields and black pine forests, but now I was returning with completely different feelings to my home town through the first glory of the early summer. My mother greeted me with affectionate pride; my strict father softened his usual stern demeanour and let me accompany him to the convivial meetings of his hunting friends.

I must confess that I behaved quite ridiculously as I let a number of the older members of the student society invite me into the homes of some of my old dance lesson flames. I didn't really care what impression I made, for I had quite different things on my mind. I'd abandoned myself to the "golden student society life" and enjoyed it to the full.

The Letts retained an old heathen mid summer festival which, by long tradition, was celebrated in Riga with great vigour. Since the church couldn't suppress it, it had simply been converted into the feast of Saint John. On the evening before, there was the so-called "Herb Evening" which was a huge market to which the farmers brought marjoram, thyme and other herbs as well as traditional wicker ware made of rushes which were gifts for the children and were then hung in the hallway. Groups of Lett women and girls with wreaths in their hair went from house to house singing an old traditional song and dancing for a small gift. Late in the afternoon the whole of Riga went to the bank of the Duna to cavort in decorated boats on the broad peaceful river. Numerous choirs and other clubs sailed on larger vessels decorated in the most fanciful manner and they let their tunes ring out over the water. We students from the "Fraternitas Rigensis" had our own raft, on which we sang our student songs and gazed in delight at the numerous little boats filled with pretty girls in light dresses that surrounded us. Then came the sunset and the broad expanse of water turned to gold. The lanterns were lit and only slowly did the "herb evening" come to a close.

During the summer every citizen of Riga who could afford it, went to the "beach"—the flat coast of Riga's bay which is composed of the finest sand. Many of the well off had summer houses there, the others rented simple one storey wooden houses which the fishermen had built for this purpose. The beach dwellings started near the mouth of the Duna and, as the population and wealth of the city increased, so the houses kept pace and stretched in a long row for many kilometres to the south.

Beach life was immensely easy. The otherwise very strict rules of social etiquette were relaxed, the interactions of young people of both sexes were freer and less self-conscious and we students with our coloured caps were welcomed everywhere as the future hope of the country. When an old member of the student society met a young one he'd straight away invite him home and give him a place to stay if things had got late and the last steamer to Riga had left. This was important because social life in Riga was confined to peoples' houses. There was scarcely any public dancing, and in many of the larger resorts there wasn't even an inn which a lady could enter even if accompanied by a gentleman.

Settling in as a member of the inner circle. The summer holidays flew by. The autumn journey to Dorpat showed yet another side of the countryside. This time the odour of glue associated with flax-retting gave the characteristic smell to this time of the year. I now had before me my next task, which was to gain a deeper understanding of the nature and purpose of the Fraternitas. I went through the completely unsorted archive of the society so as to acquaint myself with the history of our organisation and tried hard to get in touch with as many old members as possible. The fiftieth anniversary of the society's inauguration was set for January 1873. A large number of older members from Riga, from the country and from the Empire were to be expected because a demonstration of support for the Fraternitas Rigensis was seen by the older members as a protest against the russification program, and thus as both a political and patriotic duty.

A huge hall constructed many years ago by a quixotic speculator was rented for the celebration. It had stood empty for many years and hence had fallen into disrepair. It now had to be properly decorated for the celebration. An older member of the society called Poelchau who lived in Riga and worked as a drawing teacher and artist took charge of overseeing this and asked if there were any artistically inclined students amongst the current members. It turned out that only Lagorio and I understood anything about using pencil and brush. We designed and painted some huge banners which depicted monumental figures of students at crucial places and stages of their academic life. When we were finished I had to acknowledge that Lagorios's landscapes were artistically much better than my student figures. However since the enormous earnestness with which I'd approached the work had automatically transferred itself to my stiff figures they were praised by master Poelchau and were also either praised or criticised by my peers who understood nothing of the matter.

The winter holidays were completely occupied with preparations for the celebration which was set for 23rd of January so that I only went home to sleep and saw little of my parents. I did however notice that my goings on had started to worry my father. I hadn't concealed from him that I'd found little time from my society activities for study and had instead described to him the many duties of a freshman newly accepted into the inner circle. The success that my induction into the inner circle of the prestigious student society implied calmed him down for a while and he hoped that at the end of the freshman year I'd get down to serious business. Now he saw that I was still busily involved in all sorts of things which had nothing to do with chemistry and so I tried to convince him of the importance of the upcoming celebration. Shaking his head he accepted all this but made it clear that he expected that once the celebration was over I'd settle down seriously to my studies.

The celebration was as magnificent as we'd expected. Hundreds of old members came back to Dorpat although the roads were in a terrible state and the coaches crowded. In Dorpat the various "castles", depending on their size, were expected to put up a certain number of the guests. I shared the "Carusburg" with two others who like me had made the transition into the inner circle and thus become "young houses". By the rules of our society we were considered people in good standing and had to put up several old members. For the duration of the celebrations the general use of the "Du" form was introduced and the wholehearted cordiality which these old members—many of whom now held important official positions—brought to their interactions with us younger members made a great impression on me and remains till today one of my pleasantest memories. In many cases they invited us to stay with the familiar "Du" form for the rest of our lives.

The waves of general enthusiasm rolled high because we Baltic Germans knew that dark days were ahead due to the ever wilder russification measures demanded by Moscow. Apart from a degree of provincialism we regarded our resistance as a form of cultural duty since the ousted German culture was to be replaced with something we all regarded as second rate. Because of this there was, behind the usual celebratory exuberance, a great deal of genuine feeling. The recent foundation of the German Empire had so provoked the Moscow pan-slavists that they at once instituted measures to suppress German influence in the Baltic provinces as fast as the lumbering bureaucracy of the Russian civil service would allow.

My university studies. The dismantling of the celebration decorations required yet more work and as one of the few artistic experts much of it fell to me. However my father's warning saw to it that I began to attend lectures and since I didn't have much joy with them I got hold of and read books on the subjects I had to study. What annoyed me about the lectures was that the speed of my thoughts was being controlled by the lecturer. I preferred to have the possibility of spending more time on this and running ahead on that. A book in contrast is patient. It waits until one has digested a thought and sets no limits on the speed of progress. My earlier experience had left me with a preference for books and this didn't change now. Throughout my studies I hardly ever regularly attended any lecture course.

In Dorpat there was a rather well developed system of academic grades. The normal course ended with the exams and a piece of practical scientific work which, however, did not need to be published. Whoever managed this was accorded the degree of "Candidate". Most students were happy with this degree which qualified them for the practical professions like those of teachers, advocates, judges and so on. Those who wished for a career in science had to continue for several semesters on a dissertation which had to satisfy higher academic standards. If it was accepted by the faculty then its accompanying theses had to be printed and publically defended in the university's central auditorium. This ended with the degree of a "Master" which was the precondition for employment as a lecturer. A further and even more critically scrutinised dissertation together with a public defence led to the title of "Doctor" which was the requirement to be considered for a full professorship. These were the old customs from Germany as one can see in Goethe's Faust, "I'm not just a Master, but a Doctor too". Only in the medical faculty was the first level considered sufficient for the title "doctor" and because of this the title "Dr. med." was considered to be worth much less than the doctorates of other faculties.

In my case the first problem was to pass the exams for a "Candidate". These covered fifteen subjects in all, and the exams were split into three parts each of five subjects. There was no restriction on how long one left between the three exams and the choice of subjects was more or less open though with some restrictions on the order in which they were taken. The first exam for chemistry students normally included mathematical subjects (but excluding higher maths), experimental chemistry, physics, mineralogy and crystallography. Some of these were completely new to me. I'd started in a leisurely way to study, but since the captivatingly sudden and beautiful northern spring had started I spent more time in the countryside than at my books. Soon enough it was time to go home for the holidays again.

At home my father asked me how far I'd got. I answered truthfully—but he was not satisfied. He pointed out that the official length of the course was 3 years and that I'd used up more than half of that without even starting to study seriously. In vain I argued that in reality nobody finished in six semesters and that eight or more were the norm. He replied that this was only because of the stupid way we wasted our time. One word led to another and I said that if it meant so much to him I'd sit the first third of the exams, which were held only at the start and end of each semester, at the start of the next semester. My father took me at my word, and so once again I'd set the "moral flywheel" in motion.

The summer holidays were almost 2 months long and so I was able to get started on making good my sudden promise. At home in the Alexander Street we had a large garden with a pleasant garden house and a shady veranda. There amidst birdsong and flowers I was able to work my way criss-cross through the subjects I needed—taking a rest from one to study the next. I also found that the difficulties of analytical geometry were best dealt with by only staying with it until I felt the first signs of that mental mill stone so well known to everyone preparing for an exam. At that point I'd take a day or two off, usually at the beach, after which I'd find that the things which had seemed so strange 2 days before were now clear and ordered in my mind. It doesn't pay to try to force understanding, because that only involves more work and leads to exhaustion which should be avoided. It's far better to let the mind work at its own digestion rate for in that way one gets ahead faster. It's like the experienced mountaineer who starts off with a slow measured tread that seems so ridiculous to the beginner, but in this way one saves one's breath for when it's really needed in the steep places. This practical philosophy is to be found in the Swiss greeting, "Zeit lossn", (Take your time).

The exam is passed. I went confidently back to Dorpat in the autumn, though my father just shook his head because he hadn't had the feeling that I'd really applied myself. I registered blithely for the exam and passed it with a little help from some of my professors who clearly showed me the holes in my swiftly acquired knowledge, without using them to fail me. I was so ashamed about this that despite having passed I immediately went back to my books to learn all I should have known. This was easy because now the door to the scientific paradise, for which I'd come to Dorpat and which I myself had left shut, was opened to me. This paradise was the chemistry lab. It was part of the very sensible regulations of the chemistry course that one could not start practical work until one had passed the exam in inorganic chemistry. Now I had fulfilled this requirement and so there was nothing to prevent me getting started. At the appointed time one morning, I presented myself together with the few other aspiring chemists at the laboratory to start real chemistry.

The chemical labs were in a wing of the university building and they consisted of a lecture theatre, three large rooms which contained the steam baths, the balance and the work bench for beginners and three smaller rooms which were for the professor, the library, and for the advanced students. In our room there was place for around twelve beginners and the two assistants also worked there. Four to six advanced students shared the smaller lab.

Compared to the huge labs one sees today these were very small and modest rooms. The same was true for the equipment. There was no gas in the Dorpat labs until after my studies there were finished. To heat things up or do flame tests we had to use petroleum lamps—either the simple ones or those designed by Berzelius. Oxidation of organic material was done as in Liebig's laboratory with coke. To get high temperatures for incandescence we used turpentine as fuel and an air blower designed by Deville. This apparatus made such a noise that we couldn't use it when the lecture theatre was being used. A single technician, a gloomy taciturn Estonian, was more than sufficient to keep the boiler for the steam baths working, to replace used chemicals, to clean the lamps, keep the lecture room in order and in winter to heat the rooms with the large tiled stoves and light the oil lamps.

Despite the scantiness of the facilities the education was excellent. This was to the credit both of the professor and of his assistants.

Karl Schmidt. When I think of my chemistry teacher Karl Schmidt, I'm overcome with a warm feeling for him for both as a person and as a teacher he gave me a great deal. He was born in Courland in 1822. To begin with he was going to be a pharmacist, but he showed such talent that he was helped to higher things by being provided with the means to study in Berlin, Giessen and Göttingen under Justus Liebig, Friederich Wöhler and H. Rose. In all of these places he outshone his peers. Wöhler wrote to Liebig on the 17th December 1844, "Dr. Karl Schmidt has completed a piece of work in physiological chemistry entitled: 'Characterisation of invertebrate animals'—he is indeed a really splendid and clever fellow—the best brain that we have here". And Liebig answered on March 14th 1845, "Tell Dr. Schmidt that I find his work excellent, but that it would in no way have suffered if he'd left out the high flown speculations at the end—. He is very competent and once he's a little older he'll see some of these things a bit differently. When we were young we also had such stuff in our heads"

The work they were discussing had to do with the chemical composition of the exo-skeleton of lower animals. Karl Schmidt had discovered that in Salpen—a relatively highly organised class of animals—the scaffold was made of cellulose which forms plant cell walls but is found nowhere else in the animal kingdom.

Because of this work and of the warm recommendation of his teacher Wöhler who wrote among other things, "He is better than all of us in the precision and certainty of his analyses", Schmidt was appointed first as an assistant professor and then later as a full professor in Dorpat. There he enthusiastically continued his research in physiological chemistry and published in rapid succession several monographs in this area each of which was a landmark.

For example he first introduced micro analysis, which only today is really coming into its own. In work he published with the physician Bidder "On the physiological chemistry of metabolism" he was the first to be able to use precise analysis to describe the steps in digestion in higher animals and man and he made the fundamental discovery of free hydrochloric acid in gastric juice. All in all an imposing corpus of work was produced in just a few years and this is all the more remarkable since the analytical methods available then were quite primitive and very laborious. In fact, in order to follow the daily metabolism of his experimental animals Schmidt worked 18–20 h a day for several months and, to save time, he took his meagre allotment of sleep on a mattress in the lab.

The Professor. When I came into his lab Karl Schmidt was just over fifty and he'd long given up his physiological studies. He'd dedicated himself in the second part of his career to determining the mineral content of seawater from across the whole world using a standard assay system. Just as he'd once investigated the metabolism of animals, now he wanted to define the metabolism of the earth's surface. Rain water slowly leached chemicals from the earth and at the same time promoted the chemical dissolution of rocks so that new solutes were generated and set free. In the streams and rivers these aqueous extracts were mixed and averaged. The world's oceans received these chemicals and concentrated them because evaporation over the sea transfers only water to the air.

The experimental work that this involved was very monotonous, since it always required just the same simple analysis steps, and yet Schmidt was tireless in his work. He was the first one in the lab in the morning and the last to leave at night. Later, as I got to know his family, his wife complained that his diligence was getting worse the older he got.

Karl Schmidt was a tall thin man with a small head, prominent nose, grey hair and a thin beard. He moved his long limbs with astonishing speed and elegance. His speech was loud and rapid; the students claimed he could say "Limonade gazeuse" ("fizzy lemonade") as a single syllable. In his lectures he tried to avoid overwhelming his audience with his flood of words by delivering short sentences at high speed—each sentence being followed by a long pause. In personal interactions he was kindliness itself and polite to everybody. Once I saw him when his youngest daughter, then around ten, together with a bevy of her school friends met him on the street. He raised his hat with a wide arm movement called out a friendly greeting and hopped from the pavement into the snow to make way for them. He was always available to his students, always obliging, and when he was scientifically interested he was prepared to make any sacrifice. I will have occasion later on to recount instances of his kindliness.

The assistant. Although he was always around we beginners naturally didn't have much to do with him. Our lab was right next to his office and the door was always open unless the professor had guests. The routine lessons, however, were in the hands of Johann Lemberg who was the chief assistant. He was a remarkable man to whom I owe more than anyone else for my scientific education.

Lemberg came from a poor and obscure background. It was said that he supported his aged mother, though since he never invited anybody home no one ever got to see her. His lifestyle and his clothes were the simplest possible, but he kept up appearances even if his coat was faded and his cap now formless. He was rather small, bent over and, when standing next to the professor, looked more like a servant than a co-worker. His face with its strong cheekbones and broad nose reminded one of the local peasantry. By the time I got to know him he'd lost most of his blond hair so that his widow's peak now extended from either side towards the centre of his head, leaving just a strip of hair in the middle sticking straight up and running from his forehead back to his neck. It looked for all the world like a cock's comb. His chin was covered with a "democrat's" beard.

As far as I know, Lemberg's entire scientific career had been spent in Dorpat as Karl Schmidt's assistant and his published work—and in this respect he was scarcely less busy than his teacher—also dealt with inorganic chemical processes on the earth's surface. While Schmidt concentrated on the final result of the degradation of minerals in the water leached from the ground, Lemberg studied the manifold chemical processes taking place in the minerals themselves during their degradation. His leader and model was the German researcher Gustav Bischoff who had been professor in Bonn and was the founder of chemical geology. Bischoff's fundamental publications were Lemberg's bible. The experimental side of Lemberg's work consisted also in countless analyses—on the one side of naturally occurring minerals from areas where transformations were taking place and on the other of experimental studies in which he followed the long term effects of exposing well characterised starting materials to the action of various solutions. The reliability and accuracy of his results were as good as those of his teacher. However his lab bench looked very different. While Schmidt's glassware all had to be clean and crystal clear, Lemberg's maxim was that glassware had to be completely clean on the inside where it came into contact with the chemical reactions involved in his analyses, but that the outer surface could be left covered with the grey coating that all glass developed in the laboratory atmosphere.

The instruction. The laboratory instruction that Lemberg gave was unusually good. Since we were only about half a dozen beginners each of us benefited a lot from him. His lab space was in the same room as ours but at a different table so that we could always ask him for advice when we needed it and at the same time he kept an eye on us and corrected every breach of the rules before we'd even committed them. To interpret his results he'd taken over from his role model Bischoff the concepts of atomic mass, stoichiometry and reaction kinetics which at that time—and even later—were completely unknown to most chemists. He taught us right from the beginning that there is no such thing as a completely insoluble material, no such thing as a complete reaction and nothing absolute in the entire universe. In this way he laid the foundation stone for my way of thinking about chemistry which made possible all my future research work.

In contrast to the other students who'd come from schools specialising in classics and hence had neither knowledge nor inclination and talent to practical work, I brought with me a little knowledge. I was so happy with the work that Lemberg soon noticed me. My enthusiasm for science prompted him to open up a little, for he was otherwise taciturn with the young students. For someone who didn't know him he seemed even to be grumpy.

Progress. Under his inspiring guidance I analysed the 120 samples of the qualitative course in no time at all. I also did all the extra samples he gave me and before the semester was over I was allowed to leave the work bench of the chemical babies, was promoted to quantitative analyses and given a special place by a window in the library. Here I was allowed to use the chemical balance. Although it was at least 30 years old and had obviously been through a lot I treated it with awe, for I considered this transition into the empire of measurements as a promotion of the highest importance for me. I was delighted to be able to learn all the little tricks which assured the accuracy of a measurement and also in this I managed to make my way up the usual steps in an unusually short time. The fact that my work place was in the library was another piece of luck—or maybe it was a deliberate good deed that the professor quietly did for his ambitious pupil. The books were trusted to the care and honesty of the people in the laboratory; they could be taken out overnight but they had to be back in their places the next morning. You didn't need to fill out a form for them and there wasn't anybody to act as librarian. Only if one wished to borrow a book for a longer period then one had to talk it over with the assistant first and agree when the book should be back in the library. Despite this I know of no book which disappeared during the many years of my work in the laboratory.

Literature studies. This ready access to the books of my science opened up a whole new world for me and I soon immersed myself in it. Already at school I'd read a lot—I'd once even read a three volume novel on a Sunday afternoon without getting mental indigestion. Now here I saw the paradise of my wishes opened up to me and I could enter it where and when I wanted. Guided by Lemberg's short but authoritative hints, my trembling hands seized the treasures. How I had to struggle at the beginning to take my first steps in this new world for everywhere I lacked the basic knowledge. However the two bookcases which contained our library—this was in 1873 and the chemical literature was just a tiny fraction of what it is today—held the solution and slowly I found out in which order it made most sense to read the books.

Earlier I'd already seen that in studying something it isn't necessary, as some school teachers insist, to build the new only on a foundation of things which have been completely understood. First of all it's impossible to do things this way because one can never completely understand an area of science which is by its very nature inexhaustible. In addition the new knowledge makes it easier to understand the old, so that it would be inexpedient to dispense with this shortcut.

To begin with I took this shortcut with a bad conscience, because the help it gave me seemed to have been obtained almost by fraud. At that time teaching seemed to be ruled to a large extent by the same dogma that applied to medicine: the worse the potion tastes the healthier it must be. One believed the more disagreeable something was for the pupils, the better it must be for them. Only much later was I able to justify my method to myself. This was long after my conscience, though not satisfied, had at least been calmed. This was the first practical foreshadow of the "Energy Imperative"—don't waste your energy, improve it!

Beginnings of the doctrine of chemical affinity. As I was being given ever more important chemical challenges so the lessons I got from Lemberg increased in depth. His study of chemical geology had forced him to adopt an outlook on chemical equilibria which was not dissimilar from the way we view things today even if at that time a simple mathematical description was still far in the future. This was because the silicates which he was almost exclusively using are just about the most unsuitable materials for this sort of analysis. Their associations are complex, their reactions slow and they have a tendency to form colloids. Because of this he had to relinquish the search for these laws of reaction but he fully understood its importance and he seized the opportunity to place the problem in younger hands. Since I was the only one of his pupils that he considered up to the task he spared no effort to explain the problem to me and to help me along the way towards a solution. His teaching ability was soon evident in the way he taught me to see the details of quantitative analysis through the prism of this general chemical standpoint. Analysts then—and also a good deal later—considered barium sulphate to be "absolutely insoluble". As a geologist he knew that the appearance of barium sulphate in the form of barite clearly implies its crystallisation from an aqueous solution and so he taught us to regard this compound as soluble and backed this up with convincing experiments. To get us used to thinking along these lines he held regular seminars in which I so readily volunteered answers that he had to ask me to give the others, who'd been happy to leave it all to me, a chance to answer as well.

I finished the analytical course one semester later and was promoted to the preparative course. A chemistry cookbook like those many which are now available did not exist at that time and Karl Schmidt assigned me the task of reading O. L. Erdmann's papers on the oxidation products of indigo and then to make the relevant products.

Organic chemistry. This foray into organic chemistry was a little unusual in the Dorpat lab. Some of Schmidt's pupils had quite some time previously done some organic chemistry in connection with his physiological work. However they'd given this up long ago and Schmidt, like his venerated teachers Liebig and Wöhler, regarded morosely and with little respect, the rapidly rising flood of publications on preparative organic chemistry which had unexpectedly resulted from Kekule's discovery of the tetra valency of carbon and especially from the six membered ring structure of benzene. Both Schmidt and Lemberg knew that their research was directed at large and important issues and were even more convinced that they should stick to them as they saw how in the shortest time their inorganic colleagues were replaced by organic chemists so that soon German chemistry professorships were almost entirely in the hands of organic chemists. For analytical and inorganic chemists an assistant professorship was good enough.

Naturally we pupils noticed our teacher's point of view. Once when quite by chance an elderly German chemist came to Dorpat and wanted—or for some reason had to—complete his half finished doctoral work there, we natives watched his attempts to separate the isomers of chlor-dinitrobenzene with derision and considered ourselves with our old fashioned analytical skills the better chemists.

I should add that Schmidt did not neglect the new developments in organic chemistry in his lectures but rather gave us, as was his duty, a clear and as far as possible complete picture of this rapidly developing field. For us the completely foreign way of thinking of modern research presented quite some problems and when I had to attend the usually very boring student society meetings I took my chemistry textbook with me to mug up formulae.

However I was impressed with the organic chemists' idea that the number of possible different isomers of a compound could be calculated in advance from combinatorial principles. Years later I sometimes had success in using this idea in various other areas of chemistry.

And the colour theory, the last product of my scientific output, depended to a large extent on the decisive application of a combinatorial way of thinking.

Indigo. And so I couldn't resist contriving structural formulae for the isatins which I synthesised using the methods described by Erdmann, and I hoped that I'd be able to use them to achieve a synthesis of indigo—a problem which was considered of the first magnitude at the time. I wrote an outline of my preliminary thoughts to A. von Beyer[1] who'd just completed his important research in this area. He was kind enough to answer me at length and to point out the various facts which were not commensurable with my concept. In my usual excessive way I'd assumed that I was right and had informed all my fellow students. When it turned out that it was all nonsense I didn't need to wait long for ridicule, because everyone had taken it for granted that a mere beginner could never discover anything important. The view of my countrymen was that one should devotedly assimilate all the scientific and artistic advances that were made, but that the generation of new ideas was a matter of such great difficulty that a normal person couldn't possibly aim so high. I, on the other hand, having studied the work of the giants in the field, was so accustomed to seeing scientific accomplishment that I fully expected to be able to contribute to it. This was made even more natural by the fact that I was daily a witness to the quiet matter of course way in which Lemberg pursued his research and that my friend Lagorio assured me that in geological circles it was considered to be of the highest value.

The immediate result of this first failed attempt was that at the beer table my opponent, should he wish to counter one of my usually rather far fetched assertions only needed to say the word "indigo" to close the matter to his and everybody else's satisfaction.

The biology of a researcher. One day while reading the literature I decided to look in the indexes to see how much my revered professor had published. It turned out

[1]Ostwald misspelled the name of A. von Baeyer.

that after his physiological period, during which he'd achieved truly unexpected things in a very short period, there had been a long pause which was only occasionally interrupted by short and not terribly earth shaking publications. There then followed the long series of water analyses which I referred to above together with analyses of earth samples. At the same time I heard remarks about the sudden end of Karl Schmidt's productive period. All these things made me think hard and this formed the basis of my later work on the biology of genius—especially of scientific genius.

My first scientific work. I'd applied to sit the second of the three parts of the exams to become a "Candidate" at the end of my sixth semester (end of 1874). My father was happy with this decision because he could now expect that I'd need just over a year to be ready for the third exam and thus complete my studies. In the laboratory once the preparative practical was completed there would be plenty of time for my dissertation. Lemberg had given me the challenge of chemical affinity as a thesis topic and I was afire to get started on it using some better model than his minerals offered. In looking at the various possibilities I decided on the partial degradation of bismuth trichloride into the insoluble oxychloride and free hydrochloric acid. This was my first independently planned project. I asked my professor whether this would be a suitable subject for my Candidate's thesis. He was a little surprised because, since I'd only done the first third of the exam, the question scarcely arose, yet he agreed that if properly carried out the project would surely suffice. I therefore carefully prepared pure bismuth chloride and distributed it into a set of flasks that contained increasing amounts of water and put all of them in the dark so that chemical equilibrium could be established. I didn't know how long this would take and Lemberg's experience with the very slow conversion of his silicates had made me rather wary. I analysed some of the samples after a few days and left the rest to sit over the approaching Christmas holidays.

The second exam. In the meantime the exam date was approaching. Since I had a splendid memory and since during the course of the semester I'd made summaries of the difficult parts of the text books—a marvellous way to fix the material in one's mind—I easily convinced myself that I'd have no difficulties.

I clearly remembered the cases of nervousness and breakdowns that I'd seen in those facing the school leaving exam in Riga who'd swotted to the last minute without sleep or rest. A cousin of mine who I'd particularly liked suffered a serious breakdown directly after her teachers' exams, and when I met her later this well built sturdy girl had become pitifully weak.

This experience protected me from any similar folly and I decided to stop studying 2 weeks before the exam date. When I announced this decision in the pub it was met with a wave of astonishment and protest and well intentioned older friends took me aside to warn me in a fatherly way against any such foolishness. No matter how highly devil-may-care flippancy was regarded, the strongest men tended to tremble like children in the face of an exam.

As I'd expected I passed the exam. Since scientific aspirations were not common amongst the Dorpat students every single ambitious student was soon known to the professors who viewed him with a friendly curiosity which made the exam easier by turning it almost into a cosy chat. The exam was rounded off by a written essay and I rapidly covered page after page with a description of the thermo-chemical studies of Julius Thomsen in Copenhagen which had caused quite a stir on their recent publication. The choice of this topic—which had chemical affinity as its central element—demonstrates how much the examiners had been prepared to accommodate my chemical interests which were centred on the question of chemical affinity. At some point the supervising dean had had enough. After all the others had gone to lunch he went too, kindly suggesting that I put my essay on his desk when I was finished. The next day I learned that I had easily passed the exam.

The moral flywheel. The happy outcome of the exam had, of course, to be celebrated in a binge with my friends. That evening we were all in a happy mood because my success was judged to be a distinction for the student society which laid great weight on examination results. As always there were a few there whose jealousy made them speak of "luck", "professorial partiality" and so on. Under the twin influences of success and alcohol I talked expansively and declared that it would be no trouble for me to sit the last part of the exam at the beginning of the next semester. Since there was only the short Christmas holiday of 1 month before that, my boasting was greeted with general derision. I insisted that it would be possible and so to punish me my principal opponent, who so far hadn't managed to do any of the exams, challenged me to a large bet of a hamper of champagne. I immediately accepted and this was witnessed by all the others.

When I woke the next morning with a hangover from the alcohol and tobacco smoke and remembered the bet I decided to get busy. I realised that I'd already prepared two of the subjects but hadn't been able to sit them yet because of the restrictions on the number of subjects permitted per exam and in the remaining subjects there wasn't really anything dauntingly difficult so that the insane idea, born in a moment of bravado, was probably actually not so bad. If I managed it then I was sure my father would let me have the money to stay at the university for a further two to three semesters. After all, he was already resigned to this. In that time I'd be free of the need to study for exams and free from duties for the student society. I'd be able to set out unhindered on the open sea of scientific research. I therefore decided to go on with this enterprise that nobody else took seriously and do whatever was needed to achieve it.

I packed the books I'd need and went home to be congratulated by my happy parents on having passed the second part of the exams. In addition, that Christmas my elder brother was getting married and for this I, as the poet in residence, had to do my bit, which meant that time was short since I also had to celebrate the exam results with my friends in Riga by sharing a glass or two. Nevertheless, I managed to get the preparations finished. The professors were a bit surprised when I applied to sit the final part of the exam, but since I'd done well in the previous subjects they let it be, and so in January 1875, having passed the exam, I could ex-matriculate. I'd been a student for six semesters and since I'd idled away the first three I'd only really studied for three semesters. Since they had been sufficient it seems fair to conclude that the one and a half year fallow period that I'd given the ploughed field of my mind, had been extraordinarily fruitful.

I never did get the basket of champagne, because the fellow I'd made the bet with claimed that my behaviour had been immodest. I didn't press the matter.

Now the preparations I'd made prior to the candidate exams paid off. I analysed the samples and found that the equilibrium hadn't changed which meant that it had been established already in the short period before the exam. I finished off the analyses and wrote the results up as a small paper which was judged to be good and was entered in the university files. A brief summary of it can be found in volume 120 of the Journal of Practical Chemistry (Journal für praktische Chemie) of 1875 under the title "On the chemical mass action of water".[2] This was the first of the numerous papers, which I published.

[2]Ostwald W (1875) J Prakt Chem 12:264–270.

Chapter 5
The Start of My Scientific Career

My studies in physics. My surprised and happy father gave me permission to spend some further semesters in Dorpat. This was particularly worthwhile for me because I'd sensed that my knowledge of physics, though enough to get me through the exams, was not sufficient for the research I planned to do. I therefore asked the professor of physics, Arthur von Öttingen, for permission to work in his institute and to this he readily agreed. The space available in the physics institute was the same as that in the chemistry institute on the floor below, but except for me and Öttingen's assistant Grönberg nobody worked in physics, so I got a room all to myself.

Arthur von Öttingen was the youngest of the three Dorpat professor brothers who I mentioned already. After his studies in Dorpat he had gone abroad and to begin with had studied for a few semesters in Berlin where he entered into the circle of those who made up the newly formed Physical Society. He was thus in close contact with Magnus, Kirchhoff, Helmholtz, Poggendorf, Kundt and so on. There he managed to demonstrate the residual negative charge remaining when a Leyden jar is discharged and this served as the crucial proof of the concept of oscillations in electrical discharge which at that time was still contentious. Later he went to Regnault's laboratory in Paris, where he picked up a love of the French way of life. After that he always dressed so that one might have taken for a small, wiry French Marquis.

A wild life as a student had left him with a ruined stomach and other problems which limited his physical capabilities but didn't appreciably affect his lively intellect. His interests went far beyond the limits of his profession. He was an accomplished musician and had been stimulated by Helmholtz's then recently published work "On the Sensation of Music" ("Die Lehre von den Tonempfindungen als physiologische Grundlage für die Theorie der Musik") to write a book about the theory of musical harmony which was written in a most

R.S. Jack and F. Scholz (eds.), *Wilhelm Ostwald*, Springer Biographies,
DOI 10.1007/978-3-319-46955-3_5

ingenious way but in such an abstruse style that in all probability the only two people on God's earth who had ever read it all the way through were himself and the typesetter. Even I, despite my interest in the subject, never managed to get through it all despite several attempts.

I now divided my time between further chemical studies and the practical classes in physics which were supervised by Öttingen. I also took part in the physics colloquia which he had copied from Magnus and instituted in Dorpat. In these the participants had to describe old or new publications in physics after which there was a discussion in which of course the professor played the central role. These meetings were very useful to me, because through them the enormous gaps in my knowledge and competence became clear to me—gaps which my early successes had led me all too easily to ignore. A couple of gross calculation errors which I made in public convinced me to undertake a rigorous inspection and extension of my mathematical armamentarium.

My mathematical studies. In the course of this it turned out that I wouldn't get far without calculus which had not been part of the chemistry curriculum. Since going to lectures would have taken up too much time I, as usual, got hold of a book and was lucky enough to stumble across the excellent text from Karl Snell[1] who was a mathematician in Jena. This was not the usual sort of dry textbook but instead went into epistemological and methodical questions which in a study of calculus push themselves to the forefront and demand attention. These were happy hours that I spent with this brilliant book which not only provided me with the major part of my modest mathematical abilities but also stimulated me to think for the first time about philosophical questions.

My position as an assistant. I'd scarcely settled down in the physics institute before something happened that would affect my whole life. Öttingen's assistant Grönberg was offered, and accepted, a professorship at the Polytechnic in Riga. He had to move there immediately and so the only assistantship in the physics institute was open. I didn't think I had any chance to get it, but I applied all the same. Öttingen was in two minds because just then an older Dorpat chemist, Wilhelm Schröder,[2] had returned from Germany where he'd studied for some time. Like Öttingen he was a member of the Livländer student society which, under the leadership of Adolf Harnack, who later became well known in Berlin as a church historian and science administrator, had been reoriented towards science. Schröder had followed the tradition of Karl Schmidt and studied physiological chemistry while I instead saw my future in physical chemistry. This, together with the fact that I had formally already completed my studies and had one publication to my name clinched it for me, though Öttingen told me that he'd personally rather have had Schröder. He

[1]Snell K (1846-1851) Einleitung in die Differential- und Integralrechnung. Brockhaus, Leipzig 1846-1851

[2]Most probably Ostwald is mistaken and meant Paul Woldemar Viktor von Schröder.

kindly invited me to lunch at his house and there gave me the good news. My hands were shaking with excitement so that I cut myself with the knife and had to hide my hand under the table while trying to stop the bleeding with my handkerchief. I'm afraid that the social incompetence which I'd brought with me to Dorpat hadn't been much softened by the 10 years of student life—quite the contrary.

However I still couldn't quite believe my luck. Here I was, just over twenty one, and now financially independent because, though the salary was little enough, it would suffice in an inexpensive place like Dorpat. But this was not so important because I could rely for a while on the support of my father, who was now well off and had no trouble supporting me, especially since he was so happy at his son's rapid rise.

The really important point was that I had now irrevocably started down the path of research and teaching and so I was no longer threatened with having to spend my days at some monotonous technical job. Now I could go to the lab as early as I liked and leave it as late as I wanted. The thought that I could now invest my whole energy in science drove me almost mad with happiness. Not the least of it was that I could now look forward to many years of daily contact with Öttingen for I now knew him well enough to realise how much he was able to widen my perceptions and sharpen my intellect.

My official duties were few: they were restricted to the preparation of the experiments which were an integral part of the lectures. I didn't find this burdensome because I was interested in the experiments and so I tried here and there to improve them. I was terribly ashamed of the occasional accident which was usually due to my rashness though Öttingen was kind enough to elegantly ignore them. In one of the "Prologs" with which the students opened their theatre performances, one of them sang:

> Ostwald tends his reputation
> And he works all night and day
> The odd thermometer gets broken
> But with a genius that's OK.

The start of an independent life. And so there began a happy and productive period for me. The meetings and binges with the senior members of the student society, who were mostly rather older than the new "philistine", as those who'd finished their university course were known, became few and far between because I found little sympathy there for what had become the centre of my life. In addition the meetings robbed me of precious time for what brought me the greatest joy. Only a certain sense of duty for the student society that had so rapidly admitted me, and to which I had briefly belonged, brought me now and then back into these circles. At the same time I developed a critical attitude to the ideals of the student societies which I, as a freshman from a simple background had accepted without question. I saw some of those whom I'd admired as great figures in our circle, undergo an alcohol induced physical and intellectual decay, and I found the arrogant self confidence which many paraded in the pubs and in the fencing bouts hard to square with their childish fear of exams. I even came to the heretical conclusion that it might have been better for me to have joined the "Livland" rather than the "Riga" fraternity, for "Livland" was much more supportive of emerging talents. The members of "Riga", on the other hand, had only contempt for a student society that they thought was too academic and insufficiently devil-may-care. At the fiftieth anniversary celebration of the Livland, which took place shortly before ours, the Livland "philistines" had collected an appreciable sum of money and the interest from this was used to finance travel grants so that talented members of the society could study elsewhere. We in "Riga" had, in contrast, spent so much on booze that we ended up with a massive deficit that ate up all the money that our "philistines" contributed.

Patriotism and science. On top of all that it turned out that my scientific ambitions were not approved of by the most influential and politically thoughtful members of our society. One of them, with whom I was on friendly terms, told me that the small circle of those who, behind the scenes, governed my home town thought that it was my duty to devote my special talents to the service of my country. Pure research, which was for the benefit of the whole world, was simply a loss for the city of my birth. I wasn't convinced and in any case was so happy with my pure research work, which was just about to bear fruit, that I wouldn't have given it up for anything.

New paths. On looking back at the problem of chemical affinity during long walks in the extended park of the Dorpat cathedral I'd tried to think of some means of actually measuring it. J. Thomsen had come the furthest by using a thermochemical approach by which he'd been able, in a few cases, to measure the distribution of a base between two acids. There was, however, no way for me to follow this up in Dorpat because I had no access to the necessary equipment.

In that over-long essay I'd written for my Candidate exam, I'd developed the idea that Thomsen's heat determinations were not in fact direct measurements of the chemical affinities but rather just a means of determining the chemical status of a dilute aqueous solution which was not accessible by other means. Like a stroke of lightning I realised that instead of determining the generation of heat, one could use any other property of the solution which was affected by the chemical reaction involved. If this were so, then why not use the most obvious property which I could measure easily and accurately—the density?

No sooner thought than done. I had practiced so much glass blowing that I could make myself a Sprengel pycnometer to measure density and didn't have to wait weeks for a manufactured one to arrive. To determine whether the approach would really work I decided to repeat Thomsen's experiments using my method. Since I only needed easily available chemicals, which Karl Schmidt readily gave me, I was soon ready to start. It only took three days to do the central experiments and these showed the expected results. I told Schmidt and Öttingen my results and both of them were enthusiastic and strongly advised me to publish. Writing up the results took a lot longer than doing the experiments largely because I felt a bit hesitant at publishing such easy and short experiments in the leading physics journal of

Germany. On Öttingen's advice I timidly sent the manuscript to Poggendorff who was editor of the "Annals of Physics" ("Annalen der Physik"). It was accepted and, after what seemed like an age to me, the paper was published.

With that my life acquired a new perspective. Applying the new method to as many suitable chemicals as possible, analysing the results in the context of the concept of chemical affinity and developing parallel methodologies to examine other physical properties—all of this provided so many experimental options that I saw a field opening up before me that was certainly no smaller than that on which my revered teacher Lemberg had so patiently and so successfully worked. I saw myself spending my life in the lab every day—including Sundays (because in this I had already copied Lemberg)—and there enjoying the ever new joys of research. It was what he had done over the last 10 years without ever getting tired of it.

Friends. My old friend Fritz Seeck had in the meantime disappeared from my circle. He was not well and didn't take proper care of himself. So long as he lived at home his mother kept an eye on him. However the family had gone for a while to Germany and Fritz was left alone in Riga in his bachelor flat under the supervision of relations from whom he wouldn't accept advice. He once visited me in winter—the snow was melting at the time—and my mother was horrified to see that the soles of his boots were hanging open and his feet were soaking wet from melting snow. He said he was used to it and didn't bother about it. From this he developed tuberculosis—which he also tried to ignore.

As a student in Dorpat he was even more reckless of his health; he was the one who lasted longest at the binges and was the most adventurous in the sledge races. All the time he bubbled over with wit and intelligence and was the very best of company. He survived a severe bout of pneumonia but the doctors told him that if he continued with this life style any longer in Dorpat he would die. So he travelled first to South Germany and later to Italy where he died soon after all alone in Rome.

The wrong attitude. Ignoring the simplest rules of good health was looked on as being devil-may-care. Taking care of ones health in contrast was viewed as being small minded and even cowardly. This was a practical consequence of the education of Baltic youth in the tradition of the humanities. The school I went to in Riga was for many years the only one of its kind, for the others all specialised in Latin and Greek and there the sciences played a small and scorned role. In these schools the first years of the syllabus required that natural history be taught but nobody took much notice of it. Chemistry wasn't taught at all. Physics was given for 2 h per week, but only in the final year and the course was usually taught by the maths teacher who tended to either screw up the experiments or simply left them out entirely. Because of this our educated youth was completely ignorant of all physical and physiological processes. In addition all practical matters were considered beneath contempt and the ideal attitude preached by those in authority was based on the pernicious ideology of the Platonists. From all this one can begin to understand how such a suicidal approach to life could become so widely accepted in academic life.

Fritz Seeck was by no means the only one I saw being destroyed in this manner. Wilhelm Schwartz, known as "Bitze" was one of the older members of the student society and the son of a distinguished patrician family in Riga. He was well regarded for his intelligence and knowledge. He was not only a sharp minded lawyer and captivating speaker but was also an above average poet—as we had all seen at the fiftieth anniversary celebrations. This first rate mind was housed in a small, frail and slightly crooked body. His face was dominated by shining brown eyes. He too had weak lungs and when I as a freshman first met him, his frequent fits of coughing should have urged him to caution. Instead he was always one of the happiest and most enduring drinkers and never took the slightest heed of his health. After a few semesters he had to leave Dorpat and soon after that he was dead.

Other friends. After Seeck left I became friends with a medical student called Arved Faehlmann who came from Reval. He was a quiet amiable man with a calm but keen mind. I lost him as well, not to alcohol of which he was no friend, but to an inherited disease which led to his death during his student years. Then I came close to an older student called Karl Henko. He was small but well proportioned, had a sweet face that every girl wanted to kiss, blond curly hair and an enchanting moustache. Though he was well aware of these advantages he wasn't vain in any way. Our relationship was based on music which we both loved. He introduced me to his family in Riga which apart from the parents consisted of his two brothers and a sister. His father was a merchant and he was the eldest child.

All of his family loved and played music. Karl, like his sister, was an accomplished pianist while one of his brothers played the violin and the youngest one the cello. I am indebted to them for introducing me to Beethoven. Of course I'd played some of his piano sonatas as best I could and at school I'd even copied out some that I couldn't afford but wanted to have. I'd also heard some of his music in concerts—the Coriolan overture made a strong impression on me. But when one Sunday the four of them played the beautiful fifth symphony, adapted for four handed piano, violin and cello, I experienced for the first time that irresistible depth of feeling that I found with no other composer.

I also lost this friend to alcohol. He'd devoted himself in Dorpat to the rakish student life to such an extent that he didn't manage to pass any exams. Finally his family took him back to Riga where he tried to accustom himself to a regulated life as a civil servant. It didn't work out; he became ever more careless of his appearance and his choice of drinks. Even before I returned to Riga I had news of his death.

The only one of my student friends that I didn't lose early was the mineralogist Alexander Lagorio whom I mentioned earlier. He'd arrived in Dorpat a year before me. His father, a landscape painter in southern Russia, had died early and his mother had then married a protestant minister in Kischenew who came from Riga. In this way he ended up in Dorpat and in the Fraternitas where he was soon admitted to the inner circle. Both because of his mixed background and because of his pronounced interest in science Lagorio was, like me later, rather opposed to the prevailing views of the Fraternitas. This discharged itself without too much friction in constant banter that he accepted with good humour just as he did the nickname "stallion of the steppes" which he'd been given because of his birthplace. He won an academic prize for a careful analysis of thin sections of Baltic rocks using the methods developed by Zirkel in Leipzig. The prologue-poet I mentioned before celebrated this with the verse

> Lagorio is seized by whims
> And he gallops in all directions.
> Livland is not enough for him
> It's Russia he's going to section.

At about the same time as me he was appointed to an assistantship in the mineralogy institute and since we shared interests and ambitions we rented a flat together. The poet turned out to have prophetic powers because Lagorio was soon offered a professorship in Warsaw where he did so well both in research and administration that he was then given a position in Petersburg. There he really made a contribution to both technical and craft education in the whole of the Russian empire. By the time of the revolution of 1917 he held a leading position as "State Secretary" and to avoid being murdered like so many others he had to drop everything and flee empty handed. In Finland he and his family scraped by in impoverished circumstances until he finally managed to get to Germany. There he devoted much time to the development of the theory of colours and he contributed significantly to it. Although he is older than me he is still in lusty good health and seems all set to outlive me.

Music. Throughout my time at Dorpat music provided a powerful antidote to the excess of nicotine and alcohol in the fraternity. During my first semester I came across four leather bound volumes in a dark corner of the student society library which to my surprise turned out to be the scores of the 83 string quartets of Hayden. I took them home and copied some of them in order to be able to study the structure of these splendid works. While I was involved in this I was surprised by the society's singing master (Magister cantandi). When he asked what on earth I was doing I explained what it was all about and he then sentenced me to a large drink in the pub because he thought I was just showing off. However this public punishment had a silver lining because others in our circle who were interested in music heard about it and got in touch with me. When I told them I could play the viola they told me that the society owned one. It turned out to be in good condition and had probably survived the risk of casual destruction by drunken students thanks to its enormously sturdy case. Two violinists and a cellist were soon found and in those evenings when "nothing much was going on" we met to play string quartets—sometimes till dawn.

For a while we were viewed with some suspicion because the building of any form of clique within the society was frowned upon. However when we said that we'd be happy to take part in the "Freshmen's Theatre" which took place once a semester, the string quartet evenings were deemed to be preparation for this and nobody objected anymore. For us four they were however a source of heartfelt joy.

The contrast between the wonderful tenderness of a Hayden adagio and the rough student life led us to enjoy the first to the depths and saved us from losing ourselves too deeply in the student life which destroyed so many of our peers.

In this way I managed to be a member of a string quartet for the 10 years of my time in Dorpat. The other musicians turned over as they left the university, but there was always replacements in the new intake of students and though we extended our repertoire to Mozart, Beethoven and others I did eventually get to learn all 83 of the Hayden quartets. What fascinated me about them was that he introduced more daring harmonic novelties in these quartets than I have come across in all the rest of his compositions which I have heard or read.

Scientific development. The friendly acceptance of my publication was due in no small measure to the fact that I'd given the new phenomenon a new name. I'd titled the work "volumetric-chemical studies" (Volum-chemische Studien) in a deliberate reference to Thomsen's word "thermo-chemical". In this matter I had ignored lots of the philological advice I'd been offered and I used this word in later work without coming across any objection from chemists. This encouraged me, and later when the situation seemed to demand it, I never hesitated to invent and use new words. Almost all of them were accepted without any fuss and this would certainly not have been the case if I'd shown uncertainty or sought help from the Society for the German Language. In the satirical magazine "Fliegende Blätter", where a lot of common sense wisdom is to be found, I came across a story that offers the theoretical basis of my point of view. A group of people visited a monastery and after being shown lots of remarkable things they were taken to the refectory. One of the group asked the monk in charge if smoking was permitted. "No. You can't smoke here" was the reply. When the guest then asked how come all the ash trays were full, he got the answer, "That's from all the people who didn't ask".

Yes, indeed.

I had a lot of free time between my official duties and used it to sort out the problems that always crop up when one starts in a new area. I started with conceptual problems because the density values I'd measured didn't give an absolutely clear picture. Whether it was a leftover I'd inherited from Lemberg's way of doing things or whether it was on my own impetus I can't any longer say, but this small clouding of the results gave me no rest till I'd cleared it up. This I achieved by using the volume rather than the density. I'd anticipated this result in the title of my publication "Volumetric-chemical studies" ("Volum-chemische Studien").

In addition to the conceptual work I also increased the physical accuracy of the measurements so that the error was reduced below the fifth decimal point. A young professor of mathematical physics who saw me struggling with seven digit logarithms wanted to make jokes about my unscientific decimal addiction using Bessel's remark that you can't more surely recognise mathematical illiteracy than in the excessive use of numbers behind the decimal point. I answered that I'd happily use six digit logs if he could find them for me but five digit tables would be an arbitrary reduction of the real accuracy of the measurements. This was certainly a fruit of Öttingen's teaching that one should always bear in mind the sources of error in one's measurements. I never had to use so many numbers behind the decimal point in my later career.

Thereafter I extended the range of acids and bases studied and carried out a wide analysis with a large number of data points that cost me several months hard work. Although this involved what might seem to be monotonous repetitions of the measurements I never got bored with the work. With each new combination new questions were thrown up. I awaited the answers in suspense—and then documented them with satisfaction. In this way I shifted the balance between the work input and the results, which had been so one-sided in the first round of experiments, to a more satisfactory level and was able to contentedly undertake each long days work.

My work style. Already at this point I'd established the style in which I carried out most of my later work and which my colleague and friend J. H. van't Hoff so subtly described. It consisted of choosing and testing a methodology with which an important question could be answered. Once this was achieved the methodology would then be applied to a large number of individual instances so as to generate a broad overview of the topic. In this way I managed perhaps to publish more carefully recorded numerical data than any of my colleagues with the possible exception of my pupil and friend P. Walden who was clearly ahead of me in this respect. Where my results have been independently checked they have been shown to be free of any grave errors. For this I thank my teachers Schmidt, Lemberg and Öttingen.

Progress. By means of a subsequent analysis of the thermal expansion of the solutions under study I was able to extend the determination of the approach to equilibrium to other temperatures and in this way gained a broader view of the operation of chemical affinity. It turned out that hydrochloric acid and nitric acid are equally strong, and this is independent of the base that they react with. I'd been greatly excited to see if this remained true over all temperatures from 0 °C to 60 °C. It took many individual experiments and weeks of work to get the answer, and then one afternoon I was able to start the calculations. For me it was like for Newton when he was able to test his theory of gravity using the new French goniometer. As is well known the significance of the moment was so enormous that he didn't feel able to finish the calculations himself and had to ask a friend to do it for him. I was young and strong enough to subdue my excitement sufficiently to continue the calculations myself and in any case I had no helpful friend at hand. Eventually the mass of logarithms yielded the final results which had to be identical within the error limits of the measurements when my hypothesis was correct. The values were the same. My heart leapt into my mouth and I experienced for the first time the joy of discovery. This time it was not a case of repeating with different materials something that somebody else had thought of but rather a new and independent question which, if it turned out as expected, would open up many further possibilities. The thrust into unknown territory had ended in victory. With that I felt myself dedicated to pure fundamental research for all time and I also felt that this work was my badge of scientific maturity.

My master's degree. It was only natural that I now started to think about fulfilling the formal requirements for a scientific career. Both Schmidt and Öttingen assured me that the experiments described above were quite sufficient to satisfy the deliberately high standards set by the faculty for a master's dissertation. I therefore carefully wrote the results up as a single piece of work and went through the official steps to apply for the degree. This was in 1876, about 2 years after I'd got the degree of candidate. I was 23 years old.

The degree involved an oral examination which however covered only a few of the main subjects and whose purpose was to establish the candidate's maturity and capacity for independent work. This took the form of a pleasant conversation. Karl Schmidt as representative of my main subject had the right to the first question and asked, "Were you to write a chemistry textbook—which I hope you won't—how would you start?" My goodness ! Perhaps he had vaguely guessed how much I'd disappoint his hope for I sensed immediately that writing textbooks was going to be part of my life.

The last requirement for the master's degree consisted of a defence of the dissertation which had to include the presentation of five theses from other areas of science. In the main auditorium there was a large double lectern in the upper part of which the dean throned while the candidate sweated it out in the lower part. The three official examiners, who the candidate had chosen from among the professors, engaged him in a serious discussion after which the floor was open to the audience. Since these thesis defences were at that time not common in the faculty, they hadn't yet degraded into empty formalism but instead were regarded as being of considerable significance for the university. My examination lasted 2 h instead of the usual one. I've forgotten the details so I assume that I basked still in general goodwill and that everybody was anxious to do his bit to support a promising young man and that the sort of envy, which is always around, was reigned in. It was not always to be so.

The lecturer. At this time I was in the seventh heaven and everything was going well. By passing the masters exam I now had the rights and privileges of a lecturer. That had two consequences. First, I had to announce and prepare a lecture course for the coming semester. Second, from now on I was regularly invited to the lecturers' evenings and this had a significant influence on my scientific development. These evenings took place in the house of one of the older members who had reasonable accommodation, and there lecturers from the physics, mathematics and medical faculties came together to discuss their research results with an educated audience. We met once a month and listened to the presentation which was followed by a discussion and then we sat down to a simple meal which was such that even the poorest members could easily afford it. After the meal we stayed on for a drink afterwards and then the discussion became more relaxed and one got to know the people better. The tone of these discussions was free of any official stiffness because the academic titles and dignities—which in Germany was "Confidential Government Advisor" and in Austria "Court Advisor" and for us "State Advisor" or "Excellency"—were awarded by the government in Petersburg and so we despised them completely and never used them to address our colleagues. They were only used by the janitors and others at lower levels of the university administration. Even the students were asked to restrict themselves to "Professor" as a form of address. On these lecturers' evenings I presented my research results piece by piece as I got them and our host began to greet me with the words "What have you got for us this evening". I learnt a lot from the discussions which followed.

I enjoyed these meetings a lot. A few years before my personal circle had consisted of no more than my family and had only occasionally been extended by interactions with the families of school friends. My student years had catapulted me into the far bigger circle of the fraternity, however I'd only occasionally had interactions with other students. Now I was "pares inter pares"—an equal member of the scientific circle of the leading scientists of the country some of whom had a European reputation. The surgeon Ernst von Bergmann taught in Dorpat at that time before he left to pursue his career so successfully in Berlin. Then there was the physiologist Alexander Schmidt who worked on blood coagulation and who was distinguished from the chemist Schmidt by being referred to as "Blood Schmidt". Karl Schmidt and Öttingen have to be mentioned here as well.

From the many meetings over the course of the 6 years I was there one stays prominently in my memory, though not for any scientific reasons. We'd gathered at the house of the young astronomer Backlund, who afterwards made a name for himself at the observatory in Pulkowa and those attending this evening meeting were mostly mathematicians. Mathematics was represented in Dorpat by the two aging professors Helming and Minding. They were quite different personalities for while Helming was a good teacher but not of any significance in research, Minding was a brilliant and distinguished mathematician. Apart from that Helming was a well liked man who was always smiling. He was of middle height with a round bald head, round cheeks and eyes, round spectacles and was what was known in student circles as a "crake"—someone who liked his drink and who was not at his intellectual best when he sat in the pub. He regularly presided at the ceremonial drinking sessions with which the fraternities opened and closed the semesters. He opened the meeting by taking a glass of schnapps, drinking it down in one gulp and then saying in a satisfied tone "Drink a schnapps and you become a completely different person", and then after a short pause added, "And why shouldn't that different person also have one?" and poured himself a second schnapps. This happened with the regularity of an eclipse so that the new freshmen crowded into witness the famous ritual.

Minding was a completely different person. He was small and thin with a red face and snow white hair, the face framed by an enormous wing collar whose tips were as sharp as his nose. Reserved in demeanour, he seemed to be almost as abstract as the science which was his only interest. He was in essence a good hearted and helpful person who just had some difficulties coming out of his shell.

On this evening Backlund, who was from Sweden, had got ready a large bowl of Swedish punch which was tasted right from the start by the two old gentlemen and seemed to be very much to their taste. After the lecture and discussion Helming was in top form and didn't hesitate to tell even the oldest story, while Minding overpowered by the unaccustomed drink was dozing quietly. Helming shouted "Now for something mathematical" and at these words Minding opened his eyes, but as soon as he saw that it was Helming talking, he closed them again. Helming continued, "There was a ship which was eighty five feet long and twenty one feet wide: how old was the captain?" This was met with a storm of laughter and the old fellow looked quite astonished, for he hadn't reckoned on such a success with such an old joke.

Harmonic dualism. My dedication to scientific research didn't stop me from continuing my love affair with music. To be honest it was the sort of unhappy affair that one meets in novels—I loved her, but she didn't love me. I couldn't hide the fact that in the string quartet I was the best informed at the theoretical level but technically the worst player and that my performance did not noticeably improve with time even when I found the time to practice daily the exercises in a viola tutor. My sense of hearing, which is so important for this instrument, was simply not acute enough.

To begin with the theorist found a rich area of employment. I've already related how Öttingen had written in his youth a book called "The dual system of harmony" in which he'd more hidden than explained his fundamental discoveries. After I'd been in daily contact with him as his assistant, he learnt of my love of music, he lent me the book and explained the passages that I hadn't really understood. His thoughts on the subject seemed to me to be an expression of a pure ideal science and I immersed myself in it. His ideas on harmony seemed to be a great advance on the conventional views which were nothing more than a collection of rules whose scientific basis was thin to say the least. Some of these rules were based on experience while others were merely arbitrary conventions.

Öttingen's teaching showed in contrast that the important rules are all applications of a few simple and obvious musical axioms.

Before me nobody had shown any interest in the book and this spurred him on to fresh activity. I hadn't tried to conceal how difficult the text was for the reader and 15 years after completing the first version he himself saw some of the relationships more clearly. He therefore decided to offer a public lecture course on his views of harmony to which, given the wide interest in music in Dorpat, many interested listeners—largely female—attended. He asked me to put together a set of summary diagrams to illustrate the most important relationships. Apart from the normal lectures, which were part of my official duties, this was one of the few occasions where he asked for my help. Years later when we met again in Leipzig he still had these posters.

The orchestra. Öttingen also involved himself with music in a practical way by founding an academic orchestra which met every Sunday in the main auditorium. The central cadre was provided by the little town band of Dorpat which performed at official occasions such as funerals. This band was expanded by a quite large number of music loving professors and students. Öttingen was a good conductor and was able to make the best of the rather mediocre material. He managed, for example, to present a good performance of Beethoven's third symphony; only the difficult passages for French horns in the trio of the third movement didn't always succeed.

Since there was no bassoon player around he got hold of the instrument and a book on how to play it and then told me to learn it by myself. I was happy to do so and practiced in the afternoons when no living soul but me was in the physics institute and the chemistry institute on the floor below where Schmidt and Lemberg quietly worked was also deserted. Soon afterwards Schmidt asked me if I'd abandoned my previous work to study acoustics since day in day out he now heard the strangest howling sounds. I explained it all to him, but he just shook his head and said I should stick to chemistry.

Still, I did manage to conquer the basics of this strange and difficult instrument sufficiently for Öttingen to let me play it in the orchestra. However I once ruined the beautiful wind chord just before the Allegro con brio in the Egmont overture and after that whenever it really mattered he usually said to me, "Why don't you let the cello play that".

The start of a text book. Once I had my master's degree I was expected to lecture at the university. I announced a weekly 2 h lecture course on physical chemistry and started my preparations.

The name of the subject already existed and was well known because of the title of the oldest textbook in this area—"Physical and Theoretical Chemistry" which was published in 1857 by Buff, Kopp and Zamminer as an introductory chapter to Otto´s German edition of Grahams "Textbook of Chemistry". However the part that had to do with "physical" was nothing more than a simple text on physics and crystallography for chemistry students. The part which had been written by Hermann Kopp, which covers what we today would call "physical chemistry", was called "theoretical chemistry". It is 287 pages long and deals with what we would today call stoichiometry and chemical affinity together with questions of chemical structure which would later be regarded as the central issue in organic chemistry.

Apart from these solid and well written texts, that were at this time already 20 years old and hence out of date, there was no contemporary summary of the subject because the first text describing general and physical chemistry was published only later in 1877 by A. Naumann as a volume in the compendium of Gmelin-Kraut.

I therefore had to put the material for my lecture together myself and so I decided to work through every relevant book that I could get hold of. This was possible because I was at home both in the chemistry and in the physics institutes and these had a complete set of all the necessary journals on the subjects. I looked through the indexes of the possible journals, made a note of the titles of the articles I needed, and after a few months of constant work I had a card index in my hands with which I could organise the material and which gave me a reasonable overview of the published material in the area I wanted to cover.

Naturally, like every young lecturer, I began to outline the lectures on paper. But I soon stopped doing that because it was so slow and I was confident that I'd easily be able to formulate things once I started to speak. For me it was much more important and also more interesting to establish what the lectures were going to cover. And so I went back to my card index and then I used little strips of paper to mark articles of interest, which allowed me to find the relevant literature easily. Then I started to read, more or less haphazardly, trusting my excellent memory to retain everything that was of interest. And that wonderful mental mechanism in the scientific mind which builds connections between related issues saw to it that, though I read my way higgledy-piggledy through the topics, the information nevertheless gelled into distinct areas. And so I started off with my first lecture with a description of the arrangement and organic structure of the subject which I, and others too, regard as one of my most important contributions to my scientific discipline.

The half dozen listeners who held out to the end were the external witnesses of my success as a teacher. The inner witness was my certainty that I had a good overview of the entire subject and that I could now delve deeper into the various areas at my leisure.

My doctorate. Since there was now no doubt that that my future lay in academia my next step was to complete the last precondition for this by satisfying the requirements for a doctorate. The extension of my volumetric-chemical work and some analyses of the refractive properties of the liquids provided me with the necessary material. Given that this was for a higher degree the submitted work had to be more substantial than the Master thesis. It contained more than 600 precise analyses of density and refractive index and, as its main result, a table giving the quantitative chemical affinities of all the acids I'd examined—this was twelve in all. Since this was the first quantitative table of its sort and since some of the laws which underlay these values could be discerned here it has a certain historical significance. At the close of 1878, 1 year after my Master exam this second disputation took place and I was able to celebrate my elevation to the status of Doctor of Chemistry. It was with a degree of sadness that I saw the possibility of my sitting any further exams disappear. Ever since I'd left school I'd more and more enjoyed these occasions which most others so feared. The fact that I had now taken care of all the preconditions for the shaping of my fortune was now more important to me than I had supposed possible even a short time before. However the importance had not primarily to do with my scientific development.

Analysis of works of art. Öttingen's lectures on his harmonic dualism had aroused wide interest and he was asked to hold special courses for those who wished to get to know the subject better. I had already applied my usual approach of organised massive effort in this case to, among other things, the analysis of harmony in all of Beethoven's piano sonatas. I'd discussed dubious cases with my teacher and we'd sometimes heatedly discussed the interpretation—and not always come to an agreed conclusion even though in most cases the two independent analyses agreed. I am indebted to these discussions for giving me a superb insight into these treasures, one which I could otherwise never have achieved.

A second important result of these discussions was the recognition that a much larger part of the work of even the greatest and most passionate artist is amenable to scientific analysis than had hitherto been believed. Up till then the majority of the representatives of the scientific analysis of art (or at least of those who passed themselves off as such) tended to support the notion that art was not amenable to quantitative analysis and could only be damaged by exposure to the methods of scientific analysis. This is certainly true up till now because the analysis of works of art has almost entirely remained at the primitive state of art history. One brief glance at architecture and art in the second half of the nineteenth century with its emphasis on a circus of "historical" styles should be enough to make this point. But the fact that this impairment is restricted to these areas and does not extend to music shows us the origins of the problem. Music is in scientific terms much further advanced than the other fine arts and architecture. Because of the enormous advances in the understanding of the rules of harmony, structure, counterpoint, acoustics and so on, the study of the history of music has faded into the background and so it can't exert the same pernicious effects that historical analysis has brought to other areas. Here one can see a similarity to the natural sciences where, once again, the history of a subject is only of relatively little importance in comparison to its factual content. One often hears this criticised as being due to the scientist's lack of idealism but this is completely wrong. When one thinks of the idealism involved in the enormous mental and physical effort—sometimes even in the face of real dangers—that scientists are prepared to accept to reach meaningful results, one should not doubt that they would be more than happy to accept the smaller challenges involved in historical research so long as they had the feeling that something useful would emerge from it. In areas where the extent of scientific accomplishment is low, one usually finds that the contribution of history is highest. This is particularly obvious in politics where a contribution from rational scientific thought is particularly missing.

I count it a piece of great good luck that due to Öttingen's influence I was brought into contact with the scientific side of music, which is the most scientific of all art forms. I have already described how I studied harmony and counterpoint even before my student years and some of the song arrangements I did then show that I was able to apply this knowledge. During my student years I also wrote an erotic sinfonia version of a well known student song with a very frivolous content which was performed to stormy applause. But it was only after Öttingen had made me aware of the unsolved problems in this area that I could make the large step from a pupil-like acceptance of conventional rules to an understanding of their basis, and in this way begin to understand the enormous implications of science for art.

I've already mentioned that all the seemingly aimless skylarking of my youth later turned out to provide useful contributions to the formation of those mental constructs that I feel sure were the greatest challenges of my life. That the odd collection of fragments I instinctively gathered in my youth could later be put to such serious use suggests that even then I didn't regard them just as playthings but had rather viewed them seriously and as scientifically as it was possible to do at that age.

Chapter 6
Teaching and Marriage

Art lectures and their consequences. At some point Öttingen found that he didn't have the time to carry out all his manifold teaching duties and so he asked me to take over some of the courses he'd been asked to give.

Those most thirsty for knowledge were mostly women, both young and old, who gathered in small groups and studied the new courses with great dedication. In this way I gained access to a number of households in which a rich mental life was not restricted to music, though this was often in the foreground. I learnt the appeal of charming and cheerful company and found this a welcome occasional distraction from my ever more intensive scientific work.

Mostly I was in the household of Dr. Gustav Reyher who lectured occasionally at the university though his main profession was that of a doctor with a large practice and for this, as a good looking man with a rosy face, black beard, curly hair and a friendly winning way, he was well suited. His wife in contrast was grotesquely ugly; small, short legged with a face like the former Julia Pastrana[1] (without the beard) and every stranger was astounded at the sight of her. However this first impression soon faded because she was not only an extremely kind and goodhearted person but was also highly educated and receptive to the interests of those she met. She was well versed in music and poetry and kept abreast of new developments. Some of the many artists on their way to Petersburg loved to stop in this enthusiastic town of music by the banks of the Embach, and she often put many of them up in her hospitable house. The concerts they gave livened up the otherwise rather humdrum life in our provincial town and brought us a breath of the big wide world outside to which we felt that we culturally belonged.

Because of this the Reyher house was a centre of music and the students who met there were musically educated. They were known as the "house children"—the couple's own children were still small. Sometimes there were concerts in the house at which I could contribute with my viola. However when I once tried to play a

[1]A Mexican woman (1834–1860) suffering from hypotrichosis who was exhibited in freak shows world wide. She died on an exhibition tour in Moscow.

R.S. Jack and F. Scholz (eds.), *Wilhelm Ostwald*, Springer Biographies,
DOI 10.1007/978-3-319-46955-3_6

Chopin nocturne (Opus 9, Nr 2 in E minor) in an arrangement for bassoon with piano accompaniment I didn't get further than the first few bars before the unusual sound led to an irresistible activation of my audience's laughter muscles.

I was in the meantime quite accustomed to this sort of reaction to things I said or did and I didn't mind because I considered myself second to none in our circle and it had long been a source of satisfaction to me to show my antagonism to everything that was generally accepted without having being thought about. Quite naturally laughter was the response to my absurdities which I took as an acknowledgement that I'd managed to get my point of view across.

However in this case I was sensitive to the laughter. Partly this was because it showed that my superiority did not extend to the field of music and that I would have to accept the criticism with good grace. The main reason, however, was that someone was present in whose eyes I did not wish to seem ridiculous.

My wife. For some months Reyher's niece Helene von Reyher, had been living in the house. Like me she was from Riga. Her brother, Carl von Reyher, was a somewhat older member of the same student society as me and, though still young, had made himself a considerable reputation as a surgeon. He'd been the favourite pupil of Ernst von Bergmann and went on to make a brilliant career in Petersburg which was sadly brought to a sudden end by his early death. We'd got to know each other through our shared interest in music. Professor Reyher's wife was one of those who wanted to know more about Öttingen's harmony theory and she brought together a small circle of similarly minded people. I was the teacher of this group. And so I had a chance to see the aunt and niece together and couldn't stop being astounded by the contrast in their appearance. Occasionally the aunt was absent because of one or another of her countless other commitments and for that I was very glad. I thanked fate and the aunt that these absences became more frequent as time went by. As everybody else had long foreseen—I think I was the last to grasp the point—this led to a betrothal at the end of April 1879 and I now experienced the full blast of the storms of emotion by which the human species assures its propagation.

Household worries. My betrothed's parents—her father was an official in the offices of the ruling nobility—learnt about me from their Dorpat cousins and gave their consent. However there now emerged the question of how the new household was to be financed. My earnings were sufficient for my simple needs but would not sustain a family. The next possibility was a position as a paid lecturer. The government made available to the university salaries for around a dozen lecturers who were to be made available by the university council to full professors should the need arise. Just recently one of these positions had become free and I hoped to get it. However I was disappointed in this. Maybe I'd have got it if Öttingen had supported me as strongly as Carl Schmidt had. However, as he soon told me, he viewed my engagement as a serious mistake which would destroy all my chances of a scientific career. He'd concluded that, in my best interests, he should avoid doing anything that would make a rapid marriage possible.

At this time I also suffered a disappointment in another area. The university had a substantial endowment—The Heimbürger Trust—from whose income an annual prize for excellent scientific work was awarded. In addition, it provided scholarships to support a period of work and study in another university. Since the number of possible candidates wasn't that great I'd really hoped to get a scholarship with which I could spend a year "in Europe", as Carl Schmidt expressed it, where I'd be exposed to a rather different scientific atmosphere from that in Dorpat. The only other serious candidate was the highly talented physiologist Gustav Bunge. As the elder he was the stronger candidate, however, since he'd already had a travel stipend from the Livland student society with which he'd spent an extended time studying in Germany, I'd hoped that I, for whom no such student society grant was possible, might get the Heimbürger grant as a sort of compensation. As it turned out the travel grant was awarded to Bunge. My friends thought that this was really unjust, though I must admit it didn't trouble me greatly. I had uncovered in my own work so many unsolved problems that the flood of new challenges that were bound to follow from a period abroad in Germany—I'd not have considered any other country—would have been more of a hindrance than a help.

Later I came to see the decision as being probably right. Dorpat had gained such a reputation in the field of physiological chemistry through the extraordinary early work of Carl Schmidt and later from that of his namesake Alexander "Blood Schmidt" that it was considered to be an absolute peak of natural science by those professors from other areas, among whom there were many who did no research at all. In contrast my area of physical chemistry was not highly regarded either in Dorpat or indeed anywhere else. The only full professorship in the subject was Gustav Wiedemann's in Leipzig and the reputation of this first rate scientist lay more in the direction of physics to which he later returned. And Lemberg, who was more a geologist, was known for his radical political views so that his association with me did nothing to improve my candidature in the academic circles that decided such matters. Even I was not considered quite reliable in these matters.

From all of this it was clear that the conclave of professors would hardly give their support to the academic parvenu and I can't really hold their decision against them.

When I look back 50 years later and see the effect that this decision had on my personal fate and on my ability to make contributions to the development of science then I must count it a tremendous blessing—though of course I had a very different view of it at the time. If I'd got the scholarship and gone to Germany I'd almost certainly have fallen under the sway of one of the leading chemists of the time such as Baeyer and since all of these were without exception organic chemists—an area which at that time at the beginning of the 1880s was developing rapidly—I'd certainly have become an organic chemist who'd have regarded his youthful foray into the doctrine of chemical affinity as an aberration to be forgotten as soon as possible. After all, the same thing had happened to Baeyer who as a young man was a student of Bunsen where he published a (not very good) paper about "idio-chemical induction" after which he left the field for ever though, as he later told me, he had at that time much more interest in mathematical and physical questions than in chemical ones. His was by no means the only case like this because the attraction of organic chemistry at that time was very strong and I'd certainly have fallen under its sway if I'd gone to Germany then. I don't for a moment doubt that in this area I'd have done above average work, but I'd just have been one of many and would not have been able to give the whole field a new direction. When one thinks that those excellent colleagues that I later met in this field, though talented scientists, lacked any special bent for organisation, so one comes to the strange conclusion that the travel grant committee by turning me down actually had more profound and positive consequences for the development of science than either they or I could have dreamed of at the time. Looking back I think that also at the personal level I have every reason to be satisfied.

Wishes and hopes. However, all of that lay deep in the mists of the future and so my immediate problem was to get sufficient income to settle down, since neither I nor my fiancé was interested in submitting to the scourge of a long engagement.

I'd already successfully given private lessons, but this was not a steady source of income. The kind efforts of my teacher, Karl Schmidt, to find me a well paid secondary appointment which would leave me enough time for research did not work out. Then the Professor of Agricultural Chemistry and director of the public chemical analysis laboratory at the Riga Polytechnic[2] offered me a position as an assistant in his institute.

The salary was sufficient to finance a modest household, but my time would have been wholly taken up with the repetitive daily work of paid analyses. On top of that the director had made a number of scientific mistakes which would have reflected negatively on any work I did under his direction in the institute. The following story—even if apocryphal—will give an idea of his reputation in academic circles. A bottle was sent to the institute so that its contents could be analysed. The director himself (as he had no assistant at this time) read on the label the word WINE. His report read: "Specific gravity 1.02, alcohol content less than 1 %, sugar 0.5 %, traces of acid, the taste is insipid. It is astonishing that such a concoction which tastes nothing like wine can nevertheless be sold as wine". The astonished customer wrote back that the honoured director had seemingly read the label too quickly because the first two letters, which he had misread as a W, had in fact been UR.

I had little difficulty in persuading my fiancé, who was as interested in my scientific career as I was, that our long desired home would be bought too dearly with this offer. And so we agreed to keep on hoping and wait for a better offer.

I felt it was legitimate to wait for, after all, my publications had excited interest amongst the small circle of researchers working in this area. This circle was international because since only a small number of people in any one country worked on this rather peripheral subject it was essential to be in contact with one's colleagues in other countries. This was particularly the case for England and America. In France one had always believed that chemistry was a French science and so no one there was interested in what people outside France or, which is more

[2]Tactfully, without naming the person, Ostwald refers here to Georg Thoms.

or less the same thing, outside Paris did. Thus it took 10 years after the great discovery of the conservation of energy by Julius Robert Mayer and long after the work in this area in England and Germany by Helmholtz, Clausius and W. Thomson[3] were underway before this idea began to appear in the French scientific literature. It was much the same story with the development of physical chemistry.

A ray of light. At that time M.M. Pattison Muir was working as a fellow of Gonville and Caius College in Cambridge. He was one of that little band of researchers interested in the problem of chemical affinity in which, following the failures of the Swede Bergmann and the Frenchman Berthollet at the turn of the eighteenth to nineteenth century, there had been little progress.

He'd read my publications together with the groundbreaking work of Guldberg and Waage and wrote a wonderfully clear review which he published in the leading English scientific journal (Philosophical Magazine, September 1879, pages 181–203). In the introduction he emphasised that these were the first papers to make a significant advance since Berthollets work in 1803. The last sentence of his summary read: Ostwald has given chemistry a new procedure with which to solve some of the most difficult problems; Guldberg and Waage have led the way in applying mathematical procedures in chemical science.

I've recorded this in such detail in order to make clear the enormous impression which Muir's paper and the letter he sent me at the same time made on me. The leap from being a mere laboratory assistant in the Chemical Institute of the Dorpat University (in meanwhile I had got in exchange a position at the chemical institute instead of the one at the physical institute) who'd never seen another lab and who'd quietly carried out his work in this backwater, to the status of an important internationally accepted researcher was so great that I was at first unable to grasp its full extent. My teachers, K. Schmidt, Öttingen and Lemberg were better able to grasp the significance and wished their favourite pupil luck. Colleagues of my own age were much less happy.

The school teacher. Finally, in the winter of 1879–1880, my hope of getting an appointment with a salary sufficient to support a household was fulfilled. The position of a teacher of maths and science fell vacant in the district school (Kreisschule) in Dorpat and those who were to make the appointment, in particular the town mayor, were in favour of my candidature. Of course the small salary on its own would not have been enough but Professor Karl Schmidt kindly let me retain the assistantship in his institute, for otherwise I'd have had no chance of carrying on my scientific work which he held to be so important. My application was accepted and the date of my marriage was set for the Easter holidays.

I count this teaching experience as one of the happy twists of fate in my life. It gave me a chance to develop and let flower my teaching capabilities in a way that would have been impossible in a narrow academic environment.

---

[3]William Thomsen is perhaps better known as Lord Kelvin.

This was crucial, for teaching turned out to be an important aspect of much of my later work and what now began was more challenging than the usual run of a school curriculum. Teaching in my country was in the hands of two quite different authorities. One group of schools—the crown schools—was run by the government, the other by the cities and public corporations. Though the university and the grammar schools in the provincial capitals Riga, Mitau and Reval and also a number of the primary and middle level crown schools were government run, they had a lower reputation than the schools run by the cities and provinces. This was doubtless due to the general muddle of the Russian bureaucracy. Except in those cases where the school director made an extra effort, crown schools quickly slipped back to the level of the schools in the rest of the Russian Empire.

This was more or less the state of the Dorpat district school whose task it was to educate the pupils for the middle class job market. It was not connected in any way to the university and was housed in an old dilapidated building. The school was short of resources. The director was a kindly elderly man who had troubles enough looking out for his large family and was content to simply follow the administrative regulations. My new colleagues were mostly unaspiring school teachers worn down by their humdrum jobs who nevertheless welcomed the strange newcomer with a certain degree of surprise but with lots of good will. They did their best to help me and I'm thankful to them for that. I was to teach physics, chemistry and some special fields of mathematics, such as descriptive geometry. For physics and chemistry there were just the pitiful remains of ancient equipment which my predecessors had never used or brought up to date. I had no idea what descriptive geometry was but since I'd studied perspective for my sketching and painting I had little difficulty in picking it up as soon as I started to read Monge's textbook. The lack of equipment for chemistry and physics was also no major problem given my contacts in the university institutes.

And so in 1880 I began my career as a teacher. The pupils were youths of 14–18 who were all rather coarse and cloddish in character. Nevertheless, I never had any problem with discipline for with my own great enthusiasm for the subjects I easily managed to stimulate their lively participation. The lack of equipment forced me to be graphic in my description and the pupils happily took part in the demonstrations in the physics institute which I organised on Sunday mornings with the permission of Öttinger, even though these were in their free time.

I got a lot out of this teaching experience. Since the pupils didn't have much background in the subjects I was forced to restrict myself to a more basic level of explanation than I used in the university lectures and the experience of doing things this way in the school in turn flowed back to improve my university lectures. I learned here the importance of simplicity and clarity in presentation and when later I wrote textbooks these were the qualities for which they were praised.

The new household. The experimental work on chemical affinity was going steadily forward and on the lecturers' evenings I was regularly able to present new results. In addition, through the kindness of Karl Schmidt, I was able to gather experience in another area which was important for my future career as a university teacher for now and again he would assign one of his students to me. I could then propose a topic and direct the work which would then be used for the student's final assessment. This also worked out to everyone's satisfaction. I was thus involved in many different fruitful activities and had not to suffer the boredom of mere repetitive work. Both Schmidt and Öttinger many times pointed out that my life was actually ideal for someone with my interests and aptitudes. "You'll never have it as good as this again", said Öttinger's, "I envy you your freedom from peripheral issues". I laughed, only half believing him, because the work of a full professor seemed to me bolder and more effective since there was a broader circle which one could influence. And in fact the future showed that he was only half right for later as a full professor and director of a lab I was able to keep myself largely free of unwelcome distractions.

The previous work in the university thus carried on side by side with my new activities in the school and in parallel it was now time to get our new household ready. Given the limited funds available this was humble enough—a one time student flat near the school which consisted of two living rooms, a bedroom and a kitchen which we managed to furnish tolerably. Our wedding was in Riga during the Easter holidays and a 30 h journey on the post coach from Riga to Dorpat over sometimes terrible roads had to stand in for a honeymoon. So in the spring of 1880 we began our life together and in the meantime it has extended over 50 years and endured all sorts of twists of fate.

In a different context (in the introduction to my book "Daily Challenges" ("Die Forderungen des Tages") which was published in 1910) I described our situation then and can't do any better now than to repeat that description here. "As we started our life together in the scantily furnished Dorpater student flat, my wife was astonished at the variety of different tasks I undertook every day. In the morning I had a few hours teaching at the school, which I did to finance our life together. Then till midday and into the afternoon I had my job as assistant to my honoured teacher Karl Schmidt. He has repeatedly emphasised that this was to take the form of high quality experiments and was to be carried out without regard to any administrative demands. In the evenings I worked on my first book "Textbook of General Chemistry" ("Lehrbuch der Allgemeinen Chemie") though the first volume was only ready some years later when we lived in Riga. Wedged in between all this I had my private pupils and, of course, the lectures which I had to hold at the university and which required a certain amount of both practical and theoretical preparation. This was all rather different from the easy attitude to daily work in our home town of Riga, where lots of time was always left over for all sorts of domestic social occasions. In the spirit of early marriage she wanted to match her style to mine and since she, as a result of her mother's good example, was soon finished with the housework that our two and a half rooms necessitated, she felt unhappy that she had so little to do. Given the sweeping plans for my scientific future which I elaborated to my enthusiastic wife, it was no surprise that she too wished to do something important. But soon the first signs of her future very demanding duties appeared at the horizon of our life (in the meantime she has raised five children) and with that she had to give up her other plans and make ready for the coming duties.

In this situation nothing brought her more comfort than the words which stand at the front of Goethe's "Maxims and Reflections". They read, "How can one learn to know oneself? Never by contemplation, but rather through one's actions. Try to do your duty and you'll soon know what you are worth."

"And what is your duty? It is to face the daily challenges". It was a momentous experience for me to see how through these words the whirlpool of her undefined but nevertheless pressing wishes and goals was diverted into a quieter course where they moved in a clear direction and brought forth manifold blessings. I didn't feel any need to apply Goethe's words to myself, since the daily challenges facing me at that time fitted exactly to my innermost inclinations and desires. Even the school teaching, which at the beginning I'd only taken on for financial reasons, was interesting and satisfying because, within broad limits, I was free to frame it around my own thoughts and interests. Because of my incipient scientific reputation (which in Dorpat was something of a rarity and hence known to everybody) my immediate superiors all left me a free hand. It is from the experience I gained then that I claim for myself the right to contribute to the debate about school reform as an expert and not merely as a dilettante which is what the schoolmasters, who feel attacked by my proposals, call me.

Textbook of General Chemistry (Lehrbuch der Allgemeinen Chemie). The most important event from these years was the start of my work on The Textbook of General Chemistry which I mentioned above. The material I'd prepared for the lectures cried out to be properly presented. The oral presentation of the lectures was no problem and even brought me approval from the audience so that, despite the inflated respect in my Baltic environment for publishers, the idea of writing a weighty book, far from frightening me off, actually appealed to me. On the advice of Karl Schmidt I turned to Professor H. Kolbe in Leipzig who was publisher of the Journal of Practical Chemistry—the journal in which my first publication had appeared—and asked him if he could recommend a suitable publisher. He in turn persuaded the astronomer Dr. Rudolf Engelmann, the owner of the prestigious science publishing house Wilhelm Engelmann to take on the publication of the work which was still to be written. And so I began in 1880 the preliminary draft from which a few years later the textbook developed.

To begin with I was just happy at the prospect of having a publisher should the book ever be written. I had no idea when I'd start with the final version because the collection of the material and its organisation was a lot of work and it was impossible to judge how long it would all take. As it turned out, it would indeed take a long time till I reached the final draft.

Chapter 7
My First Appointment

Prospects in Riga. My wife fell seriously ill following the birth of our first child in the winter of 1880–1881 and this weighed both of us down. However in the summer there came not only her recovery but also the hope of a considerable improvement in our situation. The Professor of Chemistry at the Riga Polytechnic (Riga Polytechnical Institute) had died and, despite my relatively young age, I hoped I might be appointed because the number of possible Baltic candidates was not large. Initially the appointment was offered to Johann Lemberg. He declined because he didn't want to give up his research activities for the onerous teaching duties in an area which lay far from his speciality as a chemical geologist. The next choice was Gustav Bunge who turned the offer down for similar reasons because there was no place at the Polytechnic for physiological chemistry, which was his speciality. Finally Professor Karl Schmidt was asked to suggest a suitable candidate. He responded by recommending me in the warmest terms. He wrote the following to the director of the Polytechnic.

Dorpat 8./20. November 1881.

"Highly Honoured colleague!"

The warm interest with which not only I, but all of our peers follow the work of Dr. Ostwald, our youngest colleague here, justifies the wish to see him appropriately placed. Had we a vacant chair of Chemistry I would not hesitate for a moment to give this in all respects excellent scientist my warmest approval and support in the hope of retaining him here. To my great regret we have no resources available here with which we could offer Ostwald a suitable position. We have only one full professorship for chemistry and one for physics with none of the associate or honorary professorships which are usual at our German sister universities. The lectureships are all filled and the university's special funds are all so strained that even with the best will in the world no special position can be created for him. Weber's death changes the situation. Amongst the younger university staff I know

R.S. Jack and F. Scholz (eds.), *Wilhelm Ostwald*, Springer Biographies,
DOI 10.1007/978-3-319-46955-3_7

none, neither here nor abroad, whom I would so unhesitatingly recommend—either as my own successor or to any other equivalent position. The last word is yours!

"Ostwald is a child of Riga and his home town can already be proud of his scientific achievements and he will justify the most audacious expectations if he can be found a suitable position. You know his work. The excited attention with which its publication is received by experts in the field like Guldberg and Waage in Christiania[1] and others requires no further comment. Ostwald comes from the same C-H-N-O-S-P-combination[2] as did Bunsen, Helmholtz and Kirchhoff. Put him in the right environment and success will be assured. I don't doubt for a moment that he will have your support—should you wish to make use of this recommendation in the appointments committee please feel free to do so.

Should you do so then I would ask you to emphasise that the author is not related to Ostwald in any way. Ostwald has been my assistant for many years and before that held a similar appointment in the physics institute. He will be a major star in the border area between chemistry and physics where careful and thorough work will lead to great success.

Ostwald is also a skilled and experienced experimentalist, mechanic, glass blower etc. who constructs his own equipment in the most ingenious ways and he is a tireless worker whose oral and written descriptions are clear, concise, strictly logical and accessible not only to specialists in the field. We have often had the privilege of having him as speaker at our meetings attended by the staff and research associates of all five faculties".

What my teacher out of the kindness of his heart had set down imposed on me the duty of proving his words true. Besides the joy at seeing myself so lustrously described by someone I honoured so highly, I also gained from this letter (which Schmidt later showed me) a precise understanding of which of my abilities he laid the greatest weight and which I must therefore take care to develop. This clear and cogent program is doubtless the strongest and portentous influence that I had from my teacher.

The appointment. Karl Schmidt's letter finally removed the reservations to my candidature. The appointment was made and at the beginning of 1882 we moved to Riga. As everyone knows, the development of a German professor is similar to the development of a butterfly and involves a series of strictly separated metamorphic phases. The larval lecturer pupates into the form of an associate professor who eventually, when the time is right, ruptures his narrow envelope and emerges in the glorious form of a full professor. I had leapt over the middle stage: from being a Dorpater lecturer I sprang straight into a full professorship in Riga. I was 28 years old.

When I try to remember my feelings, thoughts and actions at that time then I have to admit that I did not live through this episode with the calmness and aplomb that I would have wished for. There was a lot at stake.

If I count in my time as a student then I'd spent almost 10 years in Dorpat, five of them as a lecturer. I hadn't needed Schmidt's letter to realise that there wasn't the

[1]Now Oslo.

[2]Symbols for the six most abundant elements in biological systems.

slightest chance of a professorship for me there. There was equally no possibility of a scientific position for someone with my interests in Germany where no one was interested in physical chemistry and, since I'd never been there, I had no personal contacts in the academic community. A Russian university was equally impossible even in the unlikely event that I'd been offered a position. The occasional visitor from Russia had made clear to me how low the level of scientific work in the provincial universities was, and in any case the language was for me a hopeless obstacle. For these reasons there wasn't any other chance for me to be promoted than to get the full professorship in Riga. I'm rather afraid that my laying this out in conversations with Lemberg, with whom I was in the meantime on friendly terms, caused him to absolutely oppose my appointment. On top of that was perhaps a clear fondness for my wife which he demonstrated in his rather clumsy fashion by, for example, bringing her a wooden block as a footstool when she came to collect me from the institute and had to wait till I'd finished my work. And yet I convinced myself later that he wouldn't have been happy in Riga so that his foregoing the position was really no loss to him. The question of the salary was of no concern to him.

At the end of my time in Dorpat I'd become friendly with the astronomer Dr. Lindstedt who was director of the observatory and, like me, newly married. Our wives too got on well. Mrs. Lindstedt was Swedish and found it difficult to organise her household in a foreign country. My wife was happy to help with these problems. It was our custom to meet once each week alternately in their flat or in ours. Lindstedt and I studied Kirchhoff's "Mechanics" while the women chatted or read, after which we had a simple meal. The momentous letter from Riga was delivered just as we were getting ready to go to the Lindstedts. We took the portentous letter with us and bought cake and wine on the way so that we could celebrate the great event with our friends who were as happy about it as we were. Thirty years later we four met again in Stockholm where Lindstedt had in the meantime solidly established himself with his groundbreaking work in sociology and political science while I together with my wife had come to receive the Nobel Prize. The memories of our younger days in Dorpat had remained as lively with them as they had with us.

Accounting. If I try to render an account of my achievements which underlay my claim to the professorship in Riga I would list the following positive factors. I had at that time started to successfully work on the greatest challenge facing —me then – the laws of chemical affinity. I had developed two methods for determining homogenous-solution equilibria neither of which involved altering the equilibrium value. By using these methods on as many materials (acids) as I could get hold of at the time I had quantitatively determined their chemical affinities which turned out to be the product of two factors each of which was associated with one of the two materials which were in equilibrium.

After that I had turned to the analysis of heterogeneous-solution equilibria (to use a term introduced much later) and for these had shown first that they were independent of the relative and absolute amounts of the two phases.[3] This was in complete contradiction to the then prevailing opinion. In a separate set of experiments I had also shown that the crystal form and water content of the solid phase had a decisive effect on the equilibrium and that this was independent of the amount of the solid phase. For these heterogeneous equilibria I found the same chemical affinity values as in the homogenous reactions. This demonstrated that the chemical affinity of a substance depended on its chemical nature and that the affinity determines the equilibrium constant much as atomic weights determine the empirical formula.

All of these things were completely new at the time and lay in an area far from the centre of contemporary scientific research with its emphasis on the investigation of organic chemistry. The handful of researchers interested in the problem of chemical affinity, struggled to solve individual problems but their results did not permit the elaboration of general rules. The only other scientists who like me pursued the general problems were J. H. van't Hoff in Amsterdam and S. Arrhenius in Stockholm and they had either not started their work, or at least not published their results, so that neither I nor the rest of the world was aware of what they were doing.

The new position. I'd not even tried to imagine the difficulties which would face me when I took up the professorship in Riga. I'd rather ignored the fact that I'd seen nothing of the world except my own country of Livland, Estland and Courland; I'd never even visited Petersburg which was quite close. I forgot that Dorpat was the only university which I had experienced and that even there I knew very little about its administration and in any case the situation at the Riga Polytechnic was quite different. Of course I'd occasionally met some of the professors from Riga and I knew Grönberg, who'd been my predecessor in Dorpat, quite well. But I now had to join a completely different circle of people and face very different challenges—and that as by far the youngest Riga professor both in age and in experience.

I ignored all this and, fired by the prospect of my impending independent teaching duties, threw myself with youthful enthusiasm into my new job. My predecessor in office, hampered at the end by a long drawn out illness, had neglected the department so that I was forced almost from the first day to start renewing everything. I have to thank my superiors and colleagues that they placed no hurdles in the way of the young firebrand. In particular I was readily granted the resources needed for the teaching facilities and lab equipment so that in a short time I was able to organise the education of the Riga chemistry students along the lines of the Dorpat model which was the only one I knew.

The Director. The decisive element in all this was the attitude of the director, who because of the organisation form of the Polytechnic had special powers.

The Polytechnic was a joint undertaking of the city of Riga and of the province of Livland. Its external relations were handled by a board of governors which was

[3]This is now well understood. The activity of a pure phase, such as a solid precipitate, is unity (1.0) and thus does not affect the equilibrium.

formed of representatives of the city and province. The internal administration was in the hands of the department heads and of the conclave of professors. Only the Director, who was chosen from amongst the professors by the board of governors, was a member of both bodies. He was the one who very largely determined the fate of the requests which the conclave of professors sent to the board of governors and, in case of conflict between the director and the other professors, the latter had no formal access to the board of governors.

At the time that I started in Riga the Director was an elderly mathematician called Kieseritsky who year after year had been reappointed at the statutory election. He was a member of the Riga student's fraternity. We'd met at the association's Jubilee celebrations in 1873 and had kept up ever since the friendly and informal relationship established then. He had exercised great circumspection during the discussions leading to my appointment so as to avoid any hint that he was favouring in any way a member the student association to which he belonged. Now that the appointment was through he treated me with genuine goodwill which was not necessarily to be expected because we were complete opposites not only in terms of age but also in temperament and outlook. He was an excellent teacher who did not, however, look much beyond the narrow borders of his field. He'd never undertaken any scientific work. Since this was true of most of the other professors (the splendid physicist Toepler who had held a chair in chemistry for some time in Riga had long left) the Polytechnic had become something like a school which, given the imperatives of the times, was perhaps no bad thing.

Chapter 8
The Professorship in Riga

Working conditions. I now sprang straight into this rather docile institution as someone with completely different goals and character and, as is the way of youth, I didn't care whether I caused offence or ruffled other people's feathers.

The fact that I was given my head and didn't bruise any egos can, I think fairly be put down to the scientific reputation which I'd already achieved as a researcher.

I soon won over the students. I was told of the following conversation between two Polish students (there were many of those in Riga). First student: "You heard new professor already?" Second student: "No, what's with him?" First student: "You must hear him. He shovels chemistry into head with a spade".

As a result there were no protests, neither from my colleagues nor from the students, when I raised substantially the standards of the final exam. Till then this final practical exam which had to be passed in order to earn a diploma, had consisted simply of the analysis of a random mixture of salts. I soon managed to change that so that this exam involved addressing a real scientific question.

Because of the uninspiring work of my predecessor, the number of chemistry students was initially quite small but that soon changed and this final practical exam became an ever increasing load on me as the responsible professor. For a start I had to think up the projects to be worked on. They had to be real scientific questions and should involve some technical challenge yet, on the other hand, they shouldn't be too difficult. The problem was that the scientific atmosphere, which is so very effective as a teaching support, first had to be established. The first students had it particularly hard because the main part of their studies had been before my time. However, theses difficulties were overcome and I never had any complaints about lack of good will on the part of the students.

Partly this may have been due to the fact that I made every effort to remove unnecessary burdens wherever possible. Thus, for the obligatory courses, there were interim exams which served to monitor a student's grasp of the lecture material. Depending on the length of the lecture course these exams took place either once a term or once a year and they were a significant strain both on lecturer and, of course, on the students. On the one hand the students had to prepare for

R.S. Jack and F. Scholz (eds.), *Wilhelm Ostwald*, Springer Biographies,
DOI 10.1007/978-3-319-46955-3_8

exams in a whole set of subjects all at once and on the other the lecturers were so stressed by the weeks-long run of exams that they quickly grew surly and even unjust in their marking. I persuaded my colleagues, that the only thing that mattered was that the students should demonstrate that they had understood the material—when they did this was less important. On my suggestion it was left up to the students when they should register for the exam. When a student felt sufficiently well prepared he could register and the professor decided on a day and hour that suited him. Since the rules made the participation in the advanced lecture courses dependent on successfully having passed the exam, there was no risk of procrastination. This proposal was gladly accepted both by faculty and by the students and remained in force even after I left.

The Assistants. I was helped a lot in my teaching duties by my assistants. Two of them had, like me, studied in Dorpat and were members of the "Fraternitas Rigensis" student association. Both of them were older than me and one of them had been a classmate in secondary school. After my arrival in Riga, they turned up on the next Sunday, dressed up to the nines. When I laughed at them for this, they bashfully explained that they'd had no idea whether I would want them to carry over our old easy relationship into the new situation of laboratory head and subordinate. I left them in no doubt that the old relationship should be retained under the new conditions. My colleagues on the faculty shook their heads when they heard that the new professor conversed like an old friend with his assistants even in the presence of the students, but I had no disadvantage from the arrangement. On the contrary, I think I awoke in them a greater readiness for effort and an acceptance of my high expectations. In addition, I had the satisfaction of seeing that, without any direct demand from me, they soon showed an eagerness to join in my scientific work, despite the fact that under my predecessor they'd for years had no interest at all in this side of things.

Apart from these two teaching assistants I had also taken over a Pole, who turned out to be adept and reliable. He'd been educated at the Polytechnic and served me well as a channel to the students of whom less than half were German. Of course he tried, as all people from Poland do, to involve me in their nationalistic activities but he found no echo in me, for I wasn't interested in political issues at the time and in any case was more worried about my own land and people in view of the increasing threat of russification.

Scientific work. Despite the numerous demands on my time the experimental work soon moved ahead. My last experiments in Dorpat had involved a thermo-chemical method; reciprocal transformation of salts which had been melted and then rapidly cooled. It is, I might say in passing, my only publication in the field of thermo-chemistry. These experiments later brought me a most instructive experience. For a long time there had been only two well established researchers on this field: the Dane, Julius Thomsen, and the Frenchman, Marcellin Berthelot. The latter was fired by the ambition to be recognised as the most important thermo-chemist in the world, and he'd easily managed to persuade at least the French. Since the procedure I'd introduced was something new, he felt the need to take it over. He repeated my experiments and extended them by the use of other salts which were treated in the same way and then he published the work as the fruits of his own genius. To protect himself from the accusation of plagiarism he took the precaution of mentioning my name in a footnote but in such a way that no one would come to the conclusion that the basic idea had been mine.

It took me a while to recognise the base nature of this tactic. Berthelot was at that time at the height of his fame and had no need to steal the only thermo-chemical results of a young and more or less unknown researcher to spruce up his own reputation. However, I'd seen in my study of the relevant literature how he'd treated his elder rival Thomsen from whom he'd stolen a law (which turned out to be wrong!), and so I was not surprised at how he treated me. I didn't follow up with a public protest, since at that time my new position involved me in quite different things. However, this little adventure was instructive and helped me to understand how Parisian science and its personalities have worked for hundreds of years—and probably still do.

The Thermostat. My experimental work in Riga made considerable headway when, instead of just measuring equilibrium states, I now turned to the determination of the rates of chemical reactions. Here I had even fewer forerunners than in my previous work for in order to make the measurements one had to be able to hold the temperature constant for a long time. That was a real challenge at the time because there were no thermostats available that would function reliably for weeks or even months.

I'd begun to think about the thermostat problem soon after settling in Riga. On looking through the various solutions which had been tried up till then, it seemed to me that those which were based on regulation of the gas input by the expansion of some material had the best possibility of being refined. In any case they appealed to my frugality because they used no more gas than was required to balance the loss of heat by radiation. I quickly showed in an initial investigation that the oblique or slotted inlet tubes, which were generally used at the time, strongly reduced the sensitivity of the device. After several weeks of practical work during which my abilities as a glass blower—imperfect though they were—served me well, I had invented the aqueous calcium chloride filled regulator. Apart from the minor improvement of using toluol in place of the calcium chloride solution, which I made later in Leipzig, this regulator served has served a generation of chemists—and it is still used all over the world wherever gas heating has not yet been replaced by electricity. A regulator for electric heating elements would need certain modifications but I didn't wish to get involved in this.

At this time I also invented a windmill driven stirrer which was powered by a small flame. The idea for this came from a Christmas toy from my childhood which consisted of a cardboard cylinder in whose wall all sorts of ghostly figures had been cut. The covering had been formed into windmill sails and the whole thing was suspended from a wire so that it could easily turn. If a burning candle was placed underneath the cylinder then it was set in motion and the ghostly figures scurried as shadows over the dark wall.

To begin with the thermostat was used only when I was present in the laboratory. Then I dared once to leave it running while I was out to lunch. Only once I had convinced myself that nothing unexpected was going to happen did I leave it running overnight, though I did ask the night watchman to keep an eye on it.

I normally sleep soundly but that night I woke up a dozen times and looked in the direction of the Polytechnic for signs or sounds of fire. Early the next morning I was relieved to find that everything was working perfectly and that the thermostat temperature hadn't changed by as much as a tenth of a degree. The apparatus never caused an accident in my lab during the following years in Riga or later in Leipzig and I never heard of problems from any of the other labs which took it over.

I write about this at such length because the plain fact of the matter is that the whole development of chemical kinetics would not have been possible without a robust and dependable thermostat.

Chemical kinetics. My thermostat made it possible for me to carry out and publish a series of experiments on the theme of "Studies on Chemical Dynamics" ("Studien zur Chemischen Dynamik") by which means additional areas of chemical affinity were developed. These later formed the basis of my terminological and scientific research on catalysis which became the pinnacle of my work in chemistry.

For this work it turned out to be of the greatest value to me that the terms of my appointment required me to lecture on the entire field of chemistry, which of course included organic chemistry. I've already related how very much organic chemistry had been pushed into the background during my Dorpat years.

There my research work had concentrated on inorganic chemistry, though my desire to test as many acids as possible for their chemical affinity had forced me to also use organic acids. At that time there were practically no slow inorganic reactions known which would have been amenable to kinetic analysis. In contrast, slow organic reactions were the rule rather than the exception. In order to get myself ready for the upcoming lectures I, as was my custom, read a vast number of original papers and, while doing so, kept an eye open for reactions that might be suitable candidates for kinetic analysis.

It also turned out to be an advantage that I'd learned the basics of calculus during my student years. I'd studied this more or less out of a general thirst for knowledge and my later flirt with physics had convinced me of its value. I must admit that I never got far in higher mathematics, and several attempts to learn it merely showed me that here my limits were rather strictly drawn. However, I was at least sufficiently at home with maths that I could easily handle the rather simple problems that chemical kinetics posed without being hindered by terminological or technical problems.

The first reaction that I studied was the acid hydrolysis of acetamide.[1] This was followed up by the catalytic hydrolysis of esters which I discovered by studying the hydrolysis of methylacetate by dilute aqueous acids.[2] This process, which could so readily be followed by titration, was used a lot later on. Then I turned to the

[1]Ostwald W (1883) J Prakt Chem N.F. 27:1–39.

[2]Ostwald W (1883) J Prakt Chem N.F. 28:449–495.

classical problem of sugar inversion.[3] All of these reactions turned out to dependent on the same property of the acids used, since the rate constants for any given acid were always in the same order and this order was the same as that which had been determined from the earlier equilibrium analyses. Thus, for the first time it was possible to demonstrate the existence of general affinity values which were decisive for all sorts of reactions involving these acids, though of course we are talking hear only of reactions in dilute aqueous solution.

The discovery of these relationships marked the start of research in the area of chemical affinity and led to the discovery of simple quantitative laws. At the beginning of the nineteenth century Berthollet had declared his conviction that mechanics would turn out to be not just the basis of astronomy and physics but also of chemistry and he was convinced that the application of mechanics to chemistry would lead to a flowering of this science. His prediction was, however, premature because chemistry developed along very different lines from those envisaged in his "Statique chimique". Nowadays we can understand the inner logic of this development. At first the diversity of individual chemical substances had to be described in terms of their content of chemical energy. Only then did the laws describing the dependence of the conversion of chemical energy on time and concentration become accessible to analysis. Because of this, nearly a century would pass before Berthollet's dream of quantitative chemistry would become reality. During all this time chemists had become accustomed to immersing themselves in the old problems and few were able to reorient their work to the new goals.

Chemical Thermodynmics. An important contribution to the now rapid development of thermodynamics—or as I prefer to call it "energetics"—was the ground breaking discovery by J.R. Mayer of the mechanical heat equivalent in 1842. Because of the mathematical nature of the framework of this marvellous construct, the initial experimental work in this area was carried out by mathematicians and physicists who did not think in chemical terms and a long time passed before those working in this area turned their attention to chemical matters. Helmholtz, for example in his groundbreaking publication on the conservation of energy considered the various areas of physics, mechanics, heat, electricity, electrical stimulation of muscles, magnetism and electromagnetism and then mentioned at the end: "Of the known natural processes there remains just those associated with organic life". Though he properly emphasised that these were chemical processes he seemed unaware that he had just failed to develop a theory of chemistry from the standpoint of the new thermodynamics. Only in the case of the electrical stimulation of muscles did he briefly note that electrical processes are here associated with chemical ones. He did also point out that the chemical induction of heat or electricity must also be governed by the law of the conservation of energy.

A quite similar attitude was shown by the next great researchers in this area; R. Clausius who enunciated the second law of thermodynamics and William Thomson who developed it. Clausius chose to use steam engines as the subject to

[3]Ostwald W (1884) J Prakt Chem N.F. 29:385–408.

which he would devote his attention. Thomson, equipped with a broader and more fluid mind, applied the new ideas fruitfully to a range of problems in physics—but he never turned his attention to chemistry. It seems that he was not attracted by its beauty. Only much later did two scientists independently show what a decisive influence the second law was going to have on chemistry. These were Horstmann in Germany and some 10 years later but quite independently, Willard Gibbs in America. Horstmann's[4] work was published in the "Annals of Chemistry" ("Annalen der Chemie") whose readers were not used to calculus and so they simply skipped over the paper. Gibbs for his part had so successfully buried the results of his monumental research in the inaccessible journal of a provincial scientific academy that it required no small effort (in which I was later involved) to bring them back into the light of the scientific record.[5]

Thanks to Öttinger's suggestion (to whom I am also indebted for mentioning Gibbs's work to me) I was aware of the work on thermodynamics early on. Öttinger himself had time and again tried to clarify the terminology, for thermodynamics—like Sleeping Beauty—was surrounded by hedgerows bristling with impenetrable mathematical thorns. It was therefore no surprise that I became convinced that this powerful new way of thinking would help me solve the problems that interested me. The first precondition was that I equip myself with the necessary mathematical tools and this was a difficult task. How often back then did I take long walks alone so as to have a chance to think undisturbed about the second law.

As I wrote in my Introductory Summary to my "Philosophy of Values" ("Philosophie der Werte") in (1913): "I can still remember the inner struggle I underwent as I tried to grasp Thermodynamics as it was called then, or Energetics as I now refer to it today. I tried because I had to. In order to be able to teach the generally applicable results of the laws of thermodynamics which were widely ignored at the time, it was clear to me that I'd have to make it absolutely clear how the postulates could lead to such important results in so many areas. It was at that time extremely difficult to realise what Clausius's law of the increase of entropy or William Thomson's law of the dissipation of energy had to do with the calculation of the latent heat of vaporisation from the temperature coefficients of steam pressure. The first is a broad generalisation which can, and has been, readily understood and is graphically illustrated by the case of a waterfall which will never flow back up the hill. The second, in contrast, describes situations which can be subject to the test of experience and so lead to confirmed results that bring together seemingly disparate physical phenomena. This was a bit of a headache and so there was nothing for it but to fight my way through the thickets of the various formulations of the second law and to try to grasp why, having shown the variables to be

---

[4]Horstmann A (1872) Ueber den zweiten Hauptsatz der mechanischen Thermodynamik und dessen Anwendung auf einige Zersetzungserscheinungen: Annalen der Chemie und Pharmazie. 8th supplement volume, 112–132.

[5]Transactions of the Connecticut Academy (1873) 309–342, 382–404, (1876) 108–248, (1878) 343–524.

independent, the second derivatives all politely show each other the door, leaving a simple and readily understood first order relationship as the result."

The work I put into understanding the concepts in calculus didn't produce any concrete results in terms of publications, but they did prepare the way for the later development of Energetics.

Interlude. In the first years of my time in Riga I got involved in a literary effort quite different from the usual reports on my scientific work. At that time the Leipzig chemist H. Kolbe was waging a war against the first flowering of structural chemistry and a young author called A. Rau tried to support his point of view with a farfetched philosophical rationale.[6] In doing so he made a rather ridiculous mistake. He started off by following Kolbe's line and arguing that the practitioners of structural chemistry had lost the capacity for logical thought and to prove this he gave a definition of the Dalton's law of multiple proportions as used by the structural chemists. After a long diatribe about the inadequacy of this formulation he gave, as an example of logical rigour, Kolbe's definition of the law. He'd obviously forgotten in the meantime the start of his argument because both definitions were exactly the same.

This reminds me of F. Reuters story "Ut mine Stromtid" in which Mining corrects his twin sister who had referred to their grandfathers wig as "Puck" instead of "Perucke" by telling her to say "Puck"—because neither of them could pronounce the "r". Though this crusade against modern chemistry had nothing to do with me I thought it would be fun to present this ludicrous mistake for the amusement of my chemical colleagues. So I wrote a polemical pamphlet against Rau. It was called "In the Matter of Modern Chemistry" and I had it printed in Riga. I don't believe that it was widely read. However Rau wrote a response which he sent to E. von Meyer for publication in the Journal of Practical Chemistry. Von Meyer later told me that this response was so abysmally coarse that he'd had to turn it down. The only result of this incident was that H. Kolbe began to take an interest in me. What consequences that had when I first met him will be related in due course.

Problems with the laboratories. The low profile of the chemistry courses at the Riga Polytechnic could be judged by the fact that the labs were in the basement where there was little daylight or fresh air—two things which are essential for chemical work. As, after a few semesters, the number of students had multiplied I had no option but to inform the governing body that the problem could only be solved by the erection of a new building.

The energy and generosity with which the patriotic enterprise of the Polytechnic was supported was soon apparent to all. Despite the expense of the undertaking the

[6]Rau A (1879) J prakt Chem 20:209–242. Albrecht Rau [1843–1918 (some sources indicate 1920)] was a German philosopher who published work on the history and philosophy of science: (1) Die Entwicklung der modernen Chemie. Braunschweig, Vieweg 1879, (2) Die Grundlage der modernen Chemie: eine historisch-philosophische Analyse. Braunschweig, Vieweg, 1877, (3) Die Theorien der modernen Chemie. Braunschweig, Vieweg, 1877–1884.

new building was approved and I was assigned the task of drawing up the plans together with the Polytechnic's architect. I made no secret of the fact that the only labs I really knew were the ancient buildings in Dorpat and that I only had anecdotal knowledge of all the new developments that had taken place since 1870–1871 in lab design. Because of this I was told to first make a tour through Germany and visit the most important chemical laboratories so that I could ensure that the money approved for the new lab building was being well invested.

For me this was a far reaching decision for it gave me the opportunity to broaden my perspectives both in general and in chemistry in particular beyond what could be found within the narrow borders of my home country. I knew the names of all the leading German chemists from the literature and I'd tried to build pictures of their personalities from the way they conducted themselves in disputes with their peers. However, by coming into direct contact with them and by asking them about the organisation of their teaching duties, I'd have the chance to get to know them much better.

Chapter 9
Germany

First visit to Germany. So as not to disrupt the regular teaching courses, I decided to sacrifice the winter holidays of 1882–1883 for my visit to other chemical laboratories. By travelling as far as possible on night trains I could use the days to the full and so was able to visit all the important university and polytechnic institutes in Germany and Switzerland. Meeting the leaders of these institutes led to helpful suggestions and many a friendship was started then.

My route ran via Königsberg[1] to Berlin, Dresden, Leipzig, Halle, Braunschweig, Hannover, Aachen, Bonn, Darmstadt, Heidelberg, Karlsruhe, Stuttgart, Tübingen, Zürich and Munich from where I travelled back home again via Berlin.

Just before the Russian Christmas (which is 12 days later than in Europe), I excitedly boarded the train which would take me through the long stretch of bleak countryside to the border station at Wirballen.[2] The train was late. The contrast between the run down villages on the Russian side of the border with the well cared for impression given by the village of Eydtkuhnen[3] on the Prussian side made a great impression on me. After that there was a long journey to Königsberg, the first of the German universities on my list. I got there on New Years day and so I had to wait till the next day to meet my colleague there. This was W. Lossen who'd discovered hydroxylamine and the curious isomers of its organic derivatives. I found a sick and aging man who was nevertheless happy to show me his teaching curriculum. The labs were old and offered nothing off interest. I wandered through the city with its narrow twisting streets which reminded me so much of Riga that I began to think that travelling in Germany was not going to be as difficult as I had feared it might be.

[1]Now Kaliningrad, Russia.
[2]Now Virbalis, Lithuania.
[3]Now Chernyshevskoye, Russia.

R.S. Jack and F. Scholz (eds.), *Wilhelm Ostwald*, Springer Biographies,
DOI 10.1007/978-3-319-46955-3_9

Nevertheless, after a long day's journey I arrived with a pounding heart in Berlin, where I took a room at the Central Hotel,[4] which at that time was the biggest in the city, and set off to admire the metropolis. Because of the New Year holidays I had time to first of all visit the museums and other places of interest. The first chemist I met was Hans Landolt who held a professorship at the agricultural university. He welcomed me with great kindness and showed me his labs. I learned a lot of things which would be useful in teaching courses and this was the start of a long and deep friendship which remained unshaken even in difficult times and continued until his death.

To find out when I might speak to A.W. Hofmann, the reigning king of the Berlin chemists, I visited his assistants, W. Will and C. Schotten, both of whom lived in the laboratory building. We were the same age and as we got on well we spent the afternoon and evening together. Here as well a friendship formed, this time with Will, who as I'll relate in due course later supported me at a critical turning point of my life. Will was one of the most likeable and reliable people I have ever met. Schotten, who was his close friend, died young.

On the next day I listened to one of A.W. von Hofmann's famous lectures. He talked without a break for 90 min, which was a considerable challenge both for him as well as for his audience. The whole thing was very dramatically, even theatrically, staged and when the audience seemed to be losing concentration they'd be woken up with a joke which was usually at the expense of the lecturer's assistants. Altogether, I must say I admired his performance.

After the lecture I introduced myself to the professor and told him why I'd come. He didn't find the time to talk to me himself but referred me to his assistants and to the laboratory manager Tiemann. He in turn suggested that I attend a meeting of the Chemical Society on the following Monday and give them an overview of my research. I was happy to do this. To fill in the time till then I visited some other chemists and also studied the beauty and the oddities of the town, which I unreservedly admired. These wanderings led to a quite unexpected broadening of my horizons. At the appointed time I held the lecture which had been requested. Since I was rather sure that most of my colleagues in Berlin, including Hofmann, knew nothing of my work I presented the old volumetric story in the simplest possible terms and with the use of explanatory diagrams.

Hofmann had watched with some reserve as I approached the lectern, but now he called out in relief, "Why, even the youngest student could understand that", to show that contrary to his expectations he'd understood me. But the scepsis remained. I later often read from his behaviour and from that of his associates that their view was: "We don't like this approach".

I naturally didn't want to miss the opportunity of also hearing a lecture by Helmholtz, Germany's greatest physicist. He was an average sized, stocky man with a bald head and a grey moustache. He arrived rather late. With his massive

[4]The hotel was centrally situated opposite to the Friedrichstrasse train station.

oddly formed head, his slow and measured movements and his staring eyes he almost looked like a monumental statue.

The lecture was a masterpiece of brevity and accuracy; it could have been published as it stood. Nevertheless one could see how enormously bored he was with the elementary things he presented and one felt happy for him when the bell to end it finally sounded.

After that I was taken to a meeting of the Physics Society of which he was chairman. Once business was completed I asked if I might make a short presentation and, after some discussion, I was given permission. I gave them a brief synopsis of my recent work on chemical dynamics. I didn't get the impression that the chairman was particularly interested, though in his reserved manner he paid the usual formal compliments. However, he did remember me as I found out later when I met him during my period in Leipzig.

I see from the protocol of the meeting (January 5th, 1883) that just before me Helmholtz's greatest pupil Heinrich Hertz presented his data. Nobody, except perhaps his mentor, seemed to have guessed his future greatness. I too remember nothing of his lecture for it was a mathematical analysis of certain aspects of tidal movement without any implications of general importance.

Full of the many strong impressions from my days in Berlin I travelled on to Dresden, where Schmitt, a pupil of Kolbe and the inventor of the process to manufacture salicylic acid, was professor at the technical university. He received me warmly and answered all my questions. A.Toepler, one of my forerunners in Riga was a physicist in Dresden. He was an ingenious experimenter and he too was friendly and I learned a lot from him. Finally there was a third worthy colleague, the excellent gas analyst Walter Hempel who as a young man (he was only 2 years older than me) had successfully dared to substitute new, simpler and faster methodologies for the classical methods of R. Bunsen. His work laid the basis for the development of gas analysis which since then has become so important. As we shared many interests we soon came closer and he was pleased to hear that the opportunity to see his methods at first hand was one of the highpoints of my journey. This acquaintanceship too led to a cordial relationship which never dimmed.

R. Schmitt's assistant at the time was Dr. Willibald Hentschel who looked after me with great hospitality. He turned out to be a chemist and man of great originality of thought whom I expected to have a much more eminent career than he later had. Our ways met once more 5 years later in Leipzig after which we lost touch with one another.

At that time the professors of the technical university would, if they were free, meet in the evenings at the Bohemian Station (where the main station now stands). Since it was conveniently close to the Polytechnic I met a lot more colleagues there and though I no longer recall them individually I well remember their open and carefree way of interacting.

Of course I didn't miss visiting the Dresden museums. On my visit to the Art Gallery I went straight to the corner room which houses Raffael`s "Sistine Madonna" ("Sixtinische Madonna"). I expected this to be a deeply moving experience and was shocked when all attempts to build a relationship to the picture failed. I felt I ought to be ashamed of myself, not realising that I had just experienced the first step along the road towards a proper appraisal of this ridiculously overrated period of art.

From Dresden I went on to Leipzig, the only place in Germany where I already knew people. Hermann Kolbe and Ernst von Meyer, publishers of the Journal of Practical Chemistry in which I'd published my first papers, lived there as did Gustav Wiedemann from the Annals of Physics as well as my publisher Dr. R. Engelmann. The reception was even more cordial than in Dresden so that already on this first visit I felt really at home in the town in which I would later spend the nineteen most productive years of my life.

At that time Hermann Kolbe considered himself the most important chemist not just in Germany but in the whole world. Not surprisingly he oozed gravitas. He was of less than average height with a large, clean shaven face furrowed with dignified creases which would have fitted perfectly to a senator from Hamburg or Bremen.

His early success, which was due to his excellent experimental work, had turned his head and led him to believe that his theoretical opinions were unfailingly correct. These were based on the development of the radical theory of Berzelius and in consequence he considered the emerging structural chemistry of Kekulé as a terrible aberration. Because of this he soon came into conflict with the leading chemists in Berlin, though his crusade was not against Berlin alone for he found something to complain about more or less everywhere. In the journal he published he had reserved space where he put across his point of view and did not shy away from expressing himself in a pretty rough and polemical style. Since his damning verdicts, which to begin with were not entirely groundless, gradually lost touch with reality, his influence had considerably eroded. Perhaps it was an unconscious recognition of this that made him particularly appreciative of my contributions to the journal especially as they had nothing to do with the structural chemistry he so hated.

Whatever the reason both he and his son in law E. v. Meyer received me warmly. I'd dined at Meyer's house and after the meal there was a musical interlude in which Meyer and his charming wife, who was Kolbe's daughter, played violin and piano. I was invited by Kolbe to a celebratory lunch the next day and he asked me to meet with him an hour earlier so that we could talk undisturbed. The conversation turned out to be most interesting. I showed him A. Raus's silly mistake, which he didn't want to believe at first, and he told me about his plan to bring out a short textbook on organic chemistry. For this he had read the papers of O. and E. Fischer on fuchsin chemistry and he expressed the greatest surprise that though the work itself was solid and competent the authors had made the mistake of interpreting it in terms of structural chemistry. He added that all that needed to be done was to re-express it in real chemical terms.

By now lunch was upon us and I was introduced to a large circle of Leipzig professors, who I treated with awe. Kolbe waved me to the place next to himself. On his other side sat the famous mineralogist F. Zirkel, who had just turned down a professorship in Munich and who was therefore greeted with special heartiness.

In the middle of the lively conversation Kolbe tapped on his glass and made a chemically tinged address. He said that he saw at the table two volatile bodies, namely Zirkel and myself. It had turned out to be possible to secure the first and he wanted to express his wish that this would also work with me and that both of us could be held in Leipzig. I was utterly astonished, in fact flabbergasted, for I'd never for a moment thought that Kolbe's friendly attitude towards me would go so far. The others at the table didn't object to this suggestion and the meal continued with innumerable further toasts and increasing liveliness.

Apart from Kolbe, Gustav Wiedemann was of particular interest since he held the only professorship for physical chemistry not only in Germany but in the entire world. He had been a pupil of Gustav Magnus in Berlin who'd been as good a physicist as he was a chemist. The subject of Wiedemann's doctoral thesis had been chemical but he later turned more to physics and so he'd been appointed to the chair of physical chemistry which was inaugurated in 1871.

Since Kolbe and Wiedemann did not get on I had to visit them separately.

Wiedemann turned out to be a lightly built man of less than average height who stooped a bit. He had light blond hair, a closely shaved face and an agile expression. He looked rather like how one would imagine a worldly French bishop who was at home in numerous salons and had a finger in every pie. Next to him his wife, who was a daughter of Eilhard Mitscherlich, made an imposing impression both through her stature and her personality. Their son, named after his grandfather, was also a chemist and acted as his father's assistant though he had already made a name for himself through several competent publications.

The family was just back from a holiday in Egypt where the other son, who was an Egyptologist, had acted as their guide. Since this sort of journey was a rarity in these days, all the participants made an effort to make reference to it through all sorts of details of dress and décor. I was also kindly received in this circle and spent an evening with them, though it never came to the sort of cordial atmosphere I'd experienced with the family of Kolbe.

The discussions with the publisher Dr. Engelmann went satisfactorily. He gave me some technical tips about publishing and strongly advised me to send him a finished manuscript as soon as possible. To this I agreed and I did deliver the manuscript punctually.

My visit to Leipzig came to an end with an excellent production of Wagner's "Tannhäuser". A lot of threads had been spun which would later have decisive impacts on the fabric of my life. I left this friendly city full of impressions, hopes and plans.

The next morning I travelled on to Halle, where I met my extraordinarily tall colleague J. Volhard who told me the tragic tale of the building of his new laboratory. Every couple of years only a small amount of money would be authorised so that the actual building work had not yet even begun and it was unclear when it would ever be finished. The dull weather made the city with its smoke laden air even more dismal so that I got a very poor impression of the town. Later on I had the chance to go back and my impression of the place was much brighter.

My next stop was Braunschweig which made a very different impression. The cheerful old city centre with the carved wooden balconies on the half-timbered houses which were surrounded by well designed new buildings raised my spirits, for here old and new stood in such a friendly contrast to each other. On top of that my colleague Robert Otto greeted me with spontaneous friendliness. His name is associated with the "Graham-Otto" chemistry book which at that time was the most detailed chemistry textbook in German and had been used by generations of chemists. He was a small agile middle aged man who was instantly recognisable by a terrible scar on one side of his face which was deep enough to lay a finger in. It was the result of an attack by a deranged laboratory janitor who had struck him down one night and left him for dead. Had not his assistant Beckurts (who I got to know on this visit and who went on to become Otto's successor as professor) gone late to the lab, which was not his normal habit, to collect something he had forgotten and chanced to hear Otto's groans, he would certainly have died from loss of blood. This astonishing experience had not in any way dimmed the professor's happy nature and he did everything he could to strengthen the positive impression which the town had made on me. A visit to the ancient "Collegium Carolinum[5]", in which a part of the university was housed, only increased my appreciation of the town.

The next city, Hannover, made a very different impression on me for there flowered a passion for resuscitated gothic which a then famous professor of architecture had applied to secular buildings. I can well remember the inconvenient stairs and uncomfortable rooms in my hotel which had been built in this style.

The professor of chemistry was Karl Kraut who is best remembered for the Gmelin-Kraut handbook.[6] He was a tall thin man with a short full beard and spectacles for his short sight. He spent an astonishing amount of energy getting to know me. Once he'd done that to his satisfaction he let me know that he intended to do his utmost to see to it that I was offered the chair of analytical and physical chemistry at the university. When he asked, I assured him that I did not feel permanently attached to Riga. Nothing ever came of this.

In my conversation with him I ventured to ask why the "Gmelin's Handbook" which he was writing came out so slowly in dribs and drabs. He made a face as if he'd bitten on a pepper corn and with a degree of self irony said, "I'll be damned if I finish the book so long as I'm living. At the moment people like you ask me about it from time to time. If it was once finished nobody would be interested in me anymore. In fact he never completed the book.

His laboratory was on the ground floor of the old Hannoverian castle and contained nothing of any interest.

After Hannover came Aachen, where I visited the Technical University's chemistry laboratory which had been built after the 1870/1871 war by Landolt

[5]The *Collegium Carolinum*, founded in 1745, was the forerunner of the Technical University Braunschweig (Brunswick). Ostwald refers here to the old building *Collegium Carolinum* which was destroyed in WW II.

[6]This is now the *Gmelin Database* of organometallic and inorganic compounds.

who'd been given unlimited funding. Each work place was equipped with a special fume cupboard equipped with seven taps: one each for gas, water, low and high vacuum, compressed air, and steam.[7] It was unclear where amongst all these appliances one was supposed to work and I, for one, decided not to copy this idea.

In striking contrast, Alexander Classen's laboratory for electrogravimetry was the first groundbreaking lab of its kind. Its designer, now in his late eighties, still follows the developments in the field.

In Aachen the chair of physics was held by Adolf Wüllner, the author of the standard four volume textbook of physics who, because of his pale, well formed face, dark blond full beard and upright bearing, was known as "handsome Adolf". I wanted to meet him, because I'd recalculated his results concerning the reduction of vapour pressure over aqueous solutions by various salts for my textbook and found that equivalent concentrations of salts of similar composition resulted in the same effect. This was the precursor to the main part of Raoult's work on the relationship of aqueous solutions to reduction of vapour pressure, for which he alone is given the credit. That is almost always the case when a new discovery is only published inside a large book. The scientific community is so used to seeing new things published as papers that it regards everything in a textbook as being old hat. It wasn't the only time this happened to me.

When I told Wüllner of the discovery I'd made using his results, he wasn't particularly interested. I'd thought that it would be a real pleasure to him but obviously his annoyance at not having seen for himself the full significance of his results overrode the pleasure in recognising that in this way science had been advanced. It took a long time before I recognised that this sort of psychological reaction to such situations is almost automatic.

The next stop was Bonn where I hoped to meet Kekulé—the man who had regenerated organic chemistry. My official request to be shown the laboratories had been turned over to his assistant Otto Wallach, at that time a wispy young man with blond hair, but now, after a glorious career as the successor to Wöhler in Göttingen, a white haired old man laden with honours.

At that time Kekulé suffered from depression brought on by private problems. He had more or less withdrawn from scientific work and published only sporadically. Perhaps on that day a shaft of sunlight beamed on him because he joined us and we began to talk. He held a wonderful short and humorous lecture on the advantages of his demonstration bench. At that time he was 54 years old with a good figure and a remarkably attractive head with a full beard. Max Klinger had used him as a model for his Zeus on the large canvas of "Christ on Olympus". Finally the talk turned to my own work and he expressed his approval that one had had the courage to take this on. "But, my dear young friend", he finally said and laid a friendly arm on my shoulder, "I can only advise you to give it up. Years ago I spent 3 days and nights continuously thinking about that and convinced myself that it would be a waste of time".

---

[7]Ostwald gives six functions for seven taps. Perhaps he miscounted.

From Bonn I went to Darmstadt where I met professor Staedel, who showed me curious multi coloured isomers, and then I travelled on to Heidelberg where I arrived with a pounding heart, for there Robert Bunsen—my scientific idol—still worked though he was now in his seventies. Despite the winter the journey along the Bergstrasse towards Heidelberg was soothing and once in the city I climbed up to the castle to take in the wonderful panorama over the surrounding countryside. First of all I visited August Horstmann the founder of thermodynamics[8] with whom I'd corresponded for some time. He was a haggard stooping blond man with a pale face, bent nose and straggling beard who was sadly almost blind. One eye was completely blind and with the other he was so short-sighted that that he had to peer at the book he was reading held just a few centimetres from his face. He'd discovered a mistake in Bunsen's experiments in which a mixture of hydrogen and carbon monoxide had been burned in the presence of limiting amounts of oxygen. Bunsen interpreted the results as being due to the stoichiometric relationships of the reactants but he'd ignored one important component—water vapour. Horstmann pointed this out and, in consequence, he was not popular amongst Bunsen's fans who were, of course, in the majority in Heidelberg and they did everything they could to make life and teaching hard for him. Luckily he was economically independent. We had a lot to talk about and during the evening that I spent with him and his family we became good friends. That evening was one of the most pleasant of my entire journey. Horstmann also talked about the possibility of my being offered a chair in Germany because his near blindness ruled him out as a candidate.

The next morning I went to Bunsen's lab. I first approached his assistants who showed me round. I learnt a lot of things which had never been published and saw the old master himself. He was a burly stooping figure with a red face which was framed by a short beard. He had no moustache and his hair was nearly white. I was warned not to speak to him because when he was working he pretended to be deaf so as not to be disturbed. The next morning I listened to his lecture of which I have no memory, and afterwards I introduced myself to him. He was courteous and didn't pretend to be deaf. I'd found out that he lunched at the "Badische Hof". When I arrived there he greeted me as if we were old acquaintances and good naturedly let me ask him about the history of spectral analysis. He gave me lots of useful details.

Kopp, who I'd also wanted to see was unfortunately not in town.

From Heidelberg I went on to Karlsruhe where I got to know Engler and Bunte—though I can't recall any details of the meeting. From my next stop—Stuttgart—I can't recall anything except for my visit to the rather odd Swiss chemist Urech. He too had started to study the kinetics of chemical reactions and used the inversion of sugar by acid which was the classical model system introduced by

[8]Ostwald refers to the fact that Horstmann applied the Second Law of Thermodynamics to chemical equilibria. Horstmann studied in Zurich under Rudolf Clausius, the discoverer of the Second Law of Thermodynamics.

Wilhelmy. It had the advantage that one could follow the reaction without interfering with it simply by measuring the optical activity.

I met him in a little room which contained his solutions, the polarimeter, a constant temperature oven and a bed. The unhappy man only left his prison for a few minutes each day because he was constantly engaged in reading the thermometer and adjusting the gas input to the oven so as to keep the temperature more or less constant—in other words he had converted his entire room into a thermostat. I predicted that he'd find the same order for the reaction rates of the acids as I'd already shown for the distribution of a base between two acids—and that is how it turned out.

My next stop was Tubingen, where Lothar Meyer, whose scientific interests were similar to mine, worked. He was at least 20 years older than me. I presented myself in the late afternoon and from the unfriendly mien and the surly look on his face I'd obviously disturbed him in the middle of some work. He asked my name which his servant had obviously not understood and looked me up and down suspiciously for after the long journey I did not present the picture one expects of a full professor. However once he'd heard my name and confirmed my identity with a quick question, his face lit up, and when I look back on that my heart melts once again. He took me home and introduced me to his wife—the famous "Lotharia"—and I spent one of the pleasantest evenings of my journey with them.

A second happy memory from Tübingen was my meeting with the physiological chemist Otto Hüfner. He was a pupil of R. Bunsen and had often managed to solve his physiological problems using physical methods. I visited him in his laboratory which was then in the old castle with its meter thick walls. He was of average size and looked the part of the typical German professor: short red-blond hair, a beard of the same hue, spectacles and a shade careless in his bearing and dress. He was happy to meet me. We had a lot to talk about so we had lunch together—he was single—and drank light Rhine wine to a lasting relationship which indeed held until his much too early death.

From Tübingen I went on to Zürich which was the penultimate goal of my pilgrimage. I was looking forward to getting to know Victor Meyer who was then involved in sensational studies of the density of vaporized halogens at the highest attainable temperatures. In fact I found him in front of his white hot furnace wearing an overcoat and hat because the windows had to be wide open to make it possible to remain in the room. He was a slim man with dark hair and beard, a typical, very handsome Germanic face which was lit up by two bright blue eyes. He knew my work and engaged me in an interesting chat, though during it his nervous temper, which resulted from overwork, was obvious and he tended to impassioned arguments which aimed at a dramatic effect and often ran rather out of control.

The physicist Friedrich Weber who I also got to know there was a completely different personality. He was averse to company and struck me as being gaunt and taciturn. He lived for his work to such an extent that, as his wife told me, he neglected her and the children. However, he thawed out a bit when I asked him a question concerning a detail from his last publication. He was obviously happy to meet someone who had studied the paper in detail and so then he showed me his lab which contained much interesting equipment, and then he took me home with him. His wife was astonished at this, but assured me that it was a most pleasant surprise.

The last city I visited was Munich, where I went after Zürich. Although I was rather exhausted from the rushed schedule of my journey I didn't want to miss the chance to see the art treasures of the city. At the technical University I met Erlenmeyer, at the university A. von Baeyer. From this conversation I remember his animated warning never to mention anything about work which had not yet been published. "None of my assistants and doctoral students is allowed to know what the other is doing". I must confess that this warning merely confirmed me in my contrary view of the matter.

When I visited Baeyer's laboratory I didn't find anyone there because it was too early and in any case a Saturday, but I was told that at lunch time one could find them at the House of Arts and Crafts. I went there and had just started to eat when a voice from a noisy nearby table asked, "Have you seen this guy Ostwald yet?" "Who's this Ostwald" asked a second. "The guy with the volumetric chemistry" shouted the first speaker in reply. I introduced myself and soon was part of their happy circle of ambitious young people of my own age with whom I spent the afternoon and evening. I don't really remember who all took part but von Pechmann, Knorr and Bamberger were probably there.

From Munich I travelled straight home and arrived there four weeks after my departure. With the resilience of youth I soon put the exhaustion of the rushed journey with its countless impressions behind me and was left enriched with the many pleasant personal relationships I'd made and the numerous stimulating scientific discussions of many different points that I'd had. These certainly had a major impact on the shaping of my career since without personal contacts it would be almost impossible to get appointed to a professorship in Germany.

However I didn't experience any change in myself as a result of this first journey to "Europe" as Karl Schmidt liked to refer to his visits to Germany. I was about 30 years old, had followed my own way for the last 10 years and this had led me ever onwards and certainly closer to my goal. I was more on the lookout for new means of solving the problems that interested me rather than having my goals defined for me by others. And so I regarded the journey more as welcome chance to waken the interest of my peers in the areas which interested me, and which till then had been largely ignored by most active researchers.

Chapter 10
Back in Riga

Building the Laboratory. Having now seen almost all the important labs, I resumed my teaching duties again—slightly tired but very excited by all the contacts I'd made and the things I'd experienced. I also took part in the planning of the new laboratory building. It was designed for 150 students. There had been a rapid increase in the number of chemistry students since my appointment and currently there were 120. Those in power in the administration reckoned that the number had now plateaued but when the new building was finished in 1885, 195 students had to be accommodated, and in the following year there were 210. As a result, despite the new building, the problem of space had not been solved, and the necessary modifications to make bigger student laboratories made a mess of the original design.

The textbook. The second great challenge that the journey had brought me was the drafting of the text book on general chemistry.

Though I was in the unusually fortunate situation of already having a publisher before the book was written, I'd taken my time with the actual writing. I'd already started to collect material in Dorpat and had to a large extent got it all together and, in addition, I had most of my lectures in written form. The publisher had asked me to send them some sample pages to Leipzig so that they could decide on the technicalities of printing. I gazed at the printed pages they sent me with all the excitement of an author-to-be and showed them to all my friends. But I hesitated to give the final go-ahead because I was still struggling with the problem of finding the most rational way of ordering the material. I tried to get some clarity in the matter from Wilhelm Wundt's "Logic of the Exact Sciences" ("Logik der exakten Wissenschaften") and discovered that this far-seeing thinker had already predicted the development of chemical kinetics with whose experimental implementation I was now engaged. This encouraged me to write to him and send him a copy of my last paper to show him how right his prediction had been and to ask his advice on my problem with the order of the presentation of general chemistry in the textbook. He answered fully and in a very friendly fashion, and even if I did not take all his suggestions they nevertheless helped me to see the problem with new found clarity.

R.S. Jack and F. Scholz (eds.), *Wilhelm Ostwald*, Springer Biographies,
DOI 10.1007/978-3-319-46955-3_10

Part of my hesitation stemmed from the attitude of my teachers Karl Schmidt and Lemberg. Both of them despised the writing of books and saw the heart of science firmly placed in experimental work which was to be described as tersely and dispassionately as possible. Neither of them recognised the value for the advance of knowledge of ordering and summarising what was known, although there were numerous examples where the publication of a textbook had brought the results from disparate areas into context and had often made the development of the field possible. I did recognise this and the success of my textbook soon became a further example of this rule.

Nevertheless it had still needed the final prod of my personal encounter with the publisher Dr. Engelmann before I overcame these various causes for delay. Soon after my return home I set to work to finalise the organisation of the text book and little by little got the manuscript ready so that the first part (the first half of the first volume) was printed during 1883. As is the publishing custom, it carried the date of the following year, 1884.[1]

From then on, year in year out, a similar amount was finished so that by the end of 1886 the whole two volume work was completed; it was dated 1887.[2] My desire to meet the regular deadlines kept me going over several periods when I was tired out and once again the principle of the moral flywheel came into play for otherwise these periods would not have been so quickly overcome.

The book was quite a challenge to me. A few years previously a new edition of Gmelin's "General and Physical Chemistry" had been brought out by A. Naumann which disappointed me for it consisted to a large extent simply of articles which he'd put together over the years for an annual report[3] on his area of chemistry and I missed the broad integrating vision which seemed to me to be the essence of such an undertaking. Because of this I felt it necessary to do more in this area than otherwise might have been the case. To develop these generalisations it was necessary to summarise the ways in which the work of many scientists over many years had led to the solution of a problem. In addition many of the results had to be calculated anew or otherwise worked over to permit comparisons from which a consistent picture emerged. This often required considerable work on the already published data. Nevertheless I look back on this period of long hard work with pleasure because the consistency of the final results was a source of real lasting happiness. I'm convinced that the positive impression which the first edition made on many readers should be, and must be, considered a reflection of the author's feelings of happiness while writing it, for experience has shown me that the mood of an author—even of something as comparatively dry as a chemistry textbook—will manage somehow to communicate itself to the reader. I think the reason that

[1]"Lehrbuch der allgemeinen Chemie" Volume 1 "Stöichiometrie".

[2]"Lehrbuch der allgemeinen Chemie" Volume 2 "Verwandtschaftslehre".

[3]Ostwald refers to "Jahresbericht über die Fortschritte der Chemie und verwandter Theile anderer Wissenschaften", an *annual review* founded in 1822 by J. J. Berzelius and edited later by various chemists, among them H. Kopp, J. Liebig, A. Strecker, A. Naumann, etc. The series ended in 1910.

most of my books are read eagerly and with pleasure by so many is because I wrote them all gladly and with pleasure. I never had any problem putting my thoughts into readable sentences and so there was always some spare energy left over which could be used to maximise the clarity both of the overall design and of the details of the text. On the other hand I've also come across scientific authors of the highest standing who struggled with every sentence and still never got it quite right. Their books are inexhaustible sources of knowledge which those who work in the field must frequently consult—but you can't read them, you can only use them for reference.

A temptation. Life at home had changed because we now had two children and a third was on the way. My salary had seemed enormous to us in comparison with what we'd had in Dorpat, but nevertheless things soon got tight what with the increase in the family and the fact that life in the rich trading city was more expensive than in the province. The money I'd got for the textbook made little difference so I started to look around for some way of improving my financial situation. Just at this time I was offered a well paid job as the director of a flourishing company for medicinal waters in Riga. I was ready to take this on as an adjunct to my work as a chemistry professor just as others in the polytechnic directed engineering works and other technical companies, but in my case I would have been required to resign my chair. Neither my wife nor I considered this for a moment and so we let this offer go without regret in the hope that we'd find some other means of improving our financial position which did not involve me giving up a position that allowed me to exercise those abilities which I prized most. For the time being I supplemented my income by holding popular lectures for which I found many appreciative audiences in Riga.

Friction. I'd begun to think it was time for a change also for other reasons. At the beginning the great difference in age and character between me and the director of the Polytechnic had been easily bridged by his genuine goodwill. However the chief administrator in his office, who'd managed to worm his way from a position as a servile scribe to become the director's right hand man, was a good deal less friendly. He probably missed in me the degree of respect which he felt was due to him through his position, and began to intrigue against me. The bills for all the equipment and materials ordered by the section heads went through his office. To put the laboratory into a state suitable for the new situation I'd had to purchase a lot of equipment, as I did not even have a still to make pure water. This involved funds which I had the right to use without having to consult the director who, as a mathematician, didn't understand the needs of a chemistry laboratory. This sometimes led to a degree of annoyance which was nurtured by the servile administrator who examined each bill searching for some irregularity. He was soon able to triumphantly present the director with a serious violation for amongst bills for Bunsen burners, tripods and clamps there was itemised 28 marks for a parlour chair, (Salonstuhl). Bubbling over with anger the director came down to the laboratory and in front of everyone else gave me a lecture on the use of institute funds which were to be used solely for equipping the labs and not for my private living quarters.

I tried in vain to interrupt him, but he grew ever more enraged until, in a devastating finale he showed me the offending entry in the bill. I was naturally very curious and burst out laughing when he pointed out the entry. It wasn't a chair (Salonstuhl) but rather a "Ballonstuhl"—something which is used to hold and manoeuvre the large glass vessels in which acids are kept. I had one of these brought and demonstrated to the enraged despot. He departed, not without repeating his warning that I must adhere strictly to the regulations, and of course he blamed me for this embarrassing scene.

In order to have access to the necessary literature for the work on my textbook I'd got, with the permission of the director, a key to the well stocked Polytechnic library where I could work undisturbed on Sunday mornings. I was sitting there one morning when the door was opened. A glance showed me that the intruder was that servile administrator who lived on campus and who, now dressed in leisure clothes not intended for general inspection had been on a morning walk. From the sound of things he felt perfectly at home and I could well understand his outrage at my unexpected presence for which he wasn't prepared.

I'd almost forgotten this ridiculous incident when I received a decree from the director demanding the return of the library key on the grounds that I could simply borrow the books I needed and take them home with me. It was in vain that I explained that I had to chase information from book to book—the man who had the director's ear looked all set to defend the place where he paused on his Sunday morning walk. It was only after a personal meeting with the chairman of the board of governors, who at that time was also mayor of Riga, at which I emphasised the serious disruption of my scientific work of which there was little enough at the Polytechnic, that I finally got free entry to the library again.

Nevertheless this burlesque did have the advantage that my serious efforts in science had been registered and welcomed at the highest level. From then on I was supported from there, for example with readily granted support for scientific visits to Germany about which I'll give an account a little later.

Feeding the mind. The days sped by, fired with such large and small events. In me the overflowing energy of youth had not been exhausted as it had been in so many of my countrymen by the alcoholic abuses of the student years. Not, I should hasten to add, through any early insight on my part—that came later. It had been an entirely automatic consequence of my passion for chemistry which in Dorpat had often driven me from student revels back to the laboratory or to study and with my growing success I'd gradually lost all interest in the joys of the beer halls. On top of that the others considered my view, that there was no point in extending student drinking rituals into real life, somewhat unsociable and as a result there was a pretty rapid separation before our differing views could lead to any alarming sorts of disagreement.

I found my relaxation in the rotation of my work. Research, teaching and writing—the three pillars of my scientific life—were all daily challenges and gave me triple joy when I got permission from two to work for a while on the third.

Private entertainment was an important part of life in my hometown but I had no time for it. I annoyed my kindly father in law by telling him that, though I enjoyed the company, I'd nevertheless no longer be coming to his Friday evening get-togethers. I managed this diplomatically by telling him that I was allergic to the tobacco smoke, which was much in evidence at these events, and which made me feel ill so that the next day I wasn't fit to work.

My father was a heavy smoker and perhaps because of that I disliked tobacco and in my youth didn't join in when the other boys started to smoke.

I passed my student years in a thick cloud of tobacco smoke and often enough hated it. Well meaning friends told me that I'd stop being annoyed with it as soon as I started to smoke myself. I didn't take them up on this for the advantage they promised seemed to me to involve too high a price. And when I saw how unhappy the revellers became when late in the evening the tobacco ran out before the beer, I felt my dislike of the tobacco devil quite justified. I haven't changed my views on this since then and there isn't the slightest chance that I will do so in the few years remaining to me. I never had the feeling that I'd missed out on one of the joys of life by not smoking.

On the contrary I think that by not smoking I won a great deal. I'm not talking about the money saved because I was never in such financial dire straights that the cost of tobacco would have been a problem. What I mean can be illustrated by the following example. There arise in everybody's life through the necessary dealings with other people numerous empty gaps of 15 or 30 min. The smoker passes the time by lighting up a cigarette. I, on the other hand, found it necessary to fill these gaps some other way. If I was at home or in the lab I'd open a book, if books were not at hand then I start to turn over some scientific problem in my head. I can't even begin to count the number of unexpected discoveries I made in these quarter hour gaps, but they were many.

My necessary physical regeneration took place in the long summer recess. In Riga, as everywhere in the north, the academic year was divided into the autumn and spring semesters so that the holidays were at Christmas and in the summer. That is a lot more sensible than the system in Germany where the holidays are at Easter and in the autumn between the winter and summer semesters because these holidays fall in times of the year that are better suited to work than the semesters and it's all made more complicated by the stupidity of Easter being a moveable feast. In the Polytechnic, lectures and exams came to an end in the middle of June. The hot months that followed were spent "on the beach" which is to say on the Baltic coast which lay about ten kilometres from the city. Here, midway between the sand dunes and the woods behind them lay the summer houses of the well to do citizens of Riga and the apartment houses which the coastal farmers had built and rented out. In the middle were the houses of the rich while closer to the city or farther away the buildings were smaller, cheaper and simpler so that there was something for every purse. Because there was only a tiny garden at our house in the city, I brought the family out to the summer house as soon as the weather permitted. I then had to stay alone in the city until the end of term when I could move out as well.

In these summer weeks the days were comfortably filled with swimming, eating, chatting and going for walks. I'd made it a rule for myself that in the holidays I'd answer all the children's questions. The children soon noticed this and exploited it to the full. I can see myself at my most common occupation, swinging slowly in a hammock slung between two pine trees (no other trees would grow in the dunes) while the children sat on my stomach and asked question after question till we were all overcome with the heat of the day and drifted off into sleep.

At the beginning I had some misgivings about spending my time like this, but on the one hand I owed it to the children (and they were later grateful for this), while on the other I made the following reassuring observation. After about 4 weeks of this lazy life without any need or longing for scientific work I'd begin slowly to turn over in my mind my next research project and this grew ever more until I'd be impatient to start and couldn't wait to test my latest thoughts experimentally. Once the short northern summer was over and I had some free time in the lab before the start of the semester I was able to immediately carry out the long planned experiments so that I could enter the lazy summer days as a profit rather than a loss. In those years of frenetic work these holidays made it possible for me not only to maintain my physiological balance but also to collect excess energy which was available later and which substantially increased my performance. Since the value of this performance grew faster than the output of the extra energy I'd tanked, the holidays provided me at the end of the day with a significant excess of achievement.

Sadly It was only much later that I developed this theory of holidays and so I often failed later on in my life when the challenge were even greater, to give myself this extra boost.

Chapter 11
My Colleague

Arrhenius. Till now I'd worked in my special field more or less on my own, but now this under populated field area started to attract independent minds and other researchers coming from different areas began to move in. The first of these was Svante Arrhenius.

I'll never forget the day—it was in June 1884—on which I heard his name for the first time. All on that 1 day I'd got a painful gum infection, a wonderful new daughter and a manuscript from Svante Arrhenius entitled "Études sur la conductibilité des électrolytes".[1] All that together was too much for 1 day and I had a feverish night with bad dreams.

The gum infection soon went away and my daughter caused no great problems, since the birth was easy and her mother recovered surprisingly quickly and I only had to take my role as father seriously in the later stages of her development. But the paper caused me headaches and more than one fretful night—and this was most unusual for me. What he'd written was so contrary to everything that was known and accepted that at first I thought it must be nonsense. But then I saw that the obviously very young author had presented some calculations of the chemical affinity of the acids which agreed with the values that I'd already reached from a completely different starting point. Finally, having read the paper in detail I convinced myself that this young man had approached the problem of the relationship of acids and bases—a problem I'd expected to spend my life working on and for which I'd

[1]This is the Ph.D. thesis of Arrhenius submitted to the University of Uppsala. The correct title is "Recherches sur la conductibilité galvanique des électrolytes. Première partie: La conductibilité des solution aqueuses extrêmement diluées déterminée au moyen du dépolarisateur. Bihang till Kongl. Svenska vetenskaps-akademiens handlingar 8 (1884) No. 13. Seconde partie: Théorie chimique des électrolytes. Bihang till Kongl. Svenska vetenskaps-akademiens handlingar 8 (1884) No. 14. Stockholm 1884. Kongl. Boktryckeriet.

R.S. Jack and F. Scholz (eds.), *Wilhelm Ostwald*, Springer Biographies,
DOI 10.1007/978-3-319-46955-3_11

managed by hard work to clarify a few points—the most important of which was the demonstration that there is a fixed quantity for the affinity which is irrespective of the type of process involved. He had looked at the problem in a much more generally applicable way and had, to some extent, even solved parts of it.

You may well imagine the jumble of feelings which this caused in a young researcher trying to build his future in a field he believed to be under populated but who now found that he had a forceful and energetic competitor. His work had its weaknesses (which were soon to be seized on by other critics in a rather excessive way), so that I could still believe, if I wanted to, that he'd hit on the right answer simply by chance.

I struggled inwardly for a few days just as the the black and white retainers struggled for the soul of the huntsman in Bürgers ballad "The wild huntsman" (Der wilde Jäger). Since not very many people were interested in this field, it wouldn't have been difficult to simply ignore this interloper. One simply had to damn the whole work on the basis of the errors it contained and in any case the publication was in the Annals of the Swedish Academy of Sciences which automatically restricted its impact because few chemists read this journal. I only had to ignore the paper and I'd have got rid of my competitor, maybe not forever, but at least for the immediate future.

I must add that these thoughts were never formulated then with the clarity and precision I give them now. It was more in the form of waves of feeling that now and then managed to wash over the borders of consciousness.

The details of the techniques for dealing with unwelcome colleagues and competitors were things I learned only later, initially by witnessing the intrigues of colleagues against me and then, once I'd recognised what was going on, by following the developments in the chemical literature. On the other side I held to the scientific ideals which I'd learned from my teachers Karl Schmidt, Johann Lemberg and Arthur von Öttingen, as the prerequisites for work in this highest region of human endeavour. The selfless and heartfelt letter of recommendation which Karl Schmidt had sent to the Polytechnic and which was doubtless decisive in their decision to offer me the professorship was an ever present example for me. In addition, it is always a joy to find a new colleague with whom to open up a new field, especially when there is room for everyone and particularly when such new colleagues are equipped with those intellectual tools which I lacked and which together with mine would guarantee that we would both succeed in progressing.

The electrical conductivity of acids. In a few days it was clear to me what direction I must follow. I wrote to the author of the manuscript in Uppsala and tried to get a clear detailed view over one major issue. This was the relationship between the electrical conductivity of acids and their relative chemical affinities which I had established.

Already in Dorpat I'd noted the work of F. Kohlrausch who'd invented the measurement of the resistance of electrolyte solutions using alternating current.[2]

---

[2]Kohlrausch F (1876) Ann Physik 235:233–275.

For the few acids he'd investigated the order of their electrical conductivity fitted with the order of their chemical affinity. However, I wasn't able to make any sense of this and the measurement of conductivity was at that time a complicated business so that I wasn't in a position to determine it for all the acids I'd investigated.

But now Arrhenius presented in his paper an interpretation which not only explained the parallel between the two sets of values but actually demanded it. In the meantime Kohlrausch's method for determining conductivity had been simplified to such an extent that I could set it up myself.

I hadn't procured any electrical measurement equipment in Riga but now my love of tinkering which I'd retained from my boyhood came to the rescue. I'd found an old but adroit mechanic in the Polytechnic who was just hanging about because no one seemed to need his services. I at once took lessons from him at the lathe and bench vice which surprised my elder colleagues and caused a certain amount of amusement amongst them. In this way I was able to make modified versions of the necessary equipment and was able to borrow for just a few days a resistor box from the local post office which we copied so that in a short time I was able to determine the electrical resistance of solutions with more than enough precision. I measured the whole collection of acids which I'd saved from my earlier work and, with a pounding heart, found that one value after another fitted exactly into the prediction. Since all the solutions were to hand and each measurement only took a few minutes, the results poured in faster than I have ever otherwise experienced. The end result was that I had here a means of determining in just a few minutes the chemical affinities which, using my old methods, had taken as many days. I quickly wrote up a short communication[3] of my confirmation of Arrhenius's idea and sent it to the editor of the Journal für praktische Chemie (Journal of Practical Chemistry). By a happy chance it could be published right away. In it I expressed my firm opinion that the work of Arrhenius was one of the most significant results to be published recently in the field of chemical affinity.

The second journey. On my first visit to Germany, which I'd made because of the construction of the new laboratory, I'd been under such time pressure that I'd only been able to gather an impression of a tiny part of the country's vast treasures in the arts and sciences. Because of this I wanted to make my second visit more relaxed. In addition it now turned out that the library in the Polytechnic was insufficient for my now well advanced work on the text book, so that it was now desirable, indeed necessary, that I visit other larger and more complete libraries. I therefore applied to the Board of Governors for a travel stipend which was given me for the summer holidays of 1884. After these events were all completed I decided to follow up my written contacts with the strange Swede with a personal visit and to travel back to

[3]Ostwald W (1884) J Prakt Chem 30:93–95.

Germany via Sweden. Since there was a direct steamer connection between Riga and Stockholm the matter was not difficult to arrange.

Visit to Uppsala. After I'd visited the museums in Stockholm and had seen the wonderful landscapes which make it one of Europe's most beautiful cities, I travelled on to Uppsala where Arrhenius was waiting for me.

So that we'd recognise each other he came to meet the train holding up like a flag a reprint of my paper which I'd sent him.

This meeting was the start of a long friendship which has continued ever since. He insisted that I stay with him for there was naturally a lot to discuss and we planned how we would together study the extensive new area which had opened up before us.

On the first evening we went to a beer garden where as the sun set the waiter brought us felt blankets to ward off the cold evening fog. From time to time Arrhenius would be called over by young people who greeted him warmly and shook his hand. He told me later that these were student friends who wanted to congratulate him on the special honour done him by a well known foreign full professor who had come all the way to Uppsala to visit him and who treated him as an equal. It turned out that he had used the work from his publication to produce a thesis for his accreditation as a university teacher. He'd submitted this to the faculty but there had been some problems concerning its contents. Earlier he'd run up against the physicist Thalén whose lowering and unfriendly manner I would later get to know. I should say right away that this was the only Swede of this type I met for otherwise they are about the friendliest nation I have come across. Thalen had either refused to allow Arrhenius to work in the physics institute or at least made his life there so unpleasant that he'd gone instead to Stockholm to the academy member Edlund who worked in the same laboratories in which Berzelius had spent the last 10 years of his life. I visited Edlund later with Arrhenius and was met by a strange little bent over man whose goodness of heart shone out of his eyes. He showed me the remains of Berzelius's chemicals and equipment which lay all higgledy piggeldy in a cupboard. I gazed in astonishment at the balance with which Berzelius had made more than one extremely exact measurement because it was a very primitive thing that even back then one would hardly have offered to a beginner. It was immediately clear to me that the equipment is less important than the man who uses it. I saw here a justification for my habit—inherited from my rather impoverished childhood—of being satisfied with the simplest equipment.

Arrhenius had carried out the experimental part of his project in Edlund's lab and had measured the conductance of a number of electrolytes which Kohlrausch had not looked at. However the procedure which Edlund had suggested he use turned out to be difficult and not terribly effective. He later brought this apparatus with him when he came to Riga to work with me. However once he saw the simplified version of the Kohlrausch equipment which I had built, he used it instead and took the unopened box containing his equipment with him when he left.

Of course in Uppsala I made the usual round of visits. I was received in a particularly friendly manner by the respected chemist Cleve, who could not hide his surprise that I was so impressed by the strange ideas of Arrhenius. Nevertheless, he was prepared to hear me out on the subject. This discussion took place some 2 years before the theory of the dissociation of electrolytes had been formulated and yet Cleve, with rigorous logic, drew one conclusion after another from Arrhenius's hypothesis and at the end asked me, "So do you really believe that in this beaker with the solution of sodium chloride the sodium atoms swim about on their own?" I answered, "Yes" and he glanced at me sideways as if to signify that he now had reason to doubt my chemical common sense. That didn't in any way change his friendly attitude and he invited us both to lunch the following Sunday. We got there a little late because as Arrhenius put on his best clothes the trousers ripped and it took a while to repair the damage. Instead of sitting down at table we collected plates from a well arranged set and went to the lady of the house who distributed our soup and other food. One went off with the booty and—standing—ate as well as one could. I was told later that this was an old Swedish custom that nationally minded people were trying to revive. I do so hope that in the meantime they have given up the attempt. All the other cheerful lunches and dinners I had in Sweden were eaten at table as in the rest of Europe.

I also remember an excursion to the baroque castle of Skokloster, though plans for the future were more at the front of our minds than old weapons and furniture. We agreed that Arrhenius should come to Riga as soon as possible so that we could work together on the solution of our problems. For that it was necessary that his thesis be approved and that he be given a travel grant. These difficulties had been much reduced by my personal appearance in Uppsala and in fact were soon overcome. Later Arrhenius wrote me that none of it would have been possible had it not been for my visit.

Apart from these things which were important for me, I hadn't forgotten the reason for which I'd been given the grant by the Board of Governors. I turned to the library there to look for the references I needed only to find that this was a matter which lay in the hands of Professor Thalén. This was the situation mentioned above, in which I met the exceptional Swede who was unfriendly. It turned out that he just had a rough outer shell, for my request was granted. After that Arrhenius and I went to Stockholm for the meeting with Edlund which I described above.

In addition I met a number of Swedish chemists with whom I thereafter stayed in contact.

Chief amongst these was Oskar Petterson, a small wiry figure with a square head, short light moustache, a sunburned face which was darker than his hair. He moved liked a sailor and was in fact a passionate yachtsman. His friends told me that he considered a summer wasted if he had not at least once been shipwrecked, though he always managed to get out unhurt. He had recently published remarkable papers in the field of physical chemistry and I valued the independence of his mind. He introduced me to his friend Nilsson[4] who as Professor of Agriculture lived in a marvellous official villa outside the city. He didn't have much in the way of

[4]The spelling is incorrect. Ostwald refers here to Lars Fredrick Nilson.

teaching duties and so could devote himself to science. He was blessed with a sweet wife and a house full of children and in general lived a life which seemed to me to be ideal. Seeing that this sort of thing was possible awoke in me the desire to achieve something similar for myself. Through all the changes of my later life I kept this goal in mind until I was finally able to realise it with my country house "Energy".

From Stockholm I went on to Göteburg. Close at hand lay the beautiful Aspen lake which I tried capture in watercolour paintings. I'd already bought a paint box and paper in Stockholm and had tried to paint some of the wonderful scenery I'd seen all around there. The results were not terribly impressive and I was convinced that this technique was inadequate for my purposes. When later, in Norway, I had to leave one marvellous scene after another unpainted because the colours would not dry in the damp air; it became clear to me that the only thing that would work would be oil paint. After I returned home I tried them out and afterwards enjoyed many happy hours painting.

From Göteburg I took a ship through the Schären islands to Christiania. There a great treat was waiting for me—a meeting with Guldberg and Waage.

Christiania. We entered the Christiania fiord on a misty morning. To begin with we sailed between rocks bare of vegetation on which the waves broke, later came larger islands with the odd weather beaten pine tree. The further we went the more the vegetation increased and with it came the first signs of civilisation—fishermen's huts and farmhouses. From time to time we could see the green banks. Suddenly a ray of sunshine broke through the clouds and the silvery light illuminated a glittering city at a bend in the fiord. Then a cloud and a bank of mist covered it all up again. The experience was like a well constructed symphonic movement. It was this that sparked my later thoughts on the art of motion pictures.

Once I reached the city I immediately went to visit my two colleagues whose solid work had smoothed my first entry into the field. I'd established from their addresses that Waage was a chemist, but that Guldberg was a mathematician, and so I went first to Waage. I was met by an elderly man whose unkempt hair and beard almost covered his face. He was of short stubby stature and looked more like a farmer than a professor. He glared at me suspiciously as I introduced myself. When he'd finally grasped who I was he was overcome with joy. He danced around me shouting "So young. No! He's so young". He'd imagined that I must be a dignified old man like himself and his brother in law Guldberg and had a hard time convincing himself that it was really me.

Waage insisted that I stay to lunch and sent to invite Guldberg too. At home he had a little flock of children, some grown up some still small—most of them were daughters. In addition there were several older ladies of the sort one sees around pastors. It turned out that he had a strong interest in youth clubs, church homes, abstinence societies and the like, all of which had a strong Christian bent, though that didn't stop him being an enthusiastic hunter and mountaineer. The main part of the meal consisted of snow grouse which he had shot himself and he recounted the details of the hunt in a way that reminded me of my father's hunting stories.

In the meantime Guldberg had arrived. He was outwardly the opposite of his brother in law. He was tall and slim, with an aristocratic-intellectual face, protruding nose and white hair and beard both of which were cut short. He looked more like a senior military officer than a professor. It soon turned out that the idea behind the work had been his. Waage seemed not to have done much more than deliver the chemical analyses. Nevertheless in questions of hunting and mountaineering he was not a whit less enthusiastic than Waage and insisted that the next day I should come and taste his snow grouse. These were hunted high up in the mountains, immediately roasted, tightly packed and dipped in melted butter so that they were preserved sterile and could be kept for a whole year.

Although it was already late in the year my Norwegian friends wanted to show me some of the beauties of their country. Since they were held up by their lectures they worked out an itinerary for me up to Hönefos by train, car and on foot which I undertook over the next few days. Since I'd grown up in flat country the wild and magnificent Norwegian scenery shook me with an intensity I'd not expected.

The weather was almost always dull and it often rained so that the distant scenery was usually hidden and yet the 3 days of wandering left me with unforgettable memories. It was the first such experience I'd had—and the strongest. What particularly impressed me was that I experienced water not as a peaceful element but as a living crashing cascade, which was often enough of enormous power. At Högsund I could see within the breadth of one glance three huge waterfalls. I returned to Christiania shaken and soaked to the skin.

From there I took a ship to Copenhagen to visit Julius Thomsen the thermo-chemist whose work had been an ideal for me and to whom I wished to pay my respects. My Swedish and Norwegian colleagues had warned me that in him I would be meeting a self-assured and rather unapproachable colleague. However he had recently publically praised my work so that I had no hesitation in going to see him. I indeed met a very distinguished closely shaved man—every inch of whom might have been a privy councillor, but who had an unusual growth on his left temple. Nevertheless there soon developed an intensive and fruitful discussion and after an hour as I was leaving he retreated far enough from the formal plane to suggest that I visit the Tivoli Park in the evening where there would be a garden party. I went there and it was very nice.

During a visit to the Thorwaldsen museum, I experienced for the first—and almost the last time—a sense of artistic elation from looking at sculptures. A visit to the Church of the Lady with its paintings from the late pious period was by comparison a disappointment.

A steamer brought me to Lübeck which I was glad to see for merchants from this city had been the founders of Riga. Lübeck is really very pretty. From there I went on to Leipzig to discuss the success of the first part of my book and to say hello to all my acquaintances there. I was also going to meet up with Arrhenius in the city. My publisher couldn't tell me much but he was obliging and friendly. I met Kolbe and von Meyer and was once again convivially greeted as was Arrhenius whom I introduced to them. However the plans to offer me a professorship, which a year ago had seemed so firm, had not got anywhere and Kolbe pointedly refused to be drawn on the matter. My other Leipzig acquaintances were all away on holiday. In his lab Kolbe showed me the start of his work on indigo. He'd challenged his rival von Baeyer, who'd made significant progress in this work but then given it up for a while, and bet him that he'd solve the problem within a year. However he had not forseen that his early death would free him from his duty in this respect. About a year later he'd attended a meeting in the best of health at which decisions were to be made about the future of a Society of which he'd long been a member and in which he expected to have a senior position. However the majority voted against him and he was so enraged that on the way home he suffered an apoplexy and soon died.

Arrhenius and I arrived in Magdeburg in time to attend the Natural Scientists Meeting there at which I was slated to hold a lecture. It was the first meeting of this sort that we'd attended. We were lodged as guests with the family of a young businessman where we were kindly looked after and I learnt the blithe goings-on of this science fair. I knew quite a few of my colleagues from my previous visit a year and a half ago. Once again there were hints that I might be appointed to a professorship, though after my recent experience on Leipzig I regarded them all with polite scepsis. In particular there seemed to be some interest in Göttingen. I don't remember whether my lecture was a success or not. I conclude from that that it wasn't a success otherwise my memory would have worked a little better. From there we went home, which I was longing to do, and in any case the holidays were coming to an end.

The only other thing I can remember from the Magdeburg meeting is that at the city's one-time mayor—Otto von Guericke of air pump fame—was smuggled into each of the innumerable speeches and toasts as a means of making some sort of connection between the Natural Scientists Meeting and the city.

Chapter 12
Progress

Work on conductivity. I got back to Riga with my head overflowing with research plans but first I had to deal with the move to the new laboratory building and the installation of the various new pieces of equipment. This challenge was made more difficult by the fact I mentioned already that the administration had substantially underestimated the growth in the student numbers so that by the time it was opened the new laboratory was already overcrowded. This situation remained unchanged for the rest of my time in Riga.

I needed a new assistant and since there were no suitable candidates amongst my Riga students I asked Lothar Meyer if he could suggest a suitable candidate. He suggested the Swiss Ulrich Schoop who later on made a name for himself by inventing his metal spraying process. In Riga, however, he never really seemed to find his feet.

My experimental work was now directed to questions of electrochemistry, in particular to the conductance of electrolytes which, since Arrhenius's work, had become of such importance. First of all we had to modify Kohlrausch's methodologies to fit our particular problems. In Stockholm the newly invented telephone communications system had been rapidly developed and there I'd found suitable telephones. The tubular rheostat and the induction device needed to generate roughly sinusoidal current waves could be replaced with a resistance wire coiled around a 1 m long ruler plus the smallest commercially available induction coil. The measurement of electrical resistance was carried out using a measurement cell which Arrhenius suggested and the thermostat I'd developed earlier turned out to be indispensible because of the temperature sensitivity of the conductance we were measuring. Slowly our equipment took form and it was later used by numerous other researchers in thousands of experiments.

However, before I could interpret the results I first had to deal with an objection which came from an authoritative source, namely from G. Wiedemann in Leipzig. He doubted Kohlrausch's assumption that the effect of polarisation was completely

R.S. Jack and F. Scholz (eds.), *Wilhelm Ostwald*, Springer Biographies,
DOI 10.1007/978-3-319-46955-3_12

excluded by the use of alternating current and this drove me to test this point experimentally. The results fully substantiated Kohlrausch's view: there was no doubt that the procedure was generally applicable and it has proved its usefulness down to the present day.

I then started work and from a series of separate determinations established the relationships in the conductance of acids. The largest and most diverse variable was the effect of dilution. While the conductance per equal equivalents or moles of a strong, well conducting acid changed little on dilution, the effect was large for weak acids. Kohlrausch had already shown that for acetic acid, the only weak acid he examined, the conductance increased with the square root of the dilution.

It turned out that these two types of acid defined the two ends of a continuous series so that for all monobasic acids there was just one function describing the relation of dilution and dissociation. At a given dilution the different acids differ only in the position at which they would appear on the curve. However from that point, dilutions of all of these acids all follow the same curve. This was the first step to the establishment of the "Dilution Law" which my chemical colleagues were so kind as to refer to by using my name.

I of course looked for a mathematical formulation of this law which was later found by use of the dissociation theory developed by Arrhenius. I remember that I tried this possibility out on my data since it was almost suggested by Kohlrausch's remark that the conductance of acetic acid varied with the square root of the dilution. In fact it fitted more or less—but not closely enough to be accepted as an experimentally demonstrated result. A different mathematical model based on tangents fitted better but I was unable to find a rational explanation for its use. Later it turned out that my data contained a systematic error which had been caused by the ammonia content of the water in Riga. Ammonia was present only in trace amounts and was hard to demonstrate but it had an appreciable effect on the small quantities of the acids which were being measured. At that time all the later discoveries about the effects of trace contaminants on conductance lay in the future.

Joint work. In the meantime Arrhenius had as planned put his affairs at home in order and come to Riga. He arrived at the beginning of 1886 and we then had the joy of working and thinking together. We thought it best to divide the practical work between us. He wasn't familiar with the physical chemical measurement methods which I'd developed and so he started to work on individual problems such as reaction rates, conductivity and internal friction which could be approached with these methods. I, in the meantime, had progressed from conductivity to electromotive force and to measure this in absolute terms I'd built a dropping mercury electrode which had been described by Helmholtz. I'd already read in Dorpat the fundamental work of G. Lippmann concerning the relationship between the surface tension of the mercury and the potential jump on the surface and, not without some considerable difficulties, had managed to construct a proper capillary electrometer. Only many years later did the fruit of this preliminary work become apparent.

J.H. van't Hoff. Winter, spring and summer passed in uninterrupted work, and on top of that I was getting the last part of the textbook ready. This was the most difficult section of the book because it should describe the current state of knowledge about the field of chemical affinity. Shortly before this, I'd read a publication that caused me even more headaches than had that of Arrhenius. It was entitled "Études de dynamique chimique" and had been written by a completely unknown researcher whose name according to the title page, was J.H. van't Hoff. It described both theoretical and experimental studies of the laws governing the rates of chemical reactions. It deferred neither to past not present researchers and ended with some confusingly written paragraphs which suggested that the author had made considerably greater advances in the application of thermodynamics to chemistry than had Horstmann—or I.

Third journey. As the summer holidays approached, those around me pointed out that I seemed to be over worked and in need of a rest. I therefore decided to go with Arrhenius to Germany. He went to Würzberg and Graz while I first spent a few weeks on Rügen. Later there was going to be a glittering meeting of scientists in Berlin where we would meet again.

On Rügen I spent pleasant days in Saßnitz, Göhren and Binz.

In Göhren I met the philosopher Vaihinger who at that time was already professor in Halle. I look back with pleasure on the walks we had together and on our conversations though I fear my unrepentant naturalism and my obvious lack of respect for the details of philological work sometimes annoyed him. If I remember correctly, however, his "as if" philosophy was not mentioned.[1]

However the main thing this time was painting from nature. I'd started some years ago diffidently trying to paint landscapes. At the beginning I'd used water colours but after my experiences in Sweden where I'd learned that, especially outdoors, this could easily go wrong, I'd switched to oils. Since I didn't have a real teacher or guide in Riga and as there were only few I could learn from, I had to find my own way. In my scientific work I was used to having to put together or invent the necessary equipment and protocols, and so for this journey I'd put together a light, functional paint box which I could easily carry on walks which did not require an easel. It was the third or fourth version of the basic design which I'd originally tried out on the beaches around Riga and then improved. The exceptionally diverse landscapes on Rügen were completely new to me and spurred me on to a certain freedom in the conception and execution of the works. The idea for this came from my school drawing teacher Clark who'd suggested I paint not on canvas but rather on paper prepared for oil paint by soaking it in diluted glue. This made it possible for me to keep several wet sheets in the cover of the paint box without them damaging each other. It allowed me to paint on a sheet for just one to 2 h, which is how I preferred to work.

[1]Vaihinger proposed the idea that we behave "as if" the world matches our models.

On this visit to Rügen I experienced for the first time the beneficial effects that painting had on my tired brain. Much later my colleague in Leipzig Flechsig, the famous investigator of the human brain, provided me with the theory. Every type of mental activity is associated with a particular set of cells. If such a set is more than normally excited by overuse then it can be stimulated even by extraneous stimuli so that it never gets rest and cannot regenerate. However if some other group of cells is strongly stimulated then they call for energy and the circulation directs the energy stores there. The tired out cells in the first group are ignored and can now come back into equilibrium. I was not a great painter but I had a certain talent, in particular a good colour sense, so that the stimulus of painting was sufficiently strong to have the healing effect he'd described. Over the years I took advantage of this beneficial effect. Painting helped me to recover from the health threatening effects of working and at the same time it provided me with many very happy hours.

No less important was the fact that I could take my painting cure to the ends of the earth and never had to fear that I'd be bored. I didn't have to look for company in the holidays, which in any case would have meant that science would be discussed, but could go wherever I wanted on my own.

I think that what I've just described is generally true. Everybody who is involved in mental work needs a hobby he loves and the further the hobby is from his main work the better.

The Natural Scientists' Meeting in Berlin. Indeed I was fully recovered when I went on to Berlin so that I could take part in all the activities there without any adverse effects.

In Berlin there was much to be done. To begin with I saw again most of those colleagues whom I'd met on my hurried trip round the laboratories in 1883 and could use this opportunity to get to know them better. From the friendly looks of most of them this seemed welcome to them too. In the sessions I was able to present some of my results and was able to note an interesting tactical manoeuvre. While those chemists whose work was closer to mine greeted me warmly there was a group of organic chemistry "purists" who made it clear that anything outside of organic chemistry was unwelcome. And, although we were a tiny minority, the organic "purists", perhaps subconsciously, reacted jealously to the budding competition that we represented.

For example, I presented the results of my conductivity measurements on organic acids and had got the audience sufficiently interested that one of the "purists" considered the situation ominous. We were scheduled to leave in 10 min to take part in a trip to the lime works in Rüdersdorf and I'd timed my presentation accordingly. Suddenly, far too early and in the middle of my presentation, Professor Tiemann, a leader of the organic "purists" shouted out, "We have to leave now" and by doing so saw to it that nobody heard the end of my presentation. At the time I was so naïve that I thought that this had happened just by chance and only later did others explain to me the reason behind it.

On this trip to Rüdersdorf I experienced another aspect of the atmosphere in Berlin. We'd been well fed in Rüdersdorf and went from there to Erkner where there was a chemical factory for us to see. As we reached the place we could see a long line of tables set out for us which promised to challenge our well filled stomachs. It was indescribably astonishing to see that those colleagues who were in the front started going faster and faster and those behind hurried to overtake them until a serious race developed to be first at the coffee tables. What drove these educated and respected people, who all had had more than enough to eat, to lose all self control in this way? It was an involuntary expression of the competition and envy that was the daily fare of people in that city.

The question of the journal. The Berlin meeting gave me the opportunity to discuss a new project which I and my publisher Engelmann had been thinking about for a while. Once the fourth and last part of my textbook had come out and had been well received we'd wondered if it would not be a good idea to provide a forum for newly emerging work in general chemistry. The publisher was enthusiastic and agreed to produce a journal initiated and headed by me. Though my self confidence at this point was at its height I nevertheless felt a little diffident at the idea of officially taking on the leadership of the new chemistry and representing it to the rest of science not only in Germany but in the entire world. After all I was only thirty three and knew most of the established people in the field only fleetingly.

Because of this I'd written my publisher that while I personally had great interest in accepting the challenge, I wanted to use the opportunity presented by the Berlin meeting to get the views of my peers in the field.

The results of my survey were completely negative. The first person I asked said that there were far too many journals as it was and a new one would find no subscribers. The next said that since papers on physical chemistry in the existing journals were scarcely read, publishing them in a separate journal would simply ensure that no one at all would read them and that was not in our interests. Even people like Landolt and Lothar Meyer, whose good will was not to be doubted, advised me not to go ahead with such a risky project whose chance of success was so uncertain.

I noted these opinions and passed them on to the publisher, but we didn't give the idea up and instead quietly got on with our preparations.

Art. Apart from the scientific experiences the visit to Berlin brought many artistic ones. A wonderful performance of Wagner's "Valkyrie" was given at the Royal Opera for those attending the meeting and there was a glittering artistic festival at the Lehrter train station. This was the first time I'd experienced a large modern exhibition which contained some 3,000 paintings. Now that I'd made my first attempt at painting myself I looked forward with burning interest to the multifaceted statements of contemporary art from which I could learn so much. Sometimes when I was faced with a choice between science and art the art won and I spent long hours at the exhibition.

Journey home. I travelled back to Riga feeling completely rested and excited by all I'd seen. I was determined to come back to Europe again as soon as possible. Once again I got hints from several sides indicating that there was a serious possibility of my being offered a professorship at a University in Germany in the near future. Where and when were, however, still unclear. Nevertheless, now that general or physical chemistry had been presented in my textbook as a well ordered rich branch of science with clearly defined goals, many thought that it should be properly represented in the universities.

I found the idea of teaching at a university in German more than welcome, because the situation in Riga continued to deteriorate. Kieseritzky, who'd been the director for many years had retired because he felt that partly due to me the atmosphere at the Polytechnic had become too heated. Under his rather young successor there developed also now in Riga a rift between the professors who saw themselves as Balts and the German appointees.[2] This was the same problem that had overshadowed my final years in Dorpat and it was fought out over the proper response to the politics of the government in St. Petersburg. While the foreigners were ready to go along with the russification measures so as not to risk losing their jobs, the Balts were ready to oppose the measures in any way possible even at the risk of being fired.

To begin with this rift was just latent because the measures were postponed. During this period of waiting I'd invited a circle of like-minded colleagues to a dinner party at my house and the talk quite naturally turned to the measures which St. Petersburg was threatening to impose on us but which we still hoped might be avoided. Since we were all quite young there sprang up that evening some gallows humour and somebody suggested that we should think over how we would manage if we were all fired. I'd entertained my guests with all sorts of chemical tricks and so we decided that we'd make our living as wandering street performers. Everybody had to show what he could do. The first could swallow knives. The second laid the newspaper on the piano and melodramatically sang the adverts in the style of a Wagner opera. A third could fall backwards off a chair without doing himself any harm. Only the newly elected Director had nothing to offer and since he was an economist he was sentenced to look after the cash box.

Turbidity. This happy evening was about the last time that we innocently enjoyed our lives because soon thereafter, when the new rules became reality, conflicts between the two groups started. Although a Balt, I had good personal contacts with my German expatriate colleagues and in the beginning I was able to prevent the threatened conflicts developing. However nobody could be in any doubt that the points of conflict could not be avoided for long and that they would, and indeed must, constantly threaten to break out again. This in fact is what happened and it led to increasing bitterness. On top of this a German member of the university was accused of currency irregularities.

---

[2]Ostwald distinguishes between the Balts who where state residents of the Russian Empire, and the state residents of the German Empire.

On the other hand my increased standing in science—the newly finished textbook had already begun to have an effect on the development of chemistry—made it clear to me that there was a basic incongruity in the situation. True, the number of chemistry students had more than doubled during my brief tenure of the chair, but the students themselves were just the same as before. They were not really up to the standards of university students. Only a small fraction were from the German speaking Baltic states, where a level of culture was practised that was in no way inferior to that in Germany itself. However most of these went if at all possible to the university and that left only the less talented for the polytechnic. A lot of our students were Poles and Jews from the furthest corners of the Russian empire and they'd mostly had an inferior education. Then there was a lot of Letts and Estonians who were almost all fervent nationalists who only picked up German culture in order to be able to oppose it. They all had the simple goal in mind of passing the exams as soon as possible and going off to the sort of practical jobs which were easy to find as Russia's industrialisation was getting underway.

There was only one student from my time at the Polytechnic who later gained a scientific reputation and high positions first in Riga and then in St. Petersburg. This was Paul Walden, who is currently professor in Rostock. But Walden was already then an assistant in the physics institute and, though he was working along the same lines as us, he did so rather independently.

I have already related that I wanted to raise the standards by getting the students to carry out a proper scientific project. I soon found out, however, that the projects would have to be very simple and it wasn't that easy to come up with three or four dozen such topics per year and then to supervise the students and make sure they could complete them in the few months available. Nevertheless I put in this work because in their future jobs my students would often enough have to move into unexplored territory and so the experience—no matter how restricted it might be—of dealing with new things was for them of enormous benefit.

The journal of physical chemistry. The discrepancy between my official position and my actual work became even greater as the question of a journal for physical chemistry, which had received such a poor reception at the Berlin meeting, now became acute. My discussion in Berlin had set off a wave of resonance which was picked up by an enterprising Hamburg publisher. He wrote to me at the end of 1886 asking how I would feel about such an undertaking. In reply I related the negative response to the idea in Berlin and at the same time told my publisher in Leipzig that the idea was now being taken up by someone else. I got a very confident reply from Hamburg and a preliminary mock up of the new journal's cover. It turned out that the editor of the journal was to be a young chemist in Berlin, who had no reputation either as an experimentalist or as a writer. This was presumably his reward for having told the publisher the story of my discussions in Berlin. I was expected to be happy with the job of refereeing the papers submitted by others.

Had one of the leaders in the field been nominated as editor I would have had no problem with the arrangement, but to give the top job to this fellow who I felt superior to in every way was not something that I was going to accept. I informed

Dr. Engelmann in Leipzig of what was going on and wrote to Hamburg telling them that I'd long since planned to set up a journal but that I would only work with him if I was given the editorship, for I did not want, so to speak, to rent a place in my own house. There followed a lively correspondence. While Dr. Engelmann asked me to establish a new journal under my leadership to be published by him, the Hamburger publisher tried to persuade me that being referee on his journal was actually ideal for me. I wrote to all my colleagues from whom I hoped for contributions and asked them to publish in the journal I would edit.

On Christmas Eve a telegram came from Hamburg. It read, "Are you prepared to accept our conditions?" The boy from the post office took my reply back with him, "No".

Since most of the chemists I'd written to had replied affirmatively I was soon able to get the new journal organised in Engelmann's publishing house. The list of contributors in the first year included: M. Berthelot, J.W. Brühl, Th. Carnelley, H.L. Chatelier, C.M. Guldberg, A. Horstmann, H. Landolt, O. Lehmann, D, Mendeleew, N. Menshutkin, Loth. Meyer, Viktor Meyer, L.F. Nilson, O. Pettersson, L. Pfaundler, W. Ramsay, F.M. Raoult, R. Schiff, W. Spring, J. Thomsen, F.E. Thorpe, P. Waage. As one can see this includes the most important foreign scientists and all of these men, many of them world figures, had accepted the invitation of the young academic who just a few years previously had been completely unknown. We may assume that a good part of the surprising success of the project lay in the reputation of the publishing house.

It was exceedingly important to get the support of the Dutchman van't Hoff who had recently become so famous. He'd given a sort of conditional half acceptance to the other side. In reply to my letter of invitation he gave a positive reply but demanded to be named as co-editor on the title page, adding that he had no wish to interfere in the running of the journal but would be satisfied with a formal appointment. As it turned out he held strictly to this agreement, despite my best efforts to persuade him to take a more active part. He didn´t want to do more. However, he did immediately submit several important papers from his laboratory in Amsterdam which were the pride of the new enterprise. It seemed to me to be so important to link the journal to the name of this brilliant researcher that I unreservedly agreed to his conditions. I never regretted this decision and am convinced that the rapid success of the journal was to a large extent due to having van't Hoff on board.

The first issue of the Journal was published on the 15th of February 1887. Further issues followed at monthly intervals. The final double-sized issue of that year came out on the 27th December and brought the page count for the year to 678.

The fear that difficulties would be caused by having the leadership of the journal in Riga while the printing was done in Leipzig turned out to be unfounded, because the journal was not tied to fixed dates of publication. However when June came and I moved to the beach I did have to undertake some long walks because the post office which handled registered mail was some five kilometres away.

The actual editorial work—that is to say the review of the incoming manuscripts to check their suitability—was in general not a great problem because in most cases the answer was clearly "yes" or "no". Manuscripts that fell in between were rare. However I also took on some manuscripts that caused me more work. Translating the manuscripts which had been submitted in a foreign language was not a big problem and in the first years I did all that myself. Linguistic formulation had never been a problem and gave me the same sort of pleasant feeling one gets from skating where there is an immediate sense of clear achievement. I never had to search for the appropriate words but could translate freely and simply write the sentences down. Writing short reviews of papers summarizing the work which appeared in other journals was more challenging. However, this was the core competence needed to write the textbook and so I was used to this sort of work. Since the textbook had required that I carefully assess the literature—and this I had done with great earnestness—I only had to continue this habit so as to develop this aspect of the journal which gave me a useful channel with which to influence the views of the scientific community.

The same applied to a section devoted to book reviews which I did myself for many years. This was the means I used to rapidly project the new ideas which were springing up in neighbouring areas of science and the constant repetition of the importance of the work under discussion made this transfer of ideas faster than would otherwise have been possible. My natural predilection for using lively language, which I exercised to the full here, assured me many readers and lent my point of view a certain vigour which even dedicated critics were not able to stand up to for long.

Style of working. Although this new work was of course a new challenge, particularly at the beginning, I think it's fair to say that I did not allow it to interfere with my university duties. The ability to do so many things simultaneously was due to the speed at which my mind then worked. While I chatted in the office with colleagues I'd be skimming through the journals laid out there and noting anything I'd need to review so that later I could just write the review straight down. This way of working on different issues in parallel could be used to fill odd gaps and my brain was able to switch between tasks without any difficulty. I count the loss of this capability as the greatest drawback of old age.

I did all my writing alone using a pen. I never wanted a secretary or typist because the idea of being dependent on someone else was intolerable. At that time even in Germany typewriters were seldom seen outside trading circles and journalists liked to show off by saying that they had "typed" their reports. Wilhelm Wundt was one of the few academics who used a typewriter and his friends all hastened to explain that this was only because of his poor sight. In any case the enormous costs involved put most people off this idea. And so I was dependent on pen and ink because the ball point pen was only invented much later.

The wish to get all my work done with the minimum fuss—my instinctive anticipation of the energetic imperative—had led me already during my years in

Dorpat to conduct comparisons of various pens, inks and paper so as to find the best combination for writing. I invented all sorts of designs which maximised the amount of ink taken up in one dip, chose nibs with broad or rounded tips, replaced the iron based ink which corroded the nib with coloured solutions, prevented deposition on the nib by adding glycerol, used thick smooth paper and by doing so made writing such a smooth process that it afforded me real immediate pleasure.

Chapter 13
The Appointment in Leipzig

Letting go. Though I did not permit my work as editor to interfere with my official duties, I still had the feeling that this, together with my other scientific work, was loosening my ties to the professorship in Riga. Though I'd managed to communicate my urge to work to my assistants and they began to produce useful results, there was little chance of getting any useful help from the students. However, there were no "Privat dozenten"[1] and the half educated students went straight from their exams to conventional jobs. Only one, J. Spohr, the son of well off parents, who was therefore not dependent on earning a salary, worked for a longer time as a volunteer with me and produced quite a lot of work which brought us forward.

Nevertheless, it became ever clearer that the Polytechnic in my home town was not the right place for me to conduct the many tasks which had more or less by chance fallen into my hands. It was inconceivable that the situation at the Polytechnic could be improved particularly at a time when the entire institute was embarking on a fight for its existence against political influences—a fight from which it would not emerge without having suffered some wounds.

Because of this I set my sights on getting a position at a university in Germany. There had been several rumours and hints about possible positions but till now nothing had come of them. And then in 1887 there appeared on the horizon a chance so glorious that, though I secretly wished for it from the bottom of my heart, I couldn't really expect to get it.

The situation in Leipzig. Hankel, the professor of physics in Leipzig, had in his old age retired and been replaced by G. Wiedemann. In this way the only professorship for physical chemistry in Germany was now vacant and I naturally eyed this possibility which had opend up. At first it was without hope for both Landolt and Lothar Meyer were in the field and would certainly be considered first. Both of them were full professors who taught the entire spectrum of chemistry and I assumed that they would jump at the chance to be able to concentrate on their own

[1] "Privatdozenten" are senior scientists who had the formal right to teach.

R.S. Jack and F. Scholz (eds.), *Wilhelm Ostwald*, Springer Biographies,
DOI 10.1007/978-3-319-46955-3_13

speciality. In fact both of them were considered and both were attracted by the possibility but after a closer look at the situation in Leipzig both of them declined.

The agricultural chemist W. Knop had retired at the same time as Hankel and this confused the situation considerably. The honorary professor F. Stohmann who covered this area had long complained that his laboratory space was in the "Cow tower" which belonged to the Department of Agriculture and lay outside Leipzig. He'd successfully concentrated on thermo-chemistry for several years and was at that time the only researcher in this area in Germany. He was a very precise researcher and had uncovered a methodical error in the work of the distinguished thermo-chemist Julius Thomsen in Copenhagen which threw doubt on many of his results. He viewed Wiedemann's now vacated physical chemical lab space as ideal for his important experiments. This laboratory had been built by O. L. Erdmann and was considered one of the best in Germany.

Space for the physical chemist who was to be appointed would therefore be available in the institute where W. Knop had hitherto worked. It was on the ground floor of the agricultural department and was controlled by the professor of agriculture. Its only asset was that it lay close to the other scientific and medical institutes. The laboratory was out of date and could not be redesigned because that would have interfered with the intended purpose of the entire institute. In addition the professor of chemistry, J. Wislicenus, had used the opportunity to clear some space on his own overcrowded department which housed both a chemical and a pharmaceutical department by assigning the pharmaceutical unit to the space of the new department which was to be known as the "Second Chemical Institute".

Under these circumstances Landolt, L. Meyer and Cl. Winkler from the Freiberg Mining Academy all turned the offer down. With that the opportunity for me became tangible because no one else except me and van't Hoff really fitted the bill. However van't Hoff had the great advantage that the decisive voice in Leipzig was J. Wislicenus's who had for years energetically supported van't Hoff's doctrine of 3-dimensional chemical structural formulae and his tetrahedral structure for carbon. In addition Wislicenus had recently published the results of experiments which were based on these concepts and which provided strong support for them. The publication had been widely read and well received.

As co-editor of the journal I was in constant contact with van't Hoff and I'd asked him if he would be accepting the position in Leipzig which would certainly be offered him, though as a full professor in Amsterdam this would not actually be a promotion for him. He had answered evasively and so I concluded that he would accept the offer and buried my hopes of getting the position.

The "Acid-journey". This was the situation in the summer of 1887 as I started out on my fourth journey which I undertook in order to help bring my scientific work forward. The remarkable and interesting relationship I'd noted between the structure and electrical conductivity of organic acids made me want to investigate this in the broadest possible way. The thousands of acids described in the literature were not commercially available and synthesising them would have cost me years of preparative work. The only possibility to get them was to ask their discoverers for a small sample for my measurements. The amount needed would be less than that required for a structural analysis and the university collections might contain some interesting samples which I could get—so long as the professor in charge was willing. And so I set off in the summer holidays on my "acid-begging journey". This took me first to Austria where I'd not been so far.

The journey to Vienna via Warsaw introduced me to the unbelievable filth of Poland. In my student years I sometimes joined in the canon:

> The Pole says the Russian
> Is not free of vermin
> To which the reply comes
> The Russian says the Pole
> Is not free of vermin

And so on ad infinitum. On this journey I was able to convince myself that the Poles are absolutely unbeatable in this regard.

In Vienna I was given a friendly reception by my colleagues there. I got to know the physicists Stefan and Victor von Lang, the chemist Barth, the botanist Wiesner, the physiologist Fleischl and other well known professors as well as a large number of younger faculty including Goldschmied, Zeisel, Wegscheider, Weidel, Exner and Herzig with whom I spent most of my time. Many of the connections I made then have lasted my whole life.

I soon had the goal of the journey fulfilled because I was given all of the preparations I wanted and all of my colleagues there were happy to help me in my work. All of them had heard of it and they knew of the textbook and the journal. The only exception was the physics professor Exner who seemed to me to lack the necessary scientific impartiality which is the basis of all research. I soon came into conflict with him and, although I have since then avoided giving him any cause for offence, the repercussions seem not yet to have been forgotten.[2]

Vienna seemed to me to be a city of contradictions. The newly built university shone forth in excessive luxury, while nearby the physics institute was housed in rented accommodation and was in a worse state than that in Dorpat or Riga—and with no prospect of any improvement. In a similar way the Polytechnic building looked magnificent from the outside but the labs in it were unbelievably inadequate. The institute budgets were all completely inadequate, while millions were thrown out to finance the prestigious looks of the buildings. Because of all this there was an oppressive air to the place which did not let a happy scientific culture develop. Nevertheless I was happy to be given the hint that in four years Professor

[2]For the details of this conflict see Part II, Chapter 3.

Loschmidt, would retire and that then every effort would be made to appoint me to his position. The city had certainly impressed me as it impresses so many from the north with its elegance and cheerfulness. Already on the first evening I'd noticed that almost every cabman had a carnation or some other brightly coloured flower stuck between his teeth or behind his ears. The harmless merriment of the people in the Wurstelprater[3] contrasted favourably with the attitude of a similar crowd in Berlin.

Graz and Innsbruck. From Vienna I went on to Graz where I met Arrhenius who was working at the physics institute there. L. Boltzmann, one of the cleverest and most original researchers of his time, was the professor of physics and he received me warmly. We had a long talk and met several times thereafter. Thus began a relationship of affection and trust that brought us often together and ended only with Boltzmann's tragic death.

H. Z. Skraup was the chemistry professor in Graz. He was a splendid researcher and I liked him right from the start. He gave me many acids from his collection and both he and his lively and amiable wife were most hospitable. Once again this was the start of a long and happy association which was only terminated by Skraup's untimely death.

Arrhenius introduced me to Dr. Walter Nernst a chemist of our age whom he had introduced to our way of thinking. As we described our joint work in Riga he expressed the wish to work with me as well. I of course agreed at once and we arranged to start in the next autumn. Up till then he had carried out research on magnetism at the Technical University with the physicist von Ettingshausen who turned out to be a dignified and interesting scientist and a close friend of my teacher von Öttinger who had given him many positive reports of me and that helped no end in the development of our relationship.

While I found the institutes in Graz to be first rate—the chemistry institute had been built by Pebal and was rightly held to be the best in Austria, perhaps one of the best anywhere—The situation in Innsbruck where I went next was simply unbelievable. Chemistry and Physics were housed in a dark and dirty building which had at one time been a Jesuit college. They had almost no equipment.

The physics professor was Leopold Pfaundler a well known researcher for whose original and elegant work I had the greatest respect. When I met him he was depressed because he lacked the funds needed to pursue his research. His colleague[4] in chemistry was more interested in hunting chamois than in running his lab and was understandably less discontented.

Apart from the scientific high points I also had a chance on this journey to admire the wonderful scenery through which I journeyed and managed to use some days to visit the especially beautiful places. Although I had brought my paint box,

[3]An amusement park.

[4]The chemist, whose name Ostwald discreetly conceals, was Carl Senhofer.

I scarcely had time to use it because I never stayed anywhere for long enough. The time of year—end of June and the beginning of July—could not have been better and I was all the more in the mood to appreciate it because the discomforting memories from Riga had been driven into the background by the many congenial interactions with my Austrian colleagues. Even thoughts of Leipzig did not trouble me because I no longer had any realistic hopes in that direction.

Getting closer. From Innsbruck I travelled on to Munich where I renewed my acquaintance with A. von Baeyer. He was at that time at the height of his most productive phase and received me in a friendly fashion and invited me to dinner.

Of the physicists the one that interested me most was Sohncke at the Technical University because of his work on crystal structure. He, for his part, was glad to meet a chemist who had such a lively interest in his research which was not widely accorded the merit it deserved. I spent a number of instructive hours with him. I also got to know the leading crystallographer P. Groth with whom I also had a long discussion and we have remained on friendly terms since then.

From Munich I travelled across Lake Constance to Zurich. I'd travelled as usual by night so that I watched form the boat as the first rays of the sun lit up the snow covered Swiss alpine peaks—an unforgettable experience.

Victor Meyer, whom I'd met on my first journey, was no longer in Zurich because he'd moved to Göttingen. His position had been taken by A. Hantzsch whom I'd met before and who had just come as a very young researcher from Leipzig where he had been G. Wiedemann's assistant. He'd been recruited by the famous educator Kappeler. He also gave me a friendly reception and made me a gift of lots of acids. He radiated happiness at being the possessor of such a famous professorship and of such a beautiful wife. She was a daughter of the Dresden sculptor J. Schilling who had created the national monument at Niederwald for which she had served as the model for the figure of Germania.[5] Many years later we met again in Leipzig, though unfortunately his wife had died early.

From Zurich I went on to Basel where Nietzky gave me a number of very interesting preparations. There I met again my old Dorpat colleague Gustav Bunge who was professor of physiological chemistry. In the meantime he devoted his life to the war against alcohol and tried his best to convince me. I must admit that I wasn't quite ready for that since I'd not been too abstemious in Munich and thanks to my strong constitution I'd survived without any obvious harm. Nevertheless the earnestness with which Bunge treated the problem left its mark, and in the longer term was not without effect.

From Basel I moved on to Strasbourg. A. Kundt was working there at the Institute of Physics and since his excellent experimental work had often been physical chemical in nature I was anxious to get to know him. In the morning as I went towards the physics palace—now that it was part of Germany the university had been rebuilt at enormous cost—I was greeted very politely by a number of

[5]Germania is a personification of the German nation.

students which surprised me quite a bit, for despite my self confidence I really didn't think I was so famous that the students in Strasbourg would recognise me. Kundt was giving a lecture and so I waited in the pleasant gardens near the institute. Once again I was respectfully greeted by students. Once the lecture was over and I saw Kundt I understood the students. We were both of similar size and posture, both had a red beard (though Kundt's was a lot redder than mine) and our hair was done in the same way so that the students had thought I was their professor.

Kundt turned out to be a lively, friendly and obliging colleague who gladly showed me his institute and was also interested in my work. He was amused by the similarities between us. When I accepted his invitation to dinner he got us both to stand together in front of his wife and asked, "Who am I?"

Of the younger physicists I met there I particularly remember E. Cohn. He had a fine mind but was shy and diffident in everything and he even kept his scientific achievements more under wraps than one would have expected from his ability and astuteness.

Fittig was professor of chemistry in Strasbourg. He was a gaunt, elderly, intense man who didn't say much and who wasn't interested in giving me any of his collection of acids of which he must have had a large number. He had to go off to a lecture and made an appointment with me for the afternoon and at this he thawed a bit and we became friends. But he didn't give me any samples of his acids.

During our discussion he asked me where I was going next. I mentioned Tübingen, Würzburg and Leipzig. He made a strange face and said, "In your situation I'd avoid Leipzig". Surprised, I asked, "Why? I have to go there to pick up travelling funds from my publisher" "Don't you know that your appointment there is being discussed?" he replied. He couldn't give me any more precise information but said that I'd probably hear more in Würzburg from F. Kohlrausch who was a friend of Wislicenus who was in charge of making the appointment.

Already in Munich I had news from home that van't Hoff had written to say rather elliptically that he would not be standing in the way of my appointment in Leipzig. I hadn't given much thought to Leipzig because I'd more or less given up on the whole idea but Fittig's remark brought the idea back. Nevertheless it wasn't strong enough to make me change my travel plans which led me to Tübingen where I was glad to see Lothar Meyer and Hüfner once again. Hüfner had in the meantime left the antiquated laboratory in the castle and was now well established in a smart new building.

The Tübingen botanist Wilhelm Pfeffer interested me very much. The strange but momentous use which van't Hoff had recently made of Pfeffer's discovery of osmotic pressure prompted me to enquire into the historical background of his work which had been done some 10 years previously. He told me that he'd been a lecturer in Bonn and had done the experiments in his lodgings because he'd had no lab space. He'd been pretty sure that his discovery was of great general importance but he hadn't been able to pursue this line in the direction of chemistry because he had to stick to his botanical problems. He'd therefore invited R. Clausius, who was an important physics professor in Bonn to see his experiments. Clausius came, read on the manometer the astonishingly high pressure generated by a dilute solution of sucrose, shook his head and departed without saying a word. Once he was in Tübingen Pfeffer had repeatedly asked Lothar Meyer to join him in investigating the phenomenon but Meyer had seen no way to attack the problem.

F. Braun was a physicist in Tübingen with whom I had common interests. I'd criticised a conclusion of one of his experiments in my textbook and he'd accepted this very well. He told me personally, "The only serious criticism of that piece of work is the one you mentioned and you were absolutely right". Since one doesn't get to hear things like that very often, this little episode stuck in my memory. A quarter of a century later we met again in Stockholm where we both received the Nobel Prize: he got the physics prize for his work on radio telegraphy while I got the chemistry prize for catalysis.

I learnt nothing in Tübingen about the business in Leipzig, though everyone seemed to agree that I was the best candidate. L. Meyer told me that according to a friend of his in Leipzig Wiedemann had opposed his candidature and when he asked his friend "Then why don't you take Ostwald?" the friend had replied "He would be even more inconvenient to Wiedemann".

The great news. From Tübingen I journeyed on to Würzburg where I finally learnt some details of the situation in Leipzig. Quite apart from the situation unfolding in Leipzig, here in Würzburg Kohlrausch was of particular importance to me because he was the inventor of the method to measure conductivity and the leading expert in this field. He'd written to me with the urgent request that I express the results, which up until now I had presented in arbitrary units, in absolute terms, and I had accepted his suggestion with thanks.

He was a tall thin man with a grey blond beard, cool and noncommittal who spoke little and was uninterested in things that one could not prove.

After we had finished our scientific discussion the talk turned to Leipzig. I mentioned Fittig's remark and he said that Wislicenus had spoken approvingly of me. More he didn't know. He took me home for coffee and invited me to an academic bowling evening. This was a cheerful affair in the company of the Würzburg colleagues. There Kohlrausch thawed a little and offered to telegraph to Wislicenus the next morning. The answer should come around midday and I should meet him for lunch. This was done but it didn't really solve the problem because the answer was something like, "Where is Ostwald going? I support his application and hope he will be appointed". Kohlrausch sent another message asking if I should go on to Leipzig. He had in the meantime decided to help me and for his support I am eternally grateful. I was lucky to have his wholehearted backing despite the fact that because of the closeness of our areas of work and given the difference in our temperament—he was a pure scientific classicist while I was a romantic—it was clear that we would have our differences of opinion. These, however, could always be sorted out by friendly agreement. As evening fell, Wislicenus's answer arrived. I should go on to Leipzig to personally discuss matters. I left at once.

Events in Leipzig. In Leipzig the situation was as follows. Once the attempts to appoint a German professor had all failed, Wislicenus had pushed for van't Hoff's name to be put on the list. However the offer was accompanied by the condition that without any further negotiations he must answer simply yes or no. Van't Hoff had quietly gone to Leipzig and looked the situation over. He considered the situation to be unsatisfactory and replied with a list of changes which he wanted. Wislicenus had to tell him that since he had not respected their conditions the matter would not be further pursued.

As a result of this chain of failures the philosophical faculty in Leipzig refused to continue looking for a suitable candidate. However Wislicenus wanted at all costs to avoid losing the second professorship for chemistry and appealed to Gerber who was the education minister of Saxony and whose confidence he had, to grasp the last possibility and offer the job to me.

The faculty made clear its reservations because of my youth and reputation for being impetuous, but had to admit that there were now no other candidates. And so an offer of the professorship was sent north from the ministry while I, all unawares, was roaming about in the south.

After journeying all night I arrived in Leipzig in the early morning, collected my travelling funds from my publisher and went as early as was seemly to visit Wislicenus. It was an odd feeling to find him established in the same room in which a year ago on my previous journey to Germany I had met Kolbe and had been received by him with quite unexpected and overwhelming friendliness. In fact Kolbe's old lab technician Schumann was still at work there. Wislicenus received me. He was an imposing figure, with a broad chest, almost white wavy hair and a strong beard. In short, a head like Zeus. Nevertheless he was very cordial and we had a long conversation. He told me that the first candidate for the position had been Landolt who was at that time professor at the agricultural university in Berlin. However his demands (a new building etc.) were out of the question. They had similarly failed to reach an agreement with Winkler in Freiberg. He said nothing about the van't Hoff episode. Then my name had been brought up but Wiedemann had been against my candidature and gave as his reason that I published too quickly and proclaimed laws on the basis of insufficient evidence. Wislicenus had accepted this to some extent but pointed out that this was due to my youth, that this would change as I matured, and that I had more ideas than any of the other candidates. W. Wundt had strongly supported me.

Wislicenus told me that in order to resolve these contradictions the faculty had asked Clausius and Kohlrausch to give their opinions and both of them had written very much in my favour. They had both said in almost the same words that I was at times somewhat bold in drawing conclusions but that my experimental work was absolutely solid and that they therefore strongly supported my appointment. Wislicenus had presented these letters to the minister who, since the faculty remained silent in the matter, had personally decided to support my appointment. Since he had the right to make the appointment without the formal agreement of the faculty Wislicenus believed that this is what had been done and that the official offer had by this time been sent to Riga.

He then gave me a summary of my future duties and the salary I could expect which was several times more than I had in Riga. As the talk developed he became ever warmer and by the time we parted he was as affectionate as an old friend.
The appointment. A few hours later he appeared in my hotel just as I was packing my bags. "Ah, there you are—everything has worked out. You are now a Leipzig professor". Immediately after our conversation he had telegraphed the ministry in Dresden and been told that I should go immediately to Dresden: the letter of appointment was already in Riga. I was flabbergasted. I had difficulty grasping the fact that the goal around which my thoughts had circled these last few days was now reality. Wislicenus was, if anything, even happier. He admitted that he'd had the feeling that he was taking a great risk in pushing my candidature but that our personal talk in the morning had removed his doubts. He was now convinced that I would fulfil their expectations and in adjusting to the new situation would soon lose the rough edges that his colleagues had criticised.

Wislicenus thought it important that I should meet personally with Wiedemann. He brought me there and then left on pressing official business. Wiedemann for his part assured me that he was looking forward to wonderful results for science from our work together. I didn't realise at the time that he had wanted the position I had just got for his son.

As I was on the point of leaving Wiedemann, we were met by a little old man with an angular face, a red wig and indescribably shy eyes. We were introduced: it was Carl Ludwig the greatest and cleverest physiologist of his time. He told me how happy he was to be able to greet me as a colleague and added, "I learnt a lot from your book". I didn't know what to say because of all the happy moments on this eventful day this was the most intense.

In the evening I went to Dresden so as to present myself in the ministry the next day. I was met by Petzold who was in charge of university affairs. He was a dignified and kindly official and after some friendly remarks led me, despite my travelling garb, to meet the education minister Gerber who had previously been a law professor in Leipzig. Gerber told me that the letter of appointment had been sent ten days previously to Riga and he asked me for a formal declaration as to whether I would accept the offer. I replied, "It's as if you were to ask a corporal whether he wanted to be a general. Yes".

And so I became a professor in Leipzig before my 34th birthday and saw scientific challenges stretch out in front of me which, if I could meet them, would have their effects across the entire world.

Part II
Leipzig

Ostwald during his experimental lecture at the official opening of the Institute of Physical Chemistry in Leipzig on January 3 1898

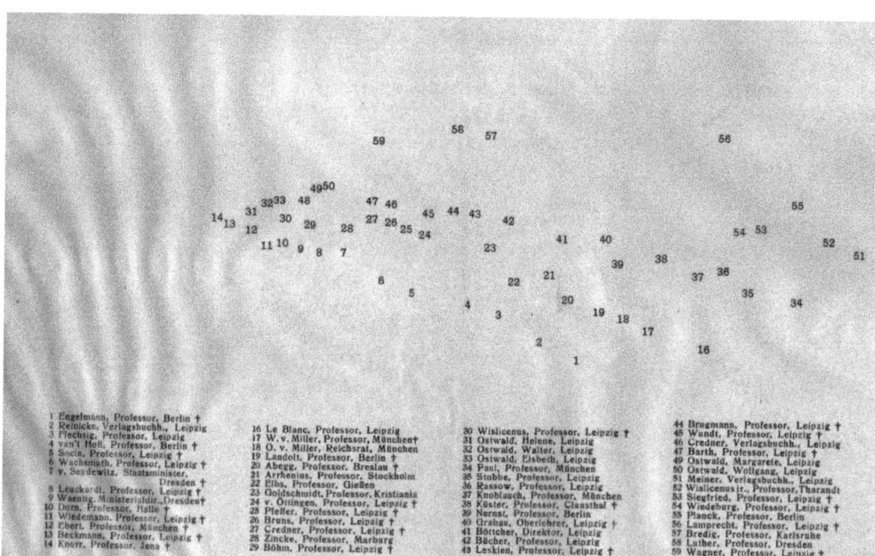

Ostwald with his wife and co-workers in front of his living appartment in the Leipzig institute

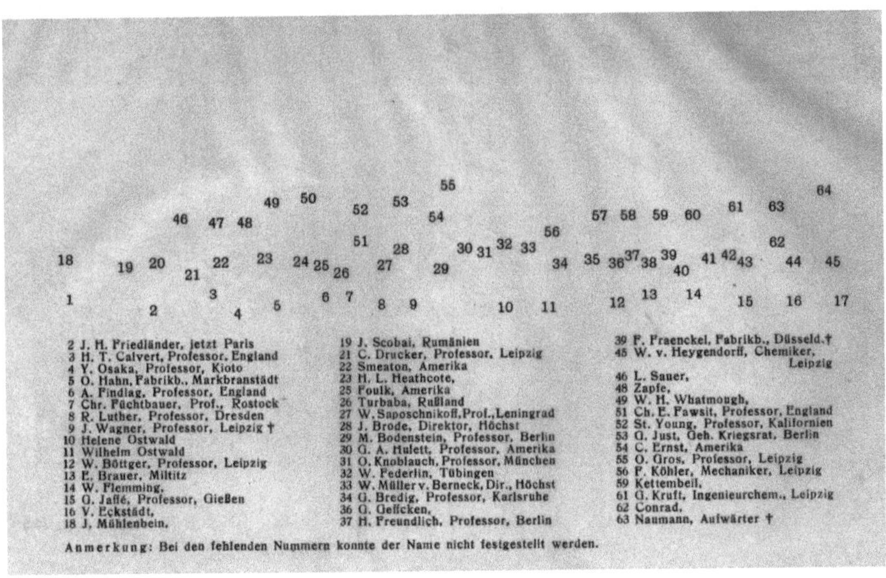

2 J. H. Friedländer, jetzt Paris
3 H. T. Calvert, Professor, England
4 Y. Osaka, Professor, Kioto
5 O. Hahn, Fabrikb., Markbranstädt
6 A. Findlag, Professor, England
7 Chr. Füchtbauer, Prof., Rostock
8 R. Luther, Professor, Dresden
9 J. Wagner, Professor, Leipzig †
10 Helene Ostwald
11 Wilhelm Ostwald
12 W. Böttger, Professor, Leipzig
13 E. Brauer, Miltitz
14 W. Flemming,
15 G. Jaffé, Professor, Gießen
16 V. Eckstädt,
18 J. Mühlenbein,
19 J. Scobai, Rumänien
21 C. Drucker, Professor, Leipzig
22 Smeaton, Amerika
23 H. L. Heathcote,
25 Foulk, Amerika
26 Turbaba, Rußland
27 W. Saposchnikoff, Prof., Leningrad
28 J. Brode, Direktor, Höchst
29 M. Bodenstein, Professor, Berlin
30 G. A. Hulett, Professor, Amerika
31 O. Knoblauch, Professor, München
32 W. Federlin, Tübingen
33 W. Müller v. Berneck, Dir., Höchst
34 G. Bredig, Professor, Karlsruhe
36 G. Geffcken,
37 H. Freundlich, Professor, Berlin
39 F. Fraenckel, Fabrikb., Düsseld.†
45 W. v. Heygendorff, Chemiker, Leipzig
46 L. Sauer,
48 Zapfe,
49 W. H. Whatmough,
51 Ch. E. Fawsit, Professor, England
52 St. Young, Professor, Kalifornien
53 G. Just, Geh. Kriegsrat, Berlin
54 C. Ernst, Amerika
55 O. Gros, Professor, Leipzig
56 F. Köhler, Mechaniker, Leipzig
59 Kettembeil,
61 G. Kruft, Ingenieurchem., Leipzig
62 Conrad,
63 Naumann, Aufwärter †

Anmerkung: Bei den fehlenden Nummern konnte der Name nicht festgestellt werden.

Chapter 14
Leaving Home

Return. The 2 days following my appointment to the professorship in Leipzig were taken up with the journey back to Riga and the beach where my family was staying. This was hard for I was exuberated at my promotion from a mere teaching position at a mediocre technical college to a full professorship at one of the most important universities in Germany and was bursting to let off steam, but I was on my own and so the news could not be shared. It was only on the rail journey from Riga to the beach that I met my colleague Grönberg and could spill the beans. He'd already caught wind of it but, like the others, had been unable to believe the news. He cross questioned me to make sure that it was really a full professorship and not just an honorary professorship which was the maximum that most of the experts felt was possible.

The news of my appointment had already reached my family on the beach where they were together with my parents in law. By chance my father in law told an acquaintance of his from the neighbourhood who had a relation from Leipzig staying with him. The next day a little old lady visited my wife and with many apologies asked her to please not spread the news any further. The visitor from Leipzig, who was a professor of law and knew the workings of the university, had explained to her that my appointment as a full professor in Leipzig was completely out of the question. A career jump of this magnitude was impossible. At the very outside it might be an appointment as a replacement for the honorary professor Carstanjen who had died a few years previously. There must have been some misunderstanding and in our best interests my wife should avoid spreading the impossible news that I had been appointed as a successor to privy counsellor Wiedemann. Although my wife could reply that it did indeed involve Wiedemann's position, she naturally felt a bit unsure of herself and impatiently awaited my arrival. The enormously sensitive gossip circle of the Riga beach had however

R.S. Jack and F. Scholz (eds.), *Wilhelm Ostwald*, Springer Biographies,
DOI 10.1007/978-3-319-46955-3_14

quickly picked up the news and its interpretation and passed it on to my colleagues at the technical college which explained Grönberg's doubts and his questions.

The successor. In order to push through my resignation quickly I had turned to the chairman of the administrative board, the minister (*Landmarschall*) von Öttingen.[1] He was the fourth of the Öttingen brothers the other three were all professors at Dorpat and he was a powerful politician with a great say in the running of the Baltic country. He met me with the remark, "You come to tell me something I don't want to hear", but he didn't turn down my request to put aside the formal period of notice. He rightly thought that I'd only half heartedly do my teaching duties if I had to stay an extra semester in Riga, and so he only demanded that I find my own replacement. I turned to Wislicenus with this problem, and he so warmly recommended his pupil and assistant C. Bischoff that he was soon elected. This was especially helpful to my Leipzig colleague and patron because though he trained a very large number of pupils year in year out he managed to get only a very few of them so far that they were appointed to academic positions. I didn't realise this at the time but learnt it much later when I studied the development of research careers.

Because of the semester holidays in Leipzig Professor Bischoff was able to move right away to Riga so that I got to know him and could personally hand over the Institute. He turned out to be a completely different person from me. In his scientific interests he'd been brought up in the arrow circle of organic chemists and the spatial organisation of the atoms in an organic molecule was for him the ultimate problem. I suppose he threw up his hands in horror when he found that there was no preparative organic chemistry in Riga. He knew nothing about physical chemistry and probably, like the majority of his organic colleagues, didn't consider it to be chemistry at all.

During my appointment in Riga I'd often enough wondered if we shouldn't have at least a departmental section devoted to organic chemistry—which after all was the central area of chemistry then. It would surely have been possible to get such an assistant professorship financed. But I always ended up with the thought that there was no point because there was no industry established in the Baltic provinces or in Russia which would have provided employment to our graduates. Any attempt to compete with German industry, particularly in the area of dyes, would have been a waste of time. In contrast, there was a continuously increasing industrial requirement for analytical inorganic chemists for whom an understanding of the basics of the doctrine of chemical affinity was enormously important and this gave our graduates an advantage when seeking employment. For this reason I had deliberately decided against establishing organic chemistry in Riga. My successor saw things very differently.

[1] August Georg Friedrich von Oettingen.

Just as in scientific matters, so we were completely different in our approach to social occasions. He was a fine figure of a man with a good voice who could render Wagner's "Seid mir gegrüßt in diesem edlen Kreise"[2] so as to at once conquer all the ladies' hearts. And so as soon as I left, the rudder was swung over to point the Riga laboratory in a completely new direction. Soon almost all traces of my work there had disappeared and only Paul Walden remained as a fully fledged representative of the new science which he carried on and expanded quietly and tenaciously. Once the new direction of teaching that I instituted in Leipzig began to achieve public approval he managed to restore its prestige in Riga. Walden was granted a leave of absence to come and work in Leipzig and afterwards in Riga his lectureship for physical chemistry was, in due course, converted into a full professorship. Because of the great influence he exerted in the Polytechnic in Riga and in the Petersburg Academy of Sciences, he finally managed to establish the science we loved in the place where its physical representations—the textbook and the journal—had been started. He himself added many important results to the new chemistry.

Farewell. Because of the different rhythm of semester starts in Germany and Russia I had a few months free before I had to move to Leipzig. They passed in a flash because of the numerous things which had to be done to close the tents in Riga and get ready to open them again in Leipzig. The events I described above had played out at the start of August while the new semester in Leipzig didn't start till the middle of October.

My leave taking from Riga was not an emotional wrench. In the five and a half years as professor at the Polytechnic I hadn't really established myself in society circles in my old home town. The development of a salon at home, which was usual in Riga society, had not been possible because of my limited salary and because social occasions in the well off business community in Riga were considerably more opulent than in academic and democratic Dorpat. In addition, the rapidly growing family made such demands on my wife as mother that she had little time or energy left over for other things. However the real cause lay in my personal attitude. The many scientific demands, be it research, teaching or writing, left me neither time nor any great interest for the sort of social occasions which were usual in my home town. At that time I simply wasn't interested in anything outside of pure science. In my own country there was little interest in this but things were different in Germany where the roots of my being had—at first subconsciously but later quite openly—always driven me. Because the publishing house for the text book and the journal was in Leipzig I'd found a sort of spiritual home there which gave me everything that my home town couldn't offer. Since I'd made no secret of my views, I aroused without wishing to—but also without trying to counter it—an opposing response from my peers in the Polytechnic who in the meantime were all rising in the ranks of the appropriate social circles. If instead of making a career in chemistry I'd applied myself to the humanities, perhaps to philology or even

[2]The reference is to Act 2 of Wagner's opera Tannhäuser, though the quotation is inaccurate.

theology, then I'd have been widely accorded respect and acceptance, but nobody was going to show appreciation for special achievements in a strange subject like physical chemistry.

I'll discuss the general causes of my inner separation in greater detail below.

The farewells to my colleagues at the Polytechnic were made in sincere friendship but without any great heartache. They were close enough to appreciate what I'd achieved, both in research and in teaching, and yet not so close that—with one or two exceptions—there was any jealousy or envy. They didn't take it amiss that I'd jumped at the chance to leave them and they viewed the chance I'd been given to go to Leipzig as a real stroke of luck.

Problems. A dark shadow had fallen over my parents' lives during the last years in Riga. Once my father had reached sixty he had accumulated what was for him a considerable fortune and had decided to leave his business so as to have more time to devote to charitable activities in town and to the guild. So as not to be economically completely inactive he granted money to a relative to start a lumber yard and he joined this venture as a partner. The business went well for a year after which my father more and more left the running of the business to his partner. This man, however, became ambitious and, wanting to make a killing, let himself be seduced by the Jewish lumber syndicate[3] without whom nothing could be achieved. In this way he got involved in a risky venture through which he lost not only the entire fortune but a great deal more besides.

This hit my parents so hard that it was almost unbearable. All his life my father had made sure that he didn't give out a single penny that wasn't covered. In this sense he'd have got over the loss of his fortune even though he'd worked hard his whole life to amass it, but what was for him quite inacceptable was the feeling that he was saddled with debts which there seemed no way to pay off.

My small savings were just a drop in this ocean, what little my brother Eugen had was not much more—and the debt had to be paid. At this point my father in law filled the breech with a loan which my father accepted with the determination not to rest until he'd paid it back.

I can't relate all the details of how my father, with a degree of mental flexibility and intense hard work which we normally associate with America, managed to pay off the debts despite his advanced age. In fact he managed to do this in a surprisingly short time and he amassed a second fortune—smaller than the first but sufficient to ensure that he could live out the last 10 years of his life in peace and in the sort of circumstances which he'd longed for all his life. He bought a small estate which was charmingly situated between the sea and the forest and enjoyed long walks in the woods on which he'd watch the antics of hares, deer, foxes and other forest creatures which he'd earlier always shot. From Leipzig my family and I visited my parents who were still fit on their golden wedding anniversary and there he died when he was nearly ninety. My mother survived him by many years.

---

[3]Ostwald uses the term "Holzjuden", which has a clearly pejorative and anti-Semitic smack.

These events also had not been helpful in aiding my entry into polite society in Riga.

The students. What hit me most was the separation from the students. I'd challenged them much more than my predecessor had done and yet had found a readiness on their part to accept this. This had filled me with real joy, however, I have to say that the leave taking was overdue.

I've already related (Part 1, Chap. 13, p. 131) that due to the nature of the existing conditions, there was no chance to bring on a new generation of scientists dedicated to my special area. Instead I had to take into account the students' interests in finding jobs in local industry and so I had to avoid letting them probe too deep into my research work. This was not only completely opposed to what I wanted, but also to their desires as well. The chance that I'd be able to assemble a cohort of foreign students who were not locked into the normal curriculum was something that seemed to be extremely difficult if not impossible. Even the simple question of space worked against any such scheme. Because of the overcrowding every last corner was packed with regular students doing their practical courses and it was unthinkable that they should be pushed aside for the sake of foreigners. In fact Arrhenius's work with me had only been possible because, since I had no other space, I gave him the bench in my own office which then had to serve both as a lab and as an administrative office.

National customs. To leave Russia one required a pile of paperwork which had to be filled out by the police and for which one normally had to wait for ages unless one used the usual Russian lubricant of a ten rouble note. In my case the responsible official refused the money without however being in the slightest degree offended. "Professor", he said. "I'm an honest man. I can't swing this for you. A week ago it would have been alright and in a couple of weeks it will be alright again. The trouble is that the day before yesterday a new chief of police was appointed and he doesn't permit it yet. It won't be long and then everything will be as it always was, but for the moment it can't be done. I'm sorry but I can't accept your money. I'm an honest man". I therefore had no option but to follow the official route. Despite this the paperwork was completed in time.

This was my last impression of the Russian Empire. In Leipzig I completed the papers renouncing my Russian citizenship and this was quickly accepted and at the same time had myself registered as a citizen of the German Empire.

The family's departure. I and my family left Riga for Leipzig in September 1887. In addition to my wife there were my two sons, Wolfgang and Walter, and two daughters, Margarete and Elsbeth (Elisabeth). Margarete at five was the eldest and she was already a sensible girl. Walter, the youngest, was just learning to walk. For the journey we'd taken along a nanny, a neglected orphan whom my wife had taken up out of pity. She proved herself and the journey passed without incident.

It had pleased me that our sons' first names, just like mine, began with a W. I hadn't reckoned with the fact that all of us would make our names as authors so that there was bound to be a bit of confusion with the work of the three

W. Ostwalds, Only once it was too late did this possibility become apparent and that was earlier than I'd expected. This was of course more of a disadvantage for my sons. They helped themselves by adding to the common consonant W their own personal vowels so that Wolfgang became Wo. Ostwald, Walter Wa. Ostwald and both were able in this way to establish their places in the scientific and technical literature. The short form W. Ostwald was left to me. However, as they gradually became better known I began to see that, to avoid any confusion, I was being referred to by others as Wi. Ostwald.

Accounting. When I try to render an account of the scientific goals I'd achieved by the time I moved from Riga to Leipzig then I can list the following:

1. By reorganising the chemical syllabus in Riga, which had scarcely existed when I started, and achieving a sharp growth in the numbers of students I'd demonstrated to myself and others my ability to build a scientific culture even in an environment which had been almost sterile up till then.
2. There had been two important scientific advances in my time in Riga. First of all the development both of chemical kinetics and of the necessary thermostat and the demonstration that with these methodologies the same values for the chemical affinity of acids were obtained as from static methods. Then came the first steps in electro-chemistry. First of all there were the analyses of conductivity measurements where my work joined that of Arrhenius who had postulated the direct proportionality between conductivity and chemical reactivity. Second was the first exploration of electromotive forces, though the further work in this area came only years later.
3. By writing the textbook I had methodically worked through the entire area of general chemistry and put it into a clear framework. It was now pretty straightforward to insert new insights at any place in the text because the basic ideas had been described and put into context.
4. The establishment of the journal had secured the future of general chemistry as a branch of science in its own right. Now the current researchers had access to a common platform of their own and no longer had to try to publish as barely tolerated guests in other frequently unfriendly environments.
5. Despite my unusually heavy load of teaching duties—a few years after I left my responsibilities were divided between three full professors—I'd nevertheless carried out a great deal of research work as well as a great deal of writing which was an assurance that my powers were up to still greater challenges.

These could be booked on the positive side of the balance. How about the negative side?

After thinking this over carefully I could only come up with the fact that I had failed to establish myself adequately in the social sphere either amongst my colleagues at the Polytechnic or in Riga society.

My colleagues had not lacked confidence in me nor was there any lack of respect from society in the town. It was merely that in the years that I was there no real relationship had developed.

When the long serving director Kieseritzky retired and my economist colleague Lieventhal was elected to be his successor he told me that in view of my scientific success the post should actually have been mine. The members of the board of governors had agreed but had thought that it would be a pity to rob me of time and energy for my scientific work by burdening me with the daily round of administrative duties which this position entailed. After all, in comparison to professors like himself, who only held lectures, I was already over burdened by the enormous amount of course work in the laboratory. I agreed entirely. I could never have taken on the office, partly for the reasons he gave and partly because I was personally quite unsuited for it. From my scientific work I was so used to abstract thought that I'd never have been able to interest myself in the daily problems of individual faculty members.

Similar considerations had governed my attitude to social intercourse.

When I now ask myself how my departure from my home town seemed so easy and involved no sense of loss then I think the reasons are clearer today than they were then. First of all my family was not old Riga stock since we'd been there only for two generations. This alone ensured that I did not really belong to the inner circles and brought with it the subconscious but nevertheless real feeling of distance in the way I was viewed and treated. I could see this clearly in my wife's family who were all down to earth and whose members formed a broad net stretching out from Riga and who owned land in Livland and Courland. They were friendly and courteous but nevertheless there was an unmistakeable undertone of condescension towards the upstart. Only my father in law had a genuine interest in my scientific efforts and he knew from his eldest son Karl (Vol 1, Chap. 13, p. 3)[4] the volatile possibilities of a scientific career. Since I didn't find anyone else who shared my enthusiasms I didn't put much effort into pushing myself into the circle and they responded by considering this unseemly arrogance on my part. In this way both sides were content to keep our contacts to the minimum. My wife quite naturally found all this painful but she understood the need for it and in any case she took seriously her increasing duties towards our rapidly growing family.

My scientific activities also isolated me in another important area. I've already mentioned that in the leading political circles of my home town it was considered nothing less than robbery that I failed to put my work and my talent, which they well recognised, in the service of my country as did almost everyone who was mentally a bit above average. On various occasions I was taken to task by the leading political figures who I all knew from my time as a member of the inner circle of the Fraternitas Rigensis. I was, however, so immersed in my scientific activities that there was no time left for narrow patriotic matters.

I can well imagine that these people considered my attitude as a serious fault which verged on treason. After all the destructive wave of pan Slavism had, as I recounted, begun and one could daily expect further pressure. It wasn't just that foreign people were intent on forcing their language and customs on us, it was

[4]Reyher, Carl Dietrich Christoph von.

much more that a clearly more primitive culture was attempting to destroy our higher values—as in fact later happened, though the attack came from a rather different source. Because of this I appeared to them to be someone who ignored the cry, "All hands on deck", simply for personal interests.

For my part a feeling of responsibility for the future of the home country had never been able to develop. The ordinary citizens like my parents had no say in the way the town was administered. My father was part of the so-called small-guild. The council and the two guilds formed the city council and though he had achieved the rank of alderman he had to recognise that he and those like him had no real power. A few years previously Petersburg had imposed a new city constitution which introduced an electoral roll based on a person's assets and my father became a member of the city council. However in reality the power was retained in the hands of same old crew because they were the only ones who understood the art of political infighting and who were therefore in a position to swing the elections. Of course this could only last so long as these political techniques were not learned by the other side.

From all of this I had grown up without any feeling of loyalty to the ruling circles in Riga or any sympathy for their continued existence. The most important breeding ground for high office in the administration had been the student Fraternitas and I'd cut my time there short by quickly ending my student years. As a result I'd never been long enough in the student society to be given any of its offices which were the training ground for a later career in the town's administration. Once again it had been my scientific ambitions which had nipped this potential opportunity in the bud.

As a result I was not accepted into the leading circles in Riga. My relationship to my class mates who were now working or in office quite quickly cooled and on the few occasions we met it became clear that I was now regarded as an outsider. Since this was happening parallel to my growing scientific success I did not consider it any great loss.

Chapter 15
The New Work Place and the First Fruits

The first days in Leipzig. In my haste to get established in the new place as soon as possible I'd gone to Leipzig much sooner than necessary and had taken my family with me. There was an apartment in the same building which housed my future institute in which Professor Knop had lived. After he'd left, the apartment was renovated and the workers were still busy when I arrived so that we had to stay in a hotel to start with. Because of the holidays only a few of my new colleagues were around but luckily Wislicenus was there and he gave me much needed helpful advice as to how to make contact within the large and famous Leipzig university.

The difficulties here were by no means trivial. I'd never, neither as student nor as lecturer ever been part of a German university so that I had no idea of the modes of conduct and I constantly ran the danger of assuming that the customs and usage appropriate to Dorpat would also be appropriate in Leipzig. I've little doubt that because of this I caused a fair amount of astonishment, umbrage and disaffection. As is always the case with these things the stories quickly spread round the university and only the person involved never gets to hear of them and so never learns to mend his ways. I think that this is one of the reasons why even later I never really totally fitted into the life of Leipzig University.

Teaching. The department which Wiedemann[1] had headed was called the Physical Chemistry Institute. The department which I was now to take over had, under the influence of Professor Wislicenus, been named The Second Chemical Laboratory. In this way it was made clear that although it was run by an independent professor it was nevertheless to be seen as an adjunct to Wislicenus's First Chemical Laboratory. This is probably the main reason that the other candidates had turned the professorship down but I, in my ignorance of the ways things were done in German universities, hadn't even registered this subordination. I'd learnt in Dorpat that teaching arrangements were immutable and hence that this sort of thing must

[1]Gustav Heinrich Wiedemann.

R.S. Jack and F. Scholz (eds.), *Wilhelm Ostwald*, Springer Biographies,
DOI 10.1007/978-3-319-46955-3_15

simply be accepted. I must add that even if I'd understood, I'd still have accepted, so great was the difference to the position I'd held in Riga and also because of the great respect that I had for Wislicenus. He was nearly 20 years older than me and had long since achieved a reputation as one of the principle representatives of his branch of science. His dignified and friendly manner had given him a sort of fatherly authority and so it was not in the least difficult for me to take on the subordinate role he'd prepared for me.

This resulted in the following situation. So that the newly matriculated students in each semester could attend the basic lecture in inorganic chemistry, he and I held this course in alternating semesters. I held the physical chemistry lectures in the semester in which I didn't have the inorganic course. This was fine by me and we kept to this scheme until 7 years later with my complete approval a badly needed third full professor of chemistry was appointed.

To begin with my laboratory course was in physical chemistry and I had to organise this from scratch. In addition there was to be a course for the beginners in analytical and preparative chemistry which was to be modelled on the course given in the First Chemistry Institute so as to ensure that the contents were uniform. This too I welcomed and this sort of course had already been offered in Wiedemann's time. Thirdly, however, I was to take over the laboratory course for pharmacists. This had nothing to do with physical chemistry but the space in the First Chemistry Institute was so packed with chemistry students and doctoral students that Wislicenus wanted to get rid of the pharmacists to make space.

In vain I argued that I knew nothing about pharmacy. Wislicenus told me that he'd give me his assistant who had up till now looked after the pharmacists and since Dr. Beckmann was himself a pharmacist and was both capable and reliable I need only consider myself formally responsible for this section. I trusted Wislicenus to be able to judge the situation better than I and agreed to this organisational curiosity without further ado. I had luck in this game because the relationship that developed to Ernst Beckmann turned out to be friendly and profitable for both of us.

For the rest, particularly the remuneration for the examinations of the medical students, pharmacists and teachers, the responsibilities and duties were divided equally between Wislicenus and me. There will soon be much to tell about the assistants but first we need a brief description of the state of the art and the problems which the new department was there to solve.

A watershed in science. Others have already noted that 1887 was a most decisive and extraordinarily fruitful year for physical chemistry. To start with there was the completion of the "Textbook of General Chemistry" (Lehrbuch der Allgemeinen Chemie) whose title page carried this date though, as I already explained the book was completed in 1886. In addition the Journal of Physical Chemistry (Zeitschrift für physikalische Chemie) began in this year. When one considers that an objective branch of science requires not only an existence in the heads of individual researchers but also a written basis then one will certainly admit that these two events were crucial for the emergence of physical chemistry as an independent entity within the group of sister sciences and permitted its further and ever more rapid development. While the Professorship in Leipzig which Wiedemann had held and which was now mine was at that time the only department of its kind in the world there is now, one generation later, probably no university anywhere without a physical chemistry department. Nowadays the division of the subject into specialities shows that this science has now grown so large that no single person can cover it all.

In addition there was an essay written by my co-editor J. H. van't Hoff which was published in the middle of that year with the at that time oddly sounding title "The role of osmotic pressure in liquids in analogy to gases",[2] whose contents were to be of enormous importance for the development of physical chemistry which as a result became in certain large areas a truly new science. Alongside these creative thoughts of van't Hoff there appeared also in 1887 the no less creative idea of electrolytic dissociation which was brought to light by Arhennius[3] and this too rapidly led to the development of a large new area of science.

The organisation of physical chemistry. It is customary in discussions of the history of science of this period to bring together with the names of van't Hoff and Arrhenius also that of Wilhelm Ostwald, though he did not at this time make any equally important discovery. The reason for this is that I was the organisational factor without which the broad and rapid establishment of a new area of science cannot be brought about.

The new subject won by my appointment in Leipzig a geographical and scientific centre point. Had Wiedemann still been the professor there then the publications of the great discoveries of 1887 would have remained without effect for a long time because Wiedemann was soon opposed to these new advances. True to his temperament he strove to avoid giving any forceful expression to his opinion in public but rather spread his view through his broad network—naturally always maintaining the friendliest manner in his personal interactions with me. And once the time seemed right his son Eilhard, who shared his disapproval of the new ideas, also argued against them publically, as will be related in due course.

With that Germany would have been closed to us for there was no other professorship of physical chemistry here and even Lothar Meyer, who was otherwise close to us, was amongst those who opposed in particular the ideas of van't Hoff. Though van't Hoff was a full professor in Amsterdam and director of a large laboratory he did not feel called upon to lead an active movement and in any case he was so tied up with official duties of a non scientific nature that he simply didn't have the time. Arrhenius, for his part, was still something of a scientific apprentice and it was many years before he got a professorship in his own country.

[2]van't Hoff JH (1887) Z Phys Chem (Leipzig) 1: 481–508.

[3]Arrhenius S (1887) Z Phys Chem (Leipzig) 1: 632–648.

I on the other hand was ready and willing to throw everything I had into the service of the new science. Animated by inexhaustible joy at the new ideas and armed with the necessary prerequisites of a good teacher I could easily make the Leipzig institute into the headquarters of the new work and from there organise the battles. The new journal gave us a forum in which to publish the results so that I think it's fair to say that never before had a new branch on the tree of science enjoyed such favourable conditions as did physical chemistry in its youth.

The article from Arrhenius concerning the dissociation of salts in aqueous solution and van't Hoff's contribution, both of which appeared in the journal's first year, formed the basis for the subsequent development of the new branch of science. These two articles complemented each other in the most fruitful way possible and laid the basis for the extraordinary number of diverse discoveries which soon flowed from them.

Osmotic pressure. The advance which van't Hoff had made in the article just mentioned can be summarised as follows. Horstmann (Part 1, Chap. 8, p. 6) had largely discovered and described the laws which govern the relationship between the rates of reaction and the equilibrium of the reactants achieved, in terms of thermodynamics, which was the best available scientific method. In practical terms however these laws were of little significance for they were limited to reactions in the gas phase and there were only a few gaseous interactions that could be studied. The vast majority of chemical reactions involve liquids, in particular aqueous solutions.

Some researchers in particular J. Thomsen (Part 1, Chap. 11, p. 8) had recognised a certain similarity between gases and dilute solutions, though nobody knew how close this similarity was nor how it came about.

This was where van't Hoff's idea, which gave precise answers to these questions, came into its own for he showed that the well known gas laws also applied to solutions and hence that the thermodynamic theory of chemical reactions could be applied to reactions in dilute solution. Suddenly the way was open to solve all sorts of problems which till now had seemed beyond our reach.

And the key to all this had been osmotic pressure.

The name of this phenomenon came from botany. Osmosis is the name given to the free diffusion of solutes in the cells through the cell envelope so that they spread equally out through the cell fluid, just like a gas diffuses to fill the space available to it, and the process does not stop till the entire available space has reached the same concentration. The cell walls however often hinder the diffusion and this results in a force which may even break open the walls. These were things which were well known to botanists.

W. Pfeffer (Part 1, Chap. 13, p. 7) had decided to investigate more closely the forces involved. He set up synthetic cells whose boundaries he cleverly constructed of semi permeable membrane which were strong enough to withstand considerable pressure and through which water could readily pass though many solutes could not. He filled the cell with solutions of substances which could not move through the membrane and then placed the closed cell in pure water. The result was that the pressure inside the synthetic cell rose and this was the same pressure which sometimes led to the destruction of plant cell walls. With the aid of a manometer he determined the pressure, showed that it was unexpectedly high and worked out the law governing its formation. In this way he had established the physical basis for his work in plant physiology. However in trying to understand the physical laws underlying this effect he had attempted in vain to interest the brilliant researcher Clausius (Part 1, Chap. 13, p. 7).

This was where van't Hoff set to work. He showed that the osmotic pressure which a solute exhibited followed the same laws as the pressure which a gas exerts. This went so far that under similar experimental conditions both forms of pressure were quantitatively the same.

Now equilibria and chemical reactions between gases are usually determined by their pressures and the laws of thermodynamics are concerned with how these pressures are affected by volume and temperature. One therefore only had to substitute osmotic pressure for pressure in the equations and one had established the laws governing the reactivity of solutes.

When these equations were compared with those derived from the few experimental studies which had been carried out using solutions then it became clear that they gave similar results. However the new equations provided many more details and more penetrating answers to general questions.

My personal view. One can easily imagine what a powerful effect this revelation had on me. For the last part of the textbook I'd recently brought together and compared all the published results about chemical equilibria and reactions and seen that they all led to the same laws. My own experimental work was directed solely to the same questions. And now all of these results turned out to be individual cases of a general law. Chemical kinetics thus achieved the same scientific status as, for example, the mechanics of the cosmos. The central tendency of my entire scientific thinking had always been to tease out the generally applicable—and hence informative basic principles and laws—and it found here an unexpectedly rich fulfilment.

To be frank this advance came at an unfortunate time for me. The paper was published in one of the last issues of the journal that I edited from Riga and the distractions of the move to Leipzig and our adaptation to the new environment delayed my proper assimilation of it. However once these obstacles had been overcome, this new idea turned out to have a fruitful affect on my own work.

Electrolytic dissociation. van't Hoff's brilliant idea had now seen the light of day, but it had been born with a defect that was almost fatal. Though it could be generally applied to the large group of neutral molecules it did not apply to the more important group of salts including acids and bases. In order to encompass them van't Hoff had had to add into the equations a mysterious factor i[4] which was greater than one and which seemed to obey laws for which no rational explanation was available.

This was the situation when, in Leipzig, I received the other manuscript. It was from Arrhenius and carried the title: "Concerning the dissociation of chemicals in water".

Dissociation means splitting or disintegration. With this term one described situations in chemistry whereby complex substances were decomposed into simpler components, usually under the influence of heat. If one of the resulting components is a gas then the reaction follows certain simple laws which I had summarised and shown in the textbook.

The study of the conductivity of electrolytes, that is to say of salts, acids and bases with which Arrhenius had started his career (Part 1, Chap. 11, p. 1) had led him to the conclusion that the conductivity was supported by only a fraction of the electrolyte which like a ship carried the electricity along, the positive charges going in one direction the negative ones in the other. That was one fraction of the electrolyte—the other fraction was inactive. At that time he was unable to come up with an explanation for the difference between the two fractions and none of the possibilities he considered provided a solution. Even during our work together in Riga in 1886 no progress was made on this front. However, he found the solution in the 1887 by making the radical proposition that the conducting fraction of the electrolyte was completely ionised while the non conducting fraction was not.

The term ion had already been introduced by Faraday. He had shown that in electrolytes the movement of the electricity was always accompanied by the movement of a part of the electrolyte that he termed "ions". He had, however, believed as did his successors that the dissociation was induced by the applied current. Clausius, for his part, had serious reservations about this and suggested that at least for a small number of electrolytes the application of current was not necessary for their dissociation. In contrast he believed that they were already ionised and that it was these ions which transported the current.

Arrhenius now showed that the restriction to a small fraction of the electrolytes was inappropriate. Rather one had to assume that the majority of salts, strong acids and bases in solution were already to large extent dissociated so that these solutions were in fact solutions of the ions which resulted from the dissociation of the salt together with a certain small fraction of the non dissociated material.

He cited a mass of irrefutable data in support of his revolutionary idea. The most important for us is that the mysterious factor *i,* which had defaced the doctrine of osmotic pressure, now turned out to be simply the number of ions into which the dissolved material dissociated. This could be either read off the chemical formula or alternatively it could be deduced using other equations which van't Hoff had derived from certain properties of the solution such as its freezing point. A considerable amount of data to the freezing points of solutions had already been

[4]This quantity was later named "van't Hoff factor".

published by the French chemist F. M. Raoult and Arrhenius showed that the values of i obtained by both methods agreed with the values predicted by his theory.

This was a dramatic development, for now a problem was transformed into a triumph. I have no doubt that those who were attracted by the growing importance of the matter and who quietly followed the developments, there were many who were most impressed and some perhaps were already convinced. Little of this leaked out to the broad public. Instead the novelty of Arrhenius's thoughts was so startling that initially it provoked an instinctive movement of rejection.

My contribution. I myself did not doubt for a moment. From earlier conversations I was aware of the basic idea which fitted well to my own thoughts on the matter which I'd started to develop a decade previously but which had never matured. In my Master's dissertation of 1877 the third thesis I proposed had read, "Water decomposes all salts".

This short sentence was the result of a long thought process developed on lonely walks in which I wondered what happens when a salt and water are united as a solution. In this process all the indications of a chemical reaction are present: heat exchange, volume change and effects on all measureable properties, and these increased with increasing dilution. Yet nevertheless in most cases the salt could be recovered simply by letting the water evaporate. This must have been an unusual form of chemical interaction but I couldn't work out what this special form might be. And so I left this problem in one of the theses in the part of the dissertation reserved for unsolved problems.

Now there was an answer to this old problem. And so there was a reason for me to join in and make the synthesis or symbiosis of the two new concepts, which had so far been restricted to providing an explanation for the irrational factor i, deeper and more complete. This made use of the experiments started in Riga on the conductivity of organic acids which had been interrupted by the move to Leipzig.

The dilution law. I found little equipment in the laboratory which W. Knop had left, as he'd had few students and in the last years had done little experimental work. This suited me because it meant that I had to re-equip the laboratory and the ministry made sufficient funds available.

Since at the beginning I didn't need to do much for the practical courses, I was able to restart my interrupted research work. A workshop with a lathe was set up and a thermostat was constructed in my office. The equipment to measure electrical conductivity was built initially with a resistor box borrowed from the neighbouring physics institute. I soon found by re-measuring some of the organic acids I'd assayed in Riga that the water in Leipzig was much purer so that the analysis done at higher dilutions gave much more precise results. This small improvement had considerable consequences.

In my mind the two concepts described above led to the conclusion that the gas laws could be used to investigate equilibria between ionic species in much the same way as Horstmann in his time had used them to describe equilibria in the gas phase (Part 1, Chap. 8, p. 6). Equilibria of this class are found particularly with weak—that is to say incompletely dissociated acids of which acetic acid is a typical example. I made the calculations on the results of the new measurements which had been carried out under more favourable conditions than had been possible in Riga and found a complete confirmation of the expected relationship. I announced the results to the scientific world in a brief publication dated January 1888 in which I emphasised that in particular the laws I proposed governing the relationship of conductivity and dilution fitted exactly with this premise.[5] With that the "Ostwald dilution law" was established on which an entire thermodynamic analysis of ions has been based. This premise was naturally the starting point for a large number of different applications which were soon addressed, partly by my colleagues and partly by me. The proof for the existence of "free ions" had now been made and it had been shown that they were individual entities whose only restriction was that the sum of the positive and negative ionic charges had to be equal.

Resistance and help. Arrhenius's theory of the dissociation of electrolytes gave rise to considerable concern in the field and I was repeatedly warned not to support it so unconditionally. My request that the objections be spelled out was usually declined. Only G. Wiedemann's son Eilhard, who was in the meantime professor of physics in Erlangen, decided to address this point and sent me an essay for publication in the "Journal" ("Zeitschrift"). This reached me in May of 1888 and I made haste to publish it.[6] The objections it listed were all easily refuted and even M. Planck joined in the discussion by showing that one of the main objections was based on a misapprehension.[7] E. Wiedemann abandoned the idea of publishing a reply and even the detailed discussion he'd originally proposed to write under the title "Concerning the reliability, if any, of certain physical-chemical conclusions" was not published.

Unexpected help came from the brilliant mathematical physicist Max Planck. From a completely different starting point he had arrived at general laws governing chemical equilibria whereby he had reached the same results as earlier workers but had made progress in important points. He compared the ability of salts to reduce the freezing point of a liquid with the theory and demonstrated that this required that more molecules be present than the normal chemical formulae allowed. He had, however, made no suggestion as to where the extra molecules might come from. He joined in the discussion and showed that E. Wiedemann's explanation of polymerised water would not serve, because the molecular size of water was not an element in the equation and hence was irrelevant.[8]

---

[5]Ostwald W (1888) Z Phys Chem (Leipzig) 2:36–37.

[6]Wiedemann E (1888) Z Phys Chem (Leipzig) 2:241–242.

[7]Planck M (1888) Z Phys Chem (Leipzig) 2:323.

[8]Planck M (1888) Ann Phys 270:139–154.

This was the only public attack made by our opponents, but they were nevertheless not convinced. They continued their agitation in private and in their influence on students many of whom were doubtless warned off the new teaching. This had the advantage for us that the large majority of mediocrities who normally seize on a new field avoided this one and so didn't waste our time. That meant that only the independent thinkers joined the field and they naturally sought out the few laboratories where the new teaching was established and where new truths could be discovered by experiments driven by theory.

Chapter 16
The Laboratory

Overview. The events I've just described all took place without interrupting my teaching duties. I had right away held the lecture course in inorganic chemistry in the winter semester of 1887/1888, the lectures in physical chemistry in the following summer semester, and had step by step started my duties as examiner and as a member of the faulty. I had also begun the endless round of introductory visits to my new colleagues and had shown my great willingness to become a typical cell in the vast organism of Leipzig University.

This was at that time a very cumbersome and even difficult business. The original university building was at the edge of the old town, squeezed into the space which the old city wall had enclosed. The wall had long since given way to wide tree lined boulevards on the far side of which lay the suburbs that slowly diffused out into the surroundings and thus irresistibly united the surrounding villages with the city. The space available in the old university buildings had for long been inadequate. Instead of doing the sensible thing and building a complete new campus suitable for the next hundred years or more out on the edge of the suburbs, "historical" sentimentality dictated that the revered old buildings could not be abandoned and that land be bought piecemeal in the south east of the city on which the new buildings needed for the medical and scientific institutes were constructed. The humanities now had their space in the old buildings.

Thus it came about that only the professors and lecturers of the humanities were able to meet in the central "professors' room" in the quarter hour breaks between lectures in the old auditoria. The various outlying institutes each had its own lecture room so that the members of the faculties of medicine and science had no place where they could meet at leisure.

This, together with my dislike of the usual social round, ensured that in Leipzig, just as in Riga, I never really became integrated into the academic scene.

The old laboratory. As I related above I had been given space for the Second Chemical Institute in rooms which had previously been used by the agricultural chemist Wilhelm Knop. He had been a pupil of Wöhler who considered him a particularly thoughtful scientist. By introducing hydroculture to investigate the nutrient requirements of plants he'd made a major advance in his field.

R.S. Jack and F. Scholz (eds.), *Wilhelm Ostwald*, Springer Biographies,
DOI 10.1007/978-3-319-46955-3_16

By the time I got to know him he was an old and rather odd bachelor who lived with his similar sister in the rambling official apartment in which they only used a few rooms. He'd been excited about the geometric forms of atoms which van't Hoff's tetrahedral structure for carbon had started and which in Wislicenus's hands had proved to be so fruitful. The only problem was that he had assigned the tetrahedral structure to hydrogen and given carbon an octahedral structure.

When I visited him I met a little man with an owl-like face and droll behaviour which was partly deliberate and partly subconscious. His head was essentially bald with just a few remaining tufts of red hair and his chin was beardless. He was very friendly and he highly recommended his laboratory assistant Naumann who'd been with him for many years and who subsequently turned out to be first rate. He gave me a present of a manuscript about his theory together with his collection of paintings which he'd had himself photographed next to holding the theory manuscript. On the lecture hall table, which was given to me, he had arranged numerous molecular models and the blackboard behind was completely covered with explanatory notes. One of the pictures showed himself with an expressive look on his face pointing with his stick at a model of benzene and on the blackboard behind was written:

There you see it
That's how benzene is constructed
Six tetrahedral hydrogens
Six octahedral carbons

When I met him once later on a social occasion he lifted up a box of cigars, held it up against his bald head and asked, "What's that?" Naturally no one knew. "Moonshine on the coast of Havana" was the answer. There was a large dose of self irony in the odd things he did. One can well imagine that the laboratory I inherited from him was not in the best shape. As I mentioned earlier it had been built for the professor of agriculture Blomeyer who was not outstanding either as a scientist or as a teacher. The architect had used the same scheme to divide the space on the ground floor for the laboratories, on the upper floors for the specimen collection on the top floor for the apartments and in the basement for the cellars. The building stood on a corner and was made up of two wings each of the same size and between which was the central staircase. Though there was an equal division between agriculture and chemistry—each got a complete floor for the institute and half a floor for the apartment—the overall responsibility for the building remained with the agriculturists. This was a completely unnatural arrangement and the unavoidable problems were not long in coming despite my readiness to subordinate myself to my much older colleague. The difficulties became much greater when a few years later Blomeyer died and his successor,[1] who wanted to put the sleepy agriculture department back on its feet, found himself hemmed in and incommoded by the needs of the expanding physical chemistry department. In the ministry the

[1]The successor was Wilhelm Kirchner.

impossibility of the situation was quickly recognised but my agricultural colleague managed to present the problems of his department as being the most pressing and these were then solved by the construction of a roomy new building for him. I too was promised a new building but the funds for it would only become available once the agricultural building was completed. For this reason I had to spend the major part of my teaching time in Leipzig in a quite unsuitable building.

The assistants. There was one assistant assigned to each of the three divisions of the institute: physical chemistry, analytical chemistry and pharmaceutical chemistry. The latter two positions were held by the men who had run these units for Wiedemann and Wislicenus. There was however no one available for the physical chemistry position until I remembered Dr. Walter Nernst whom I'd recently got to know in Graz and whom Arrhenius had warmly recommended for his talent and knowledge. Since in any case he'd wanted to come and work with me in Riga I offered him the position in Leipzig and he accepted right away. To begin with there were only two students studying physical chemistry so Nernst had plenty of time to get to know the special methods which we'd begun to develop. Up till then he'd only worked in physics and was not practiced in the usual routines of chemistry. However, he picked them up quickly and managed to publish his first results already in the second volume of the Journal of Physical Chemistry. A few years earlier Helmholtz had developed a thermodynamic theory of the voltaic pile which described the relationship between the heat produced by the chemical reaction and the electromotive force and their temperature dependence. When Czapski checked this in practice he found several situations in which the equations fitted very well, but also some in which the results were very different from those predicted and these he couldn't explain. They were only apparent in voltaic piles containing mercury whose thermochemical properties had been determined by J. Thomsen and they appeared to be just as beyond question as his other oft reproduced measurements. However, I'd begun to suspect Thomsen's measurements for mercury already back in Riga and had got a particularly talented undergraduate to check them using an independent method. The student's results were quite different from Thomsen's and agreed well with Helmholtz's equations. Because this work was done during my final period in Riga I had no opportunity to check it in the way necessary before such an important result could be published. I was therefore glad to have the chance to check it once again. Nernst's results agreed with those from Riga and so showed both that Helmholtz's thermodynamic theory was correct and at the same time that Thomsen's measurements were in error. I wrote about this to Thomsen with the request that he comment on it because I didn't want to get into a polemical battle with this respected researcher after the dispute with Stohmann had taken on such an unfortunate form. He at once set about doing experiments which checked the values again, this time using a third independent method. The results were in agreement with those of Nernst. I published his results together with those of Nernst so that it was clear that there was complete agreement and thus avoided any conflict.

Soon after Nernst found the line of thought in which his enormous special talents were to bloom.

The theory of electromotive force. Within the small group of the laboratory we constantly discussed the theories of van't Hoff and Arrhenius whose full value became apparent when they were combined. It was hard for most to grasp that indeed dissolved materials and the ions they gave rise to behaved like gases within the volume of the solute and in particular that they could exert the strong pressure which had been calculated. At that time none of us had seen Pfeffer's experiments which directly demonstrated this pressure and permitted its measurement. Later on Pfeffer built an osmotic cell for our edification and let us read the manometer for ourselves. Why don't the molecules fly out of the solution when they hit the surface? As we were aware the Viennese physicist Stefan had provided the answer: on leaving the surface strong forces amounting to several atmospheres drive them back. But, said Nernst, if one got around this force by forming a layer of pure solvent over the salt solution, then the molecules in the salt solution must immediately fly into the solvent layer. That's what they do, I replied, because diffusion starts right away. The objection was that a gas fills the empty space in just a few moments while diffusion takes weeks and months. I countered that the empty space provides no hindrance to movement while the liquid solvent gives rise to a considerable frictional resistance which will slow the solutes movement.

I left it at that, but in Nernst's mind the phenomenon grew into a picture which he analysed further and which ended with his discovery of the electromotive force of ions.

(Footnote in original) Of course I can't claim that this conversation in the corner room of the old institute, which is still fresh in my memory, provided the first seeds of this idea. Yet I had the impression that at that time Nernst found these thoughts hard to accept.

After a few months he published the basic idea which soon led to his theory of the electromotive force of a voltaic cell.

Ernst Beckmann. The work of the assistant for pharmaceutical chemistry turned out to be no less important than that of the assistant for physical chemistry. I've already told how I was forced against my will to take over the pharmaceutical section and how this annoyance was more than made up for by the personality of the assistant who came with it. In fact Ernst Beckmann developed into one of my best and most successful co-workers whose friendly and appreciative attitude towards me never faltered. He was the staunchest of all my co-workers till then and I know that, were he still alive, he would be happy at being so described.

I, on the other hand, must reproach myself because later after he had gone his own way and become the director of a Kaiser Wilhelm Institute in Berlin, I put his friendship severely to the test through my negligence. It was not deliberate and not because of ill will or because I'd changed my attitude to him but simply that I had so much else to do and because I was more and more losing interest in chemistry which he practiced so successfully. He for his part used every opportunity to convince me of his continued good will.

Beckmann's research. To begin with we were not close outside of the lab. He was already well started on his scientific career and, like almost all chemists at that time, had turned his interest and his considerable experimental ability to organic chemistry. This was a field where he couldn't learn anything from me because he knew it much better than I did. In his research he'd come across very odd cases of isomers the explanation for which depended on knowing whether they had the same molecular weights. Since the material was not volatile there seemed to be no experimental way to the answer. In discussing the matter I pointed out that F. M. Raoult had recently shown the relationship between molecular weight of a substance and its ability to reduce the freezing point of a solvent. Thus the possibility of determining molecular weight, which till then had been restricted to volatile substances whose vapour density could be determined, was now available for all soluble chemicals. Since the molecular weight determination was central to the analysis of chemical reactions this discovery had attracted a great deal of interest and many groups were now working on it. I myself, stimulated by Walden's conductance measurements of aqueous chromic acid, had shown by the reduction of the freezing point that the solution contained not chromic acid but rather dichromic acid. Dr. W. Hentschel (Part 1, Chap. 9, p. 99), who in the meantime had moved from Dresden to Leipzig and was now Wislicenus's assistant, had been persuaded by me to apply his great experimental expertise to setting up the method and had then used it to discover relationships which must have seemed astonishing to the normal chemist but not to the physical chemist to whom they seemed both natural and necessary.[2] I felt it was my duty to make this more widely known. This annoyed Hentschel so much that he terminated our friendly relationship without letting me know what it was that had so wounded him. He soon left Leipzig to do completely different things and I haven't come across him since then.

Beckmann also went in this direction and it is a shining testimonial to his sure touch in developing highly practical experimental protocols that the apparatus he built and described in 1888 is still in use practically unchanged 40 years later.[3] In particular his first description of the "Beckmann thermometer" described how by allowing for a change in the amount of mercury being used, it could cover any desired temperature range.

At the end of his manuscript Beckmann noted that at my suggestion he had also employed the alternative method of Raoult for the determination of molecular weight which involved the determination of the reduction of vapour pressure of a solution. This was of particular interest to me because I'd already seen a part of this relationship in the data of A. Wüllner (Part 1, Chap. 9, p. 7) and therefore had not shared the widespread doubt about Raoult's results. Beckmann had immediately wanted to attack the problem experimentally and we discussed the two possibilities—either measuring the vapour pressure or the boiling point of the solution.

---

[2]Hentschel W (1888) Z Phys Chem (Leipzig) 2:306.

[3]Beckmann E (1888) Z Phys Chem (Leipzig) 2:638, 715.

I knew from Öttinger's lectures that the temperature of a boiling liquid can only be approximately determined because of a small degree of superheating which at that time was considered unavoidable. In the vapour one does not measure the temperature of the solution but only that of the pure solvent which immediately condenses on the thermometer and prevents a further rise in temperature. I therefore suggested that he go for the "static" solution of measuring the vapour pressure.

Beckmann got started on this right away but found that because of the extremely small volumes of liquid which were present as vapour, the error in the determinations was even larger. However, instead of giving up the idea and satisfying himself with the determination of the molecular weights by measuring the reduction of the freezing point which he had carried out, he was sufficiently tenacious and unbiased to ignore my warning and to try the boiling point approach by applying the various methods to suppress superheating and by doing so he managed to establish a highly sensitive procedure.

In this way he completed his conversion to physical chemistry. His apparatus and the method to determine molecular weights of solutes are used throughout the world and there can scarcely be a chemistry laboratory in which they are not used.

Beckmann was the same age as me, but because he'd turned to science rather late he found himself in a subordinate position as we worked together in Leipzig, although he had already an established reputation for his research in organic chemistry. His rise was not long in coming. After a few years he was appointed to a professorship in Erlangen and a few years after that he returned to Leipzig as a full professor of applied chemistry and thus became my colleague at the same level in the hierarchy.

J. Wagner. I have much less to relate about the third (actually the second) assistant Dr. Julius Wagner. He had been educated by Wiedemann, and became only slowly and incompletely enthusiastic about the flood of new ideas and research in which the others were engaged. He staunchly carried out his duties in instructing the beginners and he also did some research in areas related to his teaching responsibilities. With his help I finally managed to get rid of the foolishness that teacher trainees were being put through a course designed to make them analytical chemists which was of no value at all for their future work in the class room. I therefore told him to organise a course which would teach them to carry out safe and instructive experiments suitable for use in a school. This was a challenge that suited him and he presented this course for many years to the benefit of the coming generation of chemists. Later on he was given a separate position devoted entirely to this.

Of all my assistants he was the one who stayed longest for he remained in the institute until his failing health forced him into retirement.

Progress. I and my colleagues watched with suspense the development of the physical chemistry course, because this was the point which would decide whether and how well I was mastering my assignment. In the first semester two students had joined us, but one of them had to leave Leipzig unexpectedly and there was not a single applicant at the beginning of the next semester, so that then the institute had only one single student.

That was more or less the situation in which this course had laboured under Wiedemann who had not managed to attract students and I began to worry that the hope I'd had when I came to Leipzig of expanding my sphere of influence was going to be disappointed.

However in the third semester there were eight students and nearly double as many in the following one. The practical courses for the analytical chemists and the pharmacists left only a rather cramped space free and this was soon completely filled. Thus the lack of adequate space for all the students who wished to study under me, which I'd suffered for most of my time in Riga, now reappeared in Leipzig. And so, after just a few years I had to inform the authorities that more space, in particular more properly equipped space, was now urgently needed.

Crucial to this rapid progress was the fact that the institute in Leipzig had, as I described, quickly become associated with a fundamental advance and hence was seen as being a centre where the fruits of the new advances could be harvested. Since in the next year, 1888, Arrhenius himself came to Leipzig, first as a post doctoral fellow and then, when Nernst for health reasons left for a while to Heidelberg, as assistant in the institute there arose a strong new attraction for ambitious young students. At that time there were no other professorships for physical chemistry available nor was it taught as a subject in its own right, and therefore we in Leipzig could only attract those who were convinced of the importance and the enormous scope of the new area of research. Seldom has an academic teacher been blessed with such a flock of talented young co-workers as I was in the old and unsuitable laboratories at 34 Brüderstrasse. Our work was collaborative and the discussions between the individual students and the professor took place in the presence of all the others who readily joined in and contributed their opinions. The success of the one spurred the others on to greater efforts which almost always were crowned with success.

Finding time for writing. It seemed to me that the many doubts and misunderstandings about the new concepts should be countered by an essay about the chemical and other consequences which stemmed directly from the theory of the dissociation of electrolytes. In particular I showed that the chemical affinities of the acids which I'd studied now for some years, and which could be demonstrated by various different methods could now be completely explained. In each case the reactions were determined directly by the number of available protons and this number was dependent on the nature of the acid and its degree of dilution. This is so clear and obvious today that we can see no other explanation and find it almost impossible to remember what a long hard struggle it was to demonstrate chemically the affinity values that the conductance measurements so accurately reflected. The conductance values themselves, however, tell us nothing of the chemical relationships. Only once it was realised that the phenomena which they reflected had been independently discovered and measured was their general applicability clear.

In this essay I considered in detail the various objections which Arrhenius and I had come across in our discussions with colleagues and students and showed that by rigorous application of the new insights these objections could be swept aside and that it was now possible to bring together various experimental results which had so far been merely unexplained observations and show that they fitted into a uniform theory and, by doing so, we were now able to explain them properly. In addition, I added some examples of the dilution law which showed that it applied over such a broad range of concentrations that could never have been investigated chemically. I think it is fair to say that since the publication of this essay no further informed objections have been raised to the concept which was from then on accepted as a legitimate part of scientific thought.

Conductivity of organic acids. In addition to this work there was one further duty which had to be completed and that was to finish the analysis of the conductance of the many acids which I'd been given during the course of my begging tour in the summer of 1887. In addition to them there were a number of preparations from the old collection of O. L. Erdmann which had belonged to Wiedemann's physical chemistry department and which I had now inherited. All in all this amounted to some 150 probes. Once again it gave me the greatest pleasure to go through these repetitive determinations which I mainly did in the evenings when it was nice and quiet in the laboratory. The many interesting and informative relationships to the chemical structures which these measurements revealed, kept my enthusiasm at a high pitch. The hours spent doing this work were happy and comfortable. On top of that they provided a vast data set in support of the dilution law.

Before I published these data I waited for the results from van't Hoff's laboratory for I'd heard that he was also checking the law. From a co-worker from the Amsterdam laboratory I later heard how van't Hoff had set his assistant Reicher to produce extremely pure acetic acid. Time and again Reicher did his best and each time van't Hoff demanded an even higher degree of purity but without telling him what it was all about. The same thing happened with several other acids. The results were published in a short paper and they fully supported the dilution law[4] "No single case of ordinary dissociation has been checked across such a broad range". The sorely tried assistant was recompensed with his name on the paper even though the professor had had the idea and had carried out most of the measurements.

Among the acids investigated was an exotic one from Japan called "Shikimic acid" which had been recently discovered by professor Eykman in Tokyo. Its outlandish birthplace did not prevent it from obeying precisely the dilution law established in Europe.

I now set to work writing up the results of the many measurements through which the outlines of one chapter of the new science was established, even if it did not receive the degree of general recognition which it deserved. This was in part due to the fact that my interest in the questions which the method was there to solve or at least to more precisely formulate was not sufficiently great and, in addition, other questions were more pressing. Nevertheless the work was not for nothing because Professor Wegscheider in Vienna took it up eagerly and pushed the matter very considerably forward.

[4]van't Hoff JH, Reicher LTh (1888) Z Phys Chem (Leipzig) 2:777–781.

In order to calculate the results it was necessary to know a particular constant (the diffusion coefficient of the anions) and here also ways were found that also resulted in the formulation of a number of new laws. Working in this new area was like walking through a garden where the fruit was just ready and one only had to touch the low hanging apples for them to fall ripe and sweet into one's hand. That's not just the way I saw it, all my co-workers felt the same way. The first volumes of the Journal of Physical Chemistry contain many papers from our group which provided the starting points of substantial scientific developments. One can see this most clearly when one compares our publications with those from other laboratories where the boundlessly fruitful ideas of van't Hoff and Arrhenius had not yet taken root.

Not a chemist. The conductance determinations of so many organic acids clarified so many structural questions that the relationship of the new area to "real" chemistry, which most chemists considered to involve the synthesis and structural analysis of new substances, could not be doubted. Nevertheless even later there were those who asserted that I wasn't really a chemist because I had never synthesised a new chemical.

I have to say that in this sense I have to be described as an "anti-chemist" because though I never increased the length of the list of known substances I did in one instance reduce it—sadly only by one entry.

The story was as follows: In 1863 H. Kämmerer recovered from a silver bath used to make photographic plates an acid which he considered to be different from all of the then known acids. The elementary analysis yielded numbers which fitted closely to the values for malic acid, however since the acid was clearly different he thought it must be an isomer and therefore named it "isomal acid". He referred to it in a number of subsequent publications but was never able produce it again. Despite that it survived unchallenged and even made it into the first edition of Beilstein's fundamental reference book.

In the collection of the old institute I found a glass tube with a tiny sample of this weird acid which was signed by its discoverer and had apparently been given by him as a gift to Erdmann. It was far too little for an elementary analysis at that time but it was ten times more than was required for a determination of the conductance of its dilute solution. The measurements were carried out both on the free acid and on its sodium salt, and the result was that isomal acid turned out to be nothing more than citric acid. I wrote to Kämmerer and asked him for a sample of his acid. He still possessed some and shared it with me. It turned out to be identical to the preparation from the Leipzig collection. Under his experimental conditions the citric acid which normally forms crystals that contain water, had been crystallised without water and this had prevented him from recognising the product. I published the result in the Bulletin of the German Chemical Society, and in the next edition of Beilstein's hand book isomal acid was no longer to be found.

Although there was at that time no other method known by which such clear results could be obtained with such tiny amounts, I don't remember ever being congratulated for having exorcised this ghost. Admittedly, there were no objections and even the discoverer of isomal acid was satisfied.

German and foreign students. After the first few lean semesters the number of co-workers in the institute rose rapidly, and in addition to Germans there soon came a number of foreigners, especially Americans and Britons. These two groups remained in the majority though almost all advanced nations were represented in the Leipzig institute at one time or another.

One of my first foreign students was the Scotsman James Walker who later was a successful proponent of the new concepts and did much to introduce them into Britain. He is now professor at the University of Edinburgh which is one of the most influential chemical chairs in the country. Of the Americans I must first mention A.A. Noyes who soon made the Technical University in Boston into a centre for research in physical chemistry. Later his health problems forced him to move to a more southerly climate and to restrict his research activities. Both of these scientists are not only excellent teachers and researchers but are among the most splendid examples of the diverse human species.

My German students were of two quite different types. The first was a small group and consisted of those who were attracted by the new concepts which we were developing and worked for the joy of it without worrying too much about the future. Of course they were just a handful and this was not surprising when one considers the mass of silent opponents of these new ideas who had a strong influence on the views of the young students.

In addition to them, there would appear every semester a group of casual students in the beginner's lab who'd signed up for the course because other practical courses were full or because their friends wanted to study here or for other secondary reasons. They were usually mediocre but afterwards lots of them found employment in the highly developed chemical industry where they rendered sterling service. Once they'd finished the preliminary courses they would apply for a position as a doctoral student with us because they knew they wouldn't be welcome in other departments.

It is a happy memory for me that even these students were so swept along by the sense of excitement for research and discovery in the institute that they usually managed to produce results which were far beyond what I would have initially thought they could achieve. And if one sees what they achieved afterwards in the struggle for existence in comparison to their mediocre chemical contemporaries, then one can fairly judge them to have been clearly better.

In this regard I learnt a curious feature of Germans which I'd never seen in my Baltic compatriots. It is the ability, despite a meagre level of general interest or education, to suddenly discover some small area in which the most extraordinary accomplishments will be achieved. Schiller's injunction, "concentrate quietly and ceaselessly the greatest effort on the smallest point", was followed exactly by these unimposing minds and often enough I could see that the achievements reached in the chosen small niche then fed back to raise the entire level of mental accomplishment. This applied only to these casual students. As the success of the new concepts became known, the number of students who joined us because they wanted to learn more, increased and I got to know many splendid minds amongst them. Their number was always so large that I was able to fill all of the increasing number of assistant positions with Germans (without exception they all later became professors) and many of the rest were able to find similar positions elsewhere.

Chapter 17
At the Writing Desk

The "Outlines"[1]: The physical chemistry lectures which I held as usual in the summer semester of 1888 came just after the fruitfulness of the synthesis of the ideas of van't Hoff and Arrhenius had been made apparent by the discovery of the law of dilution. I'd worked these new thoughts with ever increasing pleasure into the fabric of science and could see from the shining eyes of my audience that they eagerly accepted the insights that flowed from them which explained so much of the world. Since many of them were students who'd chosen other areas such as organic or physiological chemistry as their main subjects, they asked me if there was a textbook in which they would always be able to find the content of the lectures.

I myself felt a sort of creative urge to make out of the rich and beautiful new concepts an impressive overall picture. In this way I'd also be able to satisfy my deeply held wish to extend the circle of the "ions", as we were known by the others because of our constant use of the word, beyond the limited numbers who could be directly reached by word of mouth. Of course the textbook was available, but it couldn't contain all the new discoveries from 1887 since it had been completed in 1886 and in any case it was too detailed for this sort of audience.

And so I quite easily reached the decision to write a short book in which the central tenets of physical chemistry would be laid out in a way accessible to all chemists but without any excessive detail or concentration on methods. Fired by the joy of creation I completed the "Outlines of General Chemistry" in 1888 and it appeared with the official publication date 1889.

This book was an immediate success. A second larger print run had to be brought out in the following year and a third followed in 1899. After that the book was out of print for quite a while because I wanted to revise it but didn't find the time. In 1909 a fourth edition and later on a fifth were published. In total roughly

---

[1]Ostwald refers here to his book "Grundriss der allgemeinen Chemie" Leipzig, Engelmann 1889, which was translated into English by James Walker and published as "Outlines of general chemistry", London, McMillan, 1890.

R.S. Jack and F. Scholz (eds.), *Wilhelm Ostwald*, Springer Biographies,
DOI 10.1007/978-3-319-46955-3_17

1,200 copies of the German edition were sold. It would have been a lot more if I'd attended to it with the care that it really deserved.

Beyond the Germans, the book brought the news of the new science to other peoples. Already in 1890 J. Walker prepared an English translation. There followed Russian, French, Japanese, Spanish, Hungarian and other translations so that the "Outlines" turned out to be one of the most effective means of spreading the new concepts. I also noticed that in public libraries which held my work the "Outlines" always looked more used and worn than the other books.

The Classics. In the same year I made a start on the "Classics of Exact Science". In getting together the material for my textbook I had been struck by the enormous discrepancy between the vast number of published papers and the tiny fraction which was of real long standing significance. The diligence of editors ensured that the amount of completely trivial publications was and is extremely low, yet nevertheless I'd noted that for example Poggendorf who had made the "Annals of Physics" (Annalen der Physik) into a leading journal had in his last years lost his critical sense and let through a number of papers which it would have been no loss to science had they never been published. And yet the vast majority of good and useful published papers have served their purpose once their contents reach the text books. There are however a tiny number of masterpieces which tower like mountains over all the rest and whose value is not exhausted once their message enters textbooks and whose brilliance and significance should be made available to future generations who will only rarely have access to the old volumes of the journals. From them one can learn how such timeless contributions to science are generated. Furthermore such papers will frequently contain hints towards the seeds of new ideas which only await planting in a receptive mind to burst forth into new fruit.

My publisher, Dr. Engelmann was soon happy to undertake the project. A number of excellent colleagues undertook to advise us on all the main areas of the exact sciences from mathematics to physiology in which I was not competent. The "Classics" soon developed nicely and over two hundred volumes came out, quite a few of which were reprinted in several editions of several thousand. Later they were edited by my teacher A. von Oettingen and, after his death, by my eldest son Wolfgang Ostwald. Today they are printed by the "Akademische Verlagsgesellschaft" in Leipzig.

The idea of the "Classics", that masterpieces of science should be made available to the general reader in special editions, caught on. Similar reprints concentrating on various speciality topics sprang up and this German example was soon followed for French and English readers. Here lay for me the seed of the thought that developed much later into the question of the technical organisation of science and that in turn led to the foundation of the "Bridge" and to various other enterprises—but these are things that will be related in the third and last volume of this work.

Criticism of History. For me the main purpose of the "Classics" was the simple practical effect outlined above. But there was also a secondary emotional reason for me. Karl Schmidt had taught me to treasure the history of science and in fact his two hour per week course on the history of chemistry was, as far as I remember, the only lecture course that I regularly attended as a student. On top of that my own studies of lots of old papers in chemistry and physics went in the same direction. I didn't do it merely to learn physical chemistry from its sources but also because I derived real pleasure from this intimate contact with the great and less great names of science. This enthusiast's love of immersion in the historical details is often associated with an awe of everything historical and this has been elevated by the philologically oriented high priests of scholasticism to represent the epitome of "culture". As a student I followed this lead uncritically and accepted it as something one did not have to discuss.

Later on I was proud of myself that right from the start of the "Classics" project I concentrated only on the practical aspects outlined above and didn't waste any time on the usual waffle about the "ideals" of history. This of itself shows that even then I found such turns of phrase suspect. During my time as a professor in Riga I sometimes had long conversations about history with the historian Otto Seeck, the older brother of my friend Fritz. He strongly defended the standard position that science must be pursued "for its own sake" and that every historical detail is therefore important because one never knows how it may help solve a problem in the future. In fact, the brilliance of his own work lay in showing how apparently secondary notes in the original sources could be used to open up surprising and important perspectives. I thought all this went too far and wasn't prepared to accept it as a general rule and I exclaimed, "What sort of science is this is in which any blockhead who out of ignorance destroys an irreplaceable manuscript produces a hole which can never be filled?"

I don't remember anymore how that conversation ended, but I was left with a subconscious distrust of the over estimation of the value of history as a science while nevertheless paying lip service to the usual public reverence for the subject. Only in 1895, after I'd written a major historical work, "Electrochemistry: its History and Concepts",[2] did I realise that history is not a science but merely a method or tool which may help in reaching a scientific understanding of certain questions. There is a history of politics, of economics, of sociology, a history of mathematics, of chemistry or of any other branch of science, there is a history of the world in a geological sense, and so on; but there is no such thing as History on its own.

I would go further and say that the historical aspects of any subject are not, as some would have us believe, the epitome of scientific investigation but rather they are its most primitive form. This became clear to me when the urgently needed professorship for the theory of teaching, which had been opposed in the faculty by the philologists who felt that their monopoly on the subject was being threatened, was finally approved and filled. The candidate, who was chosen for peculiar reasons, didn't really know much about the theory of teaching and so limited himself

[2]"Elektrochemie. Ihre Geschichte und Lehre, Veit & Comp. Leipzig, 1896.

to lecturing on the history of teaching.[3] I was surprised at first until it became clear to me that he did this because to lecture about the history of a subject requires much less detailed knowledge than to lecture on the subject itself.

The textbook. Scarcely were the "Outlines" and the "Classics" underway than my publisher brought me the happy news that the two volumes of my detailed textbook would soon be out of print and that I should arrange for a new edition. This suited me well because the few short years since the book's completion had seen so many major advances both in terms of real results and new concepts as well as in the development of my own deeper understanding of the field that I now regarded it as absolutely essential that the field be reviewed and presented anew. I eagerly set to work. It turned out that the advances of the last few years had resulted in such a growth of the science that I had to rewrite almost everything and in the process the book grew enormously. The first volume—"Stoichiometry" ("Stöchiometrie")—which had amounted to about 500 pages was now no less than 1163 pages long. I started work on this in 1889 and completed it in 1890. It gave me the opportunity to present a complete picture of the new theory of solutes both in terms of content and history. This part was translated into English and it was influential in Britain and even more so in America in attracting many new adherents to the new concepts of physical chemistry.

The second volume—"General Chemistry" ("Allgemeine Chemie")—which dealt with chemical affinity had grown even more than "stoichiometry". It came out as two volumes. The first was 1104 pages long and came out in 1893. The second, which was 1188 pages, came out between 1896 and 1902. The problem was that new material flowed in at an ever faster rate; after all the "Journal"[4] on its own produced four thick volumes every year. Though there seemed to be no limit to the new data there was a limit to my capacity to handle it and so the second volume of the textbook was left unfinished. After the second edition was sold out there could be no thought of a third edition.

Thus my textbook suffered the same fate as many other similarly successful projects. In my literary excavations in Dorpat I'd noticed that the Kekulé's textbook of organic chemistry, on which the entire development of organic chemistry at that time was based, broke off in the middle of the most important part and couldn't be bound because one had waited in vain for the rest. Erlenmeyer's ground breaking textbook had met a similar fate. Back then I'd postponed my exam in analytical chemistry from the second to the third dead-line in order to wait for the final part of the classical textbook of Fresenius, but that last part didn't appear until long after I had become a professor.

[3]Ostwald refers to Emil Jungmann (1846–1927), a teacher and professor at the *Thomasschule* (a boarding school founded in 1212, and attended by all members of the boys Thomanerchor) in Leipzig. In 1901 he was appointed professor of secondary school pedagogics ("Gymnasialpädagogik") at the University of Leipzig.

[4]*Zeitschrift für physikalische Chemie (Z Phys Chem (Leipzig)).*

The reason for this—or at least one of the reasons—is given in Goethe's ballad about the sorcerer's apprentice. "I can't get rid of the ghosts I summoned". The stimulation given to an area of science by a successful and original textbook attracts many competent researchers who produce a flood of experimental and theoretical research in the new area. Once the field had been divided up into drawers by the definition of generally applicable concepts it becomes relatively easy to fill them and so a mass of new material is generated which must then be digested and put into context. At the same time the author of course becomes exhausted and so these two elements inevitably lead to the result that a continuation becomes ever more difficult if not impossible.

In my case there was something else which added very considerably to these two factors and that was the development of a new scientific philosophy[5]—Energetics. However, we will come to that later because I haven't yet finished with the description of the first phase of development of physical chemistry.

J. Willard Gibbs. The next book that I brought out with my still venturesome publisher Engelmann was not an original work of mine but simply a translation. Back in Dorpat Oettinger had told me about the work on thermodynamics of J. Willard Gibbs which, he said, was of great importance—but hard to get hold of. Once it became ever clearer to me how important this approach was for explaining chemical affinity, I managed, not without considerable difficulty, to obtain these papers and study them. For me the experience was like it had been for Oettinger. I found them hard to read but recognised without any doubt their great importance. Not many had come this far. The physicists Maxwell and Lorenz had been there before me but they had only occasionally mentioned or used Gibbs's ideas.

I found no better way to get to the heart of the matter than to translate the publications word for word. They certainly couldn't be summarised because they were written in such a condensed form. I thought that by bringing out a German edition I would do my part to rescue these long overlooked treasures and make them available as common currency to the current research effort.

This work had a huge effect on my own development because although he didn't particularly stress it, Gibbs worked exclusively with terms of energy and their factors and completely avoided kinetic hypotheses. By doing this he reached conclusions which for their truth and generality must be considered among the highest products of human understanding. In fact, so far no one has found a single error either in his equations or in his conclusion or—which is the most remarkable of all—in his premises. There are many scientific works in which the logic and the mathematics are all impeccable but which are nevertheless completely useless because the premises do not reflect reality. In this respect Gibbs was completely faultless.

---

[5]Ostwald used the word "Weltanschauung".

The papers which had originally been published in English were brought out in German by me in 1892 under the title "Studies in Thermodynamics" (Thermodynamische Studien).[6] For a long time it was the only form in which this extremely important work was available to the scientific community, for Gibbs had published the original in the "Transactions of the Connecticut Academy of Arts and Sciences" where no one would look for it and which in any case was long out of print. So, although they were written in their own language, even Britons and Americans had to read the papers in German until finally, after Gibb's death, a desperately needed new edition was brought out in the original language.

Willard Gibbs was an excessively modest and diffident scholar who, apart from a few years of study elsewhere, spent his entire life in his birthplace New Haven, Connecticut where his father had also been a professor. No one in his hometown or in America was aware of his greatness—and he is without doubt the greatest scientific genius that the United States has ever produced. He had to be discovered in Europe; first by the physicist Lorenz in Holland and by me in Germany once Oettinger had brought him to my attention. In Holland a whole school of researchers developed from this, beginning with Lorenz's pupil Bakhuis Roozeboom whose work concentrated in exploiting the phase law which is often absurdly referred to as the phase rule. This phase law, one should bear in mind, was just one of the many laws that Willard Gibbs had discovered and described.

In this way the scientific world slowly came to understand that Gibbs was a genius whose mental depth and fruitfulness places him alongside the giants of thermodynamics like Helmholtz, Clausius and W. Thomson. However he was so little known in his own country that the following strange story really happened. The news that Gibbs was regarded in Europe as a great genius crossed the Atlantic and spread across the country. Now, at this time there was a chemist called Wolcott Gibbs who, while he had done some nice work, was far from being a genius. Nevertheless, in America his name was much better known. Without any further questions being asked he was proclaimed to be the new star and presumably no one was more astonished than he over the congratulations which poured in from his enthusiastic countrymen from every corner of the United States. Only later was the mistake discovered and then Willard Gibbs successfully countered all attempts to make him into a popular hero.

The close scrutiny of these publications, which was needed for the translation, had enormous consequences for me. Although I could not follow his mathematics in all its details I experienced an education in thinking by following the straightforward objectivity with which he attacked the different problems and the exhaustive care with which he teased every last possible implication out of the equations. I also couldn't help noting that that the 200 equations in the main part of the work were almost all concerned with relationships between energy terms. This

[6]The story of the contacts between Ostwald and Gibbs has been described in: Moore CE, von Smolinski A, Jaselskis B (2002) Bull Hist Chem 27:114–127.

was initially just a formal observation but it became of great importance to me because it showed that this fundamental work could be described as chemical energetics.

The handbook. [7]Because at the beginning the students in the physical chemical laboratory were all qualified chemists who wished to get a practical introduction to the new area, I just let them get started right away on whatever research topic appealed to them and which seemed to me to be worth looking at. It soon became clear, however, that this resulted in a rather one sided education since they were deeply knowledgeable about their own little area and largely ignorant of all the other parts of the subject. After a few semesters I therefore decided that before starting with the research topic which was usually used for a doctoral dissertation that each student should take part in an introductory practical course in which he had to learn the most important methodologies. At the same time this involved a seminar in the use of mathematics which naturally went a little deeper than the usual rule of three which had so far been enough for almost all chemical problems.

Strangely enough a group of my students who were just finishing the usual general analytical and preparative courses did not take kindly to this decision because they had thought that they would be able to start right away with their doctoral thesis work. They let me know through my assistants that this sort of thing was not done in the other laboratories and that they would leave my group unless I rescinded my decision. For me there was no question of what should be done. I wished them luck for the future and increased the introductory program which had been designed for 6–8 weeks so as to accommodate some important new methods. Eight to ten students did in fact leave the department, though as far as I know none of them did anything worthwhile later so it was no great loss to me. However, I must admit that I reproached myself that because of a deep aversion to such things I had avoided a personal confrontation with the students.

Nevertheless I did realise the need to relieve myself and my assistants of the necessary but repetitive work involved in the introductory course and came up with no better solution than to write another book. I called it "Handbook and aid to perform physico-chemical measurements" and worked into it all the little tips and tricks that I'd picked up in nearly 20 years work in this area. I tried to coax into my students the same love of working with their hands and putting equipment together that makes the sometimes rather mechanical work we have to do in this area of research not just acceptable but a real joy.

The book was published in 1893 and was translated into several languages. Initially subsequent editions were prepared with the help of my co-workers and later on they took over the entire work because as I later turned most of my energies

[7]Ostwald refers to the "Handbook and aid to perform physico-chemical measurements" ("Hand- und Hülfsbuch zur Ausführung physiko-chemischer Messungen"), Engelmann, Leipzig 1893. The fifth German edition was reprinted by Dover Publication, New York, 1943 together with an English translation of the table of contents and a German-English glossary.

to other problems I no longer felt that I had sufficient knowledge of the daily problems in the lab to be a teacher of this material.

Legwork. In order to facilitate the use of the new methods and concepts I wrote in 1888 and 1889 a number of popular essays for the "Journal"[8] in which I described the details of conductance measurements, and since then this is how such measurements are done all over the world. I also described in detail the most important chemical reactions of acids, bases and salts in the light of the dissociation theory and these have since found general acceptance. In both cases the basic ideas were from others—the methodology from Kohlrausch and the concept from Arrhenius—and I made that very clear in the essays. Only the outline of the applications and the presentation were due to me though these were very important and effective in reducing the hesitation that chemists initially had to apply these new methods.

In particular, it turned out that the whole field of the thermo-chemistry of "salt formation", from Hess's "thermo-neutrality" of 1841 to the most recent research of Thomsen and Berthelot, could be explained in a clear and readily understandable fashion with the help of the dissociation theory. Beyond that I examined the reactions used in analytical chemistry and showed that here too a large number of facts, that had up till now been unexplained and had had to be learnt off by heart, turned out to be simple consequences of the degree of dissociation. In particular the concept of normal and abnormal reactions was suddenly made clear.

Analytical chemistry. The concept of free ions found many applications in terms of physical and chemical relationships and Arrhenius had developed these ideas largely in terms of general and physical chemistry. Because I had to lecture on inorganic chemistry and organise practical work in analytical chemistry I felt myself driven to apply the new concepts to these two areas and soon found important generally applicable results.

Arrhenius had already in his first publication emphasised that the ions of a salt were independent and this independence applied to all of their properties (except of course that the necessary equivalence of positive and negative charges had to be maintained). The properties of a salt solution must therefore simply be the sum of the properties of the different ionic species—they must, to use the term I introduced, be "additive". He showed, using a whole set of examples, that this additive property had been noted by earlier researchers who, however, had been at a loss to explain it.

For analytical chemistry it followed that the analytical properties of a salt solution were simply the analytical properties of their ions. Nowadays that is a platitude, but back then it was a real discovery. It threw the spotlight of science onto an area which till then had consisted of empirical results that lacked any integrating concepts. This was reflected in the general view that those who settled for a career in analytical chemistry were mentally not quite up to the level of those who were also able to handle organic synthesis and structural chemistry.

---

[8]Zeitschrift für physikalische Chemie (Z Phys Chem (Leipzig)).

Wilhelm Hittorf. My analysis of these questions was given depth by the fact that I had carefully studied Wilhelm Hittorf's classical paper on the migration of ions which I'd selected for the "Classics". In it Hittorf had already anticipated a large part of the theory of free ions but had failed to make the last radical step. Nevertheless, what he saw as the unavoidable conclusion from generally known phenomena and from his own measurements was considered by the leading electro-chemists of the time—Magnus and Wiedemann—to be so heretical that they attacked him with such polemics that the fundamental significance of his work was not recognised. On top of this Wiedemann's textbook summarising the then current knowledge of electrophysiology mentioned Hittorf's work only briefly and in a negative tone while numerous insignificant papers were listed and considered at length.

It seemed to me that it was a necessary duty to give this underestimated and unfairly judged researcher the recognition he deserved by reprinting his paper in the "Classics", particularly as he was still alive and held a professorship in Münster. I wrote to him to get his permission to reprint the paper and he replied with a moving letter of thanks for the recognition that his work was finally receiving. In order to spare me any inconvenience that might arise by reprinting his published defence against Wiedemann's unjustified condemnation of his work, he asked me to edit out any polemical passages and reduce it to the presentation of the facts and their interpretation.

This I did and it was only after the death of both protagonists that I took the opportunity to arrange for a new edition which included the uncensored exchanges. I did this because it seemed to me to be important that the reader be shown not only the factual arguments but also the personal difficulties that almost always accompany the establishment of important new ideas—especially when they are simple and clear.

For me the most important aspect of this work was the clear distinction between simple and complex ions. When a metal is present in ionic form it carries a positive charge and will migrate with the electric current. However if it is part of a composite or complex, then negatively charged ions will migrate against the electric current. Hittorf had shown this very clearly using a number of examples. In an analytical analysis, however, complex ions of this sort do not show the usual properties of the metal they contain, because the simple metal ions will not have been formed. This resulted, as a brief examination soon showed, in a full theory of the so-called abnormal reactions which had been well known but for which no one had managed or even tried to give an explanation.

The Scientific Basis of Analytical Chemistry. I formulated these thoughts in 1893 but I was so tied up with other things that I had no time to bring them to paper. I took them subconsciously with me in the Easter holidays which I spent on the Riviera. On the homeward journey I had to sit for 10 h in an uncomfortable seat in an overcrowded train from Munich to Leipzig and I used this sleepless night to work out in my head the entire plan for a new book which was to be called "The scientific basis of Analytical Chemistry". I can still remember the pedagogic and organisational problems that had to be solved, but that got done so successfully that afterwards the book just had to be written down without any further alteration to the plan.

By chance my older colleague Lothar Meyer visited me at this time and though we'd had some scientific arguments he was friendly towards me. I mentioned that I was writing a book about analytical chemistry. He burst out laughing and said, "That too? You've never published a paper in this area!" I told him that the book would describe an important advance through which analytical chemistry would be placed on a scientific basis. This only increased his merriment and he said he wouldn't believe it till he'd read the book—and made clear that he probably wouldn't believe it even then.

The book was dedicated to Johannes Wislicenus. This was an expression of my thanks to him for his friendly and fatherly courtesy that he'd shown me and my family. It was also an attempt to heal the schism which I could see was now quietly developing between us. His fatherly attitude was predicated on the "obvious" condition that in all matters of chemistry, including those that affected principally me and my teaching responsibilities, I should not initiate anything which had not been previously submitted to and approved by him. He even expected that I reach an understanding with him concerning my scientific views. I did not meet this expectation because I felt that his understanding of physical chemistry did not go beyond that of the average organic chemist—which wasn't very far. It was also in part thoughtlessness on my part because I hadn't fully realised the extent of his expectations in this regard. On the other hand I had so many problems with the misunderstandings, mental laziness and arrogance of the people all around me that it seemed nothing less than my duty to strongly back the newly won scientific advances. In the Leipzig Chemical Society I even dared at times in public to teach the essence of these advances to the dignified older colleagues whose knowledge of them could not be taken for granted.

Because of all this I was glad to have an opportunity to show my respect by making the dedication.

As was to be expected from the nature of such relationships the book dedication slowed up the inevitable split but was not able to stop it. This necessary separation, which was driven by the laws of nature, is one of the most painful experiences of my time as a professor in Leipzig.

Success. The effect of the book on my contemporaries was ambivalent. Some recognised its value immediately as was shown by the fact that it was soon translated and spread the new concepts across the civilised world. In fact in a short time there were eight to ten foreign language editions.

An effect on the German textbooks of analytical chemistry was not immediately obvious. When a few years later an extended second edition came out I felt it necessary to make this point in the preamble. I was also tireless in pointing out this defect in the various textbooks in the book review section of the "Journal of Physical Chemistry" (Zeitschrift für physikalische Chemie) and to demand that things be improved. About 5 years after its publication the new views began to find their way into the textbooks and once the ice was broken no one wanted to be accused of being out of date. Of course to begin with some of them just used it as a sort of intellectual varnish but with time even this was overcome. Today, as far as I know, the views developed then are now generally accepted and considered as something about which there is no argument.

Electrochemistry. The next major book that I wrote was called "Electrochemistry: its History and Concepts" (Elektrochemie. Ihre Geschichte und Lehre). The preparative work and the first drafts were done in 1893. Since the book was going to be of considerable size it came out in sections the first of which was ready at the beginning of 1894 but the publication of the whole book took 20 months and wasn't finished till the end of 1895. For this reason the book carries the publication date 1896.

The motivation behind writing this comprehensive work—it is 1151 printed pages long—was spurred on by the feeling that the redevelopment of electrochemistry which was taking place then should be coupled as closely as possible to the roots of the subject. The best way to do this was to tease out the details of the subject's development. Of course there was a volume in Wiedemann's work on galvanism (later electricity)[9] which dealt with electrochemistry but the new book differed both in its point of view and in its method of presentation. I missed in his account a general understanding of the development of the subject, which I had grasped even before I had studied the matter in detail.

This was the first book in which I abandoned my till then usual approach of concentrating on the immediate questions facing the field and instead also showed how the personalities of the researchers affected the outcome. With that I struck a note that my more intelligent readers warmly welcomed but which caused astonished reservation amongst the usual circles. This book, which I consider one of my best, can be measured against the histories of mechanics from Mach[10] and from Dührung[11] but it had no second edition and was not translated and so in this sense has an unusual position amongst my older works.

Other texts. Alongside the writing of these major works, the flow of reports about new papers and books which I prepared for the "Journal"[12] continued unabated. I piled everything which had to be dealt with on one side of my table so that the pile became ever more unstable as it grew and that forced me not to put off what had to be done. Since this sort of writing was well adapted to the method I described already (Part 1, Chap. 12, p. 129) of filling the odd quarter of an hour which would otherwise have been wasted, I did manage for a long time to hold the balance

[9]Ostwald refers to: Wiedemann G (1861) Die Lehre vom Galvanismus und Elektromagentismus nebst ihren technischen Anwendungen. Vieweg, Braunschweig. Later editions were entitled "Die Lehre von der Elektricität" 4th edition 1894, Vieweg Braunschweig.

[10]Ostwald refers most probably to Mach E (1883) Die Mechanik in ihrer Entwicklung historisch-kritisch dargestellt. Brockhaus, Leipzig.

[11]Ostwald misspelt the name. He refers to Eugen Karl Dühring and his book "Kritische Geschichte der allgemeinen Prinzipien der Mechanik". Grieben, Berlin 1872.

[12]Zeitschrift für physikalische Chemie (Z Phys Chem (Leipzig)).

between what came in and what went out and so to deal with the never ending inflow of new work.

I soon got a chance to see how useful and indeed essential this all was for the development of the subject. I had in the now established concepts and relationships a means of judging the contribution of each newly submitted paper. I constantly urged the authors to place their work in the context of these new concepts and often enough I was able to pinpoint things which had been left unexplained but which made sense and became important in the light of the new work. I am absolutely sure that I always acted objectively and was never influenced in any way by the personality of the author—most of whom I did not know. Nevertheless it was largely due to these referee reports of mine that I gained a reputation for being excessively quarrelsome and always looking for a fight. This sort of reaction is so deeply ingrained in the human heart that it is not sensible to take it too seriously. And even if I scarcely ever got an admission that my objections and demands were reasonable I could nevertheless see in the way a line of work developed or in the changes in new editions of books I'd reviewed that my comments did have a certain success.

For example in 1889 I reviewed a book by Tiemann and Gärtner[13] on water analysis in which the authors noted that, like all analytical chemists before them, they'd had problems in grouping the analysis results for salts for there was no means of ordering acids and bases. I wrote, "The idea that things that one knows nothing about should be left unmentioned seems not to have occurred to them. The rational approach would be to give the number of positive and negative charges of the species[14] (Na, K, NH_4 etc., Cl, NO_3, SO_4 etc.) separately for by doing so one has made clear everything that the analysis has given and all that we need to know".

The point I made here in passing was one of the sources of my decision to later write the book on analytical chemistry. On the other hand the Hungarian chemist C. von Thann[15] rightly pointed out that he had made the same point previously. The result of all this was that the calculation of the results in terms of ions was accepted as the only scientifically acceptable way. Thus, the Imperial Health Authority (Reichsgesundheitsamt), which was headed by my assistant of many years Th. Paul, carried out innumerable analyses of German medicinal springs and published the results in this form in their exemplary "Spa Book".[16]

Struggles. Apart from the battles that I as leader of the new concepts had to fight with the conservative representatives of the old, I also had to defend myself on a new and quite different front. Already back in Riga I'd taken an interest in the preliminary work of Lippmann on the relationship between surface tension and

[13]Tiemann F, Gärtner A. (1889) Die chemische und mikroskopisch-bakteriologische Untersuchung des Wassers. Vieweg, Braunschweig.

[14]Ostwald uses the term "Teilmolekeln", (parts of the molecules).

[15]Ostwald misspelled the name of C. von Than.

[16]Himstedt F, Hintz E, Grünhut E, Jacobj C, Kauffmann H, Keilhack K, Kionka H, Kraus F, Kremser V, Nicolas P, Paul Th, Röchling F, Scherrer A, Schütze C, Winckler A, Rost E, Sonntag G, Auerbach F (1907) Deutsches Bäderbuch. Weber Leipzig.

polarisation of mercury and Helmholtz's work on the properties of a dropping mercury electrode because I wanted to know how best to measure the individual potential differences between metals and electrolytes. I'd adapted the dropping mercury electrode for this purpose and told the English physicist O. Lodge about it in a letter in 1886 which he then made public.[17] A year later in Vienna (Part 1, Chap. 13, p. 133) I met the physicist Franz Exner who was working with dropping mercury electrodes which were, as I knew from my own experiments, not useful for his purposes because they were being operated with far too low pressure. I pointed this out to him. When he tried to argue that his electrodes worked well because the sums of certain voltages were given correctly, I showed him that any electrode of any design was bound to give these results. However, I didn't manage to convince him and I parted from him with the realisation that in this case, to paraphrase Schopenhauer, the will was stronger than the demands of the intellect.

Some time after this meeting, when I had published my results in the first volume of the "Journal",[18] there appeared from him and one of his pupils a publication that contained a strong attack on my work.[19] I replied in the pages of the journal by showing experimentally that their claims were false and put some of their errors right.[20] They replied with a vituperative publication that was devoid of factual matter and which questioned my scientific honesty.[21] I pointed out in a second reply that the attack was entirely baseless and finished with the words, "To polemically attack the scientific morality of one's opponent, as these gentlemen repeatedly do, is not my way of doing things. I will therefore now and in the future content myself with searching for the sources of the errors they make not in the nature of their character but in the deficits of their intellect".[22]

This hint seemed to work. In a third publication my opponents did not withdraw their arguments, though they had been shown to be untenable, but they did desist from personal attacks. I ignored it as it had no influence on the way the scientific community judged the issue.

However, my fatherly friend Ludwig said to me, "You've got better things to do with your time than writing things like that", and pointed out that it was a waste of time arguing with an opponent who is determined not to be convinced. The onward flow of science corrects these things on its own and usually very quickly and one avoids giving one's readers a disagreeable impression if one avoids a polemic right from the start.

I took these warnings of my wise friend to heart and from then on there were no more such polemics in the Journal. Instead, when my work on the dropping

[17]Ostwald W. (1886) Phil Mag Series 5, 22:134, 70–71.

[18]Ostwald W (1887) Z Phys Chem (Leipzig) 1:583–610.

[19]Exner F, Tuma J (1888) Sitzungsber Wiener Akad 97:917–975.

[20]Ostwald W (1889) Z Phys Chem (Leipzig) 3:354–358.

[21]Exner F, Tuma J (1889) Z Phys Chem (Leipzig) Repertorium Physik 25:597–614.

[22]Ostwald W (1889) Z Phys Chem (Leipzig) 4:570–574.

electrodes was the subject of an unjustified criticism from F. Paschen,[23] now president of the Imperial Physical-Technical Institute,[24] who however backed it up with good experimental data, I noted in my comment on the paper, "Finally I wish to acknowledge that the author's proposed modification of the drop electrode is helpful and completely solves a problem associated with earlier designs. The joy of the author for the advance he has made is no doubt the cause of his not entirely fair criticisms of my previous work".[25]

However not even this prevented me getting the reputation of a being a particularly argumentative fellow. I can show how inappropriate his was with one special example. Lothar Meyer, who had always been most friendly to me, though he was not convinced of everything in the new concepts, said to me once in a fatherly way, "It would be much better if you weren't so polemical. For example can't you make peace in the disagreement with Professor L[26] (one of his previous pupils). I pointed out that my opponent had quite unjustifiably attacked me,[27] by casting doubt on a statement I'd made and he did so without providing any experimental evidence for his point of view. I'd been satisfied by rejecting his doubts in four short lines,[28] but he responded to this by publishing three or four lengthy pamphlets directed against me.[29] I however, had not publically responded to these. Meyer had to admit I was right when I showed him the correspondence but I fear that the incident left him with an even stronger belief than before that I was incorrigibly self centred and dogmatic.

Technical matters. The increasing volume of writing I was doing, forced me to try to reduce the amount of work involved. My brain still worked willingly so that when writing my books and papers I always kept a sheet of paper handy to write down ideas that sprang up but which were not appropriate to what was actually being written at the time. I needed these notes because I'd sometimes search in vain for an idea that I'd just committed to memory. This gave me all the more reason to look for ways to improve the technical side of writing.

I mentioned earlier that I found it impossible to be reliant on somebody else and so dictating was out of the question. In any case in the chaos of writing, working in the lab, teaching and all the incidental skylarking which was squeezed into odd moments between times, the hours for writing were so irregular that I couldn't expect anyone to be on duty for these odd hours or quarter hours.

---

[23]Paschen F (1890) Ann Phys 277:42–70, and the preceding papers: Paschen F (1890) Ann Phys 276:36–52; Paschen F (1890) Ann Phys 275:43–66.

[24]"Physikalisch-Technische Reichsanstalt". This was the National Metrology Institute of Germany. It is now the "Physikalisch-Technische Bundesanstalt".

[25]Ostwald W (1890) Z Phys Chem (Leipzig) 6:369–370.

[26]Most probably Eugen Lellmann.

[27]Lellmann E (1889) Ber Dt Chem Ges 22:2101–2103.

[28]Ostwald W (1889) Z Phys Chem (Leipzig) 4:575.

[29]Lellmann E, Gross H (1890) Justus Liebigs Annal Chem 260, 269–289.

And so there was nothing for it but the typewriter. My colleague Wundt kindly told me his experiences and it turned out that he loved this technology and followed closely all the advances in typewriter design. Soon I too was bitten by the bug. The initial difficulties were soon overcome and, despite the fact that I wrote by hand at an unusually rapid rate, I noted a three fold increase in my writing output. As an example, I remember that I managed to translate S. Carnots comments on the power of fire which I prepared for the "Classics" on a quiet Sunday. It resulted in 67 narrowly printed pages in the Classics. It was of course only a translation. But that was the only time I wrote at this rate for I felt the fatigue in my wrist for several days thereafter.

For many years I did all my writing work with a typewriter till it was eventually superseded by a voice recorder.

Chapter 18
The Leipzig Circle

General. Entry into the circle of new colleagues in Leipzig was not easy for me or my family.[1] We'd been used to quite different social manners and customs and we had no German relatives or friends who might have introduced us to the mores of life in a German university. The short acquaintanceships I'd made on my journeys had been merely ephemeral personal interactions and so we were rather foreign when we came to Leipzig. In many ways we stayed foreigners for the entire 19 years of our stay.

Part of the trouble was that the number of new colleagues was so great that it was impossible to develop a closer relationship with even a minority of them.[2] At the beginning I went diligently to all the usual introductory meetings at the start of the semesters, but as the work increased I started to neglect these social duties and never afterwards took them sufficiently seriously.

The situation was much the same in the smaller circle of the faculty. In Dorpat there had been five faculties, but in Leipzig as in most German universities it was normal that both historical-philological and physical-mathematical subjects, which are completely different in nature, are all lumped together in a philosophical faculty. In Leipzig the question of separating the two groups into separate faculties had been considered now and then but the philologists were anxious to maintain the status quo.

And so both objective and subjective grounds conspired to ensure that beyond the rich and varied circle in the laboratory and in medicine and the natural sciences there were only weak connections to the university as a whole. To this was added

[1]The semester started on 17 October 1887, and Ostwald gave his introductory lecture entitled "Energy and its transformation" ("Die Energie und ihre Wandlung") on 23 November 1887.
[2]In 1887/88 the Leipzig University had 68 full Professors, 37 of whom were members of the Philosophical Faculty.

R.S. Jack and F. Scholz (eds.), *Wilhelm Ostwald*, Springer Biographies,
DOI 10.1007/978-3-319-46955-3_18

the fact that the difficulties of integration were not eased by my aversion to wasting my precious and well filled time on the sorts of social round which Leipzig had to offer.

The social round. Social contacts were made mainly at occasions which were referred to by one and all as "Feedings". Once or twice in a semester, as many victims would be invited as could be squeezed into the available room. After standing around for a while, hindered by the tea cup one was holding, one looked for the lady who one had been told on entering was to be one's partner for the occasion, and led her to table. The food, the wine and the waiters were always the same: so were the ladies. Since one didn't know what to do afterwards, one tended to sit for a long time even though it wasn't particularly pleasant. Then coffee was served and at this stage the ladies retired. Now one began to keep an eye open for the moment when the most distinguished civil servant (Geheimrat)[3] present would give the signal to leave. There was a collective sigh of relief when that happened and it was usually pretty early for the "Geheimrat" also wanted to get home to bed. Sometimes, however, he'd get bogged down in a lecture on one of his hobby horses and, since that usually required at least three quarters of an hour, our departure would be delayed by that amount.

I don't wish to be unjust to my colleagues and suggest that all these evenings followed this pattern and there were occasions when the company fitted well together and the evening was carefree and pleasant, but such evenings were few and far between because the huge numbers of university members could only be accommodated by making these events massive.

My wife and I only tried for a few years to attend these things. After that we gave them up as a hopeless form of Sisyphus work and limited our social intercourse to a small number of families we'd befriended. There were other more useful forms of social interaction, like bowling evenings, walks and so on, that I regularly took part in for many years and I got a number of useful and important suggestions from them.

Karl Ludwig. One of the most important acquaintances that my appointment in Leipzig made possible was that with the great physiologist Karl Ludwig. I have already (Part 1, Chap. 13, p. 139) related my first meeting with him and the kindness with which he received me. This kindness he extended to me and my family for our entire time together up till his death in 1895.

Ludwig was born in 1816 and so when I arrived in Leipzig and got to know him he was 71 years old. Soon after I arrived he invited me to visit him in his laboratory whenever I had time free. I often made use of this privilege and I am much indebted to him. He is one of the most successful teachers of physiology who was able

[3]"Geheimrat" (Privy councillor) was a title given to distinguished officials, including Professors. It was abandoned after the fall of the German Empire in 1918.

through an international circle of his pupils in Marburg, Zürich, Vienna, and also Leipzig to maintain and extend the leading position of German physiology which had been established by Johannes Müller. The story goes that one Russian physiologist who had been a pupil of Ludwig claimed that physiology was a Russian science and gave as his reason that in Ludwig's institute there were more Russian students than students of any other nation.

Ludwig was appointed in Leipzig in 1865, the year before the Prussian—Austrian war. Though Saxony had fought on the wrong side, Bismarck's wisdom saw to it that this did not entail any loss of territory. King Johann of Saxony decided to do what Prussia had done after its defeat in 1806 and to make up for the moral loss by gathering scientific laurels to replace those which had been squandered in politics. Though at that time Leipzig University led a humble existence, the excellent minister von Falkenstein seized the chance and by appointing good people managed in just 10 years to develop Leipzig into a scientific centre. Amongst other things the world's first professorship for physical chemistry was established there in 1871 and given to G. Wiedemann. The minister, who had recognised Karl Ludwig's patriotism and his ability to reach clever and objective decisions, used him as his advisor for medicine and science. In this way Leipzig was converted in an astonishingly short time from a provincial to a world class university whose student body soon outnumbered that of all other German universities except Berlin.

King Johann of Saxony, who had played a large role in this transformation, had always taken a friendly interest in the university. He liked to come unannounced and in ordinary clothes to lectures which interested him and the lecturers were asked to take no special notice of his presence. He'd just sit down next to the students wherever he found place. One can easily imagine the beneficial effect this had on the professors for the award of the usual honours was certainly influenced by the impression made on the king on these occasions.

At the time when I came to Leipzig these things were all in the past, for King Johann had died in 1873. His successor, Albert, was also interested in the university which he visited for a few days each year to hear the newly appointed professors. However these were well prepared, ceremonial occasions. The direct interest in science of the erudite Johann was missing in his successor who was more interested in military and state affairs. Then Falkenstein died and under the new minister, Gerber, who had been professor of law in Leipzig, the determination to maintain the highest scientific standards was relaxed. Ludwig's beneficial influence no longer worked. The neglect soon became obvious even to the king who summoned Ludwig to Dresden. There he received him with what was supposed to be a friendly remark "Welcome, my dear advisor we haven't seen each other in a long time". To this Ludwig dryly answered "Your majesty had only to call".

As a teacher Ludwig was wonderful. He gave his students not just the concepts but personally helped them with any difficulties they might have. He had given instructions that when necessary he could be called from a discussion by the institute's mechanic Salvenmoser, when an important experiment reached a critical point, so that if necessary he could personally help. I myself saw this happen more than once.

His care even extended to the formulation of the publications of his pupils. He always allowed them to publish the experiments which he had suggested and guided under the student's name alone which was unusual at the time. Since many of the students were foreigners he went so far in some cases as to write the major part of the text. The story goes that one Russian student congratulated another on his publication in the "Sitzungsbericht"[4] and asked the "author" whether he'd read it yet.

It is no surprise that such a teacher, who was also a first rate researcher and thinker, attracted many students. At the beginning of his time in Leipzig, when this reached its zenith, he had to take some days off in the week in order to hide away and deal with the enormous amount of work that the other days had brought, for otherwise the load of work would have destroyed him.

This situation was no longer the case when I started in 1887. In the doctoral students' lab (the student courses were organised by assistants) there were around half a dozen pupils and their number dropped with each semester. I don't know what the reason was. When I asked people in the field they told me that physiology had moved on and that Ludwig's area of research and his methods were no longer considered very interesting.

When I visited his laboratory the discussions covered wide areas and his contributions were always thoughtful and original. They did however become progressively more pessimistic particularly with respect to his judgement of human relations which were never personal except when he required an example to illustrate a general rule. Then humanity would be treated as if it were a chemical sample—and that made it all, once again, not personal.

Thus, for example, he sourly remarked once at the beginning of a semester, "When I consider that humanity has for several 1000 years put up with two books like Homer and the Bible without being totally corrupted, then I have to believe that there must be an indestructible sound core in human nature". There then followed a series of similarly acid remarks between which he seemed to be lost in thought. I'd never seen him like this and worriedly asked his assistants what the trouble was. "Don't you know", they replied, "in this semester not a single student has applied for his laboratory course. When he looks back on earlier times its no wonder that he gets bitter".

This incident shocked me. No one could be held responsible for what had happened. It was simply the way things go, and therein lay the tragedy. Since I'd copied Ludwig's way of interacting with his students in almost all respects (though mine had to write their own papers or, if necessary, find an assistant to help them) I promised myself that when the time came I wouldn't let things go so far. I realised that the only way to achieve this would be to recognise when the time had come and then retire from my position as teacher in the laboratory courses.

[4]"Sitzungsberichte der Königlich-sächsischen Gesellschaft der Wissenschaften" was published from 1849 to 1917.

Ludwig never really felt at home in the medical faculty which is mostly made up of people with a practical bent because he focussed strongly on theoretical aspects of science, an attitude, that despite the veneration I felt for him, I did not share. Once he remarked to me, "When one of these practical people polishes off a couple of dozen patients in a morning and then sees how I struggle for weeks with a single frog, he must think that he or I am an idiot. And since he'd never believe the first option, he plumps for the second".

In contrast, he felt an affinity with the natural scientists whose institutes lay near physiology. His 75 birthday in 1890 was celebrated by a torchlight procession of all the students and, although his best times lay behind him, the philosophical faculty honoured the part he'd played in making Leipzig into a world class university by awarding him an honorary doctorate. That pleased Ludwig very much. He repeatedly said that he was a scientist rather than a physician and that because of this he'd found his spiritual home with us.

Some years later death took him from the work which had given his life meaning and joy. As he'd always wished, he didn't have to suffer the agony of a worthless senility, but died after a short illness.

None of his many pupils ever did their great master the service of writing a full biography and now many of them are dead as well. Perhaps these lines will find a place where this appeal will germinate and bear fruit.

Wilhelm Wundt. I've already told (Part 1, Chap. 10, p. 107) that my relationship to the great founder of physiological psychology and renovator of philosophy had been formed early and this was one of the threads linking me to Leipzig. The knowledge which he'd got from our discussions of my goals had led him, when the time came, to support my appointment and he was always most helpful later when we were colleagues.

In some ways Wundt had had the same experiences that met me in Leipzig. His appointment had been made so as to get a representative of the new direction in philosophy which was taking place at that time as a result of the new ideas which were pouring out of the newly flowering sciences. Still, the matter dragged on as it was not considered terribly important. Only when an old acquaintance from Leipzig met him by chance and asked "What are you doing here in the middle of the semester?" did Wundt, who had till then been professor in Zurich, learn to his surprise that he was now a professor in Leipzig. When he presented his certificates it turned out that he didn't satisfy the requirements for a doctorate since he'd earned his title in the medical faculty rather than in a philosophical faculty. The problem was solved by making him an honorary Dr. Phil. Because of all that—just as in my case—no one expected his rapid rise. When I got to know Wundt this first phase was long past. Physiological psychology was now an established science and one university after another were creating professorships in the subject, most of which were filled with Wundt's pupils. In this area America soon took the lead. Wundt himself put aside his experimental work on particular problems and turned instead to general questions from which he built his personal philosophy and his monumental work published in the "Sitzungsbericht".[5]

Since Wundt's institute was in the old university area the considerable distance ensured that we didn't see much of each other to begin with. But once I'd visited him a couple of times to get his advice in various scientific and personal matters I was given such a friendly reception and such good advice that I began almost automatically to turn to him whenever I had a problem and never left without having been given substantive help.

Even little matters were attended to as, for example, when he readily helped me, among other things, with finding suitable terminology for new concepts. In fact the very appropriate term "colligative" describes those properties that depend on the number of molecules but not on their chemical nature.

When later our views of philosophy drifted ever further apart—he became ever more "idealistic" if one can use that term for so complex a development while I was becoming ever more practically and scientifically oriented—this did not have the slightest effect on our personal relationship. Sometimes he'd joke with mild irony about my radical views while I thought his views symptomatic of his old hang to theological philosophy. I met him from time to time in a small informal circle that for a number of years met for an hour or so once per week after dinner in the theatre café. He was one of the regulars there and his quiet nature reflected the tone of our discussions, though he was quite able to firmly defend his point of view. Later when I too held lectures about philosophy of nature I was told by some of my audience that Wundt had in his lectures warned them to avoid mine. Nevertheless I was so convinced of the sincerity of his view that I suffered not the least unease at hearing this and it had no effect at all on our personal relations.

His gentle wife became a trusted advisor to my wife so that Wundt's house was one of the few in Leipzig where we really felt at home.

I got to know Wundt when he was 55. He was thin, with a pale face, dark beard and hair and he wore large dark glasses. A far too intensive series of experiments about after-images had damaged his eyes so that he could only see imperfectly and had to move cautiously on the street. He couldn't deal with modern traffic. He led a very regulated life which let him reach a healthy old age. It was his daily custom after a meal to take a walk along the broad avenues around the old part of the city and his stooped form under his broad brimmed black hat was a Leipzig hallmark of which the city could be proud. Once I'd long left Leipzig to make my home in the village of Großbothen he decided by a happy coincidence also to settle there for his final years and so I had the opportunity of close personal contact with him till his death after a short illness in 1920 at the age of 88. He'd continued the work on his books which he loved right up to the end and once he'd finished the final corrections of the last edition he calmly awaited death as a natural event about which one should not make too much fuss because it was—natural.

---

[5]Ostwald refers here primarily to Wundt's 10-volumes series entitled *Völkerpsychologie*, Engelmann, Leipzig, 1904. Later editions were from Alfred Kröner Verlag, Stuttgart.

Wilhelm Pfeffer. There was more than one reason for my close association with the excellent botanist W. Pfeffer. I have already related how his groundbreaking work on osmotic pressure was developed by van't Hoff into one of the pillars of physical chemistry and he must therefore be seen as one of it's the most important springs.

When I visited Pfeffer in Tübingen in the summer of 1887 (Part 1, Chap. 13, p. 136) he told me among other things that he'd been offered a professorship in Leipzig which he planned to take up at the start of the autumn semester. Since at that time I'd given up all hope of getting a position there myself, I didn't pay much attention to this. Later I remembered what he'd told me and looked forward to joining an excellent colleague.

Pfeffer was married and had one son who was the same age as my children and so there started a friendly relationship between the two families which soon became closer. Later on, after the new physical chemistry building was finished we became direct neighbours, and our interactions became more frequent. We met turn about quite regularly every week or second week on Sunday afternoons and discussed mainly our scientific work. I learnt a lot about his field in these stimulating discussions. I think he benefitted too, because when I told him my ideas about energetics, which we'll come to soon, he wrote a paper on the energetics of plants in which he laid out in his own words the effect of this basic idea on fundamental questions in biology.

This close relationship lasted all through the time I was an active member of the faculty in Leipzig. When I came into conflict with the faculty which, as I'll relate shortly, led to my resignation, he sided with the faculty and without giving me a chance to make my case expressed his disapproval of my position in such an inconsiderate way that sadly our personal relationship came to an abrupt end. In any case after my move to Großbothen there was no further chance of contact with him which might have led to reconciliation.

Pfeffer was about 10 years older than me. He was tall and thin with a pale face and a black beard. He covered his almost bald head with a simple wig. He was a classical scientist being extremely careful in his work and in his writing and if one was eager and talented one could get a lot from his teaching. He was not an inspiring teacher like Liebig or Ludwig but he was extremely conscientious and critical, so that he did have a number of good pupils but did not manage to found a "school" in the sense of a coherent scientific community.

Because he was slow to make up his mind and reach a decision he did not readily admit in conversation that he might change his point of view. And yet he responded to factually based arguments by thinking them over at leisure and if necessary quietly changing his opinion. I soon gave up any attempt to change his mind during a discussion about something that we disagreed about, but I could be sure that he'd remember and think about anything useful I'd said.

Pfeffer gave me some good advice to help me settle in Leipzig. He knew the situation in German universities well for he'd spent his entire development in them, particularly those years as a lecturer waiting for a professorship. I am indebted to him for useful tips at times when I was about to make some terrible mistake, and yet his cautious, almost fearful approach to social exigencies, prevented him from venturing further than seemed absolutely safe. Because of this we disagreed about many such things and therefore restricted our discussions to impersonal matters which suited us both—though for different reasons.

Heinrich Bruns. Professor H. Bruns ran the university observatory which was just a few hundred meters from the laboratory. I knew him already before I came to Leipzig because we'd overlapped for a few years in Dorpat where he worked at the observatory. However he'd moved away before we could develop a closer relationship and, on top of this, he was 5 years older than me which at that age is a huge difference.

Bruns was of middle height and powerfully built. His healthy complexion and brown hair and beard gave no indication that he would die relatively young. He was quiet and not gregarious so that he didn't either look for or find many friends amongst the other professors. In science he was of the classical type: an acute mind with a marvelous critical talent, from whom I gained much clarity and support even though he was not particularly interested in my work. He was all the more happy to discuss general questions where again and again his critical faculty drove my thoughts to a clear focus on issues where they had been nebulous.

We talked about these things on long walks which we took together. He was a practiced and willing walker, always ready to join me when I knocked on his door on a Sunday afternoon.

Another area in which we exchanged views was in technical matters. At the observatory he had a well equipped workshop and a good mechanic whose work he closely supervised. He gave me some helpful tips for the construction of apparatus, in particular his insistence that one should always start out by defining what the problems to be solved are.

The friendship with this capable man was essentially intellectual and after I moved to Großbothen and we no longer had the opportunity to talk regularly our ways parted without our really noticing it. Neither of us had any great desire to write long letters merely to keep the connection going.

Friedrich Ratzel. I should mention just a few of the many pleasant colleagues I came in contact with through common interests, so as to give an impression of the situation in Leipzig.

One of the best of these was the geographer Friedrich Ratzel. He was a tall well built figure of a man whose bearing suggested more the gymnast than the soldier. A carefree and amiable expression lit up his well formed face which was framed in a flowing beard. Most memorable were his bright blue eyes which gave him the look of a sailor or mountaineer. There was something, in the best sense of the word, childish in his character. He liked to tell how he'd grown up in modest circumstances in the country and had come more or less by chance into science in which his fertile and independent mind had quickly brought him to a leading position. Nothing in his demeanor suggested this high achievement. He hated the dark subterranean side of a professor's business. He had always regretted having swapped his earlier position at the technical university in Munich for the seemingly more brilliant appointment in Leipzig where the downside of a professorial existence was more obvious. Not surprisingly he retired at the first opportunity and fulfilled a long held wish by buying himself a country house by a lake in Bavaria. Sadly he didn't get to enjoy this for long for he died on a walk soon after. We'd quickly become friends so his leaving was a great loss to me.

G. Th. Fechner. I regard it as a great piece of luck that I got to know personally Gustav Theodor Fechner, the father of quantitative psychology. I'd read a lot of his publications and respected this unusual personality and so I gladly used the opportunity presented by the duty of inaugural visits to introduce myself. At the entrance to his apartment the floor was covered with white sand as was usual in my home country so I immediately felt at ease. I was introduced to a kindly old man whose almost blind eyes seemed to peer into eternity. Despite his age he was as lively as a boy. He'd heard of me—presumably through Wundt—and he asked me right away whether among all my measurements there were some where the same values occurred repeatedly. This was because he was interested in the theory of frequency functions (Kollektivmaßlehre) and therefore was looking for all sorts of collections of numbers. Unfortunately I wasn't able to help though I'd have been happy to do so because it would have given me the opportunity of getting to know him better. He engaged me in a lively conversation which I sadly had to interrupt when it was time for me to leave.

I didn't see him again for he died just a few weeks later at the age of 86.

Older professors. I remember fondly the economist W. Roscher who was another of the famous old men of the university. When I'd been a lecturer I'd interested myself for a while in economics but had given it up because I couldn't see anything solid in it. I'd also read Roscher's thick volumes but didn't feel that his historical approach brought me any further than the theoretical treatments of the other authors. At that time I thought this failure was my fault.

When I got to know Roscher in Leipzig he was already 70 years old. He was a little white haired man with a rosy laughing face who gave the impression of being always friendly and happy. I remember him because of his application of practical common sense to his interactions with the university. He once appeared at a faculty meeting and announced that at his age he felt that the duty of holding lectures which properly covered the entire range of the subject was too much work particularly as he was working on several publications which required his entire effort. On the other hand, he couldn't imagine leaving the university with which he had been associated for 40 years. He therefore suggested that the faculty elect a new full professor of political economics who would be responsible for the main lectures and seminars and who would be paid accordingly. He for his part wished to formally retain his position which he expected his work to justify. The faculty agreed as did the ministry and I had the opportunity for many years to see in Roscher a man who was entirely happy with his fate. I decided then to remember his example.

Quite apart from the personal aspect I thought that this solution to the problem of "old professors" was ideal. It was a procedure which could easily be made general and I later published recommendations along these lines. Nobody took any notice of them.

Even in our faculty this case did not really set a precedent. Later there would be several cases where professors held on to their positions beyond their mental and physiological limits and I saw the damaging results that this had not only for the university but also for the professor involved who was locked in his narrow egocentricity.

Minerologist and Zoologist. The mineralogist Zirkel and the zoologist Leuckart both lived close to me and both were of an advanced age. Zirkel was unmarried and lived with a group of like minded people of his age. There was no scientific contact between us because he belonged to the descriptive side of his subject in which, however, he held a leading position. Because of this no closer relationship developed.

Leuckart was married and had a son my age, who was a lecturer in chemistry in Göttingen,[6] and several daughters. He was a small lively and agile man who got into a rage several times a day. Then we'd hear his voice echoing over the yard which separated our houses. His scientific work was characterised by his broad outlook and his original mind. His quiet and rather subdued wife took a motherly interest in mine and our children trusted and loved her.

When my oldest boy[7] reached that age at which beetles are collected he got so carried away with this that my wife felt it her duty to give his far flung efforts a scientific frame. When I was away on vacation she went over to see the old professor and asked for his advice. He said, "Send the boy to me". He examined him, found that he knew more than he'd expected and soon showed his teaching abilities. Instead of giving him books he handed him over to the curator of the zoology museum where he was set to polish bones, put skeletons together and do other practical things because, he claimed, that the endless number of different forms could only find places in the memory of someone young—and there he was completely right. That didn't stop the boy going his own way. The children were used to regarding the laboratory as an extended part of their home and they were indulged by the students and assistants. Without even realising it he picked up the idea of scientific investigation through his daily encounters in the laboratory. Half in play he'd already thought out an experiment to see if the trichoptern larvae could use other materials than the normal plant fragments to build their pupal cases. This resulted in a proper scientific report which was submitted to Leuckart. He considered it not only good but declared that it should be published.[8] The author of the report was at this time 15 years old and I was astonished that the phenotype of scientific writing which had appeared first in me in the Ostwald line should have passed so completely into the next generation. This led me to regard Weismann's denial of the inheritance of acquired characteristics as disproved. My only problem was to imagine how this characteristic had appeared in me. The idea later developed

---

[6]Carl Louis Rudolf Alexander Leuckart.

[7]Carl Wilhelm Wolfgang Ostwald.

[8]Ostwald Wo (1898/99) Z Naturwiss 72:49–86. "Experimental studies of the larvae of phryganeidae" ("Experimentaluntersuchungen über den Körperbau der Phryganeidenlarven").

by de Vries that mutations can suddenly alter inherited characteristics seemed attractive and made me consider this much criticised point of view to be correct.

Mathematician. The mathematician I was closest to was Adolf Mayer. He was the son of a rich Leipzig family and because he was financially independent he'd turned down the offer of a full professorship and taken instead an honorary professorship so that the money saved on his salary could be used to finance a new position. By this means it became possible to attract the famous Norwegian mathematician Sophus Lie to Leipzig.

I liked Mayer particularly because of his broad interests. He was thin and quite small with scanty black hair and a darkish yellow complexion which made him look a bit Japanese. In his manner he was kind and always courteous. His wife was of a similar nature and they frequently invited small groups of people who fitted well together so that his house was one of the few to which one readily accepted an invitation.

In science Mayer worked on problems that interested me so that here our relationship became closer. In the later stormy days he proved himself a real friend and I often remember him with gratitude.

The Norwegian mathematician Sophus Lie, whom I mentioned, was a strange personality. He'd come rather late to mathematics but developed such resourcefulness and independence that he quickly became one of the leading mathematicians of his time. When I got to know him in Leipzig he was a recognised master of the area he had opened up and was constantly busy developing and expanding it. He didn't look like a scholar. Massive in build with heavy limbs and a face to match, there was something primeval about him that put one in mind of a mammoth. In his character as well there was in the background something wild and Nordic. He wasn't good at dealing with the daily routine because his science, which he loved and for which he felt a fervid veneration and devotion, fulfilled him so completely that there was no room for other things. His wife was quiet and likeable. We, that is to say the two married couples felt attracted to each other since we were all children of the north who had difficulties to feel at home in Leipzig. We visited each others house on a regular basis for a long time and these visits were filled with pleasure not least because they were in sharp contrast to the usual "feedings" which we all thought were insufferable.

Slowly but surely, however, the usual illness of the mathematician became apparent in Lie. It strikes those who work in the highest and most abstract areas where the ways of thinking are very different from the normal ones. It seems to be reinforced in mathematical research by the fact that there is no possibility of recovery through the distraction of mechanical aspects to the work, because after all, writing a few equations doesn't really count. A chemist or a physicist is in a much better position because the effects of his research work on the brain, that highest and most easily damaged organ, are alleviated by mechanical tasks which are directed by lower centres and thus provide the necessary chance of recovery. In addition, the mathematician who opens up a new field works hard to gather in the fruits of his labour and ignores the symptoms of exhaustion which are telling him to take a rest. By doing so he easily destroys the delicate protective mechanisms and a neurological disease is the inevitable consequence. Luckily such conditions are to some extent treatable because they are not caused by physical change to the organ but rather by exhaustion.

This is what happened to Lie and he had to stop work and go for treatment from which he returned mistrustful and irritable. That naturally had a bad effect on our relationship.

Later on he suffered from that uncanny disease which involves a progressive loss of red blood cells. There is no treatment and so the patient is fully conscious of his inexorable descent into eternal night. After all that he had already suffered, this cruel fate must have been very hard for him.

Karl Lamprecht. When I arrived in Leipzig W. Maurenbrecher was the Professor of Recent History. He was a large man with a strong voice who loved to tell how as a professor in Bonn he had to hold lectures in history for the crown prince Wilhelm. It hadn't been difficult for him to inspire his pupil with his fiery adoration of Bismarck and that together they planted an oak tree in Bismarck's memory somewhere, though I forget where.

After Maurenbrecher's early death Lamprecht, who had already made his mark, was appointed. Since in contrast to most of his colleagues in the field, he was interested in uncovering laws of historical development, I felt attracted to him right away, especially as he too took up a position contrary to certain of our Leipzig colleagues whose ways of behaving I also disapproved. He soon joined the coffee round I mentioned (Part 2. Chap. 5, p. 196) in which Wundt, he and I soon formed the core.

Lamprecht had been born in 1856 and so was a few years younger than me. He was of middle height with dark brown hair and a short full beard of the same colour. He was a lively person whose face readily broke into smiles and laughter and his eyes sparkled through his spectacles. We two shared in common the ability to carry out large amounts of work in the shortest time. He completed his 19 volume work on German history in Leipzig in 11 years and at the same time he was constantly engaged in writing responses to his many critics and in the huge amount of organisational work directed to the extension of his branch of science at the university. He did all of this at the cost of his health. From my own experience I warned him what would happen but he ignored this even when clear indications of an impending breakdown were evident and had forced him to interrupt his feverish activity. As a humanities man he did not understand the inexorable nature of natural relationships and believed that he could drain inexhaustible amounts of energy from his body by sheer will power. As a result he died in the middle of his work in 1915 at the age of only 59.

In contrast to me Lamprecht was an expert in handling people. He well knew how to make the decisive people receptive to his organisational ideas and so, despite his ceaseless literary battles, he managed to achieve a considerable influence over his Leipzig colleagues. In this way he was able to extend his history department into a huge structure with many sub departments which he knew how to inspire with the spirit of his many—though not necessarily terribly deep—interests and his network. This structure was his personal construct and its life as usual hung on that of its initiator.

Every time we met we turned out to have opposing views and immediately got into an argument. However, this never led to personal animosity and we both enjoyed these duels. From his humanistic heights he rather looked down on the natural scientist and I mocked him when he expounded and boasted about systematic discoveries in his subject which we in science had made long before. Our main bone of contention lay in the place of history within the structure of science. Though his attitude towards the reactionary historians was welcome I could not hide the fact that I was unable to discern any science in history. I pointed out that his own writings came to very different conclusions from other works about German history, suggesting that history is merely a technique with which one tries to reconstruct relationships in the past from whatever has survived. But in order to understand the content of these relationships one needs special knowledge of the subject—and this the historian does not and cannot have. To illustrate my point of view I challenged him to write a history of recent progress in physics in Germany.

Lamprecht argued instead that the historian was essential in order to merge the many single stories into a coherent whole. The argument ended with the question of what can be more easily acquired as an auxiliary science: historical techniques or expertise. We never did settle that.

Since both of us were operating at the front line of our subjects we did have useful things to say to each other and these positive aspects of our discussions were what brought us together time and again. Now and then I was able to help him with technical expertise and he kindly mentioned this somewhere in a preamble. However once when he gave me his handwritten version of a complete volume and asked me to check it through for scientific errors. I had to return it to him at once. The whole volume had been written in longhand. Of course, just as in my manuscripts, there were only a few corrections but the letters were narrow and as tall and thin as spiders' legs so that I was quite unable to make them out with certainty and was certainly unable to read them quickly.

All in all I consider my relationship with Karl Lamprecht to have been a worthwhile enrichment of my life and I found his early death a real loss.

Science and scholastics. At the time I arrived in Leipzig I had been perfectly happy to regard the humanities as knowledge and to defer to them because of their long history. I was, however, not prepared to regard my science and the other natural sciences as being of less value than the humanities which could make no claim to be scientific. At that time I'd just read Dührings remark that the only visible function of classical philologists at the university was to churn out teachers who in turn would pass the information on to new teachers in an endless circle without anyone ever contributing in a meaningful way to solving life's problems. The inbred veneration of philology made me resist this description, though I had no objective arguments against it.

Sometimes even in Dorpat I did notice the uselessness of philological work. In my time as a lecturer there I was sent a university publication written by a respected historian who later became well known as a professor in Bonn.[9] He'd come across a relief in a collection of ancient art works and he was not in agreement with the description that the author had given it. He proved that it must certainly show a scene from a drama by Thyestes.[10] It was known that there were two dramas with the same name. Something was known about one of them and nothing about the other. And now using "philological perspicacity" he showed not only that the relief showed part of the second, unknown drama, but he also deduced which act and scene it must represent. The energetic imperative which even then slumbered in me rebelled against this sort of "work" which seemed to me to be entirely childish.

Once later when I related this to a young philologist he turned red and said, "We consider it one of the most brilliant achievements of our esteemed colleague".

I was reminded of this story in Leipzig when at some university occasion I had to listen to an address by the philologist Lipsius. A copy of a code of laws had been discovered that contained a precise version of an ancient text that had been till now only available in corrupted form. Lipsius proudly emphasised that in comparing this new text to that which the philologists had reconstructed from the corrupted version it turned out that in nearly half the cases the corrections were correct; in the other cases the corrections were wide of the mark. I thought that the 50 % that were correct were doubtless the easiest cases so that the real success rate was not really 50 % but only at best 25 or 20 %. If that was the case then the correction work had been a complete waste of time. Shaking my head in disbelief I returned to my own work where I was sure that the success rate was very considerably higher.

Because of this I was irresistibly driven to view critically the dominant influence which colleagues from the humanities exercised in the faculty. Since I expressed my doubts quite openly and tried to discuss the matter with them I was soon regarded with suspicion as someone who lacked the proper degree of respect for these great "cultural achievements".

At that time classical philology was represented by the professors O. Ribbeck, J. Lipsius and R. Wachsmuth. While Lipsius was a typical head teacher type (he'd been a school teacher for many years), Wachsmuth was a man of broader horizons and had a finer mind. Later, after I'd got to know him and had explained my viewpoint to him, I learned that in his youth he'd studied science and had only later turned to philology.

The Heidelberg Statement. Very direct differences of opinion brought me into open conflict with the prevailing views of the majority of my colleagues and this soon became apparent. At the time of my arrival the rector of the university was the classical philologist Otto Ribbeck. In his time he was very influential in his field and was determined to hold off the danger that the natural sciences might threaten

[9]It is highly probable that Ostwald probably refers to Georg Loeschke, who published several interpretations of Greek reliefs and vase paintings while in Dorpat.

[10]Ostwald is mistaken. "Thyestes" is a drama by Seneca.

the continued dominance of the philologists. In the way of all priesthoods, whose power derives not from the facts of a subject but rather from usurpation and tradition, he was not interested in real values about which one could discuss, but was only interested in questions of "conviction" that is to say in emotional points of view which it was one's "duty" to defend. There was no duty to examine the facts of an issue in the light of one's own experiences or those of others and indeed to have done so would have been considered equivalent to treason. Because of this it became a duty to fight any and all opponents and in this, as is the way of priesthoods, one wasn't choosy about the weapons employed.

Just at that time a petition from the University of Heidelberg had been sent round in which it was proposed that secondary schools teaching the classics were the only acceptable preparation for university entrance. At some point in a dark corridor by the medical faculty's examination hall I was met by a university official with a file in his hand. He opened the file and passed it to me with the words, "His Magnificence the rector, requests that you sign this". I was cautious or maybe curious enough to first read what I was to sign and saw that it was the Heidelberg petition. I declined to sign it. That such a young fellow should so brusquely defy the official point of view was bad enough, but I went on to heap yet more guilt on myself by further delinquencies.

During the examinations for the medical students I'd often meet Carl Ludwig who held his exams at the same time as I did and we'd often walk home together. His remarkable friendliness on our first meeting had led me to trust him and he was the only one of my new colleagues who would sometimes in a fatherly way point out errors or clumsiness in my way of behaving. Since I'd more than once heard him say to the young medical students "Once again you knew almost nothing. However I must accept that that's not entirely your fault. After all you spent 9 years of your short life in the unsuitable environment of a secondary school specialising in classics. I can scarcely blame you for your inability to learn to think scientifically".

Ludwigs view was that we should not let public opinion be misled by the Heidelberg petition. Since I immediately offered to help, he asked me to sound out those of our colleagues who might be of the same mind. In this way we formed a small committee to which, apart from us two, the astronomer Heinrich Bruns and the physician Albin Hoffman belonged. We put together a statement to the effect that non-classical secondary schools were a better preparation for the study of medicine or the natural sciences. We sent this to all members of the faculty with the request for a response even if they were not prepared to support the statement. I, as the youngest, was put in charge of the administrative aspects and I did this gladly. The result was that a majority of those asked supported us. We of course made sure that the result became public knowledge and as I'd been in charge of the correspondence I naturally was somewhat in the footlights and became the target of the dissatisfaction of many in the faculty and the object of the hatred of the philological priesthood.

This set the stage for the way I came to be regarded in the faculty in the future. The rapidly increasing success of my teaching activities increased the antagonism as well as the dangers to be expected from those colleagues who now felt somewhat insecure. My opponents used the clever trick of defining anything which worked against their traditional hegemony as being "uncooperative" I acquired this reputation and as a result was never admitted to those higher circles which run things in Leipzig as in every other university. That was all right by me because I'd not have wanted to lose the time involved in these things. However, I had not for a moment thought that there would be directed against me the huge amount of anger that became obvious at the end of my time in Leipzig.

Chapter 19
The Spread of the Concept in Germany

Heidelberg. At the Natural Scientists Meeting in Heidelberg in 1889, 2 years after my move to Leipzig, I was able to get an idea of how our work was being received. I had been invited by A. Horstmann to stay with him and we celebrated with lively emotions our first reunion since our meeting six years earlier (Part 1, chap. 9, p. 104). What a lot had happened in the meantime! The progress, however, had been restricted to our joint interests and to me personally, for his problem with his eyes was now such that he'd been unable to contribute to the new work.

The first impressions were not auspicious. I fell in with a swarm of organic chemists who all regarded Emil Fischer as the future leader of our science: anything that was not organic chemistry was for them not chemistry at all. When Fischer made a disparaging remark about the new direction I replied that the organic chemists had us to thank for the means of determining the molecular weight of non-volatile substances. That had been impossible till then, but through our work it had become easy once Beckmann had technically refined his procedure. Fischer replied, "That was completely unnecessary. I can see just by looking at a new substance what its molecular weight is. I don't need your methods".

I assumed this was just a smoke screen to cover his escape and hoped things would improve with time. This never happened, however, for I have since then repeatedly experienced this sort of rejection though it seemed most often to be directed against me rather than against the use of our discoveries.

Even now, (1926) a whole generation later things haven't changed much. Though the technical knowledge from physical chemistry becomes daily more useful and important, the chemists from the "organic" schools know almost nothing about it so that there is a vast surplus of organic chemists and a dearth of physical chemists.

The way in which Victor Meyer expressed it was in striking and noble contrast to this quite inappropriate opposition. He was in the process of leaving Göttingen to

R.S. Jack and F. Scholz (eds.), *Wilhelm Ostwald*, Springer Biographies,
DOI 10.1007/978-3-319-46955-3_19

take up the position which had been held in Heidelberg by his teacher Bunsen. He considered this the crowning moment of his brilliant career. True, he took up his new post as a sick, nervous worn out man who was paying for his days in the sun with severe exhaustion. But this wasn't obvious when he held one of the plenary lectures which were always considered to be a highpoint of the congress. The second such lecture was held by Heinrich Hertz and it caused much wider ranging repercussions.

In his lecture on "Current Problems in Chemistry", V. Meyer described the difference between the organic chemists whose research work was inspired by feeling and fantasy and the physical chemists whose research was driven by strict logic. He made no secret of his preference for the former which he described in all honesty as being a more childlike and artistic activity than unemotional science. However he did not deny the enormous importance of the new direction which he rightly described as a rebirth of chemistry and he warmly supported it. By doing so he exercised great influence on his colleagues and did a lot to ensure an increasing acceptance of physical chemistry in his circle.

On other occasions I also had cause to be thankful for his generous behaviour which was free of all rivalry. This was particularly true for the meeting in Munich to discuss the state examinations for chemists, about which we'll come to later. That was the last time I met him for he died suddenly a few weeks later.

Other experiences. These matters were of great importance for me but the Heidelberg meeting was also interesting for other reasons. I've already mentioned the very impressive plenary lecture by Heinrich Hertz in which he described the detailed similarities between electrical oscillations and light. This work has since become fundamental in more than one area.

In addition Edison's recently invented phonograph was demonstrated in the presence of the inventor. Edison was a big man with a face like a Caesar who towered motionless beside his invention. He spoke no German and even his native English could only escape in snatches through the fence of his teeth. He had a German–American secretary who spoke for him. The secretary did this with all the arrogance of a little man who has the chance to represent a great one.

The Grand Duke of Baden who took a lively interest in the scientific activities of the state's three universities, had announced his intention of attending this demonstration of the phonograph. For some reason his arrival was delayed for more than an hour and the organising committee had to find a way to keep the large and tightly packed audience occupied for this period. Rudolf Virchow took on this task. He played a major role in the committee and perhaps because of his assiduous political activities was the one best able to speak without having anything to say. He didn't consider us as more than just another town meeting and I have to say that I have seldom suffered such embarrassment as when listening to the pointless chatter of this eminent man. We all breathed a sigh of relief when the Grand Duke finally entered and Virchow quickly brought his talk to an end. Our sentiments were different but no better as the secretary made a mess of his presentation and tried to be funny by making jokes about professors—who of course made up the majority of the audience.

Helmholtz and Kopp. Helmholtz also attended the meeting because he liked to revisit the beautiful city where in a circle of talented and lively colleagues he'd carried out a large part of his important research work. He held a lecture on the subject of standing waves in the atmosphere which he had discovered by examining the differential equations governing the movement of large masses of air and he pointed out that the regular formation that is so often seen in fluffy clouds is a consequence of these waves. This explanation was a relief to me for I'd often seen such regularities when I was painting but deliberately suppressed them because I was trapped in the myth that irregularity was freedom and hence artistic. It now turned out that I was twice wrong. Since then I've painted the clouds as I saw them, much to the benefit of the results.

I now met Hermann Kopp, the author of stoichiometry and the peerless historian of chemistry, whom I hadn't met on my first visit to Heidelberg though I'd corresponded with him about bringing out one of J. Liebig's papers in the "Classics". He invited me to a large dinner party that he was holding on one of the following days.

When I arrived in my dinner jacket at the appointed time and greeted my host and hostess I had to suppress an outburst of laughter. Kopp was an unusually small man with skinny limbs and a curiously pointed stomach. The expression on his face was accurately described by his old friend Wöhler as being as if something was not quite right in his intestines. His wife next to him was one and a half times as tall and I guess three to four times as heavy and understandably far outstripped him in strength and presence.

Just as I'd finished greetings them my colleague Wiedemann from Leipzig arrived and was most warmly welcomed. "We have to make do the best we can", said Kopp's wife, "We invited the top people, Helmholtz, Hertz and so on, but they all had previous engagements". Somewhat put out but nevertheless with perfect manners Wiedemann replied, "Well, never mind, there are still quite a few tolerably well known people here". I'd just found out how I had managed to be given one of the coveted invitations.

I still have a lively memory of my conversation with Hermann Kopp. He complained about the discomforts of getting old, especially of the loneliness that accompanies it. "Look at this", he said as we stood beside his writing desk. "It's like a graveyard. The inkwell is from Liebig, The pen is made out of the first batch of technically produced aluminium and was a gift from Deville, the medal there is from Berzelius and is made of selenium which he discovered—they're all dead."

A few years later Kopp too joined his departed friends.

Bremen and Halle. In the following year, 1890, the Natural Scientists Meeting was in Bremen and there physical chemistry was honoured with a lecture in one of the plenary sessions instead of having to be satisfied with being mentioned in a friendly way by others. As the semi official representative of the new science I was invited to talk and I did this with joy and pride.

The introductory words characterise the situation as it was then so well that I can do no better than repeat them here.

"Who has not experienced the marvellous sensation of the mountaineer who, after a vigorously begun climb in the early morning, stops for the first pause? Of course the goal has not been reached; seemingly insurmountable rocks and ice are piled up before him but he knows his strength and can trust it. Now he can really enjoy what was denied him for as long as he had struggled and kept his eyes fixed on each obstacle that had to be overcome. Now he gazes forward and back. His starting point lies hidden in the mist far below. Buoyantly he looks along the path he has taken rejoicing over the difficulties which have been surmounted and the panorama stretched out before him. Now he can see that in some places he could have taken a short cut and some of the rocks he tackled head on could have been easily got round. But he doesn't regret the work he put in, for he had the fun of doing it and the experience he gained will serve him well on the way further up. This he now examines with a cool eye. Though the difficulties may increase the higher he climbs so too does the wonder of the panorama laid out before him. The reward one gets is always closely linked to the effort invested.

I'll be talking today from this perspective. I don't represent here merely myself nor will I be reporting on my modest contribution to the progress we have made. Indeed I did not for one moment think that the honour of holding this lecture was offered to me but rather to the scientific field—physical chemistry—that I represent. The saga of the unexpected revolution in the way we regard large areas of chemistry, a revolution which in its own small way is no less radical than the paradigm change form the phlogiston theory to the oxygen theory, has long since trickled out of the laboratories and seminars. Many of those whose areas of work are affected may worry what of all that he has relied on so far remains useful, while others, indignant and scowling refuse to accept any changes in what they have till now held to be the unquestionable fundamental basis of their science. In view of all this we have been invited to this largest German science meeting to render an account of what we believe we have achieved and to show where we want to go from here. We, the physical chemists in whose name I speak, are glad to accept this challenge. This invitation is an indication that our work has progressed so far that those who work in other areas now begin to believe that our path does lead onwards and upwards rather than into the wastelands.

The description of our work that I then gave was given a friendly reception and led to the sort of personal discussions which are the hallmark of this series of meetings. Since I'd just returned from the meeting of the British Association, about which I'll report shortly, where we'd won after a hard struggle, I was feeling pretty confident and that probably made an impression on the doubters.

Physical chemistry had now won so much ground that at the next meeting, in 1891 in Halle, I was asked to report on the advances to the combined physics and chemistry sections. The main thrust of my lecture was as follows: The concept of chemical equilibria in the broadest sense, that is to say including vaporisation, dissolution and freezing had been developed fundamentally by W. Gibbs. However his equations contained numerous functions which were unknown given the state of research at that time. Many of these have now been clarified by the dissociation theory so that quantitative values could be given to the constants and this permitted experimental tests and where these have been carried out and so far this has been done only to a very limited extent, they have confirmed the concepts. There is now therefore a wide field open to analysis where the low hanging fruit needs only to be gathered in.

In this year physical chemistry was finally properly established, the Journal[1] had developed well, several textbooks had appeared and were selling well and being read while here and there researchers who were not connected in any way to the centre in Leipzig began to work on problems in the new area.

The new results were reported in the appropriate sections at scientific meetings and soon they were so numerous that they were grouped together in their own section together with the physicists who could usually get through their program faster than the chemists.

Wilhelm Hittorf resuscitated. There was a happy surprise for the physical chemistry community at the Natural Scientists Meeting in Nuremberg in 1894. It was the custom that at the beginning of the meeting the attendees would stand up and give their names. This was because names were more widely known from the literature than faces and in this way one not only learned who was present but also knew who to approach to discuss a particular matter. This was how things were done in Nuremberg and near the end of this quite extended procedure, for there were a lot of chemists, a short tubby bald headed man wearing a distinctive pair of spectacles on his clean shaven face and looking for all the world like a catholic priest of the better sort, rose and said, "Wilhelm Hittorf". We were all astonished and asked ourselves if it could possibly be true, for since he hadn't published in a long time he was for us an honoured but almost mythical person and many were not sure whether he was still alive.

It was indeed Hittorf and he turned out to be a charming old man who unselfconsciously enjoyed his late fame like an unexpected glass of fine wine. It gave us all a warm sense of wellbeing to be able to express our respect and thanks to him.

After all he had suffered much in life, principally at the hands of his colleagues who'd considered it a crime that he'd tried to clear up existing ambiguities and

[1] Zeitschrift für physikalische Chemie (Z Phys Chem (Leipzig)).

errors with his own clear and coherent ideas. I have already related in a different context (Part 2, Chap. 17, p. 183), how I'd tried to accord him a degree of delayed justice by including his work in the "Classics" series. I'd also corresponded with him and I knew that he had suffered a great deal from the disdainful rejection of his work and that as a result he had been emotionally disturbed for some time. Because of this I'd assumed that as a broken man he preferred to live in seclusion now and I was most happy to see a robust old man with a healthy complexion and cheerful nature who, despite his 70 years did not turn down the chance to join the social get-togethers. At these he attached himself in particular to my wife who had accompanied me to the meeting and she was proud of the knight she had acquired.

The meeting worked like a youth cure for Hittorf. The sight of all these hard working young people spurred on his scientific creativity and not much later he published in the Journal[2] his remarkable discovery concerning the electrochemical properties of metallic chromium.

Against the flywheel. At the anecdotal level the opposition continued for years. I was told that the important physicist August Kundt (Part 1, Chap. 13, p. 136) who'd moved in the meantime from Strasbourg to Berlin, did not permit his students to refer to the new doctrines as being scientific. He threatened to fail anyone he examined if they used the word "ion". Since he'd been friendly and courteous when we had met in Strasbourg and I had noticed no change in later chance encounters, I used the opportunity when in Berlin to visit him together with Dr. Nernst and talk the matter over. It turned out that he had taken great exception to my description of a positively charged electrolyte as having an excess of positively charged ions, as for example protons in the case of an acid. He considered this to be impossible and said, "If you can demonstrate that experimentally then I'll believe you". We returned to Leipzig that same night during which we discussed how to carry out a convincing experiment. The next morning we could telegraph Kundt that the experiment had worked. We wrote this up together—"About free ions" ("Über freie Ionen") in passing I might add that this was the only paper that I published together with a colleague—and described in it the experiment and discussed its implications.[3]

---

[2]Hittorf JW (1898) Z Phys Chem (Leipzig) 25:729–749; (1900) Z Phys Chem (Leipzig)30:481–507.

[3]This refers to the paper: Ostwald W, Nernst W (1889) Z Phys Chem (Leipzig) 3:120–130: Ostwald and Nernst performed the following experiment: A round-bottom glass flask was filled with dilute sulphuric acid and the outer glass surface was covered with a layer of tin foil. The flask was placed on a piece of rubber for electrical isolation. Using a piece of moist cotton, the sulphuric acid was connected with a sulphuric acid solution adjacent to the mercury column of a Lippmann electrometer. The mercury was earthed. When the positive pole of an electrostatic generator was connected with the tin foil, there was a strong movement of the mercury column and hydrogen bubbles at the interface between sulphuric acid and mercury were observed. Ostwald and Nernst interpreted the results as follows: The positive charge on tin prompts a negative charge of the sulphuric acid in the round-bottom flask. The latter "expels" the positive charge (hydrogen ions) towards the mercury where it is discharged to form hydrogen. Their conclusion, that a solution can have a positive excess charge is of course wrong from modern point of view: charge separation occurred only in the double layer regions, and not in the bulk, but the experiment indeed proved

Kundt, however, was not convinced by this speedy fulfilment of his demands and remained to the end an opponent of the new concepts—and indeed these concepts did not flourish in the air in Berlin.

Thus it was that this bet ended with the same result as the one I'd made before my final exams (Part 1, Chap. 4, p. 57): I won brilliantly, but my opponent refused to keep his side of the bargain. In this case also I didn't demand the loser's stake because I knew that I had won.

Naturally the experiment was later declared to be worthless. People said that it was just a common or garden electrolysis, because it makes no difference whether one uses static or normal electricity. As is so often the case with these things the original starting point of the argument has simply been moved. The question had been whether the small current that would be induced from static electricity could only flow if ions flowed at the same time. It's well known that Faraday himself had believed that in addition to the conduction dependent on ions there was in addition a small "metallic" conductance that was ion independent. I had argued that free ions in the electrolyte were essential by means of the following logically correct syllogism:

Electricity moves freely in the electrolyte
Electricity moves in the electrolyte only together with ions
Consequently the ions move freely in the electrolyte

However my opponents tried to overturn this logic by reference to the "metallic" conductance. The experiment had been directed against exactly this point and the result showed that the "metallic" conductance did not exist.

Göttingen. The first university after Leipzig that made room for physical chemistry was Göttingen. W. Nernst, who'd been my assistant till then, moved there in 1890, gained the qualification needed to become a professor and was appointed to a new chair for physical chemistry. In order to support this appointment I was invited to go there and discuss the matter personally and this, of course, I gladly did. I considered some friendly advice that I should not promote a future competitor ridiculous. The chance to spread my science further was so important to me that such considerations played no part for me. In any case I was so satisfied with the

(Footnote 3 continued)

that positive charge carriers must be present in the solution, as otherwise no charge transport is possible. So, finally, Kundt was right with his criticism of a positively charged solution, but Ostwald (Arrhenius, Nernst) were also right with postulating the existence of ions in solutions. Nernst gave a correct interpretation of the experiment in his book "Theoretische Chemie", 2nd edition, Stuttgart, Enke, 1898, on p. 357. He understood, in agreement with the double layer structure published by Helmholtz in 1879, that excess charges are located at the interfaces. Concerning the separation of charges, Nernst writes clearly: "Such charge separation is observed only to a vanishingly small degree by electrostatic induction of electrolytes [Ostwald W, Nernst W (1889) Z Phys Chem (Leipzig) 3:120–130], where the free electricity of ions is accumulated on the surface; a measurable amount of "free ions", not accompanied by oppositely charged ions has never been seen. This will never be possible, as an accumulation of free ions of the same charge is counteracted by enormous electrical repulsion forces."

results of a free and wholehearted cooperation like I'd had with Arrhenius (Part 1, Chap. 11, p. 115) that I saw no reason to follow any other path.

The meeting was entirely satisfactory and resulted in the decision to build a new institute of physical chemistry to be directed by W. Nernst. On this occasion I got to know Althoff, the leader of the Prussian government's university department, and gained an idea of the energy and views with which he dealt with the manifold challenges of his post. The fact that he frequently acted against the interests of the professors is no reproach for when those unedifying professorial qualities, envy, pettiness and avarice were evident, he experienced them in undiluted form.

The further development of physical chemistry in Germany was for a while asymmetric in that part of it—electrochemistry—was well looked after because of pressure from powerful technical and economic interests. This is a matter that will be treated in a separate chapter later. For the moment suffice it to say that sooner or later the other universities all accepted the new subject so that it is now represented everywhere. Munich held out the longest because there A. von Baeyer wanted no other Gods set up alongside himself.

Chapter 20
Impact at a Distance

The British contribution. With van't Hoff in Holland, Arrhenius from Sweden and myself in Germany, the new science had an international flavour from the very first and this was considerably strengthened by the numerous foreign students who flocked to the Leipzig laboratory. Because of this it didn't take long for the movement to transfer out into countries other than Germany particularly where efforts in a similar direction were already underway.

This was especially the case in Britain. There, the "British Association for the Advancement of Science" had been founded in 1831. It was modelled on the German organisation[1] but differed in that it played a direct role in organising scientific advance. In Germany this was considered to be the province of the universities while in Britain the old universities were not set up to do this. As a result, it was customary that questions which were considered important for progress would be considered in special sub-committees. The sub-committee appointed an expert who was responsible for preparing a summary of the subject and where possible made recommendations as to the directions in which progress was necessary and where it could most likely be expected. Where necessary experimental work could be carried out and it was not difficult to get this financed by the British Association.

In the area we are interested in there were two such committees: one for electrolysis and one for solutions. The first was headed by the physicist O. Lodge, who was very active and wrote to virtually all the scientists in the world from whom he could expect support. There weren't many. He had the answers printed and sent copies to all his correspondents. In this way he'd been in touch both with me and with Arrhenius, who in 1887 had sent his thoughts on electrolytic dissociation from Germany so that their first publication was in a report of this "Electrolysis Committee". Some years previously the great physicist Lord Rayleigh, had made

[1]Die Deutsche Naturforscherversammlung.

R.S. Jack and F. Scholz (eds.), *Wilhelm Ostwald*, Springer Biographies,
DOI 10.1007/978-3-319-46955-3_20

clear his conviction that the next large advance in chemistry and physics would come from a deeper understanding of the processes involved in electrolysis, and with the publication of this report this prophesy was fulfilled. My first publication on the dropping mercury electrodes was also reported to this committee.

In 1886 a similar committee had been formed to consider the problem of solutions, though no equivalently active chairman had been found. The committee contented itself largely with collecting the literature and since a few experimental approaches had not led anywhere this line of work was abandoned.

The concept of kinks. Things were stirred up by the intervention of the famous Russian chemist D. Mendeleev. Because of his serendipitous treatment of the relationship between the properties of the elements and their equivalent mass, he had enormous standing, which was helped along by the fact that he spoke no English and only very few people in Britain spoke Russian. He could speak some broken German as I found out when we met occasionally in England. However, what is ignored because of his great discovery is that he made no other similarly important contributions; in fact no other contribution from him has become part of the corpus of science.

Mendeleew was encouraged by the committee to consider the question of solutions. Like most chemists of his time he assumed that a chemical reaction takes place between the solvent and solute and he'd thought up a way to show this which was both ingenious and wrong. Whereas all observers till then were agreed that all of the properties of a solution varied continuously as the concentration of solute was varied, he proposed that the changes in the first derivative would be non-continuous. If one drew these properties in the usual manner as lines along a scale representing their composition, then these lines would not appear as smooth curves as had so far been assumed but rather as broken lines of varying slope whose breakpoints or kinks showed where the two components were in a simple stoichiometric proportion and this would reflect the composition of the chemical species which were assumed to be formed. Mendeleev had published a number of papers on the behaviour of the density of solutions together with the associated graphs which were, to begin with, received in Britain with the respect due to a great discoverer. However once his data was examined it turned out to be false: the kinks couldn't be found and the curves were confirmed to be continuous.

Nevertheless the impact of this idea was great enough that it took root in the mind of a British chemist. His name was Sp. U. Pickering and he had convinced himself that the kinks of Mendeleev were not present in the density curves for solutions of sulphuric acid. However, he thought that the basic idea that solutions are composed of chemical associations of solvent and solute must be right and so he came to the idea that if the curve is smooth then the second derivative might be kinked. Search and ye shall find!—and so he found the kinks. Yet even if the assumption that solutions were composed of stoichiometric amounts of the reaction products of solvent and solute were true, the laws of chemical equilibria require continuity of the curves and their derivatives and hence rule out the kinks. He was unaware of this—and when it was pointed out to him he didn't believe it.

Because of the enthusiasm he invested in the kinks, Sp. M. Pickering (we called him Kinkering) became a leading member of the solution committee and had acquired a following amongst those who knew nothing about the subject. In this way something like an English theory of solutions was born and its adherents considered our physical chemistry as an unwarranted interference in national affairs.

On the other side stood my pupil, J. Walker, who was in the meantime an assistant to Crum Brown who held the chair of chemistry in Edinburgh. Walker convinced Crum Brown of the great virtues of the new concepts. In addition Professor William Ramsey from University College in London was also a member of the committee and he too was convinced of the scientific value of the new concepts. They exerted their influence and, as a result, the British Association decided to invite van't Hoff, Arrhenius and me to the next meeting which was to take place in 1890 in Leeds. The idea was that we would debate the issue with the supporters of the other point of view. The discussion was not to be restricted to the question of the theory of solutions but was to extend particularly to Arrhenius's dissociation theory which seemed to conservative British minds to be nothing short of scandalous.

First meeting with van't Hoff. I thought the invitation was a great opportunity to carry the new gospel to the heathens. The ground had been prepared because a translation of my "Outlines" had just appeared and since competence in foreign languages was not common amongst British scholars, it provided the first accessible comprehensive description of the new concepts and it had called forth general opposition. It turned out that Arrhenius who was by this time back in Stockholm couldn't make it, but van't Hoff could. I arranged with him that we travel together to Leeds and I thought I'd visit him in Amsterdam first for a few days in order to get to know him personally, for till then we had not met.

It was with a real feeling of pleasure that I greeted my valued colleague in the station in Amsterdam. He was a slim young man (in fact he was almost exactly 1 year older than me) of medium height with the typical long skull of his countrymen and a pale face in which the eyes under the puckered brows suggested the thinker. A personal relationship was soon established since we'd known each other for several years through our lively correspondence. In the years that followed this developed into a scientific friendship that never clouded. Several times we made the same discovery at the same time without ever getting into a quarrel about priority. He knew that I acknowledged him as the greater thinker in our joint area, and I knew that he was happy to cede me the leadership in organisational and didactic matters. In this way we complemented each other and since Arrhenius had an excellent relationship with both of us we formed that band of three—van't Hoff, Ostwald and Arrhenius—(the names are arranged in order of decreasing age) who have left ineradicable traces of their joint work on science.

Van't Hoff introduced me to his wife and children. Our children were in the same age group; he had four and I had five, and so began an enduring friendship between the families. Once he'd moved to Berlin so that we could all get to know one another this friendship developed and extended beyond his all too early death.

I viewed the city and its harbour with great interest, for it was the first non German town I'd got to know. I remember a procession of quaint half dressed orphan girls on the street and the servant girls washing the outside walls of the houses.

Of course I visited the laboratory and heard to my astonishment that my colleague had to spend an inordinate mount of time with all sorts of analyses for official control bodies. Because he'd been offered the professorship in Leipzig he'd been given a newly built institute to replace the inadequate old building. We climbed all over the walls of the new building which had just been roofed until I saw that he had no head for heights and was having difficulty keeping up with the climbing.

He took me along to visit his parents in Rotterdam. His father was a doctor and spry despite his old age. He was small as was his wife who looked as if she'd just stepped out of an old Dutch painting. Both of them were very friendly to me, since they knew my relationship to their son, and we sat down to lunch. The roast was kept warm on a rack under which some pieces of glowing peat were covered with white ash.

On a visit round the town I noticed over some doors carved and painted heads with grimacing faces and their tongues sticking out. Van't Hoff explained in answer to my question that this was an old symbol of an apothecary. In the old day customers were given a diagnosis and medical advice after they'd stuck out their tongue for examination. The head with the typical gesture was put up so that people from the countryside who couldn't read would recognise an apothecary's shop, rather in the same way as our inns are decorated with a blue angel or a golden sun, and this is still the tradition today.

Leeds. It was with a degree of worry that we took ship to England. Our knowledge of English was restricted to reading and we weren't sure if that would be enough for what stood before us. In the end everything was all right. We didn't waste any time in London because we assumed that all our colleagues there would already be on the way to Leeds. Once our eyes had got accustomed to the green and fertile countryside of Yorkshire, our train entered the black and smoky industrial town. We'd arrived one day early and we had some difficulties to find the houses where we were to be put up by our colleagues.

Leeds is an industrial town with large textile and dye works and at that time was home to one third of a university. It housed, together with Manchester and Liverpool, the Victoria University which had a single administration for the three institutions. Even then efforts were being made, which if I remember rightly, soon led to a separation of the three.

My host was the professor of chemistry there Arthur Smithells, a slim, good looking young man of my age who lived with his lovely, lively wife on the outskirts of town in a villa with a garden. It wasn't easy for me to get used to the ways of life in the new surroundings especially as I found it very much harder to understand spoken English than to read it or speak it myself. But Ramsey, who I soon got to know personally and who spoke tolerably good German, kindly helped me over the initial difficulties. He'd studied for some time in Germany and so he well knew the difference between German and British customs and he quickly gave me tips when I ran into difficulties.

The British Association. I found it very interesting to compare the British Association with the German Scientists' Meetings. The main difference was the absence of physicians for in Germany they make up half of the attendees. On the other hand Sociology was represented by section F which included economics and statistics which were not represented at our meetings.

Altogether the organisation is more constrained and consistent than in Germany. After a meeting the German Society more or less hibernates and during this time only the managing committee shows occasional weak signs of life. Sometime before the next meeting, the preparations start and they are largely the work of the organising committee which is elected from among the scientists at the meeting location. The result is that the only continuity is provided by the longer term of office of the managing committee. In the British Association, the numerous sub-committees, which are called into being whenever one is needed, and which are active throughout the year, ensure that there is continuity of effort quite apart from the general meetings. In addition they have accumulated for these meetings an organisational tradition and a collection of technical equipment which makes life much easier for the local organising committee.

The form of lecture is very different from ours. The proper term for it is, "He reads a paper" and that is to be understood literally. A written version is almost always read out word for word and afterwards given to the secretary for publication.

One can understand the utility though not perhaps the elegance of this procedure and I missed both of these properties in the major lecture held by the President before the assembly, half of which were ladies. This time the chairman was the excellent chemist Abel, who worked on explosives. He held an address of over 2 h which he meticulously read out of a notebook in a monotone an overview of his subject, though only a few of his audience could follow the technical details. The strangest thing of all, however, was that all of his listeners had found a printed copy of the talk they had to listen to on their seat. The whole thing was carried off with that air of being a matter of course which this nation always brings to its official ceremonies. Not the slightest restlessness in the huge hall betrayed the stupendous boredom which I assumed these thousand people must be suffering. Apparently they felt none. For me, it was a welcome opportunity to learn to understand spoken English. I listened carefully and when I didn't understand a word I looked into the printed text. By doing so I quite quickly found my way into the sound of the language and this was of great importance for my participation in the discussions.

The discussion. A good deal of time had been set aside for the discussion of solutions and electrolysis. It started with a long lecture by Mr. Pickering who was determined to place himself as the leader of the discussion and to distribute praise and criticism as he saw fit.[2] The lecture was in two parts: the first consisted of a description of his procedure to discover hydrates and the second of a refutation of the theories of van't Hoff and Arrhenius though it turned out that he had understood them only in a very superficial way.

This all took up the entire first session which took place on a Thursday. Only the opposition was represented on the next 2 days—apart from us there were only opponents—so that by the end of the week it appeared that the continental theories were completely disposed of. We'd had to limit ourselves to trying, as far as we could, in personal encounters with the more influential participants to counter errors and misapprehensions, during social events. In particular I remember an occasion on the Sunday afternoon when Smithells had invited a number of important chemists and as the discussion became ever livelier van't Hoff and I finally managed to convince some of our opponents. These were, of course, the smartest.

On the Monday things changed. Arrhenius had recalculated the results of a large number of Pickering's own data on the reduction of the freezing point of solutions of sulphuric acid using the equations of the dissociation theory and had found an astonishingly good agreement which showed that Pickering's experiments were a good deal better than his theories. He'd sent this work to Walker who read it out calmly and soberly and this made a great impression. Then I reported on a phenomenon, the deposition of metallic copper by current passing a semi permeable membrane, which gave a very concrete illumination of an unexpected consequence of the theory. For the first time in my life I had to give a scientific presentation in a foreign language and once my friends had told me that it had gone very well I was encouraged to take part also in the discussion and to dispose of the arguments raised against us, most of which were due to misunderstandings. Van't Hoff had also held a lecture and took part in the discussion. And so the "hydrate theory" completely disappeared from the field of view and the discussion concentrated entirely on the question of the feasibility of the new concepts. The excellent physicist and mathematician Professor Fitzgerald, had an erroneous notion of them and so thought them all totally impossible. Ramsey who was a close friend was able to convince him that this was not just some nonsense dreamed up by people who had no idea of physics, but rather a carefully thought out and well founded theory and that in terms of thermodynamics there was nothing to be said against it. This, it must be said, did not take place in the open sessions but rather during extended personal discussions.

Kinetics and energetics. Typical of the sort of difficulties which even well inclined critics of the new concepts had with the doctrine of osmotic pressure, was the

---

[2]Pickering SU (1891). The present position of the hydrate theory of solutions. In: Report of the British Association for the Advancement of Science (60, 1890, Leeds). London, Murray, 311–322. Ostwald published a translation: Ostwald W. (1891) Z Phys Chem (Leipzig) 7:378–416 (Ostwald's commentary: 416–426).

frequently repeated objection, that while one could well imagine in view of the kinetic theory of gasses how pressure would result from the constant bombardment of rapidly moving molecules, it was quite another matter to imagine how such a thing could take place in a solution where the solute molecules were constantly colliding with those of the solvent and hence an effect on the vessel wall would be insignificant. Van't Hoff's counter, that Pfeffer's experiments demonstrated the existence of osmotic pressure beyond doubt and so if a kinetic theory had difficulties finding an explanation for it, then that was a problem of the kinetic theory. The trouble was that the respect for the kinetic theory which was largely based on the complex mathematics required for the analysis of individual experimental situations was so great that people were automatically leery of anything which didn't fit—even if it was true.

For me this sort of discussion in which I had to take part on numerous occasions, was a reason to distrust the kinetic theory. At that time this was completely against the trend. I convinced myself that the yield of real scientific results emerging from the theory was modest and was largely without predictive value being limited to converting what one already knew into laws. Apart from the independence of the molecular friction from the pressure, very few new insights were generated by the concept and these insights had little effect on the progress of science.

I expressed these views at first hesitantly but later with ever more emphasis. Once I'd understood the general methodology of energetics and had seen how quickly and with what certainty they led to clear quantitative results, where kinetics required pages of calculation with the consequent high probability of errors, my mind was made up. From then on I expected from myself and from others that in order to understand phenomena one required the energy calculations and I mistrusted kinetics as being too uncertain and not very helpful. However these things came later and should be related when their time has come.

When we finally set off for home, tired but satisfied, we knew that the ice of prejudice had now been broken and that the intrinsic value of things would now drive their further development. Of course we couldn't expect that opponents like Pickering and Professor Armstrong had been convinced and indeed they both later defended their positions. We were, however, sure that we didn't need to take them seriously because in both cases their opposition was based just on gut feeling rather than on any scientific insight of which neither had sufficient to allow them to reach an objective conclusion. In any case they both soon lost the influence they once had in Britain.

William Ramsey. My friendship with the distinguished British chemist William Ramsey covered the whole period when chemical matters formed the centre of my work. We were of the same age and had similar views on many subjects and yet were sufficiently different in the conception and execution of our work to find each other interesting. We developed a good relationship based on mutual trust that lasted till the outbreak of the world war. Then, Probably under the influence of his wife, who'd always had a soft spot for the French, he allowed himself to be carried away. He not only agreed with the anti German propaganda which was an essential part of the British war effort but even increased it with his own diatribes.[3] He died during the war.

Our relationship had begun with the Journal.[4] He'd carried out extensive research on vaporisation and condensation and had submitted some of this to me for publication. I asked him to write a review of his work for the Journal and this he gladly did. I was also involved in publishing work from him in other areas.

At about this time he applied for the position of full professor of chemistry at University College in London, which was the most productive of the London universities. In Britain candidates are not selected by a conclave of professors or by the government, rather free positions are advertised and candidates apply themselves. It's the custom to collect testimonial from as many well known experts as possible because the appointment committee will often not contain a specialist in the field. Ramsey had asked me as well for a letter of recommendation and I was actually rather proud to be able to do this for him.

I no longer remember when I met him first. Since Ramsey had studied in Germany under Bunsen and Fittig (at that time in Tübingen) he often visited Germany and it's possible that he visited me at some point in Leipzig or somewhere else. In any case we already knew each other before we met frequently during the week-long meeting of the British Association in Leeds. Ramsey helped me a lot to find my way in the strange surroundings and we became so close that he invited me to accompany him on his holiday which he was going to spend with his parents in law on the Scottish coast. He was Scots by birth and explained that there were great difference between the Scots and the English and that this was obvious even just in the language, for the Scottish dialect is much closer to German than the London one. Beyond this, he said, the Scots are much more talented in science than the English are, as one could see from the fact that so many of the leading chemists in Britain were from Scotland. The constitutions of the Scottish universities were more similar to those in Germany in contrast to the clerical organisation of the old English universities in Cambridge and Oxford.

I think he was right on all counts.

I took Ramsey up on the invitation and he showed me something of his home country. We started in Edinburgh which would be one of the loveliest cities in Europe were it not for the dismal climate. There was a national enthusiasm for Mary Stuart despite—or maybe because of—the fact that the character of that dazzling personality was so totally different from the earnest and down to earth nature of the Scots who only rarely have handsome faces.

[3]See, for example: Ramsay W (1914) Nature 94 (No. 2345) 137–139. This paper was a reaction to the so-called Manifesto of the Ninety-Three (Aufruf an die Kulturwelt, 1914), a proclamation by 93 prominent German scientists, scholars and artists supporting German military actions in World War 1 and denying German responsibility for the outbreak of the war. This manifesto was also signed by Wilhelm Ostwald.

[4]Zeitschrift für physikalische Chemie (Z Phys Chem (Leipzig)).

Then, at my request, we went to look at the Forth Bridge which at that time was the most advanced construction of its sort. I was impressed. Its great height was best appreciated from a steamer from whose deck a train crossing the bridge looked like a toy. I took a number of photographs of it home with me, because I was trying to persuade my eldest son to become an engineer. However, I didn't have any luck with that, because by this time he was in and out of the lab and had made friends with many of the students so that nothing was going to stop him also becoming a natural scientist.

I spent a happy week with Ramsey's family. His wife had the face and figure of Queen Victoria and vainly tried all the time to get her increasing girth under control. They had two nice children, one boy and one girl who were about the same age as mine. His mother in law was a sweet old lady exuding goodness and friendliness. His father in law was silent and kept himself in the background. Life there was as pastoral as near the beach where I grew up—only that one spent more time messing about in boats which was something Ramsey particularly liked to do. I enjoyed it as well. What was surprising was the demonstrative religiosity. I was not spared the services on Sunday, one in the morning and one in the afternoon, nor the daily prayers at home. I didn't mind fitting into the family's customs.

Towards home. On the way home I briefly visited Glasgow where William Thomson (Lord Kelvin) was professor of physics. I'd been introduced to him in Leeds and had been impressed by his lively youthful spirit, though he'd reached the considerable age of 67. He'd particularly won my heart during a scientific discussion when he agreed without the slightest hesitation that the opinion he had expressed was wrong after somebody—I think it was G. Stokes—had raised a well-founded objection to it. However we didn't get to know one another personally at that time. Ramsey, had been his pupil and encouraged me to visit him, but it turned out that he too was on holiday.

Incidentally, Glasgow turned out to be an indescribably filthy city whose hazy smoke-filled atmosphere was particularly repellent after my interlude in the clean sea air. I went back to Edinburgh and from there took ship to Germany leaving the other parts of Britain for later. I'd been elected as a foreign member of the British Association and they'd made me promise to attend many of their meetings which in fact I later did several times.

Edinburgh. I didn't attend the Nottingham meeting in the following year but I did go to the one which was held in Edinburgh in 1892 and it was a real high point.

To get there I went, as I always preferred, by ship. I left from Hamburg on a British steamer which was making straight for Edinburgh's port of Leith and on board I met some of the meeting's German participants. There was going to be a decision on electrical units and representatives from the Imperial Physical and Technical Institute (Physikalisch-technische Reichsanstalt) had been sent along as experts. I remember the extremely tall Dutch physicist du Bois who as a rich heir could run his life as he wanted. He had installed himself in a private laboratory in Berlin and had engaged himself in lively contact with the scientific circles interested in magnetism.

The crossing was quite stormy though I wasn't sea sick and the small circle of us who dined had made friends with the captain. From the captain's bridge we were able to admire a wonderful sunset as we sailed into the picturesque port. We were so engaged in talk that the captain forgot to keep an eye on the course and would have brought us straight onto a rock if he hadn't been warned at the last moment. He did manage to turn the ship in the nick of time and then asked us not to disturb him any more at his work. That surely would never have happened to a German captain.

The status of the new concepts. A completely different situation awaited us at this meeting than at the one in Leeds 2 years previously because the assessment of the new concepts had radically altered in the 2 years since 1890.

They had become so to speak, fully acceptable. There was no question now of their complete rejection for they were generally considered to be an integral part of science. There were divergent opinions as to their consequences but everyone agreed that they were without doubt a major step forward. The influential professor Crum Brown had contributed significantly to this success. After one of my best British pupils, James Walker, had become his assistant and had told him about the new work, he had started to think about it himself and to promote it. At this meeting he and Walker presented an elegant synthesis of organic acids by electrolysis and this opened up a fruitful new area of the concept.

Influenced by this atmosphere the Electrolysis Committee, which included such illustrious British chemists and physicists as Lord Kelvin, Lord Rayleigh, J. J. Thomson, A. Schuster, J.H. Poynting, A. Crum Brown, W. Ramsay, E. Frankland, H.B. Dixon, J. Larmor and many others, presented its seventh and last report in which was included a summary of the relevant material in tabular form. After this the committee ended its activities principally because the matter had now taken root and no longer needed special support. Perhaps there was also the feeling that the focus of the work was in Germany and that this made establishing the field in Britain very difficult. One could take it for granted that the work would be well done in Germany—and this was exactly what happened.

British celebrities. This time my host was Alexander Crum Brown who was professor of chemistry at Edinburgh University. He was a man of unusually wide interests. In addition to his excellent chemical research he was interested in geometry and in problems of the physiology of the senses. He was one of the simultaneous discoverers of the importance of the semicircular canals in the human inner ear for balance. In character he was a brisk man of medium height, strongly built, with white hair, a short beard but lively dark eyes. His behaviour was more that of the man of the world than of the scholar. He lived in a large magnificently furnished house in which he had enough rooms to put up several other visitors to the meeting as well. In this way I got to know Sir George Stokes, the distinguished mathematical physicist who fascinated me with his mixture of personal meekness and scientific rigour. He looked pretty much like a British version of Kohlrausch, though he was a lot older. He asked me if the dissociation theory provided an explanation for the strange effect of chlorides on the fluorescence of chinin salts.[5] I had to admit I wasn't aware of this work, but promised to look into it. Later I came across an exhaustive study of this question by another British scientist but he seemed not to have got much further.[6] The completely changed status of the new concepts was made most obvious at the first ceremonial session. As I mentioned, important decisions concerning electrical norms were to be made and for this there had been sent not only some officials of the Imperial Physical and Technical Institute but also Helmholtz, the Institutes president, had come in person to Edinburgh where, together with his old friend Lord Kelvin, he was hosted by Lord Kelvin's friend and colleague P. Tait. When the meeting's officials—the President, the local committee and the section heads—had taken their place on the platform, Helmholtz and Lord Kelvin were invited to join them as guests of honour. A similar invitation was made to me and I felt like a small mouse between two lions as I hesitantly accepted, for both of them were undisputed masters of mathematical physics, a subject which they handled inventively, while I knew only too well my limits in this direction.

It was refreshing, almost touching to see the friendship of these two great men for it is hard to imagine two more different people. Helmholtz was short, stocky and grey blond. He moved little and his face was almost immobile. Lord Kelvin was tall and thin with lanky limbs which were almost as constantly in motion as were his face muscles. The one was a classical scientist, the other a romantic—each of them being a most extreme example of his type. They had often worked on the same or similar problems but never got into quarrels. On leaving the meeting the stocky Helmholtz carefully guided his lanky friend who had had an accident and was limping badly.

At a small dinner party in his house my host Crum Brown had placed me next to Lord Kelvin and I had an hour's animated conversation with him. In the friendliest possible manner he pointed out an error I'd made in discussing his work in my textbook and which I'd corrected in the next edition. On the other side I'd pointed out in a lecture in the physics section a case where his hypothesis that the complete enthalpy change accompanying reactions in galvanic cells could be converted into electricity was certainly wrong since it contravened the second law.

To amuse me Lord Kelvin tried to recall his German memories and he recited word for word quite long passages from "Faust" though his enunciation was a bit strange. All in all he was in every way a charming and energetic dinner companion.

He acted in exactly the same way in the meeting. An Edinburgh newspaper published cartoons of the most interesting participants and Lord Kelvin appeared as a Jack in the box who shot up every time the button was pressed.

---

[5]This refers to Stokes publication Stokes GG (1869) J Chem Soc 7:174–181, in which he describes the fluorescence quenching of quinine by chloride ions.

[6]This refers to the work of Edgar Buckingham, who was born in U.S.A. Buckingham published the results in; Buckingham E (1894) Z Phys Chem (Leipzig) 14:129–148. See also: Stock JT (2002) Bull Hist Chem 27:57–61, on Buckingham's life and research.

Aftermath. I left Edinburgh certain that the further development of our ideas in Britain needed no further help. To be sure, H. Armstrong argued heatedly against us from time to time but no one took him seriously and we could simply ignore him. On the other hand J. Walker, who was a born teacher, and soon had an independent position—he is now Crum Brown's successor in Edinburgh—trained a whole flock of dedicated "ions" and in the Leipzig laboratory there were always several people from Britain many of whom were very talented and who went on to become important professors and they established physical chemistry firmly on the island.

After the meeting I accompanied Ramsey for a few days to his summer resort on the Isle of Arran where he had a fisherman's cottage that was even more primitive than the summer house on the beach by Riga. One of his relatives of the same age took us out on a small sailing boat in which we spent most of the day; often enough in showers and gusts of wind for the weather was mostly dull and stormy. From a hill on our island we could see a neighbouring island to which we had to sail when we needed fresh vegetables. There, there was a herb garden protected by high stone walls. On unprotected sites there grew only a hardy sand grass and low thorn bushes which formed impenetrable masses through which a few small paths had been hacked. There were no fields that might have been worth fertilising but only a few patches from which the bushes had been cleared. This simple life was very welcome after all the eating and drinking in the crowds at the meeting and I took leave of my friend and his family much refreshed and set off straight for home.

I was often in Britain later, to be awarded honorary doctorates of which I was given more there than in any other country. However these journeys came later and I shouldn't record them here for other more important things are waiting in the wings.

Chapter 21
Energetics

The beginnings of energetics. Already during my student days Öttingen had made me aware of the term energy. He'd taken a great interest in the first exciting developments of thermodynamics when as a young man at an impressionable age he'd taken his first steps as a scientific journeyman in Berlin. As his assistant I listened to his lectures which were largely concerned with getting to the core of the first law, which states that energy can be transformed from one form to another but can neither be created nor destroyed, and the difficulties in understanding the second law were a major concern to him. I found all this very stimulating and have already described with how much interest I followed the development of thermodynamics during my time as a professor in Riga.

By the time I moved to Leipzig these matters were so at the centre of my mind that for my inaugural lecture before the assembled faculty and students in the main university lecture hall I used the title "Energy and its conversion".

The lecture began with a historical review of the optimistic idea that the till now disparate areas of science might increasingly converge. Physical chemistry is a good example of this. Some aspects of its development led to the following comparison. "One can visualise the development of human knowledge quite easily when one represents it as the formation of a continent from the ocean by the slow rise of the sea bed or by a reduction in sea level. To begin with just a few of the highest pinnacles protrude as islands which don't seem to have anything to do with each other: here the humanities (or sciences of the will as I like to call them), there

---

Ostwald's first papers on Energetics were: Ostwald W (1891) Berichte über die Verhandlungen der Sächsischen Akademie der Wissenschaften zu Leipzig. 43:271–288. Reprinted in Z Phys Chem (Leipzig) (1892) 9:563–578; and Ostwald W (1892) Berichte über die Verhandlungen der Sächsischen Akademie der Wissenschaften zu Leipzig. 44:211–237. Reprinted in Z Phys Chem (Leipzig) (1892) 10:363–386. For a detailed commentary on Ostwald's concept of energetics see: Deltete RJ (2010) Thermodynamics in Wilhelm Ostwald's Physical Chemistry. Phil Sc 77:888–899; Deltete RJ (2007) Wilhelm Ostwald's Energetics 1: Origins and Motivations. Found Chem 9:3–56; Deltete RJ (2007) Wilhelm Ostwald's energetics 2: energetic theory and applications, part I. Found Chem 9: 265–316; Deltete RJ (2008) Wilhelm Ostwald's energetics 3: energetic theory and applications, part II. Found Chem 10:187–221.

R.S. Jack and F. Scholz (eds.), *Wilhelm Ostwald*, Springer Biographies,
DOI 10.1007/978-3-319-46955-3_21

the natural sciences, and in between them a vast sea of ignorance on which with the boldness of youth the boats of philosophical systems sail. Most of these boats disappear into the infinite or come to grief on the hard rocks of reality. Only rarely does one find a safe anchorage.

Slowly, as the water level drops ever more small islands appear, fuse and are added to the few main peaks. When so many islands and rocks appear we of course know that under the surface they are all part of one area even though we can't yet see that directly. And when more and more islands appear then we can be sure that the ground that holds them all together cannot be far below the surface.

Later on Wilhelm Wundt liked to tease me about the "sea of ignorance" and jokingly accused me of consigning his scientific work and that of his pupils which he'd put together under the overall title "Philosophical Studies", to a boat on this sea. He knew of course that I was convinced that he was one of those who had found a safe anchorage.

The following relationship could be drawn between energy and material. In the beginning elements were not materials but rather properties. Those of Aristotle were earth, water, air and fire and they represented the properties solid, liquid and gas while fire is equivalent to heat which one might more generally describe as energy. Similarly the alchemists' elements mercury, sulphur and salt represent properties of matter to wit; metallic, inflammable and soluble and these "philosophical" elements should of course not be confused with the chemical elements of the same name. As time passed these abstract elements became ever more concrete until today they are defined as the ultimate tangible components of all physical materials.

The different forms of energy, in contrast, were initially described in terms of materials. In the eighteenth century one was constantly hearing about the "fiery material" the "electrical liquid" and so on. As time passed these materials began to disappear and were replaced by the concept of force. However, when one considers that according to the law of the conservation of energy the different "forces", or better energies, and their amounts are indestructible and can only be changed from one form to another—just like matter—so one sees that in this sense energy and matter are clearly comparable.

"If one asks what property defines real objects then the answer can only be that no human or natural power is able to create or destroy them. I don't want at this juncture to get into the argument of whether this can serve as an absolutely objective criterion or indeed if there is such a thing. Here it suffices to say that real objects can only be imagined as those whose existence is independent of human will.

There are only two such types of object so far known: physical material and energy. Only these are substances in the sense that they will continue to exist no matter what happens. I have no doubt that in 50 years the reality and substantial nature of energy will be part of the mental world of every educated person just as the idea of the reality of physical matter is today. It is the job of science to now reach the necessary conclusions for it has the task of reaching out beyond the limits of what is generally understood today rather than merely being content with what we already know.

Those who know the history of energetics will realise that the basic idea was much the same as that which the discoverer of the revolutionary law of the conservation of energy tried to convince his contemporaries nearly 50 years ago. And they will also know that the idea was completely lost so that I myself, and even more so my audience, thought it was an audacious step to describe material and energy as being comparable. The reason for this was that because of the mathematical nature of the development of thermodynamics, energy was viewed merely as a function which had the curious, but mathematically very useful property that in a closed system it remained constant. And so when 8 years later I took the next step and placed material conceptually below energy, which now became the only "reality" because it alone explained force and outcome, this was a point of view which, with very few exceptions, all of the experts in the field vigorously opposed.

For the moment, however, I put these ideas aside because the immediate application of the new concepts of van't Hoff and Arrhenius took up my entire energy and that of my colleagues. Nevertheless from occasional oblique references to it in my publications one can see that these ideas were hovering at the edge of my consciousness.

Further development. In large measure my views of energetics crystallised as a result of my daily discussions with students and co-workers in the laboratory. Since the young people, who at that time had the courage to subject themselves to several semesters of physical chemistry, were of necessity independent thinkers and born researchers, our discussions were naturally particularly fruitful. These students did not simply consume and quietly digest whatever the teacher put before them but rather responded with arguments of their own. In particular they were all influenced by the widespread tendency to view things through the prism of kinetics and could only slowly and reluctantly be persuaded to consider what they thought of as my abstract and impenetrable approach. The result was a lively exchange of views which forced me to constantly sharpen my view. In such a critical atmosphere any idea of parallelism between matter and energy would immediately have had its weaknesses exposed.

If one wants to explain something clearly to someone else then it must first of all be clear to oneself. This is probably the most important benefit of teaching because the necessity of achieving clarity is the best possible means of getting progress in one's own lines of thought. Although I don´t remember the day and year, I still have vivid memories of the environment and the atmosphere in the laboratory where the radical idea of pure energetics became clear to me for the first time. I can see in my mind's eye the dark and none too clean rooms in the old laboratory in the Brüderstrasse where I used to go from one lively discussion with a student to the next. First we'd talk about the project, the student's point of view—and where he wanted to go. After these points were settled the discussion would often easily move on to general matters because I have always found that putting a project into the more general context of the field is the best way to visualise how the work should be developed. Quite naturally the discussion was not restricted to the individual student at whose bench I was standing but often involved everybody in the room.

One day after such a discussion I'd just managed to disengage myself from the swarm of students and was on my way through the small library to my office where I had to clear up some of the official business that I always had to force myself to do, when I all of a sudden stood stock still. The discussion we'd been having had continued subconsciously in my mind and had reached a completely new level. The difficulties, indeed the impossibility of a dualism or parallelism of matter and energy had been inspired by the discussion and had reached a point at which I had to mentally "stop for breath" and find another solution. How would it be if energy alone had a primary existence and matter was merely a secondary product of energy or perhaps formed as a result of the mixing together of different forms of energy?

Fundamentals. As far as I remember the question arose when someone asked how energy can be a reality when an object moved at different velocities must have different energies. The speed is simply a number relationship between distance and time, so how can a number relationship force a new reality into an object which posses a completely different form of reality, namely mass.

The clear contradiction that this objection brought out drove me to respond in the opposite direction. The speed and with it the greater or lesser kinetic energy must represent reality because this number relationship is determined only by its value and is independent of other properties such as, for example, direction. Physics implacably demands that giving speed to an object requires the input of work—there is no way to avoid this. The necessary work, or energy, cannot be conjured up out of nothing it has to come from somewhere where it was already present. Consequently speed is undoubtedly a part of reality because if one wants to describe its properties then it follows that speed, like energy, not only cannot be created, it also cannot be destroyed.

The relationship between distance and time can be simply an abstract concept, but the actual speed of an actual object is part of reality which is only subsumed in the concept by an act of abstraction. Exactly the same distinction must be made with energy. Every scrap of energy in a piece of coal, in a charged accumulator, in a planet orbiting the sun or in the spark of static electricity that converts the electrical energy in a Leyden jar into heat, is a form of objective reality and all these forms can be placed within the concept of energy.

The realisation that here we are being forced into error by the language which assigns the same term to the general concept and to the individual case without making clear the huge difference between them—this realisation that an real object does not lose its reality when it is describe in terms of an abstract concept, hit me like bolt of lightening. I had an almost physical sensation in my brain as if an umbrella had been turned inside out in a storm. From the relative equilibrium of my mind which till now had satisfied itself with the parallelism of matter and energy, my mind now suddenly switched to an alternative stable equilibrium in which energy was the leading factor. Bulk and weight, the principle properties of "matter" were to be regarded as merely factors or aspects of particular forms of energy.

The breakthrough. I can't pretend that much was changed by this new mental approach. In the following period I suffered from exhaustion caused by the birth of the idea and so I carried on in the same old way as if nothing much had happened except that I often told myself and my pupils that we'd maybe progress a lot faster if we made use of this thought. However, as with every really new idea, it was as necessary as it was difficult to carefully think through all of the consequences, before the new viewpoint could be applied. We know, for example, that Julius Robert Mayer thought of his energy law in a sudden flash of enlightenment on the quay in Surabaya, but that he needed several years' hard work to convert it into a solid scientific concept which could be applied to any case.

This is what happened to me with my first thought on pure energetics. The first idea, which as far as I am aware had not occurred to anyone else before, was and remained the most far reaching: J.R. Mayer had already regarded energy as being as real as matter but he hadn't ventured any further. Since then things had gone backwards because all later researchers in the area of thermodynamics used energy merely as a mathematical abstraction which they regarded as a useful function of physical state variables rather in the way that potential functions are used. And even someone thinking along the same lines as me at the time—the mathematician Georg Helm—absolutely refused to take this radical step and almost petulantly refused to accept any attempt to regard energy as a substance or to accord it a reality equivalent to that of matter. In this sense he was even behind Mayer. For me, however, the real advance lay in conceptually dividing matter into its energetic components and in the realisation that that everything we experience is based on the energetic relations between our sense organs and the environment. Every such interaction is effected by changes in the energy balance and no sense organ will record a sensation unless such an energetic change takes place.

A metaphysical experience. [1] The development of these thoughts took quite some time, the details of which I no longer remember. However, an experience I had 6 months or a year later remains clearly fixed in my memory. It must have been in the spring of 1890, though I'm no longer exactly certain.

I'd earlier had the chance to get to know the physicist E. Budde who was the first to recalculate the old data from Regnault on the compressibility of hydrogen gas and show that the total volume was made up of two components: the first strictly obeyed Boyle's law of the inverse relationship between volume and pressure, and a second which was independent of pressure or which was not compressible. In my textbook I'd stressed the importance of this result which up till then had been largely ignored and Budde had been grateful for this. Since he was, as he'd shown in his writings, a brilliant, experienced and determined man with a happy go lucky approach to life, a close friendship had sprung up between us.

In thinking about energetics it became clear to me that the whole of physics, which till now had been understood as a doctrine of forces would have to be rewritten as a doctrine of energies. Aspects of this which I tried to elaborate did give very good results but I couldn't possibly work over the entire area for I could

[1]Ostwald uses the phrase 'effusion of the Holy Spirit' a metaphor used in German for Pentecost.

neither spare sufficient time from lecturing and research nor was I sufficiently sure of the mathematics to be able to easily rewrite the whole of physics.

I'd heard that Budde was looking for some remunerative work to support his household (he had married a Turkish lady)[2] and my experiences with the economic performance of scientific textbooks were so positive that I wanted to suggest to him that we should come to an agreement about a basic outline and then that he should write a textbook recasting physics in terms of energetics.

In order to get things moving I went to Berlin, where he lived, to talk it over with him. When I arrived he had both hands deep in a pot in which he was energetically kneading a white mass. He told me that this was a new and particularly useful insulation material which he was hoping to sell to the large electro-technical firm of Siemens and Halske. When I told him what I'd come about he left his dough and, having not without some considerable effort managed to wash his hands, went with me to the Spatenbräu restaurant where he said we'd be sure to meet other Berlin physicists when they left the Physics Society meeting.

He was open to my new views and we'd already covered a great deal when our physicist colleagues arrived; I no longer recall who all was present. I wasn't in the least shy about presenting my plans but the others found them so absurd that they didn't take them seriously at all but instead heaped scorn on them and tried hard to put me off my energetics. In view of the discussion I'd just had with Budde this didn't put me off at all but rather reinforced my conviction that a radical rethinking of physics was now necessary. The conversation became very lively; we went home late. I found a room in the Central Hotel, where I usually stayed, and slept for a few hours before waking, still in the middle of the same train of thought, and couldn't get back to sleep again. The sun had already risen.

In the early morning I left the hotel and went into the Tiergarten Park. There in the sunlight of a wonderful spring morning, I experienced a real Pentecost, an effusion of the holy spirit. The birds were singing in all the trees, golden-green leaves shimmered against a light blue sky, butterflies opened and shut their wings while sunning themselves on the flowers and I wandered through this springtime landscape in an exalted state. I saw everything with new eyes and felt that I'd never experienced delight and wonder like this before. I can only compare my feelings then with those of my first love which lay 10 years behind me. In my mind the plan for a complete energetic concept of the world unfolded effortlessly and joyfully. Everything around me seemed to be waiting, as in the biblical creation story in paradise, for me to give them their true names.

This was the real birth of energetics. That rather strange and even eerie thought which had suddenly entered my mind a year ago now turned out to be an essential part of my innermost being. It had been assimilated and had developed half unconsciously until, like the sudden bursting open of a bud, everything was there

[2]This is incorrect: Budde worked as a journalist for a German newspaper in Constantinopel where he married the daughter of the German consul Reiser.

and my enraptured gaze had only to dart from here to there to encompass the new creation in all its glorious perfection.

This wonderful feeling stayed with me all morning and I didn't tire of wandering through the lustrous spring and seeing with my inner eye the endlessly wonderful infinitely broad vistas which had suddenly appeared.

Then slowly the great town awoke and took me back into its din and dust. As soon as was reasonably possible I visited some of the experts and tried to explain my views to them. They told me later that I had acted like a prophet or one inspired. They'd been used to my oddities but had never seen me in such a state. I should add that never again did I undergo such a thing and never again experienced such concentrated joy—even though I was lucky enough to be granted many exhilarating —and also some harrowing—births of substantial new ideas.

A futile sermon. My attempts to awaken a similar response to the new ideas in my colleges failed completely. True, I had managed to gain general acceptance for the ideas of van't Hoff and Arrhenius against the broad opposition of my colleagues, but only because these had not been my ideas. However, they refused to believe that I would be able to come up with similarly revolutionary ideas for, after all, I was neither a Swede not a Dutchman but merely a German. To begin with they ignored my new insights, and then later, when I discussed them in public lectures, they eagerly attacked them.

I can't pretend that their lack of interest particularly annoyed me and I didn't consider that it was due either to stupidity or malice. I could easily understand how impossible it would be to even begin to excite sentiments similar to those which I had experienced in the strange but glorious spiritual elation of my Pentecostal vision, merely by relating the bare facts of the matter.

The energetic re-interpretation of physics never got off the ground because Budde soon afterwards was given a leading position at Siemens and Halske and this claimed his entire energy and time.

Creative work. I returned to my routine work mentally strengthened by this experience and tried to pin the ideas down in the daily round of discussion with my co-workers. Here as well I was met with both passive and active resistance. This spurred me on to apply the energetic prism to all of the problems being dealt with in the lab so as to try to show its practical usefulness.

It was lucky that just at that time the problems we were dealing with were all completely new so that no one had considered them up till then. Because of this I had a clear and as yet undefined field for the application of pure energetics. In this way it was much easier to apply the energetic idea to individual projects than if these areas had already been defined and given a more or less rigid form by previous workers. "First come, first served" applies probably even more in science than anywhere else. He who has opened up a new area is free to define and form it as he personally sees it. Everybody else who comes in later is more or less forced to accept the constraints the subject had been given, either because the constraints are taken over or because their rejection results in their mirror image re-enactment. In every case the personal imprint of the discoverer can be seen in the work of those who follow.

Because of this, whenever one works with new general ideas on an established field, one always has to work at two levels. First of all the old rigid form must be melted down and rendered malleable. All the "obvious" ideas, which are so very useful because one doesn't have to think about them, have to be thrown overboard, for only then can the work of re-design start. If, however, one is lucky enough to find a virgin field then this first part of clearing up, which is usually extremely difficult, is not necessary and one can get right on with the real creative part.

Electrochemistry. In physical chemistry I felt myself to be in this happy situation. The order that had been imposed on the subject had largely been imposed very recently by me and hence was not yet sclerotic. There was, however, one area that had been compiled by someone else and then recently reworked so that it almost represented current knowledge. This was the field of electrochemistry which was laid out in G. Wiedemann's comprehensive textbook on electricity. But what had here been put together had been sharply but fairly criticised by the British physicist Oliver Lodge, who was at that time the very active chairman of the Electrolysis Committee of the British Association. He wrote that Professor Wiedemann had formulated the problems associated with the electromotive forces operating in an electrochemical chain under three headings. If an angel came down from heaven and, with all its superhuman powers, answered all three questions then we would not be one whit wiser than we are at present.

Wiedemann resented this very much and said, "One sees, that he was a blacksmith in his youth". This was a reference to the fact that O. Lodge had come from a very simple background before working his way up to become a respected scientist.

The second edition of the textbook. A special opportunity now arose to carefully tease out and formulate the energetic basis of chemistry. The first edition of my textbook had been sold out in just a few years and, shortly after my move to Leipzig, my publisher suggested that I should start to prepare a new edition. This was going to be a large amount of work because since the publication of the first edition so many new results and basic concepts had come out that large parts of the book would have to be rewritten. The first part—stoichiometry—which I worked on during the course of 1890 doubled in size. It contained no mention of my new way of thinking as there had been no reason to recast the contents in this way. However things were different with the second part which dealt with chemical affinity.

I worked on this part from the end of 1891 till the end of 1892 and I regard 1892 as the year in which my thoughts on energetics really developed. One can see that in the preface to the new edition in which is written:

"One will find that parts of the treatment of energetics and electrochemistry are rather different from the usual way of presentation. I did not shy away from this way of doing things because I think that it offers the best way of making clear to the reader parts of the subject which are important but, as normally presented, difficult to explain. In any case the current development of the quantitative sciences drives one ineluctably to the basic idea which lies behind this textbook, namely that everything that happens in the world is merely a change of energy in space and time and that hence these three parameters are the fundamental concepts on which all quantifiable phenomena are based.

"In following this line of thought I had to change a considerable amount of material and in some places the argument remains to be rounded off and made more systematic. This drawback arose because writing the text induced a rapid development of my own views on the subject. I hope that the many traces of my struggles with outdated and unsatisfactory concepts will provide the reader with a better guide to the basic ideas than if I'd waited until the new concept had fully ripened so that the contents were more complete, albeit at the cost of making them less accessible and less vivid. Something of the excitement of the first understanding of the new way of looking at things will be obvious in some parts of the text. I hope that what it misses in terms of polish, it has gained in terms of forcefulness.

"A good example of the copiousness and usefulness of this approach can be seen in the principle of changes in virtual energy as applied to electrochemistry, which can be considered as a general practical tool. In the chapter on electromotive forces I have drawn heavily on the fundamental work of Helmholtz, van't Hoff, Arrhenius and Nernst and by doing so this century old problem can now be laid out in its entirety which, I must admit, I had not thought would be possible when I started this work. I have the strong impression that much of what I wrote might be improved but also that in the future, when the complete theory of the voltaic cell is developed, it will be found to contain the principle elements presented here".

This prediction was borne out. The textbook provided the first complete theory of voltaic cells and it was based on all published observations available at that time. Later it has been developed in depth and breadth but no fundamental changes have been found necessary in order to encompass results from particular experimental set ups.

It was particularly rigorously tested because many new experiments in my laboratory were initiated after applying the new viewpoint to old data. This was because since previous researchers had not realised what was really important they had often not noted important factors whose significance they had not understood. In these cases we had to think up and try out new experimental designs which would yield quantitative data on the variables required by the new formulae. The fruits of this work resulted in rows of numbers which everywhere supported and fulfilled the predictions.

I therefore think it's fair to say that by applying energetics to the organisation and codification of electrochemistry I achieved a result that has proven to have been of lasting value down to this day.

The nature of energetics. Were I asked how I see the main idea of energetics then I would answer that it is the recognition that together with the general concepts of order, number, size, time and space one must add above them the term energy as a general concept. Whether one wished to ascribe reality to energy is a meaningless question as one can see by trying to answer the question of how one distinguishes whether anything is real or not. If one satisfies oneself with the literal interpretation and considers everything that acts to be real, then one must come to the conclusion that energy—and only energy—is real. This is because energy is the only thing that must be part of every act and affects both the acted-upon as well as the actor.

One generally thinks of matter as real and when one hears this word one generally thinks one knows exactly what is meant. The pencil in my hand, the paper on which I write, the house in which I live, the planet on which my house stands: all these things are palpable and measurable matter, reality in its simplest form. But now it's got dark and I switch on the electric light; the tungsten filament glows and I can resume reading and writing. This light is no less real! I have to pay for it and it would not exist if the machines in the power station were not driven by coal. And the thoughts that originate in my mind and which, through the actions of my hand are brought onto the paper—are they not also real? They would not originate if a particular person had not had particular experiences which he'd then thought about. But they are clearly not palpable, measureable matter, just as light and electricity are not. But they can be conceived in thought, and thinking is also part of reality, indeed for a human being it is the most important part of reality.

In other words, quite apart from matter there are a large number of very important realities which do not fit into the general definition. However one can subsume them in the term energy.

In 1842 I.R. Mayer discovered the natural law of the conservation of energy which is often misleadingly referred to as the law of the conservation of force, though force is not always involved in our definition of reality. Energy, however, always is. The law states that in everything that happens, without exception, something is involved which does not change some aspect of its properties though the amount of the thing remains unaltered. One can mould a piece of clay into all sorts of shapes without changing its mass and this is a metaphor for energy. There is of course a law of the conservation of the weight and mass of matter which has been known for much longer, but now we recognise that this reflects only a part of reality. In contrast, the concept of energy covers the whole of reality, that is to say everything which we objectively and subjectively experience. Of course this includes matter which one can also define in energetic terms.

When one is asked what properties does energy have, then the answer is that it is everywhere, that it can be converted from one form to another but it can never be created or destroyed. Apart from that it can have any property you name for everything can be described as the result of the conversion of one form of energy or another.

What do we know about the world around us? We only know that which our sense organs experience and which we interpret by thought. However, if one of our senses is to deliver information to us then it is necessary and sufficient that it receives or transfers energy and in this sense is in energetic communication with the environment. The sound that we hear, the light we see, the pressure or warmth we feel, all of these are energies which alter the state of sense organs and hence can be perceived. In order to think our brain requires chemical energy delivered by the blood and this energy is used to fuel thought; that is to say it is converted into another form. The moment the blood flow ceases, thought is no longer possible.

The thoughts we think are not blood, just as the noise we hear is not vibration in the air. The blood and the vibrations merely provide a means of transferring the energy required for perception and mental processes. Since thoughts can never arise without the input of energy and since energy can neither be created nor destroyed but only converted from one form to another, it follows that perception and thought must be considered part of the cycle of energy conversion which is the source of all that happens in the world.

Difficulties of acceptance. Nowadays these ideas are, if not generally accepted, at least no longer considered strange or silly, but they were regarded as strange and silly thirty years ago when I first dared to conceive them and then faced the much greater peril of asking other people to consider them. At that time natural scientists considered themselves to be quite distinct from philosophers who they indulgently looked down on as useless but basically harmless people. This was because natural scientists had not forgotten the defeat which German science had suffered as a result of the errors of the "natural philosophers" at the beginning of the nineteenth century and they were determined to prevent a repetition of this ignominy by keeping philosophy strictly separated from science. Even such a rational thinker as Ernst Mach was considered a dreamer and no one could understand how someone who carried out such good experimental work could involve himself with what seemed to them to be mere philosophical nonsense (though Mach strenuously denied this description of it for as long as he lived). Indeed just before his death this accusation had blocked any possibility of his moving from Prague to other universities. I too was warned from all sides, even those who were scientifically close to me, that my attempts to define generally applicable concepts were regarded as an aberration which was all the more regrettable because I was credited with the ability to generate subtle and interesting experimental results.

General energetics. Once I'd demonstrated the usefulness of energetics in general chemistry, particularly in electrochemistry, I naturally moved on, to deepen and to strengthen the concept of general energetic which I'd only briefly sketched in the introduction to my textbook and to apply the concept more generally.

At this point I ventured into a dangerous area which normally would only be approached by researchers who could readily deal with higher mathematics. I knew I couldn't do this and apologised in advance to my colleagues for having ventured into an area in which I was not familiar. But since at that time no one else was willing to take this path, which I'd personally experienced to be so important for the future of science, I thought it my duty to go as far as possible even at the risk of being accused by the experts of fumbling dilettantism.

The first important conceptual step worked well. Gauß and Weber had shown long ago that the mechanical force of magnetism could be expressed in absolute terms and is dependent only on the units of space, time and mass which are arbitrarily chosen. Later on Maxwell went far beyond the German researchers' self imposed cautious constraints and proposed that all physical quantities could be described in terms of these three basic components. He had, in particular, done this for the units of electrical and magnetic parameters whereby the basic units were expressed in strange "dimensions" which involved fractional exponential powers that didn't make any obvious sense. Despite this, no one questioned the authority of Maxwell's unproven claim. Shortly before my work on energetics there had been a curious discussion amongst the leading mathematical physicists, particularly between Clausius and Hertz which came to the conclusion that if one applied Maxwell's premises then there was not just one such system of dimensions for electrical and magnetic values but four, all of which were different from each other and all of which were equally correct and equally applicable.

One gets round this problem by simply using whichever one of these systems one wants and when necessary transforms the data appropriately if one wants to change system. However in this way the huge advantage of an absolute system, namely its unambiguousness, is lost.

To me this ambiguity was an indication that the premises for Maxwell's system must be flawed, and that time, space and mass were inadequate to define the other parameters because, if they had been adequate, then the electrical dimension would have been uniquely defined rather than in four different ways as experience showed.

Furthermore, in every group of physical parameters there are those which belong only to this group and appear in no other. Temperature occurs only in thermodynamics, measures of electricity only in electric systems and so on. Such parameters can be expressed in independent units without disturbing the choice of units used for the other parameters. This too is contrary to Maxwell's premises.

The final conclusion was that there is only one parameter which is expressed in the same units in all areas and this is not space, or time or mass, but rather energy alone. If the mass units are chosen such that the law of the conservation of energy applies—and this Maxwell had tacitly done—then one can and must determine in each situation one specific parameter, whereby together, if necessary, with the units for time and space all other subsidiary parameters in the field are unambiguously defined.

So far as I am aware no objections were ever raised to this way of dealing with the fundamental problem of absolute values, though on the other hand a clear of acceptance of it was also missing. The internationally accepted definitions which were made on the basis of Maxwell's mistaken doctrine have not been reconsidered, though for all practical purposes the requirements of energetics were thus tacitly satisfied. This has resulted in the happy situation that there is no reason why anyone should publically declare that an error has been made. At some point after my death a suitable reason will be found to put the matter properly to rights.

The law of events. A second paper centered on the extension of the second law of thermodynamics, which till now has been restricted to events in which heat is exchanged. The second law states that heat never spontaneously moves from lower to higher temperatures and the extension of this law states that for each form of energy there is an equivalent parameter which will never spontaneously rise from lower to higher values. This value may be referred to as the "intensity" of the energy form. In the case of electrical energy it is the potential, for kinetic energy it is speed and so on. It was of particular interest to me to consider the intensity value of a chemical reaction. It turned out that W. Gibbs in his fundamental work had termed this centrally important value "chemical potential". The choice of this term which till then had been used only in the context of electrical or gravitational energy, shows that Gibbs understood that the these terms were fundamentally the same, though it seems that he did not expand on this idea.

This provided the possibility of extending the results of research on the second law, which so far had been restricted to questions of heat exchange, to the whole of physics, that is to say to all events, and thus to define the criteria that must be met in order that anything can happen. This would require that there are intensity differences in some forms of energy.

A close analysis shows that these conditions are necessary but not sufficient. There are (at least apparently) quiescent conditions in which nothing happens even though differences in intensities are present. For example, as everyone knows, diamond consists of carbon whose chemical potential is much higher than that of the oxygen in the surrounding air with which it could react so long as the chemical prerequisites are met. It doesn't happen at room temperature but will happen at red heat and goes faster the higher the temperature. Here we are dealing with questions of time and one could reasonably argue that the diamond does indeed oxidise even at room temperature but that this happens so very slowly that one would never be able to see a change in its weight over the course of an entire human life. Not even centuries or millennia would be enough and in any case balances capable of registering small differences in weight have only been available for around 100 years.

The two forms of perpetual motion machine. Perhaps the easiest way to view this is as follows. The "first law" or the law of the conservation of energy can be defined as: "a perpetual motion machine, which creates labour, or in general energy, out of nothing, is impossible".

On the other hand one can also imagine a perpetual motion machine, that is to say a machine that works for nothing, which does not extract the energy it requires from nothing. It could, for example, take its energy from the heat in the oceans for just a tiny fraction of that heat would be enough to run all the machines in the world. Why wouldn't that work? Because the temperature of the ocean would have to sink "by itself" or alternatively the heat would have to spontaneously move up to a higher temperature which is ruled out by the second law. Thus there is conceptually, in addition to the first sort of perpetual motion machine, which would contravene the first law, a second type which would contravene the second law. This can be simply expressed as: "a perpetual motion machine of the second kind is impossible". In this way the impossibility is not restricted to machines involving the flow of heat but rather to all imaginable forms of machines.

In the publication in 1892 these and other fundamental relationships were developed and presented and after all this time I can't see anything of significance that should be altered, though some of it might be rephrased in a briefer and more readily accessible form. Only one section should be removed and that is the ninth which deals with the derivation of the intensity value of heat where an error cropped up in the calculation. I particularly emphasise this point because it had considerable consequences.

Exhaustion. The development of the energetic concept had made great demands on my brain. This was because it involved only mental work and the welcome interruption with manual manipulations, which had characterised my previous experimental work, was entirely missing. As a result clear signs of exhaustion became evident during the course of the semester. I had the feeling that my head was full of cotton wool so that the usual automatic ignition of my mental facilities as soon as a new challenge arose no longer took place. The machine no longer produced solutions without any effort on my part. It either failed to work, or ran idle without producing anything, so that I had to wait until a productive period came round once more.

Though my mind at that time was full of the concept of energetics, I had ignored the energetic conditions of mental work and did not realise that these symptoms were physiological, as indeed they were, but instead thought that they were due to a moral failing and I made the mistake of blaming myself for my poor performance. My father used to tell us that anything could be achieved by will power and that lack of will power was "naturally" considered to be an ethical failing. Because of this I thought it was my duty to boost my will power, whereas my brain had failed for physiological reasons and the clear feelings of exhaustion should have warned me that it had been pushed beyond its normal limits. I think it must have been due to this overwork that my normally well functioning self criticism failed and let the error I referred to above (This chapter, p. 16) slip through—an error which my healthy brain would surely have detected. In scientific work one develops an instinct which permits one to clearly distinguish viable results from those which are not yet fully ripe.

Against atomism. The time that followed was filled with many activities which at times took me away from pure mental work but which also used my energy in other ways. The "Handbook", the "Analytical Chemistry" and the "Electrochemistry", whose contents and purpose I described earlier (Part 2, Chap 17, p. 181) were written and published in the years between 1892 and 1895, and on top of that came the routine work for the Journal and the new edition of the textbook. A new challenge came with the formation and leadership of the Electrochemical Society which I'll soon describe.

All this work prevented me from writing any more papers on energetics, but it didn't stop me continuing and expanding my efforts in this direction. In particular it became clear to me how much more productive energetics was than the almost sterile kinetic and atomism doctrines which, after a short flowering in the eighteen sixties, had fallen into the problem zone of mathematical thickets which removed its flexibility and prevented its adherents from pursuing new experimental paths. The many new insights, which atomic theory has established today beyond the mere level of hypotheses, were at that time not visible on the scientific horizon. And although the concepts of osmotic pressure and that of electrolyte dissociation were certainly viewed in atomic terms by their discoverers, there was in their entire field of application not a single case which centred on the single atom but only cases where atoms were considered as a general conglomeration. In other words there was no problem with discontinuities in the molecular range of 10^{-10} cm which were theoretically predicted, so that it was not thought necessary to take this seriously. One should bear in mind that X-rays, which initiated the new atomism had only been discovered in 1896.

Then again it seemed to me that the tendency to sweep real scientific problems under the carpet by proposing arbitrary assumptions about the status of atoms and their oscillations did science enormous damage. I saw it is my duty to do everything possible to reduce this damage by taking every opportunity to point out the barrenness of these so-called explanations.

Just as I had earlier been a convinced supporter of the kinetic concept of atoms, I now pursued with the zeal of the recent convert my new insight and tried to win supporters for it. This worked only to a small extent. Even close scientific friends like W. Ramsey refused to follow. They agreed that I might be right but explained that they would be unable to continue working on their problems without the current graphic concept of atoms and they considered my alternative concept too abstract, even if it was indeed correct.

The Natural Scientists Meeting in Lübeck. This back and forth argument drove things to a debate which took place in the autumn of 1895 at the Natural Scientists' Meeting in Lübeck. I proposed to hold a general lecture with the title "Beyond scientific materialism". When Wislicenus, who was chairman of the organising committee and who'd put my lecture on the program, got from me an outline of the contents, he felt his scientific beliefs to be so threatened that he looked around for ways to counter me. He managed to persuade one of my most illustrious opponents, Victor Meyer, to hold a lecture in reply. Meyer lectured to the title "Problems of Atomism" and it was a great success, though at this time he was struggling with serious symptoms of exhaustion due to overwork.

In order to provide enough time to properly discuss energetics, there was, in addition to my lecture, going to be an extended discussion round in the combined physics and chemistry sections. Naturally the lecture had to be first in order to lay out the fundamental ideas. Wislicenus, however, had changed the normal order of events and set the lecture on one of the last days, hoping that in this way it would do the least damage and would make a serious discussion of the contents impossible. I don't doubt that he believed he was doing this in the best interests of science.

In the discussion I faced a solid block of opponents. My only supporter was G. Helm who was professor at the Technical University in Dresden and who prior to me had attempted an energetic view of science which he had presented in a publication that underscored his great independence of thought. However he differed from me in that he found the idea that energy was part of reality distasteful and, as a result, we were only partial allies who had to be constantly on our guard against each other.

This was the first occasion that I'd met with such unanimous opposition, though later on this happened several times again. I never felt any depression or worry or fear during these various discussions and, as far as I remember, I left no question unanswered. However the fact that the lecture was after the discussions was not helpful and, as was to be expected, we parted without either side having convinced the other. Nevertheless I do believe that something was left hanging with some of the members of the audience from which they later developed their own lines of thought.

The lecture. The basic idea behind the lecture was to show that the mechanistic view of natural phenomena is inadequate and that the inadequacies can be removed by substituting an energetic concept for the mechanistic one.

I justified the first law on the fact that in all the mechanistic equations time (t) is always in the form of t^2 to ensure that the results are unaffected by whether time is given a positive or a negative value since $(+t)^2 = (-t)^2$. The result is that each event described by these equations can be run equally well forwards as backwards. However real events are undoubtedly not reversible because a man never becomes a child, and an oak tree never becomes an acorn. In other words these equations do not satisfy the simplest description of reality.

The second is based on the fact that everything we experience about the outer world is transmitted through our sense organs. A transfer of energy from the environment to a sense organ is necessary and sufficient for this to happen. In other words all we actually get is an energy transfer from the environment and everything else is due to our addition of subjective ingredients. And these subjective mental ingredients are, at the end of the day, nothing less than an application of energy in the brain that, in contrast to what the mechanistic view requires, does not necessarily involve changes in the motion of molecules in the brain.

None of the older terms such as matter, motion or power have the general and precise property of energy, and only energy has the enormous advantage that it requires no hypotheses and thus represents a scientific ideal.

Perhaps the clearest description of my viewpoint can be taken from the lecture, in which I said.

"Yet I can hear the objection that if the concept of atoms in motion is taken from us what is left to make a picture of reality? In reply to this sort of question I want to cry, "Thou shalt not make unto thee any graven image". It is not our job to construct a more or less distorted view of the world but rather to make as precise a description of it as our mental powers will permit. The job of science is to determine the relationships between aspects of reality, in the form of demonstrable and quantifiable parameters, so that when some of the parameters are known the others can be calculated. This goal cannot be achieved by setting up some hypothetical picture of the world but only by demonstrating the relationships between quantifiable parameters".

The centre of nature. The two side of the argument can best be illustrated as follows. In the discussion I said, "imagine an empty box from which two levers protrude at different positions. When one is moved the other also moves but at a different speed, e.g. three times faster. Energetics says that the power with which the second lever moves must be three times less than that of the first. The first law of energetics requires this result so long as no energy remains stored inside the box. One can think of a vast number of different mechanisms by which the two levers might be connected but for as long as the two levers are all that we have access to we can't say anything about which mechanism is actually present and speculating about it is a waste of time because we can't learn anything more about the levers than we already know".

My opponents protested that it might be an important scientific challenge to understand the various mechanistic possibilities because maybe one would indeed be able to learn more about the mechanism.

I replied that this was equivalent to the discovery of other levers which would have to protrude out of the "centre of nature" or the box and which one could and should then do experiments with.

And so the matter polarised into a contradiction which one might describe under the heading of science politics. And since politics is well known to be a field in which the weaker the rational basis for an argument the more passionately it will be pursued, it is no wonder that the discussions in Lübeck were not free of such feelings.

My comrade in arms G. Helm suffered from the tone of the other side to such an extent that he protested against the way he had been treated. He said he'd come to take part in a scientific discussion and not to the sort of execution which had been meted out to him.

For my part I saw no need to protest. It was for me an everyday experience that an opponent would found himself quite unable to even consider my point of view and so the fact that there were here a large number of opponents instead of just one did not particularly impress me. In any case many of the objections were about things that I'd disposed of—or believed I'd disposed of in my main lecture and so I needed only to refer to it. The whole business would have been much more fruitful if the normal sequence of events had not been stood on its head by the "tactical" games of the chairman.

For me this experience did indeed show up a few mistakes in my presentation of the general idea, yet at the same time it reinforced my conviction—which I retain down to this day—that in the main I was and am on the right track. And when I consider the way that science has developed since then I can see that it has swung round to my way of thinking, admittedly in a different manner from that which I had envisaged.

Later developments. A few months later K.W. Röntgen discovered X-rays and there followed the development of a new branch of physics through which the granular structure of matter, the fundamental idea of the atomism, was experimentally demonstrated. The field was thus rescued from its previous fruitlessness and it has unearthed a wealth of new data. In the course of this the atomic theory lost its purely hypothetical nature and became a part of experimental physics and chemistry. I have since then never tired of pointing out that my previous reservations about the usefulness of the atomic theory had been removed and that its practical successes could not be doubted on the basis of a disputed scientific status.

Energetics was not affected by these developments because, since it is the more general concept, it is not affected by whether atoms exist or not. If one was to ask how it has survived the struggle for survival to which all concepts in science are exposed, then I'd say it's done very well. As a result of the new physics the other parameters such as mass, which had till then been regarded as invariant, have had to let go this quality but the absolute law of the conservation of energy remains, thus demonstrating that energy is the only reality on which the development of science is built. The new theories now attribute mass, that property which had always been the defining characteristic of "matter", to energy. In this sense energy has supplanted the concept of "matter".

Immediate consequences. The lecture in Lübeck was a sensation which affected circles far beyond the academic one to which it was addressed. How it was regarded in academia can be seen from a letter of Victor Meyer which has in the meantime been published. "The idea was extremely interesting and I have seldom heard anything so strange".[3] The daily press took notice because of the title "Beyond Scientific Materialism". They suspected a spiritualistic turn and since in many circles from orthodox religion through philology and on to left wing circles in the humanities there was considerable fear of a scientific world view, these people hoped that in me they'd find a welcome fellow spirit from the enemy camp. From various hints it even seemed that I ran the danger of being awarded for this "act" an honorary doctorate in theology. However a few of the smarter members of these circles discovered in time that their point of view would not be aided but rather made worse by my ideas. The reason was that the problem of the relationship between mind and body, which scientific materialism had left unsolved and which the convinced proponent of materialism Bu Bois Reymond had solemnly declared to be fundamentally irresolvable and therefore part of the eternal mysteries of the world, lost in the light of energetics its unapproachable character since both mind and body were subsumed in the higher level concept of energy and hence the two must be innately bound up with each other.

Conclusion. The Lübeck Natural Scientists' Meeting can be seen as a critical turning point for me and for the views I propounded. In the first place it resulted as I described, in a lively discussion of these new ideas which thus became part of normal life and so contributed to the start of my public recognition as a philosopher with something to say. This was not unconnected with a public rejection of

---

[3]Meyer R (1917) Viktor Meyer: Leben und Wirken eines deutschen Chemikers und Naturforschers 1848–1897. Leipzig, Akad Verlagsgesell (In the series "Große Männer: Studien zur Biologie des Genies W. Ostwald ed, p. 294).

energetics which the physicists Boltzmann and Planck published in "Annals of Physics", though their motivations for this publication were quite different.

Boltzmann was a convinced adherent of the kinetic theory of heat which had prevailed and had become a most fruitful concept. For him it was important to show that one would get further in science with the kinetic theory rather than with energetics and the quickest way to achieve this was to deny energetics the right to exist. He based this on the proof that one of the mathematical formulae I'd given contained an error and the unspoken assumption was that if the equation was wrong then so too was, in all probability, the whole idea.

Max Planck had a rather different starting point. In a remarkable piece of work which he'd done as a young man he'd come very close to energetics, however he'd seen it as just a mathematical function with very interesting properties with which one would quickly come to understand natural relationships.[4] A few years previously Planck, Boltzmann, Hertz and I had a lively discussion during the Natural Scientists' Meeting in Halle,[5] during which Planck and I had argued against Boltzmann that in special cases—we were thinking of the laws governing chemical equilibria, a subject we were then both working on—that thermodynamics would lead more quickly to experimentally testable results than would the kinetic approach. We pointed out that despite great efforts in this area, including those of Boltzmann himself, the kinetic approach had not led to the elucidation of a single new law while pure thermodynamics had delivered many. Even for the derivation of well known relationships, kinetics required a disproportionate mathematical effort while thermodynamics reached the same result in just a few lines, and for this there were examples in Boltzmann's own work.

Though he didn't concede defeat, Bolzmann had at that time no real counter argument. Instead he emphasised once more his conviction that atomism was correct and finally said, "I see no reason not to view energy as being atomically divided.[6] For the first moment I thought that this was a deliberate joke in which he exaggerated his point of view and I laughed. But in my heart I was hit by the boldness of his thought and this impression was so strong that I remember that conversation till today.

If I'd thought a little more carefully over Boltzmann's remark then I'd have had to welcome it as a fusion of atomism and energetics. However at that time the objections to this were so much in the front of my mind that I didn't even want to consider this fusion as a possibility. That the thought remained in my mind bears witness to the fact that scientific concepts have their own life and once they worm their way into the mind of a researcher they exercise a logical imperative even when

[4]Most probably Ostwald refers to: Planck M (1887) Das Prinzip der Erhaltung der Energie. Leipzig, Teubner.

[5]64th Assembly of German Natural Scientists and Physicians, Halle/Saale, 21–25 September 1891.

[6]Ostwald writes "atomistisch eingeteilt".

opposed by the will which seeks a different direction. I no longer recall whether Max Planck made any comment to this remark. However his bold and idiosyncratic concept of "quanta" which he later used to explain in a completely different context the origin of radiation also builds in its own way a fusion of energetics and atomism.

In this way science turned out to be always and everywhere the bringer of peace, and the battles which take place can be regarded as the inevitable friction which arises as the ephemeral parts of the structure of a branch of science are shaken off in order to liberate the true core of a fruitful new idea.

Chapter 22
Overload, Breakdown and Recovery

Holidays and artistic problems. The first time that I felt exhausted by my work was in the late summer of 1886 in Riga. In the previous 7 months I'd completed the last and most difficult part of my text book while at the same time I'd had the strenuous experimental work together with Arrhenius together with my lecture load and teaching duties in the laboratory. On top of that there were family claims. I was 33 years old.

I've already recounted (Part 1, Chap. 12, p. 123) that the symptoms of exhaustion were entirely gone after 3 weeks painting on the Isle of Rügen, so that I found the hard demands of the Natural Scientists' Meeting in Berlin[1] and the inception of the Journal of Physical Chemistry (Zeitschrift für physikalische Chemie) that immediately followed not only tolerable but even enjoyable.

There then followed the last year in Riga and my first in Leipzig, both of which were packed with work. However, the sense of wellbeing that came form being able to swim free worked like a catalyst and kept all feelings of exhaustion at bay—the drawbacks I had still to meet. And so I thought more of my family than of me when after the first year in my new position the question arose of where to go in the holidays. Since Leipzig smells and is muggy in the summer I was determined to get away with my wife and children in the summer, or perhaps better, autumn holidays. Lectures were finished in the first week of August. From our time in Riga the seaside had become most desirable, even essential, for holidays and so I chose Rügen of which I had happy memories from my second journey. Of the places I'd been to then, Göhren seemed the most suitable, for it is one of the few places on Rügen where one can see the sun setting over the sea. The journey with the four children and their nanny was not without its ups and downs but it was a happy time as by and large the children were well behaved. Göhren turned out indeed to be the right choice. We took lodgings in a fisherman's house with a childless couple—he was very old and she was very young and pretty. The accommodation was clean and the children were happy to be able to play all day on the beach and in the sea.

[1]59th German Natural Scientists and Physicians meeting, Berlin 18–23 September 1886.

R.S. Jack and F. Scholz (eds.), *Wilhelm Ostwald*, Springer Biographies,
DOI 10.1007/978-3-319-46955-3_22

I settled into the comfortable routine that I'd got used to on the summer holidays in Riga and gave myself the feeling I was doing something by constantly painting. Nevertheless it was clear that from ten paintings only two or three really pleased me. Why this should be was unclear and that unsolved problem plagued me.

At that time I was of the opinion that if I painted a natural scene as accurately as possible then it must make a picture. Nature is always perfect and so if I managed to translate that into a painting then it too should be perfect and thus a work of art.

Of course it was very clear to me that I was only able to put into the picture a tiny part of the perfection of nature, but the pictures which pleased me didn't have any more of nature in them than the others, and yet were clearly a lot closer to the ideal of a work of art. In fact they were often simpler than the others. There had to be some reason behind this because it was even then my belief that everything without exception can be researched even if most things take a long time. However at the time I was so deep in my chemical work, which I saw as the challenge of my life, and therefore avoided any deeper scientific analysis of things outside chemistry which might divert me from my central purpose. For this reason I didn't pursue these thoughts, though then and later they constantly resurfaced, and I never dreamed that they would at some point come to be a major concern.

Vilm.[2] We did even better the next year. On my first trip to Rügen I'd visited from Putbus the little island of Vilm which lies half an hour's trip from the town and which has beautiful beech woods and just a few houses where the forester lives and where during the summer a couple of dozen holiday makers can be put up. Because of the wonderful trees and the varied nature of the coast the island had long been a magnet for painters who brought with them an air of artistic freedom and exhilaration. In the dining room there was a large painting which depicted well the forester and his caring wife together with a number of the better known guests who'd at some point been there.

Since the university holidays began so late in the year I sent my wife ahead with our children plus two nephews we'd taken on as well. They reached Putbus without any trouble but the crossing to Vilm on a sailing boat was threatened by a storm which was raging. Only once she'd been assured that the forester who was going to sail the boat was the best sailor for miles around did my wife and the children board and they were brought quickly and safely to their goal.

The other guests on the island had awaited their arrival with mixed feelings, less because of the storm and more because they feared that this crowd of children would disturb the peace. However, when 2 weeks later I too arrived on Vilm I found that all the guests were like one large happy family. The children had not only charmed away their fears but with their gay unselfconsciousness had fitted in

[2]The island of Vilm is now a nature reserve to which there is restricted access.

so well that they were passed as darlings from hand to hand and any occasional bad manners were overlooked.

I have happy memories of meeting the painter Eschke, the architect Hoffmann and the colour chemist Lehne there.

Journey through Germany. When after 2 weeks the approaching end of the school holidays made our return necessary it seemed to me that the time had been too short for me to recover as I needed to. Since I wanted to see a bit more of Germany, I put together a pleasant journey with the aid of my Bädecker. I went over Bamberg, Nuremberg, Regensburg, Munich, Chiemsee and on to Berchtesgaden where I planned to stay for a little longer. My companion on this journey was the philosopher Oswald Külpe who afterwards achieved prominence in his field and died as professor in Munich.

O. Külpe was a Baltic countryman of mine from Courland. At that time he was still a lecturer with an assistant's position in Wundt's group. However he'd begun to develop his own philosophy so that his relationship to his director became difficult. He was a regular guest in our house on Sundays, since my wife and I appreciated his hearty appetite and his excellent piano playing. He was 9 years younger than me, of middle height with blond curly hair and a small moustache. He had a changelessly pleasant countenance and a rather feminine manner. The children didn't like him because he believed for theoretical reasons that it was necessary to get them used to unpleasant experiences from an early age and so he successfully did this by examining them on their latest pieces of school homework. When he went so far as to take a comb from his pocket and comb first his moustache and then the hair of the dolls which my daughters had shown him, they treated him thereafter with nothing but rage and contempt.

For me he was a very pleasant travelling companion, interested in nature and never obstinate when it came to deciding the details of our journey. I'd warned him that that I'd probably bore him by sitting down to paint but he armed himself against this by bringing along a pile of new philosophical books which he had to review and he read them while I painted. A book by Windelband became a hobby horse for him for he had so much against its contents that he sometimes laid out his objections for me. With the happy feeling of having just sighted new horizons in science I had little time for things that I wrongly considered at the time to be nothing more than vapid invention. I could not imagine at the time that this sort of thing would considerably occupy me in the future.

To begin with we had marvellous weather and I can still remember much of the charming and splendid scenery we passed through. In Nürnberg I was especially pleased with the continuity achieved by designing the new buildings to blend into the old townscape. On the other hand I later regretted the time wasted looking at lots of things that didn't interest me in the German Museum[3] (Germanisches Museum). At the theoretical level I was convinced by the museum's strategy so that I was a bit ashamed that I didn't actually get interested in the collection. Today I know that my reaction then was perfectly correct because there is nothing worse and more tiring than looking at things when one is not in a position to ask questions about them.

From the outside the Walhalla[4] in Regensburg looked impressive enough with its massive stairs. Once inside it struck me as being the architectural equivalent of a large barn. The reason for this lay in the fact that the original wood construction, of which Greek architecture is merely a copy in marble (in Greece wood was, and is a rare and expensive material; marble is cheap and can be found everywhere) is so clear in the interior of the Walhalla that this thought sprang immediately to mind.

In contrast I was much amused by a plaque in the town with the following message:

I'm an old and well-known house
Proud to be known as the White Lamb
I really am to be envied
Here stayed Mozart, Heine, Hayden

This folksy form of immortality seemed to me far preferable to the marble form up there in the Walhalla whose name is hopelessly at odds with the pompous style of the building.

On the Fraueninsel in Chiemsee one meets either a lady or gentleman painter at every pace. I watched them with respect and as a rank amateur didn't dare to open my paint box, though the view over the water looked marvellous. We decided not to visit the royal castles. This was not so long after the death of the mad romantic Bavarian king[5] and at the Natural Scientists Meeting in Berlin I'd got to know his doctor Gudden who met his end on the same occasion (Part 1, Chap. 12, p. 124). To be honest he hadn't made a pleasant impression on me.

In Berchtesgaden we found clean and pleasant lodgings with an attractive landlady and her violently jealous husband. On long hikes we wandered thorough the beautiful surrounding countryside and I really suffered from my inability to paint it well so that I often left the paint box at home.

Berchtesgaden failed to live up to my expectations also in another way. Before setting out on the journey I'd suffered from insomnia and had hoped that the hard physical work would let me sleep better, but this was not the case, and I have since convinced myself that the reduced air pressure at these heights is the cause. Every time I've gone to the mountains for recuperation I've had trouble sleeping, while at seaside, that is to say at the lowest level of the air, I enjoy the deepest sleep. It took me a long time to hit on this because I hadn't learnt early enough to apply the rules

[3]Now the German National Museum (Germanisches Nationalmuseum).

[4]The Walhalla is a hall containing busts of famous people in German history. It was opened in 1842.

[5]King Ludwig II of Bavaria.

of scientific observation to my own situation. I know that in comparison to most of my peers I was way ahead and am shocked when I think how unscientifically most people treat that most important matter, their own lives. The real cause of this lies in the unscientific polarity of body and soul which has been ineradicably established by our "educated" class of Platonists, in particular lawyers, theologians and philologists.

Nevertheless I did get the recuperation I needed because for the entire 3 weeks of this enjoyable holiday my thoughts were kept away from chemistry.

From Berchtesgaden we went on to Rosenheim and Wörgl and climbed the Hohe Salve which yielded me one good picture in the evening and another the next morning. On the other hand I didn't even dare try to capture the beauty of the clear moonlight shining between the silver peaks.

In Innsbruck I enjoyed looking at the paintings in which F. Defregger had captured the events of the rebellion in Tyrol. The room in which these pictured hung held nothing else apart from the six or eight paintings, and because they were undiluted this made a stronger impression than any other collection that I have seen.

In the neighbouring gallery which shows the work of Tyroler painters I was struck how little the scenic beauty of their land had affected them; the only landscapes were very recent works. Ninety percent were pictures of saints and most of these were pretty repulsive. There was only one old picture I really enjoyed—though I no longer remember who painted it. Rather in the same way that a picture of a horse suffering from every illness would be shown to stable owners for educational purposes, so the painter had attempted to depict a saint on whom all possible forms of legendary martyrdom were being practiced. He sat in a cauldron of boiling oil and a red hot iron shirt was being lowered over his head. Executioner's assistants were hard at work poking out his eyes and cutting off his ears. He'd stretched out both arms so that his finger nails could be pulled out. Right in the foreground stood an executioner who was using a winch to pull the intestines out of his body and wind them onto the axle. In order to show the saint's disregard for all these matters the artist had given him such a happy expression that it was clear that he was triumphing over all these horrors.

From Innsbruck we travelled on to St. Anton which is the highest point reached by the Arlberg railway and we walked into the wilderness of the Moosbach valley where I managed to capture the feeling of the place in a painting. It was a wonderful walk and we were quite alone. In front of us the wild snowfields vaguely silhouetted against the dark grey sky and far below in the valley the thunder of a rushing stream.

To avoid the tedium of the journey through the Arlberg tunnel we hiked up over the pass. The sky was clear to begin with and clouds were hanging only on the peaks of the mountains, but the higher we came the more the blue sky disappeared. First it was fog and then a cloud and then rain which didn't fall down on us from above but seemed to come from every direction. We were wetter than ever before and on top of that there was a cold cutting wind. We made a forced march to our goal which was Langen where we managed to protect ourselves from any consequences of the cold by drinking hot wine to a hot dinner. By the time we got to Bregenz the weather had cleared up and there was a great sunset. The next morning we climbed the Pfänder and crossed Lake Constance to Ludwigshafen[6] which at that time was completely unaware of the historical significance it would soon achieve through Graf Zeppelin's great work.

Here we separated. Külpe returned to Leipzig and I, much refreshed and ready for new challenges, set out for the Natural Scientists' Meeting in Heidelberg which I have already described (Part 2, Chap. 19, p. 207).

Meran. Since I'd quite easily regained my interest in work I embarked on an energy intensive challenge and completed in the shortest time the "Outline of General Chemistry" (Part 2, Chap. 17, p. 175). Because of that I had to go off and seek recuperation in the next holidays.

One of the many stupidities of our way of dividing up the university year is that the Easter holidays come so early that they can't be used in Germany for relaxation outdoors. The many scholars who need to recharge their batteries are therefore forced to go south. Because of this the majority know Italy better than Germany and in this way we reinforce in the best minds of the country our biggest delusion which lies in our belief in the superiority of everything foreign.

So as to be at least in contact with people having the same customs and language I'd set the goal of my Easter journey as Meran where I wanted to meet up with my friend the mathematician Adolf Mayer (Part 2, Chap. 18, p. 201) who'd gone on ahead a couple of weeks before. My expectations were more than fulfilled. The beauty of the countryside and the different scenic motives that presented themselves certainly contributed to my recovery though they lay quite beyond my very limited artistic abilities. I particularly remember one long walk along a viaduct which at times challenged my head for heights.

On one of my shorter walks I had a pleasant surprise. I heard a voice behind me saying, "That's Ostwald, isn't it. That has to be him", and when I turned round I saw a very old man accompanied by two rather old women. I was greeted warmly and it turned out that these were Mr. Fromm, my first teacher from Riga who'd introduced me to humanity's wisdom, and his two daughters. He was over 80 years old but still happy and hale and clearly had more of the joy of life than his daughters both of whom had remained unmarried. One of them had taught me the elements of French 30 years previously. Back then she'd been blond and plump. I remembered the other one as being dark and thin and that we boys were afraid of her because she put us in our places by quickly inflicting pain if she saw the slightest reason to do so.

Fromm told me that the situation in Riga had become insufferable due to the Russification program. He had some savings and these together with his pension were enough to support him and his daughters in a modest way of life in Meran. I thought about that. What would have happened to me if I hadn't got the position

[6]Ostwald is mistaken. The town is called Friedrichshafen.

in Leipzig? Would I be able to enjoy such a happy and peaceful old age after 80 years work?

I also remember some short conversations with the excellent Berlin mathematician Kronecker. He was an old dwarf-like man with an agile close shaven face who treated me with so much respect that I was surprised and abashed. He must have had exaggerated reports about me from someone—I don't know who. He was particularly interested in the "Classics" series I'd published. This meeting had a not inconsiderable influence on the recovery of my drive to live and work because it showed that my numerous activities went in the right direction and that people profited from them.

After 3 week's spa-like life in Meran I was able to return refreshed and strengthened to my work which increased ever more in the course of that year. In the autumn I had the seminal discussions with the British chemists at the British Association meeting in Leeds, which marked the start of my personal involvement in countries outside Germany, and which was immediately before the Natural Scientists' Meeting in Bremen in which I was also involved.

Riva. Though I'd liked Meran in 1890 nevertheless I had missed views across open water which had come to symbolise holidays for me ever since my time in Riga. Scanning the map using the twin criteria of a holiday in early spring and one involving open water, I settled on Lake Garda whose northern end lay in Austria where the authorities did not object to the Italian population speaking their own language. Things were very different in South Tyrol when, after 1918 the area was ceded to the Kingdom of Italy and the German inhabitants of this wonderful country suffered heavily from the barbarism of Italian rule.

It became all too clear to me on this Easter journey in 1891 that my physical powers were declining. As usual I'd planned that the journey be unbroken. It started with a night train from Leipzig to Munich. Half sleeping in an uncomfortable seat I found my mind was in a whirl and my ability to quiet it down had gone. Unpleasant thoughts from all areas of my life from science to domestic matters chased each other through my mind without my being able to stop them. I have scarcely ever felt unhappier than on that night. Even the dawn was dreary. However the sky cleared and the sun shone as we came into Munich and that finally freed me from these demons.

I stayed the night in the little town of Mori where the by-line to Riva meets the main line and viewed the strange southern landscape the next morning. The journey on to Riva in brilliant sunshine was one of the most impressive I can remember and it finally pulled me out of my sad thoughts. The little train went at first through a rocky landscape that became ever wilder and more arid and ended in a huge rock fall where house-sized grey limestone blocks lay scattered all over the place. People say that Dante had taken this place as his model for the description of hell. The monotonous silver grey of the limestone was suddenly broken by the pure blue green of a little lake which appeared out of the middle of this wasteland. We travelled through more of the wilderness and then through the meagre village of Nago, after which a further blue expanse suddenly broke into view above the beautiful lake. It looked liked a fiord hemmed in by high cliffs which in the distance lost itself in an endless plateau.

I sucked in the beauty of the scene with all my senses and it chased out the last of those dark demons that had assailed me. The railway descended in broad curves into the fruitful lowlands of the river Sarca and on past the quaint castle on a rock till it finally reached its terminus in the little town of Riva at the northern end of Lake Garda. Because of the high mountains the sun sets there at 3 pm so I'd taken lodgings a few kilometres to the West where the sun shone for 3 h longer.

Here also, hiking and painting were excellent medicine. The only problem was that I had not expected to see and be able to paint so much blue. The ultramarine I'd brought with me was soon all gone and I had to send for new supplies.

I'd edited my text book for the second edition and this had involved practically rewriting the whole thing. After a week of holidays I could report that the exhaustion I'd felt when I arrived had changed to a state of complete absence of thought which seemed to me to be a sure sign of recovery. I filled the long evenings by developing one after another the photographs I'd taken during the day. I'd concentrated on taking pictures of the local donkeys because their expressive play with their ears had so impressed me.

By the end of the second week I was completely recovered and could return home ready for work. It seemed to me to be a waste of time to be doing nothing even though the weather was perfect and the trees were all in their freshest green. But equally tempting were the buds of energetics which in that same year would open to reveal the first leaves and I was impatient to be getting on with this enormously appealing work.

For this reason I made do in the autumn with a short holiday journey through the Erzgebirge which I made in part together with my colleague Bruns (Part 2, Chap. 18, p. 198). The weather was wet and it soon drove me home.

Torbole and Shierke. Because I'd enjoyed it so much, I spent the next spring at Lake Garda again. This time I'd brought a sufficient supply of ultramarine with me as well as my eldest son who was now old enough for this. This time I stayed in Torbole, a village at the North West edge of the lake about 1 h journey from Riva. I'd come across it the last time on a walk from Riva and it had supplied me with a large number of scenes for pictures. When I now look back through the pictures from those years I can see that I made real progress there. I think it had to do with the fact that the things which were completely new the first time round had now lodged in my unconscious mind so that this time I could get right into them without any delay.

The autumn of 1892 was filled with things that distracted me from lecturing and research. At the beginning of August I travelled to Edinburgh immediately after the end of the lecture courses to take part in the British Association meeting there (Part 2, Chap. 20, p. 223). The few short days with Ramsay's family on Arran hadn't been enough time for recovery and so, on my return, I went to Schierke in the Harz Mountains so as to be at least close to Leipzig. Though the place was crowded I still managed to get some peace and quiet because the tourists got diluted out in the extensive surrounding woods so that as soon as you left the village you were on your own. My attempts at painting soon showed me how much more difficult it is to overcome the difficulties of painting a German landscape even though one had learnt to deal adequately with the easier scenery in Italy.

While I was in Schierke the last great cholera epidemic broke out in Hamburg, though it was quickly brought under control. When I checked back I realised that on my return from Edinburgh I'd passed through Hamburg on the day that the first cases were recorded. Although there was scarcely any danger for Leipzig I ended my holiday as soon as possible so as to be back with my family. Probably the short holiday had not been sufficient for my recuperation because after this I felt exhausted not only at the end of the semester but already during it.

Boltzmann and Lohengrin. In the spring of 1893 I set off for a painting holiday once again at Lake Garda, this time in Gargnano. The place was of special interest to me because the founder of electrochemistry Johann Wilhelm Ritter, whose work I was dealing with at that time, had recruited the water diviner Campetti from Gargano during his last period in Munich in 1807 and Campetti played a special role in his work. On top of that the map suggested that this would be a particularly beautiful area where the mountains started to drop off down to the plain. I could finally look forward to peace and quiet because the Bädecker travel guide said that there was only one hotel in the village.

On the way there I broke my journey in Munich to see some paintings and visit a few colleagues. I particularly wanted to see Ludwig Boltzmann who'd recently moved from Graz to the university in Munich and whom I rated highly both scientifically and as a friend. Although it was semester holidays he was still at home and he was clearly happy to see me. His wife later told me that he had often mentioned a special affection for me. I asked him where we might go to chat that evening and he replied, "Oh dear, I've arranged for this silly dinner party tonight, so I can't go out". During the course of the conversation he kept coming back to this point until finally his wife said, "Ludwig, why don't you just ask the professor to join us?" Boltzmann looked astonished and said, "I hadn't thought of that—it's so easy! Can you come?" I laughed and agreed.

The evening was very animated. I met there the old physics professor Lommel, the musician Kienzl who later composed the "Evangelimann" and a number of uncommunicative painters with their wives. Kienzl's wife sang some of his songs which pleased me very much. With Lommel I talked about the idea of bringing out together with Budde a physics textbook based on energetics. "That's a huge amount of work", he said, "but if anyone can do it then you can". That was nice to hear. The lively discussion turned to Richard Wagner and among other things I criticised the logic of the Lohengrin drama. "How could Wagner make the key to the whole story centre on Elsa asking the one question which her saviour had forbidden her to ask?" Mrs. Boltzmann replied, "I'm not sure if your right. But I do know that I would have asked the same question."

From that point on my respect for Wagner as a psychologist of women rose considerably.

The umbrellas. When I arrived in Gargnano late in the afternoon I was rather disappointed. The hotel was small and clean but the village itself was filthy with badly cobbled streets whose corners were all decorated in the best Italian manner with the residues of human metabolism. Not even dead cats were missing. It struck me that these national characteristics became evident directly beyond the political border. I hadn't seen anything like that in the villages on the Austrian side despite the fact that they were peopled by the same Italian speaking race.

At the little harbour I clumsily let my umbrella fall into the water. I promised a lira to whoever could get it out and that started a real carnival, because the village youth all came running and deafening shouts accompanied the efforts of several men to rescue my umbrella. When I opened the rescued brolly and twirled it to shake off the water all the joints of the stupid thing broke. I passed the ruins to the boys to see if that would get them to make even more noise. It worked.

This umbrella story was just one act in a longer drama. When I left Leipzig my wife had got me a cheap umbrella, because, as she said, you'll leave it lying about somewhere. She was right, and in Munich I had to buy a second one because the first hadn't left the train with me. The second one, however, quite independently journeyed on to Chiasso, while I got out in Mori. A third one, bought in Riva where I waited for the steamer to Gargnano, went the way of all the others. I actually managed to bring back to Leipzig the fourth one, which I bought in Gargnano. To be honest, it wasn't worth it.

This story had a sequel on my birthday in the autumn. My children appeared one after the other dressed up as the wild inhabitants of distant lands, all the time reciting mumbo-jumbo which was supposed to be their language but which sounded just sufficiently German to be understood. Each of them presented me with an umbrella which they said I'd left on my last journey to their country.

As I got to know the surroundings of Gargnano I changed my opinion of it completely. As soon as one left its streets one was in the most wonderful scenery of mountains and lakes. I have seen no other area composed of nothing but "scenes" (in the painting sense) than this. I busied myself with my much loved painting hobby and in this way bagged as many pictures as my father used to bag ducks on his hunting trips.

Fraunhofer. The autumn holiday of 1893 was one of the best that I can remember. It was already late in the year when at the beginning of October I travelled via Munich to Tölz and from there hiked via Benediktbeuren to Lake Kochelsee. Benediktbeuren was special for me because it was the place where the optical genius Fraunhofer had carried out his fundamental work.

In Heinrich Zschokkes "Ruminations" ("Selbstschau") he recounted his third journey to Bavaria (he had written a history of the country and was in contact with minister Montgelas). In 1817 he wrote: I went with Utzschneider to one of his properties—the one time monastery of Benediktbeuren where he showed me his drainage works, plantation, tobacco factory and glass works. In talking to the foreman of the glass works I forgot everything else for he astonished me with his experience and knowledge of material elasticity, refraction, colour dispersion and so on. Most of it was new and when I shyly expressed some doubts he immediately countered them with an experimental demonstration. In Munich I'd never heard of this amazing genius—no one had heard of him. This was the scientist Fraunhofer. I asked him to publicise his discoveries but he replied that it was just everyday stuff that came out as a by-product of his attempts to improve his optical instruments. But, I cried, this "everyday stuff" is important for science—perhaps more important than your glass work! He smiled and obviously didn't believe me. In Munich I talked excitedly in the presence of several academics about this man who would be a worthy member of any Academy. They just laughed in disbelief. Finally the famous Sömmering and Schlichtegroll suggested an outing to visit the shaman of Benediktbeuren. Later when I heard of his election to the academy and came across a paper on the determination of refraction capacity in its publications I rejoiced more than a little over my victory over the laughing disbelievers".

Bavarian mountains and lakes. I didn't know this story at that time, though I had known that Fraunhofer came from a very poor background and that he'd worked his way up. I'd made sure that his fundamental work was available to everybody by publishing it in the "Classics". Though I had the greatest respect for the clarity of his thought, I just satisfied myself with a glance at the place he'd worked because I never could accept Goethe's judgement that the place where a great man worked is somehow special. To be honest, I find that making a cult of such places is not only primitive but frankly childish.

In the afternoon I arrived hot and tired in Kochel and then saw such a wonderful sunset that I got out my paints and easel to make a picture. The days that followed were spent on the way to and at the Walchensee and they were among the best that I ever experienced. There is something slightly melancholy about the scenery there which went straight to my heart and for a long time I thought this might be the right place for me once my time at the university was over. I never saw the Walchensee again and I imagine that it lost much of its charm when the hydroelectric plant was built there.

From Walchensee I wandered through the endless beech forest to Eschenlohe and drove from there to Partenkirchen. The Badersee and the Eibsee provided me with many new motives for painting—one more beautiful than the next—and the weather provided highly dramatic backgrounds for my landscapes. Once I sat in thick fog painting a small lake beside the Eibsee called Frillensee behind which the Zugspitze[7] majestically towered, though I could see nothing of it. The fog turned to rain and then to a thunder storm. This didn't drive me from my painting because I had a good raincoat and already in Berchtesgaden had learnt to paint in the rain. The thunder storm moved on as fast as it had come and suddenly there stood glinting in the evening sun the huge form of the Zugspitze which was covered right down to the bottom in fresh snow. It shone pink against the blue black clouds of the departing storm. I never saw anything half way as imposing in Italy.

[7]The Zugspitze is the highest mountain in Germany.

With this impression locked in my heart I started the new semester in Leipzig and developed the major part of my thoughts on energetics.

The Riviera. As the long hard winter semester drew to a close the necessity of relaxation became ever stronger. This time, in 1894, my choice fell on the spa Santa Margherita near Rapallo on the Italian Riviera which some of my Leipzig acquaintances had highly recommended. I travelled there with my eldest daughter who was twelve then and old enough to look after herself if necessary.

The mix of sea and rocks and plenty of green plants indeed offered me lots of challenges to my painting skills, particularly since I'd decided to do justice to the new challenge by doubling the size of my sketch block. I soon discovered that this required a pretty large change in my technique of dealing with details and was soon happily immersed in the task. In this way I did manage to clear away the mental fog which had fallen over the normally clear horizons of my mind. There was another symptom of the exhaustion that was overtaking me and that was that I had started to complain about it, while up till then I'd kept quiet because I felt embarrassed.

From that time I remember that a young man with dramatic theatrical gestures would set his white horse to gallop or dance down the street. The ladies all worshipped him and when I laughed they said, "That is a magnificent poet. He's called Gabriele D'Annunzio." When I tried to dip into one of his books I had to laugh even more over "Gabrunzio".

The North Sea. For recuperation in the autumn I went to the island of Amrum on the North Sea, though this was not terribly successful. While the Baltic Sea always calms and strengthens me, I felt the exact opposite effect here. I became fretful and nervous and once when I'd gone for a longer swim I felt sick for days on end. Only painting slowly helped me back on my feet. I started the new semester with insufficient reserves and had to face the considerable work involved in my history of electrochemistry. The plans for the new laboratory building were also a lot of work.

Lago Maggiore. In the spring of 1895 I went for a change to Locarno at the northern end of Lago Maggiore—a place which now, 30 years later, has become so well known. It lies at the foot of high mountains and the mixture of mountains, plain and lake promised a rich booty in terms of scenic motives. In fact I found lots of marvellous scenery and painted with increasing success because the depiction of the essential elements of such scenes became ever more easy and obvious. I no longer had to really think about having to do this or that to achieve an effect but rather the whole picture more or less painted itself. When one has reached this stage painting is really a great pleasure.

I'd arrived completely exhausted and needed longer till the feeling of pressure in my head disappeared. Because of this I seriously thought about giving up my professorship and spending the rest of my life in pleasant surroundings as a private scholar working on the sort of problems where I was not dependent on collaborators, or the support or ill-will of others. I looked around in the area for some suitable spot because the variety of the scenery promised to satisfy my painting needs for many years.

In looking around I was particularly interested in an old building down by the shore which was known locally as the "casa ferrata" because of the thick iron bars on the windows. According to the Bädecker travel guide it had been used in earlier centuries as a storehouse where the equipment of the various Swiss Guard units in foreign countries was stored before being sent on and at this time only a few farmers who worked in the nearby vineyards lived there. However, I abandoned this idea because although Switzerland is politically stable enough to be counted as a constant element in life, placing oneself in a population which speaks another language and has quite different customs would be less than comfortable—particularly as one grows older.

After I'd painted everything I could see there I took the steamer to the Italian side and set myself up in Pallanza. The landscape was sufficiently different from that on the other side that my painting took off again. I didn't visit the Isola Bella, which was right in front of me, since I can't stand theatrical things. Instead I visited the granite quarries in Baveno to look for twinned feldspar whose structure is defined by the "Bavenoer law" I'd already learned about in Dorpat. I didn't find any. However what interested me there was to watch the workers splitting granite into thin boards by using a set of chisels which were methodically hammered. The granite boards were used to enclose fields and gardens. I made a lot of sketches here as well.

Preparations for Lübeck. The main lecture, to be titled "Overcoming Scientific Materialism", which I proposed to hold at the Natural Scientists' Meeting in the autumn was already keeping me busy. For me this was a step whose significance I felt rather than knew: the move from a particular field of science into what one might call general science or philosophy. I am perfectly sure that this change of direction took place as a natural result of the development of my mind for I never had the feeling that I was standing at a crossroads and had to decide to go either left or right. On the contrary, I never for a moment doubted in which direction I would go and any thoughts on the matter I might have had were concerned only with the speed at which I should, or could, move.

I wrote the first draft of the lecture in Munich on the way home from Lago Maggiore. I no longer remember were the table was at which I sat and dashed off the first half of my presentation. However I distinctly remember feeling elated, not as strongly as when I'd had the first conception of energetics, but certainly something of a similar order.

My illness. I have already related what happened at the Lübeck meeting. That would have been enough to throw any man, even one at the height of his powers, off balance. I, however, since the start of my professorship in Riga in January 1882 had been involved in creative work whose volume and complexity kept on growing and as signs of exhaustion had become apparent I had only done what was necessary to cover up the evidence of my decline. Now the reserves were completely exhausted and I suffered the inevitable breakdown. Sleepless nights, depression which I couldn't master, an inability to work and lack of concentration—in short all the usual symptoms of an overuse of the brain—surfaced at the end of 1895 and made me wretched. I thought, like everyone does in that situation, that my scientific career was at an end and I didn't know how I'd manage then to give my life some meaning.

In a consultation with my colleague, the psychiatrist Flechsig, I learned the physiological basis of this condition and gained the hope that it could be cleared up. He told me that complete rest and avoidance of all scientific work were what was necessary. A condition like mine was quite common amongst scholars and in most cases could be completely remedied.

I therefore applied to the responsible government agency for a leave of absence over the upcoming summer semester and this was promptly granted.

Bordighera. Once the winter semester lectures were finished in March I travelled to the Mediterranean, to get as much sun as possible. I stayed in Bordighera just because of the "seven palms by the seashore" mentioned in J.V. Scheffel's expressive poem. To fill the many empty hours that faced me I took my paint box and a large stock of paints as well as my photographic equipment.

I had no trouble entering into a sort of vegetable existence without any serious mental activities which had been prescribed. The weather was good and, since the doctors had at least not forbidden sunshine, this made possible long and enjoyable walks. Painting, which I readily took up again because it involved parts of the brain that were not exhausted, soon occupied me for most of the day. Once I'd either painted or lost interest in everything in the immediate neighbourhood, I took ever longer walks in search of new landscapes and this was entirely consistent with the medical advice I'd been given. Usually I'd make a photograph of the scene I painted and this allowed me to see the errors I'd made in my free hand sketches. In this way I gradually got better at laying out the special relationships of the scene and found this to be a substitute for the forbidden scientific work for which, to be honest, I felt no great desire.

I didn't see any need to involve myself with the other guests in the hotel, most of whom were Germans who made a pleasantly quiet impression. I do remember a couple of conversations with the poet and author R. von Gottschall who at that time was a very old man. However, probably because of my vigorous realism we weren't really on the same wavelength and found no common ground. He was a small plump but lively man with either dyed hair or a wig who carefully looked after his moustache and beard. Otherwise he was somewhat unkempt, with large sacks under his watery eyes. He often stared out into nothing.

A little later I met my colleague and neighbour from the institute Leuckart who'd brought his wife and sick daughter to Bordighera. Despite his advanced years he was hale and hearty and was happy to join me on my long walks. To this day I blame myself for not taking care to spare him overexertion from which, however, he seemed to quickly recover. With his effervescent manner he soon had a number of friends amongst the hotel guests and he led them out to taste the Asti spumanti a light and pleasant sparkling wine which was served in several inns in the neighbourhood. I must admit that I didn't exclude myself from this and it didn't do me any harm.

Later I met another colleague from Leipzig, the geologist Credner with his family. He too was a happy fellow not much older than me, but we weren't together much. Since he'd married well he was the richest professor in Leipzig and he lived in the appropriate style, though in personal interactions both he and his wife were down to earth. However, I remembered why I had come to Bordighera and kept to my routine of getting up early and going off on lonely walks with my paint box.

Freshwater bay. Early in May I completed this first part of my recovery. In the meantime I'd grown bored with the Italian landscape and the eternal good weather. It reminded me of the schoolboy joke, "Who laughs over Italy?" and the answer was "an eternal blue sky".

I travelled home and consulted with Flechsig who seemed pleased with me and suggested that I spend the spring on the Isle of Wight where the climate would make it possible to spend all day outdoors. I went to a little spa called Freshwater Bay on the west of the island where I put up in a clean quiet room in a simple temperance hotel.

I particularly enjoyed this visit. The weather was mostly sunny and warm but, in contrast to the dreary sameness on the Riviera, every day had different light and clouds, and a different sunset. The sea with its strong tides was much more varied than the Mediterranean with its unchanging beaches and because of this the challenges for a painter were greater and more varied. I plunged happily into this work.

Whitsun. A visit by W. Ramsay from London during the Whitsun holiday gave me a welcome break in this lonely, but well-filled routine. He knew about my illness and was surprised but happy to find me apparently completely healthy. Since he kept to the doctor's orders to avoid scientific matters we spent several happy days talking about other things and this formed the basis of a long lasting close personal relationship with this brilliant researcher. Our friendship was torn apart by the World War when feelings ran high and Ramsey developed a fervid hatred for everything German. He was one of the many victims of the lying propaganda—worse than poison gas—which our enemies used against us. Given his knowledge of German life, this inability to withstand these slanders whose lack of truth should have been obvious to him, was perhaps due to the illness which soon led to his death.

The start of recovery. After in this way staying away from science for 3 months, I made a first attempt in Freshwater Bay to get back to work. I got some manuscripts which had been submitted to the Journal and were waiting for a decision sent. The first of these was from Trey who'd been my assistant in Riga and with whom I'd been to school. At that time his handwriting had been impossibly bad and he'd been put in the care of an itinerant writing master who'd promised to get anyone writing well inside of 14 days. Since it didn't cost much, Trey's parents took up the offer and the results were amazing. His handwriting was indeed made sure and legible.

What is even more astonishing is that this remained so for the rest of his life. Unfortunately I never asked him for the secret of this success.

I started off with this well written manuscript because it was the easiest to read and hence to judge. It was a beautiful early summer day. High tide was past and the slowly receding sea laid bare a deposit of clean washed flint stone which soon dried in the sun. I made myself comfortable and read the manuscript through deliberately more slowly than normal. The usual stream of thought now filled the long dried out river bed and I was overwhelmed to see how the usual mills of thought went back to work. Not only could I grasp the content, which wasn't terribly complicated, but it was also immediately clear to me what the next step must be to push the work beyond its present limits.

I was happy with the outcome of this first attempt. Though I'd never really wanted to believe that I'd have to give up my science for ever, I'd had to take this possibility seriously. Now I could see that I was dealing with exhaustion rather than with the complete destruction of my brain and that the rest that I'd given it was bringing it once more back into service.

I avoided reading the other manuscripts right away and prescribed myself 2 days rest which I spent walking and painting. After that I undertook some more rather harder work and that too went well. Now I could look to the future more calmly.

In order to ensure the success of my recovery I visited the other beauty spots on the island in between times and at the end in the company of my wife who'd come to take me home. I felt that I was cured. The long autumn holidays now lay between me and the time when I should resume my work. Because of this extra pause I was sure that my recovery would be complete. Physically I'd been completely fit during the entire time.

After the treatment. The childrens' school holidays began just as my stay on the Isle of Wight ended and we spent them at the seashore in Heykendorf near Kiel. I was already doing real scientific work again for I took a good part of the papers that had accumulated for summary in the Journal with me. Here too I could see that not only was I once again able to swim freely in the flow of scientific developments but that my ability to effortlessly generate new thoughts based on whatever stimulated me was restored. I even started to make plans of the next books that I would write. This was the time that the idea for the "Fundamentals of Inorganic Chemistry" took place though the book was only published 3 years later. This holiday freshened me up in another direction as well. Aloys Riehl was at that time professor of philosophy at Kiel and the idea of energetics, particularly the form that I represented, had made a considerable impression on him. In our meetings he accorded me a much higher degree of respect for this work than I, after generous evaluation of my own efforts, would have thought to be appropriate. Since the emphatic rejection of this scientific idea by respected experts in the field had contributed in no small way to my depression, this quite different assessment worked like balsam on an open wound. And even if this highly esteemed colleague later considerably qualified his enthusiasm I remain indebted to him for the boost he gave me then.

Work in the laboratory. When I returned to university work in the autumn of 1896 I, together with my co-workers, had to go back into the old institute which was in part made up of the rooms in the basement which had been the janitor's living quarters. However everyone did their best despite the numerous shortcomings because we were soon to move into the new building. This was all rather hard for me to take and I seriously wondered if the new building had not perhaps come too late and whether it would be able to fulfil its function. However, my colleagues and I spent the next 10 years in it and our output was not less than in the old building.

The question of whether I would still be able to carry out experimental work worried me quite a lot, because the long pause, unlike my earlier short breaks from the laboratory, had not given me a craving to return to work at the bench. Nevertheless, it was important to me to find out whether I'd be able to do such work again. While working on the third volume of the text book I'd come across some problems which concerned the properties of solid materials. These were things which had been pushed into the background by the concepts of osmotic pressure and electrolytic dissociation—which of course apply only to liquid solutions. Solids were only considered in so far as they were associated with liquids. I'd already seen when writing the first edition of the book how poorly this area had been treated up till now. Because of this, I carried out investigations in the winter of 1896/97 into the area of the solid, that is to say crystalline state which threw up many new questions and involved the development of new experimental methods. The main result was that there is a minimal value for the amount of a solid below which the typical properties of that state, in particular the ability to induce crystallisation in super saturated or super cooled solutions were no longer evident. This limit lay between 10^{-10} and 10^{-12} grams, which is also roughly the limit at which a solid is visible under the microscope. Quite independent methods led to the same result. Beyond this there were other new data all of which added up to an extensive and interesting piece of work.

In this way I was able to show that my creative scientific streak was still intact—and this was the most important result for me. I was thus able to put aside the worry that I'd lost the ability to provide my students with projects and the means to attack them. In particular this work had started for me the new and fruitful field of catalysis in which I could employ as many co-workers as I wanted. We will come to catalysis by and by.

The loss. However something less welcome also became clear to me during this work. My earlier unlimited joy at bench work was no more. Of course, I told myself that this was a normal symptom of getting old. At that time I was 43 and I remembered an essay I'd read a long time ago in the "Gartenlaube" which was titled "The 40 year sickness" which suggested that particularly busy men suffered a first great disappointment and disgruntlement around their fortieth birthday. The things they'd been engaged in up till then now begin to seen empty and pointless. Success, which up till then had spurred them on, now no longer motivates and a general grey haze spreads over the whole of life. This is a common and almost normal part of personal development which luckily can be overcome.

In my case I assumed that this had been made worse by exhaustion from overwork and that this had led to the depression which I had now overcome. These considerations also taught me that what had happened had a physiological basis rather than being a moral lapse as so many people who lack a scientific perspective assume.

Historical examples. The historical research I'd been doing on the development of electrochemistry, which I'd just finished, provided many examples of busy and happy researchers who as the years pass lose their passion for bench work. At forty Liebig wrote, "My writing work has separated me from work at the bench for which I now lack the necessary patience". Even the prospect of a joint work with Wöhler, which a few years previously he'd have regarded as a piece of great good luck, was no longer able to keep him at the bench for any length of time. True, a couple of years later they did publish a joint piece of work—but it was the last.

When Wöhler was 46 he wrote to Liebig, "So, you are tired, tired of chemistry. I'm relieved to hear it. You can't begin to imagine how fed up I am with chemistry or how organic chemistry nauseates me and bores me so that I start to yawn when I think about it. Are we already so old, or what? This tedium must be really something special to chemistry. I think the materials we use, the vapours, smells and all the stinking chemicals are largely responsible for it. It's especially the practical work that gets you down".

At the same age Liebig wrote, "Since I got back to Gießen I've been feeling rather miserable. When I'm somewhere else I'm healthy; I sleep well and can eat whatever I want, but all that disappears as soon as I get back into the office or the lab. Then I have stomach problems, have sleepless nights even when I have no work. It might be better to be bored in Italy rather than to go to the dogs here. Sometimes I almost wish that the whole business would come to a halt and then everything would be fine. The work with the young students used to be a joy to me but now it is just a pain: answering a question or giving advice now makes me feel ill".

Wöhler answered, "You write like a hypochondriac. To be frank I'm no better off and practical work drives me to despair. What we lack is youth. Our machinery, like an old watch movement, wears out a little bit more every day.

Farewell to the bench. Similar things happened to me. After the work I mentioned above I turned to studying the rate at which crystals grew from a super cooled melt in a capillary. Since I didn't quickly come to any simple and clear results I gave it up and even later didn't go back to it.[8] This led to a nightmare which plagues me from time to time since then in which I've started something and then given it up with the feeling that I was now good for nothing. When I woke I'd find that there was some physical cause such as a blanket askew or an uncomfortable position which had masqueraded as a moral problem.

---

[8]Ostwald's failure is not surprising. Nucleation is difficult to control and the growth kinetics are complex. A theoretical background was developed only much later.

I got a little solace from a different experimental work I carried out in 1899. W. Hittorf had noticed some curious properties of a sample of metallic chromium which he'd got from H. Goldschmidt. He gave me a small sample so that I could look at it myself and I found that when it was dissolved in hydrochloric acid the release of hydrogen soon stopped and then started again vigorously without there being a change in any other properties. When I looked more closely it became clear that this behaviour was periodic and I showed using a stop watch that the periodicity was constant.[9] This spontaneous periodicity fascinated me because I'd already in another context—the periodic precipitations in gelatine discovered by R. Liesegang and known as "Liesegang's rings"—come across the question of how periodic behaviour can be generated in a situation where all the experimental conditions are constant. At that time I had a reasonable explanation but it was tailored to this particular problem and did not provide a general explanation of such events and such an explanation was now clearly required.

The first thing that had to be done was to decide on a procedure with which the necessary measurements could be carried out precisely and in the minimum possible time. I couldn't bring myself to burden an assistant with making the boring measurements and recording the results. On thinking it over I wondered if the phenomenon might not record itself along the lines of the principle introduced into physiology by C. Ludwig. I was conversant with the equipment he'd used from my visits to the physiological institute. I soon found the solution which was to use an elastic capsule whose movements could be recorded by a lever holding a pen that drew a line on a moving strip of paper. The necessary pressure difference was achieved by using a capillary to slow the outflow of the hydrogen. In a short time an apparatus was designed and built which with little effort was able to record six parallel determinations run for as long as one wanted and deliver the results in graphic form.

Once again I felt happy at having found a good solution to the technical problem. As far as the science went, however, though we found a number of regularities they did not lead to a general explanation. Part of the problem was that this property of chromium was seen only in the first sample. All later samples dissolved without periodic release of hydrogen. I asked for and got from H. Goldschmidt numerous further samples of chromium but none of these showed the phenomenon. So, once the first material was used up we had to abandon the project.

I could tell you about some other experimental work which I did before I resigned my position and left the institute and which, like my previous work, was a great success. These were the offshoots of particular projects which had to be pushed more and more into the background while more general problems required all the energy that I could give them.

Balance of account. If I sum up these far-reaching experiences then I have to say that I emerged from them a quite different person. The fire of youth with which I took on every scientific challenge that offered itself was now burned out. I had to

---

[9]Ostwald W (1900) Z Physik Chem 35:33–76; 204–256.

accept that I'd now have a finite amount of energy for my work and that I'd have to continue to make sure that it was husbanded with care. I was still capable of independently producing useful scientific ideas but no longer of carrying out extensive and challenging experimental work.

My capacities as an author had suffered the least, though in writing I'd been even more excessive than with my other work. But even here I had not got off scot free because I now no longer relished burying myself in the publications of other authors in order to sort them into the stamp collection of science. As the extent and for me the importance of the new challenges grew so my interest and ability to spend time on the ideas of others grew less. My mental muscles had perhaps grown in strength but they had certainly lost a good part of the suppleness they'd once had. Given the restricted amount of energy one has, this is no doubt a natural and necessary occurrence.

I did however note a considerable decline in my memory, which previously had been exceptionally good. When I'd finished writing my textbook in 1886 I had clear and complete in my mind not only all the facts and ideas which I'd written about but more than that even years afterwards I could still recount the trail of evidence which formed the basis of my science.

This was now no longer the case. My last great feat of memory was my history of electrochemistry where I put the vast mass of unsorted material into a harmonious and logical framework and this would never have been possible without a functioning memory. Even here I'd helped myself along by making use of some technical help. Now, however, I realised that I could no longer rely on my memory. Sometimes it failed in that I could no longer recover things that I'd once known. At times, thankfully only now and then, I'd find the wrong data stored in the right place. These unwelcome experiences were the first reason that I started to concern myself with questions of the ordering of things and ideas so as to have them immediately at hand. So long as one has access to one's complete remembered inventory of knowledge one doesn't need to bother about orderliness.

This is easy to see. For as long as my memory worked well my desk was a mess because I knew the piles inside out and could immediately find everything. Wives who when their husbands are away "clear up" his writing desk do not understand (and can only with great difficulty be made to understand) that by doing so they break the connection between the things on the desk and their husband's memorised picture of where these things should be. They force him to memorise his desk top anew just at the time when coming back from a journey work has piled up and so he is annoyed at having to perform this quite unnecessary extra task. I don't believe that there are any women who don't view this natural reaction as ingratitude and I also don't believe that any woman who chances to read this explanation will in any way diminish her need to tidy up her husband's desk.

As my unconscious spatial memory of the pile on my desk began to fail I had to more and more start to put some order into it and into my other work places. And if I sometimes have to admit that this or that is not in its correct place then at least I have the consolation of knowing that my memory has not completely gone.

Teaching. The most difficult result of my new physiological state lay in my relationship to the students.

When we moved into the new institute it was pretty well filled and as time went by it could only just hold all of my co-workers. In particular there was a considerable increase in the number of foreigners so that at times there were more foreign than German scientists working there. The majority of these were Americans[10] or Britons but there were also Russians Dutchmen, Italians, Frenchmen, Japanese, and so on. This was so despite the fact that I always tried to avoid giving preferential treatment to foreigners, as many of my German colleagues did.

During the first 10 years of my time in Leipzig I knew no greater joy than my daily round from student to student to discuss the work and to come up with or prompt new ideas. Now, however, I found to my horror that this had all become very different and this became obvious to me because of one quite trivial incident. I'd always conducted the discussion standing up and moving from one student to the next. And then 1 day—it was in the new institute—when a student asked a rather tricky question I, without realising it, looked around for a chair in order to sit down and explain the matter. That is to say, answering the question took up so much energy that I had to conserve it elsewhere.

Once I'd realised this I kept an eye on myself during the next few days and had to admit that this task, which is the most demanding one there is, was no longer accompanied by feelings of joy but instead had become a burden. Here also I overcame the painful side of this admission by recognising that it had a physiological rather than a moral cause.

In practice I helped myself by hiring, in addition to the positions financed by the university, at least two extra assistants at my own expense to look after the students. This was not a problem for me because at that time I had plenty of income, mostly from the text books I'd written. In this way I spent more than 50,000 marks in the last years of my work there. Each of these assistants had half of his time free for his own research and the students were adequately looked after without me having to be everywhere at once. These assistants all later went on to become independent professors, so there is no doubt that they had been qualified to fulfil their duties

In this way I was reassured that despite the recent illness from which I'd recovered, I could push forward all of the scientific work—research, writing and teaching—at the same level as before albeit under somewhat different conditions. Since the bright and conveniently laid out rooms of the new institute cried out for new work I could see a new flowering of my scientific activity opening up before me. But already fate was spinning the threads that would pull me in a completely different direction.

---

[10]Stock JT (2003) Ostwald's American Students. Apparatus, Techniques and Careers. Plaidswede, Concord.

Chapter 23
The Electrochemical Society

The preliminaries. The work that A. Wilke and I put in to form the Electrochemical Society,[1] now known as the Bunsen Society, took place partly before and partly after 1895 when the discussion in Lübeck about energetics and my subsequent illness put a considerable discontinuity in my life. However, since the most important part took place before 1895 it seems most appropriate to deal with this now before going on to the new period which followed the move to the new institute and the conquest of a new research field—catalysis.

The first general description of the new science of electrochemistry, which appeared in the third part of the second edition of my textbook published at the end of 1892, can be taken as the start and the basis for the practical development of this field which then forged ahead with almost American speed. On the technical side the ground had been well prepared. Siemens' invention of the dynamo had made it possible to generate cheaply as much electrical energy as one wanted. The first of the many possibilities for converting this energy was in the production of light by means of the carbon filament bulb which was developed by Swan and Edison. Siemens pursued the conversion of electricity into mechanical energy and built the first electric railway. The lead accumulator made it possible to store electrical energy by chemical means and the use of a generator for electroforming and electroplating opened the way for this new technology into the vast area of chemistry.

The development of electro-technology had of necessity required a scientific foundation, and it was clear that the development of electro-chemistry could also only take place in the wake of the establishment of sound scientific foundations. Now, at this auspicious moment, came the scientific treatment of electrochemistry

[1]Founded 22 April 1894, and renamed "*Deutsche Bunsen-Gesellschaft für Angewandte Physikalische Chemie*" in 1902, it is now known as the "*Deutsche Bunsen-Gesellschaft für Physikalische Chemie*".

R.S. Jack and F. Scholz (eds.), *Wilhelm Ostwald*, Springer Biographies,
DOI 10.1007/978-3-319-46955-3_23

and this brought light into what had hitherto been a confused field and allowed one to put all the established facts into their proper order. One could therefore move forward in solving the scientific challenges as confidently as the technical challenges of generating and distributing electricity had been solved and were being solved in ever more sophisticated ways. Because of this there was an enormous interest in the young field of electrochemistry.

The founding. I had begun to publish my history of electrochemistry at the beginning of 1894, even before this interest was obvious. Shortly afterwards in the spring of that year the electro-technologist Arthur Wilke visited me from Berlin to discuss the formation of an electrochemical society. A few years previously he'd been involved in the foundation of an electro-technology society[2] and had seen for himself what an impulse such a body would provide for future developments. He didn't understand very much about electrochemistry, but enough to grasp its future potential for the technical side of things.

To begin with I was surprised with his suggestion which opened up a new challenge for me. However we had a serious discussion and this persuaded me of the value of the project and since he declared himself willing to take care of the technical details of the founding of the society we quickly decided to do it. He kept me informed of all the details of what he was doing and so I had the chance to learn the basics of this sort of organisational work and that stood me in good stead later on.

First of all a list was made of all the men who should be invited to a preliminary meeting which was to be held in Kassel on the 21st of April 1894. The place was chosen since it lay at the centre of the railway network and hence the attendees would have the minimum distance to travel. Sixty five of those invited agreed to come and thirty turned up.

The meeting voted me to the chairmanship of the foundation discussions. Wilke had formulated a constitution and once the formation of the society had been unanimously approved the discussion of the constitution could begin. Some points were altered but there was such an air of goodwill and a desire for the success of the meeting that the discussion which started in the afternoon was completed by midnight of the same day and I was authorised to present a refined version of the constitution to the meeting the next morning. I completed this work before going to bed and the constitution was accepted the next morning.

With all but two votes I was elected to be the first chairman of the society and Böttinger, who was director of the dye manufacturing company Bayer, which at that time was located in Elberfeld and then later moved to Leverkusen, was elected vice chairman. Amongst the committee members I should mention Dr. Rathenau, the son of the director of the General Electricity Works in Berlin. He is the one whose name is associated with the organisation of the economic basis of industry during

[2]The "*Elektrotechnische Gesellschaft*" was founded in Frankfurt am Main In 1881.

the war and who is well known for his political work and for the fact that he was later murdered. At that time he had little knowledge of our science but I assume his father had told him to make contact with the new society because it was expected that electrochemistry would have a substantial technical and economic impact in the future. He'd taken a lively part in the discussions at the foundation meeting and since he demonstrated a certain skill in business and organisational matters we brought him onto the organising committee despite his youth.

Since he regularly attended the business meetings of the society for some years I got to know him and after business was complete we quite often sat and talked. I didn't think much of him to begin with because in appearance he was all too much the millionaire's son for my taste. However it soon turned out, I have to admit much to my surprise, that he had a lively interest in philosophy which naturally led to endless discussions. At that time I was mulling over the first attempts at energetics but I don't remember if I managed to convince him of the value of this line of thought.

Professorships. G. Böttinger was now a member of the Prussian parliament and he held there a very successful speech supporting the establishment of professorships in electrochemistry at Prussian technical universities so that the necessary trained specialists would soon be available for the development of this area. This suggestion was taken up by the minister of cultural affairs, Dr. Bosse, who in turn instructed Dr. Wehrenpfennig, the official in charge of matters concerning technical universities, to discuss the appropriate measures with Böttinger. The executive committee of the Electrochemical Society as well as our very active colleague Dr. Holtz as a representative of the Association of German Chemists were brought in as consultants. We all attended meetings at the ministries of culture[3] and finance and these went very well. In the meantime the professorship at the Technical University in Charlottenburg had been given to Dr. von Knorre a Baltic countryman of mine who was well disposed to the new concepts even if he had not done anything of note in this area. In Aachen A. Classen was already established as an excellent electrochemist, though his work was almost all directed to the use of electrochemistry in analysis.

All of these steps were directed to the development of electrochemistry in Prussia, for G. Böttinger was a zealous Prussian and in addition to a general wish to push the matter forward he had a personal desire to see to it that the main focus be moved from Leipzig into his home land. He continued these efforts single-mindedly thereafter and in the end he was, to a certain extent, successful.

It seemed to me as an official of the State of Saxony to inform my ministry about these events which had all happened extremely quickly and without any public discussion. At the same time I suggested that the necessary steps should be taken at the technical university in Dresden so that it did not get left behind the other institutes. Meanwhile, down in Munich W. von Miller had already established an electrochemical laboratory.

[3]In Germany educational matters were and are within the province of the "Kultusminsterium".

The next step in this matter was a little unexpected. I got a polite but waspish letter from a colleague in Dresden who, in the name of the professors there, wrote more or less that I should not stick my nose into their affairs and that Dresden professors were perfectly capable of looking out for themselves. I let this outburst of injured pride stand and did not mention that they could not possibly be acquainted with the details of my report. This came all the easier to me because in this matter we achieved a real success and since then excellent electrochemists have fostered the subject in Dresden.

Internal matters. The first years of the Electrochemical Society were a lot of work. I hadn't been part of the governing body of any society, scientific or otherwise and therefore had no notion of the techniques for leading such a body. Now I was suddenly chairman, not of an established society with well oiled forms and ways of working, but rather of something that had just been put together and which, like the homunculus in Goethe's Faust, had first of all to go through the process of becoming a useful entity. True, A. Wilke looked after the external affairs splendidly but I was responsible for the difficult management of internal affairs and had to find my own working procedures. However the good will and trust of my co-workers in this matter made it possible for us to do quite a bit of useful work.

Right at the beginning there was a difficult personal matter to be taken care of. One member of the committee was a professor at the technical university of Charlottenburg and, as is the way of many migrants to Berlin, soon began to look down from his metropolitan heights at people from the provinces who are regarded as second rate. He'd organised a group of electrochemists in Berlin and soon started to organise things in the name of the society, though without asking or informing me. The bills he sent to the society. At a meeting of the committee I said that this way of carrying on was insupportable, but he countered by saying that if this was my attitude then he would immediately resign from the society. I stuck to my guns and, assuming that the matter was now resolved and that the Berliner would now leave, moved onto the next item on the agenda. He did not leave, but for a while he didn't take any active part in the meeting. Somewhat later a matter was put to the vote and he took part in this. I pointed out that this was at odds with his declared intention of leaving the society but he huffily said that he hadn't meant his words to be taken literally. The others agreed that we should regard them as unsaid. I then told him that I would instruct the treasurer not to pay any bills that I had not approved and so he ceased his independent enterprises. Later the other members of the committee regretted their mediation because his membership in the society did not improve its prestige.

When after 1 year one third of the committee was chosen by lot to face re-election the deputy chairman made sure that the envelope with the Berliner's name in it was clearly marked and made clear that I should certainly pick it. I shook my head and refused to pick it. Till then he'd been a good friend to me and my family (he was considerably older and held a prominent position), but after this his attitude cooled considerably and from then on he was always to be found amongst my opponents. Clearly he'd been unable to forgive me for not having gone along with his "tactic".

Influence of the society. In the autumn of 1894, 6 months after the foundation, the society felt itself ready to hold its first general meeting, which, for various reasons was to be in Berlin. We now had nearly three hundred members and it seemed like a good time to decide on our future course of action. Eighty five members attended and there was a lot of discussion. We quickly elected the veterans of electro-chemistry Bunsen, Hittorf, Wiedemann and Kohlrausch as honorary members and all four graciously accepted. Then we decided to spend a substantial fraction of our reserves in organising lectures on electrochemistry in all the major German towns so as to inform wider scientific circles about recent advances in this area. This task was assigned to A. von Öttingen who in the meantime had fled from the pressures of Russification in Dorpat and had, with considerable difficulty found a position for himself in Leipzig. I had helped him as much as I could and had earned strong criticism from some of my colleagues who had little interest in any new competition and considered it an unfriendly act on my part that I had invited him.

Since Öttingen was a gifted speaker and was deeply into the material he made this project a great success for electrochemistry and for the Society.

From then on we held such meetings each year and they were always interesting and instructive. Humour too was always present not only in the discussions but quite naturally in the social round that followed.

State examinations in chemistry. The most momentous of these annual meetings was the fourth one held in Munich in June of 1897. By this time I'd recovered from the illness which had kept me out of the country for quite some time. During this time various people had been pushing for the introduction of a state examination together with an official title for chemists in industry and the Prussian government had been persuaded to take the lead in the matter. The main protagonist was Dr. Duisberg who, like Böttinger who was also actively supporting the scheme, was a member of the board of directors of one of the largest dye manufacturers in Germany.

The reason behind this movement was that the leaders of the chemical industry had noted that young chemists coming from the universities had serious deficits in their education. They lacked the skill that previous generations had shown in analytical and inorganic chemistry. The reason was not hard to find. The complete dominance of preparative organic chemistry had resulted in a situation in which professors of chemistry had little knowledge of, and less interest in, analytical or inorganic chemistry. The usual introductory courses in these subjects were retained but they weren't well done and were usually drastically shortened. The lectures in inorganic chemistry were held by the organic chemists whose thoughts were far removed from the subject and who therefore could not inspire their audience.

At the time that these complaints were being made a remedy had already been found. Physical chemistry had undergone a seismic change which particularly affected concepts and ideas in inorganic chemistry and, by doing so, had thrown up new questions and hence promising research possibilities. Although the study of these general questions was carried out using examples from both organic and inorganic chemistry the majority were inorganic studies because here the relationships were usually simpler. Because of this the long ignored area of inorganic chemistry faced a new blossoming which indeed soon occurred.

However, the leaders of the state exam movement were unaware of all this and, armed with the scholastic way of thinking that is imprinted on German minds by the classical secondary schools sought to solve the problem with the introduction of external exams rather than in supporting the inner development of chemistry. This way of thinking was strongly rooted in the "Association for the Interests of the Chemical Industry" in which the leading members of the technical side of things were associated. This very influential association had won the support of the "Association of German Chemists" which represented chemists working in industry, and together these two associations had already persuaded the government to accede to their wishes.

I hadn't paid much attention to all this to begin with because I hadn't thought that anyone would dare to make such a drastic change in the status quo which, after all, had given the chemical industry in Germany the fantastic growth that had allowed it to outperform the older and richer industries in France and Britain. In fact in both these countries attempts were underway to copy the ways that we did things and it seemed to me impossible that we should now turn back to the primitive methods which had led to a loss of competitiveness in those neighbouring countries.

The defence. I first had to convince myself that the plans were really being seriously considered and then I felt it my duty to oppose these plans as vigorously as I could because I feared that they would do great damage. The meeting in Munich gave me a welcome opportunity to do this. Our representative there was Wihelm von Miller, who'd been running an electrochemical department at the Technical University there for some time. He and his gifted brother, the engineer Oskar von Miller, had done such a good job of organisation that we could look forward to a splendid meeting which would be attended by representatives of the Bavarian government and the royal court. I could therefore reasonably expect to stir up considerable reverberations if only I managed to make the issues clear to the audience.

As it turned out, the Munich meeting was every bit as splendid as we'd hoped. This was due to a curious mixture of friendly and artistic sociability together with the solid organisational work which brought the participants together at a personal level and stimulated that spirit of sincere practical idealism which at that time was typical of that beautiful and cheerful city. The venue had attracted a large number of chemistry professors who were all welcomed by Wilhelm von Miller at a festive breakfast at his house.

I exploited the opportunity provided by this informal gathering to get my problems and worries off my chest. To my astonishment it turned out that few of those attending had taken any interest in the matter and had been happy to let the influential supporters of the state examination have their way.

It was clear that the danger was real and immediate and if we didn't manage to stop the process at the very last minute then the disaster would take its course.

I quickly thought over which argument would convince my colleagues and laid out for them the scenario which I'd prepared for the lecture I was to hold the next day.

The experience with state examinations in medicine had clearly shown that in comparison to the real examination in the form of a doctoral thesis on "independent" research, the state exams meant that scientific work was reduced to a vanishingly low level. In medicine the scientific content has shrunk to a mere formalism. The same thing would happen to us chemists if a state examination was introduced.

If that happened then the result for us professors would be that the best and most important part of our teaching efforts—the initiation of projects and mentoring of doctoral students—would disappear. We would no longer be able to carry out research on a broad front because we would no longer have available the enthusiastic and voluntary co-workers who one after another plough their way through related problems and by doing so check each other's results and further scientific advance on a grand scale. We'd have to give all this up when the supply of doctoral students dried up. That, at the same time, the considerable income that we professors got from laboratory fees would also disappear was something I didn't need to spell out. Each of them could work that out for himself.

My colleagues were all ears. It was immediately decided that we must do something to ward off the approaching danger and a small subcommittee consisting of von Baeyer, Viktor Meyer and myself was formed and entrusted with looking into the matter.

This was the first and practically only time that I managed unopposed to persuade my colleagues to accept my point of view. I can be forgiven for thinking that this unusual success was due not just due to the power of my presentation nor the scientific value of what was said but rather that the recognition of the impending economic disadvantages which would inevitably flow from the introduction of state examinations acted as a high powered catalyst on my colleagues' normally snail like reaction speed.

The critical day. The main meeting the next day was attended not only by a number of ministers and high officials but also by Princess Therese the daughter of the Prince-Regent. She had a lively interest in science. I held a long lecture on "Scientific and technical education" at which I gave my considered views on the planned state examinations. The main argument was that our present system of educating chemists through doctoral work developed them to a much higher degree of effectiveness than could be achieved by any conceivable examination. A doctoral student must learn how to deal with the unknown in order to investigate it. An exam can at best force one to deal with what is already known. This, however, is not enough for German industry which above everything else needs people trained in research. It is this that the German universities deliver and the chemists they turn out are of a quality unparalleled by any foreign institution. This source of our success must be preserved.

I developed this theme in a passionately delivered address and begged my audience to ward off the disastrous state examinations with all the force at their command.

I reminded them of Bismarck's remark about second lieutenants who made the German army superior to anything that other countries could field and concluded that for as long as other countries were unable to copy our doctoral programs we would remain the leaders.

The address clearly made a strong impression. As a practiced tactician the vice chairman Böttinger, who was in the other camp, sprang to his feet and proposed that we delay any further discussion of this matter till the afternoon and in the interests of our honoured guests go on to the next presentation. I had nothing against this and later it turned out that it worked to the disadvantage of my opponents.

In the meantime Viktor Meyer had talked with von Baeyer and made clear to him what the consequences of a state examination would be. He'd felt unwell in the morning but came in the afternoon and without wasting any time he forcefully joined the discussion and produced much the same sort of arguments as I had used. His main concern was the retention of scientific work which he considered was incompatible with the introduction of state examinations. Victor Meyer, although he was already dressed to leave, took the floor and underscored the importance of a doctoral thesis. Dr. Holtz, as the representative of industry, explained that the matter was now in the hands of the central government (*Reichsregierung*) and was as good as ready. We should simply learn to live with the consequences. Böttinger moved that we declare ourselves in agreement with the central government in this matter, but he had to accept that he had no support for this among the participants. During the further discussions it emerged that along with the state exams came a new title such as "Government Chemist" which was intended to "raise the status of chemists". This brought on loud protests. The final scene, according to the official minutes that were shown to all participants prior to their publication was as follows:

Chairman (C):	Now I must ask that we vote on Böttinger's motion
Böttinger (B):	I'm perfectly happy to modify my motion so that instead of "introduce a state examination" it read "to examine the question of the introduction of a state examination" or "to consider examining the question"
C:	"We don't need to suggest that to the central government since that is exactly what they are doing anyway"
B:	"We would do well to express the view that we support the introduction of the state examinations"
C:	That is exactly the opposite of what we want to do
B:	No! Let me put it like this: "The German Electrochemical Society expresses the view that the introduction of a state examination in chemistry should be supported. That's the first thing. And then we add that we ask the central government to consider the matter. There is no contradiction. In other words I distance myself from the "introduction" of such an exam and say instead the central government should consider the matter. In other words "The Electrochemical Society strongly supports the introduction of a

state examination for chemists and asks the government to examine the matter"

C: That is a motion supporting the introduction of a state examination. Those in favour of Mr. Böttinger's proposal please raise your hands

The vote took place

C: The number of votes cast in favour of the motion is small. I scarcely need to ask for those opposing.

With that however the matter was not closed. The education minister Landmann attended the closing dinner and was in the best of spirits. Von Baeyer attended as a guest and surprised us all with a speech of such ruthless openness that no one had expected. I don't have a written copy and so can't relate exactly what was said. However it left the impression that the question of the state examinations had now been answered in the negative. In fact the central government soon after announced that because of serious differences of opinion the matter would have to be considered again. That has remained the state of play till today.

The federation exam. We, that is to say von Baeyer, V. Meyer and I well understood that our efforts would only be successful in the long run if we managed to get rid of the real deficiencies which existed in the education of chemists. These deficiencies were largely due to the lack of any means of ensuring that the students who were admitted to a doctoral course brought with them the necessary basic knowledge of general chemistry. We could not deny that there were a number of "doctorate factories" which admitted pretty much everyone who applied. In expert circles the theses produced in these laboratories were quite rightly considered to be second rate. Our discussion led to the following plan whose central idea had been formulated by von Baeyer.

The leading chemistry professors in the universities and technical universities would form an "Association of Laboratory Directors" whose members would undertake not to accept any student for a doctoral position who had not previously passed an "association exam". In this exam the student would have to demonstrate knowledge of the fundamentals of inorganic, organic and analytical chemistry. Where the professors were experts in the necessary fields, candidates could also apply to be examined in physical chemistry, physics, mineralogy or other related fields and could then be certified as being knowledgeable in these areas.

The suggestion was sent to all relevant professors and a founding meeting took place at the Natural Scientists' Meeting in Braunschweig where the proposal was discussed and accepted.

In a short time all of the laboratory directors of German universities joined this federation. Our colleagues from the technical universities joined at first but resigned en bloc shortly afterwards for "tactical" reasons. At that time the battle over whether technical universities could award doctorates was raging and quite a number of university professors were opposed to this idea.

Since then the "association" has fulfilled its role and it is still in existence which all goes to show that these things can be successfully regulated without any need for official governmental action. I soon resigned from the executive committee when I learned that Baeyer had pushed through one of his minions who quite clearly did not satisfy the conditions. Apart from me no one had dared to raise a word of protest.

This good and successful enterprise was accompanied by a painful episode. The third member of our group, Viktor Meyer, committed suicide just before our meeting in Braunschweig. He was exhausted by overwork and in the last few years had suffered a number of personal setbacks that all resulted in him suffering from frequent agonising headaches and depression. Between times he seemed lively and happy. It was in one of these good moods that he said good night to his family on the 7th of August 1897. Later he must have been hit with a particularly painful episode under the influence of which he poisoned himself with cyanide.

Personal consequences. I know that my successful struggle against the introduction of a state examination in chemistry was of great service to both to the German people and to German chemistry. I never got any recognition for this, and never suggested that it should be recognised. However the powers that be in the chemical industry who had done everything they could to get the state examination introduced and who'd so very nearly succeeded decided that I was not sufficiently acquiescent to their wishes and that I did not fit into their circle to which quite a number of prominent professors belonged.

This had consequences, nearly always negative, for my public life because it was evident that these very powerful people had an unfriendly attitude to everything I did. For me personally I counted it as a plus point.

The conclusion. The growing pressure of much work and other commitments, together with a lessening of interest in the matters of the Society, led me to think about resigning from the executive committee. Young and ambitious people in the Society were trying to undermine my position in order to bring themselves forward and since I wasn't interested in defending this position which wasn't central to me, I was quite happy to leave the field to them. I did suffer under this treatment, not, I may say out of personal interest, for my ambitions now lay in a quite different direction but rather because I felt ashamed of those who were prepared to use devious means in the pursuit of their goals. I resigned so that I would no longer have to see and suffer theses silly intrigues.

However, I still had sufficient interest in this child of mine that I wanted to hand it over to someone who would be a good leader at least for the next couple of years. I thought about it for a long time but couldn't come up with a good proposal. Finally I decided to leave it to chance and announced my resignation at the Society's meeting in Leipzig in 1898.

The meeting was first rate for many reasons. Dr. Hans Goldschmidt demonstrated for the first time his thermite process, which afterwards became so important in industry. In this process suitable mixtures of Aluminium and the oxides of other metals could be used to generate extremely high temperatures in a small area. This could be used, on the one hand, to make ultra pure samples of metals with a high melting point and on the other to weld iron by bringing it locally to white heat. As chairman I had to sum up the meeting and I described this procedure as a pocket sized smelter and forging device.

A second surprise was the demonstration of colloidal gold and its curious properties by Dr. Zsigmondy. A lot of work had been done on colloids and even Th. Graham had tried to produce a methodological review of the field. However one could be certain that Zsigmondy's work would excite considerable interest amongst the experts because the optician Siedentopf had built for him an ultra-microscope which provided a new and powerful instrument for research in this area.

The meeting witnessed a third beginning. A young researcher reported on electrochemical analyses of organic compounds and showed many results at breathtaking speed. One of our older colleagues attacked him violently but without any good reason so that I found it necessary, as chairman of the session, to come to the speaker's defence. He was quite unknown at that time, though he later became prominent. The speaker was Fritz Haber.

At the beginning of the meeting I'd announced to the managing committee that I would not be available for re-election to the post of chairman. As I went down to the lecture hall to meet the participants I worried about how this question could be satisfactorily resolved. One of those there was van't Hoff and as soon as I saw him I knew that the problem was solved. He was born for the job. True, he'd never really been terribly interested in electrochemistry but that seemed to me to be an actual advantage because I'd believed right from the start that it was neither sensible nor possible to limit the society's work too rigorously to electrochemistry. In fact the current meeting underscored this point for the most interesting presentations had nothing to do with electrochemistry.

My good friend Beckmann, had got together with a number of likeminded souls and had planned a coup d'état which would force me to stay on as chairman. But he hadn't reckoned with this move and he had to agree that it was a good solution. So he agreed and van't Hoff successfully led the society for quite a few years. At the next meeting I was elected as an honorary member.

The Bunsen Society. My last intervention in the affairs of the society was to push for its role to be expanded so as to encompass the whole of physical chemistry. Once this had been approved in principle we discussed the question of a new name for the society. Robert Bunsen had recently died at an advanced age and so I suggested that the society be renamed "The Bunsen Society". The suggestion was supported by many but opposed by the group mentioned above whose attempt at the Munich meeting to introduce a state examination for chemists I had frustrated. This time round their tactical ploy was to push through a motion to the effect that a change in the constitution would require a two thirds majority.

The decisive meeting took place in Würzburg and both sides were expecting a fierce exchange. Luckily the discussion was scheduled for the afternoon of the day on which the town had hosted a breakfast for those attending at which a great deal of the delicious "Bocksbeutel" from the vineyard of the Juliusspital was served. The warring mood was clearly relaxed by the wine. When the discussion started our most dangerous opponent was Oskar von Miller. It wasn't really a matter that interested him but the influential opposition had persuaded him to speak and he made an impressive case against the change of name. If the vote had immediately followed then he would have had a majority. I therefore asked the chairman if the other opponents might be permitted first to make their presentations so that I could respond to them all together.

This was agreed on and so there now followed a line of ever less impressive speakers from the opposition who so obviously contradicted each other that they neutralised each others arguments. In my reply I contented myself with pointing out the self destruction of the opposition. H. Goldschmidt, the thermite man, added his own personal view which went straight to the heart and in the final vote we had, after overcoming some further problems, the necessary majority.

At the end it was not a terribly important matter and had involved more effort than it was worth, but it nevertheless pleased me as a technical experiment. However, in the course of it I lost interest in any further involvement with the Society, particularly since new challenges for me lay in a quite different direction.

The mother cat. I was criticised by some for having in this way given up something that I'd invested some of my life in and now was leaving it to its own devices. Since I'd done somewhat similar things before it seemed to me that I was dealing with a law of nature whose basis should be explored. I found the answer, though only much later.

In the country house that I've lived in since 1906 the first snow of every winter precipitates a mass migration of mice from the fields into the warmth of the house where food can easily be found. Even at other times of the year there were regular attempts at invasion. For this reason we kept a cat that did her job well but in return demanded the right to give birth twice a year. Each time we allowed two of the kittens to live and the mother seemed to be satisfied with this arrangement, and this way both sides had a satisfactory arrangement. We allowed two of the kittens to live so that each would have a playmate and we never had any trouble finding takers for them because their mother had trained them well in the art of killing mice.

As everybody knows a mother cat is the archetype of mother love. Without a moments hesitation she will attack and drive off any dog, no matter how large, that threatens her brood. She endlessly cleans her kittens and when they are a little older she forgets her grown up dignity and will play with them for hours on end. At the right moment she brings them living mice and trains them in the art of hunting.

However, at some point she changes her behaviour. She no longer plays and instead leaves them to themselves and a little later she shows her growing brood, by hissing and boxing their ears, that they can no longer rely on mama and that from now on they are on their own.

I've described all this in detail because it seems to me to be exemplary. Our intellectual children, assuming that they were born with the necessary capabilities, acquire as a matter of natural development their own independent life, and the better we look after them the faster that happens. But there is always the danger that over-mothering ends up robbing them of their initiative. If the individual is strong enough then he will of necessity shake off excessive care. That leads to inhibitions and quarrels and the mother will be unhappy. Alternatively, if he is not strong enough, then the child will suffer under the constant care and die when it is withdrawn or often enough even earlier.

From the history of chemistry I knew of the tragic situation which befell the great Berzelius with the Society which he organised. It was a case where the child won its own independence. For the last 10 years of his life Berzelius was involved in an increasingly violent battle against the new concepts in chemistry and on his deathbed saw the whole structure to which he'd devoted his life go up in flames. What burned were of course only ancillary elements, but he didn't see that because his science had grown beyond his grasp and he was no longer able to see it in its entirety.

I thank science that it showed me the objective necessity of such occurrences so that I never needed to hesitate to come to a sensible decision. When I am accused of lack of loyalty to the structures I built then I can fairly answer that it was better to give up the leadership in good time than to prolong a relationship whose inner structures had changed from what they originally were.

What I have described in detail for this particular case can be applied to some other situations during the course of my life's work. The Electrochemical Society, reborn as the Bunsen Society, was strong enough to experience and survive other leaders, some better some worse, and to do so without destroying their health.

Chapter 24
Catalysis and the New Institute

What is catalysis?. For over a century processes have been known in which a chemical is able to transform another without itself being changed in its constitution or amount. In 1811 Kirchhoff[1] had shown by careful measurement that starch could be converted into sugar by boiling it in dilute sulphuric acid. The conversion requires the acid, but at the end of the reaction all of the acid is still there. In 1836 Berzelius with his extraordinary capacity to conceptually unite a wide range of phenomena, had referred to this as catalysis though he considered the time not yet ripe to explore the phenomenon and did not try to provide a mechanistic explanation for it.

Liebig came forward with an explanation but it was one that was entirely hypothetical and provided no clear means to investigate what was going on. At that time he was in the middle of the dispute with Berzelius which led to the unfortunate breach between these two great researchers and which filled the old master's last years with bitterness. Liebig took the view that in the case of starch hydrolysis the catalyst was able to impose its own molecular motion onto the substrate and hence dispose it to disintegration. Sadly, even his friend Wöhler was unable to convince him of the infeasibility of this idea.

Since then the whole area of catalysis was considered suspect and anyone who ventured into it risked losing his reputation as a chemist. For this reason the ingenious Swabian Schönbein, who'd invented gun cotton, loved to collect examples of catalysis and to point out to his peers how helpless their science was in the face of these numerous interesting situations. He himself, of course, had no explanation to offer though he was certainly of the opinion that the catalyst was essential to cause the reaction. He didn't come up with any further ideas.

[1]Kirchhoff, G.S.C. Ostwald refers to: Kirchhoff (1811), Mémoires de l'Académie impériale des sciences de St. Pétersbourg 4:27.

R.S. Jack and F. Scholz (eds.), *Wilhelm Ostwald*, Springer Biographies,
DOI 10.1007/978-3-319-46955-3_24

As time went on ever more phenomena were discovered which fitted into this class. However those who followed Berzelius and referred to these processes as "catalytic" were rightly criticised for merely using a word rather than providing an explanation. Despite this it must be said that Berzelius had introduced the term precisely because he did not wish to get into a discussion of mechanisms.

The untenable hypothesis of Liebig did much better. Because it was graphic it was widely accepted as a true explanation. It developed into the idea that a catalyst was a material with a very special type of atomic vibration which caused the molecules of the substrate to be broken up. This idea was completely inadequate because not only degradative but also synthetic reactions can be catalysed. In addition, none of those who put forward such explanations had ever tried to demonstrate and measure these vibrations. Nevertheless this pseudo explanation was as attractive then as is the current idea that a specific catalyst is like a "key" that fits to the "lock" of a particular substrate. For the scientific understanding of what is going on both of these explanations are equally valueless.

This type of explanation reminds one of the story which Wieland related about the citizens of Abdera. On the marketplace they listened to various philosophers explaining the world. Most of them were ignored but one was listened to with interest. He explained that the world was like an onion from which one skin after another could be peeled. The audience found that quite convincing, not because it told them anything about the world but simply because they all already knew what an onion looked like.

The start of my work. I'd come across examples of catalysis early on in the course of my work on chemical dynamics. Wilhelmi, who'd pioneered the scientific analysis of chemical reaction kinetics, had in 1850 based his concept on a typical catalysed reaction—that of the conversion of raw sugar in the presence of dilute acid. My first work, (1883) demonstrated the applicability of these same laws to a second intensively studied case which also involved a catalyst, namely the hydrolysis of methyl acetate.[2]

Once I had established my by now customary affinity values for the catalytic hydrolysis of sugar.[3] I asked whether these values would also apply to a completely different class of reaction namely oxidation–reduction processes. I looked for a reaction of this type which would be sufficiently slow that it could be easily measured, and settled on the reaction of bromic acid with hydrogen iodide.[4] Once again I showed that the reaction was catalysed by acids and that they yielded the expected affinity values.

---

[2]Ostwald W (1883) J Prakt Chem 28:449–495.

[3]Ostwald W. (1984) J Prakt Chem 385–408, and Ostwald W. (1985) J Prakt Chem 31:307–317.

[4]Ostwald W (1888) Z Physic Chem 2:124–147.

However there were also reactions of a quite different nature in which small amounts of a substance were able to initiate the reaction. Here I was dealing with catalysts of the type that Schönbein had described to be acting in so many different situations. By reducing the amount of the catalyst I could reduce its influence as I wished until finally there was no effect at all and the reaction then continued at its normal slow rate.

The quantitative analysis showed that the catalyst was not able to cause the reaction to take place but was merely able to speed up an already on-going process. The question of whether this was a general description of catalysis could be answered in the affirmative because I was already familiar with the idea of chemical reactions which proceeded so slowly that one could not detect them. Already in Dorpat I'd incubated sealed tubes containing a mixture of hydrogen and oxygen at moderate temperatures to see if water would be formed.

Now however I took another tack. When one is first hit by a revolutionary idea then one is at first astonished and tends to shy away from it so as not to be tempted to go too far. This is especially so when one is young and hasn't experienced this sort of thing before. Only when the idea persists does one find the courage to start to think whether it should be taken seriously. It's much the same thing that Johannes Brahms described with his musical ideas. He'd let a new idea come and go without writing it down because, so he said, if the idea was really good then it would come again. If it doesn't come again then it probably was no loss. Nevertheless it must be said that this way of doing things only works for those who have lots of ideas so that the loss of one will not be that great a loss.

In my case I should add that this was at the time that I was fully taken up with the work, based on Arrhenius's idea, which led to my discovery of the dilution law of electrolytes.

As the number of co-workers and the range of our work increased, these thoughts returned once more. Already in 1889 I'd set two colleagues, L. Henry from Belgium and U. Collan from Finland on projects which were based on this idea of catalysis—though at that time we did not talk about it in these terms. The projects were concerned with the question of whether a chemical can catalyse its own reactions. I was able to present our results on such "autocatalysis" to the meeting of the Scientific Society in Leipzig in the spring of 1890.[5]

It was only in 1894 that in reviewing someone else's wok I defined a catalyst as a material which could change the rate of a reaction but which was not involved as a component of the products.[6]

One can immediately see the enormous difference of this purely conceptual definition to the old "descriptive" hypothesis of molecular vibration for it immediately opened up a whole new field for experimental analysis. Which materials speed up which processes? What are the laws that apply? What amounts are

[5]Ostwald W (1890) Ber Verh Kgl Sächs Ges Wiss, Math Phys Cl 42:189–191.

[6]Ostwald W (1894) Z Phys Chem (Leipzig) 15:705–706.

required? These are just some of the questions, which one would not only soon ask but which could be answered by using appropriate experimental approaches.

The relationship to energetics. The laws of energetics and in particular the special laws governing osmotic pressure and electrolysis had to a large extent explained chemical behaviour so that one might now strike out further in this area without risking losing one's way. I'd now spent the best part of ten years pioneering the most important parts of this subject and had successfully led my students to apply these concepts both to their and to science's benefit. Slowly, however, this approach started to lose its novelty. The more things fitted into the scheme the more obvious it became what had to be done and what the results would be, even if the research did occasionally throw up new and unexpected results.

Because of this I now faced the question of whether all the challenges of my science had now been solved. Of course there were still thousands of individual cases to be investigated just as the organic chemists had thousands of new compounds which had to be synthesised. But were there, beyond all this, unsolved fundamental questions?

At first sight one was tempted to answer, "No". The reason was that every reaction was governed by the general laws of energetics and the second law made very clear predictions.

But though energetics defines the conditions necessary for each reaction does it also define what conditions are sufficient? Is every reaction defined in every detail by the laws of energetics?

The proponents of a mechanical view of nature would here say, "Yes". In his strikingly theatrical manner Dubois Reymond had expounded on Laplace's "Theory of everything" — an equation in which one needed only to insert the appropriate values and everything in the past and in the present would be completely explained. Energetics had taught me that this doesn't work and so I was thrown back on experience. Are there things in chemistry that cannot be fully explained by the laws of energetics?

Through discussion with my co-workers I had already long since found the answer: energetics is always necessary but it is never sufficient. Each type of energy has its own special laws which limit the many possibilities to one single reality. On top of that come the specific values for innumerable properties which define the different chemical entities and these include the rates of chemical reactions. These are not defined by energetics since the laws permit them to have any value. Catalytic acceleration of a reaction therefore does not contradict these laws. They only forbid that catalysts make possible reactions which are energetically impossible.

In this way I was now ready to complete the chain of thought which would wake this sleeping beauty and break through the hedgerow of sterile hypothetical notions which in the meantime had grown into an impenetrable tangled mass. Energetics had taught me that catalysis could not make possible any reaction which was not compatible with its main laws. In other words catalysts could only speed up reactions which were in principle possible. In this way I'd returned to my old idea and I could now see a vast new area of work open up before me, and I now had a brand new institute where I could set my students to work on these problems.

The new laboratory. I have already described how ill suited the space in the Second Chemical Laboratory was for our purposes. In the initial years when the number of students of physical chemistry was small, the space had more or less sufficed for my co-workers. But when one considers that the majority of them were doing independent scientific research which naturally requires more space than a simple student's experiment then one will quickly see that I soon suffered the same sort of space problem that I'd had in Riga.

I had of course not forgotten to inform the responsible authorities of this problem and to point out that it was getting worse with each semester that passed. However, just at this time a very ambitious professor of agriculture[7] was appointed who had no intention of being palmed off with the old building even if he was also given the floor that housed my department. He managed to convince those in charge that his needs had priority and to secure the necessary large sums for the construction of a new building from the local government. I, for my part, explained that it would not be possible to refurbish the old building so as to make it fit the requirements of physical chemistry and that a new building was therefore required. Though this argument was accepted in principle a new building was put off for the time being. The minister of culture von Seydewitz, who always supported me, pushed his politeness in the matter to such an extent that he asked me to agree to the delay as a personal favour to him. I could only reply that my part in the decision making process was entirely passive and that the assessment of the university's needs was the principle function of his office. Nevertheless I did not hide from him the fact that my institute, because of its completely unsuitable structure, had achieved a certain reputation. The astonishing contrast between the quality of the work and the unsuitability of the building had been commented on even in foreign journals.

Finally, in 1894 funds for the new building were authorised, the building plans revised and construction began. The stress associated with the new building project certainly contributed to my breakdown which occurred at this time. During my absence my assistants directed the work on the special technical aspects of the building and here Th. Paul, who was Beckmann's successor, showed his marvellous technical abilities.

In the autumn of 1897 the construction and outfitting of the laboratory was completed and we could now move in. For me this was an important change in more than one way.

The inauguration. I look back on the inauguration of the new building as a special day, a day on which new work would start in the new rooms. I provided myself and the scientific community with a collected edition of the most important publications from myself and my co-workers in the last 10 years. It came to three volumes and around 2250 pages and was brought out by the ever helpful publishing house of

[7]Wilhelm Kirchner.

W. Engelmann.[8] The volumes were dedicated to the minister of culture von Seydewitz who had supported the construction of the new building.

At the same time the institute's name was changed. It would no longer be the Second Chemical Institute, in other words a mere pendant of Wislicenus's First Chemical Institute. Ten years previously physical chemistry had been nothing more than an uncertain experiment, but now it stood solid and full of youthful energy and had demonstrated its maturity by the fact that it had spawned daughter institutes. And so the name "Institute of Physical Chemistry" was chosen. If I'd been as careful then as I am now in avoiding the introduction of unnecessary foreign terminology into German then I'd have called it "Work place" (Werkstelle) instead of "Institute". At the same time the pharmaceutical department—that foreign component that the old institute had to house—was removed. I, together with Wislicenus, had suggested that there should be a third independent chemical department and as a result the Institute for Applied Chemistry was formed and the pharmaceutical department was transferred to it. I was happy to be able to give up having to exam the pharmacists and medical students, which had long been a pain to me, though many of my colleagues could not understand how anyone could give up such a considerable and sure source of income and once again judged my behaviour to be uncooperative.

Ernst Beckmann was appointed to head the new department and so I was able to happily welcome him back as director to the place we had started out as director and assistant. He showed himself to be as reliable and loyal in his new position as he'd been in the old.

I'd taken special care of the education of school teachers and had formed a separate unit to do this which was led and developed by J. Wagner.

To publically mention all these things I arranged a special inauguration celebration to which the minister of culture, the rector of the university, the dean of the faculty and a number of friends and colleagues from near and far were kind enough to attend. A photograph of the meeting is attached to this volume. I look at the people with a mixture of happiness and sadness. Happiness, that I was able to interact with so many splendid men, sadness because the majority of them are now dead. On the accompanying page is the list of those attending and it includes van't Hoff, Arrhenius, Nernst, Beckmann, Wiedemann, Leukardt, Lamprecht, Landolt, O. and W. von Miller, Wislicenus, Bücher, von Öttingen and many others.

My previous students had for this occasion got the distinguished Leipzig artist K. Seffner to make a relief portrait of me in marble. Perhaps in subconscious anticipation of what was to come, they had not stipulated that it be housed in the institute but rather had given it to me and my family.[9] Those close to me tell me that the relief is well done and shows me as I was when I was a chemist.

[8]Arbeiten des physikalisch-chemischen Instituts der Universität Leipzig aus den Jahren 1887 bis 1896. W. Ostwald (editor). Volume 1–4. Engelmann, Leipzig 1897.

[9]The relief is still in "*Haus Energie*" in Großbothen.

To keep me happy during the sittings for the relief, Seffner told me the technique he'd used recently to make a bust of the aging King Albert of Saxony. The old man would often fall into a light sleep whereby a disturbing similarity of the shape of his head to that of a ram became very evident. Since this in no way reflected the mental abilities of this remarkable monarch it led to certain difficulties for the scrupulous artist who did not wish to misrepresent nature. He therefore had to wake his model and did so by busily using a large pair of dividers to measure the king's head. The king quickly grasped what was going on and smilingly let it happen; the artist for his part had what he needed.

In the experimental part of my welcome speech I was able to demonstrate the use of liquid air which K. Linde[10] had managed to prepare shortly before. For the theoretical part I chose to look at the defining properties of time which allowed me to move from general considerations in a straight line to the role that time plays in the development of a chemical reaction. Here it is in the laws governing chemical kinetics that time plays such an important role. Since the general laws in this area had been established and investigated in detail by the German researchers Wenzel[11] (1777) and Wilhelmi (1850) it might seem that there was little left to do and no surprises to be expected. However catalytic processes, whose introduction into chemical kinetics had been made possible by the conceptual work outlined above, now provided a basis for important and unexpected discoveries. A few years later I was able to say: When the Leipzig Institute of Physical Chemistry was established in the splendid new building I must confess that I viewed the future with some worry. The period which had just ended had been very fruitful. Large areas such as chemical dynamics and electrochemistry had undergone fundamental changes and it seemed that in the new institute in place of a happy invasion of the new areas we would have to make do with working through the details of what had been discovered. I told myself, we have to find at least something untouched, for what we wanted at almost any price was to have the chance to move into a new and untouched area. Of the all the possibilities that offered themselves none seemed to me to be more full of hope than that of catalysis.

The implementation. We soon took on this challenge and almost all of my assistants focussed their own work and that of the students they supervised on this point. While the themes of the first 10 years had been osmotic pressure and electrolysis the theme of the second decade was catalysis. The results were if anything more

---

[10]Correct is C. Linde.

[11]In his book entitled "Lehre von der Verwandtschaft der Körper" Gerlach, Dresden 1777, Wenzel wrote: "The strength of chemical affinity is proportional to the reacting masses". This was an anticipation of the Law of Mass Action. At that time no clear distinction between thermodynamic affinity and reaction rates (kinetics) had been made. In modern understanding affinity is the negative partial derivative of Gibbs free energy with respect to the extent of reaction (reaction coordinate) at constant temperature and pressure, i.e., a purely thermodynamic quantity. Only in 1954 did I. Prigogine and R. Defay define affinity as the derivative of the rate of change of the uncompensated (irreversible) heat of a reaction over the extent of reaction, thus giving the term a clearly kinetic aspect.

numerous and even more important. As I write this, there lies before me on my desk a summary prepared by A. Mittasch, one of my former students, of the technical applications of catalysis in the large chemical factories in Germany and overseas. It is an industry which is worth millions and even billions. And this is only one side of the coin, though an important one. The other is the physiological side because one can certainly make the general statement that there are no biological processes which are not influenced or controlled catalytically. In both cases the leading researchers insist that we are only at the beginning of something which will have consequences that we cannot even guess at now.

In the decade 1887–1897 I had been able to approach the work which was based on other peoples' ideas with the energy of youth. For the exploitation of my own ideas I was now dependent on the work of my assistants and students. My capacity for work in other areas had been restored but not in that of personal teaching which is the most challenging of all. This was something that had happened to greater men before. There was something tragic in the fact that I could no longer serve the needs of my own mental child as I had previously served the mental children of others. Still, I didn't worry too much about such thoughts. I had more important things to think about than this personal problem and in any case I was convinced that the work was in the best possible hands. This was the time when G. Bredig was doing his fundamental work on the catalytic properties of colloidal platinum and other metals and when M. Bodenstein was doing his no less important work on the kinetics of reactions in the gas phase, not to speak of other important projects.

The scientific family. The new building was on the outermost outskirts of the town so that the students had a long way to go if they wanted to get lunch in the city. Given the enthusiasm that most of them had for their work this was a most unwelcome interruption. After an attempt to get the janitor to deliver meals had failed my wife took on the not inconsiderable challenge of sending over from the nice house, that had been built as the director's residence, coffee and cake for the students in the early afternoon to tide them over so that they could keep their main meal of the day for the evening. A saucer was laid out so that each student could put in the few pence which were needed to cover the costs of the ingredients.

This was something which was warmly welcomed and it continued for many years. My wife was happy to be able to look out from our flat to the stairs where the coffee was set out and watch the students sitting with their coffee mugs on the steps like sparrows on the eaves. But like all good things it was finally destroyed by the influence of malicious elements. Some of the students from the outer fringes of Europe exploited the well known honesty of German students and, since there was no control, simply failed to contribute. Following the law of the ethical minimum, which W. Busch expressed in the all too true line "The greatest blackguard makes it to the top", the coffee break had to be abandoned because the difference between the costs and the contributions could not be bridged.

In other areas too I was able to develop friendly personal relationships with my students. Our house came with a small garden in which I'd planted lots of roses. When they bloomed in summer I'd cut the flowers that had opened, so as to make way for the next buds, and send baskets filled with the sweet smelling flowers into the lab. It was a charming sight to see how—so long as it didn't interfere with their work—the students would decorate not their buttonholes but rather their equipment and test tubes with these colourful children of the summer.

The advanced students who had already begun work on their projects were invited every 2–3 weeks on Sunday afternoons to coffee and then dinner. Neither for them nor for us were these occasions ever boring. My three boys, who were always in and out of the labs, knew theses students well and my two daughters were also not afraid of them. Usually those who could play an instrument were asked join our string quartet. My eldest daughter played the violin, my eldest son the cello and I played the viola. Quite often the foreign students were already settled in jobs and had married and they could bring their wives with them. The female members of the household often helped them when they reached that stage when a woman far away from home longs for the experienced help of her mother.

On the other hand at Christmas my family and those of the assistants were invited to the laboratory where the students organised a Christmas party. In the middle of the biggest teaching lab stood a large Christmas tree decorated with lights and chemical apparatus and the students offered everything that they could contribute to the party. Usually there was a humorous newspaper in which all the happenings of the year in the lab were reported. Then the presents, which all held some insinuation, would be distributed. One year I was given a nicely carved wooden cow and when you moved its head it let out a piteous "Muh". This was because when I had been asked for advice for hours on end and had finally had to almost forcibly free myself, I'd say "Today I've really been milked down to the last drop". My occasional complaint that a physical chemistry institute should really be made of rubber so that one could expand it at will was rewarded with a rubber model of the institute. During the early phase of trying to sort out our thoughts on catalysis I'd said that the concept was so difficult that even if an angel came down from heaven and told us the answer we probably wouldn't understand it. Promptly, the next Christmas a beautiful wax angel with wings of coloured glass wool swooped down over the table carrying in its hand an envelope on which was written "Catalysis".

Later on my interests broadened to philosophy, painting and various technical problems and at one of the last Christmas parties at which I took part I was given an extremely complicated looking machine out of which lots of levers and taps and handles projected each of which had a label on it. The whole thing was supposed to represent my brain because the labels referred to the various interests I'd developed and it was meant as a warning not to spread myself too thin but rather to concentrate on my students as in the old days. To me it was an indication that it was time once again for a change in my life.

Special Christmas presents. For a number of years I was able to give my co-workers a very special gift which most of them will never have forgotten. Just by chance it turned out that at the first Christmas party in the new building William Ramsey was staying with me to get some rest from a strenuous round of lectures and celebrations in Berlin. We, that is to say I and those of my children who played, would entertain him for hours in the evening with German music of which he could never get enough. He, for his part, whistled English folk songs in a masterly fashion and accompanied himself on the piano. When he was listening he'd sit in an armchair, stretch his long legs out and rub his head against the wall. Since he used to oil his hair there was soon a hand sized oil stain on the wallpaper which my wife, half complaining half laughingly, pointed out to me once he had left. I gave her a small golden crest which she hung over it.

Naturally the students asked me to bring the famous researcher who'd discovered argon and helium to the Christmas party and I easily persuaded him to hold there a half serious half facetious lecture. This was a great success for he was a wonderful teacher whose heartfelt wish it was to encourage the young scientists.

The impression he made was so strong that ever since I made my students the gift of a great man. Van't Hoff told them how Dom Pedro, the then emperor of Brazil, visited him in Paris and asked whether it would not be a good idea to give a model of the tetrahedral carbon atom which he'd discovered, to children as a toy so as to stimulate their interest in chemistry. Later on he'd prepared a whole kilogram of dibromo-butanedioic acid but couldn't get it to give the reaction he'd expected. He sadly stood before this now useless hoard and wondered what he should do with it. Then it struck him that its decomposition in boiling water would be a slow process that could be followed by titration. That is indeed what happened and this started his interest in chemical kinetics.

At another of these parties my respected fatherly friend Hans Landolt told us about his experiments to show that during chemical transformations the weight of material does not change at least within the limits of the balances available. He told us hilariously of his annoyance that the weight known as "the rider" kept falling off the balance whenever he moved it. Some of the students made a gingerbread "rider" in the blink of an eye and when the presents were distributed shortly afterwards they gave him it with the promise that it would never fall down.

The jokes about the assistants and students were not quite as harmless as those about the professors but this never led to quarrels. On the contrary I think that some of them had a positive educational effect and the better the joke "sat" the greater was the effect.

Science as a social matter. There was always some reason behind these jokes. I'd always emphasised the basic idea that the research of the advanced students was a group enterprise so as to make them understand the underlying nature of all science. Because of this I used to give out a list of the research areas at the beginning of each semester so that every student could choose what topic to work on. I did not agree with the usual method of simply "giving" each student a topic and instead expected that the student should tell me what he wanted to do. My job then was to distil out of the usually rather vague wishes a technically feasible work plan.

When the project was about half way through then the student had to present the current state of the work plus his plan for the future to the assembled members of the institute in the lecture room, and then we all discussed how best to proceed.

When the work was thought to be completed it had to be presented again and this showed clearly whether it was a rounded up story or not.

It wasn't always easy to do this, but I think this procedure contributed a lot to the success of the institute.

When I later met students from this time each and every one of them assured me that the Leipzig years had been the happiest time of his life. I believed them because for young people there is no greater happiness than in testing their youthful strength on a project which they had themselves chosen. This feeling is even more intense if one works in the company of others who are striving to a similar end and if one is in an atmosphere which allows for no other thoughts. This too one can regard as a catalytic effect.

The picture shows our physical chemical family in the summer semester of 1900 together with the assistants who pursued their own projects. The names of those who could be identified are given.

For all that. It seems hard to understand that under these happy circumstances I should be anything but happy and satisfied, but that was not the case because I felt like the crab that has built its well fitting and hard wearing shell. The shell stays as it was, but the crab is growing inside it and the shell will not expand. It gets ever more cramped. At first it is just a little uncomfortable then more so and finally causes so much pain that the crab has no option but to break open the shell and throw it away. After that it is exhausted and defenceless until it has managed to build a new larger shell.

I was undergoing just such a period of growth and in fact growing in more directions than one. Things soon came to the stage where I had to break open my shell even though it had fitted so well to my previous situation. Now I had to build a new one. How? That will be related in the third and last volume.

Chapter 25
Nitrogen

The problem. My teacher Karl Schmidt,[1] who had been trained under the direct influence of Justus Liebig, knew all about the latter's concept of the cycle of matter. According to this the elements carbon, hydrogen, nitrogen, sulphur and so on cycled between organic and inorganic forms. Plants feed on carbonic acid, water, ammonia etc. and once they die these components are released once more by decay and decomposition and so life can go on for ever.

Liebig had asked himself how this "mill of life", as I later referred to these processes, can be kept running indefinitely.[2] However it was J.R. Mayer who first found the answer and published it in his Annals of Chemistry: it is energy, energy in the form of sunlight.

While there is no problem bringing the other elements into and out of organic molecules, nitrogen is a different matter. It is easy to break down nitrogen containing compounds but at that time it seemed doubtful whether free nitrogen, which is present in enormous amounts in the air, can be fixed at all. In fact I still remember a carefully worked out lecture by my colleague Gustax Bunge in Dorpat who came to the conclusion that the possibilities for fixing nitrogen were very limited. It was therefore a matter of the highest importance for the world economy to conserve the small amount of available fixed nitrogen and to avoid anything that might destroy it. Because of this he was opposed to the idea of cremation, which was much discussed at that time, because in this way the nitrogen content of the corpse would be converted by the high temperature into elemental nitrogen and thus removed from the cycle for ever.

These arguments made a strong impression on me though I did not wish to accept them. At that time I was radically in favour of cremation and in the meantime my views are only a little more moderate. I felt cremation was cleaner and more

[1]Ostwald refers to Carl Ernst Heinrich Schmidt.
[2]Ostwald refers here to his book "Die Mühle des Lebens. Physikalisch-chemische Grundlagen der Lebensvorgänge", Thomas Verlag, Leipzig, 1911.

R.S. Jack and F. Scholz (eds.), *Wilhelm Ostwald*, Springer Biographies,
DOI 10.1007/978-3-319-46955-3_25

practical than slow decomposition in the earth.[3] I mulled all this over. It seemed to me that wood and fossil coal also contained nitrogen and so burning them must lead to a very much greater reduction of fixed nitrogen and yet that had not in any way reduced the degree of plant coverage of the earth. Since it was known that electrical discharge in the air could result in the association of nitrogen and oxygen, I concluded that the loss was being covered from this source (and perhaps from others as yet unknown) and hence that there was no reason to try to save the small amount of fixed nitrogen which would be lost by cremation. Nevertheless because of its broad consequences this idea that fixed nitrogen must be lost had made a strong impression on me and I never quite forgot this problem.

The synthesis of ammonia. The question came back into the forefront of my mind only 25 years later in 1900. At that time I was deeply engaged in the problem of catalysis whose conceptual basis I had laid by strictly applying the principles of energetics and it was also the time at which my fatigue symptoms caused by overwork had not yet been completely cured. I'd been invited to some dinner by the Chamber of Commerce, though the company was very disgruntled that night because the bank of Leipzig had just become bankrupt due to foolish speculation by its director, and I too had nearly been sucked into this.

By chance I found myself sitting next to the mayor of Leipzig, Tröndlin, and the deputy mayor Dietrich along with several bankers. I expressed my astonishment that the chemical industry in Leipzig was so underdeveloped despite the fact that the university turned out a large number of well educated chemists every year. Their answer was that all the available capital was usefully employed elsewhere. I said that the chemical industry at this time was an Eldorado in which great discoveries were possible which would lead to enormous profits and gave as an example the synthesis of ammonia. The professor of agriculture who was sitting close by told them that in fact of the three essential plant nutrients potassium, phosphorous and nitrogen, the latter was the most expensive. Only if the price could be reduced by a factor of ten, would things once again be in equilibrium. However making nitrogen compounds from the nitrogen in the air was like looking for the philosopher's stone.

I, however, was sure it could be done. Despretz and others had reported that if ammonia was blown over gently glowing iron then it would be almost quantitatively converted into hydrogen and nitrogen. That is a case of catalysis because without the iron ammonia is not decomposed to any measurable extent at this temperature. It followed that a synthesis of ammonia must possible under the same conditions because a catalyst does nothing other than speeding up the attainment of

[3]Here Ostwald refers to an article which he wrote in the newspaper *Berliner Tageblatt* (No. 248, 17 May 1914, 2nd supplement) entitled *Die Halle des Lebens* (Hall of Life), in which he advocates cremation and also the idea of the Swede Gustaf Vilhelm Schlyter (1885–1941) to build a Hall of Life nearby the crematory, where people are inspired by art to contemplation.

equilibrium. The equilibrium was presumably such that there would be just a little ammonia and lots of hydrogen and nitrogen but since during the synthesis reaction there would be a reduction in volume, the usual laws made clear that the fraction of ammonia would increase with increasing pressure.

The conversation moved off into other areas but when I left I was thinking what consequences this line of thought would have for the German people. From the notes that I later in April 1900 dictated to my wife I quote the following sentences: "Germany would be in a position to produce the majority of its bread requirements. By stimulating agriculture the future of the farming communities and hence the ability to defend Germany would be assured for 50 years. In addition loss of a sea battle would no longer lead to loss of nitrogen which at the moment is to 90 % derived from Chilean saltpetre. Even the nitric acid which was essential in time of war could be made by oxidation of synthetic ammonia by atmospheric oxygen in the presence of a suitable catalyst." These and other thoughts made such a strong impression on me that before going to bed I decided to make a start on the experimental solution of the problem the next day.

Despite the fact that I'd been deeply moved by these thoughts I nevertheless felt unable to carry out the necessary experiments myself which before that would have been not only easy for me but also a real joy. Instead I asked my assistants, first Dr. Bodenstein and later Dr. Brauer to blow a mixture composed of three volumes of hydrogen and one volume of nitrogen over bundles of gently heated iron wire and then to analyse the emerging gas for the presence of ammonia. It didn't take long to find it. The expectation based on solid theory had thus been confirmed.

I originally thought to simply publish the results and leave it up to industry to develop them, but in that case Germany would not have gained the head start which I wished my country to have in what was a matter of great economic significance. I talked it over with my co-workers, with Beckman who had in the meantime been appointed Professor of Applied Chemistry and who was my closest colleague and finally with Wilhelm Wundt whose helpful support was not missing even in this matter which was far from his centre of interest. It turned out that turning the matter over to the government of Saxony, which was what I had initially intended to do, would not be useful because I would have lost all opportunity to guide the matter further and so the best course was to apply for a patent and then negotiate with the German chemical industry. This was the solution favoured by Wundt, and I agreed with him.

This strategy offered me the chance to fulfil a personal wish which seemed to me to be becoming ever more necessary.

By any reasonable measure the remuneration would allow me the long wished for release from my teaching duties without endangering the economic future of myself and my family.

Finally, not without some hesitation I decided to take the risk of launching out on the tricky and stormy seas of economic interests for which the recent disruptions at the Leipzig Bank provided such an awful warning. The original text of the patent application has survived and I quote it here as a historical document.

"Procedure for the synthesis of ammonia and ammonium derivatives from free nitrogen and hydrogen".

"It is established that free nitrogen and hydrogen will normally not combine to form ammonia: only with the help of electrical discharge can a slow and incomplete synthesis be achieved".

"I have discovered that the fusion of free nitrogen and hydrogen can be achieved at a measurable rate at the low temperatures of 250–300 °C in the presence of appropriate contact substances or catalysts. The reaction rate increases rapidly as the temperature is increased. Metals, such as iron or copper, which have been treated to give them a large surface area, serve as the catalysts. The synthesis is not complete but rather leads to an equilibrium and the amount of ammonia produced is therefore dependent on the ratio of the amounts of the input nitrogen and hydrogen. To achieve complete synthesis it is necessary to remove the ammonia from the reaction mixture and this can be done by dissolving it in water or acid. For this purpose the gas mixture can be recycled, if necessary after cooling and then using the recovered heat once more".

"Since the amount of ammonia in the reaction mix increases with increasing pressure it will be advantageous to carry out the synthesis under increased pressure".

"Patent claim: The recovery of ammonia and of ammonium compounds by the chemical combination of free nitrogen and hydrogen in the presence of contact substances".

As the expert will readily see already here in March 1900 the basic ideas for the synthesis of ammonia which is now so important were clearly and explicitly laid out so that I may fairly claim to be the intellectual father of this industry. I am of course not, so to say, its natural father, because all the difficult and varied work which is required in order to convert a good idea into a technically and economically viable industry was all done by those who later on took over the abandoned child.

The negotiations with the representatives of big industry went very smoothly. I'd got in touch right away with the three or four leading manufacturers because I wanted this to be a German national undertaking. The first meeting was with Brunck who was director of the "Badischen Anilin- und Sodafabrick" (BASF) and who together with Dr. Knietsch visited me in Leipzig on his homeward journey from Berlin. Then I and my wife travelled to Frankfurt am Main where M. Le Blanc who had been my student and who later became my successor in Leipzig was working at the Höchster Farbwerken (Höchst Dyeworks). The negotiations with the people in power there, vom Rat,[4] von Brüningk,[5] Laubenheimer and De Rydder

[4]Should be vom Rath.

[5]Should be von Bruning.

were promising. I told them that I was also negotiating with Ludwigshafen and Elberfeld and they thought that it would be possible to divide the production between them. A visit to Ludwigshafen and a meeting with Dr. Duisberg from Elberfeld in Wiesbaden at which we discussed how the matter should be pursued led after some further negotiations to a preliminary agreement that offered me the prospect of a more than comfortable financial future. I had no objection to a clause which restricted my income to maximally three million marks because I worried that looking after these sorts of sums would take up too much time and effort.

In the meantime my experimental protocol had been repeated in Ludwigshafen and in Höchst. At the beginning this was not successful but later ammonia was regularly made although the yield was low. In Leipzig we'd had yields of up to 8 % though this was variable and sometimes the yield was much less.

Once I had returned to Leipzig I set about organising the scientific investigation of the process and the determination of the central constants. It took some time before the necessary equipment was ready. On the morning of the day on which the experiments were to start I received a report from Ludwigshafen indicating that both there and in Leipzig the ammonia was thought to be derived from nitrogen present in technical grade iron which reacted with the hydrogen and that this happened whether or not free nitrogen was present in the gas mixture or not.

Experiments that were set up in my presence in Leipzig seemed to support this claim. I hadn't known that technical iron contained nitrogen, though this had been reported in the literature and should have been noticed. On top of that there were also reports that water vapour strongly inhibited the catalytic breakdown of ammonia and hence must also inhibit the synthesis. I therefore worked with dry gases and once again we got ammonia. However, when I used as catalyst sponge iron produced by reduction of pyrite no ammonia was forthcoming.

In the meantime similar reports to those from Ludwigshafen were coming in from Höchst, and these made it necessary to do more experiments. I tried to find out what was going on but pretty soon the endless ups and downs so exhausted me that I couldn't stand it any more. In vain my loyal assistants did everything they could but I realised that only if I gave up the business entirely could I avoid a renewed breakdown of my health. I therefore withdrew from the contract and dropped the patent application. I refused the kind offer of director Brunck from Ludwigshafen to reimburse me for my not inconsiderable expenses because after all I had initiated the whole business and set the ball rolling. To my great surprise I did not register my renunciation of all the hoped for millions as a great loss.

Nitric acid from ammonia. I came back to the nitrogen story again in the autumn of 1901 in a Sunday afternoon talk with W. Pfeffer. In contrast to me he was basically a pessimist who grasped every opportunity to pick up bad news. This was the time when once again a careless remark of the emperor had led to a disagreement with Britain and Pfeffer painted a doleful picture of the state Germany would be in if it came to a war with Britain. If the British navy prevented the import of saltpetre from Chile then we would be bound to lose because we wouldn't be able to manufacture explosives. All explosives from outdated gun powder to modern picrate explosives require saltpetre (potassium nitrate) or nitric acid and the only source of these was the saltpetre from Chile.

I had to agree with him, since I'd come up with the same arguments when thinking about the synthesis of ammonia. Nevertheless I was annoyed at the victory of his pessimism over my optimism, so I told him that it was the duty of German chemists to find an answer and that the problem could be solved.

During the negotiations about the synthesis of ammonia Dr. Duisberg had pointed out that the chemical synthesis would scarcely pay off because huge amounts of ammonia were made as a by-product during coke production and very little of it was currently exploited. I accepted this and therefore turned my thoughts to the conversion of ammonia into nitric acid. For this there were two possible routes. The first involved the direct fusion of nitrogen with oxygen as happens in an electrical arc. Alternatively one could convert a nitrogen compound into nitric acid and for this, ammonia was a good candidate. Since this latter possibility was certainly the easier I concentrated on it.

Once again my energy was insufficient to do these simple experiments myself. Initially I turned them over to one of my students who was a German–American, however he delayed starting the project. The longer I thought about it the more important it seemed to be and so I transferred the project to my trusted co-worker Dr. E. Brauer, who had already helped me a lot in the attempts to synthesise ammonia and was burning to revenge himself for that failure.

The starting point was a well known experiment used in the lectures. One poured a few drops of concentrated ammonia into a beaker and held a glowing spiral of platinum wire in the mixture of air and ammonia. The platinum continued to glow and the beaker filled with a red mist of nitrogen dioxide.

We therefore built a simple apparatus which would permit the recovery of the maximal amount of the nitrogen dioxide generated. Already in the first experiments more than half of the ammonia was converted into nitric acid. In this case the catalyst was composed of a small strip of platinum covered with platinum wool.

To increase the yield the gas flow rate was reduced in the next experiment so that it was in contact with the catalyst for a longer period. However now the yield did not improve but rather fell to around 30 %. When the flow rate was increased the yield increased to 85 %.

That all seemed very strange and I couldn't understand what was going on. Then suddenly I remembered a generalisation I'd come across in the course of my work on the question of the lower border at which materials act as solids (Part 2, Chap. 22, p. 263). When an entity moves into a state of increased free energy from which it can (and must) convert into something else, it does not immediately move into the state of lowest free energy in which it must then remain, but rather moves to the next state in which its free energy is certainly reduced, but has not been reduced to the lowest possible level. Ammonia and oxygen have the highest free energy, nitrogen oxide and water which are produced from them have a middle value and nitrogen and water have the lowest free energy. The longer the platinum was able to exert its effect the more this final state would be reached and no nitric acid would be recovered at all. To remain at the middle level as I wanted it had to be in contact with the platinum long enough to ensure maximal oxidation of the ammonia but not so long that free nitrogen was produced.

Getting the optimal yield therefore requires an optimal period of contact—anything longer and anything shorter results in a lower yield. Naturally the optimal contact time depends on the nature of the catalyst but finding it is a relatively easily solved technical matter.

This interpretation turned out to be correct. In a few days Dr. Brauer had carried out the experiments which confirmed this scenario and I now had the challenge of taking this important matter further.

The simplest thing would have been to write it up as a scientific publication, but I couldn't bring myself to do that. If I had, then the results would have been lost in the vast ocean of chemical publications and the development of the laboratory experiment into a large scale industrial undertaking, which usually takes 5–10 years would have been delayed and might not have been completed in time should the worst case scenario become reality.

Together with these general considerations there were also personal ones which weighed on me. I've already mentioned that as a result of the changes in my psychological state a change in my life style seemed to be advisable and even essential. I'd come to Leipzig so as to be able to work intensively in a conducive atmosphere and with ambitious students in a way that had not been possible in Riga. I'd achieved this, but now my excess energy was expended and those mental organs necessary to stimulate students to take on projects in ever new directions had withered away. This was not just a temporary lapse, it was a permanent state. Even if the ability to carry out creative work in the lab and while writing had recovered during my cure, the ability to direct research of students—the most difficult of all my functions—followed the general law of biology that the characteristics most recently acquired are those which get lost first.

Certainly it had turned out that these functions could be successfully taken over by my assistants and I experienced time and again the joy and surprise of seeing to what a large extent they were able to push forward the direction of our research with independent ideas and new experimental protocols. There had developed a remarkable scientific air in the laboratory which made it possible to make the best of every project, despite the fact that my contribution was now much less than it had been in the old days.

I didn't realise at the time that even if I was not directly involved I nevertheless acted somehow like a catalyst to keep the laboratory's productivity at its old level. Not that I really cared because my rapidly increasing interest in philosophical and cultural matters made me anxious to break out of my then current situation.

Now, if the discovery would guarantee me sufficient income, then I had the chance to break free without putting the economic future of my wife and five children at risk. My experience of working for several years with Brauer, who had been a teacher and friend to my sons, convinced me that all of the experiments which were planned would be safe in his hands. I was also well aware of the sorts of sums that could be expected even from much less important discoveries.

When I look back on my life at all those points where there was a sudden dislocation then I can fairly say that I was almost always happy with the way things later turned out. In this particular case, however, I am not so sure, because my involvement in the maelstrom of economic competition brought me more unpleasantness than did any other project. And as to the money, it wasn't any more than I'd made from my books and was surely less than what I'd have made had I concentrated on the literary options which were open to me then—and the books brought me not only income but scientific and literary fame as well.

This move into the technical and economic area would have to be considered an error were it not for an experience which I made over the last decade which balances things out. I had assumed that the unpleasantness I experienced was because of the money that was involved. In the meantime my efforts to spread the new quantitative concept of colour in Germany whereby any profit must of necessity lie on the receiver's side, has led to a very similar experience. I have to conclude that envy and laziness, those two enormous enemies of all progress, are more important factors than money.

Above all else it is envy, which aims to influence as many people as possible, that makes the presentation of work so joyless. Bismarck, who had had a lot of experience with this, characterised the Germans as a people who fought tooth and nail against every advance even though they'd taken no notice of it till the first successes were becoming apparent.

There is no doubt at all that envy is a characteristic that human beings have inherited from their animal forefathers. In fact humans seem to represent a pinnacle in this respect for the closer an animal is to man the more clearly developed envy is. Cows are not envious, but dogs are—and in the highest degree. In primitive peoples envy and jealousy are quite natural, and if I may generalise from my own experience, it requires a huge effort of self discipline to free oneself of this mean emotion. In our most modern social order envy was written into the law when it was decreed that workers should be paid the same amount without taking into account their different levels of achievement.

Now it is quite unlikely that this characteristic is any more strongly developed in Germans than in other people. On the contrary one would expect envy to get less as the cultural standard increases and indeed the French are decidedly more envious than we are. However other peoples do manage to minimise the effects of envy by virtue of their feelings of belonging to a nation. If a Frenchman or an Italian or whatever has managed to make his way and become famous he will be seen as a symbol of the fame and glory of his country and will be treated accordingly by his countrymen and praised—usually to a degree which is far beyond what an objective assessment of his achievements would be considered reasonable.

Germans don't have this compensatory mechanism because they suffer from a pathological lack of national consciousness. Envy becomes all the stronger when a German has not only left his national competitors far behind, but has even managed to outstrip the foreign competitors. How often I have had to witness the shameful business when a German success is thrust into the shade of some foreign competitor or forerunner who, though nobody else has taken any notice of him, is unearthed by other Germans and then praised to the skies though his achievements in no way justify it.

I have already alluded to the fact that there is a considerable difference between a successful laboratory experiment and an industrial scale production and that this difference can only really be appreciated by someone who has actually tried to do it. First of all one needs a place where the experimental plant can be built, and second the resources to do all this must be found. Neither of these preconditions can be satisfied in a scientific laboratory. Even if these preconditions have been satisfied it turns out that the invention is, on its own, of no value for dozens of other inventions are needed to get from the raw materials to the finished product. Every one of these steps has to be examined closely to see if it can't be replaced by a quicker and cheaper alternative for in these technical matters it is not just the physical and chemical imperatives that count but rather the economic factors that are decisive.

Duttenhofer. This time, in trying to make a connection with industry, I initially had a lot of luck. W. Will, who I'd already got to know on my first visit to Germany (Part 1, Chap. 9, p. 98), was in the meantime the scientific director of a research institute known as the "Zentralstelle" in Neu Babelsberg near Berlin.[6] This institute had been founded and was funded by a large concern manufacturing explosives. He would visit me from time to time to discuss particular problems that faced him in his work. Since I liked him a lot I always took his questions seriously and had now and then given him useful information. Since saltpetre was crucial for his company he soon organised a contract that regulated both the development of the process and its economic exploitation.

The leading person in the company was the privy councillor Duttenhofer. He'd been born into a poor family in Swabia and had been educated as an apothecary. By virtue of his intelligence, energy and organisational abilities he had worked his way up to become the leader of the "Köln-Rottweiler Explosives Company" where he had amassed a considerable fortune. He belonged to that quite typical class of rich people who work ever harder as their radius of action increases. He had now reached the age when nature demands that things be taken a little more easily. However whenever anyone suggested that he should slow down he'd reply, "I've tried that a couple of times. Do you know what happens? They take away all the things you like to do and leave you just with the things you don't want to do".

He had open ears for the large scale economic and political aspects of this matter which really interested me and he had an open mind. He treated me very generously and soon my income from this was more than my salary from teaching. For this reason I trusted him completely and never had any cause to regret it.

When I once arrived a few minutes before the appointed time for a meeting in the Kaiserhof Hotel in Berlin I saw him taking friendly leave of a lively white

[6]The "*Zentralstelle für wissenschaftlich-technische Untersuchungen*" (Centre for Scientific and Technical research), located at Neubabelsberg near Berlin, was founded in 1898 by Max Duttenhofer. Karl Wilhelm Will was director. It was a centre of German weapons, munitions and explosives production.

haired man who was even smaller and much thinner than the small stocky Duttenhofer. To describe him as white haired was actually an exaggeration since his hair was restricted to a thin wreath that stretched from his neck to his ears and a short moustache. The man's movements were elastic and his speech was energetic. Both of these gentlemen repeatedly shook each others hand and were clearly at the end of a conversation that had moved them both.

Afterwards Duttenhofer asked me if I'd known the man and when I said no, he told me, "That was poor old Graf Zeppelin with his steerable airship. He's run out money once again. He invested all his own money, his wife gave him all of hers and I helped out with considerable sums a couple of times. He's used it all up. He wanted more money just now but I had to tell him I just couldn't do it. He seemed to understand. But both of us burst into tears!"

E. Brauer. I turned the technical development over to Dr. Eberhard Brauer, who had already carried out the basic experiments in the laboratory. His ceaseless diligence and technical talent was what made it possible to develop from these small beginnings a large and important industry which later in the World War fulfilled the purpose for which it had been designed. When, during this largest war that humanity has ever experienced (let us hope such a thing never happens again) Germany had to fight almost the entire world to keep its territory free of enemies then our success in this is principally due to the courage of our soldiers and the prudence of our leaders. But our resistance would have been technically impossible had we not had access to the enormous amounts of explosive needed which were manufactured with our new procedure.

To begin with we first had to decide how to technically translate the laboratory experiment into a large scale production procedure. The "Zentralstelle" owned a large area near Berlin which could be used for explosive experiments and for the production of explosives. We were given space here to build the first facility. At the start it was a bit of a shot in the dark because the appropriate equipment had to be developed as we went along. As one might expect some of these things did not succeed in the struggle for the survival of the fittest and were replaced by more appropriate forms. I had to be on duty in Leipzig and was only able to go to Berlin once every 2–3 weeks. Dr. Brauer lived on the site.[7]

The technical side of things developed much more slowly than we'd hoped or expected but nevertheless in reasonable time we could look forward to regular production. But then all our plans were brought to nothing by the greatest disaster that could befall us and our project—the death of Duttenhofer. An important factory of the company had been destroyed by fire and Duttenhofer had spared no effort in organising the reconstruction. Still believing in his indestructibility he'd pushed his weakened organism too far and died of overwork.

This was not just a great personal loss for me; it was also a great loss for our project. Those who took over control of the company did not have his breadth of

[7]At that time Brauer lived in Zeuthen, near Berlin.

vision and successfully managed to have the contracts revoked. Once again our child was homeless.

After a little while we found a new helper. He was in the coal industry and as the director of a large coke works was looking for some way of exploiting the ammonia which was produced as a side product. However we'd scarcely come to an agreement when he too died. He was called Klüssener.

There is no point in describing in detail the odyssey to which we were damned. We finally landed a contract with the mining company Lothringen in Bochum. Dr. Brauer built for them a facility which from 1906 on regularly produced nitric acid and ammonium nitrate profitably and in industrial amounts.

What happened with this matter afterwards will be related later.

Chapter 26
Natural Philosophy

The lecture. At the same time as the practical and economic work on the problem of nitrogen was getting started I also took on another and very different problem. Perhaps it was an unconscious desire to balance out every enterprise in which there was no avoiding a personal financial interest with some serious efforts in a purely mental sphere.

I've already explained that philosophical questions had increasingly interested me ever since the impressive meeting in Lübeck and that my lecture at the opening of the new laboratory had a more or less philosophical content. German professors are incredibly privileged by their guaranteed freedom to teach whatever they want and so are able to exploit this opportunity to map out in lectures the form of a field that is currently in the making. This is what I decided to do with the material I had slowly collected and which now threatened to overflow. I announced in the summer of 1900 a lecture series on the subject of Natural Philosophy. My own lecture hall, which could seat around 100, turned out to be far too small as was the one that was twice as big in the neighbouring Botanic Institute which my colleague Pfeffer kindly offered me. I had to move the venue to the largest auditorium of the university and even there the audience of roughly 400 who wanted to hear the lecture on Natural Philosophy couldn't all find seats.

The title of the lecture series was more than inviting, because it held associations to the unrestricted speculations which Kant's idealistic philosophy had unleashed around 80 years ago. After a short while the victorious march of unrestricted speculation had been once more stopped by the exact sciences which regarded it with contempt. This speculative phase had been encouraged by Goethe and young people got drunk on it, seeing this possibility of unrestricted freedom of thought as a compensation for their lack of political freedom. In the end, the weak were destroyed while the stronger ones recovered from their intoxication and looked back with shock and disgust on their wasted time and energy.

In this sense it was something of a risk to be sailing under this discredited flag. However the success of the lectures showed that the title far from being

R.S. Jack and F. Scholz (eds.), *Wilhelm Ostwald*, Springer Biographies,
DOI 10.1007/978-3-319-46955-3_26

discouraging had actually attracted people. The idea that philosophy can be successfully pursued by outsiders had at this time just been shown in Leipzig by Wilhelm Wundt's wide ranging work. He was a medical doctor who had been appointed to the philosophical faculty in Leipzig and it was only some time later that the faculty realised that Wundt didn't have the necessary degree of a Doctor of Philosophy—and so they had to give him an honorary doctorate to satisfy the rules. At the same time the long delayed effects of the philosophical work of the physicist Ernst Mach were just becoming apparent and so the time seemed right for my undertaking.

My colleagues in the "humanities departments" in Leipzig viewed my efforts as an unfair encroachment on their turf, which because of our freedom of teaching, was unfortunately not actually criminal. It was, however, regarded as an unfriendly act. They would not have worried so much if the lectures had been held out in the natural sciences campus on the outskirts of town just for my students, but the large auditorium lay right in the centre of the city and the large number of the audience indicated that this was going to be unwished for competition. As a result there were a number of pointed remarks directed at me in the professors' room where I went prior to the lectures.

The book. For me the large audience was a strong motivation that spurred on my thoughts. It took about 30 min to walk from my home to the lecture hall and I used this time to sort out the content of the coming lecture. This content was of course largely determined by the general direction of the idea but it had to be presented in such a way that at each lecture the audience would be presented with a rounded up piece of thinking which would resemble in its form a work of art—even if just a small one. At the same time I tried to formulate those little witty and surprising twists which one weaves into such a lecture to keep the audience on its toes and to emphasise the most important points.

For me the way from the spoken to the written form was not far and the publisher of "Electrochemistry" was prepared to bring out the "Lectures on Natural Philosophy" which I wrote up in a few months.[1] They were read by ten times more people than had attended the lectures and in 2 or 3 years four editions had been sold out. At this point I stopped further publication because I wanted to rework the material and so it was not commercially available for quite some time until the first volume of the completely reworked version appeared under the title "Modern Natural Philosophy".[2] This new version resembled the reworked first volume of my "General Chemistry" in that the follow up volumes never got written because the flood of new material was simply overwhelming.

[1]Ostwald W (1902) Vorlesungen über Naturphilosophie, gehalten im Sommer 1901 an der Universität Leipzig. Leipzig, Veit & Comp.

[2]Ostwald W (1914) Moderne Naturphilosophie. 1. Die Ordnungswissenschaften. Leipzig, Akademische Verlagsgesellschaft.

To give some idea of the impact of the book I may quote from a postcard written in English which I received in July 1902.

> Permit a stranger to express his great pleasure in reading your lectures about Natural Philosophy. I read earlier the first volume of your "General Chemistry" but the breadth and humanity of your last sentence filled me with admiration (and envy!). This book will have an enormous influence. I think the idea that consciousness is a form of energy, needs to be further worked out and I am not quite sure if you see energy as a sort of universal entity or simply as a general term for phenomena which can be measured one way or another. In either case your work is an enormous step in the direction of providing a simple explanation of things. With many thanks, William James.

The last sentence which he referred to read: "There is no better way to work for yourself than in helping others in the widest possible way. It is here that one finds the usually unconscious springs for those great achievements by which the individual benefits the many and in the accompanying enormous extension of the self lies the feeling of sheer bliss which illuminates the one who has been given the chance to do this thing".

The postcard came form one of the foremost American philosophers, whose work on psychology is fundamental. He became even better known for the pragmatic philosophy which he developed later as a deliberate counterpoint to the standard scholastic viewpoint. As will be clear from what I described above, we'd had no previous contact and I was at that time unaware of his work. Some years later I got to know him when I was for a while his colleague at Harvard University in Cambridge Mass. But that belongs in the third volume and I'll keep a proper description of that unusual personality for later.

I think I can say that James's prophecy about the book's influence turned out to be correct. One see this simply in the fact that since 1901 the term "Natural Philosophy" has again become an accepted area of philosophy and regular lecture courses are held and books written about it and this had scarcely happened before. In the same way it is now part of any broad description of the general field of philosophy. This whole development happened quite naturally as if it were something that had been expected and that one had been actually waiting for.

For example in 1905 I was invited to write a chapter on natural philosophy in a "Festschrift" for Kuno Fischer which was intended to summarise the field of philosophy at the start of the twentieth century. I turned this offer down because I didn't want to disrupt the stylistic unity of the work. In any case I had no wish to honour the philosopher from Heidelberg, whom I'd never met and whose whole approach was anathema to me. On other occasions I have however done my bit for natural philosophy.

At the moment the situation is that natural philosophy has become a normal part of science and is considered by philosophers to be a normal part of their subject, though they do seem to fear mentioning my name. On the other hand, most of what they write on the subject is of such a form that I am happy not to be mentioned.

Some remarks on philosophy. A rather different question concerns the extent of the actual influence of the ideas laid out in my book. That the academic philosophers would disagree with it was only to be expected and is scarcely worth mentioning. The low level of development of this oldest of all sciences is shown by the fact that each new philosopher is at great pains to show how different his views are from everything that went before, while in the developed sciences progress is always depicted in the context of already established results so that it can be seen as an extension of a common structure.

The reason for this is that philosophy was always a home for those areas of science whose contents could not be reduced to a logical—or better still mathematical—basis on which everyone could and had agreed on. Thus, to begin with, all sciences were part of philosophy and this can be most clearly seen in the work of the greatest philosopher of ancient times, Aristotle. As time went by mathematics, physics together with astronomy, chemistry and biology separated from philosophy and became autonomous. Chemists were generally referred to as philosophers for as long as they plagued themselves with the search for the philosopher's stone and the elixir of life, but once they turned to more mundane things such as describing the nature of materials and developing analytical tools they lost this venerable title.

In our time psychology has left philosophy for much the same reasons and the real dyed in the wool philosophers are at pains to emphasise the difference between the two fields. When I attended the opening ceremony of the new department for the Harvard philosopher Münsterberg in 1905, he emphasised that philosophy would be housed on the lower floor and psychology on the upper and that he considered it his first duty to ensure that the two did not mix. One of his American colleagues ironically expressed the hope that he would not confine the field of logic to the lower floor but would leave a little of it over for the experimental psychologists on the upper floor.

Currently the main areas of philosophy are logic (including epistemology), aesthetics and ethics. These are three divergent topics from totally different areas of science which one tries in a makeshift way to lump together under a common designation. But neither the General Calibration Commission (Normal-Aichungs-Kommission) nor the industrial norm subcommittee ever thought to extend the work on standardisation to this area. Every area of science has its normative core but this is less developed in ethics or aesthetics than in any rational science.

This is the reason why there is no general consensus that these areas be included in sociology. The current fruitlessness of aesthetics can be seen in the assessments of leading academics in the area which was published in 1925 (Meiner, Leipzig)[3] in which they all agree that the study of the current numerous works on aesthetics had been a waste of time in terms of their personal development.

As to logic, the third area ascribed to philosophy, this is part of the fundamental organisational principle of science. It forms the lowest platform of the science

---

[3]Jahn J (1924) Die Kunstwissenschaft der Gegenwart in Selbstdarstellungen. Leipzig: Verlag Felix Meiner.

pyramid and therefore is the basis for all other sciences. In the new version of the "Lectures on Natural Philosophy" which I mentioned above I tried to describe the outlines of this much neglected field and later in my concept of form and colour I gave examples of the enormous value to be gained from applying organisational and mathematical analysis to a newly opened field and indeed to the discovery phase itself.

How is it possible that this fundamental area could be methodologically so neglected that each researcher and organiser of a branch of science has to invent the necessary mathematical tools anew? The answer is that unfortunately Aristotle's logic did not suffer the same fate as most ancient knowledge but rather survived the destruction of the ancient cultures during the migration period. Since it was, for that time, a splendid piece of work it, much like Euclid's geometry, had such an influence in later centuries that it was considered unsurpassable and this prevented any further progress. As is well known geometry played no part at all in all the stupendous advances in mathematics since the sixteenth century and it was only at the beginning of the nineteenth century with the advent of Steiner's synthetic geometry that it developed independently. Things were much the same in other areas. Painting was able to develop naturally only because practically no Greek pictures survived and luckily enough those in Pompeii were uncovered so late that the retarding influence which they soon began to exert could be overcome. Statues in contrast, because of their more solid raw material, did survive and our knowledge of them brought with it great disadvantages. First of all it completely halted the beautiful down to earth development of sculpture in the middle ages from which we have unforgettable evidence, for example in Naumburg cathedral. Beyond that it has prevented the development of an art form which reflects the ethos of our times, and this is once again due to the fact that the ancient sculptures were and are considered to be unsurpassable. Because the ancient Greek marble works survived but their colouring did not, contemporary sculpture takes as its standard not the forms produced by the ancient artists but merely the forms that survived their excavation and subsequent cleaning. This is an error which dominated art for centuries and its influence is still to be felt today.

In much the same way Aristotle's logic served as a hindrance as we can see from the fact that even Kant considered it to be the final word. It was not recognised that logic is just a small part of a larger area of science whose function is to describe the general relationships of things. Logic is merely that part which formulates the inclusion or exclusion criteria for groups of similar phenomena.

The scientific study of this general set of questions is mostly carried out by mathematicians for whom mathetics is the key tool. An important part of this is symbolic logic and Bertrand Russell is its most successful proponent.

If, in this way, logic is soon to take its rightful place in the structure of science then the same can be predicted also for the other two leftovers which at the moment form "philosophy". Aesthetics is simply that part of applied psychology which deals largely with emotion and ethics is applied sociology.

It is currently fashionable to consider intellect as being malevolent and so these remarks which place aesthetics and ethics within the frame of rational science, will find little support. However we may doubt the long term survival of this particular fashion for the representatives of these phantom areas must at some point come to realise that their undoubted success in totally excluding the application of intellect to their own work is neither praiseworthy nor useful.

Ernst Mach. The "Lectures" are dedicated to Ernst Mach who was the contemporary thinker who influenced me most at that time. He was born in Moravia in 1838 and so was 15 years older than me. He spent the largest part of his life as physics professor in Prague where his work was characterised by its originality and by superb experimental design. Through dealing with psycho-physical questions he focussed on theoretical aspects of perception and on investigations of the scientific method. These studies were original and indeed fundamental. He was one of the first to recognise the general importance of the laws of energy and to help in their development, though he drew back from involvement in energetics.

I was very much impressed by his way of thinking which held all hypotheses to be not only dispensable but indeed actually harmful. I shared this view and have already pointed out in a few places my distaste for the kinetic view of atoms. In the lectures I distinguished between hypotheses which are arbitrary assumptions that cannot be tested from "proto-theses" which are interim assumptions that are there to describe an unknown set of phenomena and are there to be tested. The first I considered disadvantageous, the second are essential.

To begin with Mach had great difficulties to get recognition for his ideas because he published them at a time when the recent victory over natural philosophy led to the situation that philosophical interests made a physicist suspect, particularly at a time when the professional philosophers had recognised the bankruptcy of the creative part of their science and considered that the history of their subject was the only remaining subject for research.

I got to know Mach at the Natural Scientists' Meeting in Halle in 1890. He was a tall thin man, untidy in his dress and movements. His limbs were too long and he had a pale complexion that was rather overgrown with brown hair and a beard. He himself told the story of how after a tiring night journey he got into a hotel shuttle bus and at the same moment another guest got on from the other side and Mach thought, "where did that down at heel school master come from", only to recognise a moment later that he was judging his own reflection in a mirror.

The story was typical of this strange man. He told this story without worrying about losing face and it showed us that he had the conceptual picture of a schoolmaster more present in his mind than his own image. He did not, however, draw the conclusion that he obviously looked so seldom in a mirror that he was not very conversant with his own image.

I almost think that this incident happened to him when he arrived in Halle because there are buses with such mirrors and night trains stop there as well. I introduced myself but we didn't interact much because he left the meeting early. He'd only come to settle a teaching matter which was to be decided there and he considered the other participants far too docile in the face of philological scholasticism for which he had no time.

When G. Wiedemann died in 1899 I wanted that Mach be appointed as his successor and so I asked him if he'd be prepared to accept the position. He wrote me a very remarkable letter in which he carefully developed in detail all of the reasons which spoke against his appointment which he would otherwise have been happy to accept. The principle objection was his age, for he was then 61 years old, and this was a point that was also raised in the faculty. The appointment never materialised.

Soon after he was appointed in Vienna as representative of the history and theory of the exact sciences, though he later, after a stroke, had to give up his lectures. I never missed an opportunity when in Vienna to go and visit him in his house far outside the city in Pötzleindorf. He was paralysed on one side but he managed, with the equanimity of a scientist, to overcome this disability and was constantly involved in research and thinking.

His basic principle was that it was not the job of a thinking person to try to construct, out of the very incomplete knowledge base of current science, a rounded view of the world by formulating hypothetical or metaphysical structures but rather to accept the incomplete picture that science presents at the moment.

The few short hours that I was able to spend with him were special for me even though I knew that that he did not agree with some of my views. We were both convinced of the honesty of our positions and were ready to accept the subjective side of things which is associated with all human endeavours.

E. Mach left Vienna at the start of the world war and moved to Haar near Munich where he died in 1916. Shortly before his death my son Walter visited him as a mark of his reverence and told me about the touching impression that the old man had made on him.

The Journal. I'd learned 15 years ago with general or physical chemistry that the establishment of an independent new branch of science is greatly helped by the publication of the necessary textbook but that it is only guaranteed by the establishment of an appropriate journal. In this case I could have persuaded one of the existing philosophical journals, most of which seemed to be teetering on the edge of extinction, to provide space for publications on natural philosophy. However, that would have meant that the current editor stayed in charge and I would have been dependent on him. This would have been intolerable for me and so I started my own journal, the "Annals of Natural Philosophy" (Annalen der Naturphilosophie)[4] and took on the job of unpaid editor myself. The first issue came out in 1901 and the first paper in this issue had been submitted by Ernst Mach.

I defined the area to be covered by the journal as the relationship between the individual sciences and philosophy as the combined science of sciences:

---

[4]The "Annals of Natural Philosophy" (Annalen der Naturphilosophie) were published as a quarterly journal in 14 volumes from 1901 to 1921. Volumes 12 and 13 were co-edited by Rudolf Goldscheid (see Name Index). In Volume 14, Ludwig Wittgenstein published his main work, the *Tractatus Logico-Philosophicus*, in German as *Logisch-philosophische Abhandlung*. Bertrand Russel wrote the preface for Wittgenstein's publication.

> One can view the broad strip of land which lies between the long cultivated fields of the individual sciences and the more than two thousand year old forest of philosophy as an area which requires power and development. These various fields were once part of the forest and in most cases their conversion into ploughed land was merely a matter of practical convenience. Nevertheless in many cases the connection of the fields to the forest has withered away. Impenetrable dialectical scrubland on the one side and heaps of native rock on the other hinders traffic between them and often leads us to forget that they are all part of the same ground and that the same sun gives them the energy which they both convert into lasting material.

As one can see these are the same ideas of forming connections between areas of thought that had formed the basis of my inaugural lecture about the connections between physics and chemistry, but now the emphasis had moved to cover the entire spectrum of science. In this sense I could regard my current efforts as a simple projection of my former work and didn't need to take seriously those who criticised me for what they regarded as aimless vacillation.

Because I was particularly hoping to get contributions from the leading scientists in the various fields, I emphasised the qualms that till now had held up this sort of approach. I readily admitted that suspicion of the old form of natural philosophy was fully justified, since it had largely been restricted to speculations about matters which could not be treated by the exact sciences. Nowadays even the generalities and all-encompassing philosophical statements in the work of people like Helmholtz or J.R. Mayer were considered to be things that might perhaps to be excused but certainly not emulated.

But there was a short window of time in which these objections had to be laid aside for in many areas neighbouring sciences were coming into contact and generating new fields, and for purely technical reasons this gave rise to a need for philosophy. "Simply through the sober necessity to be able to systematically order and store in accessible form the mass of scientific data being generated it has become a major practical challenge to define the data's common general basis".

I thought it necessary to stand up against the sort of turf battles which tend to emerge at the border between neighbouring sciences and which restrict interaction between fields. These border lines are usually placed and cemented by professional philosophers though they are occasionally set up by the different fields themselves and anybody who tries to overstep these marks will get bashed on the head with a quotation from Kant.

I wanted instead that each field of science should seek out its neighbours not in order to build a wall between them but to find areas where they could work together. Philosophy, however, has to take its raw material from the established fields of science. "Philosophy becomes more and more simply an intellectual traffic control and exchange centre which has the task of putting the incoming data into context and expressing it in a generally accessible form".

Individual "philosophers" regarded this last sentence as a debasement of their high and holy science, without realising that the organisational function I was asking for operates at a higher level than the acquisition of new facts.

The success of the "Annals" was similar to that of the "Journal" though at a much lower level. I soon managed to get a enough submitted manuscripts from excellent researchers and the number of copies sold was sufficient to make its long term economic future secure. However the circle of both potential authors as well as readers was much more limited because the international community was missing and they were the ones who would have profited from the only journal at that time which was dedicated to the new discipline.

On top of that my general condition at the time allowed me to do the work of editing manuscripts and writing discussion articles on books—but I could do no more. While in the old days my entire energy and thought had been directed to physical chemistry I now had, in addition to natural philosophy, half a dozen other major projects in my mind and that meant that there was inevitably less energy available for each individual one.

Still, I can be satisfied with the list of contributors. The first volume contained papers from Ernst Mach (physics), F. Wald (chemistry), A. Scheye (mathematics), A. von Öttingen (physics), E. Sievers (linguistics), P. Volkmann (physics), L. Natanson (mathematical physics), Ch. Pflaum[5] (psychology), H. Simroth (zoology), B. Delbrück (linguistics), F. Ratzel (geography), G. Helm (mathematics) A. Bozi (law), Wo. Ostwald (zoology), K. Lamprecht (history), G. Heymans (philosophy) and V. von Türin (physics).

As one can see the Annals were run along very democratic lines since famous names are ranged alongside complete unknowns.

Energetics. Apart from the general work which the Annals covered successfully, there lay a special task which was the introduction of energetics into philosophical thought. I've already outlined the resistance I met to the acceptance of energetics in the neighbouring fields of physics and chemistry and so one can well imagine how impossible these ideas would seem in the more distant reaches of biology and psychology. And so developed the strange situation that the professional philosophers were happy to take over the newly defined field of natural philosophy but they were not going to permit the developer to plant the important herb in the area that he'd cleared for cultivation.

Because of this it soon rained "refutations" of energetics from all sides and the book review section of the Annals provided a welcome opportunity to set things to rights.

The theatre that I had to put up with was depressing. Although the first law on the conservation of energy is really not that hard to understand, I was shocked by my critics' unbelievable inability to apply it properly. Amongst those who made the most shocking mistakes were not only the ordinary run of philosophical dilettantes but also respected professors of philosophy. It troubles me that, if they used such an inadequate level of knowledge to reach conclusions in a simple matter that I well understood, then what can one expect from their conclusions in all the other areas in

[5]Christoff D. Pflaum worked at Wilhelm Maximilian Wundt's Institute in Leipzig.

which they regularly pontificate? What actually is the point of having such philosophers in the universities?

In general the introduction of the concepts of energetics in philosophy was similar to what had happened in physics and chemistry. Despite the initial rejection, traces of the new ideas surfaced ever more often, though they were either accompanied by further objections or later were mentioned without reference to their source. Nine years after the publication of the "Lectures", the then influential idealistic philosopher P. Natorp wrote a treatise on the logical basis of the exact sciences in which he treated the inclusion of energetics to the exact sciences as being a matter which was so obvious that it required no discussion.[6] The process by which an advance becomes detached from the name of its discoverer and forthwith leads an anonymous independent existence normally takes 50–100 years but in this case it had already been completed.

The effect on the editor. My work as editor of the Annals turned out to be most fruitful for me. This work is much more arduous in a philosophical journal than in a scientific one because here sense and nonsense, preliminary and mature are much less easy to distinguish—and that of course is an indication of the low level of development of the field. Because of this many of the submitted manuscripts were not acceptable and authors always regard rejection, though based on careful examination, as a serious crime against humanity. In philosophy no one is prepared to accept that anyone else understands the matter and hence no one else is in a position to deliver an objective judgement.

When I look back on the 14 years of the Annals (which was interrupted by the world war) then I can be satisfied that I rejected no manuscript that deserved to be published and that the acceptance criteria, which I felt it my duty to leave rather lax, nevertheless did not allow things which were worthless to be published. Looking back I would now not exclude more than half a dozen of the accepted papers. That compares favourably with the record of other philosophical journals, many of which have a long tradition.

Even more work, though also more gain, came from the book reviews. The first decade of the Annals fell in a time when publishers were more than ready to publish anything they were offered. If there was no willing publisher then it wasn't that expensive to self-publish. We were a rich country and there must have been an astonishingly large number of Germans who were willing to buy almost any book on the off chance. In fact even the foreign sales of German books were good. The result was a massive flood of printed paper across the country and a strong tributary of this river emptied onto my desk and waited to be judged. No matter how determined the voluntary philosopher may be to declare every objection as fundamentally wrong he is nevertheless avid to hear the opinion of others on his work.

[6] Natorp P (1910) Die logischen Grundlagen der exakten Naturwissenschaften. Leipzig, Teubner.

So even if only 10 % of what I read was worthwhile, the inner value of this 10 % can be considered so high that all the work involved was rewarding. I got many stimulating suggestions from these books even in those with which I fundamentally disagreed. The necessity of clearly formulating my objections required that I first get my thoughts on the subject in order.

Finally the Annals provided me with an opportunity to publish thoughts which had occupied me for a long time, but had not been able to find somewhere to place them. But these are threads that were spun later and should be recounted at the proper time.

Chapter 27
First Journey to America

The starting point. Early in 1903 I received a letter from the small university town of Berkeley near San Francisco in California. It had been sent by Jaques Loeb who was Professor of physiology there to invite me, in the name of the university, to give a lecture at the inauguration of his new laboratory. His name was not unknown to me but under the pressure of so many different activities I hadn't had the chance to concern myself more closely with his research. Professor Loeb seemed to have anticipated this because he had sent at the same time a number of books and papers to give me an accurate idea of his work and intentions. He turned out to be a fervent admirer of the new physical chemistry and considered its teachings to have been a major factor in his success. He wished to underscore the debt of gratitude which he felt for our science, by having me present at his inauguration celebration.

The man. Jacques Loeb was born in 1859 in Mayen near Koblenz, studied in Berlin, Munich and Strasbourg, and had made a name for himself through his early work. In this he showed that the well known phototropic property of plants, which had been interpreted by the founder of plant physiology, Julius Sachs, as a learned response to move towards the light, could also be demonstrated in animals, and that animals obey the same rules. Thus, every living organism, which is both light sensitive and capable of movement, operates by the same principle namely that it orients itself so that analogous body parts make similar angles to the incident light. If movement is limited, as in plants, then light causes it to adopt an appropriate stance; if movement from place to place is possible, as is the case with most animals, then a seeming attraction to, or flight from, a source of light becomes evident. However these movements are not "instinctive" attraction or repulsion to the light, but rather reflect a basic process which involves the positioning of the creature symmetrically to the light source. Depending on whether the head is preferentially directed towards or away from the light the result will be an approach or withdrawal.

J. Loeb laid great emphasis in the description of these significant experiments on the fact that phototropism required neither consciousness nor instinct. This was because the behaviour of animals in this respect was just the same as that of plants,

R.S. Jack and F. Scholz (eds.), *Wilhelm Ostwald*, Springer Biographies,
DOI 10.1007/978-3-319-46955-3_27

which are not considered to have cognitive abilities, and in both cases the phenomenon can be ascribed to direct physiological effects.

Throughout his entire scientific career Loeb strongly asserted that explanations for the phenomenology of life should be sought in physical chemistry and later this standpoint made possible his most remarkable scientific discoveries. His use of the Dissociation Theory, which at that time was still very new, turned out to be particularly fruitful; it was not the innumerable different salts as such which were responsible for physiological processes, but rather the ions which, independent of the particular salt involved, were brought into solution. This led to a considerable simplification of the work.

At the time Loeb sent his letter to me his research had reached a high point with the induction of artificial parthenogenesis—the generation from unfertilised eggs of sea urchins and other lower animals of viable hatchlings—by merely treating the eggs with appropriate concentrations of particular ions. This discovery had created a sensation; Loeb seemed to have brought a solution to the puzzle of life one large step closer. As is the way in America, this impression had caused a rich sponsor to make available the resources necessary for the long overdue building of an adequate laboratory.

The fact that Loeb accorded the representative of physical chemistry such a prominent role in the inauguration of the new institute was not only an expression of thanks, but also a clear accentuation of his fundamental scientific viewpoint that life processes should be elucidated by the application of physical chemistry.

The journey. I didn't hesitate to accept this invitation which had so many attractive facets, particularly since my visit to the new world would be in the university holidays and so would not involve any interference with my official duties. I was young enough—not yet fifty—so that I could face the prospect of the exertions involved without worrying whether I'd be physically up to it or whether I'd be able to enjoy and absorb the many new impressions which would await me there. The one and a half week sea voyage at the beginning and end of the journey seemed to guarantee that I'd arrive refreshed in California and return home refreshed as well. I deliberately did not choose the fastest ships so as to lengthen the beneficial effects of the crossings. In any case, experts in the matter had told me that the company on the slower ships was usually much more agreeable than the rich upstarts who travelled on the fastest vessels.

And so I set out at the beginning of August for Bremen from where I'd sail to New York. There I would be met by my one time pupil Dr. Young who would be travelling to Palo Alto where he had just been appointed to a professorship at the second Californian university. He'd kindly offered to relieve me of the technical difficulties of the long transcontinental journey and turned out to be both a willing and adroit travel companion who was of the greatest help especially as, quite by chance, my journey was plagued with hindrances which I could scarcely have coped with alone.

On the journey to Bremen I made a midday pause in Magdeburg where I met a colleague from the Leipzig medical faculty with whom I got on well. He asked where I was off to and when I told him California he wanted to hear more. I told him the story and he remarked: You'll have a triumph there. I had to admit that I hadn't thought about that aspect of things, because planning my first voyage across the Atlantic and then the journey over continent on the other side of the ocean had been at the forefront of my mind. He turned out to be right however and afterwards I could have said like Liebig, if one could get fat from honour then I would have a paunch like a Lord Mayor.

The voyage on the north German Lloyd steamer "Weser" was up to expectations. Before the world war (perhaps it is once again the case) German steamers were well known to be the fastest, most beautiful, cleanest and in all respects most congenial of ships that sailed the seven seas. On the Weser the food was beyond praise, the organisation and cleanliness faultless, and the highly educated ships officers encouraged an atmosphere which was bright and comfortable. In the morning and at dinner in the evening the ship's orchestra played good music and in this way any feelings of boredom, which the limited space on board might have called forth, were banished. At the same time personal relationships to the other passengers quite naturally developed and became livelier.

We also experienced the whole spectrum of possible weather. Most of the days were sunny and pleasant but we also had 2 days of hefty storm which I survived without becoming sea sick.

The lecture. There was just one difficulty to be overcome. In Berkeley I was supposed to hold a major lecture on the relationship between physical chemistry and biology. Before I left there had been so many things to get cleared up that I hadn't had the chance to think about it, and because it would be published in Berkeley afterwards it had to be in written form. I'd of course told myself that I'd have plenty of time during the voyage to get it ready and imagined that it would provide a pleasant way of filling the many free hours at sea. However, since I was impatient to get on with this task, I had already got the major themes laid out on the train journey and had used the half day delay in Bremen to get started with the writing.

After the first day of the voyage had been devoted to the distractions of getting to know the ship and in following the approach to Southampton, during which I saw again the Isle of Wight, I thought I'd spend a pleasant morning on this work. To my surprise I had great difficulties to concentrate on it; my thoughts escaped strict discipline and ran off like young dogs to swarm aimlessly but happily around. Because of this the lecture made little progress and, since there was lots of time, it was put off to the next day. This was far from being my normal way of working.

On the next day things went much the same, and I soon came to the conclusion that this state of cosy laziness was a direct consequence of life on the sea. I don't know how that comes about. Perhaps the action of the waves generates a fine spray of sea water containing traces of sodium bromide which, after having been taken up by the body through the lungs, manifests its sedative effects. Perhaps it is due to the relatively high barometric pressure which predominates at sea level, for I have always found that low pressure tends to make me agitated and can even rob me of my otherwise healthy sleep. Perhaps it was also being cut off from the daily newspapers, for this was before the invention of the radio telegraph and so one led a completely cut off life on board ship. On top of all that the constant exposure to fresh air led to a healthy appetite so that the digestive tract was busier than usual and perhaps reduced the blood flow to the brain. Probably all these things worked together to bring about that unfamiliar but pleasant state.

However the text of the lecture did get written without effort, and even the work on "School of Chemistry"[1] which I'd brought along with me, was brought forward by several pages.

After this experience I can recommend nothing better than a sea voyage to those suffering from mental exhaustion. Not, to be sure, on a pleasure boat where the other passengers interfere with one's sense of well being, but rather on a cargo steamer so long as it offers at least to some degree the necessary level of comfort.

The other passengers. Among the other passengers were two wholesale merchants from Hamburg who at the beginning treated with friendly irony the Professor who studied and taught a subject which actually did not exist. Bit by bit, however, they became willing to accept me and my work. The elder one told me that it had been his idea to transport petroleum in specially built tankers rather than, as had been done up till then, in barrels. He added the following anecdote: In 1890 the demand for lamp oil suddenly slumped after which the curve of the yearly increasing demand then rose from this low point at the hitherto normal rate. It was a matter of commercial concern to him to find out the reason for the slump but he didn't hit on it for a long time. Finally he found the cause. In the same year Middle European Time was introduced in the German empire, where up till then local time had been used everywhere. The result was that in the West when people went to bed at their accustomed time by the clock then they in fact put the lamps out half an hour earlier than when they'd used local time. In contrast, in the East they stayed up for an extra half hour. However, since the West is much more highly populated, the reduction there far outweighed the increase in the East so that the overall demand for light oil fell.

I complimented the old gentleman on his shrewd analysis and admitted that I would scarcely have come upon this explanation myself.

New York. The entry into New York was anything but imposing, because disembarking passengers had to gather in the dining room and there join a seated queue of passengers waiting their turn to individually tell the customs officers exactly what they had brought with them. Since I hadn't expected this I was right at the end of the queue and had to wait so long that by the time my turn came the ship had been towed almost to the quay.

---

[1] „Die Schule der Chemie: eine Einführung in die Chemie für Jedermann" was an introductory book written by Ostwald in the form of a dialog between a teacher and a school boy. It was first published by Viehweg, Braunschweig in 1903/04. This very popular book had several follow-up editions.

Once we had disembarked we were driven like sheep into a large hall with a very dirty floor and with large letters of the alphabet around the walls. Each passenger had to go and find a place next to the starting letter of his surname where his baggage, which had also been labelled with the letter, had already been deposited. After an interminable time everything was exhaustively examined for contraband goods. And then we were finally let go.

In front of the door of the customs building—nobody could enter it from the outside—Dr. Young was waiting for me. He'd been joined by Dr. Heimrod another American former pupil and together they brought me to a travel agent on Broadway where I bought a ticket to Berkeley. The insufferable humid heat which makes life in New York almost unendurable in the late summer meant that all doors were left open and the officials worked in shirt sleeves. An overwhelming din from cars, trams and newspaper boys surged in from the street while inside half a dozen typewriters, telephone bells and other things were in constant use, so that Young had to shout his wishes directly into the official's ear. In no time at all I had a terrible headache and breathed a sigh of relief when we were able to flee into a quiet side street to get something to eat. I declined with horror the offer of a tour of New York before leaving for the West. I hoped to be able to do this some other time at a more appropriate season of the year and in fact that later became possible when I spent one winter month in the city. For the moment, however, I was simply in a hurry to get out of this noisy hell—and so we left the city that evening.

The journey. When I checked my baggage through to San Francisco, the porter suggested that it would be better to secure my trunk with a strap; this would cost one dollar. I though that was unnecessary, but Young explained to me that this was a form of secret baggage insurance which had been introduced by the porters' union and now applied across the entire United States. Baggage secured in this way would be handled carefully and not stolen, so that it would be a dollar well spent. I paid up and indeed my baggage arrived safely, although the journey turned out to be very erratic, as I will relate later.

The first leg of the journey was to Buffalo and the Niagara Falls. It was not just the spectacular view that attracted me to this immense mass of flowing water but also the technical aspects of the recently completed first power station which was able to capture a part of the enormous amount of energy that otherwise would have been uselessly converted into heat.

The Niagara Falls made a strong impression on me, not only because of their breadth and shear power, but also because of their scenic beauty. The water first has a chance to lose its sediment in the upper lake after which it runs through a short stretch of river bed but remains so clean and clear that organic material only slightly tinges the natural ice blue of pure water towards sea green. A wealth of picture motifs offered themselves and I much regretted that I'd packed my paint box in the heavy baggage for the journey to Berkeley. I decided there and then that if I ever came back to America I'd reserve a couple of days to go and paint at the falls. In fact I was able to do this the following year.

The technical aspects of the power station turned out to be highly interesting and I noted with satisfaction that the more complex machines were of German manufacture. Of particular interest to me was the plant of a company which had intended to produce nitric acid out of air by means of electrical discharge. Unfortunately I couldn't get in because the factory had been closed down. Instead I was able to see the electrolysis of melted sodium chloride for the production of chlorine and sodium which at that time was an important advance.

From Chicago to Colorado Springs. The next stop was Chicago where we stayed for 1 day. The initial impression gained in New York, that America excels first and foremost in its indescribable level of noise, was here very considerably reinforced. The "Elevated", an electric train whose track runs through the streets on girders, seems to have been built with a view to extracting the maximal possible noise from the system. The houses and streets were dirty, the road surface very bad and since the local coal comes from recent deposits it produces large amounts of black smoke. From the distance the profile of the city looked very un-European. At home one always sees a mass of houses of more or less the same height from which church spires and the roofs of grand buildings stand out. In contrast, an American city looks from a distance like a ruined wall or a row of rotten teeth. Without any regularity at all low houses of three or four stories abut ones with 12–15 floors and all naturally end in blunt stubs of roofs which are as broad as the base area. In Chicago every taller building had two plumes of smoke—a black one from the furnace and a white one from the exhaust steam from the machines which power the numerous lifts that are continuously in use. Central electrical power stations were not at that time available. Overall the city made a truly repellent impression.

The inhabitants were astonishingly uniform in appearance. Every man wore the same straw hat; the same stand up collar with folded over ends, the same necktie, the same boots. Similarly the clothing of working women and girls consisted of a straw hat, white blouse and black skirt.

We thought we'd need 3 days to reach San Francisco form Chicago. Since the journey was tiring we had booked a state room in the sleeping car and this turned out to be much more comfortable and convenient than the usual sleeping car berth. I'd taken one of these from New York to Buffalo and found it much less comfortable than German sleepers. The reason is that sleeping cars were invented and first introduced in America, where the original form had been retained unchanged, while in Europe, in particular in Germany and Sweden, their design and development has made considerable progress.

The route led through endless fields of corn between which one saw few houses and practically no people. Countless billboards had been erected all along the track. Now and then one glimpsed a few trees, but there were no woods to be seen.

In the evening there was a thunderstorm of a violence which I have never experienced in Europe. The lightning flashes came ever more quickly so that soon there was no darkness between them and the thunder crashed continuously. Since the monsoon like rain threatened to endanger the not very solidly built track, the train was driven very slowly.

The next morning was pleasantly cool as a result of the thunderstorm, and this was very welcome because we had in the meantime reached the prairie. This was a gently rolling land covered with short scrawny grass. Everywhere one looked one saw the heaps of earth which the prairie dogs—a kind of rabbit—produced while building their burrows and often one of the animals sat comically on top of the pile.

Slowly the mountains came into sight—blue at first but then of an unusually clear violet-red with a colour tone of 10 in my scheme. The country became more undulating with numerous huge strangely formed rocks the size of houses scattered around. In Denver it turned out that the train, which was to have taken us to San Francisco, had left long ago. Since there was only one train per day we only travelled the short distance to Colorado Springs hoping to catch the San Francisco express the next day.

Colorado Springs is at quite an elevation and serves as a summer and recuperation resort. Close by lies the "Garden of the Gods" which is a particularly fine group of the huge rocks we'd seen on the way to Denver. The peaks of the Rocky Mountains are quite close but here they don't offer the sort of picturesque view that we're used to from the Swiss Alps.

On to San Francisco. We left Colorado Springs American style with an hour and a half's delay which then expanded to 6 h, so that although, according to the timetable, we'd had plenty of time, we now had to give up all hope of catching the express train. Unfortunately that implied much more than just having to continue the journey on a slow train, because the line here was single track and any delayed trains were treated as extras and they had to wait on a passing point to let the regular ones pass. For us this was going to mean a delay of two whole days. However, worse was to come for just at this time a great national celebration for veterans and their families was being staged in San Francisco and vast crowds of people were streaming from all directions towards our goal. The transport system in America was set up to be just sufficient for normal circumstances so that any increased strain on the system led to chaos in which the official timetables were simply abandoned. It was a case of every man for himself. The railway companies, which in America are private enterprises, had in this particular situation pressed every last piece of equipment into service and the antediluvian locomotives, asthmatic with age, wheezed and panted as they tried to get the long lines of carriages into motion.

To begin with it wasn't so bad, but it got worse and worse as we approached our goal.

Shortly beyond Colorado Springs there is the famous Royal Gorge. One travels for nearly 1 h through a narrow river gorge with much-riven shear rock walls on either side. At the narrowest point there is no space for a railway embankment and the track runs on supports which hang down from steel brackets. It was approaching evening as we entered the gorge and soon after a thunderstorm started. The lightning flashes lit up the wild scenery making this a journey to remember.

The railway line rose quickly to the top of the Rocky Mountain pass at around 3000 m, but the night was too dark to see anything. Because of the many hair pin bends the wagon was swung around a lot and it was difficult to sleep.

In the morning we were on a barren looking high plateau. The only trees were junipers which always grow on the edge of deserts but these soon disappeared and were replaced by the most grandiose view that I saw on the whole journey. Grey spherical bushes of some desert plant were scattered here and there on the ground while all around towered mountains of unusual size and shape. They were not volcanic peaks but rather the remains of a vast rock plate which had been cut up in complicated ways by glaciers and rivers. In other words, this was the same sort of formation as the sand stone mountains along the river Elbe in Saxony, but here they were magnified beyond belief and completely devoid of vegetation. On all sides there rose up a constantly changing panorama made up of the strangest castles, walls and towers.

The most remarkable part of it all was the colours. The rock was of a lively yellow to red colour which must have been around orange level three. Above it in the clean smoke-free mountain air, the blue of the sky was the clearest I have ever seen. In bright sunlight this gave the scenery a predominantly violet-red colour (colour tone 10) which, with increasing distance, graduated down to pure light blue (colour tone 15). This last was the colour of a distant mountain that rose high over the plateau on the left and was so far away that it accompanied us for half a day. The railway ran beside a stream whose clear sea green water was in stark contrast to that of the rivers down on the plain which we'd seen before. The Missouri, which we'd crossed a few days before, had such filthy water that one wondered where the fish went to wash when they felt the need for cleanliness.

By the afternoon the ground became flatter and the junipers reappeared to indicate that we had again reached the edge of the desert. The junipers were joined by pines which had black bark unlike the red of ours and this gave the landscape a gloomy appearance.

In the evening we reached Utah, the country of the Mormons. Through the use of extensive irrigation using wooden channels, similar to those I'd seen before at Meran in South Tirol, they had managed to master the arid climate. The success was magnificent; in the middle of the desert there appeared luxuriant fruit gardens and fields which could make full use of the abundant sunlight. Particularly striking was the large numbers of Italian poplars which not only lined the streets but also divided the land into large quadrants. I didn't manage to find out what purpose these trees really served.

At the beginning we got regular meals, because in the morning a dining wagon was added to the train. However the next day we were told that breakfast would be served at one of the next stations. Once there, there was a race to a wooden shed where coffee, bread and so on was waiting on long tables.

We now reached a God forsaken part of America—the alkali desert—which lies as a high plateau between the Rocky Mountains and the Sierra Nevada. The desert's name derives from the soil's high concentration of sodium carbonate which is produced by weathering of the rocks but which is not washed away by rain. I soon noticed it in the form of intolerable biting dust that got through every crack and joint in the wagon and severely aggravated the mucosa. The land is grey to white with lots of low dune-like hills. Here one sees no animals and the only plants were isolated clumps of low gorse-like bushes with hardly a trace of green. The only settlements were along the railway track.

The next morning we had breakfast in one such settlement which seemed to be largely inhabited by Chinese. Breakfast was supposed to be at 9 am but we didn't arrive till 10:30. The ensuing race ended in a barn where the tables were boxes over which red boards had been laid. Dubious ham and astonishingly tough beef was waiting for us on plates which were on their last legs. A few unwashed women went around with pitchers and distributed chicory tea while bread and butter, both rather old, were on the tables. Everybody ate what they could grab from the tables or get from the neighbouring kitchen. Dr. Young took care of me like a mother so that I was able to share my supplies with my neighbour a trembling old man who looked like a failed gold miner and ate with his fingers.

It was astonishing that all this was accepted by everyone with good humour and without any quarrelling. I've noticed on many other occasions that on average Americans, despite their nonchalant behaviour, respect each other and immediately adapt to the current circumstances so that everybody gets their fair due. As I mentioned above, over crowding of the public transport system is an everyday occurrence, and yet I seldom saw—neither then nor on my later visits—that this gave rise to any grumbling or quarrelling. I must confess, to my embarrassment, that in this respect the Americans are culturally ahead of us. Perhaps it's a left over from the early days of land settlement when everyone was dependent on the goodwill of their few widely scattered neighbours to overcome the many difficulties and dangers that faced them. This—in the best sense of the word—democratic attitude perhaps survived not only due to long practise but also because Americans very often still find themselves facing other similar difficulties. In any case I can only hope that that this attitude will be among the many things which we learn and copy from the Americans.

There was more trouble at "dinner". The train, which by this time had grown very long, stopped for dinner at a green patch in the desert where a settlement had been made possible by boring a well for irrigation water. The place was called Humboldt, which reminded me of home. We'd had nothing to eat all day and our Negro conductor had started to turn green till we brought him back on his feet with some biscuits that Young had procured as a precautionary measure in Colorado Springs. I found some half melted chocolate from home in my holdall, and so all in all we managed to scrape by. However many of the passengers were completely without provisions. Not surprisingly, the barn where we were to eat was packed full with a rather agitated crowd which pushed its way towards the tables, albeit still with a semblance of composure, and even now there were no loud arguments. Only the Negros had lost all control and dashed yelling and howling through the room. With the help of an extra dollar Young managed to get for each of us apple pie in an iron pan which apparently came straight from the blacksmith, together with coffee which was terrible, and milk which was good.

Rather stupefied by all these experiences I slept early and in the morning tried to extend my sleep by dozing for as long as possible. Despite this I got up quite early and a glance out of the window made me breath a sigh of relief, for during the night we had crossed the Sierra and were now already in California. The countryside was pretty: charmingly hilly and rich in trees and bushes of the liveliest green. Their colour was accentuated by the contrast to the earth which was bright red and almost without grass cover.

We reached San Francisco at midday instead of as planned at 6 am. The train had not delayed itself any longer by stopping for breakfast and so we just had to scrape the rest of our provisions together. In San Francisco Professor Loeb himself met me at the station. There I thanked Dr. Young, who was going on to Palo Alto where he held a professorship, for all his help on this adventurous journey, and then went with Professor Loeb to his house in Berkeley.

My host. Jacques Loeb was a small gaunt man with thick black hair, a blue sheen on his chin and cheeks, dark eyes and a, so to speak, pointed face: pointed nose, pointed moustache, pointed chin. He was a lively and rather nervous character. To be honest, the admiration which he expressed for me seemed to be almost pathological. It was, I think, largely conditioned by his strong sensibility for public recognition of scientific achievement. This was largely because he felt that he had not been accorded the recognition he'd certainly deserved for his excellent initial work (Part II, Chap 27, p. 319). On the other hand, he gave me credit for achieving a rapid rise without considering the many lucky breaks that had made my career possible.

In the meantime his name had been made in America by his discovery of the chemical fertilisation of sea urchin eggs, that is to say the demonstration that a simple chemical trigger was sufficient to induce the development of the resting egg. As is the way of journalists, this fact, which was extraordinary in itself, had been presented to the general public in the most fantastically distorted form with the result that he was daily bombarded with letters containing the most absurd queries and request. Mostly it was childless mothers or fathers begging him to use his discovery to help them get a child. This too put him out of sorts, and in fact he seemed to have a particular talent for making himself feel unhappy. The fact that I am completely different in this respect must have increased his joy at our getting to know one another.

It was with a feeling of great contentment that I moved into the cool airy room complete with its own bathroom in his house and was able to wash away the traces of the long hot journey. After that I was introduced to his family. His wife was a tall strongly built woman of cheerful imperturbable demeanour which made her the exact opposite of her husband for whom this must have been a great blessing. Two lively adolescent boys and a 1 year old girl completed the group around the table.

The dishes were served by a strange creature whose sex I could not deduce. When I asked I was told that this was the Chinese servant who did all the kitchen and house work. Mrs. Loeb assured me that she was very satisfied with him. He was dependable and reliable, though he did insist on doing everything in his own way and would not let anybody interfere. He did all the shopping for the kitchen and so on, paying for it out of the household petty cash and settled the accounts with her every day. She thought he probably fiddled the accounts just a little but only to a very small degree and in any case this was more than compensated by his defence of the household's interests through hard bargaining with the traders. In his white clothes—the trousers were so wide that they resembled a woman's skirt—and with his plaited hair wound round his head and a completely hairless old face he did indeed look sexless. Mrs. Loeb regretted that she wouldn't be able to keep him for long, for he, like most of his countrymen, was set on saving up a small sum of money so that he could return home. Being buried in his motherland was for him not just a wish but an absolute necessity.

Receptions. Already that evening there began the first of the numerous festivities in my honour. The Californian state university at Berkeley considered itself still young and a visit from a Professor from the famous old University of Leipzig was seen as a sort of accolade. The fact that Loeb had arranged my visit had markedly increased his position and for this he repeatedly thanked me. Similar feelings were expressed by the small number of scientists in San Francisco who were naturally closely associated with the university.

I spent the first evening in this broader circle. Dr. Herzfeld, a successful and well off doctor in San Francisco had begged as a special favour from Loeb the chance to be our host. Having in fact been hungry the previous day on the train, I now faced the most luxurious banquet I'd ever seen. The table, in reference to Loeb's work, was decorated with mussels, corals, crabs, sea urchins etc. In between these were arranged wonderful flowers some of them very exotic together with little electric lamps. It was all quite exceptional, but in good taste. A band was playing in a neighbouring room and from time to time a well fed gentleman in a dinner jacket declaimed something or other. I was told that he was a well known entertainer, but my English was not sufficient to enjoy his presentations for this is a language which it is much easier to speak than to understand when someone else is speaking. I don't remember all the delicacies except for a shiny silvery fish that was as flat as a flounder but much smaller. This was an exceptional speciality which had been brought from Japanese waters and the remarkable thing about it was that they were individually put into a paper envelope before being cooked in butter. It didn't taste very different from our sprats at home and as far as I was concerned it could just as well have saved itself the journey across the Pacific.

The conversation was blithe and lively in both German and English. They pointed out to me right at the start that Californians were a completely different nation from the people over on the east coast in New England. Here there was a zest for life and people were much more attuned to the arts and sciences. They thought, rightly, that I'd far prefer their approach to life.

As a matter of fact I found this opinion confirmed in the many encounters with the inhabitants of this wonderful land. The breed of people there is handsome and this applies equally to the men and the women. Both are tall and their limbs are well formed as a result of their outdoor habits and their assiduous devotion to sport. The frequency of dark hair can be ascribed to remnants of Spanish blood derived from the early settlers of this region. However their height and build showed that the ancestors of the majority of the present population had been colonists from the east who had arrived here after overcoming unbelievable dangers and difficulties. Only the most enterprising and audacious would have decided on such a journey and of these only the strongest and boldest would have reached their goal, there to pass on their genetic make up to their progeny.

An objective confirmation of this personal impression can be seen in the fact that the majority of American artists, writers and philosophers originate from California.

The ceremony. The next morning I was sat in a car to be driven where ever I wanted. However this was not terribly useful since I knew neither the city nor its surroundings and so couldn't formulate any goals. After that there was lunch with Wheeler, who was the president of the university. He turned out to be a good looking man of middle height with pleasant manners who projected with unpretentious satisfaction the air of a well educated man of the world. He was a comparative linguist and hence not entirely free of the intellectual limitations which those in the "humanities" so rarely manage to overcome. Also here the tone of the conversation was cheerful and free of mindless formalisms.

From there we went to the "campus", the extended university grounds. Here, scattered between trees and lawns, were both the buildings where teaching took place as well as the dormitories which housed the majority of the students. I was given a black robe which is essential for all official events there and was then led to the large hall where the opening ceremony was to be staged. I took my place on a dais amongst the professors and we watched the students March in.

The student body is divided into societies whose names in general consist of two to three Greek letters. The constitutions of these societies are quite similar to those of German student "Corps" as is the solidarity of the members which extends far beyond their time at university. The American student societies have the great advantage that they repudiate binge drinking and fencing duels although, to be sure, they have their own customs some of which are so barbarous that they seem to have been taken over from the native Indians along with the war cry which each society has adopted. I got to hear the complete collection as each group marched in close order into the hall, formed up at their appointed place and let loose its war cry in a sharp rhythmic chant, before sitting down.

The ceremony was opened with a prayer from which all denominational connotations had been carefully removed. This was followed by a speech from the president, my lecture[2] and finally one from Loeb.[3] I'd said right from the beginning that I'd speak German, which I did. Afterwards I frequently heard regret expressed that I had not spoken English, because though my colleagues there could nearly all read German, they did find it difficult to follow a spoken lecture. This was naturally even more the case with the students though that didn't stop them loudly applauding my lecture which I took to be a signal of thanks for my personal attendance.

---

[2]Ostwald W (1903) The relations of biology and the neighboring sciences. Univ Cal Pub Physiol 1:11–31.

[3]Loeb J (1903) The limitations of biological research. Univ Cal Pub Physiol 1:33–37.

The day was brought to a close with a festive dinner given by Loeb at which I had to endure a number of toasts and speeches about myself. With the American love of superlatives I had been proclaimed to be "the greatest living chemist" and the speeches were all along these lines. In my replies I tried to move from personalities to facts and each time that I was obliged to talk I tried to say something original. This elicited such a friendly, indeed enthusiastic, response that those who couldn't understand my German were piqued and made clear they wanted to share in things. And so I finally tried to express myself in English and this earned me clamorous applause.

San Francisco. All of this had taken place in the small university town Berkeley which lies half an hour away from San Francisco. The next day was to be for the capital city and the representatives of its intellectual life. After a lunch at the university club I was introduced to a colleague from the other side of the intellectual divide, namely the landscape painter Keith who was considered locally to be the best in his field. I was received very kindly by him, which I was told was by no means always his way, and we had a carefree discussion about art. He complained that in order to make money he had to paint the particular places which interested his customers—landscape portraits so to speak. For the pictures that he himself liked to paint—amazing natural impressions which he executed quite expressively with fleeting strokes—he could find no buyers. I missed in these landscapes the specifically American elements which had made such an impression on me during this short visit. California in particular seemed to me to offer a myriad of the most beautiful scenery. For him these were everyday things that did not fascinate him. His ideal was, as he explained to me, to paint like Hobbema had painted in his time though I could scarcely accept this without protest. At the end he gave me a sketch he'd done in his style, as a memento of our meeting. I have it to this day.

From Keith we went to the Bohemian Club which is the artists' society of San Francisco. The rooms were decorated with many paintings by present and previous members and I could see from these that American painters are strongly influenced by those in Europe, particularly Paris. I could not recognise any specifically American style and, as a result the achievement was no more than mediocre. In fact it wasn't clear to me why these many works had been painted in the first place.

The personalities I met there were full of the allure, which was so missing from these paintings, and I had to be satisfied with that. They portrayed in lively colours the artists' festival which was held each year in the form of a fanciful gypsy camp in the woods and which lasted several days. They advised me to arrange my next trip so that I could take part. Sadly it never came to that because this visit to that country—one of the most beautiful I have ever seen and certainly much more beautiful than Italy—was the only one that fate allowed me. This made it all the more extraordinary to me that no landscape painter had even tried to capture its special beauty in his works. In later visits to other parts of America I kept my eyes skinned for the discoverer of the American landscape. I didn't find him.

In the evening there was a gala dinner given by the local Chemical Society for chemists from the surrounding area—about 60 people in all. In all my life I was never so deluged with praise and prizes as on this occasion. The tone was set by some of my former students who were in the meantime Professors at various institutions and whose happy memories of their work in our team of mostly highly talented associates were reinforced with gratitude for the hospitality they had experienced in my house. My wife was also mentioned in the warmest terms and the mouths of those who hadn't had the luck to study in the institute in Leipzig started to water at the very mention of this paradise.

Since none of the speakers could expect any benefit from me I was able, even after discounting the American love of superlatives, to take these remarks as expressions of genuine friendliness and bask in these pleasant sentiments. All of this naturally considerably increased the verve of the responses which I gave for me and my wife, partly in German, partly in English and probably also in a mishmash of both so that a friendly and enthusiastic elation developed which has remained with me, and perhaps with some of the others, ever since.

The neighbourhood. The next day was given over to leisure. It began with a drive through the beautiful city park to the Cliff House by the sea with its famous sea lions. The view there was unfortunately restricted by fog. After that we went with a little rack-and-pinion railway up the Tamelpais—a mountain from which one has a wonderful view out over San Francisco bay and its opening to the west—the so called Golden Gate. The ascent is pretty in terms of the landscape; in particular one sees many groups of Californian giant spruce which are not only enormous but also very beautiful trees. The railway company didn't seem to find this line spectacular and so they'd had some problems finding some way to apply the inevitable American superlatives. Finally they'd found it: because of the many hairpin turns they declared it to be the most twisted track in the whole world and they used this slogan in all their placards and other advertisements.

A young couple joined us and I was introduced to them. The man was a trader and in the course of the conversation it soon became clear that he had no particular interest in my scientific work or in my other interests. His wife was a typical Californian woman: tall, good looking and with a truly beautiful face. Her movements were light and elastic. She seemed to be rather bashful and, since they both only spoke English, conversation was meagre. Afterwards I asked Loeb how we'd come to be in their company. He told me that I had spoken so much about the beauty of Californian women that Dr. Herzfeld had chosen the loveliest woman from among his acquaintances and patients and persuaded her and her husband to accompany us so that I could enjoy the sight of her longer than would be possible in a brief social round.

On the return journey I learned for the first, but not the last time, the Californian love of hair-raising journeys. We were sat in a little wagon without any rack and pinion or locomotive but just with a driver at the front who worked the brake. One shove brought the wagon rolling down a precipitous track at what quickly became an enormous speed. The driver's art consisted of braking just sufficiently at each of the many curves so that the wagon just stayed on the rails. When he managed this particularly well he gave a great cheer with could be heard over the roar of the wagon and each time this increased his own enthusiasm. The furious descent was completed without accident in an astonishingly short time.

The Lick Observatory. After I'd got to know the area around San Francisco, we made some further excursions. The first of these was to the famous large telescope at the Lick Observatory on Mount Hamilton. I'd met its director, Campbell, at the gala dinner and rated him highly. He'd pressed me to come and visit. The mountain is like the Rigi in the Swiss Alps. It towers above and dominates all the surrounding countryside and since its peak is far above the level of the fog the atmospheric conditions are very favourable for astronomy.

We travelled as usual by train and steamship to the little town of San Jose which lies at the foot of the mountain. From there we went up the mountain by a coach and pair with three changes of the horses. The road lay often right on the edge of an abyss and the driver did love to use the outermost edge which was not secured by any wall or stones. After around 3 h we reached the observatory.

On the top we found a whole village where about 50 people stayed because all the scientific, technical and administrative staff together with their families lived there. In fact they'd just opened a school there for the children. Professor Campbell and his amiable wife had three sons and so we discussed a number of pedagogic issues with considerable enthusiasm and by so doing brought to light the contrasting American and European points of view.

Though I have otherwise little interest in astronomical questions, I saw many things at the observatory that enthralled me. Among a collection of beautiful photographs of the moon I noticed a curious furrow on the otherwise smooth surface and asked Campbell what it meant. He said that they interpreted this as a track caused when a large meteorite hit the moon tangentially and then bounced off again because of the low gravity.

Naturally I admired the large refracting telescope with which Saturn was shown to me in blinding luminance. Lick, who had donated the money for all this, had been a successful businessman with no scientific education. He'd devoted his foundation to astronomy simple because of the celestial nature of the subject. He considered this a pious and even holy science. This viewpoint is quite widespread and no other branch of science has so many foundations devoted to it. Lick is buried under the base of the telescope and on the side of the plinth one can read in golden letters his name and some information about the foundation. The observatory is very popular and is visited by many interested members of the general public, much to the dismay of the Director who, even though he has limited visits to certain days, still has to divert his staff to acting as tourist guides. Clearly Mr. Lick had imagined that it would be nice and cosy to lie here in his grave till the end of time and listen to the praise of the multitude for his splendid foundation that actually hadn't cost him a penny when he was alive because it was only formed after his death.

I found time there to paint some pictures. When Campbell saw them he pointed to one place and said, I see you've painted our golf canyon (not wolf Canyon). I asked him what he meant and he told me that the local people were passionate golf players. As is well known this game involves using a club to hit a small hard ball made of gutta-percha so that it travels quite a way. This requires a broad flat course which, up on the mountain, was not available. The result was that time and again a ball would be driven over the boundary of their golf course and into the inaccessible neighbouring gorge where it then lay. As a chemist I noted that gutta-percha is unusually stable to damp and carbonic acid, so that in the course of time this constant accumulation would generate a new geological layer. Future geologists would break their heads over the problem of which prehistoric animal had produced these bizarre droppings.

We travelled home in a large four in hand post coach which a group of Campbell's hated Sunday visitors had brought with them. Once again the driver flaunted his Californian dexterity by racing down at a gallop and taking the corners on the convex side at unbelievable speed. Nervous Loeb was obviously suffering and I too at times felt uneasy. However we reached the bottom without accident.

The Leland Stanford University. After leaving Mount Hamilton we did not return to Berkeley but instead turned towards California's second university—the Leland Stanford University in Palo Alto. While Berkeley is a state university funded by California (which, however, does not rule out its accepting supplementary financing from private foundations), the university in Palo Alto is entirely financed by a private foundation. It was founded as a memorial by the rich parents of a gifted only child who died young. They named it after him. At the time of my visit only the mother was still alive.

Like all universities there it is built in a broad park or campus and here everything is in local old Spanish missionary style: low buildings with flat roofs bordering large courtyards. Arcades run along the walls to provide shade from the almost tropical sunshine. Palm trees and other vegetation typical of the warm zones abound.

The president of the university was called Jordan and he was quite different from his colleague in Berkeley. Tall, aged between 50 and 60 and still addicted to sport, he seemed to represent "man of nature" in contrast to Wheeler's "man of culture". He had a pronounced pride in California and America and, despite friendliness for others, considered himself and his countrymen, to be members of a better race. He made great efforts to impress upon his contemporaries and students both the rights and duties that flowed from this. He was a zoologist who belonged to the old descriptive school. At that time he worked on Japanese fish and I watched with great interest how under his direction a Japanese painter with subtle brush strokes prepared pictures of a new species.

The university buildings were not particularly interesting. In the library the librarian proudly showed under the letter "O" a complete collection of my scientific work. Mrs. Stanford had retained the power to make all decisions concerning the use of money from the foundation and Jordan complained that it was difficult to get her to approve money to equip the scientific institutes. On the other hand, money for buildings or for representation was no problem. So, for example, there was a large life-like bronze monument to the Stanfords—mother, father and son—in the first courtyard. This was referred to by the students as the "holy trinity". In addition there was a marvellously furnished museum in which were displayed in cabinets as in a fashion shop the sumptuous robes which Mrs Stanford had worn at the various university festivities.

President Jordan then drove Loeb and me in his one horse carriage through the expansive university grounds with their many scenic attractions typical of the southern countryside. All sorts of strange animals were wandering about. A huge oak tree caught my eye on which a number of woodpeckers were busy pecking round holes in the bark. Into each hole was then hammered a suitably sized acorn until less than half of it was still visible. The tree was already covered with these acorn plugs. In answer to my question Jordan told me that these were the woodpeckers' larders where they stored up food for lean months. However, he was unable to tell me whether each bird looked after its own property and, if yes, how.

At the Pacific Ocean. After all these different people and places it was for Loeb and me a pleasant, indeed necessary, recreation to visit Pacific Grove, a small seaside resort on the Pacific coast where the Leland-Stanford University had built a small zoology laboratory. Since the university did not need it at that time, it had been equipped so that Loeb could do his research there. It was here that he'd been able to collect the sea urchins which he'd needed for his important research.

We were accommodated in a simple but clean guesthouse and particularly enjoyed the peace and quiet after the last few busy days. A wonderful blazing red sunset over the sea was one more instance of the new sort of landscapes I'd got to know in the past few days.

The next morning I examined with great interest the coast of this largest of Oceans. In that area it is made up of white-yellow sand dunes rather like those at home but much higher due to the force of the wind which blew undisturbed over thousands of kilometres. In the shallow bay lay a few large rocks some of which rose up out of the water and were covered with a snow white coating. The source of this turned out to be flocks of cormorants which live there free as sea gulls. They hunt fish from these rocks which they then cover with the end product of their digestion. Behind the dunes there was a wood of abundant trees but of uneven height and this included some splendid examples of the beautiful giant spruce whose enormous size I was able to study from close up. A walk in the woods like at home would have been hampered here by the thick undergrowth, so that one was restricted to the paths.

I painted at leisure a few particularly typical spots and visited the simple labs which appealed to me because Loeb's style of work was to keep things technically as simple as possible—and this was very much to my taste.

We saw a number of Chinese and Japanese colonies in the surrounding area. They consisted of low, grey, closely packed huts that left an outlandish impression and were not inviting at all.

After deliberately loitering a whole day in the sunshine we returned refreshed to Berkeley. Neither for me nor, so far as I could judge, for Loeb, had this long unbroken period together generated any of those feelings of repulsion which can so easily arise when people are dependent on each other for extended periods time.

New Festivities. Back in Berkeley I found an invitation to a "Smoker" to be held in the university's faculty house. This is an informal get together with beer and sandwiches which has the great advantage over a German "Kommers"[4] in that there is no ceremony and one is free to move around instead of being tied to one spot.

However, this time there was all sorts of official business. After a short warm up phase, there was an address made to me which ended with my being appointed a member of the faculty club. In reply I thanked them kindly and thought that now that was it, at least for this evening. However, President Wheeler soon rolled out the heavy ordinance. He held a charming speech peppered with wit and flattering phrases in which the 4000 miles that I'd travelled for them played a prominent role and then, with an elegant twist to more serious matters presented me with a parchment diploma which witnessed my unanimous election to honorary membership of the state university at Berkeley. This together with my short speech of thanks was received with jubilation and the evening proceeded in a joyful atmosphere.

Christy, the Professor for chemical technology, had been one of the first in America to accept and make use of the new doctrine of ions in his lectures and papers, and I had several times corresponded with him. He now asked me to give his students a short lecture and suggested as a theme my work on the basic technicalities of painting which I had discussed in my "Painter's Letters" and in some other publications. I was happy to accede to this request.

Then I learned a peculiarity of the American universities which would be well worth emulating. The solidarity of the institution within itself and with its students, even long after they have left, is much closer than we are used to. Changing a university, as German students often do, hardly happens there. Because of the long—frequently life long tenure—of the university president who is endowed with wide rights and powers there develops a close coherence of the faculty which is enhanced by housing the entire university structure within the defined bounds of a "campus".

This time I was invited to a University convention which takes place about once a month and serves to make important university events made known to the students and professors. To start with the President reported on a number of minor matters. Then the professor of history, a man called Moses, took the floor. Three years ago he had been given leave of absence to take on the post of Governor of the recently annexed Philippines and had now resigned this post to return to the university. He gave a very interesting account of his observations and experiences there and since he was at pains to put them into a historical perspective, he managed to bring out a number of most interesting things.

In the evening I had to go to a dinner in my honour given by the San Francisco University Club. This group was very different. It consisted mainly of medical practitioners and a few lawyers who together with others who had some academic

[4]A formal academic feast (commercium) of the members of student fraternities in Germany, where speeches are given, songs are sung, and usually lots of beer is consumed.

education had come along out of interest. Once again "the greatest living chemist" was trotted out. In my reply I emphasised the international and supranational nature of science, a theme that was taken and illuminated in its various aspects up by subsequent speakers. Naturally on this occasion the family feeling so warmly apparent on the evening with the Chemical Society was missing.

Chinatown. On one of the few free evenings we visited San Francisco's china town. I was guided by Dr. Taylor, a young doctor with lively scientific interests. I'd met him at Loeb's house and liked both him and his wife. They were a typical Californian couple, tall athletic and unselfconscious. As the district or police doctor he knew the Asiatic quarter well and so was able to quickly show me the interesting parts of it.

The streets were narrow and steep; the houses were even narrower and appeared to be very dirty. On the streets one saw only Chinese in their national dress. Now and then a mother clad in blue trousers proudly led her child all dressed up in a green and bright blue silk dress traipsing along on high heels. The children looked strangely puppet-like as if they had been manufactured. I witnessed a funeral procession accompanied by ghoulish cries which seemed to a European mind to lack all ceremony. Then I was taken to an opium den where the customers were all lying on a set of narrow boards that looked like a large commode—a truly disgusting sight. In pleasing contrast was a temple with a lot of pretty things, most likely sacrificial offerings. All around there stood earth filled receptacles in which the devout stuck glowing joss sticks as tokens of their prayers. We spent most of our time in the theatre where one of these plays which the Chinese much enjoy and which extend over months was being presented. The actors were all male, the female parts being played by appropriately dressed and made up men who speak in a squeaking falsetto and try, by taking rapid little steps and wagging their bottoms, to appear woman-like. There was no stage decoration and the actors' entrances and exits were made through two flag like curtains in the background.

The performance was accompanied by music provided by a small orchestra in the background. The main instrument was a sort of violin that was held and played like a cello. It repeated over and over again the same short figure. There were also a few mandolins which were scratched and small drums. The actors at times took up the notes in a sing song voice and otherwise just spoke. At each entrance or exit the orchestra produced a particular noise which seemed to be graded according to the person's importance. I could see nothing of what we Europeans would consider the performer's art—the actors often turned their backs on the audience. When they talked, the sound of the last syllable would suddenly drop an octave, which sounded very odd and which seemed to be a signal to the next one to start.

Special emphasis was placed on the resplendent costumes. We were taken to the store room to admire them, but it was so cramped and dirty and smelled so strongly that Loeb felt sick and so we quickly sought fresh air and left Chinatown behind us.

Alma mater Hearst. The days in Berkeley ended with a visit to Mrs. Hearst, a rich widow of about 60 who, with respectful irony, was referred to as the "Alma mater" of the university because she had repeatedly donated considerable sums of money for its further development. The name is also known in Europe from her son who as a press baron had played a considerable political role.

We had been invited to the "Hacienda" which was a summer house—actually a palace—set in pleasant surroundings about 3 h train journey from San Francisco. I was in California at the time of the summer drought when there is no rain and the meadows all dry out so that the land is of a uniform yellow-brown colour. We got off the train at the station which was built just for the Hacienda and were collected by a magnificent car. Up on the hill we could see the Hacienda's park, green as an emerald in the midst of the yellow brown surroundings, and at the same time we could see down in the valley the source of this marvel—a steam pump station which supplied the irrigation water from a deep well. We drove through a delightful garden where flowers so beautiful they could have come out of a fairytale bloomed under palm trees and were received in a house built in the missionary style. As guests of honour Loeb and I were assigned the rooms which were used by Mrs Hearst's son when he was at the Hacienda. The rooms were full of Chinese and Japanese treasures which today I'd be able to appreciate in a way I then could not. The rooms were furnished with extravagant luxury.

We went to dine, all dressed up in dinner jackets and white cummerbunds, and met there a selection of my new friends. Loeb had been asked with whom I'd got along well and these had received invitations. Mrs. Hearst turned out to be a handsome lady of slightly more than middle height who had a friendly and gracious mien. She asked me for my impressions of the university and involved me in a discussion of possible improvements. The table decoration and the meal exceeded in their luxury everything that I had experienced so far and once again I met and ate the Japanese fish in their paper envelopes. The conversation was lively and free of formality and I was thankfull I did not have to either listen to or give any speeches. Coffee was served in a large richly decorated hall which contained among other things an organ. In her amiable way Mrs. Hearst persuaded me to play it, though to be sure I must have disappointed her and all the others who expected from me a level of musical accomplishment equivalent to what they believed my chemical achievements to be.

The next morning was a Sunday and in the best Anglo-Saxon tradition guests were free to walk through the grounds admiring the flowers and plants which, under the care of expert gardeners, flourished in copious abundance. The contrast between this colourful foreground and the drab yellow-brown of the hills near and far was formidable. Since this monotonous colour of the distant countryside with its shade—somewhere between yellow three and orange one—was almost exactly opposite that of the strangely uniform intense blue of the sky, a remarkable harmony of colour developed which made a strong impression on me. I didn't at that time know why, because I only developed the ability to consciously analyse colours after I'd come up with the quantitative theory of colour, but feeling the urge to record this moment I went and got my paint box. With swift strokes I made a few sketches. The others looked on with interest, although some, especially the older women, seemed to feel that it was almost sacrilegious for me to be doing handiwork during the hour of church service. Though I do always try to take other peoples' feelings into consideration in this case I did not wish to deny myself the lively memory of this day which these sketches were and still are.

After lunch cars and horses were available for excursions in the surrounding area. I took my paint box with me, though I didn't get much chance to paint because, in view of my imminent departure my fellow guests bombarded me with questions concerning university structure, support of science and so on. Long tradition has made us Europeans ponderous in these matters in a way that the Americans are not. They are much more ready to try new approaches which seem to offer the chance of success. In addition, scientists in America know that science is not deeply anchored in the general consciousness and that the influence of the church is much stronger. For these reasons they are anxious, both in their own and in the people's interest, to increase the general level of knowledge to an appropriate level. How difficult this task is has been clearly demonstrated in the middle of 1925 by the Scopes monkey trial.

In the evening a dinner jacket was again a must, but this time we closed early because the train next morning was leaving at 6 am.

In Berkeley there were affectionate and touching farewells from some old and many new friends. I could look back on a series of pleasant and profitable days in the course of which not one single discordant note had fallen. On thinking back I must judge it a loss that I was not able to weave these many new threads more successfully into the fabric of my life.

Departure. The time had come to take my leave of hospitable California and of the many old friends I'd met again and new ones I'd made. Cordial and rather touching leave was taken. Both sides knew that while the exalted atmosphere of these days of celebration could not be maintained for ever, they were nevertheless to be marked up on the credit side of life's ledger.

After having been able to go through all my American experiences in the care of friends, I was now, for the first time, to travel all the way on my own. This was actually not so difficult, because the train I'd chosen, the fastest one there was, managed the journey to Chicago in three days flat and I came through the expected stress in good shape because I had taken a state room. In Chicago Professor Alexander Smith, a talented colleague who was particularly interested in the teaching aspects was waiting for me. I hadn't met him personally so far though we had corresponded. After that I was able to journey on to New York without any further stops.

These 3 days journey all the way across the continent went by with no disruption due to missed connections even if there were on occasion delays. It was hot in both of the desert stages, but no longer as hot as it had been 2–3 weeks previously. After having met so many new people in the last 2 weeks I was happy to be alone for a while to put my memories in context and try to sum up what these sunny days had brought me. There was also one purely formal reason for balancing matters, for on my departure I'd realised from the date that I'd be celebrating my fiftieth birthday on the train far from all my acquaintances. In fact I almost forgot the day when it did come round.

I realised that on this journey I hadn't come across any new ideas that might have necessitated a real change of course. Although I had seen and experienced many new and interesting things, there was nothing there that had stirred the depths of my soul. In particular I hadn't taken American citizenship nor did I have any wish to do so. Not, mind you, that there had been any lack of hints suggesting a closer future link with the circle of those I'd met. I'd given evasive answers, not for any tactical reasons, but rather because I had not got the impression that I'd find there better opportunities to pursue my endeavours than were available to me in Germany. This was quite apart from the difficulties of transplanting my family into a new environment. There was so much still to be done in America before the point was reached when the sort of creative new work which interested me could be done. I foresaw that the output and value of my work would be greater in Germany—but only if I no longer had to be a university professor.

Back in Chicago. In Chicago I was met by my colleague Smith and the astronomer Hale, who wanted to get to know me and show me his observatory—the Yerkes Observatory near Chicago. After an extensive exchange of views with Smith I was passed on to the astronomer and spent with him an extremely stimulating and informative day. Hale did not consider himself to be a run of the mill astronomer, but rather a physicist whose speciality it was to use optical means to bring the objects of his research—the sun and the stars—into his laboratory.

To this end he had conceived and built quite a number of clever, indeed ingenious, instruments. Like most observatories it had a precision engineering workshop and in many cases I'd seen that the man at the lathe was always a German. That was the case here as well. Hale told me that Americans don't have the patience to acquire the necessary dexterity and sureness of touch. A long apprenticeship seems to them to be a "waste of time", while in contrast Germans are proud of their learned dexterity which becomes for them a moral value in itself and not merely a financial stake.

In Hale I found that unusual mix of the scientific and technical ingenuity which to me represents the ideal qualities of a scientific researcher and which I have aspired to my whole life. Not surprisingly we had a lot to talk about and in this I felt myself to be the pupil rather than the teacher. There developed from this a friendly relationship which I valued greatly. Later the world war destroyed practically all my contacts to my American colleagues and friends because of a defamatory campaign against me which originated in Geneva and whose lies were widely accepted on the other side of the ocean at face value. But Hale, by sending copies of his publications from the new observatory on Mount Wilson in California which he had planned and built according to his own ingenious plans, kept our relationship intact as a sign that the trust we had built up could not be shattered in this way.

From Chicago I journeyed on without further stops to New York and from there, driven by an increasing feeling of homesickness, travelled home on the fastest transatlantic liner available—the "Kaiser Wilhelm".

Chapter 28
Taking Leave of Chemistry

Attempted flight. At the end of 1900, exhausted from the ever more onerous teaching load in the institute I wrote to the ministry to tell them that I felt unable any longer to carry out my present duties to their full extent. I therefore asked that my professorship should be given to someone else and that I should be appointed as an honorary professor with no defined teaching duties but with space and resources at the university sufficient for my experimental work.

The answer was that this sort of solution could not be considered but that instead I should decide which alleviation of my duties would be necessary to permit me to remain Director of the institute and that everything possible would be done to accommodate me in this matter.

A few years earlier, at the inauguration of the institute, minister von Seydewitz had taken me aside and entreated me to limit my work so that I would be able to continue it out for an extended period without becoming exhausted or even feeling the need for great effort. I hadn't felt able to give any promises because the urge to take on problems when they arose was irresistible and in the previous chapters I detailed the flood of new tasks that had poured over me since then.

Now, once again, the minister made every effort to help. To relieve the pressure a sub-director was assigned to me to take over the administration of the institute. Furthermore, since I'd pointed out that this might not be enough to prevent a total breakdown; I was given the right to resign whenever I felt that I was no longer able to direct the institute. To give the government something in return for these concessions I promised not to use my right to resign before a certain date and, at the same time, it was arranged that my pension would, within limits, increase the longer I stayed.

I agreed to these conditions because I felt the need to meet the minister's wishes as far as possible. Nevertheless I did not hide either from him or from myself that I had little confidence that I'd be able to hold my office in the longer term even under these favourable conditions.

R.S. Jack and F. Scholz (eds.), *Wilhelm Ostwald*, Springer Biographies,
DOI 10.1007/978-3-319-46955-3_28

The anniversary of my doctorate. To begin with everything went well. I convinced myself that the work in the institute would go forward in the direction I wanted without my having to concern myself directly in the details. This had the advantage that all those involved experienced that happy feeling which comes from doing ones own independent creative work. That gave rise to the sort of pleasant and productive working atmosphere which I described before (Part II, Chap. 24, p. 290).

Then, in autumn 1903, came the journey to California that brought so much joy and so much prestige for my previous work in chemistry that I automatically cast a friendly eye on this old work which had brought me so much happiness.

At the same time—almost as if the paradise I had wanted so stubbornly to leave was to be brought home to me from all sides—students and friends decided at the end of 1903 to organise a celebration of the 25th anniversary of my doctorate. In accord with the splendid conventions of German academia, former students contributed to the preparation of a printed festschrift. It contained in its 877 pages 34 contributions whose diversity of topics, made clear to me the breadth of challenges we had worked on in the 16 years of my appointment in Leipzig. Even H. Trey, my former assistant from my time in Riga, had joined in.

The number of my former students, who were now working as independent scientists, was 147 and of these 34 were professors.

The introduction had been written by van't Hoff who affectionately described the output of my first 25 years as a scientist. He made a clear distinction between my contributions as a researcher and as an organiser. In the first category he described how I started with solving single experimental problems and then moved on to develop a philosophy based on thermodynamics. In the organisational section he distinguished between my written works and teaching, that is to say discriminating in particular between my general versus my personal influence. He remarked that, in particular at the beginning, my central works might have brought me many enemies, "However, it is undeniable that Ostwald always got it right". This was a verdict that, though late, brought me great reassurance and the recognition that I'd been in the forefront of fighting for new concepts, irrespective of whether they had been originated by me or by others.

Van't Hoff, perhaps viewing the matter through the prism of his own personality, considered this organisational talent to be a fundamental predisposition and added, "Ostwald was not satisfied simply to develop his own notions; indeed it may well be his major desire to communicate his thoughts to others, and there is no doubt that this has, to no mean extent, contributed to the acceptance of physical chemistry as we know it today".

In the same festschrift my fellow countryman and co-worker Paul Walden (Part 1, Chap. 12, p. 127) delved even more deeply into my personal contributions. Early on he had asked my wife for information about my life, particularly about my youth. From this he constructed a kindly essay which, through the rosy glow of a friendly mind, introduced me to a broad circle. Of the many gifts of that day this was the pleasantest contribution at the time and the one which had the longest resonance in the future.

The folder with the material from my wife, which van't Hoff had taken for safe keeping, mysteriously disappeared from his room in the hotel and never turned up again.

The evening, in addition to the usual banquet, also had a comic interlude in the form of a theatrical performance in which personal matters from the life of me and my wife were hilariously portrayed. The idea was developed by a member of the lab who wanted it to be performed in the familiar style of a students' rag. The actors, and in particular the actresses, protested. Another "child of the lab"—a somewhat older medically qualified lady—entered the fray and proposed to convert the content of the piece into rhyming couplets. For this my wife made over her own room and, since poetry quickly gave her a feverish head and cold feet, she made herself comfortable in a large fur lined travel bag, while the verses were being hammered out. It took all of my wife's conciliatory diplomatic skills to forge peace once more between the dispossessed planner and his poetic successor.

As I reread the essay of van't Hoff the next day I felt a twinge at a sentence which was meant to be particularly friendly. It read, "The astonishing thing is that in the midst of all this widening circle of activities he neither loses interest in his earlier work nor his mastery of it".

That might have been the view from the outside but from my perspective things were very different. With every passing day both the interest and mastery declined. There seemed to me to be an inexorable natural process which, like a mighty current, swept me on to pastures new and I had no wish to try to force the boat against the flow. All that could be done was to steer carefully. Most importantly I only felt happy in this flow. This was the insight that I knew I could not ignore.

The Faraday Lecture. However, such insights do not act as metamorphosing resolutions overnight. I still had many things left over from my old areas of interest which I neither could nor wished to simply drop. At the same time there developed from the old work new possibilities which bound me to it. It turned out that the Californian exuberance was just the prelude to more serious (though less carefree) honours for which I was now apparently considered ripe.

One of the most significant of these came in the spring of 1904 in the form of an invitation to hold the Faraday lecture in London. This invitation is sent by the Royal Society and is considered to be in recognition of major scientific achievement. British scientists as well as foreigners may be invited, though of course the former have an advantage. Among my predecessors from the continent were J.B. Dumas, the French contemporary and rival of J. Liebig, the greatest German chemist, and H. Helmholtz the greatest German physicist. In his Faraday lecture Helmholtz had presented the first reports on the atomic concept of electricity, in which the notion of the electron (though this was not what he called it) was developed. The importance of this idea only began to be grasped some twenty years later—and in the meantime every layman has heard of it.

My ambition to present something really important was fired, and so I eagerly started to think about some curious lines of thought which I'd toyed with before but never really got anywhere with. For a long time I'd been impressed with the idea of J.B. Richter, the father of quantitative chemistry, who at the end of the 18th century concluded that the law of the conservation of reactant and product masses when acids and bases react to yield salts necessitated the conclusion that mixing solutions of neutral salts would inevitably yield a neutral solution regardless of whether they interacted or not. It seemed so obvious that neutral solutions when mixed would not result in acidic or basic solutions that one can not initially grasp how from this platitude such a definitive law could be deduced. And when one had convinced oneself that this was indeed so, then the disquiet began for there must be here a peculiarly potent line of argument that ought to be generally applicable for example to the existence of chemical stoichiometry. What then is the general chemical principle that permits this broader conclusion in the same way as the conservation of neutrality of salt solutions permitted the narrower conclusion?

Simply the formulation of this question cost me an enormous effort, because it required a similar sort of turning one's thoughts inside out as I'd experienced with thermodynamics. On that occasion it had been a voluntary process, but this time I had to force it, because I needed the idea for the lecture.

Once this first clarification had been achieved I still had a huge amount of old mental residues to clear away before the clear answer could be found.

A co-worker. In this I was much helped by the fact that I had supported another lone hand working along similar lines in his efforts to get his work published in the "Journal" and afterwards in the "Annals". This was Franz Wald who was the chemist at the Kladno ironworks in Moravia. His ideas were so strange that Emil Fischer once said to me after I had accepted the first work for publication, "If you publish any more of this stuff then I'll cancel my subscription to the Journal of Physical Chemistry". I replied, "The loss will be entirely yours", and continued to publish the work.

Wald was interested in similar basic questions of chemistry, particularly in the concept of pure materials and he approached the same question from this angle, without however being able to reach the goal. And yet once it was reached it turned out all to be extremely simple.

I should add right away that I mentioned Wald's name in my lecture in London and made clear his considerable contribution. By chance the Austrian ambassador was present and informed his government of the honour done their countryman. On my next visit to Austria I was invited to the ministry of culture to give more details about this unknown man. I repeatedly made known my appraisal and when asked told them that I thought that as a teacher he would interest only a small circle but that it would be appropriate to appoint him to a professorship so that he could continue his research unhindered. I knew and told them that Wald was an impassioned Czech nationalist, but this was not seen as an obstacle and he was soon appointed Professor at the Czech technical university in Prague.

The festive day. My paper was ready in good time and, accompanied by my two daughters, I journeyed to London where W. Ramsay, whose two children were the same age as mine, very kindly put us up in his house. I'd translated my lecture as best I could into English and he corrected the language for me. I read it out loud to him so that he could correct my errors in pronunciation and this he did with kindly patience.

On the great evening—not without a degree of nervousness—I took my place at the speaker's lectern behind the laboratory bench in the round hall of the Royal Institution. Gathered in front of me I saw England's most prominent scientists as well as many members of the upper class, for the chairman bore the historically memorable name Lord Rayleigh and was himself a world famous physicist. The place from which I spoke was that from which my highest scientific paragon Michael Faraday had through the course of his life announced his epoch making discoveries. His predecessor in this role had been Humphry Davy, the inventor of the safety lamp for miners which had saved so many lives.

The theme of my lecture was the demonstration that that it was possible to derive the laws describing the mass relationships of the elements in chemical reactions directly from the concept of the stoichiometry of reactions of pure substances—without the aid of the atomic theory. My scientific audience were to a man adherents of the atomic theory, despite the fact that at that time almost none of the currently available evidence for the atomic structure of matter was known. However the atomic theory of matter was a national issue, because it had been proposed a century earlier by John Dalton in Manchester where, shortly before my lecture, the centenary of this theory had been celebrated. Nevertheless my lecture was kindly—even cordially—received and I felt it necessary to respond to the chairman's congratulations as he presented me with the Faraday medal with an impromptu speech of thanks.

The evening closed with a convivial get-together organised by James Dewar who as Professor at the Royal Institution was the third or fourth in succession to Faraday and his name is associated with many brilliant experiments on liquid air. He press ganged me into trying a genuine old Scotch whisky of which he was the proud owner and this resulted in my life long detestation of this poisonous concoction.

The honourary doctorate. The Faraday lecture and the presentation of the Faraday medal were not the only honours heaped on me during this visit to England. About one week later I was invited to the University of Cambridge to be awarded an honorary doctorate.

It was not the first such invitation of this sort. About two years earlier I'd been sent a similar offer from the University of Princeton whose president—Wilson—was later to become an inglorious President of the Republic. The doctorate however could only be awarded in person and since Princeton was not one of the better universities in America and since none of the people there had any particular drawing power for me I thought the loss of a month of precious work time which I would have wasted on the journey there and back was too high a price. I therefore turned it down. At the same time a colleague of mine from Leipzig University was also invited and he went. I saw him on his return and his exhausted physical and confused mental state confirmed for me the rightness of my decision.

This time, however the situation was different. M.M. Pattison Muir who I regarded gratefully as an early supporter (Part I, Chap. 6, p. 79) was still working in Cambridge and the invitation had been initiated by my German colleague Ruhemann who was professor of chemistry there. Although he was an organic chemist he told me when I asked that my publications had been of such scientific value to him that he was glad to be able to show his thanks in this way. He generously hosted my daughters and me in his house and, after so much English experience, we were glad of the German atmosphere which was kept alive by his kindly wife, who was also German. Many paintings in the house turned out, to be early works of Max Liebermann who was related to the Ruhemanns. They looked completely different from his later pictures being precisely painted and as finely formed as old masters.

The solemn ceremony was based on the traditions of the theological schools of the old universities and these were observed that morning. Everybody was in a gown; over their shoulders the Doctors wore a colourful silk cowl—known as a hood—which ran out into two long ribbon-like lobes. I was told by those who understood the history of these things that this was the symbolic rendering of a monk's begging bowl. The Principal of the University performed the holy ceremony in front of a large semi circle of Professors and other dignitaries by taking my folded hands between his and murmuring some dictum. Prior to that, the graduation file had been read out. It was in Latin and was read out with English articulation; however since I'd long forgotten all my Latin that was no impediment for me. Later I was given the English translation whose highly complementary contents made me smile with thanks. The accolade which made the strongest impression on me was that in my lectures and publications I combined German profundity with French clarity.

When, 10 years later during the Great War, the barbarism of our enemies—first and foremost the French—led them to drag the battle into the pure realm of science (which had never been the case till then) most of the scientific societies of France, Britain and America expelled their German regular and honorary members. This happened to me in many cases. However I never heard of this happening with the half dozen honorary doctorates which I'd been awarded over the years, (mostly from British universities). As I recalled the ceremony in Cambridge, the reason for this became clear. The conferral of a doctorate is a sacred act like the ordination of a priest and it has the same indestructible character. As is well known catholic canon does not permit rescinding the ordination of a priest even of one who has committed the most terrible crimes. In the same way a doctorate once awarded becomes inseparable from the recipient and is extinguished only by death.

An anniversary. There was an interlude of around a week between the events in London and Cambridge and I took this time for relaxation. Although it was rather early in the year we went to a small seaside resort called Penmaenmawr in Wales. I took along my paint box and made good use of it there.

Just at this time there was a celebration staged in Manchester for Sir Henry Roscoe who as head of the university had been responsible for a remarkable flowering of the institution. Roscoe had been a student of Bunsen's in Heidelberg and had been given a challenging project on photochemistry which Bunsen had published under both their names. I had this work published in the "Classics" as an example of an unrivalled paradigm of physico-chemical research and had written an appropriately laudatory commentary to it. Though I'd mostly had Bunsen in mind, some of the praise naturally also fell on his co-worker. He in turn was very grateful for this and, when I'd met him by chance at a meeting of the British Association, whose meetings he always attended, he had expressed his heartfelt thanks. And so I used this opportunity to give him not only my personal congratulations but also those which the Bunsen Society had asked me to pass on.

Again the celebration was typically English. Numerous institutions, societies and individuals had submitted nicely written speeches of congratulation which after a careful assessment of protocol to determine the proper sequence were read out and then presented to Roscoe. The contents of the congratulatory speeches had been given to him a long time before, because he did not want to have to give the expected addresses of thanks impromptu. He had therefore prepared the responses in writing and had them printed in large type in a notebook. With dignified gravitas he read the response to each of the speeches out of this book, albeit with an occasional slip of the tongue and consequent revisions. Each time he emphasised how surprised he was at the completely unexpected honour—and so on. No one present seemed to see the humorous side of all this.

A change of course. After these English honours I looked forward in the same year (1904) to an American one, for I had accepted an invitation as speaker to an international congress of the Arts and Sciences to be held in the autumn in St. Louis. I saw it as symbolic of the turnabout which had taken place in my scientific work that this time I should speak not about the old work, which I'd rather lost interest in, but instead about the new directions which had scarcely begun. I was not invited as a representative of physical chemistry but rather of philosophy and was asked to discuss my latest thoughts on the system of sciences.

One can well surmise how much this encouraged me in the new way of thinking.

My experiences at this meeting will be related in the next chapter but first I must tell you about some matters left over from the old work many of which concerned me in the time that followed.

Mutation. From all the intellectual efforts that I threw myself into in the first ten years of my time in Leipzig, creative writing was the one that I most ruthlessly exploited. My friend Walden in his contribution to the festschrift had calculated that by the year 1903 my written work would have been equivalent to 16 volumes of the encyclopaedia. Nevertheless it was in this area that I felt the least exhaustion and, after my recovery, I found, somewhat to my surprise, no change for the worse in this area. I didn't need to force myself to my desk and the work there was as pleasant as in earlier times.

And when in my old age I now see how one faculty after another degrades and disappears, then I can at least—I hope the reader agrees—be satisfied that my ability to present a coherent argument in a lively manner is still there. A decline does make itself noticeable in the reduced daily output and by the necessity of breaks in the work when my energy meter shows that it is time to pause and gather renewed strength. However, this is a general decline of the whole body and by comparison my ability to write and my desire to do so are both relatively less affected.

I must confess that I haven't really been able to interpret this situation biologically. For a start we may take it that I was the first in the Ostwald line to show an ability to write, and the general view would be that the most recently acquired abilities will be the first to be lost. That is certainly not the case with me for I first lost the capacity for teaching in the laboratory while writing will probably be something I can do until the day I die.

Beyond that I should emphasise that this capability has clearly shown itself to be inheritable. At the time of writing all three of my sons edit a journal in his professional area, two of them produced a number of books and although the third one has not yet done so, that is only because his considerable literary abilities have taken a different direction. And although the subsequent generation has not yet reached the age for writing, I am pretty sure that they will demonstrate this same family character even in those cases where the genetic material was latently transferred through an intermediate female carrier. It all looks very much to me like a clear example of the inheritance of an acquired characteristic.

However I don't wish to interpret the matter in this way. The capacity was not arbitrarily and deliberately acquired, rather it appeared in such force in me, despite a very inauspicious environment (one of my brothers had it to a very low degree and the other not at all). It seems like something which suddenly blossoms in a defined genetic background which has otherwise been preserved through many generations. It rather looks as if here we are dealing here with that strange phenomenon which the botanist de Vries called "mutation". The hallmark of this is that normal parents may produce alongside normal progeny also an individual who expresses strongly different inheritable characteristics. In this view the ability to write would have arisen in me by mutation rather than being a gradually acquired character.

This whole area is far from being scientifically explained but with the presentation of the phenomenon in the context of scientific categories the task has been formulated though not solved. Already previously I'd developed the thought that every living form resembles its parents and passes this form on to its progeny because this form is best suited to its way of life and so any change would be for the worse. When I watch the titmice which come to my window for their daily food and notice how similar they all are in their colouring then I am driven to the conclusion that there is a long acting equilibrium which maintains this pattern, even if the various elements of it may vary within the limits tolerated by their environment. In a similar way, the course of a river is always determined by the topology and remains constant independent of the ever changing amount of water in the river.

Of course, it is fundamentally the case that in addition to the explanations we have for the situations we have examined there may be other situations which may call for other interpretations, as for example when a small stream develops by chance in the neighbouring valley. Normally a river holds its course but a massive dislocation, caused perhaps by an earthquake, may push it in a new direction so that it now permanently flows into and fills a new valley with the same persistency as it had its old bed. The closer the two possible courses are in terms of the difference in elevation between them, the easier this change will happen.

In the same way, one might think that the threshold between two neighbouring species could be so low that it might be overcome without risking destruction. The new form then enters into a new equilibrium which once again represents a stable and inheritable form.

The question as to why mutations only arise in particular individuals may be answered by assuming that mutations are associated with a certain instability in the mutant organism. A river with a deeply cut course will not readily change direction even in the face of a major dislocation, while on flat land even a small disruption will suffice to effect such a change. Thus a reduction in stability, and hence an approach to a labile condition, will be required before such an event can happen. One sees this in families when the ways of life, which have been maintained more or less intact for generations, are suddenly no longer considered acceptable and there ensues a striving for change. My father was undoubtedly an example of this.

The mutation does not always lead to a viable entity, because the new characteristics may be incompatible with life. De Vries observed quiet a few such cases. Also here one finds similarities with the development of families. So, for example, the great physiologist J. Müller had a brother who like him represented a new type of personality. However this brother turned out to be a weak and frivolous person who caused heartache and worry for his relatives; luckily he had no children.

After all that, one can understand that as a grandfather I wait with a certain scientific interest to see if in my case we are dealing with a mutation of unlimited heritability.

Chemistry books. The books I wrote since 1895 fall into two categories: the professional ones containing the conclusions of my work up till then and the general science books which were the start of my new occupation.

From the second group I have already related the beginnings in another context. These were "Lectures in Natural Philosophy"[1] and the accompanying journal "Annals of Natural Philosophy" ("Annalen der Naturphilosopie"). These soon generated so much interest among their readers that I found myself encouraged to work further in this direction.

Books of the first group somehow represent end points of my work in that they were written with a view to integrating the new areas of research into the well

---

[1]Ostwald W (1902) Vorlesungen über Naturphilosophie. Leipzig, Veit & Comp.

known generalities of science. They describe the sum of my teaching experience in chemistry and show how the structure of the science looked when viewed through this prism.

For example, the analytical chemistry book (Part 2, Chap. 17, p. 182) published in 1894 was written with this goal in mind. However, in that case so little was known about the subject up till then that the book was made up largely of new material. In the last 30 years it has turned out to be an enduring text which can fairly be described as the foundation of all of the analytical literature which has seen the light of day.

The next step along the same path was to describe the whole of inorganic chemistry in the light of the new results in the span of what might be covered by 5 or 6 hours of lectures. In this case the amount of older information to be taken over from pre-existing work was of course much greater. However, that simply increased the work involved for it required a translation of the existing literature into the language of the new chemistry. In the literature some of this had already been done when research results were presented in the new light, and a textbook-like summary had been attempted by Bodländer (1896).[2] But the repeated annual presentation of the material in my lectures had brought out so much that was new and clearer that I felt I should prepare my own description of the subject. The reorganisation of teaching in the Leipzig institute as a result of the appointment of Beckmann to the chair of applied chemistry, and the building of the new laboratories meant that I would no longer hold these lectures and so it seemed a good time to summarise and publish the state of the work up to this point.

After several years work on this extended text (800 pages of small type) "Fundamentals of inorganic chemistry"[3] appeared in 1900. It was soon translated into English, Russian, French, Japanese and so on, and in German repeated editions resulted in several thousand copies being sold. Just like the book on analytical chemistry it served as the model and basis for the later textbooks. There was naturally in this case a much greater breadth of personal viewpoints and particularly in this area there have been such rapid advances that a recasting of the material in up to date form has become necessary.

Some years later I faced a more basic challenge for, in addition to the many requests from publishers which I had to turn down, there came an invitation from Vieweg and Son which I could not refuse. They had originally published Stöckhardt's "School of Chemistry" which had played such a major role in my personal development (Part 1, Chap. 2, p. 23). After the author's death in 1886 it had been revised by someone else who had however misjudged the nature of the work and had ruined it by simply converting it into something like the usual run of school books. The publishers asked me to write a new version of the "School of

---

[2]Bodländer G (1896) Lehrbuch der Chemie für Studierende und zum Selbstunterricht. Vol. 1: Anorganische Chemie (volume 2 was never published), Stuttgart, Ferdinand Enke.

[3]Ostwald W (1900) Grundlinien der anorganischen Chemie. Leipzig, Wilhelm Engelmann.

Chemistry". It should be a based on Stöckhardt's concept, but rooted in modern chemistry.

I found the challenge enticing. Chemistry had been permeated not only with ideas like osmotic pressure and electrolytic separation but also with my own conceptual and systematic thoughts, and formulating this in the form of a textbook seemed to me to provide a decisive test of their scientific value.

The literary side of the challenge also appealed to me. I wanted to write something vivid, something that a 13 year old would immediately grasp and at the same time bring a smile of pleasant surprise to the face of the experienced scientist. For this there is a particularly effective means—the poetic form. Way back in Dorpat I'd had the "Reactionary in the Waistcoat Pocket" from the poetic chemist Jacobsen to hand and the verse,

Alkalis are tough as sin,
H two S can't do them in

had shimmered over the start of my analytical chemistry career. However I didn't dare to try that for a complete textbook, even if it was only to be a short one.

On the other hand I knew from editing the classics that another form, which is also very effective, is that of the conversation. Galileo's, in many ways epoch making, "Discourses" ("Discorsi") in which he describes his discoveries in mechanics was particularly well known to me because my old teacher von Öttingen had got hold of the German translation and had often discussed it with me. It seemed to me to be an attractive challenge to write an elementary chemistry textbook in the form of a discussion between a teacher and a pupil. A quick experiment soon showed that I had here a means to present important things to the reader in a shorter and more striking form than is possible using continuous text. I therefore proposed this to the publishers who agreed, although with some reservations at what they worried might be an ineffective old fashioned form of presentation.

I, however, threw myself with delight into the work. The first part came out in 1903[4] and as I journeyed via Bremen to America I had the satisfaction of seeing the little volume in the orange cover of the Vieweg Press displayed in the town's book shops. Both on the journey there and on the way back I filled those boring hours in station waiting rooms by working on the next parts. I really enjoyed leaving lots of things between the lines and I heard later that these hints were quickly picked up by the readers who enjoyed this way of doing things. The book was soon a great success, being reprinted several times in substantial numbers and it was translated into many languages. The English translation[5] was done by W. Ramsay's daughter who—much to the distress of her grandmother—I had swung through the air when she was little. Her father worked the schoolboy idiom and slang in.

---

[4]Ostwald W (1903 Die Schule der Chemie. Erste Einführung in die Chemie für Jedermann. Vol 1: Allgemeines. Braunschweig, Vieweg & Sohn.

[5]Ostwald W (1905) Conversations on Chemistry Part 1: General Chemistry. New York, Wiley & Sons.

Last dance. The final separation from chemistry came with the publication of three books. Although this was at a later time I'll mention them here for the sake of coherence.

The first was an account of the historical development of chemical concepts and was largely based on lectures I'd given in America in 1905–1906. Though I had eagerly studied H. Kopps's seminal book[6] in Dorpat, I'd become increasingly dissatisfied with the various accounts of the history of chemistry which followed it, because I began to see ever more clearly that the heart of any branch of science is based on the construction of appropriate concepts. In this sense the history of a science is basically an account of the development of these concepts. The trouble here is that in textbooks these concepts are treated as self-evident and so no attempt is made to define and analyse them. And that, in turn, means that the very important development and changes in the concepts—which are the real life of a science—are not described or discussed.

For the lectures in Boston my former student and colleague A.A. Noyes had asked me to put the philosophical aspects of chemistry in the centre of things because the audience would consist more of lecturers than of students. I was glad to take on this challenge. The lectures were then written up and published in English in a somewhat unpolished form.[7] Later I reworked them in some detail and they were published in German under the title "Guidelines of Chemistry".[8] Because this failed to make clear the historical slant of its contents I changed the title of the second edition to "The Development of a Science".[9] This book was translated into English and French.

The goal of this conceptual work was the achievement of a "Chemistry without Chemicals"—in other words a work based on concepts and laws of nature which would be applicable to all chemical species. Thus the properties of individual chemical entities would be considered as special cases of the universal laws and could be derived from them by inserting the appropriate values into the fundamental equations. At that time this seemed a rather fantastical idea but now it is beginning to look like a real possibility. Since I'd been able without using the atomic theory to describe the laws of stoichiometry as being dependent on the concept of "pure chemical entities" this work was particularly attractive to me. In particular it seemed to me to be well worth the effort to determine to what extent these conceptual aspects form the heart of chemistry which many casual observers (and many "pure" chemists) consider to be a merely experimental science.

---

[6]Kopp H (1843–47) Geschichte der Chemie, 4 volumes, Braunschweig, Vieweg.

[7]Ostwald W (1904) Historical development of general chemistry. School of Mines Quarterly 27:87–117, 313–339, 388–413.

[8]Ostwald W. (1906) Leitlinien der Chemie: sieben gemeinverständliche Vorträge aus der Geschichte der Chemie. Leipzig, Akadem. Verlagsanstalt.

[9]Ostwald W (1908) Der Werdegang einer Wissenschaft: sieben gemeinverständliche Vorträge aus der Geschichte der Chemie. Leipzig, Akadem. Verlagsanstalt.

This ended up as a 540 page long book entitled "Principles of Chemistry".[10] It was so widely read that a second print run was necessary and, despite my warning, a Dutch translation was prepared. However, the ideas that were elaborated there have never really found there way into the mainstream of chemical thought.

I'll just briefly mention that in 1909 I wrote "Introduction to Chemistry"[11] as a school text book. I'd wanted to write something in discourse form that would be more easily approachable than my "School of Chemistry" and hence applicable as text for chemistry at the school level. I'd thought that a young and dedicated teacher would be able to perform the dialogue with a chosen pupil before the class. I had to abandon this extravagant hope after I was severely criticised in a book review written by a secondary school teacher because I had had the pupil address the teacher in the informal "Du" form. I wrote this useful book out of a sense of love for German youth and it found a pretty wide use, though by far not that which it had earned. I have to tell you in all honesty that this is by far the best school textbook in chemistry for beginners. At least it's the best such textbook of which I am aware.

[10]Ostwald W. (1907) Prinzipien der Chemie. Leipzig, Akadem. Verlagsanatalt.

[11]Ostwald W. 1910) Einführung in die Chemie: ein Lehrbuch zum Selbstunterricht und für höhere Lehranstalten. Stuttgart, Franckhs Technischer Verlag.

Chapter 29
An International Congress of All the Arts and Sciences

The occasion. A world exhibition was to be organised 1904 in St. Louis to celebrate the centenary of the purchase by the United States of Louisiana from France. At this the German Empire was to be significantly represented. Among the many events associated with the exhibition there was going to be an international congress of arts and sciences. There had been meetings of this sort at previous world exhibitions but at these the choice of speakers and the subjects of their lectures had been made on a voluntary basis. Now, however, there was going to be an attempt to properly organise the entire complex of human mental endeavour and to appoint for each field within it two leading scholars as speakers. These scholars were to be chosen from throughout the civilised world. So as to make the meeting independent of the financial situation of these representatives of human thought and knowledge, each invited speaker was to be given a travel grant of 500 dollars which would easily cover the costs of attending.

The museum director F.J.V. Skiff is said to have had the original idea and a committee, which consisted mainly of the presidents of the leading American universities, was formed to work on it. Mr. F.W. Holls from New York put forward the significant suggestion that the speakers be remunerated and Hugo Münsterberg, professor of psychology at Harvard University developed, in a letter to Holls, a scheme to systematise science which would provide the basis for the choice of speakers. In this way the rather general initial thoughts were fleshed out and made practicable. The doyen of American astronomy, S. Newcomb, was elected president of the congress and Hugo Münsterberg and Albion W. Small, a sociologist at the University of Chicago were elected as vice presidents. In addition James Bryce, Gaston Darboux, Wilhelm Waldeyer, Oscar Backlund, Theodor Escherich, Attilio Brunialti and N. Hozumi were elected as honorary vice presidents representing Britain, France, Germany, Russia, Italy and Japan respectively. The meeting took place in the exhibition grounds in St. Louis from 19th to 25th September 1904.

R.S. Jack and F. Scholz (eds.), *Wilhelm Ostwald*, Springer Biographies,
DOI 10.1007/978-3-319-46955-3_29

The systematisation of knowledge. The congress in St. Louis was supposed to present a well ordered summary of the entirety of human knowledge and ability. This then was a practical example of the old problem of methodically organising the sciences and one can ask whether and to what extent this challenge was met.

The answer has to be: not very well. What was missing was a general comprehensive concept. Chance relationships were granted too much weight and because of this it was not possible to put each field into its proper place. This became objectively obvious when one and the same area turned up in more than place as for example aesthetics which was assigned both to philosophy and to the history of art.

The various fields were first of all assigned to one of seven groups: (A) normative, (B) historical, (C) physical, (D) intellectual, (E) applied sciences, (F) social regulations, (G) social culture. The groups gave rise to 24 divisions which in turn were subdivided into sections as shown below:

A. Normative sciences

1. Philosophy

   (a) metaphysics, (b) philosophy of religion, (c) logic, (d) the scientific method, (e) ethics, (f) aesthetics

2. Mathematics

   (a) algebra and analysis, (b) geometry, (c) applied mathematics

B. History

3. Political and economic history

   (a) history of Asia, (b) history of Greece and Rome, (c) middle ages, (d) recent history of Europe, (e) history of America, (f) history of economics

4. History of law

   (a) history of roman law, (b) Common law, (c) Comparative history of law

5. Linguistic history

   (a) Comparative language analysis, (b) Semitic languages, (c) Indo-Iranian Languages, (d) Greek, (e) Latin, (f) English, (g) Romance languages, (h) Germanic languages

6. History of literature

   (a) Indo-Iranian literature, (b) Classical, (c) English, (d) Romance, (e) German, (f) Slavic, (g) Fiction

7. History of art

   (a) Classical art, (b) Modern architecture, (c) Modern painting

8. History of religion

   (a) Brahmanism and Buddhism, (b) Islam, (c) Old testament, (d) New Testament, (e) History of Christianity

C. Physical sciences

9. Physics

   (a) Physics of matter, (b) physics of ether, (c) physics of the electron

10. Chemistry

   (a) inorganic, (b) organic, (c) physical, (d) physiological chemistry

11. Astronomy

   (a) astrometry, (b) astrophysics

12. Earth sciences

   (a) Geophysics, (b) geology, (c) Palaeontology, (d) petrology and mineralogy, (e) physiography, (f) geography, (g) oceanography, (h) cosmic physics.

13. Biology

   (a) Phylogeny, (b) plant morphology, (c) plant physiology, (d) plant Pathology, (e) ecology, (f) bacteriology, (g) animal morphology, (h) embryology, (i) comparative anatomy, (k) human anatomy, (l) physiology

14. Anthropology

   (a) Physical anthropology, (b) archaeology, (c) ethnology

D. Humanities

15. Psychology

   (a) general, (b) experimental, (c) comparative and genetic, (d) abnormal psychology

16. Sociology

   (a) Social structures, (b) social psychology

E. Applied sciences

17. Medicine

   (a) public health, (b) hygiene, (c) pathology, (d) therapeutics and pharmacology, (e) internal medicine, (f) neurology, (g) psychiatry, (h) surgery, (i) gynaecology, (j) ophthalmology, (k) otology and laryngology, (l) paediatrics

18. Technology

    (a) construction, (b) mechanical technology, (c) electrical, (d) mining, (e) technical chemistry, (f) agriculture

19. Economics

    (a) economic theory, (b) transport, (c) trade, (d) money and credit, (e) public finance, (f) insurance

F. Social rules

20. Politics

    (a) political theory, (b) diplomacy, (c) national administration, (d) colonial Administration, (e) Town administration

21. Law

    (a) international law, (b) constitutional law, (c) civil law

22. Social sciences

    (a) the family, (b) rural communities, (c) urban communities, (d) industry, (e) wage earners, (f) crime

G. Social culture

23. Education

    (a) theory, (b) the school, (c) the college, (d) the university, (e) the library

24. Religion

    (a) Religious education, (b) qualification to religious office, (c) determinants of religion, (d) religious work, (e) influence of religion on the individual, (f) influence of religion on society

Criticism. It is worth taking a closer look at this attempt to organise the total of human knowledge, even if it only teaches us how not to do it.

The most obvious thing is that it displays ignorance or perhaps disregard of the simplest laws of organisation. A sensible division must follow a unified concept so that the parts fit naturally to make the whole. This however is completely missing. Normative, historical, physical, intellectual sciences and so on are not parts of the same whole, because in the historical group are found the same sciences which later form groups of their own. This demonstrates that history is not a science in itself but simply a methodology or viewpoint which can be applied to any science.

Sociology is particularly torn apart. Ethics belongs to it but this is placed under A1 as does sociology which is split between D16 and the entire divisions F and G comprising sections 20–24. On the other hand a section on the history of sociology ought to have been present in Group B.

Physiological chemistry 10d does not belong in chemistry but rather in physiology in section 13k and its separation from plant physiology in section 13c is completely illogical.

One could go on for a long time in this strain, and a discussion of this sort is by no means fruitless because with the ever more rapid acquisition of knowledge the question of its fundamental organisation becomes ever more urgent. One only needs to think of how primitively all this material is organised for daily use in encyclopaedias where all that is done is to list the keywords in alphabetical order while completely ignoring any objective form of organisation. Even within the different sciences this is the crude approach that tends to be used—even in cases like dictionaries of chemistry or physics or mathematics etc. where a form of coherent inner organisation already exists or could be produced.

The invitation. The congress's working committee with its input from the entire scientific community of America decided to personally invite the European scholars who had been selected. Newcomb took on the French, Münsterberg the Germans and Small had to work on those in Britain, Russia and Italy. For this reason Professor Münsterberg, whom I'd never met and in fact had never even heard of, visited me in 1903 one year before the congress and, after some initial small talk, told me the plan.

Münsterberg was rather large and a bit plump. He had a round almost bald head but there was as yet no grey in his moustache. His posture was that of a writing man. He was quite at home in Leipzig because he had been a student of Wundt. In addition he felt himself part of the German philosophical scene and was close to the South German idealists who, under the leadership of Windelband were working hard and deliberately to conquer all the philosophical professorships in the country. I had the impression that Münsterberg wanted to come back to Germany, preferably to an influential professorship such as, for example, that in Leipzig.

Since the empirical natural philosophy which I represented was not his cup of tea, one must regard it as a sacrifice on his part that he had to invite me not as a representative of physical chemistry but rather as a philosopher. The physical chemists were to be van't Hoff and Aurthur A. Noyes who was one of my oldest and best American students who now held a position at the technical university in Boston.

I had been invited to speak in division 1 (philosophy) in section d (the scientific method). The second speaker in this section was Benno Erdmann who at that time was professor of philosophy in Bonn and who later moved to Berlin. I was to speak about the theory of science and Erdmann about the nature and applicability of the law of causality. The fact that the Americans were treating me as a leading light amongst international philosophers miffed many of my German colleagues. For me it was a justification for the change of tack I'd made 5 or 6 years previously, though I have to say quite honestly that I personally didn't feel the need for any such justification. For my colleagues in Leipzig it was just one more reason to draw negative conclusions about my suitability for the position I held in the university.

Travelling companions. About a week earlier than necessary I travelled from Bremen to New York on the "Kaiser Wilhelm der Große". I'd planned to spend the extra week largely at the Niagara Falls and there to paint as much as possible and for this I'd brought plenty of supplies. On the ship I met other people who were also on their way to the congress. There was the anatomist Wilhelm Waldeyer, the sociologists F. Tönnies and G. Ratzenhofer, the paediatrician Escherich and the astronomer O. Buckland.

I talked a lot with Escherich who was from Vienna, knew my name and who greeted me in a very friendly way. He turned out to be a man of broad interests who had a lively interest in the sciences that bordered his area so that I learnt a quite a lot from him and the ideas which formed then blossomed only recently and led to the concept of "over-cure" which I tried to introduce into biology in 1924. In personal terms he had a most amiable appearance—tall and thin with a pale complexion, dark hair and beard and a pleasant and lively voice.

I also have pleasant memories of the sociologist Ferdinand Tönnies. In appearance he was more or less the opposite of Escherich, being small and so stooped that he almost seemed to be deformed. His face was pale and his brow was hairless. He tended to be quiet and reserved in conversation which marked him out as the lonely intellectual who finds it easier to come up with new and useful ideas but has difficulty explaining them to others. However, he seemed more to seek out than to avoid conversations with me. I had to admit that I hadn't thought much about sociology. Although I was, at a theoretical level, completely convinced of its importance I hadn't actually come across any book which had shown any concrete scientific results from this area. As an excuse I might add that at this time there was not a single full professorship for sociology at a German university. What had been done in the field was largely the work of scattered individual economists. On top of that, the similarity of the term "sociology" to "social democrat" made the whole business suspect, just as those political economists known as "Kathedersozialisten" (academic socialists) had a rather dubious reputation. At that time I followed the political line of Bismarck who accurately forecast the enormous damage that the social democrats would inflict on the German Empire.

These talks with Ferdinand Tönnies were my introduction to sociological thought for he wasn't put out by my lack of knowledge or my one sided arguments and he convinced me that his science held the key to ideas and challenges of incalculable importance. When later on I was able to tell my peers some useful things about the social aspects of science and when I myself began to see science as a social structure which it was enormously important to come to terms with, then I have to say that all this initially sprang from my conversations with F. Tönnies on board the "Kaiser Wilhelm".

I had much less contact with Gustav Ratzenhofer. He'd come from a poor background and had with his energy and talent risen in the Austrian army to the rank of divisional commander (Feldmarschall-Leutnant). He resigned his commission because of a disagreement about some fundamental issue and from then on pursued his scientific interests as an autodidact. In St. Louis he was, along with

F. Tönnies, a speaker in the Sociology section which was concerned with social structures.

Ratzenhofer was 62 years old when I met him on the ship and he looked like a typical senior officer in the Austrian army who trimmed his hair and beard like his Emperor. We didn't see much of him because he seemed to suffer from sea sickness and in general he seemed not to be in the best of health. He was accompanied by his son. He held his lecture in St. Louis but it must have been with his last reserves of strength because he died on the return journey on the 4th of October.[1]

I later learned from his papers that he had been much interested in Energetics. He'd tried to grapple with it but he lacked sufficient knowledge of physics and chemistry so that he was unable to reach an objective conclusion. He was astonished that I neither knew his name nor his publications and this may be the reason that the few conversations I had with him did not lead anywhere.

It was a real joy for me to meet on the ship my acquaintance from Dorpat Oskar Backland. He didn't know any of the others and was glad to be able to join our buoyant group. From the observatory in Dorpat he'd managed by dint of his competence to work his way up to become director of the main observatory in Russia at Pulkowa near Petersburg and it was in this position that he represented Russian science at the congress. With his quiet friendly nature he was soon accepted into our little group.

Other colleagues on board were Sir Felix Semon, the brother of the author of the term "Mneme"[2] and personal physician to the King of England as well as the Oxford professor W.A. Sorley. Apart from these I also spent some time with my neighbour at dinner the engineer Gerdes from the company Pintsch which is one of the leading manufacturers of lighting technology. They had just managed to produce large pieces of metallic Tantalum which because of its high melting point and its chemical stability was going to be important in the industry and he was on his way to America to negotiate about its economic exploitation. Later the hopes seem not to have been fulfilled — I don't know why.

Since Escherich was officially representing Austrian science and Waldeyer German science in St. Louis, those on the ship were a considerable part of the upcoming congress. Despite the great difference in age I got along well with Waldeyer and so the journey across the ocean passed quickly for all of us. Partly of course this had to do with the fact that we were on the fastest transatlantic liner which had just brought the "Blue Riband" back to Germany once and for all. This angered many Britons and we often had to suffer their bitterness which was all the greater because they had only themselves to blame.

Leaving the ship. The end of voyage dinner at the captain's table was most instructive. On one of the first days the captain had expressed his great joy at finally being able to meet me personally. I was astonished and when I asked him how he

[1]Ratzenhofer died on the 8th October.

[2]Ostwald refers to Richard Wolfgang Semon (1859–1918), a German biologist who advocated the heritability of acquired characteristics for which he proposed the term 'mneme'.

knew my name he looked piqued and said that of course he'd heard of me and had read some of my work. Doubtless its part of his duties to tell the passengers on his lovely ship things they want to hear. On this evening I was asked by the others to give the usual short speech of thanks to the captain. I was happy to do this and it seemed to satisfy the others.

After me an American also held a speech of thanks that was mostly composed of tired jokes. The best one had to do with the smuggling of goods liable to customs duty such as furs and jewels which the women of his country do with gusto. He quoted the famous signal of Admiral Lord Nelson before the Battle of the Nile: "England expects that every man will do his duty" and said that nowadays the American slogan was "America expects that everybody will pay his duty".

The captain gave a speech in reply, but to the outrage of the German passengers, he spoke English. We asked Waldeyer as the most senior member of the group to make our objection clear to the captain but I don't think he did it. This example of a German inferiority complex was all the more unacceptable coming from the captain of the ship which had just magnificently demonstrated the German superiority in shipbuilding over the British who had been world leaders up till then.

At the end there was the usual speech asking for contributions to the fund for widows and orphans of sailors. This one was held by an American woman who claimed she was regarded as the "cleverest woman speaker in America" by all her friends. And then she let loose a real Niagara of verbiage that knew no end, so that soon the people were standing up, letting off crackers, talking to each other loudly and giving other signs of their impatience. Finally her speech degenerated into a real sermon but by that time nobody was listening. This incident was instructive as a symptom of the exaggerated esteem for women in American society which is a holdover from the colonial times.

New York. Our arrival in New York went smoothly this time because the congressional leaders had arranged that we be met by officials who were able to circumvent the usual formalities. I was met by a former student of mine, Dr. Morgan, who now held a position at Columbia University in New York and who introduced me to his companion and fellow chemist Professor Charles F. Chandler. Chandler had been a student of Wöhler who he still remembered fondly and despite his 68 years he was still spritely. He was a tall stringy man with a closely shaved long red face and little hair who, after the death of his first wife, had recently married a much younger woman. When I later visited him at his home he showed me the various gymnastic and boxing appliances which he used to keep fit and in this way he got to be about 90 years old, for I learned of his death only at the end of 1925.

They had put me up at the Manhattan Hotel and there I learned the ins and outs of their huge inns. The building took up most of one block and the ground floor was one enormous hall that was so crowded that it seemed to be an extension of the street's pavement. With Dr. Morgan as guide I got to know New York better. The weather was warm but not humid and I spent pleasant hours in the broad expanse of Central Park which is the most beautiful park in the city. The traffic on Broadway, the principle business centre, was deafening; the other streets were quieter but they were also all less clean than I was used to in Germany.

Naturally I was taken to Chandler's institute, though I didn't really want to see it, but there he had a large collection of strange and funny things. Morgan also showed me his physical chemistry department and that calmed me when I thought that I would soon give up my laboratory in Leipzig because he clearly had all the necessary instruments and material with which to carry on research in my old area of science.

At the same time as the congress in St. Louis the international chemistry congress was being held in the United States under the chairmanship of my friend William Ramsey. The result was that in New York and then later in other towns I came in contact with the international chemists who considered me one of their own and in this respect Ramsey was the first. He'd arrived in New York at the same time as me and visited me in the hotel in order to invite me to a reception to be held by the New York chemists. We were both happy at our reunion and agreed to try to arrange our return journeys on the same ship—and that is how it turned out.

That evening there was a large reception in the "Chemical Club" to which I was invited as a guest. There I had to meet a lot of people including a Mr. Mallinckrodt from St Louis who turned out to be a rich manufacturer and who asked for the "privilege" of putting me and van't Hoff up during the congress. After Ramsey had told me who he was I accepted the offer with thanks. Then came the "reception". This consisted of four or five people, Ramsey at the front, then me, then an animated millionaire Nichols and one or two others who I don't remember. We formed a line before which the ladies and gentlemen present queued up and passed along. Stewards were in charge of the introductions, there followed a hearty handshake and then the next in line was on. This went on for about one and a half hours and at the end was so tiring that I slept badly.

The next day there was the usual tours and in the evening there was a "chemical" dinner where I met lots of new people and listened to lots of speeches etc. This was all too much for me so that I left for the Niagara Falls the next morning and spent several quiet and happy days there.

At the Niagara Falls. It was with real joy that I saw again the beautiful waterfall that I'd briefly visited the previous year and all my expectations were more than fulfilled. I stayed in a small German inn close to the Falls where I met nobody I knew and could leave every morning early to paint. For that, all I had to do was sit down and start to paint for from any spot I only had to turn around in my chair and there would be a new motif.

The work went faster and better than ever before, so that I managed eight to ten paintings per day while previously I'd never managed more than four or five, and this time an unusually high percentage of them were successful. In terms of landscape painting this was the climax of my artistic efforts. Needless to say that I felt truly happy and gave myself unreservedly to the work because I knew it would be the best possible preparation for the exhausting days that lay ahead in St. Louis.

Once or twice on my solitary walks I met O. Buckland, who'd retreated here to prepare his talk for St. Louis. I'd already written mine on the ship and, as I wrote home afterwards, I thought it had turned out to be better than I'd expected.

After I'd done around thirty paintings in four days I left the Niagara Falls because, apart from anything else, quite a few colleagues from the congress had arrived there who were not to my taste. I remember some conversations I had with the art historian Richard Muther whose one sided and shallow defence of the French painters rubbed me up the wrong way. I also didn't manage to establish any comfortable relationship with the biologist Oskar Hertwig from Berlin though I got on well with his brother Richard in Munich with whom several years later I spent some very pleasant weeks on Tenerife.

Meeting old students. From Niagara I went first to Toronto where two of my old students, Lash Miller and B. Kendrick, had established a physical chemical department. The day passed happily in a tour of the lab in chatting with the people and these included the physiological chemist MacCallum.[3] I couldn't really see any difference between this city and those in the United States.

For much the same reason I stopped off in Ann Arbour on the way to St. Louis where a former student called Bigelow taught physical chemistry. My reception in his house and by his colleagues was very cordial and they tried to get me to promise to come back later for a longer visit for the town and the university lay right on the edge of the Wild West and so they laid special weight on making personal contact with representatives of eastern culture.

For the journey to St. Louis I was joined by some of the faculty including the emeritus Professor Preston[4] a thin old man with a long white beard who, despite a lame leg, was remarkably spry and who, together with his dear old wife looked after me like a father.

On the journey to Ann Arbour I'd taken the train to Detroit, which was the nearest railway station, and from there an electric tram that took several hours to bring me to Ann Arbour. On the way back the son of one of the professors drove us back by car. This means of travel was at that time still quite rare and I must say that I liked it a lot. Later Detroit became famous because it was here that the great organiser Henry Ford built his factory.

Arrival in St. Louis. The journey to St. Louis was long and hot. In the sleeper I'd had to be content with an upper bunk which is always much less comfortable than the lower one. In this case that had been taken by a mother with her roughly one year old child and I could look forward to a sleepless night. However I must congratulate both of them and the American nation that throughout that long night I neither heard nor smelt the young citizen.

We arrived the next morning with the usual long delay in St. Louis. The house of my host was easy to find. It turned out to be a splendid palace and my room was as luxurious as that of a prince. Mr. Mallinckrodt, his wife and his grown up son greeted me in a most friendly way. The father came from Germany and spoke fluent German, his wife was American and had a pronounced preference for French

[3]The correct spelling is Macallum.

[4]Ostwald refers here to Professor Prescott.

pictures and books and clothes. The son, however, was proud of the German blood in his veins and longed to complete his studies in Germany.

At breakfast I met van't Hoff who was staying in the same house and was accompanied by his eldest daughter Jenny. He was not feeling at his best after the journey and also later he felt under the weather.

I found a visiting card in my room from the German Imperial Commissioner Lewald who had organised the German side of the congress in the face of many almost insurmountable difficulties. I will come back to him later.

The World Exhibition. The next morning, spurred on by curiosity, I visited the extensive grounds of the exhibition. Before the entrance there was a large square that one had to cross. The way there was flanked by a double row of men armed with a typical American instrument of torture known as a megaphone that turned out to be a speaking tube of unbelievable dimensions through which they yelled out advertisements for all sorts of things. Noise in all its forms turned out time and again to be a major characteristic of the United States. This is the origin of the strangely high pitched and shrill sound of American English which is so different from the British form. Only through the use of strong overtones can the American make his voice carry through the roar of background noise in which he lives.

My first priority was to meet my scientific colleagues who, like me, had come to the congress. But now it turned out that the organisers had in an extraordinary degree lacked imagination or interest, for at such a congress the main interest is to meet and discuss with the best minds, all of whom are concentrated for one week in one small area. The lectures themselves are less important since one can read them in comfort in the congress proceedings. In other words the most important thing was to provide a place where the speakers could gather in their free time and where one could be sure of meeting really interesting people. What someone of this calibre can say in a quarter of an hour can provide food for thought for years or decades. Every one of the participants would have gone home with unforgettable memories and a feeling of thankfulness to St. Louis and the organizers of the exhibition. If the room had writing materials and postage facilities then every speaker would come there at least once per day.

None of this had been organised, however, so that the participants of the congress began to have the feeling that they were regarded by the organisers as a foreign body and this view seemed to be shared by a wider circle because the statistics show that less tickets were sold for the scientific week than for the week before or the one after. Not even the enormous curiosity of the Americans was stimulated by the presence of all these academics. The chance to see and hear some of the best minds of our time was something that did not occur to the mass of the people who otherwise will rush to gawk at any oddity.

A thought on support for culture. The idea referred to above of providing an informal setting where the thinkers of our culture could meet personally and interact, is something which I have more than once tried to initiate when I met rich and beneficent men—so far without success. I suggested that free accommodation of the simplest sort should be made available at some beautiful place, perhaps by the sea or close to a lake, where every year the most accomplished men and women (though not artists for whom this is not suitable) could meet and foster each other in an informal setting. Perhaps now, when so much in our general culture is disrupted and needs to be put in order again, this idea can be resurrected, though of course at least for the moment the number of people rich enough to finance it in Europe has become much smaller. Nevertheless the costs of such an enterprise are not that high. If one reckons on 10 marks per day per person for lodgings then with 14,000 marks one could invite 50 of the most important men and women for 4 weeks in summer for a stimulating and peace promoting exchange of views. When one thinks that of all cultural activities, science came closest to the ideal of uniting all peoples, until the barbarism of the French during the World War destroyed also this sacred sanctum of humanity, then one can see that the realisation of this plan would be a blessing for all mankind.

I had a foretaste of how this might work in the garden of the "German House" where the Munich beer attracted many colleagues—and not just Germans. I was happy to meet there J. Loeb who told me good things about my eldest son who was working as his assistant. Although he didn't drink alcohol he agreed to try a jug of beer since there were quite a lot of our colleagues at the table among whom a lively and excited conversation soon erupted. It was always a real joy to get into an interesting conversation with a complete stranger and then to learn from his name that this was one of the foremost men of his time.

The distribution of speakers by nationality. 150 invitations had been sent to foreign scientists and of these 117 had accepted. Less than twenty of these had been unable to attend for whatever reason so that on the day the congress started 96 foreign speakers were present and 4 more arrived during the course of the congress making 100 in all.

The congress report did not give a breakdown by nationality but I checked this and the result was as follows:

Germany and Austria	32 + 10 = 42
Britain and Canada	21 + 3 = 24
France	16
Italy	4
Japan	4
Holland	2
Denmark	2
Belgium	2
Russia	1
Switzerland	1
Sweden	1
Mexico	1

Since the total is 100 the individual values show the percentage. The first point to note is the enormous predominance of German science since if we take them together with the Austrians to whom they are related and who think in the same ways then we already have 42 % of the total followed by the British ahead of the French. One has to take into account that the American scientists would in general be more aware of work published in English than in German and that St. Louis was decidedly Francophile for historical reasons so that both of these countries had advantages that had nothing to do with science.

The numbers of speakers from other nations are too small to be statistically significant. For example both Holland and Belgium sent two speakers each although the level of their scientific achievement is around 100:1. The 4 speakers from Italy, however, do roughly reflect the country's level of scientific achievement in comparison to the top three: small but noteworthy.

The lectures. Of course I went to hear van't Hoff's lecture. The room assigned him was not very large so that it was packed full. The master clearly explained that physical chemistry had developed along two lines: the atom centred view of matter and the theoretical view which was based on affinity and he characterised these two views with the names of those of the most important researchers who had promoted them.

I	II
Lavoisier, Dalton (1808)	Berthollet, Guldberg, Waage (1867)
Gay-Lussac, Avogadro (1811)	Berzelius, Helmholtz (1887)
Dulong, Petit, Mitscherlich (1820)	Mitscherlich, Spring (1904)
Faraday (1832)	Deville, Debray, Berthelot
Bunsen, Kirchhoff (1861)	Thomsen, Berthelot (1865)
Periodic system (1869)	Horstmann, Gibbs, Helmholtz
Pasteur (1853) Stereochemistry (1874)	
Raoult, Arhennius (1886-87)	
Radioactivity (Becquerel, Curie)	

After the lecture there was loud applause after which I rose and pointed out the following.

When one follows the two ways of thinking back to there roots then for the first one must mention J.B. Richter (1792) who was the first to note the law of stoichiometry and should therefore be at the head of group I. The first researcher to try to measure chemical affinity was K.F. Wenzel in (1777) and so he should head the second group. By chance both of these were Germans, though I should naturally have named scientists from any other country had there been any historical basis for doing so. Now in our time both of these ways of thinking fused in the mind of a man who has done groundbreaking work in both areas. The complete list should therefore be seen as

Richter (1792)	Wenzel (1777)
Lavoisier, Dalton (1808)	Berthollet, Guldberg, Waage (1867)
Gay-Lussac, Avogadro (1811)	Berzelius, Helmholtz (1887)
Dulong, Petit, Mitscherlich (1820)	Mitscherlich, Spring (1904)
Faraday (1832)	Deville, Debray, Berthelot
Bunsen, Kirchhoff (1861)	Thomsen, Berthelot (1865)
Periodic system (1869)	Horstmann, Gibbs, Helmholtz
Pasteur (1853) Stereochemistry (1874)	
Raoult, Arhennius (1886-87)	
Radioactivity (Becquerel, Curie)	

J. H. van't Hoff

These impromptu remarks were strongly supported by the audience and led to renewed applause for the day's speaker.

My lecture was in the afternoon of the same day. The room was large enough but I couldn't speak undisturbed because in the middle of my talk a very energetic military band marched past. I had to stop for several minutes before I could come back to my peaceful remarks. If you will, this was almost symbolic of later events. I saw in the audience quite a number of the distinguished scientists who'd come to the congress and others who I got to know later. It was one of the highest ranking scientific audiences that I have ever spoken to and I acutely felt the wish that I had spent more time on the preparation than I had.

For my part I listened to the lectures from A. Harnack and from H. de Vries. The first of these surprised me by his complete lack of narrow dogmatism so that I tried to get him to publish it in my "Annals of Natural Philosophy" but he didn't want to. The botanist H. de Vries had just recently published his sensational research on mutations or sudden changes in the phenotype of progeny which had so interested me that I was glad of the opportunity to hear him personally. Unfortunately this wish was only partially fulfilled. Just before the start of the session someone had noticed that there weren't enough rooms and so without further ado they had made two rooms from one by inserting a simple wooden wall. No one had expected that the wall would act as a resonance board and transmit every sound from the neighbouring room. Unfortunately at the same time as de Vries a remarkably loud voiced clergyman spoke in the neighbouring room so that the scientist's voice, which was in any case rather weak, was completely drowned out by the massive trumpeting of the gentleman next door.

However I did at least have the chance to personally meet on this occasion that splendid Dutch researcher.

The organisation of science. Since my lecture dealt with the issue that was the basis for the whole congress, it is worth spending a few moments looking at my solution to this old problem which has fascinated researchers ever since the time of Francis Bacon's failed attempt to solve it. I came across the problem in 1901 when I needed a fundamental means of organising my material for the lectures in natural philosophy. The solution I came up with resembles that which A. Comte had proposed 70 years earlier but of which I was unaware of at the time. My version contains a number of crucial improvements, but the similarities of the two approaches show that the problem now comes close to a solution.

The basic idea is that the organisation of science is based on the organisation of concepts, for the operation of every science involves the formulation and integration of concepts.

The concepts are characterised by their content and amplitude. The content defines the subsidiary concepts that are contained within it, the amplitude defines the number of things which are covered by the concept. The content of the concept "man" contains all of the characteristics of man; the amplitude is around 1.5×10^9.

Content and amplitude are inversely proportional so that the poorer the content, the greater the amplitude, and vice versa. If one defines a thing negatively as the property which distinguishes it from all other things, then it has a minimal content, namely the difference, but a large amplitude since it encompasses everything that fits this criterion.

Now one can organise all concepts so that one begins with those having the greatest amplitude and the smallest content and then progressively moves through those with reduced amplitude and increased content until one ends with those with the smallest amplitude and the greatest content. In such a scheme every concept will find its place as soon as its amplitude and content have been defined. In ordinary life concepts do not fit this scheme, but scientific concepts have been so constructed that they conform to these criteria. Examples are the systematic systems of zoology and botany whereby here, as everywhere, perfection remains an unattainable ideal.

If one applies this approach to the whole of science then one ends with a pyramid composed of layers whose width is a measure of their amplitude and whose vertical position in the pyramid represents their content. The following figure broadly represents this organisation.

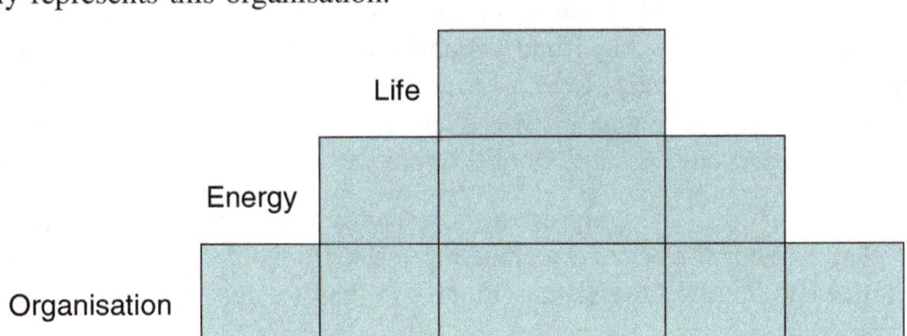

It turns out that the concepts Organisation, Energy and Life are the most important.

Sciences concerned with organisation are, in order of increasing content, didactics or special organisational sciences of which logic is a small part, mathematics, geometry and kinematics.

Sciences of energy are mechanics, physics and chemistry.

Sciences of life are physiology, psychology and sociology, to which all the humanities belong. Each of these sciences exists in a pure and an applied form. The pure form is a preparation for the applied.

It is particularly noteworthy that each general science has a larger content than the sciences above it and these are dependent on this larger content. One cannot do physics without logic, mathematics, geometry and mechanics. The reverse does not apply for one can be an excellent chemist without having any knowledge of sociology though one does need knowledge of the lower ranked general sciences.

Politics is applied sociology. If one asks whether our politicians have any knowledge of organisational sciences, or physics or physiology then one will realise how backward our age is and what a stupendous challenge faces our children and grandchildren.

This is just a small hint of the great value which can be derived from this simple scheme if one manages to bring it to life.

Threads from home. The imperial commissioner Lewald showed a particular interest in getting to know me better. At that time in Berlin the resignation of F. Kohlrausch from the chairmanship of the Imperial Institute[5] and that of Professor Landolt from the professorship of physical chemistry at the university meant that two important posts were free and naturally the question had surfaced of whether I might be suitable for one of them. It seemed that Lewald had been instructed to sound me out because he kept steering the conversation back to this topic. Already at that time I had the strong feeling that I no longer wanted to be tied down, but on the other hand these positions in the capital of the German empire would give me much more influence. In those busy days I had no time to find a means of resolving of this dilemma and made no attempt to be diplomatic in my replies. I emphasised, perhaps more than was necessary, my dislike of all those formal and official duties which Helmhotz, much like Goethe, had found rather satisfying and I also made clear that general philosophical and organisational problems now interested me more than physical chemistry. In short I clearly said everything that might lead to reservations and I feel sure that this was a success and that it led to a report (if there was one) that raised considerable doubts about my suitability for these positions.

After dinner speeches. The congress days were filled with banquets. The first was the best. The Chemical Society was meeting in St. Louis at the same time and Mr. Mallinckrodt had invited its leaders to dinner, so that we three chemists, van't Hoff,

[5]Physikalisch-Technische Reichsanstalt (Imperial Physical Technical Institute), a forerunner of the present *Physikalisch-Technische Bundesanstalt.*

Ramsey and I together with the members of the family made up a select dinner circle which we all enjoyed.

The other banquets were huge machines mostly with several hundred guests. The best of these was a breakfast given by commissioner Lewald in the exhibition's grand "German House". By now I'd met him several times and I fondly remember his mixture of polite friendliness and tenacious energy which was so characteristic of this most excellent public official.

An American "dinner" differed from a German banquet in a number of ways which, in general, were in the former's favour. To begin with there was a better functional division. Given that the mouth is required both to eat and to speak each of these functions was assigned its time. To begin with one only ate and at the most chatted with one's neighbour and—at least when I was there—the wine consumption was luckily not too high. Once this part had been completed, the toastmaster took over and the second and better part of the evening could begin.

The country has been a democracy for nearly ten generations and I remembered from my inn at the Niagara Falls where, on an afternoon when there was nothing to do, the owner and the servant sat side by side in their rocking chairs on the veranda reading their newspapers. In such a country the toastmaster represents the last vestige of absolute monarchy. He begins this part of the evening with a few words of greeting and then sketches in a few clear words the topic of the evening and closes with the phrase, "I now ask Mr. X.Y.Z. to let us know what he think about the matter". No one would dream for a moment of refusing such an invitation which is in reality no less than a command. If the toastmaster is particularly considerate then he may speak for a little while longer or introduce a further speaker so that the victim has a little time to collect his thoughts, if not then whether he likes it or not, the victim has to get to his feet and say whatever he can manage.

It's the toastmaster job to choose speakers in such a way that by his remarks between the speeches he directs their thoughts so that he creates a sort of symphonic impression. Each of the speakers tries to do his humorous best and may even manage, in the most civilised possible way, to make a joke at the expense of the toastmaster or a previous speaker. It is in other words a sort of mental football which, depending on the company, may be finer or rougher but it is usually an interesting game. I found it exciting enough even though, since I was not a native speaker, I may have missed many of the finer points.

As one can see the after dinner speech in America has developed into an important part of social intercourse and it is a natural consequence of their democratic constitution. When all important decisions are dependent on a majority then it becomes a matter of life and death to be able to put such a majority together. The usual way to do this is by convincing people and so the ability to convince a large number of people becomes the basis for all success—it is indeed essential for survival. Just as the essential movements of the muscles in the arms or legs can be developed to a high degree by exercise, so the after dinner speech is a form of mental sport. On the one hand it develops essential mental abilities like rapid thinking, plurality of ideas and dexterity, while on the other it trains the ability to see the audience's mental world so that they can be approached in such a way that they will be swept along or at least will present the least resistance.

Since at the moment we are in Germany in the same democratic situation we have every reason to develop this side of things and to methodically purse the "speech-sport".

In this respect it would be a good idea to abandon the custom we copied from the French of holding the speeches during the meal. One only needs to think of the many disruptions this brings with it—the incessant rattling of plates, the food gets cold, the suspension of serving and so on—of the present way of doing things which are completely avoided by the American division of eating and speaking. All of these advantages bring with them, as far I can see, no disadvantages at all.

After I'd been applauded for my part in this game, the toastmasters on the subsequent evenings felt justified in nominating me again so that finally I became a little impatient and decided to revenge myself. As a grateful guest I'd always looked for some way of saying something kind about our hosts. I could see that the average American (even the most gifted of them) was not only able to tolerate a vast amount of praise but was more than ready to swallow it down without complaint. Having done that he then opened his mouth and waited for more. I decided to exploit this.

I began with a long winded praise of American "can do" which went down very well. Then, as a proponent of energetics, I posed the question as to where they got all this extra energy from. I considered various possibilities and in particular noted that from my experiences at the congress they ate neither more nor more concentrated food than Europeans. Once I'd got them ever more curious about the solution I said that it must, of necessity, be something that wasn't present in anything like the same amounts in Europe and now my audience was ready to accept any flattery no matter how ridiculous.

I told them that I had finally identified the energy source which in this country everybody from millionaire to street sweeper could—and indeed had to—take up. It was in the form of vibrations in the air to which everyone was exposed all the time. Inexperienced people might call it noise and every foreigner would agree that here there was vastly more noise than in old Europe. Unlike the Americans we foreigners didn't have the necessary transformers to convert this useful energy source; however, now that we had learned so much about it we could set out to find a way of doing this.

For a moment my audience was completely surprised but then there was such a mixture of thunderous laughter and angry grunts that it took the toastmaster quite a while to restore order.

At the end of the congress there was a huge banquet for nearly a thousand people, which, as is the way with these things, was not particularly pleasant. It was in the "Tyrol Alps" which was an emulation of a village in Tyrol complete with a clever and impressive backdrop painting of mountains. Conversation with those at the table was continually interrupted by a regimental brass band which the Americans had imported at huge cost from Paris.

In between times the official final speeches were given and after each of them the band played that country's national anthem. However when it was Germany's turn the French musicians refused to play "Wacht am Rhein" (The Guard on the Rhine) and it cost a huge amount of effort to paper over this incident. I don't remember now how it was done.

This incident made a strong impression on me. For the whole week we'd heard a lot about how science was a powerful means of bringing the nations together, that in place of national jealousies and animosities it stood for the amiability of shared culture. This was the basic idea behind the congress. And yet now, after 40 years of peace, and after a never ending process of accommodation on our part, this idea of cultural unity was thrown aside by the French national hatred of the repulse of their foolish attack on us in 1870. Already then the French showed themselves to be barbarians and proved it again ever since 1914 in their unrestricted war against German science.

Washington. After the congress we bade farewell to our hosts and a large number of the scientists travelled to Washington where President Roosevelt had arranged a reception for us. Van't Hoff and I travelled together and passed the time in conversation. In the evening he showed me the rules he had developed in case of a train crash. To begin with he thought that during the night he should lie with his head in the direction of travel. If the train had to suddenly brake then his body would still have a large amount of kinetic energy with which he would be projected head first through the compartment's partition wall. His skull would be smashed, like an egg dropped on a table, unless the shock was absorbed. For this he piled all the cushions, pillows, his coat and anything else he found that was soft, as a buffer.

There were however other possible types of accident which could lead to serious injury. For these cases he had in his waistcoat pocket a thin glass tube within which was sealed around one gram of pure sodium cyanide. It was sealed to protect it from reaction with the air; he'd chosen the sodium rather than the usual potassium salt because it is more readily soluble and the tube was thin walled so that in case of an accident which would cripple him he could break it with his teeth and die instantly.

Luckily neither the one nor the other type of accident happened.

Early in the morning the train stopped in a larger city—I think it was Indianapolis—and the conductor said that we would be there for half an hour. Since there was no restaurant car I went to the restaurant in the station but couldn't get anything to eat. Ten minutes later when I wanted to get back on the train it had already left. I had to wait till the afternoon for the next train. In the meantime van't Hoff had collected my hand luggage (I'd even left my hat on the train) and since we'd luckily already decided where we would stay I had everything again that evening.

In the meantime the members of the congress had been to see the President. van't Hoff said it was such a pity that I hadn't been there, because Roosevelt had asked for me. Since I thought he was making a joke at my expense I left it at that.

There was a lot to see in Washington. For me the most impressive thing was the Library of Congress with its wonderful mechanical device consisting of a little railway into which one put the order form and the requested book was then delivered within a few minutes. There was also a lot to see in the enormous

National Museum. I skipped the "historical" sites like Washington's house and so on because I didn't have any time or interest in visiting such places of pilgrimage and didn't want to disturb the devotions of those who did. It seemed to me to be a waste of time.

In any case I was exhausted by the busy days at the congress and I compensated with days spent sleeping and doing nothing. It hit van't Hoff even harder and he was really ill for a while but recovered in a few days. Because of this I didn't join in much of the activities which were offered us even though there were lots of interesting people to meet. I remember having a few words with Graham Bell, the inventor of the telephone, and also with the excellent physicist Michelson. The first of these was a splendid old man with snow white hair and beard; the latter looked more like a military man than a professor.

Baltimore, Cambridge, Middletown. From Washington I travelled to Baltimore where one of my former students, H. Jones, was Professor of Physical Chemistry at the Johns Hopkins University. Because of the holidays most of the scientists were absent but I did have a short meeting with the university president Ira Remsen, whom I'd met in New York. He was also a chemist and told me that quite a lot of former chemists turned out to be suited to university administration. An American university president is much more than a German university rector. He is elected for a long time—often for life—and has much more power to change matters in the entire organisation than does a rector elected for just one year. Because of this the universities there are indelibly stamped by the personality of their president. One of the most influential of them was the then president of Harvard Ch. Eliot who was also originally a chemist. At that time I didn't know him though on my third journey to the USA I was brought closer to him and his university.

A drive through the beautiful town park, a short visit round the university, a small lecture to those chemistry students who were still there and a dinner with the faculty which was arranged by Professor Jones, filled the day.

With this visit to Baltimore the main part of the journey to America was now completed. Professor Münsterberg had invited the European speakers to Cambridge in Massachusetts where he held a professorship at Harvard and, insisting that it was very important for me and my interests, had made me promise to spend the evening with him. After that, if I didn't want to travel any further, I'd be able to take the liner home in the company of William Ramsey which I was looking forward to.

That is what I did. I went from Baltimore to Boston where my former student A.A. Noyes had successfully established physical chemistry at the technical university there (Massachusetts Institute of Technology) and from there went on to neighbouring Cambridge to do the rounds with Münsterberg. This turned out to be a stifling scrimmage of far too many people in far too little space. Münsterberg wanted to introduce me to the president Ch. Eliot who for reasons that I didn't know about at the time wished to meet me. Unfortunately he had an urgent prior appointment elsewhere that day and my planned visit to Middelton would brook no delay. I didn't realise that in the following year I'd spend an entire semester at Harvard.

My planned visit was to the physiologist Professor Atwater at the small university in Middletown NY. Spurred on by Pettenkofer's respiration apparatus, he'd built a highly sensitive appliance to follow the entire energy and metabolism of a human being and with it he'd already produced important results. In my Journal I'd repeatedly pointed out the importance of this work and didn't want to leave America without having personally met this researcher and his apparatus.

Middletown was a little town in the countryside with a denominational university in which this independently minded and dedicated researcher had overcome a thousand difficulties and through his enormous scientific and technical talent had managed to make his research plan reality. His results became rightly famous, but if I remember correctly he paid soon after this for his remarkable achievements with a total breakdown.

The journey home. At the appointed time I boarded in New York the steamer "Baltic" which was one of the largest ships in the British liner fleet. The comparison I inevitably made with German ships was not to the Baltic's advantage. The cabin I had was not nearly as pleasant and clean as those in the second class on our ships. It smelt of sea sickness and not just from the last occupant because there was a similar smell in the other cabins I visited. The food was terrible — tasteless and badly cooked. While on the German ships one runs the risk of eating too much of the good and tasty food which is so invitingly offered on the menu, on the Baltic one lost one's appetite completely within a couple of days because the food was so unappetising and boring.

I met Ramsey as we'd planned and he introduced me to the three or four others with whom we shared a table at dinner. These were scientists of no great note who showed no great interest in a foreigner. The other guests to whom Ramsey sometimes introduced me also made no attempt to get closer acquainted. I'm pretty sure this had nothing to do with my own personality but rather was simply that a German was not welcome on a British ship. For me it was a good excuse to keep myself to myself and get a chance to recover from all the recent exertions.

From the occasional conversation I gathered how deeply the British had been hurt by the superiority of the German passenger ships and they never missed an opportunity to emphasise that their ships were just exactly as they wanted them.

Another example of the insular narrow mindedness of the British came in a conversation with the owner of an important spinning mill. He was the chairman of an association dedicated to the abolishment of the metric system which he wished to see replaced with the British system. I agreed that the basic unit of the decimal system was not well chosen because it only had the factors 2 and 5 and that it could well be replaced by a system based on 12 which had the factors 2, 3, 4 and 6. However since the British system is not a strictly 12-based system, adopting it would in effect be a retrograde step.

A few days before the ship landed in Liverpool there was benefit afternoon to collect money for sailor's widows and orphans. Ramsey had suggested that I lend them my paintings of Niagara which were hung as a small exhibition in one of the lounges and for which a small admission fee had to be paid. However just as the passengers were beginning to get interested a concert which had been supposed to start an hour later began. A singer belted out a national song at full pitch and this immediately drew all the people into the hall and the pretty lady who had sat at the cash desk for the exhibition now had nothing to do.

I want to emphasise that this sort of unfriendly treatment was only evident in the non scientific circles of our neighbours. My colleagues in Britain have always treated me most kindly and this is shown for example in the fact that I received more scientific awards there than anywhere else. I have already described some of that.

We finally arrived in Liverpool. Although I knew people there I wasted no time but took the quickest connection to London and from there home where my family were all in good health.

Chapter 30
Free!

Biological aspects of research work. Earlier (Part 2, Chap. 23, p. 280) I described the natural and necessary process by which a mother and child, who at first form a close unit, begin separate ever more as the child begins its own life so that at the end they are quite foreign to each other. The process is not only useful and necessary for the child but also for the mother who would otherwise not be able to bear and bring up a new child.

How long this process takes depends on the nature of the animal concerned. While the love and care of a cat mother lasts no longer than 3 months, other animals look after their young for years. The time involved depends largely on the size and on the liveliness of the species since animals with a more rapid reaction time will more quickly get through their educational tasks than those of a more phlegmatic nature. Naturally, the nature both of mother and offspring are important here but since they are obviously very similar one sees this difference only in special cases.

One can transfer these observations directly to the relationship between a creative scientist and his work. Sanguineous researchers with rapid minds, whom I describe as being of the "romantic type" don't pay too much attention to their discoveries once they have been made. The discoveries are brought into the world in such a way that they soon take on their own life and the arguments for and against them can be left to others. Then again, this type of scientist soon has new ideas and discoveries and to make place for them he has to put the old ones quickly behind him. In contrast scientists of the "classical" type carry their progeny for a long time before they give birth, just like elephants do, and then keep them at their side for years on end. They spend a lot of energy re-examining the results, improving and extending them. Helmholtz who was a typical "classicist" always found it difficult to finish a paper and send it off for publication. He said explicitly that he hardly ever submitted a manuscript without already the next day thinking of a number of places where he ought to have expressed himself better or more precisely.

R.S. Jack and F. Scholz (eds.), *Wilhelm Ostwald*, Springer Biographies,
DOI 10.1007/978-3-319-46955-3_30

While the progeny of normal animals are all more or less the same so that the process always takes about the same time, products of the intellect can be very different and the amount of care they require can vary from a few weeks to an entire lifetime. The latter is in fact the most common since quite a lot of people pursue one single creative idea throughout their whole lives.

Here we see a situation in which the necessary is happily combined with the desirable. The "romantic" researchers, in whom one idea chases the next, are usually excellent teachers in their young years when their intellectual output is at its height. They train many co-workers and these in turn will then look after the ideas for them. The "classicists" on the other hand are usually not good teachers—not because they can't teach but because they don't want to. In this case there would be no one to look after the development of an idea if the originator didn't do it himself. Like this everything is well organized.

Evidence. This will become clearer with a few examples.

In his later years Berzelius explained in a letter to his student and friend Wöhler the difficulties involved in filling a professorship for chemistry at a Swedish university because there was no suitable candidate in the country. This shows the almost unbelievable situation that the country which, through Berzelius, had exercised hegemony in chemistry for a generation was not in a position to bring forward its own teachers in the subject.

The reason was that Berzelius was a researcher of the "classical" type who produced excellent work but who lacked the ability or interest to awaken this talent in others. Of course Berzellius had lots of students, many of them from Germany, but they all came to him as mature chemists who wanted to learn the methodologies he had so successfully applied in his own work. The special position that he had as an academic without teaching duties gave him the possibility of avoiding teaching altogether. Had he wanted to, he could easily have got a professorship or could have created a teaching post at the Swedish Academy.

In contrast the "romantic" Liebig was enormously successful in training an army of students who took over chairs in chemistry not only in Germany and Austria but also further afield—Carl Schmidt far away in Dorpat was one—and his influence was extended in this way also to Britain and America. Some of these students even made their mark in the French speaking world as the names Regnault and Gerhardt testify.

My own case. It was thus only natural that I, as a "romantic" introduced many young researchers into my science. Earlier (part 2, Chap. 28, p. 344) I listed the number of professorships which were filled with people from the Leipzig institute between the years 1887 and 1904. Later on this number was nearly doubled.

I always encouraged my assistants to get the "habilitation" (the prerequisite degree for becoming a professor) which allows them to give lectures. They did this and thus all became professors, some of them becoming famous in their field. The result was that in Leipzig there was in addition to my lectures on general physical chemistry also a whole set of special lectures in the same subject which extended the education of the students in particular areas. During my sick leave the main lectures were held by a deputy and later the authorities sometimes gave me leave to delegate the lecture course again.

An incident. Johannes Wislicenus died in 1904. His life which had begun under difficult circumstances had blossomed—before ending dismally. His wife, who bore him five children, fell ill and since she suffered from severe depression had to be put in an asylum. The children were by this time grown up and had all inherited their father's strong build but one son committed suicide a few years after I came to Leipzig. The second daughter, who was unmarried and looked after his household, suffered from depression like her mother. On top of that there were embarrassing scientific quarrels which he lost. His health was also not the best. He was not yet 70 and his massive build seemed to guarantee a robust old age but then he aged rapidly. Rheumatism and gout limited his movements and yet his build made exercise not only desirable but essential. Though he was a scientist and had even worked for some time in physiology he didn't apply the common rules of physiology to his own body. His father had been a subtle theologian in Halle who'd had to flee to America to avoid being imprisoned by the clerical government of Prussia for the publication of his views. The son had inherited the same stubborn sense of duty which led him to ignore reality. Instead of preserving his ability to work by going for treatment to a health resort, which would have meant being absent for half or even a full semester, he carried on and so totally exhausted himself that he died in harness.

His father's experiences, which made him responsible for the family at an early age, had left him with a dislike of the church. He made no secret of his dissidence but did not make an issue of it so that even in clerically minded Würzburg he acquired not only a scientific reputation but was so highly regarded generally that he was elected to be rector for a centennial celebration of the university, even though he had been rector the year before. When he moved to Leipzig he made no issue of his dissidence, though he did have the bible quote, "God hath arranged all things according to measure, number and weight" which his predecessor Kolbe had hung in the lecture hall, removed. However Kolbe had been thought rather maladroit in this matter by the disciples of orthodox teachings so its removal had not made any waves.

At that time the orthodox professor Luthardt[1] ruled the theology faculty with an iron fist (you get an idea of this from the bust in the Leipzig Art Museum). As every force calls forth an opposite force, people harmlessly collected stories that showed him in another light. One of these related how he greeted a small skinny theology student who had just finished his studies with the words, "My dear son, never forget that you are entering a difficult and austere profession". He hadn't noticed that a

[1]Luthardt was a lutheran.

maid had just arrived with his own usual breakfast: a large glass of sweet wine and four rolls heaped with caviar.

In the year that Wislicenus died the rector was the lawyer E. Wach who was also religiously inclined. It was the normal custom that the service for professors who'd died be held in the university church—even in the case of the catholic law professor Windscheid for whom the protestant church had been quietly re-sanctified by a catholic priest before the service. Nevertheless the rector decided that it was impossible that the coffin of the dissenter, who in the course of his life had been accorded all the honours that a professor can be given, be carried into the university church. The laying out was therefore moved to the First Chemical Institute, where the funeral service was also held.

I, as his closest scientific colleague, was chosen to prepare an obituary for this scientist as a token of the gratitude of all his colleagues and students. I did it with all the warmth I'd felt for him in the initial years of our joint work. I lauded the unusual circumstance that he was not leaving on his last journey from the usual place as an involuntary recognition of his unshakeable sense of duty in his work which made him more than worthy of special acclaim even if narrow minded religious zeal had closed the university church to him. Since I'd thought that this remark might cause misgivings I'd written out my address and at this point took the sheets in my hand and read the passage out. The congregation of mourners heard the sentence in absolute silence.

The next day there was no sign of reaction though the matter was hotly discussed. However, there then appeared a report on the incident in a daily newspaper in which my words were completely misrepresented. The writer was a colleague from the theological faculty who had more than once run into difficulties because of his uncertain and selective memory. I replied in the same forum giving the exact words I'd used and pointed out that I was fully justified in using them on this occasion. No sooner had the reply been published than the rector asked for a meeting at which he defended the position that the religious nature of the university church made it impossible to accept the body of a dissident. I replied that Leipzig University was not a religious institute and that therefore its facilities must be open equally to all its members. The discussion was carried out in a most civilised fashion. Then the theological colleague was announced and the rector asked us to promise not to carry on our disagreement in public. To this we both agreed. Despite this a new polemic from the theologian appeared in the newspaper a few days later. When I reminded him of his promise he claimed that he'd already written the piece beforehand and had "naturally" not considered that the promise extended to this article as well. I, however, kept my promise anyway.

Since then I have noticed more than once that theologians don't consider themselves bound by the normal moral constraints. Some of them seem to think that morality is a matter which they can play fast and loose with—something which applies to others but not to them. Every profession has typical diseases associated with it and this is the professional disease of priests.

The result of all this for me was that the distance between me and the colleagues from the humanities became clearly greater and they complained even more about my lack of "collegiality". This they understood as a diligent consideration of their feelings and opinions. Quite naturally they saw no need to consider my feelings and opinions for the good reason that they differed from theirs.

Self criticism. In the previous chapters I described some of the new challenges which grew organically out of my work in physical chemistry and which now took up more and more of my energy. Even when I look back after all my subsequent experiences I do not believe that going in this direction was in any way going against the ideals which I'd followed since my youth. My life was enriched by this more than it would have been by staying with the old challenges and it has also been more influential in the best sense of the word. I can't pretend that along the way I didn't make some stupid mistakes that I'd rather forget all about, but I doubt if I'd have made less of these if I'd stayed in my old position.

I should add that while I've related the most important things that I got from these new thoughts and work I haven't told everything. Perhaps there will be time later to come back to some of this and then it will become clear that all of what might seem to be just larking about was in fact directly related to the central springs of all my work. In other words I see nothing erratic in what I did, it was merely that events in external reality sometimes pushed me to change my priorities as to which of the many problems facing me should be tackled first.

The lectures. One of the many good things for which I lost all enthusiasm after my illness was lecturing. Before that I held very good lectures as my audience in Riga had made clear (Part I, Chap. 8, p. 89) as did the growing number of official attendees in Leipzig which, it must be said, is an underestimate because the students often attended without being officially enrolled. I never had any difficulties in expressing myself so that the preparation of the lectures was not a problem. I think the effectiveness of the lectures was largely due to the fact that their structure and content were clear to the audience. The observation of a creative event stimulates the audience to comparable creativity.

There was only one difficulty that annoyed me now and again. Sometimes, when I had to present a long mathematical derivation on the blackboard I lost the thread and had some difficulty deciding on the next step, which should follow. This was a strange sort of mental blindness whose cause I was never able to make out. It wasn't due to a lack of understanding of other peoples work because it sometimes even happened in the middle of derivations which I had originated and which I knew inside out. The problem was limited to the use of a blackboard and chalk because it never happened when I was explaining to a student or group of students. It got worse when I was overworked and exhausted. I began to find the lectures, which I'd previously enjoyed holding, increasingly embarrassing because of these lapses.

The explosion. During the winter semester of 1904–1905 I'd suffered several times from this and had therefore petitioned the ministry to be excused from holding lectures during the upcoming summer semester. This would not cause any disruption because in the current semester there were eight lecture series on various aspects of physical chemistry and for the next semester there was a similar number on offer. The ministry sent the request to the faculty. I had never known that this sort of thing was ever commented on; one treated such requests as a purely routine matter of form.

However, as the ministry's request was being dealt with in the usual course of faculty business someone objected. At the start the objection was guarded, but gradually more voices were added. It became more intense and ended in such a flood of criticism and accusations that I was utterly astonished. Those who were most worried about the negative effects of the lack of lectures on my students were the representatives of the philological and historical departments - in other words those who had no idea of the special nature of education in my field. They applied their own way of doing things, in which the central educational element was the lecture, to my field without understanding that here the emphasis was heavily on the lab work which I did not intend to give up.

It slowly became clearer that the psychological basis for this was the feeling that they considered that what I did and how I did it was not what was expected of a university professor and they wanted to ensure that I was granted no special dispensation and no special distinction amongst the other professors. Shortly before this I was visited by an influential member of the faculty who said that it seemed to him that I was asking for a position which would make all the others second class professors. I explained to him that such matters of hierarchy were of no interest to me, but was, however, so unwise that I added, "but of course there are in any case third class professors already among us". In reality it was true that I had no ambition to a special role amongst the Leipzig professors because my work and effort had put me in a class far beyond them. The others must have registered this as well and this was why there was so much opposition to me and why I was accused of insufferable arrogance.

I had relatively poor lines of communication with the faculty because basically all matters relating to the institute were handled directly by me with the ministry. The faculty was not involved. The faculty was however involved in all matters to do with lectures. In my teaching activities lectures were but a small and not the most important part while the practical work in the labs, which had begun in a small way and had required ever new negotiations with the ministry, quite naturally was at the centre of my attention in a way that the lectures, given my minimal interaction with the faculty were not.

I was reassured and even elated by the attitude of my closest colleagues, the physicists Wiener and Des Coudres and the chemist Beckmann. They took my part vigorously but weren't able to win against the general storm. The majority of the faculty approved a report opposing my request. This turned out to be without effect since the ministry approved my requested dispensation for the coming semester.

The faculty meeting at which this unusual business was debated was the last which I attended. I sent my resignation to the authorities and considered myself retired, though of course I continued to perform all my duties until I was formally released. I told myself that there had been no way to prevent this having been done to me once, but I'd make sure it couldn't happen again.

The friends and colleagues I mentioned above tried long and hard to make me change my mind. However, resigning the professorship was no loss for me; on the contrary it was the fulfilment of a longstanding wish which had merely been accelerated by the discussions in the faculty. I clearly remember the astonishment of my colleagues when I declared that I had no wish to end my life as a professor and that the life of a free researcher was the ideal that I intended to realise. "Our company isn't good enough for you" was their tetchy response—and my reply that I wasn't looking for company did nothing to improve things.

Wundt and other colleagues whom I regarded highly tried several times to get me to stay on at the university and asked me to name my conditions. I finally agreed to retract my resignation if I was given the right to decide whether or not I should hold lectures. All my other teaching activities would be left unchanged. To my great relief the faculty refused and so my resignation stood.

These events took place in February and March 1905. The University of Leipzig obviously did not appreciate my having in the shortest time made the physical chemistry department the best in the world. It was therefore with satisfaction that a few weeks later the press ran the story that Wilhelm Ostwald, Professor of Physical Chemistry at Leipzig University had been selected as the first German scholar to represent our scientific establishment as the first exchange professor with the United States in a program which had recently been founded by Kaiser Wilhelm II.

This seemed to me a confirmation that the challenges that faced me required a greater stage than could be provided by the professorship in Leipzig.

Part III
Großbothen and the World

Chapter 31
The Doctrine of Happiness and Its Applications

The theory of happiness. As I sensed the approach of those major changes which were going to completely change my life I felt deeply the need to study the origin, content and consequences of such a process scientifically. True to my conviction that there is nothing in heaven or earth that cannot be explained—and if necessary improved—by scientific analysis, I asked myself how something that was initially an emotional response could be rationally described. For this I had to first seek a general concept into which my particular case might be subsumed so as to be able to define as clearly as possible the natural laws governing such situations.
The situation was that I had felt joyless and yearned for a happier life. I therefore had to pose the general question: What is it that makes someone happy or unhappy? Or, in other words; I had to apply the law of happiness to my case.

Although I had read a great deal in the last 50 years, I hadn't yet come across any sort of discussion of this matter and I didn't get any help when I asked around in the circle of my friends and acquaintances. Only one of my previous students, Dr. Helmut von Öttingen, the oldest son of my teacher in Dorpat, was able to help out by bringing me a little book in which a busy collector had assembled the views of the leaders of world literature on the subject of happiness. This did not yield a scientific concept of happiness though it did supply a lot of valuable raw material and so it was left up to me to produce the necessary science. This wasn't the first time I'd faced such a challenge. I got straight to work on this task which would be decisive not only for my own future but also for that of my family. That was of course a good reason to diligently deal with the challenge.

It was clear from the start that the answer would be found in the field of psychology. Not in the sort of atomistic psychology that my respected friend Wundt had pursued along the way to developing his psycho-physics in which aspects of the mind were studied separately, but instead in the other form of psychology which treats each individual person as an organic whole and seeks in this context for laws of correlation.

I'd gathered quite a lot of material for this during my study of the history of science and I have already told (Part 2, Chap. 17, p. 177) that I'd been fascinated

R.S. Jack and F. Scholz (eds.), *Wilhelm Ostwald*, Springer Biographies,
DOI 10.1007/978-3-319-46955-3_31

right from the beginning how the personalities of the scientists affected their contribution to building the greatest achievement of mankind. Now I had the chance to see if this work, which had been started simply out of curiosity, could now be put to practical use which was of course its only moral justification.

I approached the problem with the expectation that it could be solved. Perhaps not completely, for that is not in the nature of the solutions to scientific problems. In fact I couldn't really expect that here, for I was dealing with a first attempt at an extremely complicated new problem. However, since what I was looking for would be essentially a yes/no answer I hoped that I would at least come up with a rough idea of what was important.

After a few weeks of goal-oriented thinking I had come sufficiently to terms with the material that I could start to formulate a general result and it was like a wink of fate that the path to the result once again took me by way of energetics.

The happiness equation. I assumed that all life is characterised by a steady state in which a continual energy flow circulates. The energy flowing through the body is used for all of its activities and as a result the energy loses its free work capacity or potential. This loss will be made up by energy from food.

This applies equally to the lowly single celled organism as well as to highly developed human beings; the difference is merely that in humans the process is registered by conscious feelings.

This conscious appreciation varies from case to case. Certain processes register as pleasant while others do not and one tries to avoid these latter as far as possible, though it is beyond one's power to override them completely.

The conscious experience of an activity is, in general, proportional to the total energy flow involved. This falls into two parts: the first, which we will call part A, is associated with those aspects which are welcomed and wished for, while the part W is associated with unpleasant experiences which generate resistance. The first results in feelings of happiness, the second in feelings of misfortune. Depending on whether the difference $A - W$ is positive or negative will determine whether the experience is considered in total happy or not.

To begin with I thought that was the answer and was a bit put out because all it says is that in order to be happy one should maximise welcome experiences and minimise those that are unwelcome. However, that doesn't take us very far, for we knew that already.

But then I realised that in the past merely overcoming difficulties produced feelings of happiness, which nowadays no longer happens to anything like the same extent. Some source of happiness must therefore arise by the use of energy and this has to be included in the law.

The total energy involved is the sum of the welcome and the unwelcome, that is to say $A + W$. Since there is no feeling of happiness either when $A - W$ or $A + W$ are equal to zero the two components must be viewed as factors of a product rather than as simple arithmetic elements of a sum. When one now adds in a factor k which defines the degree to which the energetic flow is converted to the psychological level then the happiness equation, in which H is the extent of happiness, becomes:

$$H = k(A - W)(A + W)$$

In other words happiness is proportional to the excess of positive feelings (A − W) and the total energy involved (A + W).

Now the equation has become much more informative. In a paper published several years after the result was obtained in the 4th volume of the Annals of Natural Philosophy (p. 459). I listed quite a large number of examples and showed that the most diverse phenomena such as alcoholism or religious belief could be explained using this equation. Here I only want to consider the point which was crucial for the decision that I faced.

Explanation. To maximise one's feelings of happiness both of the factors (A − W) and (A + W) must be as large as possible. The total energy involved (A + W) can only be influenced by looking after one's health, because of course illness is accompanied by a loss of energy. Nevertheless the energy flow is strongest in the youth or young man and inevitably falls off with increasing age. This reduction first becomes obvious between 40 and 50 and increases from then on.

In youth (A + W) is large and so one can attain a high level of happiness, though of course it can be reduced by a high level of unwelcome experiences. If (A − W) becomes negative then the high value of (A + W) ensures that the result will be a strong feeling of misfortune. This sort of impassioned oscillation between happiness and misfortune is typical of youth.

With increasing age the total energy is reduced and hence the product is bound to get smaller and in order to achieve a useful degree of happiness one has to make sure that the second factor (A − W) is as large as possible and this means that the feelings of unhappiness (W) must be reduced to the minimum.

Because of this the nature of happiness is quite different in old age from what it was in youth. The young have the happiness of heroes which requires the application of their entire energy even if great difficulties have to be overcome. The happiness of old age, in contrast, lies in avoiding unpleasant experiences and concentrating on the calm enjoyment of welcome activities. It is the sort of homely happiness that stormy youth despises.

Guidance of science. Once I had obtained this scientific result it became clear to me how my future should unfold. After my illness and recovery my energy reserves were reduced and there was no doubt that I was now beyond the peak of my life. So far as I could I must now take care that with increasing age I increasingly avoid unpleasant experiences. Since it was absolutely impossible for me to remove the unpleasantness that had piled up against me in Leipzig University—and which could only get worse as time went by—my decision was now quite clear.

I want to emphasise that science to which I had devoted my life from my earliest days did not abandon me now. Today, in 1927, I am thankful that I followed this scientific analysis which brought me two decades of useful and happy life and I have never for a moment, not even in my dreams, regretted entrusting myself to that high goddess.

Settlement of accounts. My feelings as I resigned from the professorship in Leipzig were comfortable. The only thing that was not comfortable was the antipathy, indeed hatred, which a number of my former colleagues clearly felt for me, though this did not really bother me. Since I'd never done them any harm or even been unfriendly to them I didn't consider myself responsible for their dislike and only wondered how such a harmless creature as I could call forth such a storm of disapproval. Nor was their hatred any great intellectual or personal loss since none of these people were close to me and it was a matter of indifference whether I should enter them on the credit or debit side in the book of my life. The only one whose opposition wounded me was the botanist Pfeffer. True, our relationship had cooled as my rapidly growing recognition by the international scientific community had so reduced the original large difference in our places in the hierarchy that there was a danger that I would soon take precedence. On the other hand I was rewarded by the friendly support given me by my close colleagues Wiener, Des Coudres and Beckmann.

Thus the contradiction between my great eagerness to accept the position in 1887 and my equally great eagerness to resign it in 1906 was more apparent than real. Of course I'd had every reason to view my early appointment as a professor in Leipzig as an unbelievably lucky opportunity, indeed it was the only real possibility for me to apply my energies without any hindrance as demanded by the law of the happiness of youth. The work in the new position brought more positive fruits, both in scientific terms and for me personally, than I could have imagined possible. The personal gains for me were a deeper scientific understanding and the chance to meet and interact with the best minds of the age and with the marvellous circle of students and younger colleagues who had joined my laboratory. The scientific fruits were the success of my many books and papers and a rapidly increasing international reputation as the organiser of the new science of physical chemistry for which new professorships were being created everywhere and for which few candidates apart from my students were available.

In this way I had taken charge of the field I had opened up and I and my students were responsible for a very large fraction of the fruits of this effort. When I had started there had been only a handful of colleagues in the field while now hundreds of talented and busy researchers spread across the entire civilised world were churning out so many results that not only was the journal I had founded filled to overflowing but other equally good journals had in the meantime been founded as well. I realised that in my own department my co-workers were able to push forward independent work in the broad direction I'd set without any requirement for my personal involvement. In short a broadly based field had been established which not only managed to maintain itself, but as is the way of youthful structures, grew constantly in breadth and depth.

This seemed to ensure that I would be accorded honour and a pleasant life if I simply left things more or less alone. I didn't really have to take the conflict with the faculty seriously. I could simply have ignored it and their decision would have been rendered irrelevant because their powers ceased at the door to my institute. Some sort of formal assent that would have satisfied the core of my wishes and their demands with respect to the lectures could easily have been found if only I had had any real interest in retaining the professorship. This was what experienced colleagues had repeatedly suggested as a way out. I only needed to want to retain my position and everything would have been left as it was.

The individual and science. Despite all this I was not for a moment in doubt as to what I should do. With the force and enthusiasm of youth I'd pushed the cart of the new science up the steep hill towards general acceptance. Of course I didn't do this alone. I was decisively supported at the beginning by just a few colleagues who were, however, first rate—but my role had been the most demanding. Now, at the top of the hill and on the flat plain up there on the top, it was relatively easy to keep the cart moving. And then began the descent where the cart now rolled under its own momentum.

As the leading horse I had the pleasure of seeing the increased rate of progress without having to do very much, but then I realised that the momentum and the speed were increasing all the time. Soon it was hard for me to keep up with all the new results in the field, as I'd been used to doing, and I could see the time approaching when the movement would be ever faster and when it would be irrelevant whether I still pulled or not.

The two ways. In this sort of situation one has two possibilities. Either one tries to stay at the forefront, in which case one has to move ever faster to avoid falling under the cart's wheels. The more effectively one has worked in one's youth and trained those who follow so the greater will be the chasm between the declining powers in old age and the lively progress of the movement. This is bound to end in catastrophe—not of course for the cart, which will carry on its way undisturbed, but surely for the old nag which will be mercilessly crushed by the wheels. The development of science is uninterested in the individual who should not expect any thanks or special dispensation.

There is then only one alternative possibility: one gives up the now dangerous position at the front and steps aside. If one is not then going to fall into depression at the sight of the field, to which one has devoted all one's powers and the best years of one's life, carrying on completely unaffected on its own, then it would be best to find a fresh field where that sort of danger is not yet present.

This is easier to achieve if one has earlier maintained broad interests, but if one has had nothing in one's head apart from one's work, then the prospects are poor. To use a different metaphor, one has to be able to switch from the horse to a donkey and, if the worst comes to the worst, then just hobble along on foot. To the outsider things may not seem so bad, especially if, as is often the case, reputation and fame keep on increasing. But the aging scientist feels this inexorable long descent and the situation is not improved if he tries to cover it up by emphasising his accumulated honours. Sometimes it seems to work and we have quite a few old heroes who sit there in state as their own monument and graciously accept the incense offered them. But their unavoidable recognition of their own inadequacy, even if it remains subconscious, makes this an unenviable existence.

Much better off is he who still has left over from his potent youth other challenges which he never got round to doing anything with because the problem he'd first picked up had proved so fruitful that he'd had no time for anything else. Now he can return to one of those loves of his youth and indeed developing a close relationship with his old flame is now not only allowed but necessary.

This is one of those sadly rare situations in which personal and objective needs combine to produce a benefit that results both in an increase in personal happiness as well as the generation of things of objective value.

New challenges. In my case there was a whole flock of loves from my youth that I wanted to return to. For as long as I was a professor of physical chemistry in charge of a large laboratory full of ambitious students I could only risk such an infidelity, so to speak, behind my own back—and if I'd done it I'd have been plagued with feelings of guilt. The result was that, though I did now and then take up these things, much less came out of them than I felt sure would have been possible had I been able to concentrate on them.

The things I principally tended towards were colours and painting and they later became combined with elements that originated in my career as a scientist which can all be grouped under the heading "philosophy".

At the beginning of the twentieth century when these questions became foremost in my mind, philosophy had taken on a largely historical orientation. The majority of philosophers had followed the cry "Back to Kant" and under the guise of epistemology pursued a rather sterile scholasticism which was based on Newton's principles that Kant had used. None of them ever bothered to ask whether after a hundred years of incredible scientific advance these principles still held or whether they might require some correction and amendment.

I, on the other hand realised that the enormous expansion both in extent and depth of scientific knowledge ranging from mathematics to biology provided a much better and more useful basis for concepts of knowledge and cognition than Newton's ideas which were restricted to mechanics and astronomy. The attempts of the best mathematicians to extend them to other fields of physics had some initial success but in the long run must be judged a failure. The concept of energy however with its two far reaching laws had already proved useful in philosophy—and promised to deliver even more.

In this respect energetics, which I had developed because of its usefulness in chemistry and physics, had an even greater hold on my thinking, but exploring all its possibilities required a degree of dedication which the busy institute director, journal editor and lecturing professor could not possibly supply.

Biological imperatives. This fundamental change of goals was sparked by aging. When I felt the first twinges of age I wanted, as always—but in this case most urgently—to get scientific clarity on the causes and nature of the phenomenon. This, after all, is the only way to achieve a conscious adaptation of one's lifestyle to new circumstances and by doing so to direct one's life as far as possible in a productive and happy direction. The tragic experience of my respected friend Karl Ludwig who saw his life's work crumble away before his very eyes (Chap. 8, p. 192) had shown clearly that this great physiologist had failed to apply to his own life the rules of the science to which he had contributed so much.

The fact that I wasn't a biologist could be as much an advantage as a disadvantage for me now as I faced a similar problem. In science it is necessary to concentrate on lots of details and this often results in a sort of mental short sightedness. The better and more successful a researcher is the more likely it is that he will lose touch with the large general questions facing his branch of science. It is one of the great strengths of our university system that the coupling of research with teaching works against this short sightedness because to present the field in a convincing way in lectures requires that the broad general problems be returned to again and again.

Short sightedness is, of course, not a problem for those who have never worked in the field, but they instead face the danger that they miss important things which those in the field never mention because they are "obvious".

The best means of avoiding this is to start out with a broad based taxonomy of science. If one can assign a field to its place in a general scheme of science then one has a clear framework into which the contents have to fit even if one doesn't know at the moment what these contents are. The appropriateness of the contents can be seen in the extent to which they fit into the framework.

The case of Berzelius. I can't pretend that all this was clear to me when I turned fifty and felt the need to reorganise my life in the face of advancing age. However, I did have a number of examples, some from personal colleagues like Karl Schmidt and Karl Ludwig, and others from the history of my field which I'd garnered from direct or indirect reports in the literature. One of the most instructive of these latter was the case of the great Swedish chemist (Part 2, Chap. 23, p. 281 and Part 2, Chap. 30, p. 380).

Even though his experimental work was extraordinary both in its range and significance, Berzelius had been even more important as an organiser of the field than as an experimenter. His textbook which he'd written at a young age and which he kept improving right up to his death was at that time the key script from which almost all chemists at the time were trained. Though he himself scarcely taught, and when he did it was only to small groups, his textbook made him the world's chemistry teacher. Of course this was not achieved by the original Swedish edition of the book since few people could read it. The widely effective edition of Berzelius' book was the German edition which was carefully translated by his student Friedrich Wöhler who put a lot of work into it. For later editions the Swedish original was no more than a handwritten manuscript and only Wöhler's translation was published.

Never was an intellectual empire more complete and never was it better justified than by the selfless pursuit of pure science by that king of chemists Berzelius and yet he lived to see it all fall into ruin and all his efforts to stave this off were to no avail—though he fought hard till the day he died. New areas of science develop which then require a new fundamental organisation. As is so often the case the small amount of new takes the stage at the expense of the vast amount of old material which no one talks about anymore because of course it doesn't change. As a result one has the impression of a fundamental revolution when in fact all that has happened is that an annex has been built whose integration into the whole just requires that a wall be knocked down.

Application to myself. Even as a student I'd been fascinated by this tragic tale from the history of science when I came across the first traces of it in the chemical journals which at that time I read more or less at random. Berzelius's concepts were not considered old hat in Dorpat, though they were in Germany which at that time had become the leader in chemistry once Kekulé had opened up the new field of organic chemistry. As time went by I gained more insight into this as the letters of the leading scientists of that time were published and I came to understand that what had happened was a simple matter of necessity. As always I then looked for the practical lessons from this and they were going to have a major relevance for my own life.

The result was that for me there were in fact even more pressing grounds to resign my position than there had been for Berzellius. He had dedicated himself from his youth to his chosen field and till his death his attachment to it grew in strength. In my case however, other thoughts had now sprung up. Some of them had been with me from my youth while others originated in my work as a chemist but then they took on a life of their own. These new thoughts began to infiltrate my work which not only made it easier for me to leave, but also made it my duty to science, to my students and not least to myself to do so. A duty to science and to the students so as not to do them damage, and to myself so as to allow me the opportunity to develop sides of my personality which had been held back till now.

Versatility. In the diary of my student years an older member of the student society who studied astronomy had written "If only you didn't want to be so horribly versatile!" He stands as a witness that even then, despite my unrestrained enthusiasm for chemistry, I was also interested in all sorts of other things whose only connection to chemistry was that they were also present in my brain.

I don't want to suggest that I had any plan to aim for universal knowledge. It was more a sort of instinctive response which was due to the fact that everything that interested me made my fingers itch to try it out myself. It's the sort of itch which is present in any halfway mentally active youth and what he ends up doing will depend on what chance happens to place in his hands. Because I'd read so much as a boy I had a wide collection of wishes and hopes and the limited possibilities of realising them in the situation in which I grew up only made them stronger. Some of them had brought me the joy of seeing them fulfilled, though usually only enough to whet my appetite for more. But of course that is exactly the best precondition for a successful development of the personality: enough success that curiosity is not starved to death and by far not enough so that it dies of surfeit.

Manual skill. I think it was very favourable that as the son of an artisan I was brought up to regard manual skills as something that was of special value. I've already told (Part 1, Chap. 2, p. 16), how my father, once his work lay in business and business matters, still loved to go to the workshop and do the really demanding manual tasks so that he'd know they'd been well done. I inherited this trait from him and that opened up for me many opportunities which remain closed to those who, following the Platonic way of thought, consider only literary and intellectual work as worthy pursuits and regard all manual labour as beneath them. I don't doubt for a moment that Helmholtz would have come up with quantitative colour theory half a century before me if only he'd been as used to working with paints and brushes as he was to working with differential equations. On the other hand it was a natural, almost unavoidable, development that painting, which to begin with I'd taken up just as a hobby for the purpose of refreshing my mind, led finally to the most general artistic questions and to the aim of introducing science into painting and art in general.

Relationship to the present. Looking back now, at the end of my life, I see myself as having been something of an exception in an age which was characterised by specialisation which shrinks the horizons. If I read the signs of the times correctly then this age is now coming to an end and in the future the challenge will be to synthesise our strands of knowledge. Probably this will not be in the "pure" sciences where on the contrary specialisation will remain the dominant theme but rather in the applied sciences—in technology and in economics.

I really can't name anybody of my time who managed to work on so many different issues as I did. Not, however, that I ever started work in a new area merely to provide evidence of my "versatility". On the contrary I often felt a little guilty when I set out once again on something new. I only need to remind you that the sudden resignation of my professorship in Leipzig (Part 2, Chap. 30, p. 384) was due simply to my inability or unwillingness to be satisfied with my position in physical chemistry—a position which was surely large enough to fill a life. I have to regard it as an inborn part of my being that as soon as I'd settled down somewhere in fertile soil I had to send out exploratory roots to search for new places to grow in much the same way that we discriminate between these plants which conquer their territory by sending roots down deep to drive the branches ever higher and those which instead spread out ever further.

A preamble. An example of this is in the volume "Essays and lectures on general subjects" which I wrote in the spring of 1904 one year before the end of my time in Leipzig. It brings together lectures and essays from the years 1887 to 1903 which I'd written for a broader audience. In total there were 27 pieces which were organised in five groups: general and physical chemistry, electrochemistry, energetics, philosophy and finally technology and economics. The preface of this book so clearly reflects my views at the time that I quote it here, for it shows that I have not fallen into the trap of making up a suitable construct retroactively but that the description I have given was accurate.

The preface, dated May 1904, reads as follows:

"To begin with I must admit that this collection of talks and lectures aimed at a broader audience was not, as is usual with these things, suggested by my friends.

Rather, now that the development of science is undergoing a great change, I think that I and my colleagues should take stock and I don't deny that I feel we can do that with a certain degree of satisfaction. For myself I consider that my main contribution was to establish physical chemistry as a recognised branch of science and this has been largely achieved. Naturally this does not mean that that is the end of the matter for there remains more than enough to do both here in my German fatherland as elsewhere. However I do think that the new branch of science has been well enough established that its further development is assured and it is now impossible that it will simply disappear. What the future holds for it will depend on what the field can deliver, rather than on any personality conflicts.

At the same time this collection should be an explanation for and, I hope, a justification of the change I mentioned. Through my own assiduous work in the new field and the invaluable experience of guiding so many talented young scientists at the start of their careers I have accumulated a pretty comprehensive practical view of how a scientific field develops and how this development can best be supported. It would seem sensible to apply this knowledge to other fields and that is what I intend to devote my remaining energies to. Those friends who note with concern that I leave the pursuit of results in my own field will realise from reading this collection that this is the result of a long thought out idea which so far has had to take second place to the routine demands of normal daily work.

Here one will find not just the widely read essays such as the lecture on overcoming scientific materialism or that on catalysis, but also discussions of other less well known topics. I have not aimed for completeness and in particular I have omitted many of those general points which I developed in book reviews and instead have limited myself to those which represent a rounded view of some aspect of science or life. In all cases the text is as it was originally written. The essays are not in chronological order because that would have resulted in the presentation of what might seem to be an aesthetically unpleasing jumble of random topics and I also wished to avoid giving the impression that I was bragging about my many varied interests. I've arranged the essays in five main groups in each of which the development of my thoughts on the topic becomes clear, while the common thread that runs through them all emerges naturally and does not have to be reiterated time and again.

The footnotes serve on the one hand an explanatory role and on the other permit a few necessary improvements and modifications. I have not included any of the many polemical arguments with my critics, particularly about the lecture on scientific materialism. When someone like me looks back on a long and varied period of scientific development then it is easy to convince oneself that it is more profitable to win new friends from amongst the community of unprejudiced young scientists rather than to waste one's time trying to convince dedicated opponents. No one could complain when a dedicated representative of energetics invests his energy not in convincing opponents, of whose narrow mindedness he has no illusions, but rather invests that energy more profitably elsewhere. I hope that this reprint of these essays will be received in this light. Wasting energy, whether it be by ignorance or malice is the greatest sin one can commit, because the loss can never be made good.

When here and there in this volume you come across instances where it is clear that such waste of energy is reduced or where it is redirected in some useful direction, then the author's goal will have been met.

The question of values. Well meaning preachers of self restraint considered it "obvious" that I would be more productive working in the field I knew well rather than in a new area which I'd have to learn my way about in. They forgot, however, that the work they praised me for was done as a complete beginner—as a pioneer. It wasn't obvious to me at the time, but now looking back I can see that this type of pioneer work was the sort of thing that I could do much better than the average of my peers. It was my special talent and my personal passion to provide new areas with the necessary scientific content and structure. The constant changes of tack were merely a necessary consequence of indulging this talent which required ever new challenges.

One could say that at the end of the day every single scientific problem is part of the general network of knowledge, so that without changing tack one can dig ever deeper into ever more difficult and important areas. That is without doubt correct, but it is also true that the relationships between things not only go deeper but also spread out on all sides. How one works is in the final analysis simply a question of temperament. The "classicist" will be happier digging deep, the "romantic" will by his nature prefer to explore the neighbourhood. I am a romantic.

The only thing that is important is the consistency of the work or the coherence of the set of problems attacked. In the case of the "classicist" these are easier for an outsider to discern. But the unbreakable fundamental law defines that the romantics' changes of direction are not inconsistent and I think I can claim that this sort of horizontal versatility can be very productive, for the exploration of neighbouring areas can often provide unexpected new insights.

I might add that at the time I started out in science the analysis of these sorts of relationships was neglected much more than was good for a proper and harmonious development—so I must praise the aptitude that my chromosomes brought together that made integrative thought both a necessity and a source of happiness to me. The fact that I carried out this work in relative solitude not only allowed me, but actually forced me, to bring in an extraordinarily rich harvest from what had otherwise seemed to be an empty field. This had of course the disadvantage that to begin with only a handful of people recognised the value of what I'd done. In fact the effects of much of the work I did at that time will only become apparent in the future when I won't any longer be around. When I ask myself what I feel about this loss then the answer is that it is inconsiderable. The happiness of finding and working out such general ideas is so great that recognition by others, no matter how much I sometimes enjoy it, is by comparison relatively unimportant—especially as many of my peers find it necessary to mix a little gall into the praise.

Once I'd managed to reach this degree of scientific reassurance about my future direction I was able to turn to the many challenges that the transition necessitated.

Chapter 32
The Exchange Professor

The idea behind the exchange program. Kaiser Wilhelm II had pursued a many sided, but not always terribly successful effort, to forge a close relationship between Germany and America. He had noted that of the many forms of intercourse between the two countries scientific matters were particularly strong. Admittedly this consisted largely of talented young Americans coming to a German university as doctoral students and this went so far that the American universities protested against the generally held view that a scientific career was impossible unless the candidate had a doctorate from a German university. On the other hand there was no particular interest among German students to study in America, largely because the universities there aimed at younger students and hence put more emphasis on personal relationships than was the case in Germany. Following the British tradition, on which they were modelled, these institutions were more directed to education and teaching rather than to research. Professors were frequently at the level of secondary school teachers and had nothing like the social standing of their colleagues in Germany. The average level of research was much lower than here and this meant there was little reason for German students to try to surmount the difficulties caused by distance and language.

Because of all this it was a great and somewhat one-sided compliment on the part of the Emperor to suggest a regular equal exchange of professors between the two countries' universities. From what I saw it was not really successful because, despite the great personal impact which German exchange professors later made there, the scheme never became really popular. Part of this may have been due to the low social standing of professors over there.

The implementation of the scheme. By the beginning of 1905 the negotiations had reached the point at which the first exchange professors could be nominated. Harvard University in Cambridge near Boston was chosen as the receiving institute for the German professor because it had the reputation of being the best university for science in the country. Harvard therefore had to choose a German professor while the American professor who would lecture here would be nominated by the university in Berlin and formally invited by the emperor.

R.S. Jack and F. Scholz (eds.), *Wilhelm Ostwald*, Springer Biographies,
DOI 10.1007/978-3-319-46955-3_32

To everyone's surprise Berlin selected the American church historian Peabody, who had produced nothing special of note. One suspects that this was intended to be a hint to the Americans that they in turn should invite the Berlin professor of church history who the emperor wanted to reward with a well deserved distinction. This is of course no more than a suspicion.

In any case, in spring the newspapers brought the news that Harvard University had nominated the physical chemist Wilhelm Ostwald as their choice for the exchange professorship. This came as a great surprise to me for though I knew several people in Harvard including the chemist Theodore William Richards, the philosopher William James (Part II, Chap. 26 p. 309) and the psychologist Münsterberg (Part II, Chap. 29, p. 361) neither they nor the university president Charles Eliot had contacted me about the matter. The newspaper reports were, however, only a little ahead of reality for shortly thereafter the matter was officially confirmed.

I have no idea why the choice fell on me and why I was held to be the person best suited for this purpose. Of course in the previous year Münsterberg had got me to break my journey home in Cambridge and wanted to introduce me to President Eliot though, as I described, nothing had come of this. Probably it was due to the large number of American students who had finished their education in my group many of whom were now professors in their home country. In fact quite a few former Leipzig chemistry students were now professors in Harvard and Boston. But I was by no means the only German professor in this situation because every half way prominent German chemistry professor had had a large number of American students in his laboratory and this was also true for the representatives of many other fields. The most likely reason is that William James in his impulsive way had formed a great liking for my philosophy and also perhaps for me.

This news was not welcome in Germany. Till this point the Emperor had probably never heard my name because the award of the Prussian Order for the part I'd played in founding the physical chemistry institute in Göttingen passed off without a personal meeting.[1] The discussions about Cambridge were in the hands of Althoff who didn't have much to say and he passed me on to Harnack with whom also no fruitful discussion emerged. The Emperor, who then and later regarded the exchange professorship scheme as his own project, expressed no wish to discuss with me personally his aims and so I just had to muddle through on my own.

Resignation difficulties. The whole business was complicated by the situation in Leipzig. The quarrel with the faculty, which I described at the end of the second volume, had ended with me handing in my resignation to the ministry of culture just as the invitation to Harvard arrived. The attempts of my friends, to whom I am

[1]Ostwald is mistaken. The official Kaiser Wilhelm/Roosevelt professorial exchanges, established in 1905, were between Columbia University in New York and Germany. Ostwald was not part of this prestigious program. Instead he was invited under a private professorial exchange scheme run by Harvard University. This latter scheme had no official connection to the governments of the US or Germany.

eternally grateful, to settle the matter by arranging that I be excused from holding the main lecture course were torpedoed by the faculty. On the other hand it would not look good to retire me just as I was to undertake a new challenge which would be anything but peaceful. A discussion with the ministerial official Waentig led to the agreement that my resignation should be put aside until I had returned from America and had taught one last semester in Leipzig. This was a very kind gesture on the part of the minister because in this way my pension was increased, though at that time I was so well off from the income from my books that this was not a consideration for me. However, later when after the war economic mismanagement by the young republic led me to lose my wealth, which had been invested in German state bonds, this pension provided me with most of my income. So while the ministry, despite the troubles we'd had over the last few years, felt it necessary to recognise my accomplishments at the university, the faculty and the university had a quite different attitude which was apparent even years later and sometimes breaks the surface even now.

I rather fear that the nomination for the exchange professorship, which was in complete contrast to Leipzig faculty's evaluation of my accomplishments and which came just as they had made their evaluation public, was doubtless considered yet another example of the "un-collegial" attitude they accused me of and this must have strengthened them in their attitude to me.

Departure. My new duties in Cambridge began on 1st of October. Since the semester in Leipzig ended at the end of August, I had about 4 weeks holiday which were really necessary given the strains of the spring and the work which faced me in America. However they were cut short by the developments with the saltpetre business which due to Dr. Brauer's ceaseless work was now so far advanced that the launch of the first commercially sized unit was imminent.

Since my wife felt that after all the recent agitation she didn't want to spend 6 months alone among the hostile ladies of Leipzig, I took her and my two grown up daughters with me to America. My two eldest sons were already independent and the third was being well looked after so that nothing stood in the way of the journey.

The Hamburg-America-Line had offered the Emperor to give the exchange professor free travel on one of their ships but when we looked into the possibilities it turned out that the only places still free were on the small steamer Blücher because this was the time of the year when the numerous American tourists in Europe went back home and they had all booked their passages long ago. The Blücher was going to leave Cuxhaven on the 21st of September and would not reach New York before the 1st of October so that I had to postpone the start of my lectures in Cambridge for a day or two.

We left Leipzig on the 20th of September loaded with flowers and sweets from our loyal friends Beckmann, Des Courdres and Wiener and then stayed the night in Berlin. From there a special train took the passengers direct to the harbour and a small steamer took us out to the Blücher. My wife was impressed by the unbeatable cleanliness and tidiness of the cabins and the unfailing helpfulness of the crew. She considered this a realisation of the ideal way to run a house. The captain was an older man with brown-red skin, ice-grey hair and beard. He had shining blue eyes which looked a little odd in his weather beaten face.

On the journey. Since most of the passengers were Americans with German roots who'd enjoyed their holidays in Germany and were now happily travelling home, the mood was light and my family soon made friends and got used to the social norms that they'd be living under for the next 6 months. We got to know the unbelievably strong attraction that America had for the immigrants. In this respect the Germans showed a sort of melancholy tenderness for their old home but nevertheless they were not only willing but proud to have become Americans.

For the rest, the journey was much as ever, though this time there was no storm and only a little fog. We sailed pretty close past some icebergs about which the ship had been warned by wireless telegraph, saw whales playing, lots of dolphins and several wonderful sunsets and spent the days in lazy contentment which I found very relaxing.

I was asked to hold a lecture at the end of voyage festivity which was to raise money for the orphans and widows of sailors. Shortly before this I'd begun to be interested in the idea of a general world language and was full of the idea that here would be an application of the energetic imperative. I'll describe all this in a later chapter, for the moment all I want to say is that at that time I thought that the synthetic language Esperanto was probably the best solution.

In the lecture I described in lively colours how helpful such a world language would be for all humanity. It should not attempt to replace national languages but would rather be used as an international means of communication. It would be everybody's second language and once everyone could speak it then no one would have to learn any of the other languages.

Since at that time this sort of idea was discussed only in a small circle, the lecture was received with great interest. This was increased even more when my daughter Elsbeth gave the audience a chance to hear the sound of the language by reciting some poems in Esperanto.

New York. The disembarkation routine was as laborious as always because of the customs procedures. My daughter Margaret had brought her violin along and she had to play something for the customs officials to convince them that it was for her own use. It all took so long that it was too late to travel on to Cambridge unless we wanted to arrive in the middle of the night. We stayed in New York for the night and one of my former students took me to a small dinner while my wife and daughters observed life in the enormous Manhattan hotel where we were staying. My assistant for Cambridge, Dr. Harry Morse who'd been a student with me in the best years in Leipzig, had been waiting for us as we left customs and he turned out now and in the coming months to be a clever and cheerful supporter who was of great help to me in many ways.

Lodgings in Cambridge. We left for Boston on the 2nd of October where Professor Th. W. Richards met us and took us on to Cambridge.

Although the town is not small it had no hotel at the time. With some difficulty Richards had managed to find a boarding house which was run by two rather old women, a mother and daughter. Here we were rather inadequately housed for we had only one bedroom for the parents and one for our daughters, one living room and a small dressing room for my wife. As usual over there the rooms were small and with low ceilings, but they were kept clean. Sofas served as beds and were not uncomfortable. People apologised that they had not been aware that I was bringing my family and suggested that we move to Boston which was 35 min away by tram and where there were lots of hotels. However I told them that I'd be glad to take on the discomfort of the accommodation so as to be close to my new colleagues. My family also had nothing against this as my wife felt herself transported back into our first years together in the student accommodation in Dorpat.

The food was rather like the accommodation. It was clean and made with good solid ingredients but it was cooked without imagination. As time went by meals were ever more boring and slowly we began to accumulate a collection of dips and sauces which were supposed to make the meals more interesting but they managed this only to a small extent. One of the best things when we went back home was the return to our own food, for my wife was almost unfailingly able to able to teach the domestics to cook to perfection.

The kitchen and the house were attended to by black servants. We did not practice the strict racial separation that is normal in America but accepted the blacks as they were and so they developed a great degree of gratitude and devotion which was very apparent when we left Cambridge.

The lectures. The discussion about the topics that I should lecture on in Cambridge showed that the audience would consist both of chemists and philosophers. The major lectures were on the philosophy of science and were for 4 h per week. In addition there was 1 h per week for advanced chemists on catalysis and a further 1 h of elementary chemistry for beginners so that I could demonstrate my lecturing technique. The lectures on catalysis were in German, because the advanced chemists could all at least read the language, but the other lectures were to be in English.

As you can see all this was quite a challenge because the work on the lectures was increased not only because I had a new audience but also because of the foreign language. My former student Dr. Morse who was now a lecturer at Harvard had been assigned to me as my assistant to help me out with this and he had once again turned out to be a splendid and thoughtful help. We organised it so that he sat in the audience right in front of me and when I couldn't find the right English word I used the German one and he gave me the translation which I then repeated. Since we did this unselfconsciously before the audience it was not disturbing and in fact it increased their concentration as they wondered whether the translation was accurate. Soon a few of the older ones who understood German began to offer their translations. This went on for 4–6 weeks until Dr. Morse said to me, "In the last week you didn't need help with any words. You don't need my help anymore".

Nevertheless I asked him to stay because the feeling of security he gave me had helped me a lot with my English.

The official grades. The major lecture was treated as an official part of the student course and for these the regulations laid down that around the middle of the course the students had to produce a short written account of what had been presented so that the lecturer got an idea of how well he'd managed to put his ideas across. Dr. Morse passed on to me these essays from the fifty odd students. I looked at quite a few of them and could see that most had understood the major points. I left the grading to Dr. Morse since I wasn't sure what standard was applicable in Harvard. He told me that only a few essays had had to be marked "unsatisfactory".

The lectures were my only official duties and I wasn't expected to attend any of the meetings concerning the inner workings of the university. Because of this I have only a very incomplete picture of the system there. What the American colleagues wished to know from me about university administration was discussed in the numerous unofficial parties that my family and I were soon invited to. In addition, several of the professors attended my lectures in particular the philosopher W. James who came regularly to the main lectures for as long as he was in Cambridge. Later he went off to Berkeley in California to put some life into philosophy there and I have no doubt that he managed to do that well for he was exactly the right man for the job. Professor Theodore W. Richards, who I knew well from Leipzig where he'd spent a semester in my laboratory to learn our teaching and organisation, came to the chemistry lectures. He was now full professor in Cambridge and had made a name for himself with his excellent precise determinations of the atomic weights of many of the elements.

In addition to these official duties I also held many other lectures. The old saying that one is punished in the right way for one's sins turned out to be true here. I'd given up the professorship in Leipzig because I hadn't wanted to take on the modest burden of the regular lectures but now I found myself with the much greater burden of a whole set of extra lectures that had to be held in a foreign language. But I accepted this punishment with good grace because the lectures only had to be held once rather than being repeated again and again.

My new colleagues. I start off this section with President Charles Eliot who was both the official head and the intellectual leader of Harvard. He'd been born in 1832 and so he was 72 years old when I got to know him in Cambridge. He'd studied chemistry but for a long time now he had applied his unique talents to general questions of culture. He had a huge influence on Harvard's fortunes in that he made sure that the university was based on the highest standards of independent science. He was the accepted doyen of the university presidents in America of whom I met quite a few.

President Eliot was a tall well built man who despite his years held himself erect. He put this down to his daily sports work out which he never missed. His large long face, more English than American, was close shaven except for a narrow beard on his cheeks. His face was well formed but had a large birth mark on one side. His behaviour was grave but friendly, more dignified than hearty.

His wife was rather similar. She radiated the dignity of an English bishop's wife, though once one got to know her, her good qualities shone through.

Once I'd presented my credentials to the president in his office our two families got to know one another better. This took place at the house of Professor Richards' mother in law, Mrs. Thayer who was an exceptionally likeable long-widowed lady. Richards was the only other guest. Once the initial stiffness had been overcome we had a lively and pleasant evening because our views of science and education were quite similar even if Eliot, as a practised manipulator, held back from my more radical views. However he had managed long ago to get rid of the requirement taken over from the British that every student had to learn Latin. He had made attendance at lectures voluntary in all subjects and had asked the students to report whether the lectures they had attended had been for their speciality or for their general education. The result, he told us, had been that not a single student had voluntarily attended Latin or Greek lectures out of general interest. In this respect the Americans have been ahead of us for over a quarter of a century.

I had the impression that we understood one another that evening and that we both confidently looked forward to the success of the new experiment.

This good relationship continued. One Sunday he brought us to the country house of a close relative to show us a well preserved example of New England culture from the colonial time. It was a stately property with a large well attended park and a spacious country house whose furnishings had remained more or less unaltered for a century, which for an American is an almost unimaginably long time. The rooms were indeed very uniform and pleasant.

We journeyed there in two one horse buggies. One was driven by the President himself with his wife, my wife and me as passengers. My daughters were in the second one which was driven by his granddaughter Ruth, a lively girl about 16 years old who quickly made friends with my daughters. Her grandfather explained before we started that she was every bit as good a driver as he was.

Since my wife is nervous in a carriage she once or twice pointed out to him a rapidly approaching vehicle. That wounded him in his driver's dignity and though he didn't say a word I could sense a small drop in the temperature of our relationship thereafter.

Commonsense. He gained my full respect on a somewhat later occasion. It was in the autumn when the usual football competition between the different universities had led to the deaths of 17 students. This was widely reported in the press and there followed a heated discussion of the matter fired on both sides by deeply held feelings. Now there was stretched across the whole country a network of football professionals who were involved in every competition as trainers, referees and so on. I was told that especially good players would rent themselves out to colleges who wished to secure a victory.

Because of the public outcry these people were worried about their influence and their business and therefore proposed a meeting of all the university and college presidents to discuss the matter. In an open letter which was soon widely published President Eliot refused to take part. He argued that he could not believe that those who were responsible for making the game so rough and life threatening were the right people to organise the necessary improvements. Instead he abolished the yearly competition between Harvard and the neighbouring university of Yale in New Haven which till then had been a national institution like the Oxford and Cambridge boat race in Britain. The competition which took place when I was there was the last of its kind to date.[2]

His radical statement, that there is no surer way to foil a necessary reform than by asking those who have been in charge up till now to suggest and institute the necessary improvements, was one of those unforgettable enlightenments that one experiences from time to time. When one asks why the necessary reform of secondary schools makes no progress then one only has to look at the people sitting on the committees set up to deal with the matter. These are almost entirely teachers and school directors. It is indeed too much to ask of someone that he digs his own grave. Progress will only come when one asks the people who have to introduce school leavers into the real world.

The same is even truer in politics. No one doubts that our present (1927) parliamentary form is deeply flawed. It is however absolutely certain that no parliamentary committee will ever find a cure for it.

A huge stadium which could be seen shining in all its glory from far over the other side of the Charles River had been built for the football competitions. We'd been invited to watch the spectacle and rather offended our American friends by refusing because we'd had more than enough from watching the preliminary matches. By chance I had an invitation to give a lecture in Boston, I believe it was in the "20th Century Club" on the same day and I'd gone to the tram an hour earlier because I'd been told to expect delays. However I was met by such a massive migration of people that the journey took two and a half hours instead of the usual 30 min.

Despite his advanced age President Eliot had kept himself busy for many years. He eventually resigned the post as active university president, but only so as to be able to concentrate on important cultural and political work. In particular he worked towards establishing a durable American influence in China though this idea has lost much of its importance because of intervening events. During the World War he was a bitter enemy of Germany. He died in 1925 at the age of 93.

William James. After the president the three Harvard philosophers William James, Josiah Royce and Hugo Münsterberg were of the greatest importance for my work. I'd been in touch with the psychologist William James the longest. I've already related (Part II, Chap. 26, p. 309) how he started things off with a postcard he sent me immediately after reading my "Natural Philosophy". I'd met him first on my visit to Harvard after the congress in St. Louis (Part II, Chap. 29, p. 376) when we'd

[2]Ostwald is mistaken: The Yale-Harvard football match known as The Hampden Park Bloodbath was played in 1894. Seven players were carried unconscious from the field. The match was banned for 2 years but resumed in 1897 and has been an annual event since except for breaks in World War I and II.

had a long and most interesting conversation. He'd been particularly interested in the means of distinguishing between real and pseudo problems which I'd developed further from E. Mach's work on this problem. I'd suggested starting of by assuming that the problem is real and then seeing how that would change our view of the world. If it makes no difference then the problem is not real. If it would make a difference then this difference defines the point where the problem can be attacked. This procedure is reminiscent of certain very powerful mathematical and geometrical methods (for example that of indefinite coefficients) in which the solution is formally assumed to be given and by doing so one can then define the nature of the real solution. James was already working on the idea that later became known as "pragmatism" and he found what I'd suggested to be helpful.

William James was the son of a well known writer and was 11 years older than me having been born in 1842 in New York. When I first met him he was an extremely lively man of middle height with a graceful build and an expressive face. His brown hair was thinning out and his short full beard was turning grey. His behaviour was a mixture of artist and child—they are in any case closely related. He was open and in a conversation one could clearly see how he picked up an idea, absorbed it rapidly and then would answer in his usual lively way.

In America it is normal that it's the residents who visit the new comer and visiting time is after dinner between 8 and 10 pm. James was among the first to visit us and he soon developed a friendly relationship with my family.

His wife was a very friendly lady of a quiet nature who managed to dampen down her husband's sometimes stormy nature. There were also a number of grown up sons and daughters whose genetic similarity to their parents was obvious.

In the time that followed I often sat round the fireplace in James's study and spent many interesting hours talking with him and with other guests.

He liked to tell how, without planning it, he'd ended up as an academic. After working on his own—and without ever having attending lectures—he'd published some papers on psychology. Solely on the basis those publications he'd been given a professorship. My own introductory lecture at the university, he said, was the first lecture he'd ever attended in his life—you can well imagine how nervous I was.

The fundamental two volume "Psychology",[3] developed from the lectures he gave at Harvard, made his name. Later he made many contributions to current thought, particularly by his energetic support for a practical philosophy which could be applied to normal life.

It is well known how much the pragmatic philosophy he taught affected his contemporaries. At first it was just in America but soon a broad cultural movement developed from it. The artistic side of his nature saw to it that this movement turned away from the realm of pure theory such as Mach—and also I—pursued, and instead developed a benevolent relationship with the church whose millennium long practical experience he did not wish to waste. In our times the church has come more and more to the view that its previous rigid opposition to science had been

[3]James W (1890) The principles of psychology. Vol 1 and 2, Henry Holt and Comp, N.Y.

counterproductive. Nowadays the church's policy is to combat science by seeking allies amongst scientists. These can always be found amongst the opponents of any significant advance who feel that things are being disturbed and their old rights destroyed. Another group of allies are to be found amongst failed scientists who blame their lack of success not on themselves but on the scientific establishment. In this way any turn of events that hints at a shortcoming of science is sure to produce quiet but active and often effective support for these circles that are then able to impose wide ranging changes, even if they are not always popular in the long term. With this key one can understand many phenomena of our current intellectual life where one would otherwise not have expected to find such interrelationships.

The considerable popular success of W. James can be booked at least in part under this heading, because the church was and is very strong in the United States even if, in contrast to many places in Europe, it is not supported by the state. Freedom from the chains of social tradition which Goethe had already envied —"America, you are better off than our old continent, you have no crumbling castles"—had given the church leaders powerful means of influencing everyday life by manipulating men's souls in ways that do not and did not exist in Europe.

Self destruction. This movement, which William James started, destroyed him a few years later in 1910. I was deeply shocked to read shortly before his death an essay he'd written in which he announced that human beings had vastly more powers than was generally recognised.[4] Till now one had believed that the exhaustion one felt after work defined the limit of ones capability. This was however a mistake. One only needed to set one's will to conquering the feelings of exhaustion. At the beginning this was hard but with practice it became easier and then one could get more work out of oneself. Indeed finally the exhaustion would completely disappear and one could work round the clock.

I was familiar with these sorts of ideas, but I also knew well the danger they bring with them. It simply meant that the body's warning signs of exhaustion would be raped and cast aside. These signs have evolved to protect the organism from the irreparable damage which follows from overstraining the organs and they are essential to keep the organism functioning over the longer time. If these signs are ignored then there is no further means of warning of imminent self destruction which is then associated with feelings of happiness.

It was these sorts of expressions of happiness that shone out from James's essay and I could only expect an imminent breakdown which, given my friendship with him, caused me a lot of worry. It turned out worse than I'd expected, for it was not just a mental but also a total physical breakdown that carried him off. And yet perhaps this was for him no bad thing because an existence in which a mental breakdown forced him to inactivity would have been insufferable for him.

---

[4]Most probably Ostwald refers here to: James W. (1907) The energies of men. Science, N.S. 25 (No. 635):321–332. James is regarded as one of the intellectual fathers of the Human Potential Movement.

Josiah Royce. The representative of metaphysics was Josiah Royce. He was a little older than James and the two were good friends, which is unusual amongst philosophers. Both in his looks and in his nature he seemed to be a facetious contrast to his friend and colleague. The two were of about the same size but Royce was round and corpulent with comparatively thin arms and legs so that with his protruding eyes he looked rather like a frog. His hair was thin and blond, his eyes bright and his smooth face rosy. In his behaviour he was so similar to the German Professor in the satirical magazine "Fliegende Blätter"[5] that I think this type of personality must be international—only that you find them more often in Germany than anywhere else.

I can't judge the importance, if any, of his contribution to philosophy. His reputation was also based on a two volume work whose title I no longer precisely recall but it was something like "The world and the Individual".[6] I borrowed it together with the work of other leading American philosophers from the university library so as to get an idea of the mental world of my new colleagues. I found it hard going. On the one hand I welcomed the strong emphasis on a sense of reality which permeated the work and which was in striking contrast to the fruitless scholasticism of German epistemology which was still running on the ideas of Kant—ideas which had been overtaken by the progress of the exact sciences. On the other hand it was clear that the liberation of philosophy from religion, which is the prerequisite for its progress, had not yet got very far in America, and this is quite different from the situation in Germany where, in this case, we are the winners. A clear indication of this was that all of these books had a long final chapter headed "God" or something like that whose function it was to relate the contents to theology. This was something that one found only on rare occasions in the German philosophical literature.

Professor Royce reminded one of Socrates not only in his appearance but also in his readiness to start a philosophical discussion on the marketplace or on the street, discussions which could be terminated by the other side only by the use of physical force. He'd trapped me in this way a couple of times and I'd only been able to escape at the cost of a coat button or two so that afterwards I avoided meeting him on the street. He was easy to avoid because he was normally so deep in thought that he didn't register his surroundings.

In the presence of others he was naturally not so dangerous and one could enjoy the dedication and honesty with which he followed and developed his ideas. He was also a good and inspiring teacher as I was able to see when I attended his philosophical colloquium for students. I took a lively part in the discussion of the question he'd posed in this lecture and afterwards he thanked me profusely.

Royce was also married and had grown up children. The difference in the personality of the James couple was apparent also by the Royces only this time in the opposite way. With them it was the wife who was lively and open. She had dark

[5]Humourous weekly magazine published in Munich.

[6]Royce J (1899) The world and the individual. Vol 1 (Vol 2: 1904) Macmillan, N.Y.

hair and eyes and a lively face whose changing expression accompanied every word she spoke. They too received us kindly in their home.

Hugo Münsterberg. The third philosopher in Harvard, Hugo Münsterberg, was quite different from James or Royce. I'd met him in Leipzig as one of the organisers of the congress in St. Louis and then again later in St. Louis and Cambridge. I've already given a description of him (Part II, Chap. 29, p. 361). He was from Germany and had been a student of Wundt, though he later joined the southern German group centred on von Windelband who were in opposition to Wundt's scientifically based mode of work and thinking. This dichotomy was obvious in his teaching for on the one hand he concerned himself with experimental psychophysics along the lines of Wundt but on the other he represented in his lectures and papers a very abstract metaphysics of the mind which I, despite a number of attempts, found impenetrable.

He was also of a quite different nature from his two colleagues. He was much younger than them (and younger than me as well) since he'd been born in 1863 and while the others were obviously happy to pass their lives in their pleasant surroundings, it was clear that Münsterberg was aiming for a more brilliant career than Harvard could offer him. I do not know whether he was aiming for a German professorship or whether he had some other position in America in his sights. He probably had more than one iron in the fire.

We two never developed a close relationship and neither did our families, though it wasn't for lack of trying. The central element of his personality was his enormous ambition and this prevented one from getting close to him. In addition I wasn't attracted by his work. Later friends there told me that he was the originator of some attempts to limit my activities. Our relationship ended when I left Cambridge.

Th. W. Richards. I had old acquaintances among colleagues in other disciplines. The professor of chemistry at Harvard was then, and still is, Theodore William Richards. Born in 1868 he was at that time still young being only 36. I've already recounted that he spent part of his early research years in Leipzig and that our mutual affection from that time made working together in Harvard particularly attractive. He was the son of a well known painter and had inherited a certain artistic bent which had been carefully tended and extended during his education. He was a good looking man of slender build, somewhat less than average size with a thin symmetrical face and brown wavy hair. His graceful and friendly politeness which was the undisguised expression of his personality won over everybody he met and in this sense he wasn't anything like what one normally expects from an American. He was however very definitely a scion of the cultural traditions of the old colonial families of Boston. One can recognise them from their strange belief, which they emphasise, that they are different from the later settlers of the country. Nevertheless there is in these circles a degree of real idealism which demands respect and one can see this in the generosity with which they support cultural life in Boston.

Richards was married and already had two or three children who were exceptionally pretty though they made a rather weak impression. His wife was pregnant again and this prevented any hospitality at their house. She was in any case quite neurotic and this had been to a large extent precipitated by an accident she witnessed in which several people close to her had been killed. His mother in law also lived in Cambridge and I've already related that my first social event in Harvard was at her house.

Scientifically Richards was close to me, though his area of research was independent from the clearly defined line we followed in Leipzig. In the second third of the nineteenth century the Belgian researcher J.S. Stas had managed with enormous energy to measure atomic weights and to hypnotise the chemical establishment with his results as if they were the pinnacle of what could be achieved. It seemed that this work might here or there be extended a little, but it could never be bettered. Nevertheless it was generally ignored that his measurements contained an internal discrepancy, and thus had ended with an unresolved dissonance.

Slowly this discrepancy became ever more evident as more new independent measurements were carried out. The results were internally consistent but differed from those of Stas. It took a lot of effort to break the hypnosis and restart work on atomic weights. This was done in a number of places but no one pursued it more broadly and effectively than Th. W. Richards who in doing so went his own way.

Richards was not only able to carry out extremely accurate and reliable work himself, he also had that other rarer talent of being able to guide suitable students towards a similar level of accomplishment. This was something that I'd been closely involved in for the greater part of my life and so it was a special joy when he showed me his laboratory and the way that teaching was organised in it.

All of this led to Richards being for me the closest colleague there and I think the feeling was returned. Unfortunately I followed my usual bad habit and failed to keep the relationship alive with the exchange of occasional letters. When he was an exchange professor in Berlin I failed to follow up on a miscarried attempt to meet him and then I didn't congratulate him as a colleague when he won the Nobel Prize for which I'd several times, but without success, nominated him years earlier. In one's younger years one treats friendships as if they were flowers on the meadow which don't need looking after. In old age one realises too late that they are in strictly limited supply.

Other colleagues. In the course of the 3 months I was in Cambridge I met practically all of the professors there and had many interesting and instructive conversations. However these interactions were too few and far between for them to blossom into real relationships. However, I have pleasant memories of the physicist Hall, the mathematicians Huntington and Peirce and several others, though these contacts ended when I left.

There is just one other man I'd like to mention who was at that time very old and is now long dead, Charles Norton. I met him in a club that W. James took me to, though I no longer remember what it was called. In this club the liveliest minds of Boston and Cambridge met every 2 weeks, and I found myself several times in conversation with a little old man. These conversations were spellbinding for both of us. I didn't catch his name to begin with, as sometimes happens, and so later asked James. He described the man and told me his name, but I hadn't heard of him. He'd been professor of literature and art history at Harvard and had had a strong influence on the intellectual development of the university. He was a friend of the famous British philosopher of art, Ruskin, and he'd also known Emerson and Darwin. At that time he was living in his house "Shady Hill" in Cambridge.

At the next meeting with Norton we found that we had so much to say that he said I should drop in on him in the evenings whenever I wanted for he was almost always there. I took him up on this and spent many evenings sitting with him at the fireplace and the clear tone of these conversations stays with me to this day. Two unmarried daughters, neither of whom was young, lived with him. They were calm and silent but followed our conversation intently. One of them must have been a real beauty in her youth and her face seemed to express the burden of some scarcely surmounted adverse fate.

I can't any longer remember what we talked about but I do remember that Norton was always a source of thoughtful and stimulating opinions. Our relationship reminded me a lot of the years when I had the opportunity to talk with Karl Ludwig and the two were indeed similar in build, though of course their intellectual interests and world views were very different. Norton was also milder in his assessment of people and things. Perhaps it's because of the challenge I experienced then that I later turned more and more to questions of the science of art, as one might call it, and the philosophy of art, though I couldn't show you any threads that connect these things.

My relationship to Boston. The Massachusetts Institute of Technology was a large and well organised technical university in Boston. Two previous students of mine were professors of chemistry and physics there: the chemist Arthur A. Noyes (Part II, Chap. 16, p. 172) and the physicist H. M. Goodwin. Both remembered me and my family kindly from their time in Leipzig and made every effort to give us a positive feel for their country. A conversation during one small intimate dinner with Noyes remains with me. We talked about what was needed to develop scientific work in America and I pointed out that what one needs are time and a scientific tradition. I gave as an example Clark University which had been founded about 10 years previously with generous funding and the declared aim of recruiting the best researchers no matter what the cost so as to form a large scientific potential. The attempt had failed.

Noyes was in no way cast down but said that one should learn from this so as to do it better next time and that the money could always be found. Americans were determined to achieve the highest goals here. I told him, "You will have to wait a long time till you begin to get close to what Germany has already achieved". Noyes struggled to overcome some embarrassment and then red in the face and with moist eyes he replied, "We hope eventually to move the intellectual centre of the world across the Atlantic". Since Noyes was always a quiet and restrained person this inner fire astonished me—and made me think. Whether the economic leadership, which the United States has achieved as a result of the stupid self destruction of Europe, will then lead to an intellectual leadership can't yet be judged. On thinking it over the following points should be considered.

The future of culture. Every time culture has developed it has always followed the pattern that there is an artistic flowering first, followed by a flowering of science. This was the case in ancient Greece, and had then been repeated later in Italy and in France. In Germany the period of classical literature was in the eighteenth century and science flowered in the nineteenth. At the present time one can see an artistic rise in Slavish culture, but as yet no scientific advance.

If we look at things in this way then we can see that American culture has still not experienced an artistic heyday. Certainly there have been scattered individuals who were successful in one or the other area but no new artistic movement has developed. Things look better in science. Here the number of researchers of real distinction is much greater and especially biology and applied psychology have taken on a life of their own and in some areas now threaten to overshadow what is done in Europe. This is particularly true wherever the hostile Platonic view in our humanities puts a brake on the free application of exact scientific procedures. These sorts of hindrances are much less likely to be found in America.

It is therefore quite possible that cultural development in America will reverse the normal order of art before science. The American people do not derive from a thousand years slow progress from the lowest possible cultural level, but rather is made up of immigrants from many countries who in just a few centuries have adjusted to this huge area full of natural resources and have forged political unity. Each of these immigrants brought along a considerable amount of culture which had developed long ago in their countries of origin. It was therefore not a question of slowly developing the elements of culture anew, but rather to construct a coherent whole out of the many disparate strands which they'd brought with them. This resulted in a large degree of individualism because there was no need to go through the normal tribal state.

In addition to that there had already begun what we may refer to as the scientific side of art. Music, poetry and painting all rely for their effect on the laws of psychology and hence they can all be rationalised to the extent that psychology itself develops into a rational science. Here in Europe we suffer from the Platonic mysticism that the "soul" is in a special category which is not accessible to scientific analysis and any attempt at an analysis is either ridiculous or breaks a taboo. It is especially in the field of applied psychology that the Americans' freedom from such prejudice gives them a clear advantage.

It is entirely possible that the organisation of artistic activity will be achieved much sooner on the other side of the Atlantic. In the advertisement section of a much read American monthly magazine from 1926, I came across a company which offers to quickly teach its customers to write short stories for newspapers and magazines and guarantees to provide help with placing them at no extra cost. We also have introductions of the same sort but here they are treated as secrets so that the readers of the stories don't realise that they have been churned out in factory style. Instead the readers are led to believe in the myth of inspiration (which nobody really believes) and anybody who talks of this openly will reap the fury and penal action that is the lot of every traitor.

The greatest hindrance to the development of an original American form of art is that the consumers of culture are virtually all women, for men are too busy for that sort of thing. However all great epochs in art were brought about by and for men. Naturally women were also enthusiastic partakers but they were never the leaders. I do not believe that this biologically based difference can change, not even under the special conditions in America where for men earning money is a source of joie de vivre and replaces a need for art. Boxing and betting serve the same function.

Lectures. I never tried to count the number of lectures I held in the course of these 4 months that I spent in America, but there were a lot of them. I remember once going to bed dead tired in Cambridge and before sleeping realised that in the previous 24 h I'd held three lectures in three different cities: the previous evening in New York, then after taking a sleeper in the night train I held a second one in Cambridge and the third in the afternoon in Boston. I decided never to do that again, but later in a 2 week period in New York I did hold every day two 60 min lectures with just 1 h between them—and these in English. The total number of lectures must have been between 100 and 200.

It was a good thing that at that time there was something that I wanted to bring to the attention of as many different people as possible and that was the question of an international auxiliary language. I've already related (Part III, Chap. 32, p. 404) that I'd already started this on the ship. The obvious surprise which that lecture caused convinced me of the necessity of spreading the idea. I thought I could make an impact with it and so I used the chances given me to speak about it and finally was able to start a real movement as I'll tell later on.

I held scientific lectures not only in Cambridge but also in Boston at MIT. They were concerned with the development of concepts in chemistry and thus served as an example of what I see as the real function of a scientific history of chemistry. I held the first lecture in German as Noyes had asked, because he and the others who'd been in Leipzig wanted to bask once more in that atmosphere. But then he told me that the subject had so fascinated the audience that they wanted to hear it again without the difficulties and gaps which the foreign language caused them. Because of this I held the rest of the lectures in English. For the same reason the lecture notes were in English and they were distributed without my doing much in the way of correction. This led later to the idea of a properly thought out presentation of the same material in German which appeared in two editions, the first under the title "Guidelines to Chemistry" and the second "The development of Chemistry" (Part II, Chap. 28, p. 354). Two or three lectures for the Lowell Foundation were aimed at a more general audience and were concerned with light, colour and painting. I published a little book in 1904 called "Painter's Letters" which described the physical chemical laws governing painting technique which at that time were almost unknown. The letters appeared first in a Munich newspaper and attracted many interested readers as I could see from later publications which referred to them. The book version was soon out of print and since I had no time then to rework the material this useful book has become something of a rarity.

The Lowell Foundation was dedicated to general education. The benefactor had decided that the interest from his considerable fortune should be used as follows: to begin with a director would be appointed and given a salary such that he required no other job and could concentrate fully on the Foundation. Whenever possible the director would be chosen from the members of the Lowell family which was based in Boston and sufficiently numerous that this condition would be fulfilled for a long time. It was therefore a matter of honour for each director to increase the influence of the Foundation as much as possible so as to increase his own reputation at the same time.

Another part of the income was used to attract high calibre speakers and for them the remuneration was accordingly generous. In the strongly protestant Boston of a hundred years ago theatre and concerts were considered to seduce their attendees to sin and so lectures, usually on a theological topic, were the only available public entertainment. As a result there lingered on into our times a considerable interest in this sort of event even when now more worldly forms of entertainment were regarded as permissible and were offered in profusion.

Admission to the lectures was free but it was regulated. One had to send an application in good time and include a stamped addressed envelope after which the requested number of tickets for designated seats was sent. If seats were not filled then they could be taken shortly before the lecture started by those who had no tickets, but the maximum number of the audience was no more than the total number of seats available. Five minutes before the start the doors were closed and after that nobody was allowed in so that under no circumstances would the lecturer be disturbed. From the newspaper reports and from the enquiries later sent to me, it seemed that these lectures too were a success.

Motivation for the development of a theory of colours. These lectures led to a particularly interesting acquaintanceship with a Mr. A.H. Munsell. He was an artist and painting teacher who had, for someone in this line of work, an unusually good scientific training. He'd been trained by the physicist Professor Ogden Rood who had written the best of the older books on the theory of colours. Munsell had been stimulated by this to arrange and standardise colours and had devoted a number of years to this work. He contacted me and showed me what he'd done together with a rather inadequate photometer which he'd built himself. It was already clear to him that the colours could only be methodically displayed in three dimensions and so he used the colour sphere which had been introduced by Runge in 1802. However, his attempt was doomed from the start because he'd taken over from his teacher the three unnecessary changes introduced by Helmholtz, namely tonality, purity and intensity. I was very interested to find out what criteria he had used to arrange the colours on the sphere and determine the purity (he used the photometer to determine the intensity). However he either couldn't or didn't want to tell me and instead fell back on his instinct as a painter. For the colour circle he'd used the old erroneous scheme of yellow, red and blue and by doing so had ended up with a completely wrong division of the colours. In fact in this respect it was a step back from Rood who had come much closer to a correct colour distribution.

Despite the limitations of his scheme, Munsell had set out with extraordinary energy to apply it in practice. He'd had the necessary colour pencils, and paints and a simplified colour sphere to demonstrate the colour arrangement produced and had already attracted quite a large number of adherents among teachers.

Later on he tried to introduce his system in Germany, but without much success. He died in the meantime but his son continues to spend considerable sums in spreading his father's teaching.

I was really interested in all this but a number of conversations sufficed to show that Munsell's system would not stand up to scientific criticism. However, I am indebted to this meeting for providing me with the challenge to involve myself with this large and important problem. I incubated this matter in a rather platonic manner for 10 years before I started to study it properly with experiments.

Other lectures. Apart from these lectures on methods there were also a number of other talks which I held on occasion in various places. A number of these talks were held before various scientific and charitable organisations that had elected me to honorary membership and which I had to repay with a lecture. In most cases I chose as my topic the development of an international language. At that time I felt it was particularly important to spread this message and still today I consider it most important for establishing peace in Europe and of course the contents of the lecture could be altered to suit the needs of the particular audience.

I got a number of invitations from people who wished to show me features of American life that they considered to be more advanced than in Europe. This was particularly the case with the scientific education of the female sex.

While most universities and colleges were coeducational, so that they were equally accessible for both sexes, Harvard did exclude female students. President Eliot view was apparently that the gravity and discipline necessary for education were easier to maintain when the professors did not need to pay special regard to a female audience. Instead, in parallel to Harvard, Redcliff College had been developed exclusively for women and most of the Harvard professors also taught there. I allowed my daughters to attend there lectures on topics which interested them. This was interpreted as recognition of the American method and was strongly approved of.

Wellesley and Vassar. In addition there were two large women's colleges nearby: Wellesley College near Boston and Vassar near New York. I was invited to visit and get to know both of them and I held a lecture in each. In one I described in 45 min the development of philosophical thought from ancient times till the present to the more than 1000 assembled female students and I think I did it rather well. At least I couldn't detect any inclination to sleep on any of the more or less pretty faces and was given loud applause at the end. I no longer recollect what I presented the students at the other college.

The two colleges were so similar in layout and design that a general description will be more than sufficient. They both consisted of a spacious and charming campus laid out with streams and lakes. Some of the main buildings were grouped together in a large quadrangle and numerous smaller buildings devoted to various purposes were scattered around the campus. The overall design was good and in many aspects splendid; everywhere there was lots of air and light. The dormitories were in the main building because both of these colleges, in contrast to Redcliff College in Harvard, were residential schools.

An observatory, chemistry, physics and biology labs, libraries were all there and were well equipped. There were many opportunities for sport as well as a music hall and a theatre. In short, every technical facility one could imagine was there and was regularly used.

As in the men's colleges the studies were set for 4 years and the students were 16–20 or perhaps a little older. Those who entered the college were obliged to remain for the full course.

Wellesley concentrated on scientific disciplines, while in Vassar the emphasis seemed to be more on developing social poise.

The scientific work I saw in Wellesley didn't impress me. I had the impression that most of it was treated in a rather superficial way so that the girls could put the subject behind them. However I have to give this judgement with some reservation, because I didn't have a chance to properly study the situation. However what my daughter told me of how things were done in Redcliff College tended to support my assessment.

I took away a very happy impression from Wellesley. As I was led to the assembly hall to hold my talk I had to pass through a staircase in the middle which rose up four or five floors. It was brightly illuminated from the top and on each floor it was surrounded by metal grilles. On the upper floors there were choirs of girls at the grilles who sang me a welcome to the house, first each choir alone and then all together. The fresh voices sounded marvellous in the enormous room.

From what I saw, this way of educating girls is not worth copying. The worst thing is that the daughters must leave the house in the years between 16 and 20 when they are usually at their sweetest. That's either an indication that there is no family life in which their presence would be missed or it is a means to chip away at any family life there may be. What the girls get in terms of "general education" didn't seem to me to make up for this loss because the colleges offer no real specialist knowledge of any kind.

Brooklyn. Around the middle of November I was informed by the Brooklyn Institute of Arts and Sciences[7] that I had been elected to honorary membership and was invited to a dinner to celebrate this event. The pharmacologist Professor Herter offered to put my wife and myself up for the duration of our visit. Since the German embassy was also sending a representative from Washington—this was the first sign of official respect paid me by the German government—it was impossible to turn the invitation down. I never did find out why it was the Brooklyn Institute that gave me this award.

---

[7]Now the Brooklyn Museum.

The reception at the dinner was warm. Around 100 gentlemen, many of them professors from Brooklyn, New York and its surroundings had come and it was of great importance for all of them to have the opportunity to meet me in person. As we were leaving one of them told me that he came from the countryside and as he'd missed his train had driven the 40 miles by car which at that time was an unusual achievement. He was however fully satisfied with the evening. It had been made clear that my hosts hoped for a talk from me and that they would particularly like to hear about my personal development. So as not to make myself completely ridiculous I laced my talk with a heavy dose of self irony which went down well.

The next day I was shown several other remarkable things, including the new buildings for a technical college which had been sponsored by private citizens. In a valuable private art collection I noticed next to the usual Parisian impressionist vegetables from Monet, Manet and so on, three pictures from Böcklin's early period in which he'd painted Italian landscapes. They stood out in happy contrast to the others in their lively harmony and their colours. Their owner didn't seem to understand that they were really something special.

Professor Herter was both scientist and millionaire and this is a mix which to my great pleasure I met several times over there. He'd inherited a great fortune as well as the ability to increase it at will on the stock exchange. He lived in a large house in the best part of town and he told me that he needed the interest from his fortune just for daily life, though of course a large part of this was used to attract the best teachers for his four children. When he needed money for something special he turned to business, though he never spent more than 2 days a month at this. He suffered from a chronic stomach ailment (from which he died a few years later) and because of this he wasn't able to devote much energy to his scientific work and was therefore thinking of giving it up. For his research he had a well equipped laboratory at home in which two research assistants worked.

Washington. Another lecture invitation, this time from the National Academy of Sciences, brought me to Washington where I had the opportunity to meet the President of the Union Th. Roosevelt. While the German Emperor had graciously welcomed the American exchange professor and had personally attended his inaugural lecture, there had been no official welcoming ceremony for me. I'd therefore made no attempt to make any connections there. However, once I'd received that invitation, which I couldn't refuse, it seemed to me that I couldn't visit Washington without paying my respects to the President.

My friends there took care of the organisational details and I was given a slot in the daily reception to which every American citizen had the right of entry. Dress was to be a normal suit. Roosevelt's appearance is so well known that I don't need to describe him here except to say that when he spoke he kept conspicuously showing his teeth. He started off with a friendly greeting, assumed that America pleased me very much and then continued to talk without pause, mostly about the brotherhood of nations, until our time was up. I wasn't given a chance to get a word in edgeways.

Apart from this the time in Washington was filled to the last minute. I got to know F.W. Clarke from the US Geological Survey whom I'd been in correspondence with for a long time. When I'd been working on the first edition of my textbook he'd brought out a summary and recalculation of the then available data on the atomic weights of the elements. Although the work was by no means first class, and some of the calculations were questionable, he was considered an authority on questions of atomic weights by the chemists who at that time had so little mathematical training that they were in no position to judge the matter.

In person he was a friendly old man, tall and with a long red fresh face and fading hair which had also once been red. He was an attentive host and didn't seem to expect any special degree of respect to fit his overvalued reputation. He showed me round the laboratory. It was led by Dr. Day, who was a son in law of F. Kohlrausch, and I found the new concepts and methodologies of the physical chemistry used to the fullest extent—which at that time was far from the case in official laboratories in Germany. I involved myself deeply in a discussion of their work and I think it is fair to say that one can now see here and there in their work some useful results of that discussion.

I spent a whole day at the National Bureau of Standards which was modelled on our Imperial Physical-Technical Institute and served the needs of the government in matters of physics. I saw a lot of interesting things, met and discussed with many of the staff and had an in-depth discussion with the director. In the evening many came to an informal reception—a "smoker"—with beer, Frankfurters and sauerkraut and a great deal of tobacco. When I was asked as usual to make a speech I suggested instead that they ask me questions and after a short hesitation they started and then there developed a lively to and fro. Finally there came the unavoidable speech of praise for me in which the speaker said that the Americans had once again shown themselves to be the better bargainers in the matter of the exchange professor because they'd got by far the better deal. This friendly bon mot quickly spread and I often got to hear it again later. On the other hand van't Hoff (Part II, Chap. 20, p. 217) wrote in his diary on the 17th November 1905, "M told me that that the American exchange professor can't find an audience any more".

There was more ceremony at a meeting of the National Academy in Washington which had elected me to membership and at which I gave a talk on the world language. A number of classical philologists from the nearby Johns Hopkins University in Baltimore were there and I could see that on both sides of the Atlantic these people had exactly the same prejudices.

The German ambassador Speck von Sternberg wasn't in when I called but I quickly got an invitation to breakfast the next day. He turned out to be a small thin man of sickly appearance. There I met Woodward, the leader of the recently formed and richly endowed Carnegie Institute, who questioned me about a number of colleagues who he was considering. He also asked me if I would be interested in working together with the Institute but nothing came of this.

The 3 days stay in Washington was packed with these and other social occasions so that I heaved a sigh of relief when I got into the sleeper which took me back to Cambridge in time for the lecture the next morning.

The lecture on immortality. One of the many foundations managed by Harvard was devoted to financing the yearly Ingersoll-lecture which was held by some famous speaker who had to discuss the immortality of the human soul. It was an express condition of the foundation that the lecturers were by no means only to be priests or theologians but that it could be anyone who had views on the matter irrespective of whether they were in accordance with the church's teachings or not.

The foundation's administrative committee, of which President Eliot was chairman, invited me to take on the lecture which was due to be held at the end of the current year. I pointed out to Eliot that my views on the matter were so at odds with what was generally accepted in America that if I expressed them publically this might lead to a scandal. He replied that it was the Foundation's aim to have as many different views as possible expressed because only in this way could the truth—or at least the most likely probability—be reached. I should therefore have no worries but simple present those views which were the result of careful thought.

A few days later W. James, who'd held such a lecture a few years earlier, came to see me in a state of great distress. From some of our earlier discussions he had learned to his surprise and mortification that in line with the agnosticism which was the norm in scientific circles I had firm views on such questions and that while I had found tenable objections to the idea of an immortal human soul I had found nothing to support it. He asked me to reiterate my views and then, in a state of shock, asked if I really intended to present them without caveats in such a manner, because if I did then people would take it very much amiss. He begged me, out of deference to public opinion, to at least leave the possibility of immortality open. I, for my part, was shocked that my honoured colleague could make such an unreasonable demand and I refused point blank, for any compromise would be a deliberate deception of my audience who had the right to know what my views were. In desperation James turned to my daughters who were by chance sitting in the same room (we only had one) and had heard our conversation and—hoping for an affirmative answer, asked them if they believed in immortality. Without batting an eyelid, both of them said, "No". Stunned, James left the house.

The news of my views on immortality spread like wildfire and on the evening of the lecture the great hall which was used for this occasion was filled to the last place with an excited and attentive audience. Dead silence reigned.

I published the text of the lecture in 1907 in the 6th volume of the Annals of Natural Philosophy (pp. 31–57) so I'll only give a brief summary here.

I began the lecture by pointing out that the concept of persistence was determined by the nature of our mind. Our conscious mind is constantly registering what we experience. Things which are similar are registered as "the same" and quite naturally these things can be said to persist. We expect to come across the same thing again in the future and hence consider that this thing will persist into the future. If we see no reason why this situation should change then we can consider that this thing will persist for ever. Scientific materialism had denied the existence of a soul independent of the body and hence it denied immortality. Now that this materialistic view had been replace by energetics, we have to look at the question anew because mental life can no longer be regarded as a matter of the movement of atoms which must come to an end with the death of the brain. Energetics makes no such mechanical assumptions and thus opens the way to other possibilities. What we therefore have to ask is whether the idea of immortality is, in general terms, consistent with the laws of nature in the broadest sense.

One can approach this by listing those things which stand beyond time. These are the chemical elements, mass and energy. However in recent times it turns out that the elements can change (Ramsey had recently reported the formation of helium from radium) and similar things are suggested for mass so that at the moment the only thing that stands beyond time is energy.

These more or less immortal things are characterised by the fact that one cannot make individuals of them. Two masses of water when mixed cannot be separated again so that the original distribution of the atoms is renewed—these masses retain their individuality only so long as they are separated in space. The same is true of energy. Thus individuality is incompatible with immortality and this is true wherever you look.

In particular, life involves continual interactions with the surroundings and hence results in continual changes in the nature of the living beings. Change and eternal persistence are thus mutually exclusive. There are therefore only two possibilities for the survival of the personality after the death of the body. Either it survives in a relationship to some other living being in which case it cannot be immortal. Alternatively it survives entirely independently in which case it can be immortal but it cannot have any relationship to the living or to those who died, that is to say that it is as good as dead. In both cases immortality as we like to imagine it is impossible—and we have to regard it as scientifically impossible. Survival of the individual takes place only to the extent that he altered the world and the people he interacted with. But these sorts of relationships are never eternal. As time passes they lose their individual significance and finally become merely part of the cultural history which those who have died leave to the living. The general law of increasing entropy which expresses itself in the diffusion of matter and energy applies also to the intellectual and moral world.

The loud applause before and after the lecture made it clear that many in the audience were ready to accept this line of reasoning. President Eliot who sat next to my daughters had a stern expression and Professor Münsterberg made clear that he by no means agreed. The audience had been completely quiet during the lecture so that every word echoed through the large hall as if it was empty. The whole atmosphere was one of an unusual event. The speaker too was affected by this and formed his sentences more solemnly than was his wont.

Criticism. The church in the United States still has great influence and theologians take an active part in the discussion of day to day problems and so this lecture, being contrary to the concept of immortality, which is the one thing which the different Christian churches all agree on, caused quite a stir. I heard many very different verdicts. On the one hand Professor Richard's mother in law, the very old but highly educated Mrs. Thayer (Part III, Chap. 32, p. 407), expressed her warmest thanks that the prospect of eternal peace which I'd given her made it easier for her to accept her approaching death. On the other hand I was described in the leading clerical and conservative newspaper, the "Boston Transcript" as Satan's spawn and President Eliot had to defend himself against criticism from the church for giving a forum to such a heretic and heathen, by clearly pointing out that the terms of the Ingersoll Foundation required that representatives of all opinions be given a platform.

The lecture was printed by a local publishing house which had published all the previous Ingersoll lectures in the same format. Over the years the quite large edition was, as I have seen from the receipts, completely sold out. It turned out that Münsterberg who had never been invited to give such a lecture wrote a pamphlet on immortality and saw to it that it came into the bookshops in the same format as the real Ingersoll lectures. The views expressed in it were entirely in line with the church's teachings.

It was soon obvious that the lecture had an impact on my position in Harvard. Those colleagues I didn't know well clearly distanced themselves a little from me. A public disagreement with the church was there, rather like in Britain, considered to be not just a moral failing but, more importantly, a social faux pas and this point of view was zealously and successfully propagated by the theologians. Not only I but also my wife and daughters were soon made aware of this.

Christmas far from home. The weather in Cambridge in that autumn of 1905 was particularly fine. The regular lectures were by chance so set that I had each Wednesday free and I did my best to keep this day free of any other obligations so that I could use the many possibilities offered by the trams and trains to explore the surrounding countryside. My daughters almost always came with me though my wife had increasingly to stay at home because of her frail health. My elder daughter and I usually took our paints with us and we brought home rich booty because the many strange scenes we saw demanded to be held as paintings. It was only at the beginning of December that the first snow fell overnight and lured us out the next morning. Outside we passed many a famous professor who was clearing the snow from the pavement to the front door. This was work which was not considered part of the duties of a servant and so to save time the house owners did it themselves. However the snow didn't lie for long and after some dark, stormy, rainy days we soon had blue sky and sunshine again.

As therapy for the heavy mental effort—the Ingersoll lecture had been on the 13th of December—I set up an easel at home to paint large pictures from the sketches I'd made and I finished quite a few that I gave away as gifts. In Boston I'd met the landscape painter Enneking whom I'd several times questioned about his views and creative experiences. He was quite happy to answer, but I didn't really get anything useful from him. He seemed just to start painting and then kept on correcting until he had more or less what he was looking for. Because of this he sometimes ended up with very thick layers of colour; a picture he handed me with sides of 50 cm which well reflected the mood of a late morning in autumn in the woods weighed several kilograms because of the thick layers of lead paint.[8] It seemed to me that this deliberate design of the picture offered more possibilities than the clearly more primitive unconscious momentary rendition which is favoured by painters and praised to the heavens by critics as the only truly "artistic" approach. I felt ever more strongly the wish after returning home to reorganise my life and devote it to painting. Twenty years later, aged seventy three, I have achieved this goal, though of course at this age it really doesn't matter what one does with the little bit of energy that one still has.

At Christmas I and my family were particularly looking forward to a visit from my eldest son Wolfgang who was at that time an assistant with J. Loeb (Part II, Chap. 27, p. 319) in Berkeley, California. Loeb had several times had nice things to say of him and we were all anxious to see how the move to a foreign country had affected him personally. He arrived on Christmas Eve and seemed to be more or less unchanged except for his mental development. Naturally his mother and sisters got in a female tizzy about the state of his clothes and so on so that there soon followed an extended shopping expedition.

Gifts. On Christmas evening we were brought lots of nice surprises from colleagues and friends which reflected many of the pleasant personal relationships which had developed in the course of our 3 month stay. The weightiest gift to me was a five volume work from the geologist Professor Shaler who had opened his still young heart to us. It was not, as I had thought, a scientific work but rather five dramas set in Elizabethan England which were written very much in the style of Shakespeare.[9]

Shaler had told me that he'd come from Kentucky where his parents lived out in the wilderness like settlers. For a long time he'd had no education till a German student, who'd fled his home during the persecution of democrats, chanced by. This man was a fanatical Hegelian who built everything on the triad of thesis, antithesis, synthesis and who'd managed to give the young hillbilly a very odd view of science and the big wide world. Shaler said that despite this he always looked back kindly on his teacher who, despite his strange ways, had made him develop the ability to think clearly and precisely.

I no longer remember how Shaler got from there to be professor in Cambridge. The Elizabethan dramas owed their origin to his view that all art was based on the exploitation of certain mental and technical possibilities which could in principle be learnt and applied so that it must be possible to produce a work of art quite artificially without having any inspiration at all. To test this hypothesis he'd set to work experimentally by choosing five incidents from that time and set about depicting them. The form he chose was Shakespeare's blank verse—the iambic pentameter. He told me that at that time he was so steeped in these rhythms that he wrote the text without having to consciously worry about the verse and that in

---

[8]White lead; $2PbCO_3 \cdot Pb(OH)_2$.

[9]Shaler NS (1903) Elizabeth of England: a dramatic romance, in five parts. Boston; New York: Houghton, Mifflin.

reading it over later there was scarcely anything to correct. I don't know whether these dramas had any literary success.

In person professor Shaler was a thin white haired elderly gentleman with dark eyes and lively movements. He and his wife were very kindly to me and my family and this was a reflection of his humanity because we almost never discussed science.

American philosophers. I should also mention a meeting of American philosophers which took place in Cambridge around New Year. A new university building—Emerson Hall—had just been completed. It was largely for Münsterberg who, apart from his abstract philosophy, also ran a psychophysical laboratory modelled on that of Wundt. My philosophical lecture course was moved to this building, though I only had a few lectures left. I was invited to join the philosophers' meeting and to hold a talk there. I talked about the relationship between mind and body in the light of energetics and this was received in a friendly fashion with many expressions of support. I was also able to take part in the discussion of other presentations. At that time I was treated as a colleague by the American philosophers, but in Germany this did not happen till much later. In his summing up the chairman described the national characteristics of philosophers as being depth of thinking for the Germans, clarity of expression for the French and common sense for the Americans.

Leaving Cambridge. My teaching duties at Harvard ended in the middle of January 1906, but this was not the end of my stay in America, for about a month earlier after having turned down several other invitations, I had agreed to hold two lecture courses at Columbia University in New York. Once again there was to be one course in chemistry and one in philosophy. The lectures in chemistry were a repeat, or perhaps better a new version, of the lectures I'd held in Boston on the historical development of chemical concepts. The philosophical lectures were to cover my own work on natural philosophy focussing particularly on new developments. The chemistry lectures were organised by the local professor of chemistry Chandler while the philosophical ones were organised by the psychologist J. McKeen Cattell who had been a student of Wundt and was now an influential networker in American science. The whole business was compressed into about 2 weeks so that I took tickets on the steamer home for the 6th of February. I had repeatedly felt exhausted already at the end of my time in Cambridge and expected an even more strenuous period in New York so that I was looking forward to some peace and quiet on the ship.

For the moment, however, I had to organise my goodbyes to all those I'd got to know in Cambridge and Boston. This took the form of a gentlemen's evening to which I invited 30–40 guests to whom I'd developed closer relationships. Dr. Morse was kind enough to assist me in this and he took care of all the technical side of things (organising a hall, menu and so on) and he did this perfectly. My daughters decorated the place cards with paintings and other ornaments and Goodale, the director of the botanic gardens, provided the flowers. The invitations were sent in good time and all were accepted with thanks. President Eliot, however, was unable to come because of a prior engagement, and so the dinner had to be postponed for a few days and the other guests informed. This extra difficulty also did not lead to anyone declining the invitation.

Once the meal was over I greeted my guests with a somewhat longer address in which I described the benefits I'd gained from my stay in Cambridge. To begin with I addressed the difference between professors in America and Germany in terms of their social status: Americans were appointed for a fixed period and if the occasion arose they could be fired while German professors are appointed for life and the appointment can only be terminated by the professor himself. He alone decides the contents of his lectures and in this he is completely free to describe to the students the results of his thinking and his experimental work. That is the positive side of things, the negative is that his horizons may easily become limited to the books in his office and the bottles in his laboratory, whereas the fresh air of practical reality reaches into the offices and labs of his American colleagues. A breath of this fresh air had wafted through my head and had made me more decisive than I'd hitherto been in the matter of organising my affairs.

A second thing for which I was most thankful was the willingness I found here to consider views which lay beyond the borders of accepted science.

Because of this I had spent several months in pure sunshine both in the meteorological and in the moral sense. This had involved more than the usual effort on my part but it was a happy time spent surrounded by an atmosphere of goodwill, friendly cooperation and amiable indulgence. In fact I had suffered no unpleasant experiences at all. It is hard to believe that this would be possible in our imperfect world, but in this case the unlikely had become reality.

Finally I asked my guests that instead of following the American custom of having a toastmaster, we should this time follow the German custom of letting anyone who wished say a few words.

Evaluation. The first to speak was President Eliot. He agreed with me that Americans were particularly well able to consider new ideas in an unprejudiced way and emphasised the great debt that American science owed to German universities. In this particular case the indebtedness was all the more since not only the students but also the professors had been given worthwhile motivation from my visit. He hoped that American science would now begin to pay back the debt by means of its accomplishments. The idea of exchange professors had been much strengthened by this first experiment.

Thereafter Richards (chemistry), Goodwin (philology), Shaler (geology), Wright (history), Franke[10] (German studies), Hall (physics), Noyes (chemistry), Münsterberg (philosophy) and Royce (philosophy) all spoke. Richards claimed that he had learned from me how to steer the ship of science around the rocks of fruitless hypotheses. Goodwin, a classical philologist who together with his wife had formed a friendship with me and my family, praised the hospitality of German universities.

[10]Misspelt by Ostwald: correct is Francke.

Shaler emphasised my bracing influence on all the young members of the university who were still able and willing to learn something new—he himself, however, was too old for that. He had been brought up by a German follower of Hegel whose philosophy he had in the meantime entirely forgotten and this had left such an enormous hole in his skull that he'd later been able to pack it with a huge amount of science. Trust in his own thoughts was the most important thing he'd learned from the German. Wright thanked not only me but also my family for coming. Franke said that it was well known that nothing was more difficult than walking through an open door. The open door was the general feeling of thanks and that he could not emphasise one aspect at the expense of others. He finished with Goethe's verse, which I quote here because it expresses what I would most like to think about myself.

Wide world and broad life
Long years of honest striving
Ever searching and always well basedNever finished, often rounded
Faithful to what's old
Open to what's new
Buoyant of mind and pure of purposeThat's the way one really gets ahead.

Münsterberg held a really nice speech which I hadn't expected. He said that the description of the wonderful characteristics of a German professor was readily accepted by those Americans who had studied in Germany but was scarcely believed by those who had not been abroad. They believed instead that this type of person didn't exist and had been invented as a pedagogic necessity rather like the poor but cleanly clothed goody-goody in books for children. He considered that my main contribution had been to show that this view was mistaken and by my presence had shown that the usual description of a German professor was more underplayed than exaggerated.

Noyes said that he'd got to know me 17 years ago as a teacher and researcher but now had had the chance to get to know me as a person and that he found this side of me even more impressive than the other. He laid special emphasis on the breadth of my activities all of which were pursued with undiminished thoroughness.

Royce related how a few years ago my newly published "Natural Philosophy" had been intensively discussed in his philosophical seminar. When recently I had attended this seminar and taken part in the discussion it had seemed to him to be like a wonderful continuation of that old discussion.

As Eliot suggested that we should now bring the speeches to an end I thanked them for the joy that the evening had brought me and let it finish with the thought of the power of science to bring nations and peoples closer together just as this happy occasion had. Indeed science belongs to all developed peoples who should all strive to add as much as they can to it. The only thing that is missing is a common language which would give everyone without exception access to this greatest treasure of mankind. And so the evening ended as it should, not in contemplation of the past but with a vision of the future; not with thoughts of an individual but with a view of common human notions.

Tea with students and other things. I took leave of the Cambridge students in a different way. I'd had contact with some of them beyond the lectures and so to end I arranged a college tea for a Sunday afternoon to which anyone that wanted could come. My daughters and some of their friends made tea and cakes and sandwiches and for around 2 h one wandered through the rooms which had been specially decorated for the occasion. For this I received many friendly thanks.

Most touching was the way the neighbours and in particular the black servants bade my family goodbye.

I can't count the many dinners to which I was invited in Cambridge and Boston in these last weeks, for some friendly words of farewell. They brought me once again into the company of the intellectual elite of both towns. I'd got to know them at least superficially during the last months and they left on me the impression that this society was based on powerful ambition coupled with an ideal fair-minded ethos. The distinctive self-confidence of these members of the east coast establishment, which is said to be typical of Boston and which far outweighs the considerable self-confidence of all other Americans, never struck me as embarrassing because the Bostonians lay great value on their excellent social manners.

Nevertheless in the interests of veracity it must be said that the official farewell dinner with President Eliot to which only the other officers of the university were invited, was conducted without any great warmth. I had the impression that something in my personality or manners was not to Eliot's taste. This was never expressed in words or gestures but simply in the fact that the general friendliness to which one had got accustomed did not develop with him. Probably the reason for the attitude of this man, whom I regarded highly, is already known, though as is so often the case the person most involved is the last to hear the truth. I certainly never found out.

On to New York. We left Cambridge in brilliant sunshine on the 22nd of January 1906 and after the short journey to New York were met by friends. Professor Herter had invited us to stay with him but since it was a long way from his house to Columbia University in the north of the city my wife and I stayed at a quiet hotel nearby while both of my daughters who'd met and made friends with Herter's family on a previous visit stayed with them.

We stayed in New York for not quite 2 weeks but it was by far the most strenuous part of my stay as an exchange professor. Every day I held a lecture of 2 h in English before an audience of 300–500 which was as much as the lecture room could accommodate.

These lectures, each of which was 60 min long with a 1 h pause between them, were already a considerable strain. On top of that it was clear to me that these were not just ordinary lectures but that here I had to give my American audience (many of whom were women) a strong positive impression of German science. Because of the very mixed audience these lectures could not be dry and academic. A degree of artistic presentation was needed so that each of the lectures would come across as if it were a separate stand-alone essay.

I think I managed to do that well. Vanity had driven me to employ an agency to send me all the newspaper articles on the lectures. This turned out to be a mass of paper and the whole exercise was much more expensive than I'd imagined, but the contents of the articles showed that I'd achieved what I'd set out to do.

However, I had to put all my effort into it and my reserves of energy were entirely exhausted. In one of the last lectures I had to struggle against a fainting fit without interrupting the flow. I asked a number of acquaintances in the audience afterwards but they hadn't noticed anything untoward.

In addition to the two series of lectures there was also a whole set of individual lectures which I neither wished, nor could, refuse to hold. On top of that there was almost every day a breakfast or dinner (sometimes both on the same day) with colleagues from all sorts of disciplines and at these I, as a "splendid guest", had to hold a speech and answer hundreds of questions, so that I was kept all the time under extreme mental pressure.

Worse still, my wife was now seriously ill. Already in Cambridge she'd increasingly withdrawn from social events which she no longer felt up to. The therapy ordered by the doctor who was treating her just made things worse and so she had to struggle with this as well. However in New York things got much worse so that for days on end she was confined to bed. We soon found out what was wrong—it wasn't life threatening but it did require that she take things easy.

Luckily the weather was good the whole time. Though the temperature was less than zero, the sky was clear and the short days afforded as much light as the calendar permitted. Morning walks down by the water where every twig was covered in silvery hoar frost were so refreshing that they helped me a lot to stay the course.

Journey home. On the 7th of February everything was finished and we boarded ship and found our cabins loaded with flowers and fruit and sweets as farewell gifts from our friends. As I watched New York's irregular skyline fade into the distance I had the feeling that I would not be doing this sort of thing again.

As we left harbour the weather was still fine but the ice covered ships sailing the other way showed us that rough conditions prevailed out on the ocean. Indeed we sailed into a steadily worsening storm. Half the family was immediately sea sick. I and my elder daughter held out for one more day and then too fell sick. On previous voyages I'd survived worse weather but the exhausted condition in which I left America had broken my resistance. Naturally my wife's condition worsened under these circumstances and even when 2 days later I and my stronger daughter had recovered from sea sickness, the mood remained sombre. Of my six voyages across the ocean this was the worst but at least it was soon over.

We sailed into Bremerhaven with a sense of relief at having reached the shores of home. We were met by my second son who organised our travel on to Leipzig where we found our house in perfect order. We were indescribably happy to be back home and promised ourselves not to repeat such a journey any time soon. We never did.

I first of all had 2 months holidays and then, as agreed, served out my time for the last semester. I therefore had lots of time to organise our move into my country house "Energy" which I'd had rebuilt for the purpose. The semester passed in the usual manner but without any of the drive and zest for work which had made it so enjoyable in the old days. On the other hand the synthesis of nitric acid had been pushed ahead by the tireless Dr. Brauer so it was now being regularly produced on an industrial scale. I was thus given the calming assurance that in case of war Germany would not be quickly rendered defenceless by a lack of explosives. At the end of the semester I left the University of Leipzig, without any official ceremony.

Chapter 33
Country House "Energy"

My own place. The need to own my own land was something I'd inherited from my father. I explained earlier (Part I, Chap. 1, p. 5) that my father used his first savings to buy a little house and after the financial collapse, for which he was in no way responsible and which with great courage he'd worked his way out of, he used a large part of his newly won fortune to buy the land on which he lived happily till his death.

The nomadic life of a German professor is a great hindrance to such a down to earth approach to life. Everyone in academic circles believes in the superstition that buying a house is the surest way to challenge fate to force a change of venue.

This didn't apply in my case. My years of wandering were short since Leipzig had followed soon after Riga. "Leipzig is not a place you leave", said my colleagues, "for Leipzig is a professor's heaven". This was fine by me because establishing the Laboratory as a world centre of physical chemistry was a time consuming process that would only be disturbed by moving somewhere else, and I couldn't expect to find a more understanding official authority than that in Saxony.

The minister, Gerber, who had been in charge of my appointment, was not much help in supporting the field and some time after my appointment Karl Ludwig, still red in the face from anger, told me that the minister had said that he didn't see the point of physical chemistry. The man was a lawyer and had no understanding of the subject or its possibilities, and of course his view may have been cemented by the intrigues of other professors. He died soon after and his successor, von Seydewitz, had a completely different view of the matter and did everything in his power to help me in my work. This was his objective judgement because I never got to know him at the personal or social level.

First attempts. My desire for a house of my own grew with the increasing difficulty of arranging for the necessary freedom of movement for my growing children. This got worse once they went to school because the school holidays were mostly not at

R.S. Jack and F. Scholz (eds.), *Wilhelm Ostwald*, Springer Biographies,
DOI 10.1007/978-3-319-46955-3_33

the same time as the university holidays and so I had to send my family off on its own.

My first solution was to buy an allotment garden—a so-called "Schrebergarten" in Johannistal—near the institute and my official residence. This invention which is so wonderful for those who live in big towns was due to a doctor in Leipzig called Schreber for whom a modest almost life size cast iron statue on a small pedestal was erected as a memorial in Johannistal. This was used by the many young people in the gardens as a target to throw various things at and so it always looked a little battered. However, I'm sure that Dr. Schreber would have been quite happy about this.

A lucky chance let me have an allotment nearby which one could look into from the institute windows and so it was possible to let the children run over there and if necessary to call them back with a prearranged signal.

The garden was small but it was old and densely planted with shrubs. An old garden house with a mysterious cellar increased its value for the children immensely. They explored every corner, found where the most snails were and staged races between them and, as children will, found all sorts of things to do in the little room.

I also enjoyed the garden and often enough went there when I needed an undisturbed hour to write or think, which was not so easy to get at home. I can still recall the happy feeling of having peace for serious work there which I experienced in the early years, though it was also there that I experienced the first painful inability of my overworked mind to convert the initial vague outlines of a project into a clearly defined goal.

How it happened. When in 1896 we moved into the newly built institute there was room for a little garden. It was even smaller than the allotment in Johannistal and was separated from the street only by a metal gate so that everyone passing by could look in. A large veranda took care of my requirements for in warm weather I could work there and had a view over the main part of the laboratory. The children lost out here and they did miss the hidden corners of the old garden.

One Sunday in 1901 to give my three boys something to do, I sent them on a journey of exploration. I'd seen in the newspaper an advert for a "romantically" situated plot of land in Großbothen. The railway timetable informed us that Großbothen was near the little town of Grimma which I'd visited once or twice with my paints and easel and had found lots of useful motifs. We'd also once spent a summer holiday in Grimma. Without having thought seriously about buying the plot I thought the idea was interesting enough to get the children to go and look and then report back.

They came back in the evening all excited about what they'd seen and begged us to buy this magnificent place. For me this was an old and strong wish that I'd tried more than once to satisfy. Once we'd tried to buy a plot on Lake Lucerne and a lawyer was authorised to complete the matter but at the last moment the owner withdrew from the contract and I didn't want to get involved in litigation. Now there seemed to be another possibility.

Since, despite the numerous stops, the journey by train took only an hour my wife and I finally gave into the children's demands. We found there an old house, large enough to serve as a summer house but in a terrible state. It was, however, surrounded by land that was truly charming. A broad valley ran from east to west in which the river Mulde had once flowed, though now it ran several miles further north and only a little stream was left in the valley. The plot was on the valley's north wall, on a gentle slope with the old house at the top. An orchard stretched from it down into the valley where it ended in a broad meadow. Old beech trees overshadowed the house and beside it there was a fir and oak wood. Nearby there was one of those huge porphyry rocks which are commonly found in moraine landscapes. A forest, through which it was an hour's walk to Grimma, lay on the plots west flank.

It was just the sort of country I needed; charming and varied but not heady or challenging. It seemed like a good idea to me. At that time I had a considerable income from my books and so could buy without touching my savings in any serous way.

My wife was appalled at the mess the house and garden were in. The house had been built by a widow who'd lived there reclusively for many years. After her death it had passed through many hands all of whom had let it decay more and more. An old farmer and his wife had kept an eye on the house and the room they lived in did nothing to increase the attractiveness of the place. Because of all that it took quite a bit to convince my wife to agree for she said "I'll be the one who has to clean the place up and put it in order". In this she was of course completely correct.

So we bought the house and garden and I could now feel like a landowner. The complicated business of officially registering the sale with the authorities and becoming members of the village community made it all seem to be something special—that owning land was somehow different from owning money or something else. Only years later did I change my mind and become a convinced supporter of the idea of land reform.

Energetic ideas were part of this idea that everybody, without exception, should be connected as if with an umbilical cord to some piece of land which would provide the chemical energy that he needed for his existence. This connection can be long and tangled but it must be there because nobody can survive without using energy. A people which has a direct connection to the soil is healthier and more productive, for the longer the connection the more the loss and the lower the quality of the nourishment.

The country house Energy. During the holidays I sent my older son with his teacher friend Brauer, who was later my colleague in the matter of nitrogen, out to the house to get rid of the worst consequences of long neglect. They made a primitive camp in one of the rooms and managed in a few weeks to get it to the stage where a team of cleaners could move in and make the rooms habitable again. My wife directed all this with her usual energy. An adequate supply of simple furniture was bought and finally the happy day dawned on which we could move into our own home.

The children were the happiest of all for now they had a much bigger area to explore and play in than they had been used to and they tumbled from one pleasure to the next. The impression was so strong that even today the "Energy", as I soon called the property, brings back for them that feeling of happiness. The effect can be seen with equal intensity in the grandchildren to whom the terms "Energy" and "paradise" are approximately equal. I myself feel revitalised every time I think that future generations of the family will feel the same way and that the property will serve them as an inspiration to higher accomplishments.

I too felt better in Energy. I first of all wrote my "Lectures in Natural Philosophy" there and I believe that something of the freshness and grace of the new surroundings found its way into the text, perhaps even something of the vigour and wildness of the land we'd just acquired.

Of course, in addition to writing there was plenty of manual work. With axe and pick and shovel I cleared most of the paths in the little overgrown wood and then there were plenty of things in the house that needed to be nailed or sawed or glued —all of which I happily did.

In comparison, my wife had the most work and the least joy from the new property for it involved a doubling of her house work and added all the worries of keeping it in shape. Of course all of us tried to minimise her work by settling for a very unpretentious way of life and our daughters helped out whenever they were allowed to. Nevertheless the burden was mostly on her shoulders.

And yet I do believe that some of the wonderful summer evenings with the scent of flowers and the glow of fireflies, some dewy mornings with the songs of finches and cuckoos in the nearby wood refreshed her spirit and gave a feeling of the good sides of our new circumstances.

Settling in. To begin with we used the new house only in the summer. At the beginning of the school holidays my wife and the children moved into Energy while the university semester kept me in Leipzig and I only went out there from Saturday till Monday. Then there were a few weeks of holidays together until the start of school called mother and children back to Leipzig. I then usually went off on a journey. Sometimes we went out to the new house for a short visit in late autumn or in the Christmas holidays.

The old couple I mentioned already were given the job of looking after the house, garden and meadow as best they could. They were honest and reliable but it wasn't easy to break them of their habit of using our drawing room to store eggs and corn and other supplies when we weren't there. We bought a donkey to look after the meadow and he soon turned out to be a useful and agreeable member of the household.

The old man spent his many free hours making models of farmers and artisans from wood and leather and whatever else was lying around. They were before their time, because nowadays they could easily have competed with African art or the pictures of H. Rousseau. He gave me quite a number of them but instead of keeping them I gave them to the children as toys. They were glad to have them.

Our initial experiences at living in Energy made me certain that if the house was properly extended then it would be a splendid place to live in permanently. My wife was a great deal less convinced of this.

Effect on our way of living. My behaviour during the conflict with the philosophical faculty of Leipzig University was determined to an extent by the purchase of Energy. Because I now had a roof for myself and my family I no longer needed to worry about losing the official residence, pleasant though it was. I actually found it all rather reassuring. I'd several times witnessed how after the death of one of my colleagues the family was forced to leave the official residence in which they'd lived for half a lifetime. These situations are of course necessary, but they are nevertheless cruel to the surviving dependents at a time when they are sunk in deep sorrow. Since I too might have met an early death I saw to it that the future of my family was assured and when this done I experienced a great feeling of relief.

Rounding off and extending. Soon after I'd bought the ground I was offered a neighbouring strip of land whose owner had wanted to use it also for the summer but whose resources had only been sufficient to construct a simple hut. Although the price was far too high I bought it because on this side it nicely straightened out the strangely zigzag border of my land. The addiction to rounding things off, which can often become an obsession, started to work on me. I bought one plot after another, though the farmers who were selling usually managed to cheat me. But even if it did make a bit of a hole in my bank account this was usually made good by the income from a new book or from a new edition of an old one.

This kept on for years on end till finally my land was bordered all around by roads and paths or further expansion was prohibitively expensive. Even during the war I was able to buy some fields that extended my land up to the main road. I always paid cash so that the property was not mortgaged in any way. I must admit that it wasn't clear to me why I was buying all this land and sometimes had a bad conscience as if I was indulging in a pointless and expensive all consuming passion. However since this was the only one I had I let it be.

The name. Possessing land quieted my mind and played a part in my choice of name for it. If I'd done the standard thing and called it after my wife "Helen's peace" then that would have come across as a bit ironic since, at least to begin, with it caused her nothing but trouble. On the other hand the name "Energy", which at first I'd thought of just as a joke, began to seem ever more appropriate because I had learned in the past few years what an important role energy plays not just for the outside world but for my own personal fate. The country house was for me a continually available source of new energy when I'd used up the old, and so there could hardly be a more appropriate name for it than "Energy".

I had this name hung over the entrance. Some of the villagers didn't understand the word and in their dialect it came out as "Anarchy".

Rebuilding and moving in. Once it was clear that I would be leaving Leipzig I started to rebuild and extend the house so that I and my family could live there permanently. I came across a young architect called Munde, who'd read and enjoyed some of my publications, and who was willing to take on the job which he carried out diligently and correctly. Because for other reasons (Part III, Chap 32, p. 402) there was plenty of time, the building work could be planned and carried out unhurriedly.

It was finished in good time and we moved in at the end of August 1906 after the completion of my last summer semester as professor. I still felt the exhaustion from my stay in America and the unfriendly atmosphere of the final period in Leipzig had done little to further my recovery. Because of this I was not much help in moving the furniture in, my wife and daughters had to do most of that—and they did it well.

Moving in was completed satisfactorily. The new central heating worked well in the cold days of the following winter and we could get on undisturbed with all the various things which had to be done. I was busy with finishing the publications I mentioned earlier (Part II, Chap 28, p. 351) and the ladies of the house had their hands full with putting the house in order. Before we'd moved we had sometimes discussed how we would manage to pass the long winter evenings in the loneliness of the countryside. By the time we got round to talking about that again we were sitting under the trees in our garden and it was already spring.

Morning walks. During the first years in Leipzig I'd sometimes noticed in the evening that what with the never ending work with the students and at home I almost never got out in the fresh air because the apartment and the lab were under one roof. I didn't think this was very healthy and so to avoid having to plan each day separately, I settled for a walk every morning. The lectures were every morning from 8:15 till 9:00 and so I went for a walk at a quarter to eight and had 30 min to think over what I was going to say.

I've kept to this habit ever since. In "Energy", where of course the original purpose no longer held, I have always taken a walk through the estate no matter what the weather was. And I still use these walks to think—no longer about lectures, of course, but nevertheless many useful ideas came to me on them. Walking and thinking are now so subconsciously fused that when I have something that needs some thought I automatically go for a walk to clear my head.

In this way I found "Energy" a splendid help in my work. My thoughts are not disturbed, as they had been in Leipzig, by all the sorts of trivialities which one comes across when walking through the streets. The only things which changed, and which one often noticed, were the natural results of growth and decay, the weather and the seasons. It is much easier under these conditions to form and work out new thoughts than it is in the bustle of a city.

When working on particularly complex lines of thought I was always able to find paths in the neighbourhood where I could walk without meeting anyone and on which the longest chain of ideas could be put together.

When, later on, there were numerous opportunities to hold lectures, I would, for important occasions, first get the general outline clear and then hold the talk out loud to myself on such walks. This turned out to be a very productive way of doing things. Often enough I'd find that a line of thought which seemed to fit the subject would turn out, when I talked aloud, to be not yet fully thought through. Thoughts combine to a many layered network but a talk, like a piece of writing, is one dimensional and requires that the different lines of thought be strictly regulated. Whether this has been done successfully can be tested much more easily in speaking rather than in writing. In any case writing the text for such talks would have been far too cumbersome.

Work plans. In thinking over what I should do in the upcoming last part of my life I kept in mind that Wöhler had reported that at that stage he was only able to do things which could be completed within about three days. He couldn't undertake anything which lasted longer because his memory was no longer all that it once had been. I therefore thought I'd either do some sort of experimental work which could be done quickly, as Karl Schmidt had done in his second period, or alternatively do something that required old memories which, as everyone knows, remain strong even in old age. The best solution would be to do something that was a bit of both. For this reason I took care in the planning to make sure that there would be not only a laboratory but also a library. In the discussions with the builder I wanted to keep the rooms small, which was what I'd been used to all my life. My wife, however, energetically argued for large rooms saying, "You're building this as if was for yourself alone". I backed down. As it turned out, this prediction was completely correct because every last corner of both the workshop and the library were filled and used.

The Library. In Leipzig I had always been surrounded by masses of books. The textbook and other works had to be re-worked and since this always required that I constantly check back on the original sources, I felt I had to have the most important ones to hand. For this reason I bought the entire set of the Annals of Physics, various chemical journals and the most important foreign language journals. They were at that time reasonably cheap because this was before the American market developed and drove the prices up.

In addition I bought many books which were important for their historical value. Soon enough I had assembled a library with which one could carry out proper historical research.

On top of this came books which were given as gifts and those which I was sent to review for newspapers or later for the Annals of Natural Philosophy. Another source was the generous gifts of the rapidly increasing number of publications written by my one-time students. Most of these came in the form of reprints of papers. From all of this I could see that my mania for writing had spread infectiously to my colleagues. Many of them, especially if they were native speakers of some language other than German, wrote large summaries of their work and their precision in doing this, which of course varied from author to author, convinced me that they had learned from my example.

The result was an ocean of printed paper which grew with every postal delivery especially in the period when the scientific societies of various countries elected me as an honorary member.

In this way I found I was able to settle down to the sort of work I'd envisioned for myself. I thought I'd look into various scientific questions in a much more detailed way than I had been able to do in my first historical review of electrochemistry.

This was the reason that I could now start my "Psychological Studies", the first of which was on Humphrey Davy and was published the following summer in the Annals of Natural Philosophy.

Further similar works soon followed. This material had the form of general summaries which pointed to practical consequences and before I knew what was happening I found myself in the middle of a lively debate about how the school system should be reformed. I'll come back to this later.

Other thoughts, particularly in the field of natural philosophy, had at the beginning no relation to a concrete project but soon developed along similar lines. It has always been my belief that theoretical results should be applied as soon as possible to real practical problems and so I soon found myself involved in all sorts of public issues which took up the major part of my time and energy. I'll discuss these below.

The laboratory. Next to the library I had a good sized room facing north which I used as a lab. I fitted it out with work benches and equipment and had a quite rich collection of the most important chemicals so that I could immediately carry out virtually any experiment that occurred to me without having to first order the necessary materials. It soon turned out that since I'd ordered as if for a complete teaching lab, I'd vastly overestimated what I'd need. The result is that more than 20 years later I'm still using the supplies that I purchased then.

Spending substantial sums of money for such things and on the equipment which I needed—or thought I needed—was something that I first had to get used to, for until then I'd had everything I required paid for by my office. It seemed to me to be almost a waste to be paying for it myself. Larger items became a motivation for me to undertake some literary work that I'd turned down or put off and for which the remuneration was much larger than the cost of the equipment. At that time the future of my children was not yet assured and so I considered it my duty to keep a large reserve of money available for them and was not happy at the thought of reducing this safety net. Later when all this was lost in the world war it turned out that the things I'd bought had been the better investment.

The lab was soon in working order. I already started in Leipzig to apply my chemical and physical knowledge to technical questions of painting and so now I worked along the same lines, though without any great sense of urgency. This form of occupation too turned out to have much greater consequences than I could have predicted then for it opened the way to the quantitative concept of colour which kept me busy with experimental work for a decade and which involved much difficult mental labour.

All of this was only possible because the fading energy of my later years could, thanks to my freedom from all official forms of time wasting, be focussed on these general problems.

The garden. The garden and the woods that surrounded the house begged to be ordered into a beautiful form particularly once I'd bought an old quarry in which one could have immediately staged the second act of Carl Maria von Weber's opera "Der Freischütz" complete with wolf's lair and incantations to call forth the devil. However, I must confess that I lacked the time and the desire to do all this. In terms of physical work I confined myself to pick and spade with which I worked on the paths and I built some little dams across a small stream to produce three little ponds to satisfy my need for sheets of water. However, the slow time scale of herbs and flowers and trees was so completely different from the pace of my inner life that I was never really able to reconcile the one with the other and so I left the garden to the rest of the family and enjoyed the results of their work as they appeared on the table or in the flower vases that my wife set out in the various places where I worked.

Only in recent years, since 1925, have I started to develop a closer relationship with the garden, though to be honest it is a little indirect. My efforts in painting go in the direction of depicting natural scenes as the rhythmic ensembles of form and colour through which they become works of art. Plants are the obvious subjects for this and the enormous developments in landscape design greatly increase the available motifs. Through this I develop a closer relationship to plants as part of nature. It will depend on the number of years granted me in my old age how deep this relationship becomes, though it certainly gets deeper with every year that passes. As I grow older my speed of reaction gets slower while that of the plants stays unaltered so that in this way we slowly approach one another and I can envisage a tender relationship to flowers developing in my last years.

Organisation of work. As I'd prepared for this free existence I'd sometimes asked myself how I'd mange to fill the days and whether it wouldn't be uncomfortable to have to come up with a program for each new day once I'd lost the formative constraints of my official duties. However, I'd already learned in the holidays how easy it is to develop a disciplined approach and my experience in all the years since then has confirmed this.

Usually I had a new book to write and I'd do this mostly in the morning. After that the post came and dealing with it would often require several hours. The flood of printed matter, new books and so on, often kept me busy for days and weeks on end while work in the garden and walks with my easel and paints gave me a change of scene.

These walks were an opportunity to develop new thoughts which, out of long habit, I didn't work on right away but I often returned to them later. Since I couldn't always trust my memory I kept an empty note book on my desk in which I briefly noted everything that seemed worth preserving, particularly everything that related to problems I was working on. Once a notebook was full I'd write on the cover the date of the first and last entries and put it on the pile of older notebooks. To be honest I did this more out of a sense of duty rather than from necessity because I can't remember that I ever ran out of things to do or ever had to go back to these old thoughts. These notebooks were less a real collection tank and more a means of dealing with the mental overflow.

Since I only applied myself to those things I wanted to do I was able to spare all the energy that is normally wasted when one's working life is regulated by external imperatives that require that one gets into the necessary mood or that one breaks down one's own inner resistance. That at least is the explanation I give when I'm asked how I managed to do so much. I optimised the use of my available energy to an extent which normal mortals, who have no idea of the scientific application of energetics or who simply reject its teachings, cannot achieve.

Lectures. What I've written so far related to the days I spent at home. However the daily round was often broken by journeys to scientific meetings, or to meet new people and, in particular, to hold public lectures. To start with I held a few of these for particular events and their success soon led to all sorts of invitations and I could only accept those which dealt with the various forms of cultural work that I was involved in. Energetics, schools and universities, the scientific method, organisation, world language, internationalism and pacifism were the usual subjects on which I lectured. The following chapters will show how I became involved in these things. One should bear in mind that all these things developed in the 8 years between moving into the new house (September 1906) and the outbreak of the world war (August 1914). In the description that follows each of these threads may appear to develop on its own but in reality they were all interwoven with each other, and so the texture of my life was much more colourful than the ordered presentation may suggest.

Chapter 34
Great Men and the Schools

The problem. Once I'd put the work on the books that I was involved in (Part II, Chap. 28, p. 351) behind me, the first major task that I undertook in my new surroundings was to explore in detail a chain of thoughts that stretched all the way back to the start of my scientific career in Dorpat. This had to do with the question of how the highest achievements of the great researchers are brought about.
I've related several times with what fascination I regarded the mental grasp of the various researchers with whom my studies brought me into contact. A great advance in this was made possible by the application of the energetic concept. Many details of a researcher's life, which till then had brought forth nothing more than a vague feeling of understanding but were categorised as "fate", could now be understood as the result of simple energetic relationships. It was clear to me from the start that one couldn't answer all questions in this way, but nevertheless the number of questions which could be so addressed was so large that it seemed to me that an investigation of this problem was both a scientific and an ethical necessity.

The beginnings. It was clear to me from the concept I'd developed of the general properties of a great researcher that this was not an impossible venture. Near the end of my last visit to the United States, which I discussed above, I accepted a dinner invitation to the Philosophical Club in New York which ended with the news that the others there were hoping to hear a lecture from me in the near future. I hadn't considered that (though I should have expected it) and so I had to hurriedly look for something in my mental store to tell my friendly but critical hosts; something that would neither disgrace me nor bore them. Under the pressure of events the long simmering thought that researchers fall into two very different psychological classes crystallised in my mind. There is the slow but deep and parsimonious classic scientist on the one hand and the rapid, brilliant and highly productive romantic scientist on the other. I illustrated this with a number of examples from the exact sciences and was happy to see that this line of thought was readily accepted by the audience. One of them, whose name I sadly did not note, was a short but graceful-seeming young man with a pleasant lively face who looked more like an

R.S. Jack and F. Scholz (eds.), *Wilhelm Ostwald*, Springer Biographies,
DOI 10.1007/978-3-319-46955-3_34

artist than a philosopher. He told me that he had come to exactly the same conclusion and had even used the same terms to describe the two groups.

Psychological studies. Now I used the freedom which Großbothen gave me to work as I liked and to go more deeply into the matter. After the first years of my new life I published the first result in 1907 in the form of a psychological study of the English chemist Humphrey Davy. I chose him for no better reason than that I already had a lot of material on him in my library. The paper was published in the Annals of Natural Philosophy.

I include the preface here because it well expresses the relationship between the study and my wishes and experiences in this area.

> In the first place it was personal experiences which led me to think about the life of a scientist. A completely unexpected change in my scientific interests and in particular a sudden loss of the ability to be an inspiring teacher led initially to a strong feeling of mental unease and, despite my normal optimism, made me at times quite unhappy. One of the most positive things that science gave me was the insight that this was not a condition for me alone and that one should not think of it in terms of guilt and responsibility. The automatic tendency to generalise began to make itself felt so that I was soon asking whether my experience was a unique case or whether it told us something general about the inner world of a scientist. The considerable amount of material that I had accumulated in my previous studies of the history of my special field of chemistry made it possible for me to start a preliminary investigation. The result was that there are indeed general laws which make a psychological analysis of the lives of great scientists possible. In particular, the effects of scientific discoveries on the lives of the scientists themselves were so obvious that any doubts about the operation of general laws were soon forgotten. I made the preliminary results generally available and pointed out their possible practical applications and now I feel I must bow to the methods of exact science and make the raw data available so that these special cases which served as the basis for my inductions can be re-examined. As always in the exact sciences there is no other basis for generalisation than induction from an incomplete data set so that the conclusions may be altered, improved or clarified by later research—but they will not be totally refuted.

First summary. Soon after this I accepted an offer of M. Buber who was the publisher of the collected work known as "Society" ("Die Gesellschaft") to produce a volume of around 100 pages for this collection. My inner wish to do this was coupled with an external imperative for I had chosen my new life in the expectation that I would be able to finance it by independent work. In order to test this I had decided that for the next 3 years or so I would accept every opportunity to make money that was consonant with my views as a practical idealist. That would show if it was going to work or if I would have to look for some regular job.

The question of income was not particularly urgent. Ever since moving to Leipzig I'd had a sizeable and growing income from my books and there had been no need to spend this money because the university salary and the laboratory fees were enough to cover my normal expenses. Because of this I had a reserve of several hundred thousand Marks which was more than enough to act as a pretty large reserve in case I was no longer able to earn sufficient money. However it turned out that without exerting myself or having to sacrifice my freedom my income was more than sufficient to cover my costs and even to increase the reserve.

This was the period in which writing seemed just to flow. I'd become so accomplished on the typewriter that the work seemed to go by itself. My typing speed was such that the typewriter seemed to buzz like a bee. Since I'd already thought over the content and the form of the sentences on my walks, it is easy to see that a book like this could be written in 14 days—and that included the time necessary for the walks.

This opportunity was also welcome because just as with an upcoming lecture it gave me the opportunity to put my thoughts in order. The subject matter turned out to be richer than I had expected and I began to wish to go the whole hog. The little book was consequently well done. I don't know how well it sold.

Great men. The "psychographic study" as I termed this work at the time was quickly extended to cover a number of great men. Part of this work was published in the "Annals" but soon there was so much and the conclusions and applications so numerous that I had to turn to the more complete format of a book in order to bring everything together.

It ended up as a thick book with 420 pages which was finished during the course of 1908 and was published in the spring of 1909 under the title "Great Men". Its success was immediate. The first edition of 2000 copies was sold out in 6 months and the numerous reviews in the press ranged all the way from hearty approval to bitter enmity. Further editions also sold well.

This considerable and rapid success was due to several factors. The first of these was that the biography was approached in a rather different way than was usual. Normally these things were written in the style of an obituary in which one only related good of the dead and any negative aspects were either ignored or explained away. Then there was the limitation of the usual historical approach that only the "intellectual" aspects of a life were worthy of consideration and the general or biological aspects were merely referred to obliquely as a minor matter. In short, the one sided limitations which the representatives of the humanities, by ignoring the mental rules of science, imposed on themselves meant that this type of work was generally fruitless.

In addition I had in no way restricted my historical research to Ranke's dictum of merely determining "what had actually happened". This was a necessary first step, but that was all. One reason for these studies was to use my observations and analyses to define the natural laws governing genius so far as I could identify them. From this there then flowed many practical applications particularly in questions of school education and how one might be able to identify those boys or youths who could later make great scientists.

For this reason I used "Studies of the biology of genius" as a subtitle to emphasise that I wished to project this neglected side of things into the forefront. In doing this I had gone beyond what was then known of biology and applied the new perspective of energetics to reach a new and deeper understanding. The reviews in the press clearly expressed surprise at this approach. Some approved while others did not.

Practical applications. The question of the inheritance of genius was at the forefront of the biological argument. I made the reader consider this point by describing a particular event which had made me think about this problem.

One of my Japanese students had passed on a question from his government as to how I managed to train so many talented and successful students. At that time considerable sums had been assigned to the development of the country's scientific performance and, since those in charge of this program were naturally anxious to ensure its success, they asked me for my support.

I told them first that the situation had simply developed as it was without any conscious effort on my part. I was only aware that that I'd given my young students the freedom to follow their own way and had not tried to force them down some other path. The general direction of the work in the laboratory was defined by our overall judgement of what was fundamentally important: osmotic pressure and electrolysis in the first period and catalysis in the second period of my teaching career. Since every new co-worker simply took up whichever one of the unlimited number of problems appealed to him, the result was more or less guaranteed.

Nevertheless the question—this was in my last years in Leipzig—had given me cause on long walks through the meadows by the river Pleiße to think more deeply about this. It was certainly the case that a remarkable number of people developed into highly talented scientists and this was shown by how many of them achieved academic positions and went on to become full professors. If I compared this to the situation in the First Chemical Laboratory which had at least twice as many people in it as did mine and whose director was admired and almost worshipped by his students so that he had an almost unlimited influence on them and where despite all this only mediocrities who seldom made it as far as lecturers were produced then I had to accept that there had been some special factor that positively influenced my students and was absent in the other institute.

Bad influences. At that time I only managed to solve the positive part of the puzzle. I only began to understand the negative factors in the other institute after the death of its director when his previous students described the atmosphere there and their accounts were all the more believable because of the positive terms in which they described their institute.

According to these accounts the greater part of the education and the choice and execution of the doctoral work was as follows. The director collected problems that he had come across in reading or thinking and he wrote each one down on a sheet of paper. Usually the project consisted of confirming one of his opinions or of refuting an opinion of someone else. In almost every case it was clear from the start what the expected result was to be and so each student took his piece of paper and set to work to achieve it. If the expected result was not what he found then he considered that his work had failed.

In the period when we worked together I had been the second referee on the theses of these students just as he was second referee on the theses of mine. During the course of this time I noticed occasionally some rather strained interpretations and forced conclusions that I tried to point out. The result was that the first referee strongly protested and said that the second referee had only a formal function and should not criticise work which he, as first referee, had approved. After that I held my peace and restricted myself to commenting on the usefulness of free research only when the work was carried out in my own institute.

Organisation of research work. The direction of research was indeed quite different in my institute. At the start I made a judgement of the interests and aptitudes of a student and not only left him the free choice of his thesis subject but insisted that he himself should make the choice. I often emphasised in our seminar round (Part II, Chap. 24, p. 292) that getting the expected results, so long as they were solidly based, could only be considered a success at the lowest level. Unexpected results were often much more interesting and important because, just like the bump on the bark of tree, they show where a new bud will break forth which may grow into a completely new area of science. I did this because I was acutely aware of the danger that the novice may be so blinkered by what is expected that he becomes blind to anything which fails to fall into the expected scheme.

This attitude was not due to any wish to be different from the way of work in the other institute, about which I had no idea at the start, but rather derived from my study of the history of chemistry. This had many times illustrated for me the dangers of preformed opinions and it underscored the deep impression that an experience from my first lab semester in Dorpat had made on me. I had given my instructor Lemberg silica as the result of a qualitative analysis. He said, "There's no silica in that sample". After the first analysis I'd jumped to the conclusion that it was silica and then in the follow-up analyses simply looked for confirmation of this. He, however, patiently showed point by point how I had fooled myself and warned me that in the future I should come to a conclusion not at the first step but rather wait with it until the last moment and to be always ready to change it if further results demanded.

Because of this I had, without really thinking about it, instituted a procedure which without doubt allowed the students to develop their talents and so bring them to above average scientific achievement. Now what I had to do was to bring this out of my subconscious into the light of day. This, like a birth, is a dangerous business during which the baby can be squashed, deformed or smothered just as can also happen in the generation of a work of art, though here it is always possible to put right, improve and reanimate when the first attempt has resulted in a monstrosity. In this matter I turned my thoughts over, changed and developed them until, by constantly measuring them against my experience of reality, they formed a stable and viable structure.

To reach this comparison I'd worked through the biographies of the various scientists and I was always looking for particular patterns through which I could get to the heart of the matter. It is no surprise that under these circumstances the result was portraits in sharp relief which were much more gripping for the reader than what one otherwise found in the dimly lit sketches available in the "official" biographies.

Researcher and teacher. Another point of view which emerged from this had to do with what I like to term the cultivation of genius. Those who know how our universities are run all deplore the way in which the most important resource that a nation possesses—the talents and the time of its sharpest minds—is thoughtlessly thrown away. Already in the course of my first visit to Germany I'd noted the useless demand made on the country's leading physicist at that time that he hold 5–6 h introductory lectures for students of medicine and chemistry and so on who were in their first semester. I learned from a large number of this audience that he had made no impression on them at all. In this case one cannot claim that the best was only just good enough for the country's future scientists. No matter how good the lectures were in terms of their content they were scarcely able to introduce the young students to the elements of physics since the concept of precise scientific thinking was almost never made clear in the secondary schools. For this these lectures were far too advanced even though Helmholtz tried his best to make them accessible. He didn't understand the difficulties these average students had and hence was unable to take that into account and build a bridge for them. The whole exercise would have been much more useful if the lectures had been held by a good teacher, of whom hundreds were available, even if he totally lacked Helmholtz's creative genius. The precious energy of the great scientist that was wasted in this way could have been used instead for further scientific discoveries which would have added enormous value to the cultural assets of mankind.

In the light of my studies it should have been clear that Helmholtz belonged without doubt to the "classical" type who would never have made a good teacher, especially not a teacher for the broad mass of beginners. Compelling him to do this must necessarily have reduced his capacity to make new discoveries which he alone was capable of. And if it's said that he'd made enough discoveries then one must answer that there can never be enough of them particularly as we are only just at the start of the scientific conquest of our world. One forgets all too easily how young science is—not more than about a 100 years old. It is for example scarcely a century ago that Kant was forbidden to publish anything on religion and that he had to accept this ban.

The fact that science is still in its youth explains why it is only recently that anyone has thought about organising the harvest of results. In the universities it is and has been axiomatic that the professor is there to teach—not to do research. However in complete contrast to this the appointment of a professor is made not on the basis of his ability to teach but rather on his research record: the ability to teach is a minor consideration. One he has achieved a leading position however the question of further research is a matter of choice for the professor and he retains his position even if he abandons research altogether. On the other hand the authorities make certain that he continues unabated with the secondary business of teaching. We can take my own case as an example. Although I achieved much more as a researcher than the majority of my opponents, they used the antipathy to lecturing which I developed in my later years to make it impossible for me to remain.

In this way we reach the amazing situation that the highest—and for the nation and humanity in general most important—achievement is demanded from a professor but he is not rewarded for this. He has to deliver results as a free gift and he will be punished if in order to do this he neglects his teaching duties.

The study of the psychological and physical conditions that impinge on a highly talented researcher as a result of his teaching duties may be summed up as follows. The "classic" type is from the beginning unsuited to teaching and should therefore in the common interest be excused from these duties. The "romantics", in contrast are in their youth almost always splendid teachers and one should give them as many opportunities as early as possible to do this. But because of their enthusiasm for their work, which is about the most demanding that you will find in science, they burn out rather quickly and lose both their love of science and their success.

From this one gets a means to assess these people. One should not put pressure on the "classic" type and in particular he should not be burdened with mass lectures. The "romantic" should be given every opportunity to teach so long as he is young (this type of scientist is liable to produce his masterwork in his youth) but then one should reduce the pressure on him once he has expended himself in this effort.

All this should apply to highly gifted researchers. For those talented individuals who are not quite in this class the characteristics of the two groups are less obvious and they can certainly be expected to carry a larger teaching load which they will absorb easily. In fact one often meets in these positions men who have a real talent for teaching. They are valuable members of the teaching staff and have scarcely any difficulty in the application of their talents.

The elite. Another use of the results of the study was to answer the question as to how one would be able to identify as early as possible the future great scientists. As I explained above the whole question first arose as this question was posed by the Japanese. At that time I gave a preliminary reply which had the following content.

The genius-to-be can be recognised in the school by the fact that the contents of the normal syllabus are not enough for him and that he goes beyond this and searches for knowledge, usually in books, and usually mainly in one direction.

This is bound to be the case for, as long as one continues with the severe organisational error of putting every age group into a separate class so that individual differences in the speed and the depth of uptake are methodically ignored, the normal syllabus has to be designed for the average pupil. The construction of school classes is organised on the implicit assumption that all of the pupils either have exactly the same talents, interests and diligence or that they can, and must, be forced into the same mould. In reality there is nothing more obvious than the personal differences between people and hence our schools inevitably come into an insoluble conflict with reality. It is not the only one, because such a fundamental error inevitably gives rise to a string of further errors. In order to avoid this the schools should be organised so that the talented pupils can make their way faster than the less talented ones. This is certainly possible to organise and it has been done. I'll add more to this shortly.

If we turn for the moment back to the main question then we will usually find our future genius developing a close relationship to some grown up friend who helps him. Sometimes it will be the father for quite frequently the intellectual blossoming has already started in the previous generation without yet having reached such a peak that it brings forth any very unusual achievement. More often, it may be a relation or some chance acquaintance. From this person those interested in identifying highly talented youths will be able to obtain the most precise information.

The assessment of a teacher—and this is the usual way of doing things—is to be accepted with caution. There are—frequently unrecognised—brilliant teachers who have an ideal understanding of the nature of their job and who do it well. These will usually give a correct assessment of their pupils. Understandably, however, most teachers base their assessment simply on how much trouble a pupil is. The highly talented pupil has no difficulty in assimilating the usual run of the syllabus but he will from time to time ask questions that the teacher can't answer and which therefore make him impatient or angry. On top of that, supreme talent is always a special talent. It is one sided and everything which lies outside its compass will be neglected. All of this tends to result in the assessment of an average teacher being rather negative. Usually the end result will be: when he was young we had great hopes of him, but later he caused the school great difficulties and worries so that we now are not sure if anything at all will come of him.

The ideal pupil for the average teacher is not the talented one but rather the "good pupil" who causes no problems.

The school question. These studies had started with the greatest minds to which humanity can aspire and led me to the smallest, the children who make the first steps towards their future greatness in the schools. At that time it was believed in Germany that our schools were unbeatable and perfect and the attempts of Kaiser Wilhelm II, who'd himself experienced the great deficiencies of the secondary schools which specialised in the classics, to correct the errors were blocked by the unanimous opposition of the "experts".

I'd been spared these experiences because I was educated in a secondary school which did not specialise in Latin and Greek. Now, however various matters of fact forced themselves on me and they all pointed in the same direction. In the book "Great Men" that I referred to above I devoted an entire chapter to the problem of schools and came to the conclusion that classical secondary schools are an atavistic left over of our cultural development and that for the sake of the soundness of our system they should be abolished as soon as possible.

This result did not surprise me.

The credulous reverence for classical education which I'd accepted in my youth had slowly been eroded by experience and had now been converted into its opposite. A major factor in this was the observation that the human qualities of the representatives of the humanities that I met in Dorpat and later in Leipzig showed no evidence for any elevation of their spiritual values through their studies of antiquity, and this, after all, was supposed to be the basis for the emphasis on studying Latin in secondary schools. On the contrary I often found these men to be narrow minded and not competent to understand the most important events of our time. They were the ones who displayed the worst characteristics of the German professor. When I led the faculty in Leipzig as dean for a year, there came up, among other things, the appointment of a replacement professor for classical philology. We scientists were used in such circumstances to appoint the best candidate because we believed that the better he was as a teacher and researcher the more the university would benefit and the more students would be attracted. The philologists were principally concerned to find someone who would not trespass on the turf of any of the current professors. During the discussions this was presented as the most natural thing in the world and something that was essential to maintain "collegiality"—and so it was done.

I had to acknowledge this as an additional argument against the value of classical philology. The making of new discoveries was considered to be in very bad taste and was frowned upon. This is what I had experienced.

Historical proof. This conclusion was also supported by my study of great men. Quite often, in fact almost regularly, these had rebelled against the classical schools. Liebig was expelled before completing his secondary school. Mayer was always at the bottom of the class, Davy found nothing to praise about his school except to say that it did at least leave him time for his hobbies, and the quiet and measured Darwin, whose hundredth anniversary was just then being celebrated, had in his autobiography laid out in harsh words the hatred and contempt he felt for the old classical secondary school. As an objective researcher I could only come to the conclusion that such schools are not suitable as places to educate great men. And as a scientist who must always look for the causes of things which happen repeatedly, I saw it as my duty to find out why this should be so, especially, as the general consensus in the population and in the ministries responsible for higher education was the exact opposite. At least they have always acted as if that was how they saw things.

The simple fact is that the classical ideals are unable to accept any development that goes beyond what was achieved in antiquity. Because of that this mode of thinking damns itself to sterility and the character of the majority of its proponents serves as an experimental proof for the smothering of the intellect that this brings with it.

Against the calamity of the schools. Because of the resonance which my complaints and demands brought forth in the press, I was sucked into the campaign for school reform which was developing at that time. In 1866 the word was that the Austro-Prussian war had been won by the Prussian teachers and this resulted in the feeling I mentioned above that our schools were unbeatable. This opinion remained unchanged and the result was that the schools were increasingly incapable of satisfying the requirements of our growing cultural development. A few persons and groups had protested against this, but theirs had seemed to be a hopeless battle against the ever growing pedagogic mainstream. I was welcomed by these few as an ally and was invited to many meetings and asked to give lectures. Since I was no longer hindered by holding an official position, I readily accepted most of these invitations.

My public efforts developed at first with my attendance at the annual meeting of the "Society for German Education" which was held in a wonderful summer in Weimar and at which I participated as a lecturer. The effect of the meeting was so positive that a larger congress was held in Berlin in the spring of 1909. Over 2000 listeners attended my lecture and it was clear that I spoke for them for my presentation was frequently interrupted by loud applause. I spoke without notes to the title "Against the calamity of the schools" and later I formulated the argument as a pamphlet which made it accessible to wider circles.

The main point was that the repeated attempts of Kaiser Wilhelm II to reform secondary education were bound to fail because those who were to decide the matter were the representatives of the current system. In this way a majority against any significant reform was guaranteed and the recommendations quite naturally were that the present system should be changed as little as possible. I told the story of President Eliot and the American football experts (Part III, Chap. 32, p. 407), but in this case there had been no one to point out his sensible point of view.

However the fact that the system required reform was made terrifyingly clear to the public by a rapid series of suicides by pupils which took place at that time.

Two German educative institutions had achieved worldwide recognition: the kindergarten and the university. In contrast to the secondary schools where every little detail was officially decreed, the kindergartens and universities were free to develop as they saw fit and hence the most successful methods could evolve through a process of natural selection.

In both cases the successful institutions were characterised by the joy with which both teachers and pupils took part in them. The well known bleakness of life in a secondary school is itself a mark of the failure of this type of school system.

I attacked as sharply as I could the view that a syllabus based on philology is directed towards high ideals, unlike the "Philistine utilitarianism" of the sciences. An ideal is an unreachable goal that one can only approach step by step. Antiquity can therefore never be an ideal because far from approaching it we quite naturally move ever further away from it. From this contradiction we are left with just the simple truth that pursuing the ideals of antiquity is at the end quite useless and therefore harmful.

From the classical secondary schools specialising in Latin the excessive concentration on language has been transferred to the newer school forms where modern languages again take up too much time and work. The argument that learning languages is an educational achievement is not tenable. A hotel porter knows more languages than the best educated teacher but the teacher would never consider that the porter was better educated.

I also complained that secondary education spread over 9 years was far too long because the talented youth who was already well able to take care of his own development was consigned to a state of intellectual imprisonment which would often destroy the most important thing namely the development of the thinking personality. An even if the school left something of this intact then he was threatened with destruction by the wickedest damaging element of our school system—the school leaving exam which can only be regarded as a crime against our youth.

Further reform work. I'm not sure how much support I'd have for these ideas today. At that time they were considered new. Since they were set forth in plain language they had an effect like blows with a cudgel and so the result was an enormous kerfuffle.

I was kept busy with the numerous requests to lecture over the schools issue which I gladly accepted. As a result I was in numerous cities in Germany and got to know a lot of people from all sorts of jobs who, like me, had experienced the damage done by our school system either directly or on their children or had seen its effects in their business and who warmly welcomed my efforts to improve things. I'm not in a position to judge what effect all this had but a sober assessment would probably be that not much came out of it all.

One of my most successful lectures in this regard was held in Vienna. In the meantime I'd often been in this beautiful city to give lectures on various subjects and always found its people warm, lively and readily animated. In this case the degree of approval that met my lecture exceeded by far all that I had so far experienced. The effect was so strong that the other side soon thought it necessary to found a society dedicated to saving the classical secondary school and the chairman of the society was the Austrian minister of education, the Earl of Stürgkh,[1] who was later assassinated by the socialist Adler.[2] This society immediately organised a large meeting at the university at which powerful resolutions were passed against my destructive views. The New Free Press however was friendly to me and published in my defence.

Even the worthy Mrs. von Ebner-Eschenbach was recruited into the service of the holy cause and she wrote the following epigram:

Those who understand Greek and Latin

Will soon master German

Those who want our education to sink in the mud

Will take classics out of the schools

At that time one must add she was already over 80.

My view was: By their fruits you will know them: Legal German has long been notorious and those who produced it all came from classical secondary schools.

Berthold Otto. Of all the personal relationships that developed out of this the most important for me was that to the brilliant educator Berthold Otto—the Fröbel of our time. While most of those involved dedicated themselves to fighting the deficiencies of the system Otto with fresh courage and a clear view had initiated a new form

[1]Karl Graf Stürgkh.

[2]Friedrich Wolfgang Adler.

of education and had started this with a cohort of pupils whose understanding parents had agreed. He soon demonstrated some remarkable success.

A principle idea running through his method was not to force feed a pupil new knowledge but rather to show him how to gain access to useful knowledge himself. This was done first by asking other pupils, then by asking the teacher and, when even this source failed, by recourse to books. This sort of knowledge is by definition something that the child wants for otherwise he would not have voluntarily searched for it. The questions would be inspired by experiences of daily life, such as things seen on the way to school or heard at home or on the streets. The application of his pedagogic ideas made use of the "holistic lessons" which took the form of an unstructured conversation led by the teacher.

In the course of my conversations with B. Otto there came an amusing illustration of the energetic imperative. One of the many visitors to his lessons—clearly someone of the "old school"—had complained that the pupils were not sitting in a proper disciplined way but instead lounged about. Otto agreed and said that at the beginning he'd tried to get the children to sit straight up but soon found that this was pointless because then they were less lively and less engaged. As soon as a question enthralled the children then they forgot all about how they were supposed to sit and instead applied themselves to the problem with all their limbs. It is not possible to concentrate both on an intellectual problem and on one's posture and so posture was sacrificed in the interests of intellectual progress.

At that time I did what I could to support this splendid man and pure idealist in his difficult struggle and to provide him with practical help where I could. He never lost faith and managed to gather a like-minded community around him. However, he never managed to develop his ideas in a broader context. Nevertheless I do believe that sooner or later his time will come.

The imperial school conference. A sort of conclusion was drawn to these efforts when in 1920 the new men organised a general conference on school education in Berlin. This however suffered from the same problem that had bedevilled Kaiser Wilhelm's attempts, because it was composed almost entirely of representatives of the existing school system. These were the people who drew up the agenda and they spoke in almost all the sessions. They also played the leading parts in the sub committees. In those parts of the proceedings that I took part in, it was clear that the discussions were following a predetermined course and every attempt to bring in new ideas was blocked.

The result was that the really important matters were never discussed. As is well known our school system is predicated on the notion that all of the pupils in a class come with the same basic knowledge, same skills, same interests, same talents and so on and that they all remain uniform during their entire school career. The principle function of the teacher is to bring the class all equally well through the year's syllabus. Given that these premises could not be further from the truth, the teacher is forced to waste the largest amount of his time and efforts in ironing out the obstructions and damage which this gives rise to. As a result his efforts are not directed to bringing on the best, where they would have the most beneficial effect, but of necessity he must apply himself to the weakest and here his efforts have the smallest result.

It has often been pointed out that this basic flaw in the organisation of our schools need not be accepted as inevitable. It is based on the idea of school classes and would disappear when classes are abolished and are replaced with individual teaching. It is not possible to go into details here and I can only point out as strongly as possible that this has to be the central point of any school reform and that its application must be left to a more comprehending future. This will result in a real social "class struggle"—a struggle against the class room—in contrast to the fundamentally antisocial nature of what is currently referred to as class struggle.

Chapter 35
The World Language

The beginning. The lectures on Natural Philosophy in 1900 had led me to examine more or less every major issue from the stand point of energetics and where necessary to take a closer look. In the course of this I noted the enormous waste of energy which arises from the differences between languages. This waste is so great that, even in times when the energetic principles had not been formulated, educated people had sought an explanation for it. The legend of the tower of Babel was constructed as a warning that should men ignore God and follow ideas which had not been approved by the priests then they would be punished by a confusion of languages. The authors or editors of this old myth regarded the fact that different peoples spoke different languages as so senseless and stupid that they could only explain it as resulting from a divine punishment.

Today we know that the idea that all peoples originally spoke the same language is as far from the truth as the myth of a golden age or the birth of mankind in the paradise of the Garden of Eden. Paradise and the golden age lie not behind but rather before us and it's our job to work towards them and every single one of us can do something about that, even if it's only a little. In the same way the time when all humanity speaks one language lies not behind but before us. In ancient times when the clan was the biggest group that could be built, each of them had a separate language. After all, their interactions were limited to robbery and murder, both of which are activities that can be carried out without the need for linguistic communication. It was only with the exchange of goods and the start of primitive trade that there arose a requirement for the understanding of more than just the mother tongue. From that point on the energy wasting consequences of the differences between languages began to be evident and have grown to their present monstrous dimensions.

In the light of the many new considerations that the consistent use of the energetic concept requires I suggested a few sentences which applied to the question of language and these will be repeated here:

> "Language is not only the treasure house in which the jewels of correct and useful concepts are kept, it is also the attic to which old and useless concepts are consigned. Since the connection between a word and a concept is often not obvious there is not necessarily any

R.S. Jack and F. Scholz (eds.), *Wilhelm Ostwald*, Springer Biographies,
DOI 10.1007/978-3-319-46955-3_35

conflict when the word assigned to a concept happens to sound like the word used for some other quite different concept. Because of this it often happens that when a concept is improved the old and now inadequate word for it is often not dropped immediately but hangs on, sometimes for ever. For example in German a chemist refers to oxygen as "acid stuff" (Sauerstoff) even though we now know that the sourness is a property not of oxygen but of hydrogen ions.

For this reason thoughtful minds have considered the question of whether the uncertainties of language, which are made painfully obvious to school children in the form of irregularities in the rules of grammar, should not be avoided by the development of a synthetic language which is completely regular in all respects. Many people today regard this idea as ridiculous. They hold that a language is an organically evolved structure and that trying to build one from scratch would be like trying to fabricate a tree.

However language is not an independently growing and persisting organism. It is merely a tool that humans have developed for particular purposes and as these purposes have changed over time so has language. Language is like an old house that has been lived in by many generations and each generation has altered, torn down and made additions to adapt it to their needs. Of course the old house will not be completely knocked down and destroyed because a lot of the effort put into it by our ancestors is still there. But couldn't we build a special house for special purposes nearby? When the old rooms are too dark and cramped, the old floors too uneven for the work we now need to do, then we should build a new, comfortable and suitable house nearby! Joy and sorrow, birth and death and everything that moves our feelings will still take place in the old house. However the business and work rooms can be moved to the new house which will not be built of gnarled wood and carved stone but rather of polished tiles and light but strong metal girders. We could quite easily, to return to our theme, have in addition to our mother tongue a general simple language for business and science which for the interaction between peoples would be incomparably more beneficial than the telegraph or railways.

In reality we already have several such languages. Musical notation is understood all over the world wherever European music is known and even if we can't understand anything else in a Japanese book we can at least make out the chemical formulae or mathematical equations in it. In the same way written numbers may be pronounced in quite different ways in different languages but can be immediately understood by anyone who has learned to read just his own language.

A general synthetic language is therefore not a mad fantasy but simply a technical challenge which if solved would be an enormous help to working people all around the globe".

The delegation. A few years after the publication of "Lectures on Natural Philosophy" I received a letter in which reference was made to these remarks and in which I was invited to help make a general world language a reality. The letter came from the Professor of Philosophy in Paris, Louis Couturat who'd made his name with his penetrating work on Leibniz. The letter was accompanied by a number of publications which included: a description of the current state of the work and a weighty book containing a detailed analysis of all previous attempts to construct synthetic languages that he'd written together with the mathematician Léau.

Practically all of this material was new to me. Since he had not restricted himself to write it all in a historical context of "what had actually happened" but had rather compared the various systems and their success and drawn important and wide ranging conclusions, I needed to waste no time on the lower levels of Platonic wishful thinking but could get started immediately with real practical work.

At that time the situation was as follows. Many international scientific meetings had taken place at the world exhibition in Paris in 1900. In the course of these language had been the cause of considerable difficulties. Since science had long since established itself as the human activity which was most free of national differences, the experts I mentioned above had suggested that a committee be formed to develop an international language. The committee chairman was Couturat.

The work of this "delegation" was initially to canvas support from prominent personalities and groups such as tourist organisations, trading associations and the like. Their success was quite remarkable but it was naturally largely restricted to French speaking areas. I soon joined the delegation and tried to spread the idea in Germany, though I had little success. The main problem was Wilhelm von Humboldt's mystical theory that language is a special sort of living form with its own rules and that it arises without the conscious input of humans and in some mysterious way that was never explained, it reflects the inner soul of the people. Again and again I was told that one could no more build a synthetic language than one could synthesise a man. My counter argument, that there already existed a whole set of synthetic languages some of which allow for precise gradations of expression such as musical notation or mathematical or chemical formulae, was simply not understood. These reactionary linguistic nihilists imagined themselves to be better than the adherents of creative progressiveness which one of the best minds of Germany—Gottfried Wilhelm Leibniz—had predicted, demanded and propagated.

Volapük. Then I remembered that A. von Öttingen had years ago brought the synthetic language Volapük to my attention. With his active mind he had got so far with it that he had corresponded in Volapük with a number of foreign supporters and he had assured me that they had understood each other perfectly. Of course I'd replied with the usual story of the independent life of languages which at that time I also believed, and I found his contrary experience hard to grasp.

At that time, around 1890, Volapük had an extraordinarily rapid rise. As the first true synthetic language it gathered all the progressive minds in civilised countries that understood the need for a world language and fused them into a working community. Within a few years there were 283 societies spread across the world,

312 textbooks in 25 languages had appeared and 1600 teaching certificates had been awarded. The total of those involved was estimated to be around a million. However the decline was as rapid as the rise.

Volapük was the work of one man; the clergyman Schleyer who had built quite a number of good ideas into it as well as some that were not so good. To get started everyone had just accepted the language as it was and used it. In this phase a number of errors became apparent and required correction. Congresses were held and an Academy was formed. Disputes soon arose between the Academy and Schleyer because, while he considered that he held an absolute veto over all changes, the Academy was not prepared to grant him more than the ability to delay improvements. This disagreement ended in a complete break; Schleyer held onto his original Volapük while the Academy improved it so successfully that a completely new synthetic language (idiom neutral) emerged which had very little in common with Volapük though the good ideas of Schleyer had been retained. As a consequence of this schism the whole movement fell apart and this remained a burden for the future development of such ideas.

Nevertheless the impetus led to a whole set of suggestions for new languages and this in turn brought ever more clarity about the prerequisites which must be met.

Esperanto. Couturat and Léau's book described the newest and most hopeful development in the field which was the synthetic language Esperanto invented by Dr. L. Zamenhof. It had been published in that busy year 1887, but had been largely ignored. A little later the Frenchman de Beaufort had invented a new language and was preparing his work for publication when he came across Zamenhof's proposal. He thought that Esperanto was in many ways better than his own new language and, in a way which was most unusual for people working in this area, he selflessly gave up his own project and devoted his considerable energies to working for the acceptance and dissemination of Esperanto.

That was the start of the rise of this synthetic language which was soon widely spread. However "widely" must be understood in a geographic sense because though adherents were to found in many widely separated countries, their total number was small, because most them were excited about it for only a few years after which they let it drop.

There were two quite different reasons for this. The first was that there was no real use for this new communication form. The allure of exchanging postcards with other Esperanto speakers half way round the world was initially strong but it soon palled. This is a problem that every synthetic language faces for as long as it is not required for some important international undertaking such as science.

The second reason lay in the nature of the language. Because it had been invented in a Polish-Russian area the inventor had taken over a large number of "zisch" sounds into the language. Since the normal letters of the alphabet can't handle these, he'd used symbols above the letters as is common for example in the

Czech language. However this made it impossible to write the language using a normal typewriter or to print it or to send things written in it by telegraph.

One would have thought that there was nothing simpler than the removal of these deficiencies once they'd been recognised, but now there occurred a major organisational error which seriously damaged the movement. At the first international meeting which was held in Boulogne a hastily prepared resolution was passed which declared the "Fundamento"—a short textbook of Esperanto which contained its rules of grammar and some exemplary texts—to be "untouchable" (netuchebla). It could no longer be altered in any way. In this way any development of the language was forbidden and that was its death sentence.

This situation wasn't changed by the later attempts of some ambitious and zealous fans of the old Esperanto to bring the language, now suffering from consumption, back to life. Volapük had also had just such an ephemeral period of strong growth and died for the same reasons as Esperanto because in both cases the inventor and his heirs failed to understand that the development of the language should not be suppressed but rather organised.

Esperanto in America. I hadn't known these fundamental difficulties when I went in the autumn of 1905 for a longer stay in America. My position there as an exchange professor was a clear symbol of the international nature of science and so I found the idea of being an apostle for the world language to be an organic part of my mission. After the initial period of settling in I began to recruit support for the idea of a world language and found my assistant Dr. Morse (Part III, Chap. 32, p. 416) to be both a dedicated and deft co-worker.

I held lectures in favour of it first in Cambridge and Boston and then later in many other cities in the United States, tailoring my presentation to the particular audience. I strictly separated the general idea of a synthetic language which would be everyman's second language to be used for interactions with foreigners from the special solution of this problem which existed in the form of Dr. Zamenhof's Esperanto—even though that certainly required modification. Each time I spoke before the members of a society I would end by suggesting that they agree to the general idea and join Couturat's delegation. They almost always agreed and this increased the rather short list of adherents by a large number of members.

Dr. Morse, for his part, developed a lively and successful interaction with the student societies which he knew from his own student years. The Esperanto clubs which he formed in Cambridge functioned in an astonishingly short time as models for similar clubs across the huge country all the way to California. Morse organised competitions for the best translations and I put up the money for the prizes and in this way I did more for Esperanto then than any other supporter with the obvious exception of Beaufort. The American wave rolled back on Europe and led to a revival of public interest here.

However it was just like a wave: the crest was followed by a trough and after that nothing more happened. The American movement too ran into sand after about a year because there was no real need for the new language.

Conceptual progress. The propaganda work in the States had been useful for me because the need to present the matter to very different sorts of audiences forced me to look at it from all sorts of different viewpoints and that gave me increasing clarity of the nature of the problem. I clearly remember as I was preparing one of these lectures for a meeting of lady teachers in Boston that I suddenly realised that language can be regarded simply as a means of transport, similar to the postal service, the railways, the market, the stock exchange or money, though it is of course the most important of all because it is the most general. Its function is to serve as a connection between people pursuing any sort of common aim. One therefore has to make sure that it, just like any other transportation system, satisfies the technical requirements which will allow it to do its job as well as possible.

If you consider the amount of bombast and fog which has been used for over a century to hide rather than to lay bare the nature of language then you will appreciate how I breathed a sigh of relief at this insight. I had unconsciously suffered till then from the lack of clarity on this point and this ran counter to my constant wish for clear thinking which was the basis for all my work as a practical idealist. I was accustomed to opponents and doubters coming at me with the argument that language was an organism with its own life and that one could no more construct a synthetic language than one could a homunculus. Now I could reply that a 100 years ago the synthesis of organic compounds like indigo could have been declared impossible for exactly the same reasons. Today hundreds of tons of indigo are manufactured and it differs from the natural product only in being free of the impurities which make up a large fraction of natural indigo. In exactly the same way a synthetic language would be free of the "natural" impurities of a natural language which are the inevitable leftovers of the time of their formation when ways of thought were more primitive and less logical than today.

The crucial experiment. A particularly crass example of the uselessness of Humboldt's linguistic mysticism was provided by Humboldt himself. He'd been involved in analysing language for the whole of his life and was generally regarded as being someone who understood the matter better than others. One would expect that such a person would have a better command of his own mother tongue than his peers. But in fact that was not the case. His detailed analysis of Goethe's "Hermann and Dorothy" which he considered as the summation of his views on art is so badly and vaguely written that neither Goethe not Schiller, both of whom approached the work with a positive preconception, were prepared to publish it. In addition Humboldt had been accustomed for many years to collect the thoughts and feelings which moved him in the form of sonnets and there were a great number of these. No publisher dared to bring out a full edition because they are so terrible in terms of their language, rhythm and rhyme.

Here we have the proof that the way in which Wilhelm von Humboldt approached the problem of language had not given him the capacity to write a half way decent poem, while hundreds, indeed thousands, who had no idea about his theories were able to do so much better. One would think that the originator of the theory ought to understand the matter much better than ordinary people and so one is driven ineluctably to the conclusion that the theory was not able to improve the use of the language or at least had neither a positive nor a negative effect. Perhaps the ability to produce bad poems was simply an inborn characteristic of W. von Humboldt.

I of course expect the argument that Humboldts deep linguistic philosophy was not intended to satisfy the banal purpose of bringing forth a clever use of language. To that I can only say that I know of no other or better function of a philosophy than to teach an all round understanding of the matter under discussion. If it can't do that then it isn't science but simply scholasticism, and because it involves a waste of energy its social value is less than zero.

Turning away from Esperanto. After I'd returned to Germany at the beginning of 1906 I was initially tied up with finishing off my duties as professor and in organising the move to Großbothen. Nevertheless, I kept the question of an auxiliary language in mind and used every opportunity to support it. I have to say that I had much less success in Germany than in America.

At that time there happened something which opened my eyes to a very important matter. Because of the increasing support from an ever widening circle, the German Esparantists, who up until then had been a small and sorry group, gathered up the courage to mount a general meeting which was to be held in Dresden and I was asked to hold the major lecture. I was happy to do this. So that the lecture could later be published I worked it out completely beforehand, which is something I rarely do. When, accompanied by my second daughter who'd also learnt Esperanto, I arrived in Dresden and met the Esperanto community there, I was somewhat shocked to see the company I was now keeping. When I brought up the matter of the difficulties which the strange accents over the letters caused and explained that this must be sorted out as a matter of urgency, I found myself faced by bitter resistance which was clearly discernible through the veil of politeness with which they felt it best to treat the effective new member. An old lady, who had been a follower of Esperanto right from the beginning and who now played a leading role, took me aside into a little room to explain how the Esperantists viewed this matter. In the room there was a table with a stately green satin cloth (green is the emblematic colour of Esperanto). In the middle of the table flanked by burning candles in silver candlesticks lay a green leather bound copy of the "Fundamento" signed by the master himself. The whole thing was an altar dedicated to the cult of the infallibility of the "Fundamento".

This sort of blind fanaticism, typical of religious movements, is quite common amongst the supporters of Esperanto. To me, interested in a world language principally as a means of communication, this sort of approach seemed both ridiculous and unhelpful for the movement. The object was to solve a technical problem. However, one cannot view a matter of religious adoration through the eyes of a rational technician without calling forth the fury of the congregation of adorers and, more particularly, of the high priests. And if one does not regard the matter as a technical challenge then one loses the chance of applying the energetic imperative which is the only way to achieve cultural success.

Paris. I'd written to Couturat from time to time to report on my success in America. Once I got back home I suggested that the number of supporters was now high enough to justify taking the next step laid out in the delegation's constitution.

This second step involved forming a circle of experts to design the new language and to ensure that this auxiliary language would be recognised and propagated.

After the necessary preparations the inauguration of this working committee was set for the spring of 1907 in Paris.

It was only the second time that I'd visited Paris. During my time in Leipzig there had never been any important reason to go there because in striking contrast to the willingness that the British and Americans had shown in testing and accepting the new teachings of physical chemistry, the French had ignored them entirely. There was not even any strong opposition because the French showed not the slightest interest in the new concepts, knew nothing about them, and made no attempt to learn their implications.

This was not the only time that a great and essential advance in Science was ignored in Paris. When, in the middle of the nineteenth century, the concept of energy was established in Germany and Britain and both countries competed to be leaders in the field, it was all ignored in France and there seemed to be no one there who realised the importance of what was going on. It was fully 10 years after the proposition of the law of the conservation of energy by J.R. Mayer before the leading journal in France "Annales de chemie et de physique" made any reference to it. And this all happened in a time when V. Regnault's research was directed to establishing the laws governing the performance of steam engines, that is to say in an area that achieved its success and clarity only through the application of the concept of energy in the form of thermodynamics.

Things were much the same with the new chemistry: it took 10 years before Paris realised that once again science had not waited for their approval of progress, which might have been the case 50 years earlier. Instead no one had even asked them what they thought of the matter. Finally van't Hoff who'd kept in touch with Paris since his student days there was invited to give a lecture to the Chemical Society and then the French learned to their astonishment that a strong international movement was underway which they could not silence by ignoring it. Then Arrhenius was invited to talk but it took a long time before a small stream of publications from there joined the mighty stream which flowed from sources in other countries.

I saw Paris for the first time in 1907. I was there in connection with a technical and scientific report on Claude's method for making liquid air and went to inspect his laboratories where I had the chance to meet some scientific colleagues. The principle of these was A. Haller who, though an organic chemist, had closely followed the new concepts of physical chemistry. We got on well because in addition to his scientific talents he also had an interest in organisational matters so that we found ourselves in agreement in our thoughts and plans. Later on I'll recount how these common views were applied in a practical way to the formation of an organisation representing all chemists the world over. I also became friends with the chemist Le Chatelier and with the splendid mathematician Poincaré.

Laboratories in Paris. Because I really wanted to see it I was shown the Curies' laboratory in which radium had recently been discovered. The Curies were not in Paris. The laboratory was a cross between a stable and a potato cellar and if it hadn't been for the table with chemical equipment I'd have assumed that someone was playing a trick on me.

It was already clear to me that the labs of even the most important professors in Paris were not the best. M. Berthelot (whom I never actually met) was in the habit of carrying out his thermodynamic determinations at the ambient temperature of his lab while J. Thomsen (Part 1, Chap. 11, p. 119) had set up a thermostat which kept his lab at a constant 18 °C so that his determinations were all internally comparable.

In Berthelot's work one came across temperatures as low as 6° or 7° so that I got the shivers just imagining how scientific work could be carried out in such surroundings. Clearly even at the start of the twentieth century laboratory heating was considered an unnecessary luxury just as had been the case a 100 years earlier when J. Liebig and his teacher Gay-Lussac used to wear thick clogs while they worked. Things obviously hadn't changed much and lay far behind what was standard at even the smallest university in Germany.

Incidentally, when I expressed my surprise at all this I learned that as a result of the great discoveries by the Curies a new laboratory was going to be built.

Of course in his day Liebig had held to the Paris model when he opened his laboratory in Gießen. From his correspondence with Wöhler we learn of his fear of catching a cold if he had to spend time in the unheated weighing room. However if the long cold northern winters were not to seriously disrupt work then a proper heating system had to be installed and this explains why comfortable labs for "the dirty part of physics", as chemistry was sometimes referred to, were established much sooner in Germany than in France or Britain. A major contributing factor was the introduction of gas lamps and heaters in place of the dirty coal fires. We should not forget that the great Robert Bunsen had worked for years to develop the smoke free hot gas burner in which the gas is mixed with air prior to its combustion and this design is used even to this day as the basis of all gas burners.

Progress in heating was standard in Germany, for a generation before it was adopted in Paris and then in the rest of France. This is one of the many examples of how unable the average Frenchman is to grasp that outside of Paris there may be something better. This superstition is held even by those at the highest levels and is shared by many outside France. We don't have to look far to find examples of this particular form of mental illness. It is widely spread in the German population.

I personally didn't experience much in the way of the fantastic impressions of the city that most people, under the influence of the usual propaganda, come away with. I liked best the broad area from the Tuileries garden via the Champs-Elysée all the way to the Arc de Triomphe as well as the gothic Notre Dame cathedral with its inexhaustible collection of sculptures and magnificent stained glass windows. But for the rest, the city seemed to me to be very old fashioned, like a woman who was once attractive and who still believes in the effectiveness of her aging charms. In the shop windows one could see their deference to the garish taste of their American customers. The picture postcards were vulgar in the extreme and if one found a good one then it turned out to be made in Germany (though you had to look long and hard to find this information).

The meetings of the delegation. Because of all this I knew Paris more or less when I set off soon thereafter, accompanied by my eldest daughter, for the meeting of the "Delegation". The meetings took place in a lecture theatre of the Sorbonne University where during the semester Couturat held his lectures. So as to be close by I put up at a small old hotel on the Quai Voltaire. Later I came across in Richard Wagner's memoires that this was the hotel in which he wrote the text for the most Germanic of all his operas—"Die Meistersinger von Nuremberg".

Twelve to fifteen representatives of the groups taking part in the "delegation" had turned up. The chairman was Professor Louis Couturat who together with his colleague the mathematician Léau had written the excellent history of synthetic languages which was now the basis for our discussions. Professor Léau was present for a short time at one of the first meetings but later took no further part in the matter.

Couturat was a little younger than me, tall and would have appeared slim had it not been for his bent over posture typical of one who sits all day at a desk. His hair was blond, his eyes bright and he had a pale indoors complexion that gave him a rather un-French appearance. He was married and his wife was kind to my daughter when she occasionally visited. Their household was middle class and my daughter sometimes shocked the professor's wife, who it seemed had just recently left a cloister school, with the openness of her opinions.

I made a very valuable acquaintanceship with the Danish philologist Professor Jespersen. He was at the time about the only "expert" who had really understood the nature of the problem that linguistics had to solve namely to construct a language for the scientific epoch in which synthetic and creative work would replace the mere registration and listing of facts. He told me that he had endured the strongest opposition from his colleagues even in his earliest works in which he had recognised the usefulness of the energetic imperative for the free recasting of languages over time, which was most clearly to seen in the development of English. He worked with lively diligence in the discussions and we are indebted to him for his expert input into a great deal of our results.

Jespersen was of middle height, with an open face, red-blond hair and a short pointed beard. He was lively and disarming in his interactions with people and his idealistic nature was clear in all aspects of his conduct. He was remarkably free of the ethical occupational disease of the academic and soon won respect and admiration from the meeting. A second philologist was Professor Baudouin de Courtenay from Petersburg. He had a strangely mixed up background because despite his French name he was of Nordic descent and looked like a member of one of those not yet culturally assimilated tribes from Russia. He made no attempt to hide his very radical political and social views—today we would call then Bolshevistic. He regarded it as a violation of general human rights that we discussed which word ending would be used to distinguish between male and female members of a group, for example homo for man and homino for woman. He said that women had in every area as much right as men and it was therefore inadmissible to distinguish them grammatically and thus assign them an inferior role. He didn't manage to convince the rest.

I watched Mr. de Beaufort closely. I've already (this chapter page 460) related how he gave up his own invention and helped bring Dr. Zamenhof's Esperanto to its first success. He was a well turned out person with a delicate complexion, white hair and beard and an obliging nature. He soon stopped coming to the meetings. Nevertheless, I'll come a little later to the special role he played in these discussions.

Gaston Moch whose name was well known in pacifist circles at the time, towered above the other attendees. He'd been a soldier and had resigned with the rank of captain of artillery when he found it impossible to resolve the clash between his conscious and his occupation. He was a thin wiry man, quick of movement who talked well. His speeches were accompanied by a lively and almost grotesque pantomime played out on his gaunt moustached face. He paid great attention to the female sex; his ideal here was the Parisian woman and with his ceaseless praise of them he bored his audience and drove them to inner resistance to his views. In our discussions he was courtly and devoid of fanaticism which, in a supporter of Esperanto as he was, is a rare thing. In this respect his opposite was Mr. Boirac, the rector of the Academy in Dijon, who saw it as his duty to fight with all available means for the interests of Esperanto. In this effort he was supported by several other members of the meeting who otherwise played no part.

The Italian mathematician Peano was special. He was tall, very thin, from his posture and dress a typical academic and he had no time for minor matters. With his pale yellow cavernous face, coal black scanty hair and beard he seemed almost as abstract as his science. He had his own agenda which consisted of a "latine sine flexione" in other words a highly simplified form of Latin. He advocated this with an unshakeable fervour because as an Italian he had the feeling that he was defending an ancient inheritance.

The meeting was opened by Couturat, who gave a brief account of what had happened so far and then asked that we vote for a managing committee. I was appointed to the office of Chairman, Jespersen was vice chairman and Couturat was secretary.

Since up till that point I didn't know any of the others personally I regarded this as a great honour which was due to the name I'd made for myself first as a chemist and then as a philosopher. The honour came however together with a heavy burden. The proceedings were in French because the French participants, who were in the majority, knew no other language and only Couturat could with difficulty speak adequate but stilted German. My knowledge of French was however limited to rather incomplete memories from my school years (Part I, Chap. 2, p. 28), supplemented by reading scientific papers in French which I had gradually become accustomed to doing. Nevertheless the occasional attempt to read French literature had shown me how one sided my knowledge of the language was. I had had a little practice in using the language for the daily routines of life during my first visit to Paris.

In addition to these obvious external difficulties there were also internal ones. The people at the meeting came from all sorts of countries and traditions and represented many different points of view which some of them were determined to force through come what may. There was going to be a hard struggle in front of us which brought with it the danger that differences in points of argumentation would quickly convert into personal animosities.

The others attest that I managed to avoid this danger although our meetings went on for 2 weeks and were rich in unexpected incidents. I'd first thought over how best to ensure that the meetings were peaceful and had come to the conclusion that we should avoid putting motions to the vote which were supported by only a small majority of those present for otherwise nearly half the members would be dissatisfied. I therefore took care that each point of dissension was discussed so long till there was nothing left to be said and the meeting, with the possible exception of one or two inveterate dogmatists, had come to an agreement. By this means I soon won the trust of the members and this made it much easier for me to settle the major differences of opinion that came to a head at the end off the meeting.

Babylon. We started off by letting those who had submitted synthetic languages in the competition, or their representatives, explain the line of thought and organisational plan which lay behind their entry. In this way we got the necessary overview of possible ways of doing things. There turned out to be only two. Either the vocabulary was formed using certain rules, or it was constructed from one or several existing languages. In either case the grammar was limited to the minimum necessary though the pronunciation of the phonemes could be altered from case to case.

We soon convinced ourselves that inventing new words had the sever disadvantage that it required a lot of effort of memory without having any obvious advantages. As Jespersen later put it, the future international auxiliary language should be easy to learn for the largest number of people and that therefore its vocabulary should be built of words which were as far as possible internationally known. Our conclusion was thus the same as that on which Schleyer's Volapük and Zamenhof's Esperanto were based, except that in those cases the principle had not been rigorously followed.

By this criterion the "blue language" invented and presented to us by the rich amateur Bollac[1] was excluded from further discussion. Bollac had from the beginning assured us that he would accept our decision and I am happy to be able to report that he kept his word once the decision went against him.

[1]The correct spelling is Bollack.

Ido. Once the discussion had reached this point we had a surprise. One morning, each of us found on his desk a handwritten booklet containing the complete plan for a new language that satisfied the criteria we had agreed on. Couturat told us that he had placed the booklets on the desks but that he was not the author. He assured us that he knew the author but was not permitted to reveal his identity, however it was not one of the members of the committee. He said that the author had insisted on this way of doing things because he felt that the inner value of his suggestion could be more objectively judged if the authorship was unknown.

It turned out that the booklet contained the basis for an improved version of Esperanto in which the substantial errors in the construction were removed. Of all the suggestions Esperanto had proved to be the best though it contained some major though easily remedied shortcomings which prevented its general acceptance. Now these shortcomings were being dealt with.

The Esperantists, led by Boirac, tried to mount a counter attack by interpreting the wording of the meeting's mandate to mean that we were restricted to choosing between the languages submitted to us but that we were not mandated to undertake improvements to the submissions. The view of the meeting however was that we were empowered to act autonomously and that the new project should be considered. At this our hard won unity threatened to fall apart because a group of hard core Esperantists threatened not to accept a majority decision (it was in fact almost the only one we reached for our other decisions were unanimous). G. Moch was particularly helpful in pouring oil on the troubled waters so that we could consider the anonymous submission.

The author had demonstrated his association with Esperanto by the use of the name "Ido" which as a suffix refers to a derivative form. The initial work which of course to begin with concentrated on the general questions produced a majority for the international (English) ABC system of letters without any accents or other special forms and the language was based on a simple relationship between letter and sound. Sadly Couturat saw to it that this principle was violated and that the double consonants "ch" and "sh" were accepted and assigned sounds as used in English, which is to say sounds that cannot be derived from the sounds of the individual letters alone. In vain I pointed out that if one correctly removed all sounds which were assigned to double letters (such as c = ts or x = ks) then all the necessary "zisch" sounds could be assigned to separate letters. The other way of writing was adopted because it resulted in a greater similarity to natural languages. I prophesised that later on this point would have consequences that would demand that it be discussed again and that a modification would become essential. All experience in the establishments of norms had taught me that every compromise would eventually develop into a stake in one's flesh which would be constantly painful and which sooner or later would have to be removed—usually under conditions of great difficulty. I can see the time coming when an international committee (perhaps constituted by the League of Nations) will have to take over the problem of a world language and work through it from the start and by doing so establish the principle of one letter one sound.

Conclusion. After 2 weeks in which every day we'd held two meetings lasting several hours and had agreed on the main points of the future world language (though leaving open the possibility of future corrections where necessary) we realised that we would not be able to continue the work like this. We were all exhausted and many points could only be resolved after detailed work which would better be divided up between us and done at home. We were prepared to make concessions to the Esperantists so long as these did not impinge on our basic ideas. We planned a joint action with them and they would be asked to send representatives to join us. The Esperantists on the committee agreed to this and expressed their hope that their organisation would agree to this plan. However we should give them a month to get everything organised and until then our work should not be made public. We agreed to this and kept strictly to the agreement.

A shadow. Once all this was arranged I left for home, very tired from the effort but satisfied that with the results which seemed to promise a fruitful collaboration of all those involved. Couturat was tirelessly involved in organising the next phase of the work. In order to provide regular dispatches on our work I suggested the foundation of a newspaper called "Progreso" (Progress) which was led in an excellent manner by Couturat. He did however disappoint me in one respect. I'd already developed the idea of a world format (which will be explained in greater detail in the chapter on "The Bridge") and naturally wanted that the Progreso use it. Couturat for his part had already developed a paper format of his own for his considerable body of scientific reports and this was roughly the size of normal letter format. He printed the new journal in this format and wrote me a long letter in which he undertook to prove that though his paper size was smaller than the world format, nevertheless more could be printed on it. I argued in vain that the world format was organically derived from the metric system and hence was bound sooner or later to be the accepted standard. He was not convinced by this and since the question of formats was still in its infancy I let it be.

Nevertheless this incident led to a cooling of the warm and close relationship which we'd had. This wasn't the first time that I'd noticed that alongside his international efforts, which were both assiduous and successful, he retained in his heart strong national tendencies. He hid them carefully but they often seemed to direct his actions. He tried to give his country, its language and so on, a large influence in general affairs—and that was something I could well understand. However, in addition, he not only failed to support things which were not of French origin but actively plotted against them behind the scenes. He chose the weapon of intrigue which seems to be quite common among our Western neighbours—the Paris newspapers are full of that sort of thing. I could not and would not approve of this sort of behaviour and this gradually strained our relationship.

The Esperantist war. As the month's period of grace granted the Esperantists to decide on whether to work together with us came to an end I had still received no reply from them. I wrote to Paris about this and was put off again and again without getting any concrete reply. And then suddenly there appeared in the Esperantist press a violent diatribe the essence of which was that we had attempted a traitorous attack on their holiest beliefs. They declared war with the battle cry "ni restas fidelaj" (We remain true). They declared every connection to our "delegation" to be treason and that an absolute adherence to every letter of the "Fundamento" was the only possible way forward. I turned to the men who had been part of the peaceful conference but it was in vain. War to the last, was now their slogan.

Couturat later told me that a large Parisian publishing house which printed practically all the Esperanto literature had initiated and financed this counter movement. The firm had large stocks of unsold books in Esperanto and feared that these would lose their value if the language was modified. The idea of disguising a business interest in a moral cloak is not new and the bigger the sum of money involved the more likely is it to be invoked. The explanation is by no means unlikely but I have had no chance to verify it independently and so can only fall back on what I was told.

Zamenhof. Not even a personal meeting with Zamenhof helped seal the breach. When I asked, he was good enough to break a journey home from Paris to Warsaw for a few hours in Dresden. I was waiting on the platform and met a very self effacing simply dressed old man to whom it seemed to be no sacrifice to make the long journey in a third class coach. Zamenhof was typically Jewish in appearance. He was small and thin with a pale face and spectacles and hair and beard were streaked with grey. His manner was unpretentious and disarming. However he was completely uninterested in peace talks. I had the impression that he had given a solemn promise in Paris not to agree to anything. Once he'd expressed from his side his heartfelt wish for peace I asked him how he thought that could be achieved he replied shyly that I should abandon Ido and become an Esperantist. I asked him if reform possibilities would then be open. He shook his head sadly and we parted without having achieved anything.

Schneeberger. There was nothing left for us Idists but to abandon all thought of joint work and to instead organise our own agenda. A steering committee was elected to run our general affairs and an Academy which was to deal with questions of the language. A number of Esperantists who had recognised the need to reform the language and who had been put out by the rigid opposition of the majority had joined these bodies. They were the more intelligent members of the movement. Principle amongst these was Schneeberger from Switzerland an extremely active and adroit man to whom the Ido movement has a great debt. Sadly he died in 1925.

Schneeberger was a small man with a black beard and hair who was a parson in a small village near Bern. When, on one of my many journeys, I'd arrange to meet him, he pressed me to come to his house. That I did gladly, for I respected not only his tireless unpaid efforts on behalf of his ideals but also because I liked him very much for his open ways which were free of any religious bigotry. I learnt some strange and unusual things in the manse. Outside and inside the manse looked like a farmer's house and Schneeberger made no secret of the fact that his income was small and only just enough to keep himself and his family alive. His wife, who was as simply got up as the house, took him aside to tell him a plaintive tale. He turned to me with a laugh and explained that in my honour they had slaughtered and roasted a rabbit. However, half an hour ago the cat had grabbed the roast off the table and made off with it. We'd have to wait till a replacement was ready. Shortly afterwards the woman of the house called us to table. After the custom of her country she did not eat with us but went forwards and backwards between the table and the kitchen and made sure that we ate well and lacked nothing.

Other work. The time which followed was filled with work which went in two directions. First of all Ido had to be developed out of the preliminary sketch. The original outline was changed considerably by simplifying and finding consistent forms. Couturat's expertise in formal logic and Jespersen's in linguistics were very useful here and made it possible to develop Ido to a much higher level than Esperanto whose development had been stopped in its tracks by the suicidal decision to make it unalterable. In just a few years all that was necessary had been done and all that was needed was to clear up more or less secondary problems. The language took on such form that suggestions from the Academy for its improvement became rare.

In order to combine constancy with the necessary room for improvement we reached the sensible decision that the language would be changed only once every 5 years irrespective of the continuing work of the Academy. At the beginning of each new period the collected decisions of the Academy would be brought into practice and the improved language then held constant for another 5 years. In this way we were able to meet the disparate demands of constancy on the one hand and ease of development on the other.

The second side of things—the dissemination of the new language—was not so easily achieved. Ido was soon attacked by the Esperantists and I had to conclude that their methods were not always fair.

The following incident was comparatively harmless. I met on my regular visit to the spa in Karlsbad the remarkable Austrian socialist Schulmann Glöckel whom I'd tried once or twice to interest in the question of a world language. When he asked what I thought the best solution was I'd told him about Ido. When I met him later he told me that he was more for Esperanto because the Esperantists had told him that in Ido his name would be "Closeto" and he didn't like that much. In fact Glöckel (a German diminutive for bell,) in Ido would have been "Closheto" and in any case names were not translated. From this you can see that the Esperantists were not choosey in their propaganda against Ido.

Ido in chemistry. Once the main linguistic arguments had been settled it was time to turn to more general matters. It seemed to me that simply collecting followers of Ido would not be sufficient because each individual would have little benefit from all this apart from the feeling that he'd learnt a useful language. The most that could be expected was a correspondence with people in other countries or interacting with Idists when journeying overseas. That wasn't going to be enough to form a strong movement.

I thought over what could be done to provide a practical use for a general auxiliary language and came up with the following idea. As I've mentioned several times already there was a huge language-based waste of energy in chemistry. To summarise everything which happened in chemistry in the course of a year requires one to two thousand closely spaced pages. In Germany this work is done not once but around five times for different journals and scientific societies. In English the work is summarised at least three times; twice in Britain and once in America. Then there are the French, Italian, Russian and so on summaries—or in other words around a dozen summaries of the same material.

This work could be simplified by a factor of ten or more if the summaries were not written in the national languages but rather that each country summarised its own contribution in a general auxiliary language. In this way not only would the work of making the abstracts easier but in addition the entire work would have a market which was twenty times as large and hence the costs would be considerably reduced and everybody would have access.

To test this idea in a practical way I took the index of a chemistry textbook and translated all the entries into Ido. There were only a few minor difficulties. Practically all of the names could be easily translated into Ido, so that every chemist would understand them on a first read through.

On top of that I made use of the fact that it was normal practice to provide a summary of the main findings at the end of a publication. I suggested making this summary also available in Ido so that everybody, even those who didn't understand the language of the main text, would understand enough to know whether it was worthwhile going to the effort of reading the full text. As an experiment I prepared a number of such summaries and could show that even a novice could readily learn to understand them. After all, the vocabulary used in any given branch of science is relatively small.

All these simple and necessary suggestions remain unrealised till today although they were well on the way to implementation. This was done via the International Society of all Chemists in whose establishment I was heavily involved. This effort, like so many other things, was destroyed by the world war. I'll say more about this at the appropriate place.

The world language administration. I tried another way to bring the world language into real use and this arose from the great disadvantage which stemmed from the battle between Ido and Esperanto. I looked for a situation that was free of this struggle by instituting in Bern in 1911 a movement that was dedicated to supporting the idea of a world language but which was not settled on any given solution. There was at that time a strong predisposition towards the international organisation of many technical and economic matters and some of these like the international post or railway systems and others had achieved considerable success. The administrative centres of such organisations were usually either in Switzerland or in Belgium. It wasn't that difficult to excite interest in Bern as the site of a future world language administration and to get some of the preliminary work done. A number of energetic and influential men came together and the tireless pastor Schneeberger took on the job of secretary. I held a well attended lecture and an association for the founding of a world language administration was registered with broad support. However this support was not unanimous for soon a number of Esperantists turned up and demanded that the association accept their language as the only possible candidate right from the start. They did not convince us and so they declared that they would fight against our association.

Our plan was to petition the Swiss government to make an official request to other states to found a working group to study the question of the auxiliary language and at the same time we wanted to approach a number of other governments with similar petitions. We needed a certain amount of money to get all this started and I was happy to supply it.

Efforts of this sort are always rather slow. The government officials responsible for considering this matter (supposing they were indeed prepared to consider it) were even less informed about the background than they were about any other minor questions of culture. We were condemned to patience and by the time our annual meeting came round there had been very little progress. Our association, like many other things, was then destroyed by the world war.

Recently in the League of Nations it has once again become clear that some means must be found to resolve the language muddle so that representatives of the many different language groups can understand one another. However none of those responsible seems to realise that the problem has already been solved and therefore we face the absurd spectacle that the train is ready, the locomotive warmed up, the driver ready to go, and the carriage doors are open, so that the passengers only have to climb in and they can start on their journey.

But they don't want to do this because the idea of a railway is still too novel.

The inertial law. I've often wondered where this instinctive dislike of everything new comes from. Everyone who tries to bring humanity on is faced with it and the conquest of this inertia requires ridiculously more energy than does the invention itself. Since the phenomenon is met with constantly, as if it was a law of nature, it must have a biological basis. Finally I believe that the following explanation is correct.

We evolved from lower forms which all differ from us in one special characteristic which is slowly being lost from humanity alone. This is the property of invariance. For many thousands of years the bear, the deer, the beetle have retained their structure, their way of life and their behaviour in unaltered form and one would have to go back for a very long time to see changes in these species. This sort of biological inertia is a fundamental legacy which we humans too have acquired from our ancestors.

Humans however have more recently acquired the capacity to develop new characteristics. To begin with this was, in all probability, just a small change from other species. However, this characteristic is associated with a feed forward loop so that the rate of development is much faster in individuals who acquired better genetic material at the start from their parents and hence, over time, this difference has become ever greater. Today it has achieved such proportions that the difference between two succeeding generations may be so great that they are no longer able to understand each other.

Nevertheless the original genetic propensity to inertia remains a basic part of our organisation and keeps on making itself evident despite the increase in development at higher levels which, for the vast mass of people, is in any case not very high. However, in those who are bringing humanity forward development has reached the highest level that can be achieved today. Because of this, the masses respond to the commotion caused by the demands of a new discovery not with praise for the effort involved in making it, but with inertia. The greater the consequences of the discovery, the greater the inertia. And of course it follows that this inertial resistance is as automatic and as strong as any other basic instinct.

The only way to force progress through is by constant patient repetition of the advantages of the new discovery. In this way the new idea slowly becomes accepted until it becomes regarded as an old idea and, in line with the basic biological rules of memory the oftener it has been repeated the easier it will be accepted.

Overcoming this inertia is a completely different form of work than the formulation of a new idea and this explains why it often can't be done by the inventor who has exhausted his energy in making the discovery. But everybody who is aware that progress must be made, has to accept the ethical demand to do his bit towards making sure that this period of mankind's acclimatisation to a new idea is kept as short as possible. Everyone has the possibility to do this in his own circle.

Once one has produced a scientific account of the natural imperative which drives the mental law of inertia then one is more ready to be a little more patient. Without this law humanity would be a prey to every whim of fashion. One can see hints of this in certain social circles in large cities where the antics of the followers of fashion do not contribute to culture or help their fellow men. A degree of mental inertia (in the physical sense of resistance to motion) is both desirable and necessary. The only question which remains is: How much? And there is no sense in getting morally outraged about mere quantitative differences.

Chapter 36
Festive Days

The transition. As if events had just waited for the chance, a whole slew of welcome new activities associated with the support of cultural advance crowded in on me once I'd resigned my professorship. The seeds of these activities had been planted prior to my resignation but at that time I'd been unable to do much about them. Now, as I was free, all that was changed.

I should say right away that the matters referred to in the previous chapter were by no means all and I'll tell more about the others in due course. In any case I now found my days in Großbothen just as well filled with enjoyable work as had been the days in Dorpat, Riga and Leipzig.

But my days were not only filled with work. True to Goethe's motto, "unpleasant weeks, happy feasts" the work in Großbothen (which I never found unpleasant) was often enough interrupted with happy occasions. I agree with Goethe that it is in the combination of the two that true happiness is to be found and so the "happy feasts" must also be mentioned in the account of my life.

The theory of honour and reputation. I'm not going to go into all the usual medals and titles that I got in the course of my scientific career nor those awards and decorations for my efforts which were hung around my neck or fixed in my buttonhole after retirement. These things are never really related to the value of the effort put in by the recipient as I found out myself when my all too free interpretation of my civic rights as a professor in Leipzig led to a clear delay in my receipt of such things. It is however a rather different matter with scientific honours. It is of course the case also here that who you know—and luck—play their part, but nevertheless they are awarded to the largest extent on the basis of the assessment of experts in the field and hence represent an objective value.

It is usual to damn one's pleasure in these things as mere conceitedness and some excellent researchers are allergic to being given public thanks and praise and avoid it at all costs. There is something like fear of praise just as there is fear of being enclosed in a small space. However, for most people this sort of thing is pleasurable and I certainly belong in this category.

R.S. Jack and F. Scholz (eds.), *Wilhelm Ostwald*, Springer Biographies,
DOI 10.1007/978-3-319-46955-3_36

When one thinks about it a little more then one will see that this pleasure is well founded. We are dealing here with a natural law that one has to understand in order to approve it. The fulfilment of every necessity of life must lead to feelings of happiness in every being for otherwise life would come to an end. An animal species that abhorred eating would of necessity die out. Reproductive matters, because they frequently involve great sacrifices, are associated with very strong feelings of happiness and this is necessary to preserve living beings.

Joy in work belongs in the same category and we can follow its development. The first book of Moses describes the curse place on Adam when he was driven out of the idleness of Paradise as "You shall earn your bread by the sweat of your brow". Here work is regarded as the most awful punishment that could be visited on the sins of humanity. For me, as for millions of the best, the greatest curse would be if work, and the happiness that it brings us, were to be banished from our lives. Even today we can observe wild and barbarous peoples who remain locked in the biblical view of work and the underdeveloped members of European culture have also not raised themselves much above this level. But the higher the individual is developed spiritually and socially the greater he is able to experience the joy of work. One really can't avoid the question of how people in the future will achieve this joy as technical progress increasingly reduces the work available to be done. In all probability science will help here because its challenges are limitless. And for the others there will remain games (in the widest sense of the word).

Social recognition by one's peers, or honour in the broadest sense, are essentials of life which bring happiness with them. When one observes the exhausting life of a society lady who views her social duties as imperatives to which she sacrifices without a moment's hesitation her peace of mind, a great deal of her joy in motherhood and marriage and yet more besides, then one must ask what mighty force brings forth this quite common effect. I can think of nothing other than the desire for social recognition. The precise way in which this desire is expressed depends of course on the intellectual and ethical level of the individual and also on her social environment. In any case it is without doubt one of the strongest drives experienced by people in our day and age.

That social recognition or honour is a necessity of life doesn't need to be discussed in detail. Our existence is based on the division of labour and the first prerequisite for the life and prosperity of an individual in that complex organism called society is that he finds a space amongst the other individuals where he will not be crushed but will at least be treated as well as all the others—and preferably be given a privileged position. It then becomes inevitable that this necessity brings with it feelings of happiness no matter how odd the means of achieving it may seem. After all, work, love and hunger, which are the other great sources of joy in life, are by no means free of their oddities. Especially love. From the standpoint of developmental history the desire for the happiness that honour brings with it is, within the trilogy of hunger, love and honour, the last and hence the most superficial. At the lowest level is the happiness which comes from satisfying one's hunger because this is something which relates entirely to the individual. Love is a step higher because it has to do with the survival of the species. However honour has to do with more than just the individual and his progeny because it is an element in the social cohesion of many individuals and families. Indeed honours are found as soon as groups are built so that they are present even in the most primitive peoples.

Work must be regarded as a branch of culture which developed wherever people found themselves far from those tropical islands where all one's needs are satisfied without work. Thus the need to work, and consequently the joy that work brings with it, increases with the geographical latitude. Those who live in the tropics feel this joy the least and even in relatively small areas one can see that those who live in the north work harder than those in the south, as one can easily see by comparing north Germans with southern Germans, people from north Italy with those from the south or Scotsmen with Englishmen.

I don't intend to deprive myself here of the pleasure of looking back, as I have done quite frequently before, over the honours done me by my peers. Perhaps this theoretical introduction and the following examples will help one or other of my colleagues to overcome their inner reluctance to surrender to happiness just because he'd been put off by the occasional false decision.

Naturally such a general and deeply rooted matter is of practical value in many ways. Just as an example: whenever I worry that in writing this third volume I am expressing very dry matters in a boring way, then I flip through the collection of friendly reviews of the first two volumes—which I thank my publisher for having collected—and each time I feel such exhilaration that this transmits itself to my writing and the sun lights up and illuminates the content once more.

On Liebig's trail. An Institute of Physical Chemistry was opened in the new university in Liverpool and one of my former pupils was appointed as its director.[1] I was invited to the opening and, to make the journey more tempting (which was scarcely necessary) I was told that the university wished to award me an honorary doctorate.

A rich industrialist called Muspratt[2] had financed the building of the institute and he offered to put me up in his house. I gladly accepted both invitations.

Muspratt is best known in Germany for a detailed textbook of technical chemistry which first appeared around 1854. It had been written in English and described almost entirely British methods. However since industrial chemistry was at that time only established in Britain this had been enough. A German translation was prepared by F. Stohmann who later became my colleague in Leipzig and he extended it to a compendium which became the basic text used for the education of

[1]G.F. Donnan. It is surprising that Ostwald does not give his name here.

[2]James Sheridan Muspratt.

all the older chemists. Muspratt had been a student in Liebig's laboratory in Giessen after which he founded a chemistry school in Liverpool and at the same time developed a successful process for the manufacture of sulphuric acid, soda ash and bleaching powder.

When Liebig began to utilise his results on the inorganic nutrients of plants for the preparation of a mineral fertiliser (in Holland they called it in their odd language "synthetic dung"). Muspratt offered to manufacture it in large amounts and during the course of this collaboration Liebig had stayed with him several times. Muspratt had died in 1871 and the new institute had been financed by his son who, in his impressive house outside of Liverpool, reverently looked after the mementoes of his father's great friend.[3] He seemed to want to develop a similar relationship to me and treated me with all possible friendship.

He was a thin rather sickly looking man whose complexion and behaviour seemed more that of the academic than the industrialist. He was not interested in any personal luxury. He was another example of that sort of good rich man who is peeved that he is rich, and so he was permanently busy so far as his health permitted. He didn't seem to be like any of the other Englishmen I knew—the Scotsman Ramsey was perhaps the closest. I liked him from the start and visited him on later trips to the island. He was perhaps 10 years older than me. His wife found it hard to move but she made a most friendly impression.

The doctoral ceremony had been set for the founding date of the university and it began in a festive mood. Apart from me several others were to be honoured in the same way. The ceremony was very similar to what I'd experienced in Cambridge. Once again there were lots of students in the hall's gallery but while in Cambridge they had been silent, here they joined in and loudly commented on the ceremony unfolding below them. I was told that this was an old student privilege and there was intense competition to come up with the wittiest comment. I was obviously not well enough known to them because they restricted themselves to clapping loudly as I approached the rector.

From the many people I met there I remember best Colin Roß[4] who did research on malaria. He told me that the blond northern races are much more susceptible to the infection than the dark haired peoples of the south and gave this as an explanation for the fall of both Rome and Greece in ancient times. Their intellectual leaders had been blond long-headed and, rather like the Normans, they had invaded from the north and subjugated the dark haired inhabitants. However in the course of time they had been decimated by malaria and eventually there remained only the progeny of the original dark people together with other dark immigrants who were unable to maintain the level of culture which had been achieved.

To give me a view of the English chemical industry I was taken to a factory site near by at Widnes which was unimaginably black and dirty and, to make it worse, it

---

[3]Ostwald is mistaken. The Muspratt Laboratory for Physical Chemistry was founded in 1906 by Edmund Knowles Muspratt, the brother of James Sheridan Muspratt.

[4]Correct form is Ronald Ross.

was a dark, wet, foggy day. The works in which soda ash was produced by the Leblanc process seemed strangely familiar to me and eventually I recognised that this was the works pictured in an illustration in Roscoe's chemistry text book which I'd read many years previously. The buildings were just the same which meant that the factory had not changed significantly in the last 20 years. When I asked I was told that the old works were now not actually economically competitive with the Solvay process, but that the old factory was kept going for business reasons. I thought that this was a remarkable example of the difference between progressive Germany and conservative Britain.

Aberdeen. A similar occasion brought me to the University of Aberdeen in the north of Scotland. Here the celebration was of the real founding of the university 50 years previously by the fusion of two till then independent colleges. Once again a number of honorary doctorates were to be awarded. For me they'd decided for a change to make me a doctor of law rather than a doctor of natural science. Representatives from the sister universities on the continent had been invited and I was amused at the curious costumes straight out of the middle ages which the representatives of the Sorbonne (Paris) were got up in.

The following story from the history of the university was very instructive. When the two colleges had been fused a number of professorships had to be dispensed with to avoid duplication. One of those who were considered expendable was the young J.C. Maxwell who was soon to emerge as one of the greatest physicists in Britain.

To made the celebration even more glorious King Edward VII had agreed to attend. To welcome him a ceremonial gate had been constructed and the street was richly decorated and a placard shone forth with the message "Welcome the king and queen" in huge letters. At the last moment someone noticed that it had been hung on the railings of a small graveyard.

To see the royal couple the guests had collected in a large square which had a dais at one side. Their majesties appeared and a much crippled man was carried towards them. He had sacrificed his limbs to save a number of people and the King personally pinned a medal of honour on him. There wasn't any sort of ceremony for the university, at least none to which we were invited. Once again one could see a stark contrast to Germany: there science was not of any great importance to influential men in the government while here it is highly held—as it should be.

A trip to a beautiful estate nearby showed us life in Scotland. There were all sorts of mementoes of Sir Walter Scott and we watched a sword dance performed by people wearing ancient costume to music from bagpipes. We'd already had a chance to make acquaintance with this national instrument at a banquet and had learnt that its player is readily able to produce such a loud noise with it that it could be heard through the alarums of the fiercest battle—at least until the invention of gunpowder.

On this occasion I met a number of representatives of the University of Toronto in Canada. They presented me with the doctoral certificate of their institute together with a silk hood in their colours and told me that as a special exception my earlier visit to them (Part II, Chap. 29, p. 366) would be considered as the equivalent of a personal receipt of the degree. Since it was in principle an acknowledgement of the successful establishment of physical chemistry at their university by two of my former students, I gladly accepted the honour with thanks.

Geneva. Another honorary doctorate took me to Geneva in 1909, the year of the 350th anniversary of the university. In this case my friend Ph. A. Guye was the moving force. I'd never seen the city and its beautiful lake and so I went there gladly. In any case there was someone there with whom I'd corresponded but had, so far, never met. During my research about great men I'd come across an important predecessor Alphonse de Candolle from Geneva. I'd translated his book into German and brought it out as the second volume of the series "Great Men" which in the meantime has grown by several further volumes. For permission to use the book I'd turned to his son Casimir de Candolle who looked out for the traditions and the scientific inheritance of this family which had produced two excellent botanists. He readily agreed to my request and refused to accept any remuneration.

When I visited him in his apartment I met a rather old, thin little man with a pale face and hair who, though he seemed to be rather sickly was nevertheless lively and very friendly. The fourth generation was represented by his son who closely resembled him. They lived in the old family house on the Petersplatz and showed me a window which, against all the rules of architectural symmetry one of their ancestors had installed so that he could immediately see when the citizens of Geneva were once again starting a revolution.

The whole house served the needs of the herbarium with its ancillary library which Augustin-Pyramus de Candolle had founded and each of the following owners had seen it as his duty to maintain as a centre of descriptive botany. The collection was open to every researcher who needed it for his work. I gave them my sincerest thanks for the information and left this living witness to the history of science.

Of course I also visited Guye and his laboratory which was really quite unique. There he carried out determinations of atomic weight by weighing gases which was an extremely difficult procedure that could be carried out in this scale nowhere else but which in Guyes hands yielded very good results.

I don't remember anything about the award ceremony itself which probably was no more than a solemn announcement of the recipient's name during the ceremony.

On the beautiful lakeside I saw close beside one another the Rousseau island which honours that vastly talented but morbid man and the pompous monument to the Duke of Brunswick[5] that evil bounder who, on the condition that they erect a monument to him at the prettiest spot in town, made over to Geneva the money he'd got by selling his subjects. The city agreed and inherited the millions.

I was happy to be able to thank them for the honour done me by holding at their request a lecture at the university. I chose as my theme the problem of great men

[5]This was Charles II (1804–1873) who was most unpopular in his country because of the way he treated his subjects.

and without flattering the citizens I can say that the subject interested them because many of them had bought the article at the time it was originally published. Of course I didn't mention this in the lecture.

Stockholm. The highest scientific honour that I received was the Nobel Prize. It's the highest there is because the number is limited in each speciality to just one each year and sometimes even less because the prize is not always awarded, as has happened several times in recent years.

As is well known, Alfred Nobel left his considerable fortune to the Foundation which bears his name so that it could be used to support physical, chemical and medical science. A fourth prize was instituted for those who aided the cause of peace and a fifth for literature. Because of legal problems the arrangement could only be carried out to about half the expected extent. Nevertheless the part that remained for the Foundation was so large that after the administrative and other costs have been taken care of there remains around 140,000 Marks for each winner.

The choice of prize winners is carried out with enormous care. Prize winners may be nominated by previous Nobel laureates, by the members of the Swedish Academies or by a number of highly regarded institutions and outstanding people who are invited to suggest candidates. The suggestions are then sifted by a permanent committee which passes on their recommendations to the Stockholm Academy of Science, the Academy of Literature and the Norwegian parliament who together are responsible for the final choice. The prize giving ceremony presided over by the King of Sweden takes place every year on the 10th of December, the anniversary of A. Nobel's death.

I was awarded the prize in 1909 for my research on catalysis (Part II, Chap. 24, p. 284). In a discussion of the conditions which determine the rapidity of scientific success I had pointed out that there is often a great disparity between the real value of a discovery and the way it is generally received. Of course it is clear that the more a discovery is ahead of its time the longer it will take for it to be appreciated. Then there is the question of whether we are dealing with an old or a new problem. If the discovery has to do with a well known problem then it will be readily accepted—but not otherwise. And finally much depends on the public media. Whether the inventor belongs to a circle which is able to manipulate the press will make a huge difference to the reception of the discovery. On the one hand, even a very abstract and hard to follow discovery can be made popular with the appropriate slogan while, on the other, a discovery which runs against what is generally considered to be "obvious" will run into serious difficulties. In Germany there is in addition a belief in the idea that everything from overseas is better and our recent terrible experiences have only made that worse. The most ridiculous "scientific" news from overseas is widely read and swallowed at face value while the formation of a press office reporting on advances made in Germany seems not to be considered a paying proposition.

But even within the scientific community one can see trends that owe more to emotion than to rationality. I often think back on the contempt that my old teachers Schmidt and Lemberg had for those who wrote books, particularly when I consider how much more difficult it is to come up with a conceptual advance than with a concrete experimental result. Part of this is due to the fact that the experimental result is so much easier to judge, for it can quickly be established whether something new has some unexpected property as the discoverer claims. It is much more difficult to determine whether a new concept is going to lead to new advances, which would be its only claim to be taken seriously. That is something which takes much more time.

These are the thoughts that crowd in on me when I consider the case of catalysis whose definition in conceptual terms (Part II, Chap. 24, p. 284) was my most personal and momentous contribution to chemistry. I've already related how it was only after the recognition that catalysis consists of the increase of the rate of a chemical reaction by a substance which is not a reactant, in the sense that it is not part of the product, that the scientific study of catalysis became real and possible. A huge number of catalytic reactions were known at the time because one stumbles across them continuously in chemistry. Some, such as the standard synthesis of sulphuric acid by nitrogen dioxide, were already in use in industrial reactions though nobody knew how they worked. The chemical reactions of life are everywhere regulated by enzymes which are catalysts: this was just one of the many secrets of life. Only once one realised that catalysts merely increased the rate of reactions which were indeed possible was it feasible to move on to their scientific analysis. Today both technical and physiological chemistry work all the time with this concept and the results have been an enormous increase in our understanding and in what we can now do.

But just as people are usually not able to deal with concepts—they prefer opinions—so only a few are able to grasp the practical successes of a concept even when they are held before their eyes.

For example a few years ago the German Chemical Society held a lecture series on catalysis without even considering the conceptual background of it all.

In the light of all this one can well imagine how happy I was in 1909 to hear from Stockholm that I'd been awarded the Nobel Prize for Chemistry for my work on catalysis.

The official plan was that the award should only be made public on the 10th of December. Since the prize winners had to get ready for the journey to Stockholm they were informed about 4 weeks in advance, but were sworn to secrecy and this was usually observed, though as the trip needed a fair amount of organisation it was not easy to keep the secret. In those days however one did one's best and so the newspapers which were competing to be first with the news had access only to rather doubtful sources and sometimes came up with the wrong names. In the meantime it has turned out to be better to let the real names leak out so as to spare people the embarrassment of being wrongly identified as prize winners. After all, those who might be thought worthy are all distinguished people. In the 4 weeks before setting out for Stockholm I had to apply a certain amount of sleight of hand to keep the secret without having to tell a downright lie.

Finally the day to leave came; my wife accompanied me. In Stockholm I was able to meet the many friends I'd made there on my numerous visits to this beautiful city since 1885. We stayed with my friend Arrhenius who was now director of a newly built Nobel institute for physical chemistry and was happily married. The institute was in beautiful countryside outside the city and consisted of a well designed house next to the laboratory where a few of the scientists lived. He had no teaching duties or any other duties that might have interfered with his scientific work so that everything was there that one could wish for. Things haven't changed there to this day.

The other prize winners were Marconi and F. Braun (Part I, Chap. 13, p. 137) who together shared the physics prize for their work on radio communication. Just at this moment good music is pouring from the loudspeaker into my writing and I reflect on the fact that this enrichment of my rural life is due to the efforts of those men who established the scientific and practical aspects of radio. The shorter the time my failing strength permits me to work each day the more I appreciate the chance the radio gives me to fill the empty hours without feeling bored.

The prize for medicine went to the Swiss Kocher for his fundamental work on goitre and Graves disease. I was happy to learn from his lecture that here it was a case of competition between two catalysts which regulate the thyroid in that one accelerates and the other slows down its activity and that the balance between them ensures that the proper level is achieved. In the meantime it has turned out that this is a general principle by which the organism regulates a large part of its physiological processes.

The literature prize went to the Swedish poet Selma Lagerlöf. As usual German science proved here to be the strongest.

The ceremony began with the presentation of the certificates and medals in the presence of royalty and of the intellectual elite of Stockholm. Each prize winner was introduced by a member of the Nobel committee who unfortunately spoke Swedish so that the prize winner sits as if with a full glass from which he cannot drink. However at the end he is given a short summary in his own language after which he receives the certificate and medal from the hands of the king.

In the evening there is the banquet at which each prize winner is once again praised and at which he may give a short speech of thanks. In the days that follow each of the prize winners has to give a lecture which is centred on the work for which he was awarded the prize. In between there is a banquet given by the king who talks with the prize winners. The court's ceremonial is free of formality since the Swedish throne is anointed with more than a drop of democratic oil.

All in all these were amongst the most pleasant days that I have ever experienced.

Munich. The annual meeting of the board of managers of the German Museum in Munich is always something to look forward to. They gave me a chance to meet regularly with the best organisational and creative minds in Germany in the context of a cheerful meeting at the highest artistic level. This made it easy to get to know the others, who were all at pains to be at their best. Since this was also an important cultural matter it deserves to be described in some detail.

The Electrochemical Society held its momentous meeting in Munich in 1897, which I described earlier (Part II, Chap. 23, p. 273), and there I got to know my chemical colleague Wilhelm von Miller from the Munich Technical University who welcomed me with open arms. Influenced by his brother Oskar von Miller, who was one of the leaders of the blossoming field of electrical technology, he'd tried for some years to come to grips with electrochemistry but he found the field to be in such a chaotic state that he had great difficulties. For this reason he felt greatly indebted to me for organising the field and its latest advances in a rational way in the second edition of my textbook, and he made this very clear.

Oskar von Miller. It was thanks to him that his brother Oskar organised the meeting and helped it to its great success. As a result I started a long standing relationship with this excellent man which I consider to be one of the best things in my life.

The two von Miller brothers were members of an old and honoured Munich family. Their father was the famous smelter Ferdinand von Miller. When I got to know Oskar he'd already had a brilliant career in electro-technology and as an organiser and he felt, as he later told me, a growing need to find something more fulfilling than a technical job hemmed in by economic factors. This was what brought him to his crowning triumph—the German Museum. He talked it over with me and I responded warmly so that he soon asked me for my cooperation in deciding how to convert the thought into reality.

To me the idea of being able in this way to develop the rudimentary technical thinking of Germans, particularly of German youth, was very attractive. The situation in Leipzig had shown me what an unexpectedly large effect scholasticism—learning just for learning's sake without any consideration of the value of the material learned—now had in our universities and it wasn't hard to see in this scholasticism the source of much of the backwardness in our cultural life. On the other hand I had seen in the case of my sons the remarkable educational value of direct practical science. Because of this I eagerly accepted the chance to join this enterprise whose enormous future dimensions were recognised then by no one except for its initiator.

When I got to know him Oskar von Miller was about 50 (2 years younger than me). He was somewhat over middle height, burly, with a red face surrounded by black curly hair and beard, with thick black eyebrows and protruding black eyes. His face seemed swollen and gave the impression of a boiler under steam and this underscored the man's enormous energy without which he would not have been able to do all that he had done. Nevertheless, his strong Munich sense of humour ruled out any impression of fanaticism that the strongly developed blood vessels in the whites of his eyes might have given rise to. He lived entirely for the moment whether it was a moment of work or of relaxation. He made a joke of my habit of reflection, "When Ostwald wants to drink a beer he first has to work out the relationship of the beer to the organisation of the universe".

The annual meeting. I am indebted to the German Museum and to its founder for much personal joy and support. Each year there is a meeting of the managing committee together with the museum's supporters and friends at which the business and social aspects are well mixed. At the beginning the circle was small but excellent, for Karl Linde who had invented the procedure to liquefy air and the mathematician Dyck were on the committee. On top of that there was the close relationship of Miller to the Bavarian royal family so that Prince Ludwig, the son of the Prince Regent Luitpold, was a regular attendee and he was a tireless supporter of the museum and he clearly foresaw its significance for Munich, Bavaria and Germany. Even when, after the death of his father he became King, he kept up his interest in the museum.

With his quite unusual organisational skill Miller was able to recruit every creative element in the country into the service of the museum so that it was soon a great honour to be invited to hold the scientific lecture at the annual meeting. I myself had unfortunately to decline the invitation, which Miller offered me, to hold one of the first lectures because the date clashed with a journey to America and I had to return the invitation which I had already accepted. Van't Hoff held a lecture in my place but it was rather disappointing. He had seen similar institutions, though organised in rather different ways, in Leiden, London and Paris and what he'd seen had not encouraged him to believe that the German effort would be any more effective. He told them this straight out. In this case he had underestimated the personal aspect which allowed the German museum to be much more successful.

Through the association of almost all the country's leading men, and in the absence of the usual scholastic ballast, which so lowers the value of other meetings, the Munich annual meetings became for me a real high point which I eagerly awaited and then looked back on with pleasure. There was no other opportunity to meet and interact with all the bright people in Germany in such a relaxed atmosphere.

Graf Zeppelin. One of the most interesting meetings I had there was with Graf Zeppelin whom I'd met once before in what was for him a very difficult period (Part II, Chap. 25, p. 303). Miller had managed not only to bring together the leading airship constructors, in particular Zeppelin and Parseval whose relationship was strained, at some of the annual meetings but he'd also smoothed out the friction between them so that to the great satisfaction of the other guests they publically conversed in a friendly way. After that they were both besieged by young women who all wanted an autographed postcard and who felt blessed at being the proud owners of this symbol of peace.

The social part of the meeting ended with an evening reception by Prince—later King—Ludwig and from there a few of us went on to the artists' bar "Allotria" to end the celebration in good Munich style with beer. By chance I was sitting next to Graf Zeppelin and asked him to tell me why he'd exchanged his senior position in the military for the hard life of a German entrepreneur. We should remember that at the beginning of the war of 1870/71 he'd been a young cavalry lieutenant and through his clever and bold behaviour had managed with just a few men to hold up the French vanguard in the Palatinate long enough for the German Army to move up. He was awarded the well earned medal and soon advanced to the rank of

General and then, while still relatively young, he resigned his commission. I added in excuse for my question that it was not just empty inquisitiveness, but that I was involved in the scientific study of the careers of great men so that his authentic account would be of great service to me.

Graf Zeppelin turned red and refused to be considered in this category. When I asked him to leave that to me because I had developed a particular method to define the term he relented but insisted that I should agree not to publish his comments till after his death. I accepted these conditions and have strictly observed them. What he told me was as follows:

He had developed new ideas for the training of the cavalry which the General Staff had permitted him to introduce in his part of the army. However his plans were several times disrupted by the unannounced interference of the Kaiser whose love of pomp and splendour tended to interfere with the military side of manoeuvres. He considered this to be such a grave disregard of the seriousness of his assignment that he let it develop into a conflict which ended with his forced resignation.

He then had to look around for a fresh challenge and returned to the interest he'd had in his youth of developing a steerable air ship. To my interposed question as to how he'd managed, given the usual education of German youth, to come to such an idea he replied that luckily he had never had to go to a secondary school specialising in the classics. He'd been largely educated at home and then, after passing the school leaving certificate at a technical secondary school, had spent a few semesters at the Technical University in—if I remember correctly—Stuttgart before joining the army. During his education various technical challenges had fascinated him and one of these was the idea of a steerable airship. And it was to this idea of his youth that he returned. At the end of it all he ought to have been almost grateful to the Kaiser for having forced him into this new occupation.

As is well known Graf Zeppelin made his first important overland flight from Friedrichshafen to Berlin. During the course of the flight lots of news about it flowed into Berlin via the telegraph and the Kaiser went with a large following to the expected landing place so as to be able to greet the Graf. He, however, landed at Bitterfeld about one or two hours distant from Berlin and then immediately returned home. The stories that the press printed about the technical necessity of this premature landing seemed pretty weak, but the matter was swept under the carpet.

On that remarkable day I'd by chance been in Weimar and had been sitting outside thinking about Goethe. An unusual noise surprised me and I looked up and saw for the first time in my life a steerable airship. Goethe's immortal lines sprang immediately into my mind:

Everybody is born with the ability
To let their feelings soar
When, lost in the endless blue above us,
The lark sings,
When over the craggy pine covered hills
The eagle soars
And over meadows and lakes
The cranes seek their home.

At this point he had remarked in resignation: These wings of the spirit will not so easily be joined by any real physical wings—and now, after a hundred years I saw with my own eyes the poet's dream come true. But it was not the artist that had realised it—it was a triumph of technology.

Towering figures. I'll now mention just a few of the other important men that I got to know at these meetings.

I liked Rudolph Diesel, the inventor of the extremely important engine named after him. He was a graceful and elegant man of less than average height with a pleasant lively face, a small moustache and short black hair. In conversation he was cheerful and confiding. What was particularly interesting about him to me was that he had very deliberately worked his way towards his invention. The idea came to him during his studies at the Technical University where he'd been fascinated by the second law of thermodynamics from which he'd concluded that the efficiency of a motor would be higher the higher the temperature difference in the system was. From this point on he decided to design a motor in which the combustion chamber temperature would be as high as possible. Afterwards, once he was able to build an engine to his plan, h'd had for a long time to drive it with an auxiliary motor because the combustion didn't work properly and the motor was unable to do any work. After a lot of trouble and numerous changes, which challenged his patience to the limits, one of the workers noticed that the belt which connected the both motors was no longer taut on the upper side which carried work from the auxiliary motor to the "diesel" but rather on the other side: this showed that the diesel was finally running and was doing work. Cap in hand the worker showed the inventor the evidence for the realisation of his dream. After that, the sorting out of the remaining problems was a comparatively simple matter.

Diesel was interested in many social questions and we talked about many things. I liked him very much and his strange death in the channel while on a journey to England struck me personally as a great loss.

Emil von Behring, the discoverer of the diphtheria serum was a completely different sort of personality. Through his discovery he became rich and famous overnight and to begin with this had led him to somewhat lose his self-control. When I first got to know him he told me that all that was behind him. What now interested him were new fundamental ideas in the fields of biology and medicine which he had so far not achieved. He seemed to believe that I could help him here and he sent me his latest publications and at our next meeting talked to me about their contents. I'm afraid I wasn't able to give him much in the way of help because I never really was able to work out the crux of the matter.

He made the impression of a man who had survived some huge effort and had never quite recovered from it. He spoke nervously but calmed down appreciably if one didn't interrupt him for a while.

The director of Continental Gas (Dessau) W. von Öchelhäuser was a curious mix in his mental structure. On the one hand he was a successful technologist, organiser and businessman, while on the other he was a connoisseur of the arts and literature with a strong bent to the classics and an almost lyrical love of aesthetic values and one could see this in his mode of dress and way of speaking. Despite my different opinion in these matters we got along well because his enthusiasm was honest and not at all superficial. In fact he was one of the very few men who thought like this who I wholeheartedly liked and we spent many stimulating hours together or in a small group. His highly developed technical way of thinking clearly protected him from developing the sort of vacuous opinions that one so frequently finds in aesthetes.

The last time I saw him was in Berlin where we lunched together with Krause, the director of the Borsig locomotive factory. Krause was a typical Berliner in the best sense of the word. It was in the fourth year of the war. There wasn't much to eat and the fading hope of a good end to it all made us all a bit depressed. But nevertheless the three members of the trio, who only shared in common a belief in the future, soon saw to it that after two happy hours we parted reluctantly to apply ourselves once more to the grey daily routine.

I had quite a lot of contact with Georg Hirth the founder of the magazine "Youth". He was the first to suggest that there were not just inheritable handicaps but that there would also be inheritable improvements in other words that every mental and physical improvement that an individual underwent would make things easier for his progeny. My endless optimism eagerly seized on this idea and I discussed it with Hirth and publically supported him.

My relationship with him did however become difficult when he stumbled across my old hobby horse, the concept of free ions, and on the basis of scientifically unsound arguments made them the source of enormous biological forces. He published a number of articles on this which he sent me, asking me to use my influence to support his supposed discoveries. It was not so easy to decline without hurting him and this I wanted to avoid doing because I valued him highly on account of the enormous breadth of the work he'd done on the history of art which he'd achieved without, like most art historians, ending up being incapable of understanding and supporting the artistic life of our times.

By the time I got to know him he was very obviously at the end of his career. He had a broad well formed face, lots of grey hair which stood straight up, a full moustache and lively features. His rather portly body was not very mobile and he needed a walking stick. Mentally he seemed to be tired—or even exhausted. He died a few years later.

I was often together with the excellent architect Gabriel von Seidl who drew up the plans for the museum. He was a real Munich personality full of humour and making no attempt to hide his Bavarian dialect. He was small and thin with white hair and beard. Like most people who work with their eyes he found it hard to express himself in words. I was in this respect exactly the opposite and despite the good intentions which both sides brought with them we never developed a close relationship.

I especially enjoyed the chance which the annual meeting gave me of getting to know Karl Linde who had invented the method of making liquid air and had then developed the industry which depends on it. He was of medium height with sparse brown hair and beard and he was very friendly and straightforward so that everybody liked him. He was of great service in the development of the museum.

As a weighty final point there was the, at that time all powerful, prelate and leader of the "Zentrum" party Daller who I once by chance sat next to. He obviously didn't know my name for he conversed with me in a grave but confiding fashion. Just then the museum had been presented with a huge model of a rolling mill ("Walzwerk") and it was delivered with an explanation of the order of the rollers which, one after another, would work to form the piece of metal until at the end of the line it left the rollers (roller road = "Walzenstrasse") in its proper shape. Daller said he couldn't understand it because the talk had all been about the roller road when the thing was surely a road roller (Strassenwalze)!

My sixtieth birthday. It is nice of people when, as one's life slowly nears its end, they take the opportunity to give one a boost. While my fiftieth birthday passed almost unnoticed by everybody including myself, for I was on a train somewhere between San Francisco and Chicago, a big thing was going to be made of the sixtieth.

Here the "Monistenbund" (Society of Monists) played a prominent part. The great success of the meeting in Hamburg which I'll discuss in the next chapter had not yet been forgotten, and the various activities that I'd involved myself in, in order to breathe life into the Society, had not yet resulted in an open break. But perhaps there was already then a semi conscious feeling that these happy times could not last much longer and the first hairline cracks in the edifice may already have been evident and begun to expand. In times like these, well meaning people tend to be extra careful to try to convince themselves and others that all is for the best and always will be.

I had promised not to resist the plans that were being made by my family and friends and so they had a free hand.

In the morning I was awakened by a string quartet. They played Hayden's op 76, Nr. 4 in B flat major known as "The Sunrise" in which right at the beginning the first violin comes in so beautifully on E in the middle of a B flat triad. This music had brought me many hours of joy since my student days and it got this beautiful early autumn day (it was the beginning of September) off to a good start.

I went for my usual morning walk but this time I needed much longer than normal for all along the way there were posters attached to the trees which recounted the stations, or better paths, of my life in verse. They had been made by my eldest son and because they expressed things so well and made me so happy I reproduce them here. On the posters the words were illustrated by various symbolic drawings which my eldest daughter had added. As one can see, the verses are in the manner of Arno Holz.

1. Fireworks
   Let the colourful bombs explode!
   Let the alarming crackers bang!
   Crackle you ladykillers, rush you rockets
   To fiery bouquets, to flowers of fire
   Just like you did 50 (not 60) years ago
   As you shot proudly into the sky
   As if Papa had made you himself
   Biff, baff, bummh, sch—sch—sch—knattrattatah!

2. Love
   The Helen of this story was the girl
   Who caught him in her magic net
   And there the researcher was really snug.
   His wife takes care of his daily needs
   And she likes nothing better
   Than to jump into the river for him
   Weren't our lives
   Peaceful and happy, my cockerel?

3. Music
   If you feel a sense of disharmony
   Sit yourself down with a bassoon.
   You'll get it to make a sobbing ohhhh sound.
   That's a credit to you and makes peace with the world!
   So much better rises from this tube of the Gods than from a lyre
   A wordless song—especially one from Mendelssohn

4. Chemistry
   Oh chemistry—noble science
   To you I gave the best years of my life
   I found new ways for you
   And brought you together with physics.
   He helped you up, oh physical chemistry
   With energy and art!
   You rushed up like a wild spring
   And I gave you form in my textbook.
   What didn't I all do for you?
   Thinking, writing, planning
   Teaching, speaking, organising
   How many students I led to you
   I sent the news of free ions
   Into the furthest corners of the world
   More than that, the Journal, the Institute!
   What a rich full life it was!

And then came the great “Electrochemistry”
“Inorganic Chemistry”
“Compounds and Elements”
Principles” (for those who could understand)
And “The School” for all the rest
And from it any fool could learn chemistry
“The shorter Ostwald” the “My Career”
It’s a long list
“Basics of Analysis”
And of course Catalysis
Etc, etc., etc.
Yup, for those who see it, it’s overwhelming!
But now its time to move on
And I have other ways to go
Come, wanderer, think this over with me here
And make a bow

5. Painting
You can’t go any further?
Your feet are sore?
Then try something different
Oh helpful Calliope!
So take a rabbit’s paw
And a roll
And you’ll soon
Simply paint your sorrows away

6. Philosophy
I loved you, philosophy,
And I breathed youth into you;
I gave you “energy”
And new freshness and drive.
I blew away some of your old cobwebs
But built new replacements.
So carry on further, wanderer
Carefree and Energetic

7. Internationalism
I’m an internationalist
You, wanderer, probably don’t know what that is!
An internazi is the sort of man
You won’t find in the Balkan.
World language, world law, world money
That’s what pleases him
World peace, world format, world symbols

All this is what an internazi seeks
Stamp collecting is not for him
Instead he's a close friend of Bertha

8. Technical things
Oh technical matters and industry
One never knows how it will be with you
You are sometimes too thorny to touch
And one loses sleep and nerves because of you
But anyway even hedgehogs sometimes sing love songs
And a cactus will flower now and then.

9. Monism
Monism is just a tool
To control the world in and around you
You probably thought it was just a pillow
Made for dogmatic slumbering pleasure
No, wanderer, you have to hold the plough fast
And always be ready to sharpen it again.

10. The Bridge
From islands in their solitude
Now bridges spring in graceful arch
You're still building bridges
As if your back, which has carried so much, would never tire
And every bridge is a monument to your motto
"Don't waste your energy, use it"
You give the world the finest bridges
And your greatest power is that of synthesis

The curious last lines in the second poem refer to an old family joke from Dorpat times which I don't need to go into here.

And so this morning walk turned out to be a stroll through my whole life up till now. I was never the sort to spend much time ruminating about the past, for I always had so much new in front of me. But on that day I was glad to be able to be led by dear and understanding hands through the old and new byways of my life and to have the feeling that in the future too the work would not be forgotten but would be carried on even when I was unable to do any more.

Five children, three of them married and five grandchildren assured me the continuation not only of my intellectual but also of my physical progeny. They were all there that day and they made me happy with a wealth of gifts.

Then the visits from the others started and most of these were from the Monist Society. The society had planted a memorial tree early in the morning as a symbol of their devotion. Despite being well looked after it died a few years later.

The Austrian Monists had prepared a "Festschrift". A year earlier I had visited in passing the group in Vienna which seemed then to be on its last legs. A couple of strategic changes of personnel had converted it into one of our most active and successful sections and it had expressed its thanks in this way. The Festschrift contained contributions from Rudolf Wegscheider, Ernst Haeckel, Friedrich Jodl, Paul Kammerer, Wilhelm Exner and Rudolf Goldscheid all of whom had competed in writing about me the nicest possible things which were more or less compatible with the truth. Their candour was underlined by a few passages in which they made friendly reference to points about which we were in disagreement. Sometimes when I feel the need for pleasant mental refreshment then I know I will find it by browsing through this little book.

In the afternoon I ceremonially inducted my two youngest grandsons into the ranks of the Monist Society.

In the essay to commemorate this day, which was printed for the family and close friends, there is already the news that I had started on the work of measuring colours and putting them in their proper relationships to one another.

My seventieth birthday. The autumn of 1923 was in the time that the Mark lost most of its value and this destroyed our savings. It made life difficult and the future uncertain and cast a depressive atmosphere over everything. Despite all this the day was a happy one for me because it was much simpler than my sixtieth birthday 10 years before.

On the previous evening my wife and I were asked to stand at the open window. We could here distant singing that came ever nearer: a dozen strong boys from the Leipzig secondary school, which three of my grandchildren attended, were singing under the guidance of their cantor. The choir was excellent and was the equal of the famous old Thomaner choir.[6] They stood under the window together with my sons and daughters who later told me it had been like a picture by Spitzweg: the old couple in the brightly lit window, outside the dark figures, and up above a clear starry sky. One song after another was sung; the best was "In stiller Nacht" from Brahms.

The next morning was a Sunday; the sun shone the whole day and huge white clouds in a pure blue sky cast enormous shadows over the landscape. We were wakened by a song from the boy's choir and as I was leaving for my usual morning walk my wife was asked to join me which she normally didn't do. Song sprung from the bushes by the door and as we went on our way more songs sprang up in the oddest places because the boys had run ahead and taken up new positions. Particularly impressive was the sound of Mendelssohn's "O Täler weit, o Höhen" echoing out of the old quarry on the western edge of the park.

Afterwards the youngsters visited me in the laboratory where they were rewarded with paint boxes and booklets of paper coloured in accordance with my colour doctrine. I explained some of my work to the older ones. They thanked me by singing the Brahms song again and, once the women of the house had fed them, left happy.

My five children and their spouses had all come as had the nine grandchildren of which only two were girls. We, the grandparents, were sat in the middle, the next generation sat along the wall and the nine grandchildren, clothed in the eight main

[6] St. Thomas Choir of Leipzig.

colours plus white, came forward one at a time riding a hobby horse each of which was decorated to represent one of my mental hobbyhorses and sang an appropriate verse clear and precise. They really are all quite musical. Only the last one, little 5 year old Gretel, burst into tears. Apparently the poet had, because of the artistic structure of his work, not given her a hobbyhorse and she found that terribly unfair. It took a while before she calmed down.

In the meantime the postmen and messengers had delivered all sorts of congratulations and gifts. A few old friends from the university and Professor Bodenstein from Berlin, who'd been my assistant for many years, came by train. That gave rise to a lot of greetings and speeches: Leipzig Academy, Bunsen Society, Union of German Chemists and so on. My previous students had sent me an album with their pictures in it. I was particularly happy with the remark of the representative of the Chemical Society who said they were keeping the traditional wish for dignified old age "otium cum dignitate" for my ninetieth birthday. In my reply I tried to express my thanks appropriately and yet lighten the gravity of the occasion by telling them that the first to congratulate me that day had been our cat Mia who had presented me with a litter of six little kittens. As always only two were kept.

After lunch my sons and daughters looked after the guests in the garden and in the woods while we rested for a while. Then at coffee they greeted us once again before leaving. The family members stayed on to read the many telegrams and letters and essays for the occasion. The evening drew to a close with some songs for three sopranos which my eldest son had written and set to music. They were sung by his wife, her sister and her niece. They had to sing it all once again because it was written in the modern manner which was not immediately accessible for me. Thus the concluding artistic event of the day pointed once more into the future.

The day made me think of the future also for other reasons. My three sons each of them fully involved in his own field hadn't seen each other for quite a while and they now realised that there work could be better integrated than had been the case till then. It became clear to me that in these hard times, when the trust we'd all had in the structures of the state and the community had been shaken to its foundations, that the family now was the only social unit in which each could trust the others without reservation. We were now a group of twenty people, some of whom were pretty smart, and so we could hope to support each other when society at large collapsed. I'd just recently had the happy experience of having been helped in my work for quite some time with an informal daily meeting with my second son, Walter, which had cost him a lot of time and effort.

There were no congratulations from the official side. However Dr. Hübschmann the mayor of Chemnitz came personally to thank me in the name of his town's industry which had received a considerable boost from my colour doctrine. The colour chemist Professor Ristenpart, whom I valued highly, had made Chemnitz the centre of his activities in which he, better than anyone else, applied my teaching to the manufacture of luxury goods.

In the meantime Germany had begun to recover, despite the foolish principle of voting for lists which trapped the government in bondage to the parties, and it should be able to increasingly trust in the health of its inner being.

Chapter 37
The Monist Society

Ernst Haeckel. Around Christmas 1910 I got a letter from Ernst Haeckel who asked me to visit him in Leipzig where he was staying in the house of his son in law. I hadn't met him up till then though some years previously, when I'd happened to be in Jena, I'd tried to visit him. He hadn't received me. From the date it later became clear to me that this was the time when he was writing his book "The enigma of our world" (Welträtsel) and had hidden himself away from everybody.

My view of Haeckel was somewhat ambiguous. I agreed with the general trend of his ideas, but not with many of the details. In particular I disapproved of the way he made use of physical concepts and terms because this merely showed up the large gaps in his education in this area. I'd pointed this out in a number of book reviews in the "Annals". On the other hand I was really disgusted by the way in which his opponents treated him and I mean here not only the clerics but also his colleagues, that is to say other professors, and because of this I was inclined to help him if necessary.

Haeckel's appearance with the flowing white hair and beard is so well known from photographs that I don't need to describe him here. Despite his 75 years he still held himself erect.

From his style of writing I'd expected to be meeting a wild man and so I was completely surprised by his—in the best sense of the word—childlike open friendliness. There was not the slightest trace of that abrasive self assertion which most people at the time associated with him. On the contrary, he was shy and self-effacing. The well known description of the personality of Darwin, who was his scientific ideal, could as well be applied to Haeckel.

The reason he had invited me was his wish that I take over the leadership of the Monist Society which he had formed in 1906. This had had its ups and downs and at the moment found itself at a low point and Haeckel, probably rightly, did not believe that he was the man to bring it back up. Like Liebig he was one personality in the flesh and a completely different one at his writing desk.

R.S. Jack and F. Scholz (eds.), *Wilhelm Ostwald*, Springer Biographies,
DOI 10.1007/978-3-319-46955-3_37

The idea seemed to me to be not unproblematic. From long experience I knew that I was able to sway and even inspire masses of people. But, on the other hand I also tended, as Goethe once put it, to put people off after a while and I was never able to work out what drove the one or other of these characteristics. On top of that I had a lot of irons in the fire just at that time and I worried that if I took on anything else some of them might grow cold.

Haeckel was able to dispel my worries by presenting it as a social duty which I could not refuse. He pointed out that professors were fearful of expressing any views that were contrary to the teachings of the church and that if a prominent researcher did this then the situation might be much improved.

The campaign against Ladenburg. Here he touched a nerve that convinced me. In 1903 the chemist Ladenburg had held a lecture at the Natural Scientists' Meeting in Kassel about the influence of science on our philosophy of life. In this he more or less repeated the well known mechanistic philosophy which had been presented there a number of times, for instance by Dubois-Raymond. I'd followed the talk with an indulgent smile, because he really had nothing new to say. But the times had changed. A storm of orthodox protest broke over Ladenburg. It was inspired by circles close to the Kaiserin, who was deeply religious and an inveterate builder of churches. She normally spent the summers with her children at beautiful Wilhelmshöhe near Kassel and considered the theological defilement of the place by this faithless chemist and Jew to be a personal impertinence and an affront.

At that time I was a member of the executive committee of the Society of Scientists and Doctors. When we met in the winter in Breslau to plan the next meeting a letter from some very high level was passed on to us which suggested that we might wish to officially disavow Ladenburg. The chairman of the committee at that time was van't Hoff. In the discussion which followed it became clear that quite a few influential members were quite willing to take the hint. I argued fiercely and passionately for the independence of science and managed to achieve a vote against accepting the hint—but it was a close run thing. At the next opportunity I was voted off the executive committee.

In the light of all this I accepted Haeckel's invitation. The necessary formalities of my election were soon completed, the members of the executive visited me in Großbothen and in the shortest possible time I was leading a movement which till then I'd only heard the name and the rough general goals.

Monism. I could accept only some of the views which Haeckel had expressed at the foundation of the Society. The greatest danger facing such societies is that they are dogmatically locked into a program which is defined in enormous detail and which thereafter cannot be modified in any way. In my meeting with Haeckel it became clear that this was not the case here. On the contrary the Monist movement had developed many different approaches all of which were united in their rejection of the ever stricter religious orthodoxy that the Kaiser supported. I had read Comte in the meantime and agreed with his division of the levels of cultural development into theological, metaphysical and scientific. In this sense I could see that the task of the Monist Society was to oppose the attempt to turn back the development of culture and instead to support the study and spread of a scientific view of life. Since I knew from history how differently science can impinge on a society at different times, I saw it as an important goal of the Monists to prevent a fixation on any particular set of ephemeral scientific concepts. There was a danger that something like this might happen even though this wasn't Haeckel's way of looking at things. I'd known before this that a large part of the success of his book "The Enigma of the World" rested on its dogmatism which admittedly was not theological in nature but rather based on science. However the meeting with that most unusual man set my doubts in this regard at rest. I found him free of stubborn dogmatism on the details we discussed so that I felt sure that we would quickly find agreement. In particular he was not the sort of person to use petty or underhand means to force through his opinions. This unconditional ethical idealism was what gave him the childlike qualities which won my trust and then my heart.

The meeting in Hamburg. The value of the Monist way of thinking and the effectiveness of my leadership were put the test the following summer. An international meeting of Monists was arranged which would take place in Hamburg. Though the movement's headquarters were in Munich, there was a numerous and active chapter in Hamburg. There had developed a clear difference in their ways of doing things—the people in Hamburg were used to reacting quickly and considered the Munich approach a bit too laid back. Both of them were happy to have the top executive in a third place because this lessened the contrast between their ways of doing things. However the ambition of the Hamburg group to make the meeting really successful led to an increase of effort generated by their "moral flywheel".

In Hamburg both the goals and the name of the movement had been in use before Haeckel founded the Monist Society. There Dr. Rée, a friend of Nietzsche, had already in the second half of the nineteenth century, started to try to reform the school system and free it of old ideas. In the winter of 1901 a "Society of the Friends of Free Thinking" was formed under Kahl's leadership which at the start led a tenuous existence under close police supervision. Soon after (in 1903) they adopted the name "Monist Society" and hence had been in existence for 3 years before Haeckel founded and endowed the Monist Society in Jena which soon successfully developed its activities. Once the German Monist Society was formed the Hamburg group immediately joined without insisting on any special privileges which is a monument to the organisational generosity of their ways of thinking and acting.

In fact the group in Hamburg produced an organisational master work. The scientific program was set up by the famous dermatologist Paul Unna and the technical details were taken care of by the industrialist Jakob Wolff and the businessman Carl Rieß.

Unna had produced a well designed program of lectures. Svante Arrhenius would speak about the universe, J. Loeb about life, I would talk about science and the Vienna philosopher Freidrich Jodl about culture and Monism. Then came lectures from rector Höft on the separation of schools from the church and from Professor Wahrmund on the separation of the church and state. The final lecture was from Dr. Ernst Horneffer on Monism and freedom. This well thought out program which covered all the great issues (centred around Jodl's discussion of ethics) provides the explanation for the enormous attractiveness of the Hamburg congress and the influence that it had. Because of oversubscription the lists of attendees had to be closed several days before the start.

I'd gone a day earlier to Hamburg to take part in the final preparations. There it turned out that even on the technical and economic side everything was in splendid order; our colleagues in Hamburg had collected donations which totalled around 40,000 Marks.

I suffered a bilious attack that evening which brought me several hours of pain but it didn't affect my ability to work over the next few days. That was just as well because this was the first occasion where I really had to accept the responsibility for the office I had taken on a few months previously. Up until then organisational matters had been carried out in the usual manner from Munich.

During the course of the meeting I was put up by the well known painter Mrs. De Boor who was the sister of Professor Unna.

The congress began with a welcome evening at the restaurant "Uhlenhorster Fährhaus" at the Außenalster.[1] We weren't sure if the congress would be a success when we went there—we were sure it would be when we left.

The crowds of people in the hall showed how the mood of the congress was going to develop. Some people had worried that the lectures had been aimed too "high" for the broad masses and that we would only get a handful of attendees. The opposite was true: over 2000 people were already impatiently waiting for this first evening which was only to be for the usual words of greeting. And without quite knowing why, there was a happy and expectant atmosphere before the first words were spoken.

After Carstens, the chairman of the Hamburg group, had said a few words, greetings were brought by R. Penzig (Ethics and Culture), Carus (American Monism), Schmal (Union of Free Thinkers), Weigt (Masons), Helene Stöcker (Maternity Protection), Polako (Natural Morals, Paris), Bloh (The Peace Society) and a number of other representatives of like minded movements. The unexpectedly numerous and diverse allies were greeted with ever louder applause and there develop a happy and enthusiastic atmosphere which everyone there experienced as a strong and pure feeling of goodwill. For me it was then easy at the end to get a warm and hearty reception as I announced the following: "Moltke's motto 'March separately, strike together' has been used so often in Germany that the emphasis is laid on the independent choice of the route of march and little on the united striking.

[1]Outer Alster Lake, one of two artificial lakes formed from the Alster river in Hamburg.

Today we have seen how happy a united march can be and we'll try to remember that for the future".

Haeckel had wanted to come but was prevented by illness. We therefore had decided to go and visit him in Jena after the congress. This news was met with such storms of applause that there was some worry whether we'd be able to sustain this level of elation over the whole meeting. The success of the meeting proved that it had indeed been possible. The fact that it was possible even in the absence of the revered founder of the movement was an indication of the real practical value of the meeting.

The next day started with the business meeting, the first I'd had to lead. The most important result was that the previous definition of the goals of the movement as "natural science based philosophy of life" was considered to be too restrictive and was changed to "scientific philosophy of life". This had been proposed by F. Jodl who afterwards in his important lecture provided a profound justification for the change.

Apart from that there was a personal matter which had a real and serious background. To the general public the movement was represented by the vice chairman Ernst Horneffer since the chairman Unold, who had kindly stepped aside to make my election possible, was more involved in writing than in giving speeches. Horneffer was a brilliant speaker who appealed to his audience's emotions but he was sceptical of down to earth science. Now that Jodl's change which we had just approved put science more in the forefront of things he felt that this was a negation of his activities up till now and made very clear that there was a real difference between him and the movement. The discussion was thus about the fact that the movement was composed of two streams of roughly equal size—the "emotional Monists" and the "scientific Monists"—and the great difficulties of getting them to march together. In many other movements there are similar problems which often seem to be insuperable. It's the old difference which was personified in ancient Greek culture by Aristotle and Plato.

In the evening there was the first public session of the congress. For this we'd chosen the largest hall in Hamburg, the Curio House, which had space for 2700. It wasn't just filled; it was over full with an audience of about 3500. Hundreds of late comers had to be turned away. In my introductory remarks I pointed out that since 1871 Germany had experienced extraordinarily rapid economic growth and that now there was a general feeling that man did not live from bread alone and that there was a yearning for intellectual sustenance to nourish the mind and the intellect. This was not to be found in the old church traditions, but it was to be hoped that it would be found in science and delivering it was the challenge to which the Monist movement was dedicated.

On this occasion too I was quickly able to establish rapport with the huge audience who responded to the various announcements and introductions that my position as chairman required with ecstatic applause. It's a great feeling to experience the unrestrained goodwill of such a huge mass of people.

There followed a number of greetings from allied movements from Paris, New York, Chicago, Iasi, Washington and Antwerp most of which were made in person by their representatives and this gave rise to the idea that an international organisation should be founded.

Ernst Haeckel had written an essay on the fundamentals of Monism which was read out by his pupil and friend Heinrich Schmidt and this gave rise once more to stormy applause.

The main presentation of the evening was the deeply thought out and extremely impressive lecture by S. Arrhenius on the universe in which he presented arguments—many from his own work on astrophysics—in support of the Monist philosophy of life. The audience was clearly deeply affected and I reflected their feelings in the speech of thanks which I gave for his support of the Monist idea. I recalled how 25 years previously we'd begun our scientific cooperation, how our paths had drifted apart and how now we had come together again. The applause seemed to go on forever and the success this day was no less than that of yesterday.

Once again we could rely on the remarkable organisational talent of the Hamburg group. Since an even larger crowd was expected on the next day, which was a Sunday, they organised during the course of the night a second hall, persuaded Horneffer to give an impromptu lecture, had placards prepared which were carried and driven through the streets the next morning so that around a thousand people who couldn't get into the Curio hall could be offered an alternative. Horneffer was equal to the challenge and he whipped up his audience with a lecture on Monism and the challenges of today which was received with loud applause.

Before the afternoon public session we held a committee meeting at which the formation of an international Monist organisation was approved and representatives for North and South America, France, Spain, Poland, Russia and Rumania were elected.

Then the congress went on, once again to a full house, with the lecture by the brilliant biologist J. Loeb (Part II, Chap 27, p. 319) whose discoveries concerning test tube fertilisation had recently caused such interest even amongst laymen. His talk was followed breathlessly and his conclusion, which was presented with the greatest determination, that all aspects of life were explicable by physical chemistry sent a shock wave through the audience, as I noted in my speech of thanks.

Then came my own lecture about science. It started with the description of the curious forms of independent life which science let us grasp and then went on to show that from the three great forms of cultural advance—religion, art and science—the last one was doubtless the youngest and but therefore also the highest product of human endeavour. The relationship of applied and fundamental science, their property of thrusting into the future and the organisational structure of the pyramid of science were then described.

At the end the following argument was presented and it clearly made a large impression on the audience.

Humanity had always projected its wishes and desires onto its Gods, making them symbols of the strength, intelligence and goodness which they sought but did not find in this world. If we now ask where all these ideas come together then the answer is in science, which is now the epitome of all that mankind strives towards. Thus the superhuman qualities which used to be attributed to Gods are now to be found in science, though not in the absolutist form of old tradition. Absolutes, like omnipotence, omniscience and omnipresence, do not exist.

Science is not omnipotent since it is limited by natural laws (which were often considered to apply to God), but it is the greatest power that is available to humanity. It can move mountains, multiply harvests prevent death and prolong life, in short it can do all we want and to such an extent that humanity's common work goes forward.

In the same sense science is omniscient since it encompasses all that humanity knows and increases that knowledge every day.

It is also in this sense omnipresent because everything that we do that requires more than mere animal instinct is based one way or another on science.

The impression this lecture made on the audience was increased by its contrast to Loeb's presentation. In it a pale thin dark haired pessimist had delivered, in a series of hard and unforgiving hammer blows, a cold and impersonal structure which was based on a ruthless and almost fanatical belief in the central importance of the inanimate aspects of life. Now came a broad, comfortable, blond haired man with a warm personal tone, though still practical and soberly scientific. In Loeb's talk one had almost to fear science, in mine you could love it as the repository of all that was finest, highest and best of what mankind had brought forth.

This explains why I was interrupted several times by loud applause and people shouted their agreement and support.

The day ended with a banquet for around a thousand attendees at which donations totalling some 17,000 Marks were made as an "energy treasure" to support the work of the movement.

On the third day we started off with elections and other business matters.

In the evening there was to be the last of the public lectures and I was a bit worried whether these final presentations might not be overshadowed by what had happened on the Sunday. In the afternoon there was a trip in brilliant weather round the harbour in small steamers. I joined it, although I was by this time quite exhausted. My companions were kind enough to note my tiredness and spared me from having to give further speeches. An acolyte of the Mazdaznan religion, who had latched onto us for propaganda reasons, seized this opportunity to try to convert me. It was to no avail that I explained that I wasn't interested; she could not be shaken off and finally my friends had to form a living wall between us, so that I could get some peace.

I got back home late in the afternoon rather exhausted and wondered how I should best get enough energy together to face the congress again in the evening. I wanted to avoid stimulants like coffee or alcohol and so I asked my hostess, to make me a hamburger and tried to justify my request for a meal at this rather odd time with all sorts of psychophysical reasons. She understood the problem and agreed at once and so later I was able to ascribe the success of the evening session to the energetic-chemical basis which she had laid.

And so we came to the third evening session which marked the end of the congress. It seemed impossible that on this evening the vivid experience of the previous days could be excelled, but indeed this turned out to be the case.

The central element of this meeting was the lecture by Friedrich Jodl on Monism and current cultural problems. Jodl had originally been interested in the history of science till he changed his interests and wrote a remarkable work on ethics which had led him to the same conclusions as scientists had come to, namely that ethics is a social matter and hence must and can be viewed in this light. He had accepted the invitation to Hamburg in order to use this opportunity to present this fundamental and widely applicable result. He did this despite the fact that he was not well and he paid for it with a marked deterioration of his health.

His lecture therefore was directed to the point that science and the humanities, which till then had gone their separate ways and were often in conflict with one another, could now move forward from a common position. He told the Society of Monists that it was now their duty to work on this point of common contact and not just theoretically but to turn it into a practical effort. Once ethics was scientifically shown to be a social structure then Monism, as a scientific philosophy of life, must apply itself primarily to social issues.

This idea was soon accepted by the Monists as their guiding principle and so we once again had the feeling that the congress had not just been a pleasant game of words and ideas but that it had to do with important down to earth work.

In my words of thanks I emphasised that Professor Jodl was the first representative of the humanities who had found his way to us and that in this way Monism now indeed represented all academic disciplines. For this his future in the history of human endeavour was assured.

The two talks that followed, Professor Ludwig Wahrmund form Innsbruck on the separation of church and state and Rector Gustav Höft from Hamburg on the separation of the church and schools, were both excellent reviews of current situation but of course contained nothing as new as the ideas presented by Jodl. They were both given the full attention of the audience without any of the acoustic problems we had worried about when an impatient audience starts to fidget and which increases dynamically once started.

The last speaker was Horneffer and he was warmly welcomed because the audience expected that he would speak to their emotions which had perhaps been a little neglected so far. He spoke about Monism and freedom, emphasising the Platonic idea of individuality. This was something that was dear to his heart and so his more poetic than scientific presentation more or less papered over the resistance to the change of direction which had been so forcibly put forward at this meeting.

An outstanding member of the Hamburg group sat behind the speaker. He was a lawyer and in our business meetings he had shown himself to be sharp witted. He had a bald head and a narrow agile face like Mephisto. During the lecture he caught my eye with his expressive smirks which showed his pleasure at the embarrassment I'd suffer in having to say a few words of thanks. I nodded in reply and murmured to myself—let's just wait and see.

Once the applause for the lecture had died down I said: It has been my pleasant duty in these last days to translate our thanks and applause into our dearly loved German language and at the same time to pin the well deserved laurels on the lapel of the speaker's jacket. This time however I will not pin a sprig of Laurel that but rather place a crown of roses on the head of our much admired vice chairman.

As you have seen I had made it my practice as chairman not to just address the speaker at the end of his lecture with the usual formal words but rather to try as precisely as possible to express what it was we were thanking him for. Once the audience had begun to understand this, these remarks of mine were listened to with increasing interest and were especially welcomed. As this last speaker had been nearing the end of his presentation I began to think that I should not just say something appropriate to what he had presented but instead sum up the success of the entire congress in a few memorable words. I thought it over quickly and decided that what I had to do was to emphasise the hope we all had that practical results would flow from the congress. And so, once the final speaker had been dealt with, I called out: "I now close this first international Monist congress and declare the start of the Monist century".

Never in my entire life, neither before and nor afterwards, have I experienced such a strong response from several thousand people as after uttering these words. The jubilance reached ever higher levels and I had to resist the temptation to say more for that would simply have diluted and destroyed the effect. I congratulated myself on having left it at that. This was the high point of the congress and at the same time the pinnacle of my effectiveness in the Monist Society.

The pilgrimage to Jena. Some 250 people had signed up for the trip to Haeckel in Jena which I had announced at the beginning of the congress. Since it was quite far we went there in a chartered train. On the way the "Weimar Cartel" an umbrella organisation of various societies of freethinkers, reorganised itself so that when appropriate it would be able to act with united power. However no leader emerged who was able to hold this diverse group together so the effectiveness of the cartel was minimal.

We reached Jena in the evening. Once I'd shaken off the dust of the journey I went to see Haeckel and tell him about the torchlight procession which would be coming once it got dark. He looked forward to this happily. He accepted this honour, which was modelled on the torchlight processions to Sachsenwald in honour of Bismarck, with the same childlike happiness that he'd accepted all the other honours that his acolytes accorded him and if anyone wants to call this vanity then it was vanity in its most engaging form.

The torchlight procession arrived at the right time and, standing with him on the balcony, I made an address to him in which I compared his ideas on Monism with the noble seeds which lie a long time in the earth and then suddenly sprout in all their magnificence. In his reply he ascribed our success to my efforts although I well knew the huge part played in this by the Hamburg organisation.

In the evening as we all sat down to a drink and I pointed out that very few professors and students were members of the movement and expressed the hope that in the free atmosphere of the Jena university this situation would first be changed. This hope was not fulfilled.

The next morning we visited the phyletic museum (which had been created by Haeckel. He personally showed us around and had to be almost forced to conserve his energy for the banquet which he had agreed to attend. Here too there were lots of speeches as well as a poetic homage which the poetess in person delivered and which deeply affected the 77 year old Haeckel. We parted from him in the highest spirits and each went his own way to think over what these last days had brought.

The newsletter. The Hamburger meeting had decided to develop the movement's newsletter into a decisive means of stirring up the rather lethargic circulation of ideas within the Monist Society. Despite the enormous success of the congress we had to keep in mind that only a fraction of Monists had attended and their experience in these few days would never be sufficient to breathe life into the movement. This however was essential if our movement was going to be able to change the intellectual climate in our country by convincing everyone that the solution to every problem must be left in the final analysis to science.

The society already had a monthly newsletter, "The Monist" but it had no influence because it was scarcely known outside our circle and even our members didn't pay much attention to it. It was published by a member in Berlin who ascribed to all the right views but he suffered from the delusion that he was a great but as yet undiscovered poet. When his bent to versify got hold of him then his priorities were clear and everything else, including his publishing activities, had to take second place. To begin with I thought I'd be able to get along with him but I never did manage to convince him to alter his assessment of the relative value of poetry and publishing to our movement.

Attempts to find some alternative ran into all sorts of difficulties. Finally there seemed to be no other way than to take care of the newsletter myself. My eldest son had introduced me to W. Blossfeldt and he turned out to be well suited to help me in this endeavour. I also contributed to financing the newsletter. This was not a great sacrifice at the time for I was well off but when things changed then my enemies accused me of siphoning money out of the newsletter.

The earlier newsletter had mostly published the sort of general and theoretical essays which turn up in a group where everybody on their own is struggling to throw off old ideas and come up with a useful philosophy of life—which usually turns out to be cobbled together from all sorts of weird sources. Such personal views, that are often hard to digest, are held by their instigators to be extremely important and they are seldom prepared to consider alternative solutions as being worthy of consideration. Here I thought it would be important to concentrate on the common elements of all these personal philosophies rather than arguing about their differences. I considered the common elements to be the scientific way of thought as indicated by the results of the Hamburg congress and the challenge, laid out in Jodl's lecture, was to develop theoretical and applied ethics in the form of social work. I therefore set up a special section in which I let the representatives of the different factions express their views. A little later I'll have more to say about this.

The metaphysicists. From those who are critical of the demand that we accept a leading role for science (curiously enough this includes almost all working scientists and professors) you will often hear it claimed that mankind has an innate metaphysical need to find reasonable solutions to questions that science cannot answer. Since I don't register this need in myself I can fairly doubt both its necessity and its generality. I recalled that previously doctors and laymen were convinced that everybody must suffer from measles and scarlet fever during their childhood and thought that the belief in a metaphysical need was no sounder than this medical old wives tale. This was the only time that W. Wundt, who I admired so much, was annoyed with me for he ascribed to the idea of a metaphysical need and he replied that I was fooling myself if I thought that I didn't have it. He failed, however, to show by what means he had diagnosed this metaphysical need in me.

Mach had the final word on this issue. He characterised the scientific man as one who is satisfied with the incompleteness of his world view which is due to the incompleteness of science, and doesn't try to fill the gaps with conjectures which he knows to be no more than hot air.

Reaching an agreement with these people was going to be difficult, but not impossible. However, there was a sizeable group—the emotional Monists I mentioned earlier—and the hope of reaching an agreement with them was vanishingly small. Like their paradigm Giordano Bruno they saw the poetic content of their ideology as its most important aspect. In the course of an annual meeting in Magdeburg we made an excursion to nearby Helmstedt at whose former university Bruno had been a professor for several years. That was quite normal in his time because Latin served as a common language amongst academics irrespective of where they came from. This highly desirable situation could be reintroduced at any time if only scientists would agree to a synthetic world language. The preparative work for this has all been done for the Ido language so that all that is needed is a decision to apply it. If only there were no law of mental inertia!

The poetic Monists were influenced and led by some who had developed their natural rhetorical talents and now applied them as a sort of work of art. The idea that cold intellect was in principle superior to warm feelings, was for them unbearable. Some of them were immediately opposed to the new leadership, while others simply turned their backs on a movement which no longer seemed to offer them what they were looking for and went elsewhere. Those who stayed, however, immediately started their opposition. I wasn't aware of them to begin with and once I could no longer ignore them I followed the dictates of my nature and made no effort to fight them. I left that to my colleagues because right from the beginning I'd decided not to sacrifice too much in the interests of the movement but to restrict my activities to the necessary leadership work. I considered that my resignation was just a question of time as it had been earlier with the Bunsen Society which was much closer to my heart.

Democrats. Further difficulties arose from a misunderstanding which I have learned from political events in the last years to be very widely spread. The movement was of course democratically organised and the highest legislative power lay with the annual meetings at which all the elections took place. Between these meetings the elected managing committee ran the movement. There was always however a number of members who interpreted a democratic organisation form to mean that they only had to announce their wishes to the managing committee for them to be immediately applied. They were astonished and upset when told that they would first of all have to organise a majority for their proposal at the next annual meeting for it to be successful. They considered this to be the result of the managing committee's undemocratic craving for domination.

The Monist sermons. My personal interactions with the members of the movement had convinced me how different their views of Monist ideology were. Even Haeckel himself was not free of a tendency at times to slide back from the scientific to the metaphysical viewpoint, and so the supporters of all sorts of strange ideas could invoke the Master. At that time I held the ridiculous hope that I would be able to unite perhaps not all but certainly the majority under the flag of science and was ready to put a lot of work into this effort. Remembering the good results that hard work had achieved in establishing the new concepts of physical chemistry I hoped here to achieve a similar success. What I'd failed to realise was that establishing a new idea in science is so much easier than converting laymen to a new philosophy of life. It's true of course that emotions play a part in the acceptance or rejection of new scientific ideas but, at the end of the day, all that matters is the purely logical proof. Philosophies of life however are hypothetical matters that are based almost entirely on emotion.

The first way I tried to influence matters was through the movement's newsletter. Till then it had come out once a month so, to begin with, I managed to have it published once every 2 weeks and then once a week and I had no difficulty finding lively contributions to fill it. On the contrary, there were so many contributions sent in that I had to take care to restrict the amount of space I used for my own essays, even though there were lots of things I wanted to say.

Because of this I looked for a way to get my thoughts across without impinging on the others' needs. After a few initial attempts I finally settled on the plan to print one of my essays every 2 weeks in a section titled "Monistic Sermons" each of which should be no more than half a printed page long. I'd noted on a number of occasions that I found it easy to formulate a set of ideas within a certain word limit and could get this right to within three to 5 %. I didn't find the space limitation a problem but more a challenge which I enjoyed taking on without compromising either the content or the style, just as a composer enjoys operating within the strict limits set by the rules of counterpoint.

From the technical aspect I also had no reason to fear the work. Around 1910 I'd bought a dictaphone which saved me the trouble of writing everything down and it increased my writing speed by several fold even in comparison to the typewriter. Once I'd arranged my thoughts and made a note of some key terms, a text ready for printing could be dictated in around 2 h. My secretary soon accustomed herself to the sound of my voice from the machine's cylinder so that her typed version was almost free of errors. Whenever it was necessary to change a sentence around during the dictation then I either repeated it or completed it in the new form and put the necessary corrections into the typed version. My secretary had strict instructions to type the sentences exactly as she heard them even when they sounded odd. For corrections a lot of space was left between the lines.

By doing this I noticed that the spoken form is quite different from a written text and that I had to avoid falling into the style of a speaker when I dictated. The difference is mainly a question of the repetitions which a speaker has to use to fix the important or difficult points in the minds of the audience. For a reader however it was only necessary to give the necessary hint to take things slowly at this point and by so doing I spared both the reader and myself a lot of unnecessary repetitions.

Before starting I got around five of these "sermons" ready and made notes on the contents of a dozen more and having done this I happily set about publishing them.

From the positive and in some cases moving feedback from readers it was clear that these "sermons" were an immediate success. When at the end of a year 26 "sermons" had been printed I had them published as a book and had to organise one edition after another to satisfy the demand.

The themes of the sermons were sometimes general questions while others dealt with specific current topics. For the sermons on general topics I used a scheme in which the major questions concerning the scientific background to the philosophy of life were addressed one by one. In between these came essays on other topics which might have been stimulated by current events, by readers' letters or from the wide range of material which I reviewed for the "Annals" and soon in increasing quantity also for the "Monist Century". I enjoyed writing this mix of order and chance and it was clear that they were read with interest.

These sermons were written over the course of about 3 years. The war brought them to an abrupt end.

Connections to other efforts. The challenge facing the Monist Society was to gather in all those who'd developed beyond the religious and metaphysical levels to the scientific one. They had to be encouraged to stay at this level especially when atavistic relapses threatened to cloud their thinking and judgement. Secondly it was necessary to defend our point of view against all those who at that time were propagating the old ideas with considerable effect. Under Kaiser Wilhelm II the orthodox reaction had gained considerable power. Though Wilhelm I was certainly more pious than his grandson he considered respect for the freedom of thought to be his plain duty so that he would never have considered using the state's authority to impose his own opinions on others. In this his grandson had no such reservations and in fact his thinking and his ways of doing things more resembled those of his great uncle Friedrich Wilhelm IV. He believed he enjoyed a closer relationship to "his God" than was granted to other people and from this elevated position thought it was his right and his duty to lead his "subjects" to the proper way of thinking. This tendency was increased by the orthodox views of the empress and through his love of the pomp of the Catholic Church to which he felt much attracted.

I didn't feel that our work should be limited in any way and I was very happy to accept the cooperation with the Monist Society of other movements which I felt were socially useful. Already the participation of excellent foreigners at the Hamburg congress had shown the opportunity for making international contacts. At our next annual meeting in Magdeburg Dr. Juliusburger convinced us of the need to work against the evils of drug abuse, in particular of alcohol. We also soon made contact to the land reform movement and to the feminist movement. Our member Dr. Dosenheimer was particularly active in seeking reform of the justice system. In addition we were also linked to those seeking reform of the schools, sexual reform and a number of other things.

Social democracy. All of these activities lay at the left to far left end of the political spectrum and so there naturally developed a closer relationship with social democrats some of whom suggested that I should join their party. I told them that this would not be possible until some way was found to bridge the gap in the concepts of socialism and class war. A party that advocates class war is quite clearly anything but social.

However this didn't stop me from getting together with social democrats I liked. In particular Heinrich Peus went far beyond the party template in that his views on many issues including the world language and land reform were close to mine. He was one of the first to understand the wide applicability of the energetic imperative and I can think of no one who exploited it better or more successfully. As president of the parliament of the Free State of Anhalt and leader of countless social organisations he'd been more successful than most other members of his party who'd failed completely in their attempts to socialise the economy.

Leaving the church. The inner dishonesty of many of the supporters of the church which I noted during my time with the Monist Society led me to get involved in the "Leave the Church" movement which was quickly developing at the time. This work involved giving lectures in which the outmoded nature of the church was made clear. The church, for its part also organised lectures in its defence. Often enough there would be a debate at which the chairman would use all sorts of "tactical" tricks to help the party he supported.

At the start I was eagerly involved in this but I soon lost interest largely because this work was not effective and that was not in line with the energetic imperative. Sometimes I had a strange feeling when I attended meetings at which professional speakers were involved. There were many of these on both sides and they wandered from one place to the next given the same lecture each time. These speakers seemed to deliberately avoid destroying the other side's arguments but rather always left a bolt hole open. This reminded me of the way in which battles in the middle ages were regulated by the representatives of the mercenaries from both sides meeting before the battles to agree how many dead and wounded there should be on either side. This was because it was not in their interests that a war should be ended by a decisive victory for otherwise they would be jobless. I don't believe that the speakers consciously acted in this way but I am sure that there was an unconscious impulse in this direction.

This work brought me together with the most radical wing of the Social Democratic Party and with others who shared their views. I didn't find this much to my taste and so I gave up this work soon after. In the same way the subcommittee which the Monist Society had elected to consider the matter of school books on secular morality met a few times but didn't result in any real action.

The Monist colony. I must not omit to mention an experiment I ran during this time. Like many others I was concerned at this time with the hasty re-orientation of the Germany economy to the production of goods for export because the working conditions especially in the large towns were causing a physical and moral deformation of the coming generation. I thought then, and think now, that the best remedy for this is to ensure that each worker have a direct connection with the earth in which the energy he needs for himself and for his family is collected.

Because of my habit of applying general scientific knowledge to questions of practical life it was clear to me that to survive each individual must be connected with a piece of the planet's surface just as an embryo is connected to the placenta. The umbilical cord may be long and bizarrely twisted but it must always be there for otherwise the individual must die. In the final analysis every living thing derives its energy from the sun but the sun's radiation energy can only be converted into the chemical energy from which humans and animals live through the mediation of the earth and the plants that grow on it. And therefore a little piece of the earth's surface belongs to each and every one of us and this piece of earth is what allows us to live. A people that imports a significant fraction of its food has its pieces of earth outside its own borders and the umbilical cords are always at risk of being broken. A similar danger is also within a country's borders when too much of the land is in the hands of a few. In any case it is always best to keep the umbilical cords as short and safe as possible and this would be achieved when every family had an inalienable right to its own piece of soil. This is the prerequisite for the rebirth of real family life. The difficulties which up till now have stood in the way of such a development are daily being reduced by progress in mechanisation which allows us to transcend time and space. Electrical energy, which is easy to distribute, will change our lives by helping to disperse population centres.

These considerations led me first of all to suggest that our Monist Society join the land reform movement which was so effectively led by Damaschke and to describe their goals and methods in our newsletter. And then I found the idea of setting up such a colony myself increasingly attractive. Of course by moving to the country house "Energy" I'd done this already for myself and my family. However since my readiness to make sacrifices did not go so far as to accepting outsiders in or around our house, I sent my assistant who had experience in farming and gardening to look for a suitable piece of land. He found something close to the town of Eisenberg in Saxony-Altenburg which was large enough to sustain about ten to twenty people and I bought it for a pretty stiff price. It was beautiful countryside beside a little river and it had not only extensive fields and meadows but also a mill, houses, barns and everything else one needed to live.

The hardest part of these things is always the choice of one's collaborators and I have to confess that here I was not up to the challenge. On the economic side I'd imagined that I would give the colonists free use of the ground and the living quarters, but that food, clothing and so on would have to be covered by the proceeds of their agricultural work. Naturally I had to put up money to put everything in order to begin with.

Looking back it seems to me that this was a basic error because the colonists got the impression that in this scheme money was not an issue. The major goal, which had been to make the colony an economic success, thus faded into the background and with that failure was programmed. When I looked through the accounts for the first few months it turned out that each of the colonists had bought a special toothbrush and charged this to the account which was to be used to get the farm started.

I had failed to set up suitable criteria for the choice of the colonists and had never intended to be personally involved in the development of the colony. I understood nothing about farming and in any case I had neither the time, the interest nor the talent required to settle any of the difficulties and quarrels which were bound to arise. I therefore left the administration in the hands of the man who had been up till then my secretary. He was from a poor background and had managed to educate himself. He was a gardener and in his youth had learned how to look after animals. He accepted more or less anybody who applied even when their only reason for applying was their financial plight. Since the news of the scheme was more or less limited to the members of the movement, a Monist philosophy of life was the real selection criterion. However, I have to say that some of them were indeed idealists who honestly set themselves to all the work which the operation required.

The experiment lasted about a year and a half. As always with these things there were disputes which at the beginning I tried to sort out myself but the solutions lasted only a short while so that I got tired of the wasted work and just left them to sort out their problems themselves. The result was a rapid rise in the temperature of the conflicts which culminated in a firmly worded letter to me that they would all leave the colony unless I immediately found a new manager. Their objections to him were largely of a personal nature. Since I'd now realised that the project was hopeless I took them at their word and left the letter unanswered. The colonists then left as a group for Leipzig apparently believing that their threat to close the colony would make me change my mind. This however played into my hands because it let me avoid having to get into tedious arguments with each of them. I let them know that I was happy with their decision and settled the accounts by paying some outstanding bills.

This was not the only example showing that a social project which is going to survive requires a strong personality to breathe life into it. In the middle ages it was widely believed that a cathedral, castle or similar construction would not last long unless a living being was walled into its foundations. That can be taken as a symbol for all such constructs. It doesn't take a whole life but such a project will consume at least a dozen years of one's life until it develops a life of its own which does not require unbroken nurture. I'd freely made this sacrifice for the Bunsen Society (Part II, Chap 23, p. 269), if one can call something one likes to do a sacrifice, and it was successful because its transition to a life without me at the head was achieved smoothly. If I'd known then what I know today then I would have been able to predict that the Monist colony lacked the basic necessities and was bound to fail.

Unfortunately nature is so badly organised that before the event one doesn't have the experience which one really needs. One only gets that after the event when it isn't worth much anymore. A number of other ventures I was involved in failed for this reason. Luckily most of these were isolated cases in which I tried to breathe some life into a general thought. The failure applied to the attempt while the thought remained alive and could safely wait for a new attempt to be fleshed out and set to work.

The holiday courses. A very effective project was begun in the spring of 1914 and would have been pursued and developed had it not been for the outbreak of the world war. This was to be a meeting for scientific lectures.

These sorts of "holiday courses" were at that time very popular particularly as a further education measure for certain occupations. Ours was planned around the idea of having lectures from experts covering certain important areas where science and daily life came into contact. The centre was to be in Jena which we regarded more or less as our home city because it was there that Ernst Haeckel had spent most of his life. In addition, a property there, endowed by Ernst Abbe, provided cheap suitable accommodation and lecture halls of various sizes. The meeting was set for the Whitsun holiday; the lectures were to be held from 8 am to midday. The afternoons were for discussion, outings and so on, because we felt we shouldn't burden the participants with more than four lectures per day.

The lectures were from Staudinger on cooperative societies, from Bozi on the principles of reform of the justice system, from Magnus Hirschfeld on sexuality and from me on organisation. They were attended by several hundred people. The largest audience was for Hirschfeld. The discussions in the afternoon demonstrated their interest in the material. We separated with the clear recognition that we had started something worthwhile and with the determination to repeat it. As I mentioned however the world war destroyed these good intentions.

A low point. In the first sudden wave of elation after the Hamburg congress the Monist Society showed itself prepared to follow me and trust my judgement. Slowly, however, a group of members who felt that they had been unfairly pushed out of the limelight, coalesced together into a fundamental opposition. Among them were those who had tried unsuccessfully to borrow money from me. I hadn't wanted to say no and so I'd given money to those who'd tried first, though I never saw a penny of that money again. However as ever more queued up for money I made a firm decision to overcome my shyness and they took that very badly for they truly imagined that by taking on the chairmanship I had somehow become the common property of all the members, all of whom must now have access to all my resources.

And so there soon emerged a group that was dedicated to causing trouble for me. After a while they felt so strong that at the annual meeting in Düsseldorf they accused me in the finance committee of having enriched myself at the expense of the society's newsletter. I refused to even reply to this personally and the society's secretary Bloßfeldt, who was chairman of the committee, was easily able to prove the opposite.

But, just as I'd experienced in the philosophical faculty in Leipzig, there is in every such social group a surprisingly large number of people who consider one's successes as a personal affront and can't wait to get their own back. In this case I saw how even highly placed friends in the Monist Society turned away from me and became ever more hostile. One case which hit me hard and in which I really could see no fault on my part was as follows. We had coffee together in the Hofgarten in Munich and I got him to give me a detailed explanation about the state of a project which was of common interest to us. Since I had another meeting later we left and my companion carried on his explanation on the half hour's walk back to my hotel close to the main railway station. I had a number of objections to his scheme and tried to bring them up but he wasn't finished. Again and again I tried to interrupt him but I couldn't manage it. When we finally got to the hotel I said to him with a smile, "Hmm, you haven't let me get a word in edgeways for the last half hour". He stopped short, realised that I was right, and from then on avoided all personal contact with me and on public occasions treated me with acrimony.

The end. The outbreak of the world war brought hard times also to the Monist Society. There was a serious rift within the society. We were of course all, in general, supporters of world peace and saw no reason to believe that just because there had always been wars that this automatically must mean that there always would be wars in the future. We thought, in contrast, that progress would banish the evil of war just as it had banished the plague, cholera and other terrible epidemics, at least in civilised countries. But now that we were being attacked from all sides and had to defend our lives a part of the society was of the view that we Germans must do all we could to defeat the enemy. Another faction, however, thought that now was the time to demonstrate our peaceful intentions and that each and every one of us should refuse to take part in any warlike or defensive actions.

I for my part belonged to the first group; though I gave the other side the chance to air their views in the society's newsletter. Here I made a strange discovery. Those who supported the war were prepared to respect the others' views and to argue fairly about the pros and cons. Those who opposed the war showed a real wish not to convince their opponents but to destroy them.

Under these circumstances I thought it best to resign as chairman particularly since I saw no chance for any fruitful future for the Monists after the war, irrespective of whether we won or lost.

My friends reminded me about my foolhardy call to a Monist century with which I had taken on the chairmanship and they tried to convince me that it was my duty to stay at my post. I however argued that the energetic imperative required that I reduce or even remove the current high level of friction with my resignation.

When I now look back on how the century has so far developed then the first impression can only be that the prophecy that it would be a Monist century was very far from the truth. But one should bear in mind that only about a sixth of the century has passed so far; five sixths still lie ahead of us. And since for me a Monist century is a synonym for a scientific century I think there is still room to hope that my prediction will be fulfilled.

Chapter 38
The International Union of Chemists

A small international matter. At the Natural Scientists' Meeting in Kassel (Part III, Chap. 37, p. 498). I met my Swiss colleague Ph. A. Guye from Geneva whom I'd known from the literature as an independently minded worker in the field of physical chemistry. In order to support our field in French speaking regions he had started a newsletter which I considered a help in our joint efforts and I heartily approved of it. Guyes own work was characterised by independence of thought and exactness of his experimental work and I'd been happy to remark on this more than once in the "Journal".

Our meeting increased my positive impressions. He was a tall thin almost wiry man with the short dark brown hair and beard of those from the south. He had a lively and winning way that gave the impression of great sincerity. He told me how he had come to physical chemistry. His chemical education had been like that of all chemists at the time "organic". At that time chemistry in Geneva was taught by the excellent chemist Graebe who was one of the co-discoverers of synthetic anilines and Guye had been his assistant. One day while cleaning up he'd come upon a rather dusty and as yet unread copy of Ostwald's Textbook of General Chemistry. The unusual title interested him and he took the book home. He explained to me with heartfelt emotion that he had read it through almost at one sitting and as a result was won over to the new field of physical chemistry for the rest of his life.

At this time William Ramsey had been invited to Kassel to give one of the main lectures. He already knew and had become friends with Guye and so the three of us spent time together and were egoistic enough to move away from the mass of people so as to have some peace to talk.

A little later van't Hoff joined us and was heartily welcomed. Like Ramsey he was accompanied by his two grown daughters, sadly I'd left mine at home. We arranged to have lunch together. There were flowers for the girls and there developed a feeling of absolute contentment and happy abandonment to the

R.S. Jack and F. Scholz (eds.), *Wilhelm Ostwald*, Springer Biographies,
DOI 10.1007/978-3-319-46955-3_38

moment which none of us would ever forget. Ramsey the stolid Scotsman raised his glass and said with a quaking voice and shining eyes "Long live friendship".

When I later looked back on this small but happy incident it seemed to me that it could be taken symbolically. Each of us four belonged to a different nation and yet of the thousands attending the meeting it is unlikely that any group was more closely bound to each other than we were. The union of chemists from all nations seemed to me then the most natural and necessary goal and one that all of us would support. And since the way from a theoretical idea to an attempt to bring it into practice is never very far with me, I looked around for a chance to bring it about.

An organisational plan. In thinking about this I remembered what the distinguished Belgian statesman A. Beernaert once told me. I'd met him in Brussels at a meeting of international societies and despite the great difference in age and outlook we'd got along well from the start. He'd been born in 1829 and so was almost a quarter of a century older than me and was politically close to the clerical party. But at this time his main interest had to do with organisation—he was a member of the Hague Peace Conference and a winner of the Nobel Prize for his efforts on behalf of international peace—and so his friendly courtesy after my lecture at the Brussels meeting was very welcome. During a meal at his house we had a long talk and he told me, among other things, the following: many years before he had been involved in an effort to create an international body on shipping, but it never came to anything. Later the idea did become reality because institutions of this type were formed independently in the different port cities and they had then developed into national associations. When he then tried to organise these national associations into an international federation it turned out to be easy to achieve. Since then, he told me, he never again tried to start that sort of thing at the international level. The beginning always has to be local, then it has to develop to the national level and only then can one start to consider an international association.

This valuable insight into the practice of organisation had impressed me greatly and I'd decided to make use of it if I ever faced such a challenge. Now an opportunity to apply it had appeared in the form of the idea of making an organisation for all the chemists in the world.

The imperial chemical institute. This challenge also had grown out of a much smaller and more limited problem. I'd returned from America in 1906 with the conviction that chemistry too, just like physics, ought to have a research institute which was completely free of teaching duties. In physics there was the Imperial Technical Physics Institute which Werner Siemens had founded and liberally endowed in 1887. He entrusted the organisation to his friend Helmholtz and appointed him as first president of the institute. I'd already visited this institute on my first journey through Europe (Part I, Chap. 9, p. 98) and met there a colleague who was working on a problem that I had already solved, namely the construction of a thermostat. I'd followed the rapid development of this Institute with lively interest since for a long time it was the first and only institute of this type. Only much later were physics institutes modelled on it opened in Britain and in America.

The integrative and general concepts which I'd concentrated on at the end of my time as a chemist had shown me that there were a great number of challenges which could not be taken on by a single researcher working on his own because they required an organised attack involving many different areas of work that would have to be carried out using common protocols. For this reason I began to think in terms of the formation of a suitable equivalent chemical institute. I got a lot of support from my chemical colleagues—but no practical help. There was little point in applying to the government because this would have involved endless delays even after we had got to the stage of getting a first "positive assessment" from the responsible officials.

I talked the matter over with representatives of Germany's chemical science and industry on the occasion of the International Congress of Chemistry in Rome, about which I'll have more to say later. As usual it soon became clear that the question of money would be the first hurdle that had to be taken. "Go and see rich Dr. Mond", someone called out. "You know him well enough and he lives here in Rome". That seemed like a good idea, though I learned that he was not well and that it was difficult to get an appointment with him. However, in response to my letter he agreed to see me and after listening to what I had to say he granted a considerable sum of money. If I remember correctly it was 200,000 Marks.

Ludwig Mond. I'd known this multimillionaire industrialist for more than 20 years and since he was in many ways a most remarkable man I'll give a brief outline.

The first time I met Mond was in 1889 at the Natural Scientists' Meeting in Heidelberg (Part II, Chap. 19, p. 207). On one of the excursions an undersized but powerfully built Jewish looking man with black hair and beard on a curious squint head introduced himself as Dr. Mond. He'd come to the Heidelberg meeting because he'd studied chemistry there in Bunsen's laboratory.

I knew nothing of his industrial success and treated him coolly because he made some remarks which seemed to show a lack of respect for the representatives of science. I was rather sensitive on that issue because it was at that meeting that Edison's German speaking representative, who had replied on his behalf for the award made to him, took the liberty of making tasteless jokes about academics (Part II, Chap. 19, p. 208).

I thought that Mond was doing the same thing and behaved towards him accordingly. However Mond soon made clear what huge value he laid on real scientific work and told me about his attempts to create electricity on an industrial scale by burning gas, initially using hydrogen. In this he soon had my complete attention and his unstinting way of viewing chemical processes from an economic standpoint was something that later influenced the way I would view these matters.

Later I met Mond several times because William Ramsey knew him well. Mond had recognised early on that the physical chemistry that was taking shape under his eyes was going to have not only a great chemical but also an economic significance. As a pupil of Bunsen he was well prepared for this and he was further encouraged by the positive opinion of Victor Meyer at the Heidelberg meeting. This explains his interest in me at that time. On top of that the work he'd been involved in was in the same area as mine.

Although he was from Kassel and had been educated in Germany Mond, like so many German Jews, was willing to take on the nationality of the country in which he'd made his fortune and in his case he had wholeheartedly taken on British nationality. Because of this he felt uneasy that the centre of the development of physical chemistry was in Germany and he wondered how to establish it also in Britain where it had previously been so well tended by Davy and Faraday. Their suggestion that a professorship should be founded at a university was not to his taste. In Britain the most important discoveries are made by private people and so he settled on the idea of endowing a private institute for physical chemistry that would be open to anyone who wanted and was able to carry out such research. He planned to have it managed by the "Royal Institution of Great Britain" a private society which had been formed at the beginning of the nineteenth century by a number of rich men to provide them and their friends with first class scientific entertainment and, at the same time to provide fruitful ideas for the country's industries. The main organiser was the American Benjamin Thompson who is better known by his later title of Count Rumford. The Royal Institution had always had a good eye for the scientists they appointed for the first had been Humphrey Davy and then, after a short interlude with Brande, they appointed the outstanding Michael Faraday. Another was John Tyndall who in the second half of the nineteenth century was better known in Germany than in his own country. Indeed Helmholtz had arranged for a German translation of Tyndall's book on heat.

Mond bought a house close to the Royal Institution and equipped it with everything that was needed for successful research in physical chemistry. To do this he studied the leading institutes in Germany, France and Britain and bought the best equipment that money could buy.

During one of my many visits to London he took Ramsey and me to the house which at that time was not yet quite finished to show us the equipment he'd bought. I noted with satisfaction that of the best equipment that Mond had scoured the world for, three quarters of it came from Germany. The mass of equipment was matched by the lab benches which were of massive mahogany. Mond told me that mahogany was really solid and that the price had not been that excessive. He called it the Davy-Faraday Research Laboratory and had felt a need to equip it in a way fitting to these illustrious godparents.

Like most houses in London it was rather narrow but extended over four floors. To reach the top floors we got into an electric lift that had just been installed. Mond started it. Due to some sort of failure in the mechanism the motor ran through and we shot up at such a speed we expected to be squashed flat at the top. Ramsey, as the tallest, would have been first, then me and last of all Mond who was barrel-chested but quite short. Luckily our trip to heaven was interrupted before we reached our destination for our Elijah's wagon halted just under the roof and we could step out. None of the three of us had shown any sign of fear.

Much later in 1909, when I met Mond shortly before his death in Rome, where in the cold months he lived on account of his health problems, I took the opportunity to ask him if he'd been satisfied with the success of his splendid endowment. With a degree of openness and unselfconsciousness that commanded respect he told me that he now considered it to have been a failure. They had not managed to initiate a tradition of physical chemistry there. The problem had been that though the director was competent he'd not been able to organise a group of co-operating researchers. Then he added that in the meantime he was convinced that his institute had not in any case filled a gap because there were in fact more possibilities to carry out physical chemistry research in London than were actually needed. His institute degenerated into a sort of club where a number of elderly gentlemen of leisure with a scientific bent would meet to smoke a pipe and play at being chemists.

Despite his personal dreams and hopes he had the ability to look reality squarely in the face and come to an objective judgement. I think this was one of the most important characteristics behind his extraordinary industrial success.

As to his verdict on the institute I think it may have been a little hard. He was at that time rather ill and perhaps to have seen the shadow side of things a little stronger than they were in reality.

This splendid man died at the end of 1909.

The Kaiser-Wilhelm-Institutes. Once Mond's endowment had made the founding an Imperial Chemical Institute if not certain then at least possible, a founding committee was assembled in Berlin and I was invited to join it. The result was a meeting on the topic at which I held a lecture on the background to it all. Once the organising committee had completed its work it turned out that there were so many people in Berlin who just had to be given an important post that only a subordinate role was left over for me. I turned it down.

However it never came to an Imperial Chemical Institute as we had planned. The general idea had diffused out to the royal court where the focus changed and resulted in the crystallisation of a rather different idea. This was that not just chemistry but also a range of other sciences should be given research institutes each of which would have an independent director but all the institutes would share a common administrative body. Kaiser Wilhelm II liked the idea and so the "Kaiser-Wilhelm-Institutes" were formed. Money was no problem because the Kaiser himself made the rounds of the rich industrialists with his begging bowl and they were richly rewarded for their contributions with medals and titles. This by the way is a convincing example of the value of a monarchy which our Republic, much to its detriment, has so thoughtlessly dispensed with.

I viewed this development as being somehow "unfair competition" because the Kaiser's iron hand was able to sweep all the money available for idealistic scientific aims into this one project so that there was no chance that other social causes could be funded. And I had a whole slew of projects in mind, none of which could be funded by this research institute program.

The International Union of Chemists. One of these projects was the organisational work within the field of chemistry which I alluded to above (This chapter, p. 518).

Since this concerned all chemists in the world, the idea was to combine them all and then work our way up from there.

This idea became concrete for me during a visit to the Swiss Natural Scientists' meeting in Basel. They had distinguished me by making me an honorary member and I thanked them by trying to organise the establishment of a Professorship for Natural Philosophy at the university there. It would have been the first chair in this subject and Basel would have been the leader in the field. However the idea must have seemed too fantastic to the sober Swiss and they did not pursue the matter. In the meantime others have taken it up.

The Parisian chemist A. Haller with whom I'd long been in correspondence was at that meeting. He was at the University in Nancy and had tried there to introduce German teaching methods which he rightly felt were the cause of the phenomenal rise of the German Chemical Industry. In the meantime he'd been offered a professorship in Paris where he carried on this work. He was one of those splendid people of Alsace who contributed a great deal to French academic chemistry. Our shared interest in organisational matters soon brought us closer and together we made a plan to form an international association of chemists. Remembering the wise words of Beernaert (This chapter, p. 518). I asked myself if local and national associations were already in place and found in the chemical societies of the various countries the necessary starting material. To organise all chemists all that needed to be done was to form a union of the ten to twenty national chemical societies. Thus the challenge, which had seemed impossible, became easily manageable and in my heart I thanked old Beernaert for his useful hint.

Haller suggested that he would propose the idea of an international union to the French society and at the same time I should do the same in the German society. After that we could get in touch with the sister societies elsewhere. I had to tell him that if I proposed such a thing in Berlin it would undoubtedly be immediately rejected out of hand. On the other hand, if the French society sent an invitation to Berlin then it would equally certainly be enthusiastically accepted. Haller shook his head but he did as I suggested. Not surprisingly the approach was a great success.

We agreed to contact the British society as well and that each should then send three representatives to Paris to discuss the final organisation in the spring of 1911.

The German society sent from Berlin my colleagues Jacobson and Wichelhaus. They tried to avoid sending me but Haller advised then that this would not be possible. Jakobson[1] was the general secretary of the German Chemical Society and hence it's appointed representative. Wichelhaus was one of the founders of the society and as such seemed well placed to describe its organisation. Britain was represented by Ramsey, Frankland and Meldola though the latter fell ill and could not attend. France was represented by Haller, Hanriot and Béhal. When we met with the German delegation Jakobson in particular could not disguise his displeasure at seeing me there.

[1]Ostwald misspelt the name. The correct form is Jacobson.

The line of discussion had been prepared by a subcommittee of the French society to which Gautier, Maquenne, Le Chatelier, Lindet, Bertrand and Urbain also belonged and they'd done their work well because the basic structure of the future organisation which they had prepared was later accepted.

The working areas were: Nomenclature of compounds in inorganic and organic chemistry, atomic weights, standardisation of formulae, organisation of publications, preparation of summaries of published work, general questions of the choice of language, standardisation of the format of publications, avoidance of multiple publication of the same work and preparation of a complete list of the entire chemical literature.

Once the idea of forming an association with the name "Association of Chemical Societies" had been accepted in principle, the French colleagues suggested a constitution which was accepted without much change.

National Science. The only difficulty was to decide if individual chemical societies should join the association or whether in each country in which there were several societies these should first be fused. I was in favour of the first option but the objection was made that in that case small and insignificant groups would have the same influence as the large societies with thousands of members. This was settled by requiring that a society must publish its own journal and have a certain minimal number of members before it could be accepted into the association. I'd objected to this "national" solution on the grounds that science is the most international thing there is and hence it makes little sense to define the nation as the basic unit of organisation. Then again there were soon to be difficulties: did Austria–Hungary have the right to one or to two representatives? Should Canada be considered independent of Britain?

I didn't manage to persuade a majority. Neither the representatives of Britain or France openly referred to it but it was clear that they were influenced by the fact that there were many more chemical societies in Germany than there were in their countries so that if one simply counted societies then they would be at a disadvantage. My German colleagues on the other hand liked the idea that with a national solution they would now represent all the other German societies while otherwise they would have been just one of many. And so I was alone with my suggestion, though I was far from convinced that it was not the best solution. I mention these aspects here because sooner or later it will turn out to be necessary to form other similar associations.

Voting Worries. Once this matter had been dealt with we turned to the paragraph that laid down that the chairmanship of the association would rotate annually from country to country in alphabetical order. Following standard diplomatic usage we used the French names for the countries so that Germany as "Allemagne" would have the chairmanship for the first year.

I noticed that Jakobson, my colleague from Berlin, was not all happy at the thought that I would win this election and he tried to drum up support for Wichelhaus. The discussion had lasted all morning. In the afternoon the election was to take place and it was likely that he would be elected, even if only by a thin majority, so long as I, as everybody expected, voted for him, After all I could hardly vote for myself.

Before the meeting I went with very mixed feelings for a walk alone. To begin with I was simply furious that in this way I was to be robbed of the fruits of my work. However I didn't want to let myself be carried away by these feelings which came pretty close to jealousy which I have always considered to be the lowest and stupidest of sins because it always ends up doing most damage to the one who is jealous. I worked on myself until I was able to unemotionally contemplate the thought that I might lose the election.

If, however, that happened then all of the work we had put in would be in danger because the job would be done merely formally but without initiative. Current business would be attended to in Berlin and I would as usual be excluded so that I would not be able to realise my organisational plans. Thus, objectively, it was my duty to make sure that I became chairman.

It would have been mortifying to have to vote for myself, because I would not have been able to explain my reasons for doing so. To abstain would have been even worse. Then it suddenly struck that since it was most unlikely that anyone else would abstain all I had to do was to vote for Jacobson and my election, though not certain, became much more probable.

There was tension in the air as the votes were counted. Since there were eight members present five votes would make the majority. I got five, Wichelhaus got two and Jacobson one. My worries had been baseless.

We parted tired but happy after the French society had held a banquet for us. In a dinner speech I summed the occasion up by saying that our child—it was to be called the "Association of Chemical Societies" was very modern since it had two fathers who were not even jealous of each other. In addition, the motherly duties were also split. It had been born in the lap of the French society, the German one had taken over its immediate care and the British society stood by as a careful nurse who was not entirely sure why people had to have children but was nevertheless determined to do her best to make sure that the child thrived.

The International Institute of Chemistry. During some sleepless hours in the night following the meeting I wondered how best to give the new organisation some physical form when I remembered talks I'd had with Ernest Solvay about the organisation of science. I decided to turn to him to help put the association on the map. He replied positively to my telegram asking for an appointment and, rather nervously, I journeyed from Paris to Brussels with the intention of asking him for practical help.

I hadn't made a mistake. He was in principle prepared to put up a considerable sum of money (a quarter of a million to begin with) in order to develop the idea on the condition that a close relationship be established with Brussels, perhaps in the form of a permanent office or later perhaps an institute. There seemed to me to be no objections to that for there were already many international organisations operating in Brussels. There was plenty of room in the Institute of Physiology in the

Leopold Park (Part III, Chap. 40, p. 549) and if necessary we could get more. In this way the future of the association was assured.

Organisational Work. The next challenge was to develop the association into a real world organisation, for at the start there were only the three societies in it. We hadn't invited the American society, which was the equal of ours, because there was no chance that three leading representatives would make the journey across the ocean so long as the association was merely in the planning stage.

Immediately after the Paris meeting, however, I had followed our decision and contacted the American, Italian and Russian Chemical Societies and all of them had applied to join. In addition I had informed all the chemical societies of other countries about the formation of the association and left to them to apply for membership. This was done by the following countries: The Netherlands, Switzerland, Austria, Norway, Denmark, Japan and Spain, all of whom were accepted. In this way the large majority of all chemists in the world were now represented in the association. One could soon expect that the few remaining nations would join for some of them had simply delayed until they were sure that the association would actually come into being. All in all I could be satisfied with the results of this part of my work.

The Berlin Meeting. With that, the first year of the association came to an end and in the spring of 1912 representatives of the member societies were invited to Berlin to the second general meeting. Around 20 colleagues attended.

The work went very smoothly because there were no great differences of opinion and those that did arise were soon settled by discussion. Once again, like at the world language meeting in Paris, I managed to get virtually all of the motions passed unanimously.

One of the main results which should be mentioned concerned the catalogue of chemical names which given the hundreds of thousands of compounds was a question that was as difficult to deal with as it was important. Extra subcommittees were formed for the various languages which had the job of keeping in touch with each other so that as far as possible a name would sound similar in the different languages.

To my great joy a motion I submitted that in the future the association should consider uniting the entire chemical literature in a synthetic world language was accepted.

The same was true for the motion to use a standardised format—to begin with for the publications of all the societies and after that for all publications in chemistry. At that time I had just completed the ground work for the standardisation of the format in the work for the "Bridge" which I will come to shortly. The association agreed to the "world format" of 160 × 223 mm and joined the "Bridge" as an association.

You will well understand that it was with a very satisfied feeling that I attended the final banquet which was hosted by the German Chemical Society. My Berlin colleagues, who hadn't wanted me in the chairmanship, were diluted out by the large number of foreign members who were satisfied with my leadership of the association. To be true the welcoming address given by the Berlin people was based on the idea that if only A.W. von Hofmann had lived long enough then he would surely have initiated the formation of the association and so that the merit was actually his. That, however, only increased my merriment and in my toast to the future of the association I reminded them of the Arabic tale in which the ill king could only be healed by being given the shirt of a happy man. After lots of searching they finally found a completely happy man—but he had no shirt. Today, I continued, the king could be healed because I was now completely happy and I had a shirt even if one couldn't easily see it (my waistcoat was covered with medals which I'd put on for the occasion). This was as well received as it was intended and I could happily turn over my chairmanship to my friend Ramsey as the representative of Britain.

A New Organisational Thought. The next meeting didn't take place in London but rather in Brussels. The discussions with Solvay had resulted in his agreeing to give the association in addition to the first quarter of a million, a further million plus a building when it agreed to meet once every 2 years in Brussels and to install a permanent office there. For Solvay the important thing was to support and stimulate scientific work in his country, and this was badly needed for of all the journals of the national academies, which roughly reflect their country's achievements, that of Belgium was the weakest both in terms of size and content.

I've often wondered how it came about that two neighbouring nations like Belgium and The Netherlands, whose climates are similar and whose populations are related, can nevertheless show such enormous differences in their scientific output. If one takes the Nobel Prize as a measure, then the Netherlands are not right at the top in terms of absolute numbers however, it is at the top when the population size is taken into consideration. A few years ago Belgium did not have a single scientific Nobel Prize. The only explanation I can find is that The Netherlands are protestant and Belgium is catholic. The result is that there is no illiteracy in The Netherlands, whereas there is a lot in Belgium. And since the level of achievement that can be reached is determined from the baseline it is not surprising that The Netherlands with its higher baseline has achieved so much more. By this argument the schools in The Netherlands must be better than those in Germany. I'm not an expert on the school system there but I do know that it was freed a generation ago from the smothering influence of the classical philologists and that therefore its youth now has a better chance to flower than ours does. As an example, as van't Hoff has more than once clearly said, he missed out on the "blessing of a classical education" and this may explain in large part how he was able as a young man to produce such enormous achievements.

For the organisation of his foundation Solvay had drafted a very liberal constitution which left the decision as to the details of its form largely up to the association and required Belgian representatives merely for technical administrative reasons. The most important and novel proviso was that the money would have to have been entirely spent in the following 28 years. He'd told me that he thought that a benefactor was necessary at the start of something new, but if it was going to survive then it should have learned in a reasonable time to stand on its own feet. In order to force that to happen he decreed that the capital must have been used up in the 28 years and that is more than enough time. In addition it made sure that more money was available each year than would have been the case if the association had followed the norm of living off the interest.

This piece of organisational wisdom impressed me and I stored it away for possible later use. As Professor in Leipzig I'd been involved in the administration of several charitable foundations which had become entirely silly because the circumstances for which they had been instituted no longer existed. I'd sometimes asked myself why those long dead benefactors should have felt it right that they dictate indefinitely the actions of their descendents. For books the personal intellectual property rights of heirs lasts for no more than 30 years, or in some countries for 50 years, after the death of the author. Why should money be given such an advantage over the personal achievements of great men particularly since no one checks how the money was made!

Foundations should not be permitted to exist for ever. Instead they should be examined every 30–50 years to see if they should continue or whether something else should be done with the money.

Brussels. It was with great satisfaction and even greater hope that I travelled the next year to the annual meeting in Brussels. There it was clear that the subcommittees dealing with the names of compounds had done sterling work and that the idea of dealing with problems together had made considerable advances. I myself had concerned myself with making a start on the challenge of putting together a complete collection of the entire chemical literature. I had planned, along the lines developed in the "Bridge", to prepare a large keyword based chemical register which would contain everything that had been published about a particular compound, process or concept so that this would become the complete repository of our science. By a suitable copying process it should be possible to give any searcher access to a copy of what he was looking for. Alongside this complete repository there would be a middle sized one from which all the outdated, doubtful or otherwise dispensable parts had been removed. Finally there would be a third small repository which would be a version of the middle sized one but with the entries reduced to keywords. Because of the form as a registry it would be possible to easily add in new results once they had been worked through.

This seemed to me to be the first and most important task, but it was not the only one. Because of the difficulties of language the registry would still remain closed to many. By translating for example the middle registry into a synthetic language—for example Ido—one would make the entire chemical literature available to everyone. After all, the time needed to learn Ido was a lot less than that needed to synthesise a chemical compound. One can start reading after half an hour's study of Ido and one will get through a document a lot more quickly than one would through a Latin text after 5 year's study of the language. If one restricts the vocabulary to that which is needed for chemistry then the dictionary can be very small.

These were just a few of the plans I had for the association and Solvay was fully in agreement with them because he too was an adherent of energetics.

The most important decision for the future of the institute we wished to form was of course the election of its future director. Since the institute was to be in Brussels it was felt that the director should be French speaking. From all those who might be considered no one seemed to me to be better suited than Ph. A. Guye from Geneva. I suggested this to him but he didn't want to leave Geneva. He'd just recently had an offer from Paris to take on the chair which the famous Moissan had held and he had turned it down even though his wife was from Paris. I was sorry about this but at the end I could understand him.

The question of the future institute had not held up the work which had already begun. In particular the collation of chemical nomenclature had made progress along the lines we had agreed. The British subcommittee had made a list of 18 points so as to provide structure to the upcoming discussions. Common language and format were among these points and both were referred for further consideration to committees as was the suggestion that a global address book of all association members be prepared.

The discussions took place in a friendly atmosphere in which German English and French were all spoken. However it turned out that the French colleagues had the least knowledge of the other languages; only Haller spoke German and understood English.

This however was not a situation that was going to last because the Italian chemists made it a condition of joining the association that their language also be accepted. It was only a question of time before the others made similar demands for their languages.

We don't need to go into details here. Work progressed on all fronts and by the time our meeting was over it was clear that the association was now a viable entity. As defined by the constitution the chairmanship now passed to the French. Haller of course was elected and he was certainly the best choice. I still have among my papers letters from him from the end of 1913 and the beginning of 1914 in which he described progress, particularly with respect to the establishment of the future institute.

The next meeting was set for the autumn of 1914 and of course Paris was naturally the venue of choice, though we could hope that the foundation of the institute would by that time be so far advanced that it might be preferable to hold the meeting in Brussels. The question of Paris or Brussels or perhaps both was left open for the managing committee to make the decision.

The End. The outbreak of the world war destroyed, among so many other things, the association of all the chemists in the world which we had so successfully started. As everyone knows, once military hostilities were at an end various groups got together to pursue the war against Germany by other means. One of these was a so-called International Organisation of Science whose goal was less the support of science and more the exclusion and persecution of the Germans. That organisation's chemical branch took control of Solvay's foundation for its own ends. German attempts to challenge this legally were unsuccessful. That was the end of this great and worthy effort.

Chapter 39
The Bridge

The beginning. When I got back from Paris and Brussels in 1910 full of thoughts and plans for the future of the association that had just been formed and happy at Ernest Solvay's understanding help, which had made our friendship a lot closer, I found a letter from Munich waiting for me. This soon kept me very busy, although both head and heart were already fully engaged in the matter of the chemical association. The letter came with a typescript of 177 pages and a short accompanying note in which the authors asked for a personal appointment. Their names were K.W. Bührer and Dr. Ad. Saager neither of whom I knew. In the preface it was noted that the plan was Bührer's and the manuscript had been written by Saager. The title was "The organisation of intellectual work by the Bridge". I read the manuscript through and the result was that I wrote back to Munich to say that I would be happy to receive them. The idea they had developed was so close to my plans and wishes and in particular there were a number of points at which it provided solutions, where till now I had only seen problems, so that I felt I could not refuse the hand they had extended. The visitors arrived a few days later and after a discussion lasting several hours they left with my promise of help.

A conflict. When I thought it all over the next day I began to feel a little ambiguous about the whole thing. The arguments in the manuscript had to a large extent convinced me for they concerned fundamentally new ideas which I was happy to support. Of course there were a number of points that I thought should be improved, though they weren't central and served more as starting points for further fruitful developments. In these points my guests had shown themselves to be flexible and said they regarded the manuscript as a preliminary means of gathering feedback on the scheme.

However I found it difficult to project this readiness onto the personality of my visitor. Karl Wilhelm Bührer was Swiss and came from the border area between the German speaking and French speaking cantons. His mother tongue was German. He was a little over average height, with a long and mobile face, brown hair, nervous in his movements and there was something odd in his nature and appearance. It was something that I hadn't come across before in any of the many

R.S. Jack and F. Scholz (eds.), *Wilhelm Ostwald*, Springer Biographies,
DOI 10.1007/978-3-319-46955-3_39

excellent people I'd met so far in different countries all of whom had an easily identifiable psychological focus which gave them the equilibrium so necessary to master some part of the world. I couldn't see this in Bührer.

It was odd that he hadn't written his thoughts himself but had rather left that to a "chance acquaintance" (as he himself characterised his relationship to Saager) and that had also made me a bit suspicious. Nevertheless he'd reminded me of the curious fact that I'd heard of before that German speaking Swiss have enormous difficulties in writing German. Perhaps this has something to do with the fact that their real mother tongue is "Swiss German" so that German, particularly written German, is for them like learning a foreign language. In any case I was aware of the facts of the matter and could accept them as an explanation for his unusual behaviour.

In contrast to the personal impression made by Bührer, the written text testified clearly to the presence of clear and penetrating ideas of very high organisational value and hence to an intellect that could formulate them. My judgement of the text was thus not in accord with my judgement of the man, but since the latter was based on a very shaky assessment I put it aside as probably flawed and accepted the manuscript at face value.

In the meantime I have learned that both judgements were correct and that this gaping discrepancy between the man and his manuscript was real. I still haven't managed to solve this riddle but I believe that the following explanation is not only possible but also probably correct.

When we began to work together Bührer was about 50 or perhaps a little older. From some things he let slip I gathered that in Switzerland he'd had a large and apparently successful business as a publisher, editor and advertising manager which had suddenly turned sour. After that he moved with the rump of his business to Munich. There he'd established himself in the circles of rising intellectuals in the coffee shops and as an older and self confident man he'd made quite an impression in ways that would most likely have put me off. He didn't talk about what had then happened but it had been a sudden shock and he claimed that he had been terribly slandered. Since he was still in contact with the Swiss consul in Munich I assumed that whatever it was it had not resulted in him having problems with the law or of being socially ostracised.

I think it was more a case that just like me he had worked too hard and become unable to properly look after his affairs. From his often excited and odd behaviour it would seem that his breakdown must have been serious and had perhaps resulted in a longer stay in a sanatorium and that it was not yet completely healed. Thus I had in front of me a man who in his youth had had these creative ideas but his breakdown had destroyed the best part of his youthful talents and now, as I learned later, he periodically suffered from depression and when that happened his views took on strange forms.

The bridge book. The manuscript consisted of two parts: the first was a general basis for the organisation of intellectual work and the second concerned the methods required to achieve this. One can probably ascribe the contents of the first part to his collaborator Saager while the second part contained Bührer's ideas. The line of argument was as follows.

To start with it was shown that the division between science and technology or between pure and applied science does not apply any more because science has become more technical while technology has become more scientific, and this is an inevitable development. Because of this, the differences in thought and approach and the economic imperative, as Ernst Mach had already recognised, now also applied to pure science as it does to technology even if in slightly different ways.

Then the principle of totality was stated as being that every challenge facing science and technology should be and has to be investigated completely. This was the point which I found hard to swallow. For me the totality principle in science meant that there is nothing that is not open to scientific investigation. The authors of the Bridge book took it to mean something rather different.

They considered that the totality principle demanded that nothing was too insignificant for scientific investigation because one could never know when it might become useful. From this they drew the conclusion that it doesn't matter what you look at because more or less everything should be investigated for future use. As you will see later, this was the point on which the great and hopeful project finally foundered. For this reason its worth looking at this question a little closer.

As far as I understand the workings of science nothing can be explained totally. No problem is ever explained in such depth that there is nothing to be added. On the contrary, in my own publications I placed special emphasis on those points which required further investigation. Everybody who has studied the master works of the great researchers will find a number of left over points which could be a spur to further work.

This is not just a mere academic point. If one understands that every piece of work is just a step along an endless way then one will try hard to make sure that everything was done in the best possible way but one will not try to convince others that it is the end of the story. One would merely try to make the biggest step forward that was possible and to do this one would concentrate on the central point and try to avoid secondary issues. In this sense one would apply good economic principles to a purely basic science problem.

In philological scholasticism, where only a very small number of interesting problems are available and where therefore there is always a problem of finding something useful to do, thoroughness is regarded as the highest ethical value. Every author has to know and cite and weigh up everything that other authors (even the stupidest) have ever written on the subject. It's clear that this is supported by the leaders in the field because it means that every achievement they can claim, no matter how trivial or worthless it may be, will at least be quoted by their followers in the field, even if nobody else wastes their time with it. The demand that everything should be considered of equal value is equivalent to a refusal to evaluate the various options and hence contravenes the energetic imperative which prohibits wasting energy in useless work. It is exactly here that we disagree with scholastic philologists who know nothing about the energetic imperative and constantly transgress against it. Scholasticism is to science, what bureaucracy is to normal life;

it too is characterised by an enormous discrepancy between the effort put into it and the value of what comes out—and it too derives from ignorance of the energetic imperative.

What was actually new in the idea of the Bridge was independent of all that and so to begin with I was prepared to accept it. I guessed that Dr. Saager had been educated in philosophy and philology and hence was predestined to make this mistake while Bührer knew nothing about science and so I assumed that it would not be a problem to show my future co-workers the proper way of viewing things. Sadly this turned out to be a one of the many times when I have allowed myself to be deluded with optimism.

Fundamentals of organisation. Viewing all human knowledge as an organism led to the principle of the division of work as the basis of progress. The range of work which can be undertaken is constantly extended by the development of ever more and ever more finely elaborated organs. However all the individual pieces of work have to be integrated with each other and this leads to the second principle—the linkage of work. Both principles have started to crystallise out as applicable methods and the function of the Bridge was to identify and develop the most successful protocols.

Division of labour is easy; its integration is much more difficult. In other words what was missing was an organisational scheme. Because of this every little group was like an island working on its own and what were missing were the bridges between the islands. It was to be the main job of the Bridge to form these connections.

To begin with we should see what was already available. What was initially needed was a register of all the islands both in terms of who worked there and in terms of what results had been achieved. Once this was done the Bridge would be in the position to act as a clearing house for questions that required links between the islands. To do this organisational and technical facilities are required because as each new island develops there is no way of knowing a priori what connections to other islands will be useful in the future.

If intellectual work was organised in this way then one would expect a vast increase in what could be achieved. Till now, for example, each researcher had to look into the background of his problem on his own and quite often the same work would be done several times because the individual scientists were unaware of the parallel efforts. In future the "Bridge", by virtue of the network nature of its organisation, would naturally be able to produce and make such summaries available so that the researcher could immediately get into the work of making new discoveries. By having a catalogue of all currently available information the "Bridge" would be able to connect each enquiry to the appropriate answer. It would become the central information office of all information offices.

A suggestion as to how this system would help secure world peace completed the first part of the manuscript.

Technical basis of the system. If an idea is to be incorporated into human culture it must be able to be communicated and hence not be bound by space or time. This is done by expressing the thought in a language and hence the written page becomes the technical basis of all culture, that is to say of the collected achievements of creative people of all nations and of all times. If one is going to organise intellectual work then one has to start by providing an organisational framework for these written pages.

This was Bührer's idea which I considered new and important and it was the reason why I felt it necessary to support him in any way I could.

It has long been common that notes which refer to the same thing should have the same format. Now notes referring to lots of different things were to be brought together. Clearly these notes would all have to have the same format. Of course there are not only sheets of paper but also notebooks, pictures and so on, and so they too should have the same format. In this way there emerged naturally a requirement for a standard format for paper.

At the time this was far away from being a generally understood need. A few international organisations like the world post, with its standard for postcards or the international library centre in Brussels with its catalogue cards, had established certain norms but they had done so without giving any thought to producing a radical standardisation of all formats. When he was still working in advertising in Switzerland Bührer had suggested a format called "monoformat" which he thought was a good solution and he'd seen to it that it was used throughout his business. In America too people had been concerned with the question. The result, which was suggested in the "Bridge" manuscript, was a format of 11.5×16.5 cm. Larger sheets would be able to be folded to yield this size.

The arbitrariness of this choice was excused by the argument that all such formats are in the end arbitrary. I wasn't happy with that and decided to try to find a better solution which I managed to do within a few weeks. I'll describe that a little later.

The monograph principle. Another fundamental idea of Bührer was the discovery that the book binder is one of the greatest enemies of the organisation of science. In 99 % of cases the entire contents of a book are never used. Often enough all that is used is just one chapter, or even just one page, for example for a particular table. When this is the case, then the unused rest of the contents of that book are unavailable to others. However, if one could dismantle the book into chapters and paragraphs then each user would only have to take out the part which he required and everything else would be left for others.

When new editions are prepared it is usually the case that only a few pages or words are changed while the rest remains unaltered. The owner of the old edition however has to buy the new one if he needs access to the altered parts. He then owns two copies of most of the book. Were the book to have been published in a deconstructed form then he would only have to buy the few altered pages and thus would have money available to buy other books.

It is therefore important that the results of intellectual effort be published as separate monographs or reprints. In scientific journals that does happen but then comes the book binder and binds all the reports, whose only common point is their shared year of publication, into one huge volume and by doing so he destroys the independence of the individual reports.

The advantage which would be achieved by individual publication was quite rightly compared by Bührer to Gutenberg's discovery of the printing press which had been so fundamental for the development of Europe. Before Gutenberg books had been printed but this had involved carving the entire text into wood blocks and then using the blocks to print. The advantage of Gutenberg's process thus lay not in the ability to print but rather in deconstructing the page into individual letters which could be put together in any desired combination. In this way one set of letters could be used to print any desired text. In a similar way each person will be able to have his own books by putting together monographs which reflect different aspects of a subject of interest and the book can grow and divide or can be altogether removed as his interests change with time.

All this however can only be done once the basic format has been fixed.

Two other important considerations also sprang to mind. First, we would need a common language which could only be a synthetic language with unrestricted plasticity. Second a system to categorise all human knowledge would be essential and for this the system invented by the American Dewey could be used. In this system numbers are given first to the large divisions and then to ever smaller branches and finally to very restricted fields. In this way every term is assigned a set of numbers each one of which narrows the field defined by its predecessor until one reaches the level of the term itself. This is in principle the ancient organisational principle of Aristotle which always goes from the general to the particular.

For these two problems no new solutions were required. Instead the already existing solutions could be integrated without alteration into the "Bridge" where they would serve to accept the inflow of data and direct it to the appropriate output.

A further challenge which the "Bridge" would devote itself to was the organisation and standardisation of colours.

The founding of the "Bridge". These ideas seemed to me in their breadth so important and so full of potential that I decided right away, together with Bührer and Saager to formally found the "Bridge" and to provide the money necessary to do this. It seemed to me that the money I'd recently been given as part of the Nobel Prize would find no better use and would be in line with Nobel's ideas. The money was then pretty much used up in this enterprise.

The founding meeting of the "Bridge" took place in 1911 in Munich and at it I was elected to be chairman, Bührer to vice chairman and Saager to secretary. Since the other two had no talent for public speaking, I held the main lecture which was enthusiastically received. The business centre was moved to Bührer's Munich apartment and from there he at once began a propaganda campaign which attracted a large number of excellent individuals. Some of the names in the membership list were: G. Kerschensteiner, Munich, S. Arrhenius, Stockholm, F. Bajer, Kopenhagen, P. Otlet, Brussels, W. Exner, Vienna, H. Beck, Berlin, P. Beck von Mannagetta, Vienna, E. Solvay, Brussels, K. Sigismund, Berlin, E. Rötlisberger,[1] Bern, A. Schlomann, Munich, F. Oppenheimer, Berlin, Hj. Schacht, Berlin, P. Langhans, Gotha, K. Schmidt, Hellerau, K. Oppenheimer,[2] Berlin, E. Jäckh, Berlin, St. Bauer, Bern, P. Behrens, Neubabelsberg, E. von Behring, Marburg, J. Brinckmann, Hamburg, E. Francke, Berlin, A. Gobat, Bern, E. Metschnikoff, Paris, H. Muthesius, Nikolassee, H. Poincaré, Paris.

We had thus come a long way towards our goal of making the "Bridge" into an "organisation of the organisers" because these were all people who in their different ways contributed to culture.

The world formats. For me there developed out of the general idea of the "Bridge" a number of special challenges. I'd been concerned with the question of the definition of norms on the basis of absolute values and had been led by energetic considerations to overcome the false viewpoints of current physics which had resulted from a conceptual error of J.C. Maxwell (Part II, Chap 21, p. 237). In this way I had developed a longstanding interest in these matters particularly as even today there still exists the wide spread error that all physical values can be derived in terms of time, space and mass.

In the case of the suggested monoformat it seemed to me to be necessary to put it on a solid base rather than defining it arbitrarily. For this the axiom should be applied that no new norm should be introduced where one already exists. The height and breadth of the format are lengths and the unit of length is the centimetre. Thus without any arbitrary assumptions it was clear that the centimetre would be the unit of measurement.

The second point here was that a single format would not be good enough. A whole range of formats, from the placard in the street to a postage stamp, would have to be defined and for technical reasons and to minimise waste they should be related to each other by doubling or halving.

When one does this however, then the relationship of the sides to each other changes. If one starts for example with a square with sides 1:1 then if one folds it in half there is a rectangle with sides 1:½ and if it is halved again then we have again a square with sides ½:½. This is obviously a great disadvantage. Is there no way round it?

The answer is that there is one, and only one, rectangle that when folded in half always results in a geometrically similar shape i.e. in a rectangle for which the sides have the same relationship to each other. Lichtenberg had already set himself this problem and had found the correct answer. This splendid rectangular format has sides of $1:\sqrt{2}$ which in round numbers is 10:14 (or better 14.14). A rectangle of this format is pleasing to look at and indeed it turned out from a large number of measurements that the format chosen for books usually turns out to be close to this one.

[1]Misspelt by Ostwald. Correct form is E. Röthlisberger.

[2]Misspelt by Ostwald. Correct form is C. Oppenheimer.

This then gives us a rationally derived format which is based on centimetres. The first level (level 1) has the dimensions 1:1.4 cm. The second is 1.4:2, and the next ones are 2:2.8, 2.8:4, 4:5.6, 5.6:8, 8:11.3, 11.3:16, 16:22.6 cm and so on. In each case the second value behind the decimal point is rounded off.

Luckily the format which Bührer had suggested 11.5 × 16.5 cm was so close to the rational value of 11.3 × 16 cm that Bührer immediately agreed to abandon his suggestion in favour of the rational solution. In this way we avoided the problem that different shapes would be produced by folding a sheet in two.

We now had something with which we could start on the practical work. I wrote a small pamphlet on the "world format", as we called it, which was widely read and reported in the daily newspapers so that the idea became generally known. Bührer himself worked hard to support the idea because he was well aware of the practical value of the matter from his publishing experience in Switzerland. We soon had success with this in many widely separated areas. We held to the concept that the use of the format should be completely voluntary for we were sure that once the usual inertia had been overcome the advantages would be so obvious that they would result in an increasingly rapid acceptance of the norm.

The annual meeting. In order to give the "Bridge" a public presence I suggested to Bührer that we organise a public meeting in Munich in 1912. Just before that the moral weight of the "Bridge" had been considerably increased because I had managed to persuade Wilhelm Exner to take on the position in the managing board which Dr. Saager had so kindly agreed to vacate.

The Munich meeting was most impressive. A considerable number of the leading citizens of Munich, first and foremost King Ludwig, attended the opening ceremony at which I held a lecture. I'd been told that the planned time had to be scrupulously adhered to because the King had to go on to a second celebration. I therefore faced the challenge of keeping my remarks memorable and yet keep strictly within the 35 min assigned me. It's fair to say that it was one of the best talks I ever held and it lasted exactly 34 min without my having to alter the tempo to fit the time frame. I'm not sure, but perhaps the three cola sweets that I sucked before starting had something to do with this success for they always worked well in similar circumstances. I never tried to decide whether this effect was directly physiological or whether it was psychological i.e. a placebo effect based on belief. In any case it was certainly there.

I might add that afterwards I have held good lectures without the benefit of cola sweets; then again some of these lectures were not so good.

However when the costs and the earnings were compared it turned out that the former far outweighed the latter. Since I'd decided to use the substantial Nobel Prize money for this project I readily agreed to Bührer's request for further financial support and left the day to day administration entirely to him. In any case I had no time to look after the administration myself because this was the period when I was leading the Monist Society and that required a lot of time and effort. I found the first set of accounts not terribly clear but I assumed this was due to my lack of experience in these matters because they had been certified by an official accountant.

Gathering shadows. To begin with blue sky reigned over the "Bridge" but soon clouds began to gather. The first of these came in the form of complaints about Bührer's management style from an employee who'd been fired. I took this to be an attempt to get revenge and didn't investigate further. Then there came complaints from members of the governing body in Munich. As an example of what the "Bridge" would be able to do Bührer had decided to collect a complete set of all the postcards of one city (he'd chosen Ansbach) and only by my vigorous intervention was I able to stop this project which I rightly assumed would have made the "Bridge" seem ridiculous. This was an example of the consequences of the ludicrous idea of "totality" (This chapter, p. 533).

Instead of the postcards Bührer now turned to the advertising pictures which were produced in enormous amounts and were avidly collected by children. He collected them from all over and pasted them onto cards of the correct world format. To begin with he did this himself but soon he set the staff of the "Bridge" to doing it since otherwise he'd never have been able to handle the thousands of cards. When, all too late, I heard about this it struck me as being an ominous turn of events.

However in the meantime there were some successes. A voluntary worker who was a keen skier had managed to persuade the German ski clubs to make all of their announcements in the world format and to centralise their transportation which not only led to considerable savings but also improved and speeded up their work. This had been so marvellously well done that I considered it to be the best example of the organisational advantages which we could achieve.

At the same time the complaints increased. Important letters were left unanswered for weeks on end, deliveries were not paid for and in short the organisation of the management, which by the "Bridge" should have been perfect, seemed to be terribly deficient.

Eventually the number and the seriousness of the complaints increased to such an extent that I decided I would have to personally go and see to matters. Since I knew that I wasn't in a position to judge the business side of things I accepted the help of my son Walter who looked into the complicated and unpleasant situation with eagerness and dedication for which I am much in his debt.

The collapse. The result was that the "Bridge" had to be wound up. There had been no dishonesty involved, though considerable sums had been thoughtlessly and uselessly distributed on all sides and the clarification of all this was far from easy. It seemed to us that Bührer had lost control of the business side of things and that he had been faced with matters that he was not competent to judge. He seemed to be incapable of distinguishing between important and secondary matters which were no more than silly games. We couldn't get an explanation from Bührer because he was ill and communicated with us only indirectly through a confident.

I was unable to understand Bührer's behaviour at the time. I've already related that his Swiss business had ceased to exist before the foundation of the "Bridge" and that he had hinted that that was due to a serious illness. I also mentioned that I'd guessed that this illness was in all probability a serious mental breakdown and that his incomplete recovery had left him unable to shoulder the responsibility of managing the "Bridge". His attempt to do so had led to a relapse. This would at least explain his strange behaviour. He died a few years later.

I saw to it that the "Bridges" debts were paid and in return took over the remaining assets including the innumerable pasted advertising cards. Soon after that the world war broke out.

This was the end of a very hopeful undertaking but the organisational ideas behind it, which at that time were not widely known, are now common currency. Much of what we were trying to do has in the meantime been achieved. This includes the standardisation of the format of paper, the deconstruction of books into separate monographs and the organisation of colours. I add here a few words to the first two points; the last grew into the central element of my last years.

Committee work. I tell here the further story of the world formats although this was something that developed at a much later time. After the unfortunate outcome of the world war there began zealous attempts all over the field of technology to develop better organisational forms especially in terms of defining norms. In other words the program that the "Bridge" had set itself 10 years earlier for intellectual work now became applied to the entire field of technical work. Since the time was now ripe for this, the necessary ideas sprang up in all sorts of places.

The question of the format of paper was handled by a subcommittee that was largely made up of representatives of the publishing houses, though manufacturers of presses and paper as well as government officials and the major users of paper were also involved. The whole was led by the Leipzig book industry which also sent representatives to the subcommittee. I was invited to join the discussions. The meetings took place In Leipzig and began in 1919.

It soon turned out that few of those present had given any thought to the question of the format of paper. I realised at the first meeting that I was the only one present who understood the matter and so I laid out the three prerequisites that must lie behind the formation of such a standard. These were that the formats must be rectangles which when halved gave rise to the next in the series. The relationship of the sides must be $1{:}\sqrt{2}$ because only then does the geometric shape remain the same as one progresses through the series by folding the paper in half. Finally, the basic unit of measurement should be in meters or centimetres.

In the course of the discussions it soon became clear to me that the leading men there had a visceral dislike of me. I didn't try to find out why, but I assume that it was a side effect of the atmosphere in which I'd terminated my relationship to Leipzig University. There arose a clear wish to try to find some solution other than the one I had offered and this resulted in a mass of senseless work which was in complete contradiction to the purpose of the exercise which was to avoid wasting energy. Many different suggestions were made including one that would have resulted in so many different "norms" that, as the chairman pointed out in his support of it, no change in the current situation would then be necessary.

It took a few years before the members of the committee had understood the problem sufficiently and then they came back to the suggestions I had made. Nevertheless, to demonstrate an illusory independence, the metric units were not, as

I in line with what was done in many other fields had suggested, applied to the length of the sides, but rather to the areas of the rectangles and this gave rise to some very odd numbers. In this form the norm was adopted and it has begun to be slowly accepted. From my experience with the "Bridge" I'm sure that it would have been more quickly accepted had the original "world format" been used.

The collected editions. The most original idea of those who planned the "Bridge" had been to remove the obstructive work of the bookbinder by deconstructing books into their basic parts. Of course this would not apply to large works of art like plays or novels which have to be viewed as a whole. However it would apply to scientific works, to newspapers and the like. The objective was to permit each individual to put together his own personal collection of information which would be tailored exactly to his needs and which could be updated and altered as his interests changed with time.

The idea was not really new to me because this was more or less what I'd done from the beginning of my time in Leipzig. The "Classics of the Exact Sciences" (Part II, Chap 17, p. 176) had arisen from the wish to make important parts of the scientific literature available to the public as single papers which had till then almost all been published in journals and the success of this series attested to the need for it. I myself in this final period of my work have again and again turned for some information I needed to one of the "Classics". But I had never extended the thought beyond this application to the scientific literature. I was therefore all the more ready to seize on it when it was presented to me in a much more general form and enriched by Bührer's idea of a standardised paper norm.

In the course of my organisational work for the Association of Chemical Societies (Part III, Chap 38, p. 521) it seemed to me that an opportunity had opened up in my own branch of science. One of the important points in the program for the international chemical institute which Solvay supported was the preparation of a complete handbook of chemistry by collecting and collocating reports about every chemical compound and concept. From this every chemist would have had access to the entire literature of every aspect of the subject and it would also have represented a complete history of the science. Unfortunately the world war had destroyed this hope probably for many years to come.

When later on I felt the need to make my research on the concept of colour public our scientific publishers had fallen on hard times and submitted manuscripts sometimes had to wait for years to be published. I therefore formed a new type of publication which was based on the ideas of the "Bridge". I called it the "Collected Edition" and brought it out in place of the other journals. Just like the journals the Collected Edition prints individual papers but they are published in such a form that each paper can be treated as an independent unit. Instead of a bound journal the Collected Edition comes out as a folder, which contains a number of individual papers, and these folders, just like the journals, are published at intervals and each folder contains, on average, roughly the same number of papers. Each paper in the folder has its own page numbers, a numerical identifier, the date of publication and key words which show the principle areas to which the contents relate. The technical details of the printing, which require an understanding of the publishing process, were looked after by Dr. Manitz. This collected edition was initially called "The Colour" but this soon turned out to be too limiting, because I soon had to publish my research on the concept of form in it as well. Some 40 such folders have been published and this format has in the meantime been copied by others so that this concept too has survived the failure of the "Bridge".

The third challenge, the organisation of colour, has occupied me in the last part of my life ever since 1914. It required a synthesis of my organisational, physical, chemical, physiological and psychological knowledge and this has led to the quantitative theory of colour which I consider to be the pinnacle of my scientific achievements.

The value of this new field can be seen in the fact that already new fields of science such as the theory of form and the general theory of beauty (Kalik)[3] have begun to grow out of it.

[3]This was the term Ostwald proposed for his theory of beauty.

Chapter 40
The Energetic Imperative

*The categorical imperative.*Kant's fame and influence are based to a large extent on his "Critique of Practical Reason" which, in the edition of Hartenstein is a little book of 169 pages in which he presents the concepts of God, freedom and immortality in an extremely clever manner and by doing so rescues them from the floods of "pure reason" which had threatened to drown them. He did this by constructing next to pure reason an independent form of practical reason in which they could be embedded. By doing so he won over, in addition to those few who admired his research into areas of logic and psychology, masses of people for whom this book satisfied their thirst for metaphysics.

His basic idea was that practical reason must be based on purely formal arguments divorced from any factual elements. In this sense then he formulated the proposition which has become famous as the "categorical imperative":

> Act so that the maxims of your will could be taken as the principles of a general law.

At first glance this proposition does indeed seem to be entirely formal since the demand for general applicability appears to be purely logical and it is so formulated by Kant. However, if one looks closer then the generality that is demanded is not logical but social: this law is there to regulate human interactions and is far from being merely a general concept.

Thus here one comes across the social nature of ethics in a place you would least expect to find it. This hidden social aspect is what makes the categorical imperative so fruitful.

Since I of course was not satisfied with Kant's idea of the inborn nature of moral laws, I asked where the source of ethical and social fair-mindedness might be. Here I came across a much older thought. Much earlier, when I had realised that the second law of thermodynamics applied to everything including human actions, I'd looked at it in the light of the concept developed by theological ethics of the sin against the Holy Ghost. As everyone knows this is the only sin which can never be forgiven, although, as Lessing pointed out, we don't really know what this sin consists of. This sin therefore has certain similarities to the dissipation of energy whose consequence, the reduction of energy can never be made good. This idea was

R.S. Jack and F. Scholz (eds.), *Wilhelm Ostwald*, Springer Biographies,
DOI 10.1007/978-3-319-46955-3_40

more or less just a game until I realised that all cultural activities are directed towards restricting the dissipation of energy in nature so that it can be used for human purposes. Unregulated energy dissipation is the squandering of energy and so quite naturally the categorical imperative of energetics becomes: Don't waste energy, use it.

The line of argument which is given here briefly had in fact taken 10–15 years to develop.

It clearly demonstrates the relationship of the categorical to the energetic imperative. The categorical imperative is claimed to have its source in the supposed inborn conscience and hence to be absolute. The energetic imperative has its source in the inescapable natural circumstances into which man is born and it defines his relationship to the environment and to the level of culture which he has attained. Culture, however, is a product of society and Kant's imperative turns out to be an answer to the question, "What limitations does the second law set for the prospering of society?" The answer is fairness because every unfair act results in opposition and hence in the dispersal of energy. Thus the categorical imperative turns out to be simply a special case of the energetic imperative. These brief hints will be fleshed out with a more detailed account later on.

The pyramid of the sciences. For me the way from physical chemistry to the science of culture (which is often less usefully referred to as sociology) went by way of energetics. Already in my first attempt to organise my thoughts for the lectures on natural philosophy at the beginning of the twentieth century, it had become apparent that the scientific investigation of human affairs was the most difficult, but also the most important challenge. Then I came across Comte's organisation of science which had driven him to propose the existence of a highest level of science which didn't exist at that time and which he proposed should be called sociology.

A great advance in my thinking came with the recognition of the fact that the simpler and general sciences are prerequisites for the higher specialised sciences but that the reverse is not true. In this way chemistry is essential for psychology but you don't need to know any psychology to understand chemistry. Earlier I liked to depict the totality of science as a net in which the connections between the knots are known. Now I understood that this image is imperfect because it does not reflect the essentially hierarchical nature of the relationships.

Now it seemed to me that science was more of a pyramid in which the lower levels support the higher ones but not vice versa. Though this idea is very simple it turns out to have important consequences, because it allowed an overview of the entire corpus of human knowledge. Later as we go along I'll have a number of opportunities to point out some of its applications.

The idea first struck me in 1903 during my first journey to America (Part II, Chap. 27, p. 319) as I prepared my address for the opening of J. Loeb's new institute. Since, as a chemist, I had to say something about biology I first of all had to get the relationship of these two areas of science clear in my own mind. To avoid any errors in this assessment I needed to clear up the relationship of all the sciences to each other. In doing this I constructed the pyramid of science for the first time.

I was soon able to make a significant advance on Comte by making not mathematics but the science of organisation as the most general of all sciences. However, at that time, because of the special nature of the talk I was to give, I only worked my way up to the level of psychology.

I'd already applied the concept of energy to psychology in the lectures I'd given on Natural Philosophy by defining all mental processes as energetic that is to say as processes driven by the transformation of chemical energy present in the nerves and brain. This idea was at odds with the then current dogmas as can be seen from the comments that the psychologist W. James made about it in a letter to me (Part II, Chap. 26, p. 309). What was important to me was that it got rid of Dubois-Reymond's[1] notorious "riddle of the universe" which was concerned with the question how the mechanical motion of atoms in the brain could be converted into thoughts. Leibniz had already pointed out the impossibility of this idea. However, instead of recognising that processes in the brain could not be based on atomic motion—a view that the dogmatic mechanic Dubois-Reymond had been unable to grasp—he confused his own incompetence with the unlimited possibilities of human thought and romanticised it all as an eternal "puzzle of the universe".

No such puzzle was required for the energetic explanation of mental processes, because it became merely part of the natural phenomenon of energy transformation. We cannot currently devise and test useful hypotheses over the details of these processes because our current knowledge of the chemical dynamics of the brain is so rudimentary.

I first included sociology in the pyramid of science in 1904 for my lecture in St. Louis (Part II, Chap. 29, p. 370). In the meantime I'd come across Comte when I'd started to take an interest in earlier attempts to deal with the problem of the organisation of the sciences. As the highest science which encompasses all the so-called humanities, sociology makes use of all the other sciences and these can be divided into 3 groups; organisational, energetic and physiological-psychological sciences and these in turn are further subdivided into the different branches of these sciences. For me the energetic basis of sociology was the most interesting part.

Sociology. To begin with sociology was for me was just an empty frame that contained no graphic picture. I first began to see something in it during the journey to St. Louis (Part II, Chap. 29, p. 362) when I met the sociologist F. Tönnies and heard from him some interesting arguments. Once I'd resigned the Leipzig professorship and had time to do what I wanted, I slowly began to formulate a set of interrelated thoughts on the subject.

Part of this was driven by the request from the publishing house of J.A. Barth that I write an article for the first volume of a collected edition entitled "Knowledge and Competence". I agreed provided that I might use the title "Energy" and this they were happy to accept. In this way I put together for the general reader a short description of the meaning of the term energy. This was different from everything that had gone before, firstly because I kept it free of the kinetic hypotheses which were till then

[1]Ostwald misspelt this name. The correct form is du Bois Reymont.

inseparable from this concept, and secondly because I strongly emphasised the fundamental importance here of the second law (Part II, Chap. 21, p. 238). The little work was widely read so that it had to be reprinted and it was also translated into other languages. It contributed significantly to bringing the ideas of energetics across to the average reader.

In the last part of the manuscript I included a section on "sociological energetics" in which I tried to lay out the basis of this new and very wide field of study.

Legal energetics. Among the new fields which now awaited further work was one that I had already started on, namely the law. In my last years in Leipzig the backwardness of legal concepts had been underscored by the fact that in the following case no reason for punishment could be found. The accused had secretly tapped into someone else's electricity supply and used it to power a lamp for his own use. This was clearly theft but the judge refused to refer to it as such because the Corpus juris,[2] and hence all the law books, defined a theft as the illegal appropriation of some movable thing belonging to someone else. In this case "electricity" had been stolen but since this is not something you can grasp or weigh it cannot be described as a thing. The confusion was even greater because not even electricity (in the physical sense of an amount of electricity) had been stolen because this had remained unchanged in the main circuit. What therefore had actually been stolen? Of course it was clear that not even the most serpent tongued lawyer could argue that using someone else's electricity to run a lamp or a motor or whatever, is a fundamental human right.

To me the case was very welcome, because here what had been stolen was the electrical energy which is the article of value that is produced from coal by machines in the electricity works and has to be paid for by those who use it. In this case energy, which was widely viewed merely as an abstraction was something real that was produced and sold as a product and hence could be stolen.

I wrote an explanatory essay for the German Legal Journal, but it was not well regarded by the jurists. Their comments took the quaintest forms which merely went to show the severe deficiencies of a legal education. The result of this case was that a new law was passed which provided for the punishment of those who illegally expropriated energy.

I made the following point in discussions about this with my Leipzig colleagues from the law faculty (though I could endure only the most infrequent and subtle contacts with them). As the sciences awoke at the end of the Middle Ages they had to free themselves from the earlier prejudice that the Greeks and Romans had reached an unsurpassable level of achievement. Modern physics began with Galileo's struggle against Aristotle's preconceptions in mechanics, astronomy with the rejection of the ancient belief that the earth was the centre of the universe and so on. Every modern science underwent not a renaissance but rather had to be born anew and only managed to unfold once the ancient views had been overthrown. One can see this in every branch of science—with the sole exception of

[2]The legal term Corpus Juris refers to the entire body of law of a country, jurisdiction, or court.

jurisprudence. Just as the doctors in the Middle Ages swore by Hippocrates and Galenus and the mathematicians by Euclid and considered it a sacrilege to go beyond these sages, so the lawyers considered the Corpus juris to be the epitome of all legal wisdom, though its failings were all too obvious. The senior member of the Leipzig law faculty, Windscheid, had energetically contributed to the lawyer's bible—The Imperial Civil Law Compendium—and had made it so useless that the whole thing had had to be completely revised.

I should point out here that the situation I am referring to was that of 25 years ago. I am well aware that in the meantime the essential work of removing ancient prejudices has begun and has made some headway. How much progress has been made is something I am not able to judge.

I have recounted these things because they made me think continuously about these questions and in particular because they led me to understand that they are part of sociology which was an area that was then coming into the forefront of my mind. I became a member of a society for the philosophy of law which had been founded by the Berlin lawyer Josef Kohler. I'd read about his progressive views which he was very successful at getting aired in the daily newspapers. It was this society that invited me in 1910 to hold the lecture on the topic of "Function and Value" in the context of the philosophy of law which led me for the first time to articulate the energetic imperative (see below). However, as so often in these situations, I soon run up against the determined opposition of the experts in the field. Kohler himself had no interest in getting himself involved with a line of thought which he had not initiated. Gradually it became clear to me that the society was really just a private enterprise to generate influence and prestige for Kohler and so I no longer associated myself with it. In any case I had too many other things to do at the time.

All this had however directed my attention to the cultural consequences of the second law. In the first brief sketch of energetic sociology with which I had closed the little book on energy (This chapter, p. 545). I had already gone so far as to recognise that the maximisation of the efficiency of energy transfer from a lower to a higher level is a general task of the entire culture. In this way I had enunciated a law which not only permitted a broad general understanding of these things but was also something that could be directly applied to specific problems.

This sort of law can be expressed in a number of different ways depending on the point of view. I instinctively looked for the shortest and most demonstrative version and found it in the energetic imperative.

The energetic imperative. I've already described how the idea of the energetic imperative developed (Part II, Chap. 21, p. 227). By putting together the first and second laws of energetics for the purpose of developing practical applications I came to the formulation: "Don't waste energy, use it" which I called the energetic imperative and which has in the meantime become part of the normal way of thinking of every thoughtful work right up to the highest levels.

I've tried to remember when I first used this name for it, but without much success. It certainly arose first in 1908 during discussions I took part in concerning the social significance of the second law—a matter that was of great interest to me at the time. The results of my thoughts on the matter were published in the little book "The energetic basis of Cultural Science" whose preamble is dated April 1909. However neither there not in the "Bridge" book of Bührer and Saager, which was published 2 years later and contained my thoughts on the matter, was the energetic imperative referred to as such.

On the other hand I clearly remember that during the discussion after a lecture I held in Vienna in 1910 in the Society for Legal Philosophy on the subject of "Purpose and Value", I summed up my thoughts on the matter in the form of the energetic imperative. This may have been the time that it first saw the light of day before an admittedly rather small audience.

In discussing the work of Lichtenberg, the intelligent physicist and writer of popular literature, Goethe remarked that whenever he made a joke you would be sure to find a problem lying hidden behind it. I quite often do something similar: when I formulate a new idea it is often in form that I like because it comes across as really novel and yet later I see in it new and fruitful relationships. To begin with I tried to convince those I was talking to of the importance of behaving in a way that took account of energetics by putting it on the same level as Kant's famous "categorical imperative" by referring to it as the "energetic imperative". Only later, after due consideration, did it become clear to me that the energetic imperative was not only equal but was in fact superior to the categorical imperative. The categorical imperative is restricted to the social interactions of individuals while the energetic imperative applies and governs all of the actions of an individual.

The first place I can find where I used the term before a broad public was in an essay that I published the following year (1911) in the Berlin newspaper "Tageblatt". In the same year I published a second essay under the explicit title, "The Energetic Imperative" which explained its content and breadth of application. The term soon became popular so that in 1912 I put together a collection of essays most of which had been published in newspapers and gave it the title "The Energetic Imperative". The roughly 50 essays on a whole set of different topics nevertheless could all be quite naturally incorporated under this title. From this time on it became possible to use the term without putting it between inverted commas.

Ernest Solvay. I got a new impetus in terms of social-energetic thinking when Ernest Solvay sent me the publications from the Institute of Sociology which he'd founded. They showed that this remarkable man had worked his way towards a form of energetic thinking and in doing so he had come up with some highly individual and important ideas. I had been happy to review them positively in the "Annals" and this led to the development of a personal relationship which must be counted as one of the most outstanding of the many valuable and enthralling experiences I have had.

In particular he had thought widely about the application of energetics to social questions. As always with talented but self-taught amateurs his results were a mixture of far-reaching good ideas together with some pretty primitive errors. Since superficial reading tended to make the latter stick out he had come across only polite rejection when he'd tried to interest other researchers in his work. To provide a home in which these ideas might develop he had built his own institute and funded it well from his vast fortune. It was typical of him that he knew that he shouldn't do this just to push ahead his own hobby, so he'd at the same time founded two other institutes; one was devoted to physiology and the other to business education. Each of these institutes was given a large house with all the appropriate, even magnificent, fittings and a scientific staff. The directors of the three institutes were paid a salary such that there was no need for them to earn money elsewhere in order to keep themselves and their families in comfort. They were also guaranteed in their contracts complete scientific freedom so that their jobs did not depend on their acceptance of the founder's views.

"Apart from these institutes Solvay was always ready to help finance other projects which were of general value and for one such project—a new Institute for Sociology"—he visited Berlin. I too had been invited to take part in the discussions. The idea had been initiated by an original but rather limited thinker called Pollack who had written a book on organisation with the title "Political Methods". It had been sent in for review in the "Annals" and I had read it through and written a largely positive review.

However, this time things did not work out. The problem was that the organisational ideas put forward by the sociologists who attended the meeting were primarily aimed at making sure that the new institute should provide them with the basic data which they needed for their own current work or the work they planned to do in the future. Solvay thought that this was too narrow and showed no interest in supporting their plans. As the meeting came to an end he invited me to dine with him so that we could talk about questions relating to social energetics. Though my French was very rusty and Solvay knew no German we spent an interesting evening together. At the meal he proved to be very abstemious for he neither drank alcohol nor smoked.

This was the beginning of a closer professional and personal relationship which became something like a cordial friendship based on mutual trust.

Personal matters. Solvay was a few years over 70 when I first met him. He was slightly under middle height with a thin but wiry figure and with short but full grey-blond hair and beard and blue eyes. His forehead was deeply furrowed from much thought. He didn't look his age because he was agile and moved with an air of determination. This was the result of his sporting activities which he undertook to maintain his equilibrium amongst his numerous diverse business interests. As he told me once later when I visited him in his lovely country house near Brussels, his best relaxation was the roughly 15 km walk to his apartment in the city. He said "These are the only hours in which I can think seriously about science, because only then can I be sure that I will not be disturbed". A little later he sent me some photographs of himself together with his guides on a difficult ascent of a snow-covered peak in Switzerland. I wrote back to tell him that I would have felt jealous at the sight of the card if I not much earlier worked hard to banish as far as possible this mean-spirited emotion.

Later on he survived the world war, though the occupation of Brussels by the Germans hit him hard, and he died in 1922 at the age of 84.

Because of the matter of the World Association of Chemists, which I related above, I was in constant contact with him and was several times at his house in Brussels or at the country house for days on end. Though both houses were built like palaces and richly furnished he lived very simply, he had no interest in showing off his wealth. He too belonged to that happily not so small section of the rich who treated their wealth as a duty rather than as a source of pleasure.

Typical of this was the celebration he held in 1913 for his 75th birthday, his golden wedding anniversary and the 50th anniversary of the introduction of his soda ash process. The numerous soda ash factories in Belgium, Germany, Austria, France, Britain and so on were represented by their directors and Solvay had asked them not to bring any personal gifts. Instead he suggested that any money set aside for that purpose should be used to support the workers in the factories. It pleased him a lot that three million francs were raised for this purpose.

The celebration was however more or less a state affair and Solvay just had to accept this. I attended because it amused me to see an idealist in this situation. It consisted as these things always do of speeches and banquets and so on and ended with a theatrical presentation given by around half a dozen members of the Paris Comédie Française who'd been brought to Brussels at an outrageous cost. I'd most kindly been given a seat in the first row so that I'd miss nothing.

I must confess that I have seldom seen anything so childish in all my born days. Perhaps it was due to the then current Ultramontane movement or perhaps due to Solvay's wife (which is saying much the same thing) that the one act sketches were so harmless that they bordered on fatuity and the stilted way in which the verses were recited only increased this impression. It was a bit like the school leaving play at a very pious girls' school.

However, perhaps because they'd known what was coming, the number of guests dramatically decreased, and so I was rewarded with fifteen minutes chat with the birthday boy. I left for home the next day with the satisfied feeling that here in this man external success and inner values went hand in hand.

Energetic cultural science. In the meantime so many new thoughts had come together that in my usual manner I organised them into a book so as to make a bit of space in my brain. This was in 1908 and in the following spring the "Energetic basis of Cultural Science" was published. It was dedicated to Ernest Solvay the founder of social energetics.

The content of this little book (it was 184 pages long) is clear from the chapter headings. Work, Efficiency, Raw energy, Life forms, Mankind, Mastering external energy, Conquering time and space, Socialisation, Language, Law and punishment, Value and exchange, The state and its power, Science.

As always these new thoughts received an unfriendly reception from the experts. A few of the crushing reviews by well known sociologists convinced me of the necessity of my work because these critics were simply unable to follow the arguments presented. I can't say whether in the meantime that has changed much for I haven't kept up with the newer sociological literature. In current practice one does come across numerous example of the application of energetic views, though it doesn't necessarily follow that this is a direct or indirect effect of my book. Where survival of the fittest is the rule, what's right quite naturally wins whether or not it had been previously analysed theoretically or not.

Soon after this I was given an opportunity to check the applicability of my idea. In the summer of 1909 an international sociology congress was held in Bern and Ludwig Stein, who was professor of philosophy at the university there, had seen to it that I was invited. Since this fitted into travel plans to Geneva where I was to be awarded an honorary doctorate I was happy to accept.

Stein had a large circle of pupils in Bern and though he originally worked on the history of philosophy he now tried to teach them all the recent developments in the subject. He was the first of the professors of philosophy there to repeatedly use energetics as a subject both for seminars and for dissertations. He was always looking for connections between the ancient Greek philosophers and current trends and was overjoyed to find the seeds of recent concepts in Aristotle.

He offered to put me up for the duration of the meeting and I thankfully accepted. Since he was very well off he had bought a grand house in a beautiful part of Bern and there I was given a wonderful room. Although, as organiser of the meeting, he had a lot to do he found the time to have long and detailed discussions with me and these resulted in the inception of a number of wide ranging charitable social projects. Like me he was by temperament and scientific conviction an optimist but he was even more optimistic than me and he thought that by applying his vast fortune and his influential connection he'd be able to start a movement of cultural energetics the outlines of which we formulated together. Unfortunately these plans had to be delayed because his doctors said that he should take a longer break to recover from over work, and then the world war destroyed everything.

At the congress it was easy to see how uncertain the sociologists were about where their field fitted into science. I think this was the reason that they were so ready to accept my explanation of the pyramid of science and the place of sociology at the very top being dependent on all the other sciences. I was elected to membership of the Paris based International Institute for Sociology and I later wrote a number of articles for its journal.

I got to know and respect a large number of interesting and independently minded thinkers and researchers at this meeting though, since no longer term relationship developed from them, there is no need for me to list their names here. The one exception is Ludwig Stein with whom I have remained ever since in friendly contact, particularly after his move to Berlin in 1910 which brought us geographically closer together. I am indebted to him for his untiring readiness to help and for his support in a number of academic and personal matters.

The peace movement. With the energetic imperative I was in possession of a means of judging the cultural value of the various movements that applied to me for support. For example I was unable to see any way in which the German Language Society was going to reduce the waste of energy and so I did not do anything for them, even though I'd considered it important throughout my entire career to try to make my lectures and papers as German and as expressive as possible. In fact I have enriched our language with a number of new terms which were generally accepted without any opposition and these I introduced when necessary without asking anyone's permission.

War, however, was a waste of energy on a massive scale and so I did not fail to answer the call to join the public opposition to it. My contact with the movement was through the dignified figure of Wilhelm Förster.[3]

I'd got to know him in Paris in 1907 where I was involved in the world language meeting. By chance the managing committee of the International Office for Weights and Measures was meeting there and Förster was Germany's representative. He was staying at the same hotel as I was so that I often met him and some of the other delegates. He was very interested in our work on the world language but turned out to have been indoctrinated by a number of convinced Esperantists.

Wilhelm Förster was a lively old man, short and thin with white hair and beard and blue eyes which reflected his kind-heartedness. He was already 75 but wasn't worn down by his age and with his open and confiding way made a winning impression.

What particularly attracted me to him was the fact that long before me he had reorganised his life much as I, without having heard of him, was to do later. He'd had a respected scientific position as professor of astronomy and director of the observatory at the university in Berlin and had given this up in order to be able to devote his life to practical idealism. I often took him up on the offer to drop in on him in Berlin and never left without the feeling of having breathed the fresh sunlit air of the mountains.

What led me to collaborative work with him started with certain technological questions. In the winter of 1909 I'd published an essay in the New Free Press in Vienna in which I pointed out that with the conquest of the air, which was just beginning, the third dimension of space was being opened up. For as long as mankind had been restricted to the two dimensional land surface, border lines had been enough to separate countries from each other and these could usually be controlled and closed. Now that man was learning to move through the air the border lines would have to become fences which were high enough that they could not be over-flown. That was naturally technically impossible and thus this development must inevitably lead to the abandonment of political and economic borders which were of course responsible for a huge waste of energy.

Baroness Berta von Suttner read the essay and soon wrote me to suggest that I might present these and any other similar thoughts to the Austrian Peace Society

[3]Ostwald misspelt this name. The correct form is Foerster.

which, under her leadership, was very active. I readily agreed because in any case I was often in Vienna on other business and so I got to know her. The idealism which characterised her made my dealing with her very pleasant and after that I never missed the chance to drop into the old house on the Zedlitzgasse whenever I was in Vienna.

At that time Mrs von Suttner was already almost 70 but with her slightly plump figure and average height she was a lively and amiable person. The apartment was filled with beautiful old furniture. In the middle of her drawing room stood a table on which under a glass plate was the certificate of the Nobel Prize which she had won for her services to peace. She was rightly proud of this.

At around the same time I had a long talk with the Japanese ambassador in Vienna (his name escapes me now) who had questioned me for hours for detailed information about the world language. As thanks, he invited me round for the evening and apart from me the other guests were Mrs. von Suttner and a number of other Vienna internationalists. Later we were joined by the Japanese ambassador to Petersburg who was on an official journey together with his wife. She was a graceful still young looking Japanese woman who was my dinner companion but she unfortunately was wearing European dress. We had to speak French because this was the language the Asiatic diplomats were most at home with. Mrs. von Suttner spoke it fluently while the rest of us had only a smattering. Despite this the evening passed in cheerful and interesting conversation.

In particular I had asked the Japanese if the rapid assimilation of the foreign European culture by the Japanese people would not lead to a serious reaction, comparable to digestive problems, as had happened in similar cases elsewhere. They replied that their people had managed this several times in the past as, for example, when they absorbed Chinese culture and they were sure that also now the assimilation would be painless. Till now the results seem to show that they were right.

Another city in which the peace movement was strong was Frankfurt am Main. There the peace movement was already 25 years old when I was invited to the celebrations which took the form of a general meeting and I was asked to give a presentation. I used the opportunity to initiate a number of useful and pleasant relationships with people, both in and beyond the peace movement there, who made me like this vibrant city.

I still have very lively memories of the peace conference in Stockholm in 1910 where I got to know the movement's international leaders of whom I remember best the Danish Nobel Prize winner Bajer, who in his personality resembled Wilhelm Förster. I met Gaston Moch again there whom I'd become friends with earlier in Paris (Part III, Chap. 35, p. 467). At this meeting too I held a lecture this time to the title "Culture and War" which was met with loud applause. Afterwards the Swedish philosopher of culture Ellen Key congratulated me most kindly.

I think I have never met a more mixed bag of congress participants. Here some Parisian ladies made themselves obvious by their luxurious dresses and make up, over there were long haired men looking like farmers, and in between faces which ranged from stubborn fanaticism to cherubic benevolence.

The town had arranged for a banquet so long as no alcohol was drunk. This was unhesitatingly accepted and I've never attended a banquet at which there was so much discussion, much of it at a very high standard. When we'd finished with the meal and the fruit lemonade the discussions were not nearly finished. In the garden there was a sort of natural pulpit on a rock and from there the talks carried on for some hours till the audience eventually drifted away. A Chinese who spoke good German had joined me and I tried to convince him of the necessity for his people to join world culture by learning the world language.

At some of these meetings I'd been rather embarrassed that apart from the energetic imperative I hadn't been able to add much to the well known reasons for pursuing world peace. I was therefore more than happy to seize a new and unexpected opportunity.

The French monthly magazine "Grand Revue" asked me to write something for them and left the subject matter up to me. I answered right away that I'd be happy to oblige but that I rather feared they'd send the manuscript straight back. They asked again and I wrote an essay, "The Great Step" in which I developed the following line of thought.

The political role of France on the European continent had always been to try out new political experiments on itself. For example they had invented the absolute monarchy 500 years ago and the people's revolution a 100 years ago. The French nation was now in a position to make the greatest achievement in this area by signalling the start of world peace. All that was needed was that the French unilaterally disarm. France should not fear attack from any of its neighbours and in particular Germany did not seek war with anybody because there was nowhere a goal for which the working people would be prepared to go to war. As proof of this was the fact that when in 1905 Russia lay paralysed by revolution and defenceless, Germany had not attacked her to gain control of the Baltic provinces with their German speaking populations.

The essay was very well translated and it was published. It caused a lot of debate and a number of letters from both sides of the argument were published in the following issues. Had the suggestion bourn fruit, what tragedies the French might have spared themselves and the rest of the world!

Chapter 41
World War and Revolution

Rome. There were two occasions on which I was able to sense the shadows that the world war threw out in front of it.

In the spring of 1909 I was officially requested by the imperial government to represent the German Empire at the upcoming International Congress of Chemistry in Rome. This was for me completely unexpected since I had no close relationship either to the government or to the International Congress. In fact till then I had avoided this congress because it had almost nothing to do with science. The large chemical industries had secured considerable influence over it so that the meetings were nothing more than social events at which the host countries tried to outdo each other in pomp. The last such meeting had been in Berlin and I had deliberately stayed away from it.

The obvious thing would have been to nominate one of the organisers of the Berlin meeting as the imperial emissary. This was also the view of the colleagues in Berlin, for I found them and all their friends very hostile to me when I met them in Rome. They had no idea how much I would have preferred to leave this honour to them. As it was I had my work cut out not to keep tripping over the obstacles they kept putting in my way. And though my unknown benefactor in the Imperial Office for Internal Affairs doubtless wanted to do me a favour, it turned out to be more the opposite.

In any case the invitation was welcome because I'd never been to Rome and so was given an opportunity to get to know it. I must confess that I never felt any strong wish simply to improve my "education" by going to Rome, for if I had, I'd had plenty of time and resources to do this.

As the congress approached I travelled to Rome and used the ample free time to see all the much described and enthused over treasures of the "eternal" city. The actual congress was over in 3 days but the ceremonial banquet given by the King of Italy and his wife, both of whom had attended several of the meetings, was scheduled for a few days later. Because of this there was plenty of time.

R.S. Jack and F. Scholz (eds.), *Wilhelm Ostwald*, Springer Biographies,
DOI 10.1007/978-3-319-46955-3_41

Rome made the impression on me which I had expected. I've already mentioned that my ability to enthuse over places and things of "historical importance" is underdeveloped and that I only register the current value of things. Because of this I lacked the emotional fog which most ordinary visitors bring with them to Rome, and I have to say that the eternal city didn't have much to say to me.

This was particularly the case with the ancient ruins. Apart from a few busts which I considered expressive and honest in their brutality, the rest of the classical remains couldn't dispel the air of ceremonial boredom which I knew well enough from the numerous copies from which no one can escape. Rafael's frescos looked worse than the well known copperplate engravings because not only were the compositions overly theatrical but the colour scheme was disastrous. In the Sistine Chapel I was able to enjoy Michelangelo's wonderful figures of the sibylline women, but I thought the artist had been very careless with his composition of the Last Judgement; scattered groups without any obvious connection between them had been placed on a blue sky which had been smeared on with a thick brush and there seemed to have been no attempt to devise an artistic form for the room.

There is no doubt that the painter's contemporaries considered this a great work of art even if it wasn't worth the ridiculous praise that later art historians paid it. But what new things the works of these times had to say has all been long absorbed so that the artistic needs of people today cannot be satisfied with them any more than his scientific needs could be satisfied by Euclid and Aristotle.

Unfortunately I didn't keep this criticism to myself and thus managed to hurt the feelings of many of my colleagues.

Very different was my response to meeting the famous chemist Canizzaro, who I'd imagined had long belonged only to the history of science, but who was still alive in his late eighties. He'd been professor in Rome for 30 years and still held lectures for the young students. Together with a few of my likeminded colleagues I went to one of his lectures. The very simple lecture hall had probably never contained such a collection of important chemists. When the lecture was over we went forward to pay him our respects and he was particularly happy to see me. He was thin little old man with the simplicity and naivety that one associates with great intellect. I'd sounded out the other members of the managing board of the Bunsen Society as to whether we couldn't in this case overstep the mark and make him an honorary member of the Society, but the others didn't want to take the responsibility for this move. He died soon after.

I saw on this occasion what a very minor role the university played in Rome and how much lower the social position of a professor was in Italy than in Germany.

At the official receptions and other court activities I could see how very much the King and his ministers gave preferential treatment to the English and French representatives in comparison to the way those from Germany or Austria were treated. Since at that time Italy was part of the triple alliance with Germany and Austria this was even more surprising. I pointed this out in no uncertain terms in the official report on the congress which I prepared for the imperial government, for I considered it a matter of the greatest political importance. However, I don't think they paid any attention to it especially since German politicians of the time preferred to believe what they wanted rather than paying any attention to the facts.

For me this was the first shadow of the coming world war.

The Royal Society in London. In the spring of 1914 the Royal Society of London for Improving Natural Knowledge celebrated its 250th anniversary as one of the oldest such societies. In 1664 the name had been coined not in the sense of contemporary science but as protection against the then all too common accusation of witchcraft. However the Society's activities were more or less those of such a society today with the exception that they concentrated on tangible things; mathematics and theoretical research were only later admitted.

The "Royal Society", as it is generally known, has since its inception been the centre of scientific life in England since the sister societies founded later in Edinburgh and Dublin did not provide much competition. It became all the more influential because the old English universities in Oxford and Cambridge did not assume the scientific leadership as the universities in Germany did. In Britain it was more the Scottish universities which took the lead and it is only in the last thirty odd years that the situation has changed in favour of the English universities.

The sister societies from all civilised countries as well as the universities and such like institutions were invited and they all sent representatives. I was able to calm down a one-time colleague of mine from Leipzig who wanted it made clear that I did not represent his university with the news that I wasn't representing any official body but had rather received a personal invitation from the Royal Society.

This meeting brought me together again with many old friends and acquaintances whom I'd met at similar occasions in the past. It was clear that there was a rather small group of scientists who were tacitly accepted as the international leaders of their branch of science and who repeatedly met together on such occasions. We'd got used to meeting each other with a laugh and the thought "Off we go again!" I could count myself as a member of this group and this was a source of no little happy pride to me.

From the celebrations I particularly remember the service in Westminster Abbey because of the splendid choir and because of a remarkable sermon held by a distinguished bishop of the High Church whose name I no longer recall. Instead of launching into a polemic against the inroads of science into the church's intellectual territory, he simply declared the entire progress of science as being evidence of the glory of God who had produced such eminent minds as, for example, Charles Darwin.

Apart from that the celebrations passed off as these things usually do. There was only one point which seemed a little strange to me and that was that there were absolutely no after dinner speeches and when I asked why, I was given evasive replies. I later realised that the British were worried that because of the strained international situation the French visitors might speak out of turn. This ban on after dinner speeches was strictly enforced.

Even so the atmosphere was charged. I got involved in a discussion on internationalism with a leading British statesman, though I've forgotten who it was. At that time the first flight across the channel had just succeeded and I said to him that Britain was no longer an island and must now change its century's old policy of isolation.

The war. In August 1914 the war suddenly erupted to the great surprise of me and of almost all other Germans. I had just turned down an invitation from the Australian government to the meeting of the British Association for the Advancement of Science (Part II, Chap. 20, p. 215) which was to be held in September 1914 in Melbourne. This was not because I felt that war was imminent but simply because I doubted my ability to be happy for months on end among the English, even though I knew many agreeable people and had many good friends among them. A colleague in Berlin who did go then had to spend many months out of the country. We thought so little about war that when it broke out the Kaiser was sailing around Norway and none of the necessary preparations had been made.

I've already related (Part II, Chap. 25, p. 299) that the question of producing the nitrogen compounds which are essential for making all sorts of explosives needed in war had long worried me until with the help of Dr. Bauer the catalytic production of nitric acid from ammonia was achieved not only as a laboratory experiment but also scaled up to industrial production. I'd long tried unsuccessfully to get this through to the appropriate people at the army administration. Now they turned to the leading Berlin chemist Emil Fischer. He himself had publically explained that he'd been at a loss to see how explosives could be manufactured once Britain had declared war and closed the sea route to ships bringing saltpetre from Chile. Then he studied the trade reports of companies in the German Empire and found out that the Lorraine Union close to Bochum had for years been producing nitric acid and nitrate by the catalytic oxidation of ammonia. This was the way, indeed the only way, to solve this existential problem as the battle of the world against us developed.

The factory in Bochum was very quickly considerably extended and a whole set of new factories based on it were constructed and these were indeed able to satisfy the considerable requirement for nitric acid which the authorities had till then not thought about. Without this the war for us would have been over within 3 months.

I nevertheless have to say that the discoverer of the method, as well as his chief technical collaborator, were both carefully kept away from the project. Because of my age I could not be called up and so had offered my services as a volunteer but no one was interested. Dr Bauer was drafted, but he was assigned to quite different tasks which were not of any great importance.

In this context, I want to make clear that I did no chemical or other work for the war. The only request to me was whether I could suggest ways of detecting mines at sea but here I was unable to help as I lacked any of the necessary background and experience. Apart from that I neither offered advice nor engaged in any other form of war work.

I emphasise this because during the war there was a lengthy campaign of slander against me from the enemy side which largely originated in Geneva. There were detailed reports purporting to show that in my Großbothener laboratory I was busily inventing new types of shells with which one could rapidly burn down whole villages and other equally stupid things. I don't know why I was chosen to be the target of such lies. Perhaps it was because the enemy realised the absurdity of suggesting that someone whose life had been devoted to promoting the better understanding between nations was unsuitable as an honorary or foreign member of their scientific societies.

Personal views. As the war became inevitable I hoped for a German victory even though this was bound to be a hard struggle. From the point of view of the energetic imperative I had to consider war to be the greatest waste of energy. However, I told myself that of all the wars fought in the last century the Prussian–Austrian and the German–French wars had involved a relatively small waste of energy because they'd been short and the victor had worked hard to make a fair peace and to get back to a longstanding peaceful relationship with the defeated opponent. If one compares the fact that shortly before the fall of Paris the Germans brought in masses of food so that as soon as the city surrendered the people could be fed, with what our enemies did in 1918 then one can clearly see on which side the "Huns and barbarians" were to be found. Quite apart from my patriotic feelings as a German I had to hope for the success of our arms for the sake of our cultural heritage. It is a fact that the war against our culture continues to this day.

I tried hard to forget those four hard years and in this I was largely successful. The food shortages in the last war years were not as bad for my family and me out in the country as for those in the town and, since I've long preferred a largely vegetarian diet, the lack of meat was not a major problem for us. Heating, however, was a problem even though the park provided wood for the fire but the spacious house had a central heating system which we could not use in the last winter. We got by by restricting ourselves to just a few rooms and installing some fireplaces. I could only continue my work in the lab by having hot water bottles at my feet. Still it worked out—only a few bottles containing dilute solutions were broken by ice forming in them.

My three sons were all inducted. My eldest had already served as a volunteer and was called up as soon as war was declared. To see him one last time we drove to Leipzig and found him in the barracks in Gohlis getting ready to leave that same evening. We and his wife Pia stayed there and accompanied him part of the way. He expressed the atmosphere of that time in a song which was distributed widely in various different musical versions:

We who march gravely
Through the dark streets,
Wives and children at our side,
Wives and children with us;
Wives and children in our souls;
Soldiers, soldiers, stay firm!
Even when our throats are burning,
None of us will fail

We are men, grown men,
And nothing can easily shock us,
But we are also sure:
That the Empire is more than each of us!
That our lives are nothing,
When the nation is threatened,
Nothing is more important,
No hardship harder than that of our people.
We will have to put up with a lot
And we have to leave you alone,
And in all this time
There will be nobody to help you.
You will bear my child,
You may bury one;
You perhaps will have to bring them up
And feed them by yourself!
But we want to bear all this,
Every one of us, all united,
Till we can say once again,
That freedom's sun shines.
We who march gravely
Towards the battlefield are ready:
And we will fight for
Germany's fame and glory.

He served for about one year in the trenches at the front before being recalled for scientific work.

Our second son was turned down because of an eye defect but served voluntarily as a driver before being given the job of developing a new method he'd invented for making nitrates from coal. The third son was able to apply his technical skills in the air force. All three finally came home safe and sound: a rare piece of luck in those hard times.

Work at home. As in many other places a convalescence hospital for sick and wounded soldiers was set up in Großbothen under the direction of the local doctor Dr. Panitz and funded by patriotically minded members of the community. On average there were twenty five patients there. I contributed money and furniture but left the actual work to the doctor and his nurses. My wife supplied the food for 6 months until she was so exhausted that she handed it over to other willing hands. She did however stay in charge of organising the laundry which for a hospital is at least as important as the food. My eldest daughter, who was unmarried, served as matron and helped to get the inexperienced nurses organised and to smooth out the sorts of personal animosities that always crop up in such a situation. By her unfailing friendliness she and the doctor were able, despite all the pain and discomfort, to make it such a happy place that even till today their patients' eyes light up when they look back on their time there.

In this way we all had work to help us through the hard years of the war and the equally hard ones that followed.

The famine that threatened the country in the last year of the war hit my family and me less hard than most of my former colleagues who lived in the city and who had to struggle with all the difficulties of rationing. Though we didn't have much in the way of meadows and fields these nevertheless helped us through the hardest times because I leased them in exchange for food and that kept serious want from our door. Best of all we didn't need to stand in line for hours to get a meagre ration. My wife's cooking skills, which she'd learnt from her mother and then improved further, were such that she was always able to prepare a tasty and healthy meal from simple ingredients We all lost weight especially in the "turnip time" and our clothes had to be taken in but this didn't do us any lasting harm. I for one was happy to see that the large demands that I made on my brain in the course of my work on the theory of colour were well tolerated. This all was once again a proof that in normal times we are accustomed to eat far too much.

Revolution. I lived through the cataclysm of 1918 with very mixed feelings. I was dismayed by the inability of the rulers to stand up to the often rather weak forces of revolution. A 100 years ago Goethe had written:

Why, as if with a brush,
Can such a king be swept away?
If they'd been real kings
They'd be there still.

At that time I was democratically inclined and had basically nothing against what was done. But I knew from science that every volatile change results in a considerable consumption of energy which in cases like this is always to the detriment of ordinary people. And from history I knew that a revolution never leads immediately to better conditions and that this is due to the great waste of energy that results from the endless friction that comes with the establishment of new political relationships. Once more Goethe had it right:

And when one smothers the tyrants
There's still a lot to lose.
They didn't want to give the empire to Caesar
But weren't able to rule themselves.

And that's what happened in the years that followed. The German revolution couldn't have come at a worse time than at the end of the war. The result was that the responsible people had no idea of diplomacy and that brought us the outrageous "peace" whose sole purpose was to continue the war against the German people by non military means. And the lunacy of building a new Europe dedicated to denying our people all their options is driven by such a short sighted thirst for revenge that it will require the greatest care to make sure that the shaky edifice does not collapse and lead us all into a new disaster.

At the moment there doesn't seem to be anything that can be done about it. It would be far the best if the current common border between France and Germany could be got rid of. To do this it would be necessary to neutralise Alsace, which has turned out to be indigestible for France, as well as Lorraine. Then there would be a broad band of neutral territory between Germany and France which would border with Luxemburg and Belgium on the one side and with Switzerland on the other so that a German French war in the future would be prevented. This group of countries could form a customs union and thus serve as the seed for a united Europe which is widely recognised as being needed to balance the United States.

As it was, however, things took the course one had to expect after the "peace treaty" was signed.

Short sightedness and lack of experience was not only evident in politics but also in economics and this led to the collapse of the German currency. My quite substantial savings which had been in the form of German Bonds went up in smoke and I had to start to organise my economic existence from scratch. This was made much easier by the fact that I owned the "Energy" that secured at least our daily bread. I thanked heaven that I had earlier bought more land than had seemed strictly necessary. I didn't shed any tears over the fortune I'd lost though I must admit I thought that this practical lesson in the down side of a people's government had been bought rather dearly. Nor do I wish to hide the fact that my lack of ready cash often held up my work on the theory of colours—work whose completion would be for the good of science and of the German people.

New work. When one considers the last five chapters which summarise my work from 1906 to 1914 then one will see that these hard and to some extent successful efforts were completely flooded and torn apart by the war. I told myself that the destruction was so complete that it would take years until something could be made of it again. At my age this meant that all the work which I had undertaken in the last years now had to be written off.

It was not easy to get over this. No new work had grown out of the war because, as I related, no one took up my offers to do voluntary work. It would have been impossible to survive the agitation and the burdens of the war years with nothing to do. There was no other option than to find fresh fields of endeavour which the war could not destroy and to work to bring them to fruition.

Were there such fields?

Indeed there were—in the area of pure science. To be sure the French were so driven by hatred that they carried the war even into these areas and convinced their allies to join them in this barbarism. But all that they could achieve was to destroy the already highly developed facilities of collaborative science. But for an individual researcher the infinite breadth of the imagination was open to him and he could start a new field in the limitless expanse of virgin soil. Of course the war made it difficult to get hold of supplies and equipment but since my youth I'd been used to getting by, also in science, with very little and in my later years had learnt to do a great deal simply by thinking deeply about the problem. I'd thought that once I'd moved to "Energy" my days at the bench would be over and, had it not been for the war, that is probably what would have happened. But the enormous demands now made on the German people could only be met if every one of us contributed the energy he still possessed. And since I was now cut off from the major part of the work I'd been doing, I didn't hesitate for a moment to pick up where I'd left off 8 years before.

I didn't need to look far for the scientific work I was going to do. One of the challenges which had faced the "Bridge" had been the question of how colours should be organised. I found this area attractive since for many years I'd been used to recharging my batteries by sketching and painting and, by doing so, had learnt enough so that the friends to whom I gave paintings were particularly impressed with the colours. There were a number of other factors which I'll come to a little later that made a general examination of the concept of colour attractive. Even before the war I'd started in an almost playful way to make the important questions clear to myself in an experimental setting. Now I could devote myself entirely to this line of research. As is always the case as soon as I started work new questions arose until at length this work of my old age became a completely new focus of my life and the previous broad efforts in many areas which had characterised the last years was now replaced by an investigation in depth.

Chapter 42
The Theory of Colour

The beginnings. When I try to look back on my childhood then I see pictures before me whose colours are so clear that I could paint them now. This goes to show what a large part colour plays in my mind's response to experiences.

The relative poverty of my youth forced me to make for myself the materials I needed for my attempts to work with colours. Because of this I learnt early how to produce dyes and colours for painting. This early knowledge was maintained when, later on, painting landscapes brought me not only a rich source of joy but also a means of maintaining my energetic equilibrium in the face of the effort I put into chemistry which was the centre of my life. It was perfectly natural that I sometimes made considerable outlays and applied my knowledge of chemistry to the technical aspects of my hobby.

Although I found these questions fascinating I felt for a long time that it was not proper to make them the subject of real research. It was only once I'd turned to natural philosophy and had broken through the walls surrounding my area of work that I found the courage to follow these till then suppressed goals and turn my research experience to these other areas.

On painting technique. The first approach quite naturally was closely associated with chemical questions. Towards the end of the nineteenth century the awful consequences of focussing the education of artists solely on idealistic and aesthetic matters became clear. At that time almost nobody had any idea of the chemical and technical basis of their work so that in their ignorance the paintings they completed tended to self destruct soon after leaving the studio. Even such a careful painter as Menzel prepared the canvases for his oil paintings with Florentine varnish without realising that this is more or less the worst thing you can do. That is a thing of the past because we now know better—and it's no longer used. However, his earlier works were soon covered with a network of cracks which grew with the years.

R.S. Jack and F. Scholz (eds.), *Wilhelm Ostwald*, Springer Biographies,
DOI 10.1007/978-3-319-46955-3_42

Böcklin experimented like an old style alchemist without the slightest knowledge of basic science. Instead he drew his ideas from obscure remarks of ancient writers who, he was convinced, had some supernatural access to the secrets of painting. Most of the others did much the same. Instead of taking advantage of the available chemical knowledge, which Pettenkofer had already shown to be useful, they spent their time arguing about the interpretation of ancient texts written by authors who usually had no personal experience and were only passing on things they'd picked up by hearsay. The suppliers sometimes used this ignorance of the artists to palm off on them worthless materials like the newly discovered aniline dyes.

A society was formed in Munich, to combat this mischief. I didn't really give it much chance of success because most of the members were artists who had little if any real knowledge but I was nevertheless prepared to join in. I got some of their material and did a little background investigation. It turned out that the leader of the society was the distinguished painter Lenbach who liked to use bitumen to paint although it is about the most dangerous pigment you will find because it never completely dries. Because Lenbach applied it very lightly the terrible consequences were not so evident. But for most of the other inexperienced painters—and that was the majority—making use of bitumen was suicidal. When I got the society's list of completely dependable so called "normal colours" I found to my astonishment that bitumen was among them. I looked into this and found out that Lenbach had lent his name to the society only on condition that that the reputation of the bitumen he loved would not be tarnished by the suggestion that it was not dependable. Later, once Lenbach died, bitumen was quietly removed from the list.

I therefore chose not to join.

The painter's letters. By painting during the holidays I'd picked up enough practical experience so that I could now apply my knowledge of physical chemistry to painting. I wrote a series of essays which were first published in a Munich newspaper and then subsequently in 1904 in book form under the title "A painter's letters". In these essays I invented a number of new terms without emphasising their novelty and these terms have now passed into general use so that it seems clear that the book was widely read. It soon went out of print and once again I didn't get round to preparing a new edition.

In the preface I'd written that I found the then current "philosophical" approach to art unsatisfactory and wanted to see instead an empirical scientific approach firmly based on experimental work. From then on I was considered to be beyond the pale for most art journalists and experienced every sort of effort to neutralise my work. There were only a few exceptions but they soon learned that failure to run with the pack is not good for your career.

My remarks could be summed up in the slogan: The artist should be conscious of what he's doing. He must be constantly aware of his goal and of the technical means he wished to employ to reach it. After saying that this is the way that all human work is carried out I closed as follows: even in art unconscious inspiration must yield to conscious competence.

A quarter of a century later it was a pleasant surprise for me that in looking at that old book again I could see the basic themes of my later interest in art were already clearly formulated. However I don't want to hide the fact that at that time I thought it was the task of the painter to produce a picture which would have the same psychological effect on an observer as would the original scene in nature. This is a fundamental error. I'd taken this idea over from Helmholtz who regarded it as a matter of course in his famous lecture on "Optics and Painting" and this was due to the impressionists who were the dominant artistic movement at that time. I have already (Part II, Chap. 22, p. 248) indicated the enormous difficulties I got into by uncritically accepting this axiom. It was only by inventing quantitative colour theory and the laws of colour harmony which are derived from it that I was able to free myself from it. Looking back it's easy to see why neither Helmholtz nor I was able to recognise the error. It was due to the lack of numbers and measurements in the world of colour.

These experiments led me to further develop pastel painting which is the form freest from bonding agents and this led me to make large pictures from the sketches I'd made on my journeys. These made such a good impression on those who viewed them that I gave many of them away to friends and family. One friend was so happy with his picture that he said, "Hey, if I could paint like that I'd do nothing else all day". At that time I just laughed at his remark but later I began to wonder if he wasn't right. Now I have finally decided that when I finish this book I'll do nothing else but paint and I'm already looking forward to that even though I know that it is probably too late. But maybe a little of the youth that this artist no longer has will be compensated by the youth of the ideas that I follow.

The problem of colour systems. In the next few pages I'd like to go in more detail into my work on colour theory than I did with my previous work and I do this for several reasons. My subjective opinion is that in this work one can see most clearly the special characteristics of my mental processes which make this the high point of my scientific achievement. Objectively the subject is of broad interest for colours are what we perceive with our eyes and they are by far the most important sense organ we have. The forms, figures and things that we believe we see arise from our interpretation of the blobs of colour that fill the eye's field of vision.

The reader need have no fear that what follows is going to consist of impenetrable abstract science. What can be clearer than the colours which constantly fill our eyes? And when he grasps the unexpected cultural dimensions that flow from these studies then he would be fully justified in criticising the author were he not to provide him with at least a glimpse of the how the theory arose and grew.

Finally, it is possible here in this isolated example to demonstrate how a new field of science can be opened up. Since I did it all myself, the interrelationships can be clearly seen and one can view it as a tutorial for discovery. Since we are dealing here with very elementary relations the necessary ideas and concepts are so simple that anyone can grasp them. In fact I've seen how readily children pick up the basic ideas of this new colour theory. The only problem is that artists claim that they can't understand them.

I'd probably have remained on the primitive standpoint of a landscape painter had I not received a stimulation from a completely different direction. The organisation and standardisation of colours was part of our rich "Bridge" program (Part III, Chap. 39, p. 536). This was a challenge that the German Work League[1] had tried to solve by putting before a committee all the previous attempts to address the matter. In most cases the suggested solutions were presented by the inventors or their representatives. This approach, however, did not yield a single useful result. In about 1912 I joined the Work League and started my work on the committee. Although there were things I didn't like in their approach, I suggested to the League at their meeting in Cologne that we should work together so as to avoid unnecessary doubling of the work. They agreed—but this was all in that fateful year 1914.

I hadn't till then really done anything much with these questions though I did find the whole matter very provocative. During my stay in Cambridge I'd met A. H. Munsell in Boston who showed me his work in this direction (Part III, Chap. 32 , p. 417) and I've already explained why I didn't find his solution satisfactory. However since he'd managed to come up with a real, even if unsatisfactory, colour theory I thought I could surely find the correct solution particularly since I'd already done some experimental work in this area. That work had of course been based on Helmholtz's view, which was not challenged by me or by anyone else, that hue, purity and luminosity were the three elements of colour.

I still remember how, when alone on the channel ferry on the journey back from London in the spring of 1914 I struggled to understand how the whole spectrum of colours should be generated using just these variables. One thing was very clear and that was that the different pure colours do not all have the same luminosity but instead are all quite different. Goethe had emphasised again and again that yellow had the greatest luminosity while blue was the darkest and since then many determinations of the luminosity of the different primary colours had been made and they all gave approximately the same answer though the actual values differed. Munsell had tried to express this by taking Runge's system of organising colours in a sphere but he set the circle of primary colours not at the sphere's equator but rather at an oblique angle. Methodologically this was impossible because the primary colours must all be equidistant from the axis. Without having a model available it wasn't clear to me what would come of this.

Once home I got to work on making a model but it led to such convoluted relationships that I became convinced that nothing would come of this.

My own work. Grey. Soon after this the world war broke out. As I related all the international contacts, which till then had occupied most of my time, were abruptly interrupted. I was 61 and hence too old to serve and my offers of voluntary work

[1]The German Work League (Deutscher Werkbund) was formed in 1907. It was an association of artists, architects, designers and indutrialists.

were all ignored. The only thing that was left was to get more deeply involved in this scientific problem, particularly since it soon became clear that I had much underestimated the difficulties involved.

Just before this I'd been working on the principles of organisational theory for my book "Modern Natural Philosophy" and so I saw here an opportunity to apply this powerful tool. The first result was that the work had to be divided up and the first thing to do was to sort out the simpler case of grey with its end points black and white. When one looks at how this problem is presented today, for example in my "Colour Primer", then one will have difficulties trying to imagine what problems had to be overcome to get that far. The most advanced researcher in the field of colours at that time was E. Hering and he had never heard of pure white. In his book "Concepts in Vision",[2] to which I am much indebted, the term "whiteness" (albedo) which had been introduced by Lambert more than a century earlier, is never mentioned and because of this he was unable to deal with the simple group of "non coloured" colours. At times he mixed up the terms "reflection" and "whiteness". If I hadn't, quite by chance during my time in Riga, read the work of the astrophysicist F. Zöllner in which there is a lot about the albedo of the moon I'd probably still be stuck in that morass, but in this way I managed to avoid it and thus got a lot of unexpected help from the "spiritism" which Zöllner was a slave of and which gave his books a broad readership at that time.

Another question to which I found no answer in the literature was, "What does pure grey look like?" It was well known that mixing white and black dyes gave rise to a bluish grey and the reason for this was also well known. Now the theory indicated several ways of getting optically pure grey and thus be able to compare that to the mixtures of white and black. The result was a protocol in which one mixed chalk, rust and yellow ochre to yield a neutral grey. I can well remember the surprise and joy with which an experienced colleague and friend greeted the first sheets painted with neutral grey. This was new to him and to everyone else.

Introducing measurement and quantitation. Now I had to be able to measure the different shades of grey. To begin with I did this as usual using a spinning colour disk fitted with white and black disks. But then I knew that every white contains some black and every black some white and you can't measure this with a spinning disk. I therefore invented the half-shadow photometer (Hasch) and with its help I could make pure black simply by switching off the light. Different white powders could then be compared for their relative degree of whiteness.

It turned out that all the white powders I looked at approached a limiting value which was therefore very close to white. Among them were some which were the whitest and which couldn't be distinguished from each other and these could be regarded as practically pure white or, if you will, samples lacking any trace of black. The lightest of these was pure barium sulphate which was thus pure or normal white.

---

[2]Hering E (1905) Grundzüge der Lehre vom Lichtsinn. Leipzig, Engelmann.

This now gave me the possibility of measuring every shade of grey because my Hasch Photometer let me reduce the illumination of the pure white and hence to make it appear the same as a given grey sample. Once this had been done I had a quantitative measure of the fraction of the light which was reflected from the grey and hence a quantitative and unambiguous definition of the grey shade. All grey samples which gave the same result are then equally grey, those which reflected more light were lighter and those which reflected less were darker. In this way I was able to prepare any shade of grey I needed. I therefore prepared samples with one tenth, two tenths three tenths and so on black and expected to get a simple step relationship.

That, however, is not what I got. The first samples with one tenth, two tenths or three tenths black didn't look grey at all. Instead they were white. With four or five tenths black I got a clear grey and as I went on with seven, eight and nine tenths black there was no gradual stepwise increase but rather huge leaps in the greyness. The nine tenths sample was still very far from black, incomparably further than the one tenth sample was from white.

Then I remembered Fechner's "Elements of Psychophysics" which I'd read already in Riga, for no particular reason other than that the idea of a quantitative analysis of cognition had struck me as being fascinating and I'd wanted to see what was behind it. Later I met Fechner personally in Leipzig (Part II, Chap. 18, p. 199) and later on through my association with Wundt, who'd extended Fechner's ideas, I'd again thought about problems in psychophysics. Once again a simple scientific curiosity had led to a rather random contact with a field that was far away from mine and turned out at the end to have been a very worthwhile investment.

In this case I saw right away that the law which had first been enunciated by Weber[3] and then developed by Fechner whereby a stimulus must rise or fall in a geometric progression so that the sensation is perceived as an arithmetic progression. In this case the stimulus was the fraction of white in the samples.

I then prepared a set of grey samples whose content of white varied as a geometric progression and was happy to see that this yielded the graded steps of grey that I'd been looking for. I then constructed a sort of ladder in which the steps were these grey samples fixed between two strips of cardboard and when I then laid it on a grey background then one could immediately see which of the steps was most closely matched. In this way I could now as easily measure the greyness of a sample as one could measure the length of a line with a ruler.

This was all very simple but in those days it was also very novel. Ever since Maxwell had shown 60 years ago that colours can be quantitatively determined using a spinning disk there had been thousands of attempts to quantitatively determine grey, but no one had ever thought to ask whether one shouldn't relate the results to absolute black and absolute white. Absolute black was known to physiologists and psychophysicists because one can see it by looking through a hole in the wall of a box which is painted black on the inside. What had remained to be

---

[3]Ernst Heinrich Weber.

done was to define absolute white and then fit the scale between to Fechner's law. Of course that required someone who had knowledge of organisational matters and who was thus able to formulate the steps required to break down the problem and then clearly define the ways to the solution. There weren't—and aren't—many who can do this.

Fechner's law. As we have seen the standardisation of the "non-coloured colours" required the use of Fechner's law. Fechner had built on the previous work of Weber and had formulated his law in 1859. Since then it has been accepted as one of the basic laws of psychophysics. Nevertheless its use to standardise the values of grey is, as far as I am aware, its first practical application. Up until then it had merely served physiologists and psychophysicists as something that they could discuss and try to present small deviations as a refutation—in short it had been merely the subject of sterile scholastic argument.

However this law is fundamental to all questions of perception. Currently judges are slowly starting to realise that punishments should be graded according to the nature of the convicted individual because, for example, a fine of 100 marks can be ruinous for a poor man while it will scarcely register with a rich man. This would have been obvious long ago if one had only been able to apply Fechner's law. On the other hand the Standardisation Committee of German Industry would have saved itself from a number of mistakes if only it had realised that the assignment of standardised equivalent steps should be made as a geometric progression as defined in Fechner's law rather than as steps in an arithmetic progression which is what is normally done.

It is indeed the case that an enormous amount of worldly wisdom can be derived from this law. The common saying that everything is relative acquires a new and clear meaning, and if one organised one's own life according to this law then one would avoid all sorts of difficulties which are not real but which arise only from our mistake of thinking in "absolute" terms.

The standardisation of grey. [4] In this way I'd cleared up the notion of grey because the question of which of the infinite number of possible geometric progressions should be used to determine the extent of grey was answered as soon as it was posed. Since the uniform progression of numbers is standardised by the formation of groups of ten, the geometric progression must also be based on 10, because once a norm has been defined it has to be strictly applied in all cases. Thus we start with white = 1 and go on with 1/10, 1/100, 1/1000 and so on. However since a black containing 1/100 white is hard to prepare, these steps are far too large. What we have to do is to place between 1 and 1/10 nine geometric steps and similarly nine geometric steps between 1/10 and 1/100. In this way we end up with 20 intervals between white and black and these turn out to be sufficient. Indeed, for most purposes, they are too close together so that one only really needs every second one.

[4] Ostwald referred to grey as "uncoloured colours" (unbunten Farben).

To briefly denote these steps I chose the same procedure as is used in music and gave each step a letter of the alphabet. Thus "a" was pure white, "b", "c", "d" etc. were light greys of increasing darkness, "g", "h" and "i" were middling grey and "k" and "l" were dark grey which bordered on black while "n" and "t" are basically black. On paper one usually doesn't get beyond "p" which is the value of good black printing ink. If, as I suggested, each second step is eliminated then the grey series consists of a, c, e, g, i, l, n, and p for the usual shades of grey.

That is, in simplified form, the broad outline of the results of my thoughts and experiments on the standardisation of the grey series. It wasn't all done in one step because there were alternative organisational possibilities some of which seemed initially to be more obvious. However I knew from experience that in fundamental matters it never pays to introduce extraneous elements because they always turn out in the long run to be unsatisfactory. The later they are got rid of the more difficult and expensive it becomes because everything which has flowed from them must then also be changed. Thus, for example, natural languages are so unsatisfactory because they were standardised at a time when nobody understood anything about the logical and technical demands that a good and useful language has to be able to meet. For the same reason we carry on with a calendar that is grotesquely silly because the unit of a month varies from 28 to 31 days, that is to say by 10 %, when it should of course remain a constant value.

Extraneous elements in this sense are all arbitrary assumptions which are introduced into a standardisation based on some older and fundamental system such as, in this case, the decimal system. Only once all the unnecessary assumptions had been removed could I regard my work with grey as finished. As a consequence, the result is robust and durable. This standard has been used by me and my co-workers for more than 10 years in all sorts of situations and never once have we been faced with the necessity, or even the wish, to change it. My opponents too haven't found anything to complain about.

The only possible improvement that might be undertaken in the future would be to move from a system based on ten to one based on twelve. The reason for this is that twelve has the factors 2, 3, 4 and 6, while ten has only 2 and 5. However I rather fear that it will take several centuries to get the world prepared for this change.

Quantitation of colours. Finally I was done with the greys. I must confess that I'd had to reign in my desire to plunge immediately into the more attractive world of colour, but because grey is simpler it would have been a terrible waste of energy to go straight to the more difficult task before dealing with the easier one. Since none of my predecessors in the field of grey had even tried to sort things out properly it was clear that I would have had even more difficulties with colours.

The fact that I'd managed to keep my feelings under control to such an extent and applied myself to scientific experimentation paid off enormously. Looking back on my slow climb up to the level I'd achieved then, I can see that this path via grey was the only way forward because the ways of thinking developed here were absolutely essential if one was to reach a solution to the much more difficult problem in the field of colour.

A light grey piece of paper will be recognised as such independent of the strength of the light source, that is to say independent of whether a lot or just a little light is reflected. However in order to decide that the paper is light or medium grey we have to able to see it in the context of its environment and so be able to compare its greyness with other known objects. If we view the piece of paper through a tube which is black inside and has a small field of view so that we see nothing but the piece of paper then we are uncertain and vague in our judgement. The evenly illuminated field one sees in a microscope with nothing under the objective is not grey even if the light source is turned down; instead it remains equally unrelated to white and to grey.

These and many similar observations, some of which were quite remarkable, were collected by E. Hering though he did not interpret them in terms of a unified standpoint or recognise their relatedness in terms of related and unrelated perceived colours. This was primarily due to the fact that he was unaware of Lambert's definition of white or albedo. He was unaware of a whitest white and instead believed that white could keep on getting whiter forever.

I'd had to struggle hard to free myself from this error of my distinguished colleague who was still living at the time and with whom I corresponded. I managed to do it with the help of Lambert's work on photometry and I had access to a good version of this in the form of one of my "Classics of the Exact Sciences" series (Part II, Chap. 17, p. 176). This was not the only case where the work put into writing the "Classics" paid off handsomely and we'll get to further examples a little later. Finally it became possible to find the laws which made it possible to place the whole question of colour in a proper context.

Related and unrelated perceived colours. Here we are dealing with the creative definition of a term, much as was the case with catalysis, except that here no special experiments were needed. I realised that colours fall into two groups—the related and the unrelated perceived colours. Unrelated perceived colours are those which appear in a dark field of view. These colours are seen in optical instruments and the normal spectrum of light is the best known example. They are termed "unrelated" because they are perceived without having to be seen in the context of the surrounding environment and hence one does not know how they are related to the source of light. In the case of the "related" colours all this is reversed.

The consequences are both odd and extensive. The unrelated colours have no grey or black and instead only white and colour. If one, for example makes a spectrum ever darker by reducing the intensity of the light source until it disappears then one does not see black here one simply sees nothing at all. This is a huge difference. If one now increases the intensity of the source then eventually all that one sees is blinding white. In other words: with unrelated colours there is no black, but only colour and white. Black and grey are present only in the case of related colours.

Related colours are the colours associated with objects in our environment. They are defined by the part of the input light which is reflected. If all is reflected then the object is white irrespective of whether the intensity of the light source—and hence of the reflected light—is high or low. If none of the light is reflected then the object is black. If only a fraction of the input is reflected then the object is grey so long as the same fraction of all wavelengths is reflected. If different parts of the spectrum are reflected to different extents then the object will be coloured. It may be red, yellow, green or whatever, depending on the dominant wavelength in the reflected light and the more dominant one particular wavelength is the purer is the colour. (By "colour" I mean here, and in what follows, every non-grey colour which contains yellow, red, blue or green. The other meaning of the word "colour" which implies something obtained by mixing different colours will not be used at all).

In other words: the colours of objects in our environment or the related colours are defined by the diffuse reflectance from an object's surface.

Helmholtz. Nowadays one can scarcely imagine how excited I was when this idea dawned on me. To begin with it was now clear that Helmholtz, whose work on the physiology of vision is rightly regarded as a scientific masterpiece, had only ever worked with unrelated perceived colours as one would expect from a physicist. The organisation of colours was not something that interested him much and since he had only the unrelated colours in his mind he developed a hypothesis which covered those colours that could be derived from saturated colours and white. Of course he knew that there are three types of colours, not just the two in his scheme. The missing third dimension was treated to a cursory and not very appropriate remark.

Because of this neither he, nor any of those who followed and used his terminology, was in a position to construct an organisation of colours. The American researcher O. Rood who had been Helmholtz's best pupil in this area wrote later, after many years of quite successful work, "In reality neither our knowledge of colour nor our experimental approaches are at the moment sufficiently advanced to allow us to even suggest how we might achieve a proper scientific classification of colour. And, even if we could, then a great deal of work would lie between the suggestion and any real result".

Goethe. Historically a different type of explanation came from Goethe's view of Newton and of the physicists of his time on the question of the concept of colour. He detested their experiments and started one of his poems against Newton with the line: "Friends, get out of the darkroom". His objections to the experiments were not of any interest for Goethe was no physicist. It had more to do with the fact that he was unable to see in the colour world of the physicists, which of course consisted only of unrelated perceived colours, the much richer world of colours which he daily experienced. From this it was clear to him that the challenge of colours, which filled his soul, had not been answered. In this he was perfectly correct. But since neither he nor his opponents were aware of the terms "related" and "unrelated" perceived colours the dispute had to be left to be decided a 100 years later.

The recognition of the central role played by black was what led me out of Helmholtz's labyrinth. When the presence or absence of black is the decisive factor distinguishing the two closed groups of colours then black has to be an important element. And if that is so then white must be another element. The third element is also quite clear—it consists of the pure or saturated colour which I, because this term has been applied in slightly different ways, called "full colours". In this way were defined the three components of colour and changing any one, or of combinations of them, permits the generation of all possible colours. Hence any colour must be definable in terms of these three components.

Once I'd got this far it seemed to me that the matter was settled. If full colours, black and white are the elements then the following organisation of colours emerges. Greys are composed of black and white. The unrelated colours are composed of a full colour plus white. The related colours are composed of a full colour plus white plus black. That still left the possibility of a group of colours which were composed of a full colour plus black, but we know of no examples because common experience tells us that all known colours contain white. Nevertheless, it is certainly possible that this group of colours may be made. I gave up looking for them since it seemed to me to be more important to sort out the known groups first.

I should note here that much the same analysis had been achieved earlier by E. Hering. However, his approach was very different from mine because he took a psychological line while my approach was one of scientific organisation. Given the general law of nature that one always comes to the right answer last, my approach—being simpler—was better than his. He relied on perception to claim that every (what I would refer to as related colour) can be seen to contain an element of white and a bit of black in addition to the full colour. Given the fact that there have been quite a few physiologists and psychologists who have claimed that you can see a bit of green in blue and yellow, it is not surprising that Hering's claims were not generally considered convincing.

Nevertheless the fact that the results of the two very different approaches were in agreement was a welcome indication that they were correct.

The composition of the colours of objects. The next question was how the apparently uniform colours which we see in our environment are constructed from these elements.

In the case of greys the answer is well known. Each grey is composed partly of white and partly of black and together they make grey. If one terms the elements "w" and "b" then we have the equation $w + b = 1$ whereby w and b are both real fractions. The greater w is, the lighter is the resulting grey and the smaller w is, the darker the grey. Ideal white contains no black and so here $w = 1$. Conversely ideal black contains no white and so $b = 1$.

At this point I asked myself the worrying question, "Where does the "1" come from?" None of the many hundreds of equations from chemistry and physics which had passed through my hands had ever contained such an absolute value. That sort of thing only happened in certain geometric equations like in the well known case from trigonometry where $\sin^2 x + \cos^2 x = 1$. But even these are rare. Under what conditions do they arise?

In the case of geometry one can understand it. A line or an area can be extended infinitely and so here no absolute number can arise. But an angle can be no more than four right angles. Thus there is a limit to an angle and a limit to the natural extent of any angle so that angles can be properly expressed as true fractions.

This is reminiscent of the situation with greys. As is well known one can generate a grey area by taking a white background and adding to it small black dots in a regular array. The greater the percentage of black, the darker is the grey. But this value reaches a limit at the point when the dots entirely cover the surface. The surface then is the unit whose fraction can be white or black and these fractions together can never be greater than this unit value.

That then is where the "1" comes from in the equation for grey. It is an expression of the organisational law that governs the formation of mixtures of such elements. Thus, because in any area there can be no more black dots and white dots than will entirely fill that area it follows that $w + b$ cannot be of variable size but must always yield the same value.

One should not regard this simple idea as obvious. I remember how much thought went into formulating it in this simple way. Its real power, however, only emerges when it is pursued further.

If we now turn to colours then it is immediately clear that the same sort of idea applies also here. $f + w + b = 1$. In other words each colour is made up of three elements—the full colour, white and black—and the sum of these 3 fractions is always 1.

One can illustrate this in exactly the same way as with grey. One need only imagine taking a white background and adding to it dots of a full colour together with black dots and one will immediately see why all three elements are needed and why they have to be real fractions that always add up to the value of the unit.

While one may think that the equation for grey is simple common sense, the equation for colours has to be regarded as novel. Its discovery could only be made by someone who had achieved clarity about the concept of the elements of colour. Hering had this clarity but from his psychological stand point, which lacked terms for pure white and for related colours, there was no way to measure this fraction and indeed no way in which such a measurement could be conceived. This was what he wrote me when I sent him a letter telling him that I had discovered such a possibility and emphasised how the earlier qualitative description of colour could now be expressed in quantitative terms. This was shortly before his death.

The discovery of the means of measurement. At this point it became clear to me that so far I had only uncovered the difficulty of the challenge but had not yet solved the problem. The equation $f + w + s = 1$ might look nice and simple but it would remain a mere formalism until such time as a means could be found to put numbers on the components of any colour. In fact this was by far the most difficult part of my research into colour and I was determined to find a way because all further progress depended on it.

Sometimes I'd managed to make a discovery in much the same way that one finds a beautiful flower or a good friend: one keeps one's eyes open and when something worthwhile crosses the field of view one grabs it. Only in a single case—that of my Faraday Lecture (Part II, Chap. 28, p. 345) did I make a discovery—driven more by wish than by a real problem—which I didn't just find but had to work hard to reach. As I've already related the amount of work invested in that was greater than I invested in any other discovery.

However, now things were much easier. I'd achieved a large degree of understanding of the methodology of discovery and invention and was also able to apply the combinatorial approach which Leibniz had recommended as well as the general concepts of organisation. Because of this I was pretty confident that I'd reach my goal, for under these circumstances one can more or less order a discovery much as one orders a pair of trousers from a tailor: one will certainly get them, it's just a question of how long it will take.

To get an idea as to how the question of colours could be solved I first of all recalled how they'd been solved with grey. In that case there had been the two values of white and black which together had to make one. One therefore only needed to measure one of them for the other is then simply the difference between that value and one. In the Hasch I only had to quantitatively reduce the intensity of the input light which was directed to a white field until that field looked as grey as the sample and I then had the fraction of white in the grey. Black was the rest which was required to make one.

Let us suppose that I was able in a similar way to measure the amount of white in a colour would the problem be solved? Clearly not, because now both black and the full colour would be left and these two could be present in any ratio. In an equation of the form $f + w + b = 1$ two of the components can be varied independently. Measuring only one component is not enough; one has to be able to determine two of them.

Does it really have to be two? Could one find conditions under which one of the components has no effect? In other words can one find a way to make one of the elements inoperative?

It was a considerable advance simply to have phrased this question and it was all the more difficult because I was entirely on my own and couldn't reach anyone who might be prepared to think about the matter. The war was raging all around and it consumed all of the available energy.

The phrasing of the question had however brought my prey to a familiar place—that of the special case. During my work on energetics it was a way of thinking that I had employed many times to help answer difficult general questions. If I could manage to find the appropriate special case with the problem of colours then, mathematically, one could remove one of the variables and the problem would then be no more difficult than measuring grey.

What is a special case? To understand this important term we should look at a few instructive examples. If we draw a line which bends down, like a long chain fixed to two pillars, then this line can be made of innumerable points but only one of them is the lowest. This is the special case.

In a circle there are innumerable points which all share the property that for each of them the distance to points on the circumference are all different. However there is one single point, the centre of the circle, where the distances to all points on the circumference are identical. This is the special case here.

You can draw an infinite number of lines between two points but only one, a straight line, is the shortest. Again this line is a special case.

There are lots of such examples and the question then is, what do they tell us?

There is a curious and important law which states that the special case is always simpler to describe than are the other cases.

In the case of the chain each point has a different inclination to the horizontal and to describe the point we have to know the angle of inclination. In the special case there is no inclination.

To define a point in a circle we need to know the distance from the point to two positions on the circumference. The special case of the middle point requires only one value.

To describe a given line between two points we require lots of information along its entire length. The special case of the straight line between the two points is characterised by the fact that it is the shortest possible line.

One should note that in certain circumstances there may be more than one special case as one sees in the case of an ellipse where there are two special cases, one at each focus point.

These examples are from geometry because they are the easiest to see, but the concept of a special case applies far beyond geometry alone. The general principles of physics, mechanics and other sciences are all in their own way special cases.

For research this leads to a rather general approach: if there is a complicated problem one should always look for the special case because there the complexity will be least and hence the problem will be easiest to solve at that point.

The special case for colour. To be able to apply this I therefore had to find the special case (or cases) for this particular problem. I won't describe the many false paths I started down and luckily I've forgotten them all in the meantime. The successful ideas were the following.

What is different between a colour and grey? The answer is of course the presence of the full colour. Can I somehow eliminate this so that it doesn't interfere with the measurements?

Of course I can't actually remove the colour but what I could do was to make it invisible to my eye by viewing it through an appropriately coloured piece of glass. If you look through a blue piece of glass then the whole world looks blue but it is not of a uniform blue. There is light and shadow just as there is in a picture in grey tones. Therefore if I look at a piece of coloured paper through a blue glass then it will show a blue which is exactly as light or dark as an appropriately chosen grey. Now I am in a very similar situation as when I measured grey paper using the graded shades of grey. However I still didn't know how to interpret this.

This is where we come to the concept of the special case in which all the relationships are simpler. Let us suppose that the piece of paper is red; where is the special case to be found? The answer is obvious. It is when the coloured glass or filter is also red. And there is a second special case for every colour has its complementary colour which is diametrically opposed to it. So in this case a sea-green filter would also be a special case because sea-green is the complement of red.

Let's take the second case first because it turns out to be the simpler. The sea-green filter has the property that it blocks all red light, but permits sea-green to pass. If my piece of red paper was pure red then, seen through a sea-green filter, it would look black. There is no such piece of paper because in addition to the full colour, red always contains both black and white and so, seen through the filter, it is not black but instead it is clearly lighter.

Why is that? The red light cannot pass the filter, the black component yields no light and that just leaves the white. White light is a composite of all other colours including sea-green. The amount of white in the red paper, determines the amount of sea-green that will pass the filter. The same is true for the white component in the graded shades of grey. If the two look the same then they contain the same amount of white. Since one knows the white content of the graded shades of grey one now knows the white content of the red paper.

I was indescribably happy when I got this far. This was a discovery which was not due to chance but rather by methodically approaching it with appropriate ideas until it was solved.

And of course there is more to this than the solution of one particular problem—though that is important too—because here I had shown that one could indeed organise a discovery. This was a big victory for my leading idea that everything is accessible to science and hence that the art of discovery can be taught and learnt just as one can teach and learn the art of riding a bike.

Whether this shows that others can also do this is something about which I have my reservations.

Of course I can hear an outcry from the innumerable mystics and believers in the intuition of creative people who will all argue strenuously against this. They say: Ostwald has got it wrong. We agree that he made the discovery and that it is important. But he is wrong when he claims that he got there by so-called methodical work (which we don't understand and therefore despise). The guy is simply a genius and he makes his discoveries entirely by intuition as all geniuses do. Of course he prides himself that he did it all using his limited intellect rather than acknowledging that it was a kind gift from the world spirit.

I have to admit I am pretty helpless against that argument but I do have difficulty understanding how they know all this, especially as none of them makes any discoveries despite their intimate relationship to the world spirit.

We, however, still have the second (actually the first) special case to look at. In this we look at the red paper through a red filter. Now all the red, including that derived from the white, passes the filter and if the colour of the paper contained only red and white then it would look as bright as a piece of white paper. The black component, if present, reflects no light and hence makes the paper seem to be darker. If I find a grey shade which when viewed through the red filter looks as light as the piece of red paper then I know that both the grey and the red contain the same amount of black.

Having done all this the numerical values for white and black have thus been experimentally determined and if one now subtracts them from the value one, then one knows how much of the full colour is present and so, in the equation $f + w + b = 1$, all of the values are now known and the question of measuring the elements of colour has been solved. This question had not even been posed—let alone solved—by other researchers

Once these fundamental ideas had been elucidated it was relatively easy to convert them into practical uses. To be sure I'd never till then had any experience in making optical instruments but what I needed for this task could be built from relatively simple components so that I was soon able to make a "chromometer" or instrument for measuring colour and I used it daily for years even when instruments built by fine mechanics were commercially available.

The whole new area of research that was now open to me provided work for several years just to get the main points clears and determine the relationships between them. During this period the equation $f + w + b = 1$ was used all the time and its value was frequently confirmed.

The equal hue triangle. Innumerable colours can be generated from a full colour when it is mixed in all possible combinations with white and black as given by $f + w + b = 1$. I now faced the problem of how the vast number of resulting colours should be organised. Given the "threshold" of the eye's ability to distinguish shades we can estimate the number of different colours that can be resolved as being somewhere between one and ten million. The challenge then was to organise this enormous number so that each of these million colours would have a unique identifiable position. If mathematics and organisational science was of any use then I should find the answer there.

A large part of this work had been carried out over a century ago by German researchers including Tobias Mayer, Johann Lambert and Philipp Runge. Their principle result had been that it was not possible to represent all colours on a line or on a two dimensional sheet and that instead some three dimensional representation such as a pyramid (Lambert) or a sphere (Runge) was required. Their work was rather driven into the background by the emergence of the sophisticated system of the Frenchman Chevreul which a number of people tried to use. However none of the attempts really worked because these systems are all flawed. In all of them the colours must be arranged subjectively because there was no means of defining and measuring objective criteria. Because of this none of these systems won general acceptance even though they did manage to produce some fundamental organisation.

More recently the work of Ewald Hering, who had been the successor to Karl Ludwig (Part II, Chap. 18, p. 192) and hence my colleague in Leipzig, came closest to a solution. From him we have the "equal hue" triangle which displays the totality of colours which can be derived from a full colour by the addition of white and black in all possible combinations. He had rightly recognised that every colour associated with an object has in addition to the full colour a certain degree of white and black and that these determine what the colour looks like. In addition he showed that all of the possibilities for one full colour can be represented on a triangle in which the full colour is one corner, white is the second and black the third. In the spaces between all possible transitional tones are represented.

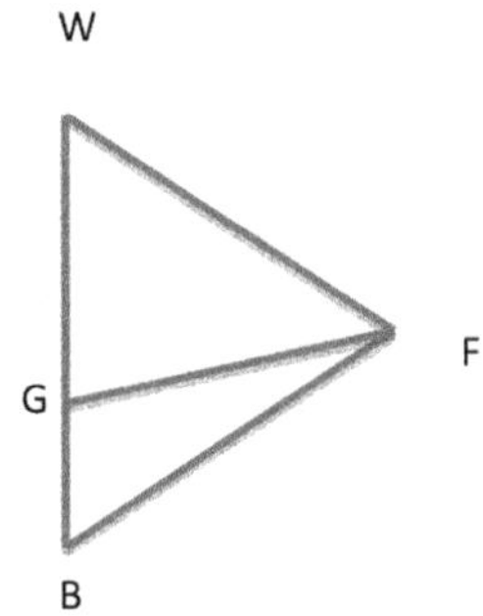

This was all correct. Unfortunately, however, he denied that one would be able to quantify these colours and, by doing so, he lost the chance to complete his excellent line of thought (This chapter, p. 569).

It was here that I, having found the means to quantify colours, could take up and extend his work. The side of the triangle from white to black covers the entire spectrum of greys. A full colour cannot lie on this line. We put it in the corner labelled F. The side extending from full colour to white represents all shades that can be obtained by adding increasing amounts of white to the full colour. Inside the triangle are all the mixtures which contain in addition to the full colour also white and black. Along a line from the full colour to the shade of grey (G) one finds all possible mixtures of that full colour with a particular grey shade. Since such lines can be drawn to any desired point on the grey line all possible mixtures of the full colour with all possible shades of grey are present. This triangle thus does indeed cover all possible variations of the full colour. Since the distances along the lines are a quantitative measure of the mixture we have here a complete description of all possible variations of the full colour.

What I now needed to do was to prepare such colour triangles for all the different full colours so as to see how they looked, for till then none of them had seen the light of day. At first I was horrified by the amount of work, but by starting with the main elements and gradually adding in the details later it turned out that the task was doable, although the war made it all more difficult. One essential prerequisite for this was of course a good working knowledge of colours and the ways that they were used. This was something that I had, while most other colour researchers did not, and that explains why I was successful in this case.

The colour circle. Before I could get into this work there was another question to be settled, one which science had up until now left unanswered, and this had to do with the proper arrangement of the hues on a circle. Hue is the general term used to describe the property of a colour that allows it to be described as yellow or blue or green. We have already seen that there are endless numbers of colours of the same hue, namely all those which are present in the same triangle. One therefore has to clearly distinguish the terms colour and hue. Colour is the appearance of a particular hue mixed with black and white. Hue is a more general and less precise term which describes that which is independent of the black and white components. It was long known that the hues could be arranged in a circle which was known as the "colour circle", or better as the "hue circle". If one starts with a particular hue then its neighbours will be those hues most similar to it. The further round the circle one goes the less similar are the hues to the starting one. This does not go on for ever because a point of least similarity—the opposite hue—is reached and it is followed by hues which are progressively more similar to the starting one. There is a way to precisely identify such pairs of opposite hues because when they are mixed they yield a neutral grey. One can then demand that all pairs of opposite hues lie opposite each other on the circle. This is the first step in arranging hues in a circle, but it is not sufficient.

In such a circle the hues are clearly arranged progressively in a particular order which is not doubted. However, only the order is clear because, just as some pearls on a string may be closer to their neighbours than others, so in the circle of hues some hues may be closer together while others are further apart. I was unable to find any scientific basis for making a decision on the distribution of hues around the circle. Once again I used the principle of the special case and in all the possible arrangements looked for the one which had unique properties. This then was the ordering which I applied to my circle of hues and I can tell you that it has become increasingly accepted as the best way of doing things.

During the course of this work I came across a problem that was a real burden on my intellectual conscience for quite some time. As is well known the spectrum shows the arrangement of light according to its wave length and this is the same as the order of hues in the circle: but with the one exception of purple which is not present in the spectrum. Red light has the longest wavelength, then comes orange, yellow, leaf-green, sea-green, ice blue, ultramarine and violet which has the shortest wavelength. In terms of wavelength red and violet must be least similar in appearance, but this is not the case. Instead red and violet are very similar, more similar than, for example, red and green. Thus while the perceived hues can be arranged in a circle, the spectral colours, to which they are undoubtedly related, can not. If the wave lengths were to be directly decisive for perception then the hues should form an arrangement like that of the greys in which there is a row with the greatest difference being between the hues at each end.

The only explanation I can find for the fact that no one else had stumbled across this problem is that even experts who know all the details of a field lack any understanding of organisational science and so they are unable to understand the problem even once it has been explained to them. This is the consequence of the neglect of organisational matters which form the basis of every science.

The "half the spectrum" colour. There was another problem which had in fact been described by Schopenhauer but since nobody had been able to solve it, it had simply been ignored and avoided for a century. It has to do with the following.

Pure saturated yellow, for example cadmium yellow, is almost as bright as white. In the spectrum, however, yellow makes up only around 5 % of the light. If the colours of objects were derived, as every text book will tell you, by the reflection of the appropriate wavelengths, then yellow should appear black because everything that reflects 10 % or less is black. For the other colours the situation is similar though less extreme.

In order to see this for myself I looked at a pure yellow solution in the spectroscope. What I saw was not the expected yellow region of the spectrum but rather the entire long wavelength part from red through orange and yellow to green. This part of the spectrum was as bright as if my sample was not in the instrument. Blue and violet, in contrast, were absent. The border ran through sea-green. I tried something else because there are lots of different yellow materials. Each time I got the same result. The largest and strongest part of the spectrum was unaffected and the border was the same as for pure yellow. With a yellow slightly tinged by red the spectroscopic analysis showed a small shift towards higher wavelength.

Hundreds, perhaps thousands of physicists, physiologists, psychologists, colour chemists and so on had seen this before me and some of them must have been a little astonished at what they saw. But none of them had decided to conclude that all the red, orange, yellow and green light was essential in order to form the related yellow colour of an object. This shows the power over adults, even over those who are used to thinking scientifically, of the views which are absorbed in school, and this makes them unable to see what is in front of their eyes. The importance of the task of school reform which I'd devoted a lot of time to, was brought back forcibly to me.

I was determined not to leave those problems lying and so I saw it as my duty not only to face the idea but to place it at the centre of the concept that for the generation of the related colours, even in their purest forms, a whole collection of spectral colours were needed. In my publications on the concept of colour one can see how this developed into the concept of "half of the spectrum" colours which not only got rid of this problem and the problem of purple but also opened up explanations for many other things. I will just give a very general resume here.

Developmental history. Goethe wrote in the introduction to his concept of colour, "Light is what makes the eye. Light is responsible for converting an indifferent animal auxiliary sense organ into an eye". Later work on the history of eye development has lent support to this view. Thus, to understand vision it is necessary to keep in mind the development of the eye.

No creature in nature has ever seen light of a single wavelength—what physicists would refer to as homogeneous light. Always and everywhere an eye picks up only mixtures of neighbouring wavelengths across a broad part of the spectrum. The primitive eye is unable to discriminate between different wavelengths and hence can not detect colour it sees only grey. At higher levels of development some rough distinctions can be made but even for the most highly developed eye there are numerous different light mixtures which it cannot distinguish. Throughout the entire period of phylogeny from primitive pigment spots all the way to the eye of an artist there was never an opportunity to adapt to homogenous light because the eye never sees it. Physicists made the assumption that the homogenous bands of light produced by a prism are not only the physical but also the psychophysical basis of vision and this was not challenged by early researchers into colour. This assumption, however, is a serious error. The eye is simply not capable of measuring the wavelengths of light with the necessary precision. At some point in the future the eye may evolve such a capability but we are light years away from that at the present. In fact the concept of "half spectrum" colours can be taken as a measure of the extent of evolution of our eyes and there is no clear indication that we have started to further evolve beyond that. A modern concept of colour must therefore be able to account for this and to examine the relations that lie behind it. That was the challenge that I faced and it was solved to a sufficiently successful degree that we can all feel comfortable with the circle of hues.

Once I'd got this far I now had to decide how to determine the hue of any given colour no matter whether it was a full colour or a mixture. For this I developed a very simple instrument—the Polarisation colour mixer which was able to determine for any colour sample its position on a properly arrayed hue circle.

The construction of the hue circle was a long and difficult work for which many of the details had first to be determined. To begin with, because of the decimal system, I divided the circle into one hundred grades of hue and at this level of resolution the neighbouring hues could scarcely be distinguished from each other. When this was ready, then every coloured sample could be assigned precise values for its hue and for its content of white and of black. The entire world of colour was now subjugated to the rule of measurement and numbers.

Implications of the measurement of colour. Culture is the intellectual capital of mankind. In order that it can accumulate there has to be a means whereby whatever the individual has worked out can be communicated through space and time to those who follow. This is what language and writing are all about. Without them there can be no culture.

However, there was up until this time no spoken or written language for colour and therefore there is no culture of colour. The Berlin psychologist von Alesch had looked into this question for many years and had come to the same result. Even though I'd been deeply involved with colours for my entire life nevertheless, until a few years ago, I was unable to judge whether a picture, an ornament or a decorative lady's hat made good or bad use of colour. And when I asked myself I got no answer. In the studies I mentioned above von Alesch showed the same coloured picture to the same person at several different times. The response was often quite different at the different times and in some cases the responses would be diametrically opposed. Everything that one might, by benevolently stretching the limits of the term, refer to as "colour culture" is limited to the achievements of individuals who were however not able to transfer their expertise to others. That is not culture. Instead it describes the situation which pertains before culture begins.

Why can the individual not pass on his knowledge of colour? Just imagine trying to describe a beautifully coloured pattern so that the person you are explaining to could, without having seen the original, produce an equally beautiful replica. It can't be done. And even if the originator of the pattern was to repeat his work without having the original in front of him then he might come up with an equally beautiful pattern—but it would be different in colour because he is not able to accurately define his colours and hence can't reproduce them. That is the reason why we place such a huge value on the master's original because it is one of a kind and will disappear forever when destroyed—and that is bound to happen in time to all oil paintings produced today. What musicians and poets produced centuries ago cannot be destroyed because it is all available as musical scores and books and can be brought back to life immediately. This isn't possible with paintings because of their colour which cannot be defined in some similar way. And why not? Because colours cannot be defined in writing. At least that was the situation until recently.

Now, of course, one can do it. One can define the hue and the content of white and of black and with these three numbers the colour is fully defined—now and forever. The colour, so defined, can be articulated, written and sent by telegraph or radio. The beautiful pattern I mentioned above can be quantified and then reproduced with the same effect when one enters the appropriate "number codes" into a sketch of its form. Thus colour becomes amenable to the same form of storage as is given by a musical score or by the letters of a poem.

One would have thought that the representatives of art history would have seized on this quantitative means of giving their subject clarity and precision. However, when in the autumn of 1919, I publically presented my results at a meeting in Stuttgart of the "Werkbund" (German Association of Craftsmen) they wrote a report claiming that my activities were highly damaging and having collected signatories they sent copies to all the German Ministries of Education begging them to ensure that the quantitative concept of colour not be allowed into the schools.

New work. Once I'd done all this it seemed for a moment as if my work might be finished. Not, of course, as if there was nothing left to do, because I have not related all the possible questions which this research answered and in any case a scientific problem can never be finally solved. Rather this new insight provided others with enough that was new, both at the practical and at the theoretical level, to keep them going for a long time.

However, there was another new field of great practical importance that needed some work. With the greys I had achieved not only a means of measuring in quantitative terms but also a means of standardisation. From the infinite number of grey shades a small number had been chosen to represent their nearest neighbours so that with just 10 or 20 such shades of grey essentially everything could be described that until then had required much more, and much less precise, information. In addition, these defined standards had been given letters which made it all very easy to use. If one said that a dress should be in grey "g" and the trimmings in "e" and "i" then we have a clear description of the shades of this important thing.

Just as greys had been organised and standardised so this should now be done with the colours. The difference—and the difficulty—was that while only one letter is required to define a shade of grey, a colour requires three. In looking for a solution I was not entirely free because the standard shades of grey had been defined and the new standards had to be compatible with them.

Fechner's law applied to colours. To begin with I started with something rather simple by taking a pure full colour, for example vermilion, and lightening it stepwise with white. At the start I made the series arithmetic 1/10, 2/10 3/10 and so on and by doing so got a result similar to that obtained with grey (This chapter, p. 571). The steps were too small at the white end and too large at the vermilion end. The red of vermilion behaved in other words much the same way as black had done. The conclusion was obvious: here also Fechner's law applied and so the white had to be added in a geometric progression.

But which? The answer was soon found because the same contributions of white must be taken as in the grey shade series i.e. cegilnp (since "a" is pure white it has no place here). What has been standardised once must be kept unchanged and here that had to be applied to the contribution of white.

With black, the story was the same. Here also Fechner's law applied but now the full colour took on the role played by white in the grey series.

How should the black be graded? Obviously it had to be done as in the grey series acegilnp. Experiments demonstrated that in both cases the ladder of graded shades was perceived as being composed of equal steps.

The "equal hue" triangle. With that the mixtures with white or with black are standardised but that still leaves the mixtures which contain both black and white. Those with white I call "light clear", those with black "dark clear" and those with grey "unclear".

Hering had already done some work on this and had shown that the derivatives of a full colour to which white and black are added can all be depicted in a triangle in the corners of which are the full colour, black and white. Then along the side f–w lie the light clear, along the side f–b the dark clear while w–b represents the now familiar grey scale. In the inner part of the triangle now lie all the unclear colours which contain at the same time full colour, white and black.

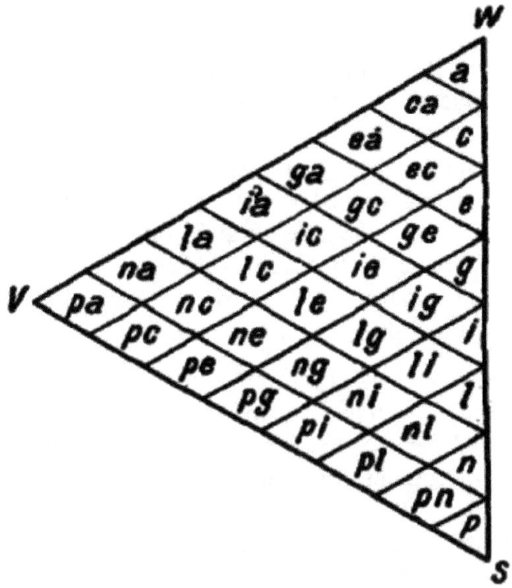

If one looks a little closer at the geometry of this arrangement then one sees all sorts of informative relationships which make the whole exercise a real pleasure. Here I only want to go into the final result that pertains to standardisation which one can see in the figure.

Every field represents a standard colour. Of the two letters in each field the first refers to the content of white and the second to the content of black as defined by the grey designated by these letters. As one can see the colours in the rows that run obliquely down from left to right all start with the same letter which means that they all have the same content of white. In the rows that run obliquely up from left to right all have the same second letter and hence the same content of black. These are "equivalent white" and the "equivalent black". The "shadow colours" are those in the vertically oriented entries.

A close look, or better still a look at a properly filled out triangle, shows that the pure colour of this hue is shown around "f", the white derivatives are clustered around "w" and the black derivatives around "b". Everywhere in the triangle neighbouring colours are related colours. Anyone who, like a painter or a decorator, has some experience of working with colour and looks for the first time at such a triangle experiences a happy revelation. This is not just my experience because I have tried this out on many people. One senses for the first time the strict inner coherence of the world of colour and recognises fundamental relationships which till then might have been sensed but were not known. For example the sudden recognition of the colour related to yellow had loosed a storm of joyous excitement in me as if a fog had suddenly cleared and shown a distant landscape in all its brilliance and beauty.

The triangle contains all of the possible mixtures of the full colour with defined amounts of white and black. These define the standards for all of the colours which can be generated by adding white and black to the full colour and hence it serves as the complete standard table of this full colour up to “f”. All of the colours in the series have the same hue and hence the triangle is referred to as the “equal hue” triangle.

It was an indescribable reassurance to my intellectual conscience when I saw this triangle in which everything worked so completely. I hadn’t been able to consult any committee of experts for at the time I was the only expert and, because of this, I had to do without collegial aid in finding possible errors and had to rely instead on my own powers of intellectual criticism. However, the rigour and the harmony of the solution argued strongly that it must be right.

Standardisation of the hue circle. Finishing off the standardisation was now relatively straight forward. All that had to be done was to construct such an “equivalent-hue triangle” for each hue. Once that was done all of the triangles were then arranged around a common axis such that their grey sides (“w”–“b”) were fixed to the axis with the tip “f”pointing out to form a double sphere. Its upper point is white, its lower point black and the full colours lie on the equator. The upper surface of the sphere shows the light-clear colours, the lower surface the dark-clear colours and the inner part of the double sphere shows the un-clear colours. The grey shades lie along the axis. Everything was so well ordered that everywhere new relationships could be perceived and this showed again that the triangles indeed truly expressed the heart of the matter.

Only one thing remained to be done. I’d started off by dividing the circle of hues into 100 levels and then quite impartially viewed the impression that the resulting arrangement made on me. This examination showed first of all the utility of Hering’s assumption of four basic colours yellow, red, blue and green but it also showed that in every case there had to be a colour in-between. The additional colours between yellow, red and blue were known for a long time as “orange” and “violet” which are both foreign words and so I use instead the German names “Kreß” (the colour of the nasturtium) and “Viel” (the colour of violets). Then it was necessary to find two colours to place between blue, green and yellow. That one can distinguish the reddish ultramarine blue from its greenish neighbour had been apparent even to Newton when, in his failed attempt to make colours fit to the musical scale, he needed seven basic colours. However, the cold sea-green is quite different from the warm leaf-green, so green too needed to be split. This resulted in the names “u”-blue, ice-blue, sea green and leaf-green for the other half of the circle of hues. In giving them their names I was careful to make sure that each had a different initial letter so that the formation of abbreviations was simple.

These eight points on the circle of hues lie too far apart to serve as standards. In fact a close examination of the question showed that three grades of each of the eight main hues were necessary and sufficient for 99 % of all uses. This gives rise to 24 hue standards which are named as first, second and third yellow, kreß, red and so on. One can easily fix these in one’s memory so that they can be learned off by heart. The only things that are necessary for this are that they are precisely defined and that they have distinctive names otherwise one will never be sure what is being referred to. Once again this is a proof of the fundamental cultural importance of language.

A short symbol was required in addition to the name. Since letters had been used as the symbols of the shades of grey, only numbers were now free. The best would have been 0–7 or 1–8 for the principle hues but then there was no simple means of defining the intervals between them which would have required infinite decimal values such as 1.333– and 1.666—. This, once again is a huge disadvantage of using a system based on ten. Because of this I had to use the numbers 1–24 for the 24 hues and in the unusual cases where finer differentiation was required then decimal values can also be used.

The colour code. Though the recognition of the necessity of a precise nomenclature of colours was a far reaching idea for the making of a culture of colour this could not begin before the idea had been fully thought through. I thus faced the practical challenge of bringing the results into a handy form such that anyone, even an art historian, could use it.

The division of the colours into the elements of the circle of hues (Part III, Chap. 42, p. 588) was of course just such a representation. It was however general in nature and did not show the standards. This was rather like the relationships one has in music. The pitch of a note is defined by the oscillation frequency but in order to determine from the oscillation value the interval that lies between two notes one has to go through a calculation whereas the musical score makes this immediately obvious. A piece of music written in the form of oscillation frequencies could not be played because the calculations can't be done fast enough even in the unlikely event that the musician was prepared to learn how to do the calculation. It's the musical score that makes it all possible.

The world of music is much simpler than the world of colour since each note can be defined by a single term—the oscillation frequency. A colour on the other hand has three: the hue, the white and the black content. It follows that a colour must be defined in terms of these three components.

Because of this I chose the short descriptions outlined above. Their combination yields the colour code much as the notes in a score yield the sounds. First comes the hue given by one of the numbers 1–24, followed by the white content by one of the letters cegilnp or further on if one manages to produce a colour with even less white. Finally the black content is defined by the letters ace—etc. White and black are clearly distinguished because white comes first. For white "a" has the largest value and the following letters, "c", "e", "g" etc. describe ever decreasing values. In the case of black "a" has the value zero and "c", "e", "g" etc. describe increasing black content. One quickly learns to read the colour code: 13ni, for example, is the first shade of u-blue with only little white and a moderate amount of black. It is a dark and slightly unclear ultramarine blue.

The colour atlas. Using the instruments I'd developed a colour can be analysed in just a few minutes (women are better at this than men), but nevertheless it was clear to me that there would be innumerable cases where such a measurement was not feasible. It was therefore necessary to have properly measured colour samples in some accessible form as a card index or an atlas for general use.

However, measuring a colour sample is a very different thing from producing a sample of standardised colour. It's roughly the same as the difference between measuring a distance with a ruler and producing the ruler. And since a colour is defined by three elements the problem here is much greater than with distance determinations where there is only one variable.

So now I faced a problem that no one had ever faced before; the preparation of quantitatively defined colours. After the hard mental work which had gone before this purely technical problem was for me now very welcome and I thoroughly enjoyed generating standardised colours. I was reminded of the biblical story of Adam in paradise when he was shown all the different creatures and gave them their names. Again and again I found myself looking forward to how one or other of the standardised groups would look and found myself happily enriched when I saw, for example, that the colour group with the white content "g" and the black content "e" was similar to what furniture makers refer to as "Marie Antoinette" with the difference that they were much better related to each other.

Despite the difficulties imposed by the war it became possible with the valuable help of my publisher Dr. Manitz to produce the first quantitative colour atlas in the years 1916 and 1917 and to make it generally available. It was strenuous work because in this first attempt to take control of the world of colour there were of course many unforeseen difficulties and often enough hundreds of colour plates had to be rejected. My eyes failed—I was 64 years old—before the last plates were finished and I had to leave these last measurements to one of my sons. Since then I have been careful not to strain my eyes more than was absolutely essential.

In planning the atlas the intermediate values abcdef—etc. were retained so that the number of colours rose to more than 2500. That was certainly a huge amount of work and effort. I did it gladly however because, in comparison to the exorbitant waste of energy in the war, I found it comforting to be doing as much valuable and constructive work as I could. I also considered it an expression of my life long close relationship to science.

Later on I restricted the standardised colours to the series acegilnp. On the triangle shown on page 396 one can see that there are 28 boxes with two letters. Since each of the 24 standard hues is formed into such a triangle the total number of colours is 28 × 24 = 672. On top of that there are the 8 grey shades so the final total is 680.

Apart from the atlas which came in the form of cards in the world format (Part III, Chap. 39, p. 537) (40 × 56 mm) other ways of displaying the standards were produced which in some ways made it easier to overview certain aspects of the whole. In none of these did we come across any difficulties or discrepancies and this practical experience shows once again that the challenge I'd taken on had indeed been properly solved.

Summary. When I look back over all this work, then I think I am justified in considering it the high point of my scientific achievements. This follows from two independent ways of making the judgement. The first is concerned solely with the amount of intellectual effort necessary to come to the solution and is independent of the cultural value of that solution. This is the abstract scientific viewpoint which is concerned with the "how" but not the "what". The value that is put on the work is, in this sense, absolute and is independent of the amount of time involved.

The other form of judgement depends on the effect of the work on the current status of culture and rises or falls with it. When, for example, it would become possible to make the surface of the moon visible to us as if it were only 100 m away then this would undoubtedly represent a huge advance in our ingenuity and skill but it is less clear what effect it would have on our culture.

Of course such a judgement is restricted to the present and one could theoretically suppose that it would be of enormous importance for us in 50 years or so. When Faraday presented his results on electromagnetic induction one "practically minded" member of the audience asked him, "What's that good for?" Faraday answered, "What's a child good for?" Today all of our electrical technology which is decisive in so many ways for our lives and will become ever more important in the future is based on the fact that Faraday's child has grown up and has brought us not only technical successes but also deeper insights into the nature of light and materials. Because of that this type of judgement is much less certain and is in any case far from being absolute.

When I look at the full extent of my work on the theory of colour in terms of these two forms of judgement then I think that it has been of great value in both senses. In terms of the intellectual effort it was concerned with a problem which neither Goethe with his enormous powers of observation not Helmholtz with his enormous mathematical intellect had been able to solve. My powers are much less than the powers of these heroes and no one knows this better than I. But also first rate minds of our time, such as Hering (I will not name those who are still alive), who have spent a lifetime working on the problem failed to solve it. Here there were many factors which had to come together. A chain, no matter how strong, will break if only one link is weak and in a similar way a whole string of different factors had to be in place for a solution to be found. The chance that all these factors would be represented in one individual was very improbable. Almost all of the many things which I'd done in the course of my long life, without feeling that they were in any way linked, turned out to be necessary to face this challenge and find the solution.

Of these factors the most important was the concept of discovery which is based on applied organisational science. For this I had not only my own experience for over the course of my life I had observed the work of my co-workers and pupils some of whom were the best in their field. Because of this I was able not only to make individual discoveries in the field of colour but was able to form these individual results into groups which contributed to the final solution. When I think this over it strikes me that even the precarious excursion into industrial matters in connection with the conversion of ammonia into nitric acid (Part II, Chap. 25, p. 299) gave me the chance to learn how to make discoveries when they were really necessary so as to move a project forward.

As to the second form of judgement, which is dependent on the cultural effect, I only need to remind you of the desire for colour which has captured all mankind after they had lived in dullness for nearly 200 years. By bringing this chaotic area under the control of numbers and measurement we can expect a development equivalent to that of the development of chemistry at the beginning of the nineteenth century which was brought about in a similar way by replacing qualitative chemistry with quantitative chemistry. Everywhere that colour matters—and where is that not the case!—will be affected and I can't even begin to think what the consequences will be. Here we are dealing with the physiology and psychology of the eye which is by far our most important sense organ.

And so far I have only considered the technical and scientific side of things. In addition to that there is an aesthetic side to the perception of colour. Here we are dealing in a deeper sense with the entry into a new epoch. I consider this aspect so important that I will consider it in a separate chapter.

Chapter 43
The Beauty of the Law

Technology and art. It was unavoidable that in connection with my work on colour I had to consider both the technical and artistic aspects.

I have described often enough in the course of this autobiography how my fascination with colourful impressions led me to try to capture them in paintings. By doing this I was faced quite naturally with general questions of aesthetics which I'd met in less pressing terms in other forms of art, particularly in music and literature. With the increasing realisation that there is nothing to which science cannot be applied I was forced to take an interest in the science of art.

Like everybody else who comes from the practical side of life to art, I found little that was satisfying in most works on aesthetics. Just the fact that the authors find it necessary to approach the problem with such exaggerated respect and treat any and all attempts to carry out a neutral analysis as petty minded carping, showed me that they didn't really have much to say in plain language, i.e. not much that was sensible or scientific.

One can say with some assurance that every time a writer treats a subject in exalted, solemn, touching, uplifting terms then he has probably abandoned logic and clarity. And equally certainly one can see that it is just these sorts of passages—or books—which make such a strong impression on the average reader. They remain stuck in his memory much more than things by the same author which are presented clearly and intelligently. What the educated person knows about Kant is largely restricted to the comparison of the moral conscience with the starry sky and includes perhaps also the moving tirade about duty. In fact both of these passages merely serve to paper over serious logical flaws in Kant's philosophy.

I therefore saw myself thrown back on my own devices and have always tried to achieve conceptual clarity about art, particularly about painting. I was helped in this effort by music which has a good though still incompletely developed practical and scientific basis. I am indebted here to the lessons in composition from Lobe which confirmed Goethe's words that the fraction of clearly enunciable and hence learnable in all arts is much greater than is generally thought.

At the same time I developed a distaste for those numerous writers on art who, without knowing anything about the basis of the subject, waste mountains of paper on senseless mystical or metaphorical drivel. Almost worse, because it comes

R.S. Jack and F. Scholz (eds.), *Wilhelm Ostwald*, Springer Biographies,
DOI 10.1007/978-3-319-46955-3_43

disguised in a scientific cloak, are the art historians who claim to be able to say something useful about current and future art forms on the basis of their historical knowledge. Such judgements are only valid in fields which have been conquered by science and whose contents have been shown to obey natural laws, for only in this sense can science predict the future on the basis of the known laws. However it's exactly this sort of work which one misses in most art historians and the few fruitful approaches, such as those of Fechner or of Morelli (Lermolieff) are left unattended.

Certainly my interest in the art historians' work was not sufficiently great that I'd been prepared to divert much attention from my various different projects to it. It was only once I'd completed my work on the theory of colour and thus provided this area with a scientific basis that was equivalent to that available for music that my old desires for some basic understanding of painting came back to the forefront of my mind. The organisational work on the colour concept led quite unexpectedly to something beautiful. I saw myself faced with a host of slowly ripened fruit which both duty and desire drove me to harvest.

The first light. In 1917 one of the many ways of presenting the colour standards had been to place the hue triangles for two opposite colours next to each other with their grey sides in contact so that they formed a diamond shape. "How beautiful!" I called out when I saw them for the first time. "How beautiful" said everyone to whom I showed them.

As will have often enough been clear from these memoirs, I am like the young man in Schiller's poem who was unable to find peace until he had peeked behind the veil of Isis. The priests who wanted to prevent that seem to us to be dull clerics with all the unlovely characteristics of that class of person, and if I had been that youth then I wouldn't have been "unconscious and pale" the next morning but rather jaunty and ready for new adventures.

And so I wasn't going to be satisfied with the beauty of the main outline and instead began to think hard. I'd put the plates together myself using the standardised procedures (Part III, Chap. 42, p. 587) without having had any intention of achieving any form of beauty and yet I had produced something beautiful just as when a chemist unexpectedly finds beautifully shaped crystals forming in his test tube. What was the source of the beauty which I had unintentionally produced? To find the answer I asked myself, what was it that I had actually made? The answer was that I had assembled colours of the same hue in a Fechner series with equivalent distance between them. This was the source of the beauty. By changing the arrangement I was able to show that the beauty disappeared if Fechner's law was not strictly observed.

In other words beauty was the result of adhering to the law!

Colour and hue. The conclusion seemed to be as unavoidable as it was absurd. I'd repeatedly read and heard in thousands of variations, that art and the artist are free and that any attempt to impose laws would destroy art, and that wherever nature happened to show apparent regularities the artist must use his freedom to replace them with irregularities. I have already related (Part II, Chap. 19, p. 209) how my own experience of painting had failed to support these notions. As I slowly conquered my fear of the absurd and developed a limitless trust in the results of scientific research, I pulled myself together, like a hunting dog that has scented a partridge, and looked around at the new possibilities.

These were not very inviting. The fact that since Pythagoras one had understood the laws of harmony in music and, using them, had been able to develop a vast repertoire of beauty, had inspired those interested in colour to formulate laws of colour harmony and there is a considerable literature on this subject. None of these attempts, including Goethe's chapter on the sensory perception of colour, had ever led to a useful "basso continuo of colour" which he had postulated and sought. In other words there is not a word in all this literature that would help to produce harmony in colours in the way that the rules generate harmony in music. It seemed that here I was in danger of falling into the same trap in which all of my predecessors had fallen.

On the other hand it was clear that the situation had now radically changed. Without exception all my predecessors had simply assumed that, just as musical notes may be defined by the frequency of vibration, so hues would define colours. None of the theoreticians had grasped that while a note has just a single dimension a colour has three and that the hue is just one of these. The hue does not define a single colour but rather a complete hue triangle which includes all variations of the hue from the lightest to the darkest, from the "clearest" to the least "clear".

My predecessors had been unable to avoid falling into this pit because they did not understand the organisation of colour. I, however, did and thus was able to avoid making the same mistake as they had.

What is beautful? I therefore set out to investigate the nature of beauty using the scientific tools available to me. On this topic there was already a vast literature on aesthetics. As I mentioned before, I had many times tried without success to extract something useful on the subject from this literature. In this I was not alone for I never found in any of the countless memoirs and letters of artists which I read that any of them had claimed to have been influenced by any of these works. On the contrary, one regularly came across the opposite claim that these works had not been of the slightest use. So once again I had to do for myself what others had failed to do, namely to open up the source of beauty.

I found two such sources. The first consists of the simulated revival of happy feelings one had in one's memory. The image of a dear friend, for example, may awaken such feelings and this is independent of whether the face is beautiful or ugly. In poetry it is the meaning of the words and sentences which call forth the welcome feelings.

The second source is a direct and spontaneous or playful use of the muscles or sense organs as one sees in its purest form in dance, in the harmony and rhythm of music or in the verse structure and rhyme of poetry. These things all involve rhythmic elements which are repeated in the same (or similar) form and without which the welcome feelings we seek and which we call beauty cannot be called forth.

If one looks more closely then the generally accepted view of things begins to appear a little differently. Objectively this has to do with those well known properties of works of art which are usually referred to as content and form. Goethe had with quiet precision observed that these appeal to different aspects of the mind in his verse:

The content in your breast
And the form in your mind

He assigned the content to feeling and the form to the intellect.

It was now no longer difficult to answer the question what law had to do with beauty. Law implies repetition because a law has the general form: always when "A" is the case then "B" is also the case. Repetition however results in rhythm and all artistic and beautiful forms are based on rhythm in one way or another, in other words, repetition. The idle talk of "artistic freedom" is just nonsense. In fact it's the other way round. The artist tries hard to make his work beautiful by creating a unique composition and he knows he will spoil it when he inserts something arbitrary in it. We are dealing here once again with the "special case" (Part III, Chap. 42, p. 577) and the special case tolerates no arbitrariness.

Access to colour harmony. Thus it is that part of the work defined by the form in which the law functions as a source of beauty. The content is not, or is only scarcely, affected by it. The beauty of form—and that is the general conclusion of my investigation—is always based on the application of the law. In this sense form is by no means restricted to spatial design but includes also temporal and organisational elements.

A work of art in its entirety is characterised by the simultaneous and reciprocal interactions of form and content. Here we are not going to have a discussion of the meaning of art but rather one on the doctrine of beauty. The general term beauty can be applied to either of the two factors. Here, however we are not concerned with the beauty of the content, instead we are concerned only with the beauty of the form which in normal language can be referred to as harmony. For this we can now say that adherence to the law brings forth harmony.

This was the key to the explanation for the observation that the plates of the hues were clearly beautiful although they had not been created by an artist. It is not without good reason that I compare the production of these plates to the growth of a crystal, for no matter what they are made of all crystals are beautiful. Their beauty too derives solely from the regularity in the way their basic components are added on to the whole. I knew from my micro chemical work that crystallisation under the microscope produces the most beautiful structures and it does so under simpler conditions and with smaller amounts of material than are required to make large structures. The law of harmony now explained the astonishing fact that all crystals are beautiful.

The law of harmony also explains the beauty of colour for all assemblages of colour will be harmonious so long as there is a relationship between them supported by the law of harmony. In other words, hue alone is not sufficient to define harmony, rather all three elements—hue, white and black components—are involved and only when all three are suitably related to each other can harmony arise. The simpler the law, the easier it is to grasp the harmony which unites the colours.

Once again I had opened up a new area which revealed a rich new world of unprecedented vastness.

Grey harmonies. Since access to the analysis of colour had started with the analysis of the various shades of grey I decided to try to apply the proposed laws of harmony to the simpler world of greys. This was all the more inviting because till then no one had suggested any sort of harmony for the shades of grey. I therefore started down my usual road for thinking which ran between Großbothen and Grimma. It ran over hills which offered wide views under whose influence it was possible to think in broad terms.

What sort of law would permit the organisation of grey? Since greys have only one variable component—the brightness or extent of the contribution from white—it is clear there must be a relationship to this. If I place two shades of grey together then they must differ to a defined extent in their lightness. Lightness is a simple unitary number and can therefore not involve a law since a law always relates two or more values. For this reason harmonies are only possible between at least three shades of grey. The simplest law would be that the relationship of the first two shades should be the same as the relationship of the second to the third. That, however, is the definition of a geometric progression.

Geometric progression! That is what I had used to define the standards acegilnp to follow Fechner's law of maintaining equivalent distance between the standards. And what was the result? It was that the grey standards set together in this way should be harmonious.

In the first moment I was astonished, but then it struck me that my investigation of special cases had shown me that in general a case which is special in one sense is also special in other senses. The standards had been chosen in a non arbitrary fashion to satisfy the simplest rule that they should be equidistant from each other and in this sense they were a special case. Because of this one could well expect that they would constitute a special case in terms of harmony.

I impatiently turned back so as to immediately test this idea experimentally, for from the work on standardisation I still had the samples of the grey standards. On the way I thought over what could be done and found a good dozen grey harmonies between 3 shades of grey.

One can well imagine with what excitement I set up the first experiment. It was a Chinese ornament out of an atlas that I set in colour in this way.

The result fully supported the theory. Not only I but everyone else in the house thought that the result was beautiful. The following experience was particularly convincing. Similar pictures which were generated deliberately with colours which were not equidistant were not rejected when I first showed them because everybody is accustomed to seeing the usual grey in grey pictures which are almost always not harmonious.

When, however, the eye had become accustomed to the impression of the grey harmonies then the viewer found the non harmonious pictures repugnant for now he could see the difference between them. Since then I have carried out this experiment hundreds of times with all sorts of people and always with the same result. To be precise I must say that on one occasion the experiment did not work because the subject considered the un-harmonious picture as beautiful as the harmonious one. The person involved was an art expert.

I can summarise my experience as follows: almost everybody, particularly women, are able and willing to sense the beauty of the equidistant grey harmonies. Those who have never experienced this learn very quickly the pleasant impression made by such pictures by looking at a few of them. After this he is soon able to detect disharmonies. In particular, I found that that disharmony in colours is not as easily detected or "understood" as with disharmonies of the simple greys. That is easy to explain because till now no one was able to produce pure harmonious colour combinations and so no one has had the chance to learn and recognise them. The situation is comparable to that of a Chinese or Japanese person coming across European music. He must first learn how the simplest tonal harmonies sound before he can recognise and enjoy them. One must be able to feel how happy I was with this result. The general formulation Harmony = Law now seemed to me to be a fundamental general law of beauty. As described above, I had soon subjected this idea to a particularly hard test by using it to generate beauty (in the field of grey) where no one would have expected to find it. The completely convincing result of this bold experiment was indeed a lucky stroke. It had, after all, been perfectly possible that people would have shown no sensitivity at all for the harmonies since they had never before been able to experience them. That the experiment worked must surely be due to the fact that in this case of grey we were dealing with the simplest possible harmonies that the field of colour offers so that the challenge was not beyond the abilities of average Europeans.

Kalik. I think that with the formulation Harmony = Law, I have discovered a fundamental feature of beauty and by doing so have made the field available to scientific investigation as a psychological discipline. I can't predict when this will happen, though I guess that it will take a long time, for we are dealing here with the most fundamental paradigm change in a field which has till now been dominated by historians with a bent to the scholasticism of the middle ages. The same hatred which the religious authorities directed at Galileo when he overturned Aristotle's errors in basic mechanics will be directed by the priests of the scholasticism of modern art to oppose the entry of science into their area should their current attempts to hush matters up, which they pursue with military precision, fail. For me this will be a confirmation that in the "humanities" it is possible to predict with a fair degree of precision at least the qualitative nature of the flow of events in these peoples' group soul.

Since I feel it is important to clearly separate my efforts from the aesthetics which has existed up till now I wish to use the word "Kalik" as a name for the new branch of science that I plan. The term is derived from the Greek word kalos (beautiful) and is structured in the same way as physics, optics and acoustics and should be pronounced with the emphasis on the first syllable. As one can see we are dealing here with the development of an "aesthetics from the bottom up" which Fechner vainly tried to achieve half a century ago. That ingenious researcher had recognised that the usual "aesthetic from the top down" started at the wrong end and that this explained its lack of success.

Coloured harmonies. Now I had the explanation for the beauty of the standard hue plates (This chapter, p. 594). The colours in the triangle were not only arranged in rows parallel to sides of the triangle in accordance with the law of closest relationship (similar extents of white, black and pure colour) but also their colour values differed by the simplest of all rules, namely that of equality. The whole plate was imbued with simple laws which were so novel that the simplicity never seemed monotonous.

As I hinted above, the organisation of the single hue triangles involves three types of regular rows and one can discern three types of harmony in these rows of colour just as I described for the row of greys. There are in other words three classes of hue harmony.

These rows had been discovered and used by artists for hundreds, indeed for thousands of years. Initially this was used to show the strength of light in relation to the position of an object with respect to the light source i.e. shadowing. Then came the association of different colours in ornamental work. To achieve this, the deepest colour would be used for the deepest shadow and white would be added to the colour for the less shaded parts. Even today it is not widely known that in this way one makes very faulty shadows for the colour is far too pure at the dark end. It was one of the most important discoveries of the great Leonardo da Vinci to detect this error and to find a solution to it. Nevertheless I'd dearly love to meet the art historian who understands this fundamental point. And how many artists understand it? Even I had not understood it and had made this mistake in my earlier pictures until I grasped the proper way of generating shadows on the basis of the quantitative theory of colour.

I set about colouring one of my grey patterns using the three laws of colour and produced plates of unimagined beauty and of completely different artistic character depending on whether the rows of equal shadow, equal white or equal black were used. The harmony of the equal shadow series could be immediately understood because this is what we see daily all around us. However their authenticity in comparison to the error laden shadowing which painters achieve resulted in a large increase in their beauty. The "equal white" harmonies seemed more novel but still I could "understand" them. The most difficult were the "equal black" series and one can see that in the near future they will take a special position in all deliberate attempts to produce harmonious images.

The harmony of equal values. The next question was what further simple rules govern the relationship of colours to each other? The answer came from looking at the sphere.

If one starts from any point on the sphere then one can make a line from it through the axis and this line will pass through the group of most related and hence most harmonious colours. By doing this we create a principle section and we have already seen how useful that can be. But one can also start from the same point and create a circle whose midpoint is the axis of the double sphere. This defines a circle of colours in which each colour has the same content of white and black. The only thing that changes is the hue. I refer to such colours as being of equal value; this circle is therefore an equal value circle and the differences between the colours are brought about by differences in the hue.

We have thus finally reached the point, which till now had been investigated by those interested in the harmony of colours but which they had never managed to explain. Now we can see why they failed. The question of whether, for example, red and sea green are harmonious is insufficiently well defined and hence cannot be answered. Even if one defines the hues precisely and takes the 1st red 7 and 1st sea green 19 there are still 28 different white and black 1st reds (if one restricts oneself to the standards—otherwise there are thousands) and just as many different sea greens which gives us 378 different pairs from which only a few will be harmonious. Only after the disposition of white and black have been settled can we define a particular pair.

The simplest law here is that of equal value. We therefore have to look at the collection of colours in an equal value circle and only here can we hope to find the first colour harmonies.

Here the results are not as simple as with the grey harmonies because of the much larger number of possibilities. We have in the colour sphere 28 different equal value circles corresponding to the 28 fields in the hue triangles and the same hue connections appear very different depending on the circle from which they are taken. Opposite colours from the pure colour circle, such as pa or na make a very "loud" impression which some people describe as "screaming" or "brutal", while the same pair taken from the "unclear" area such as ge or li are softer and yet seem lively.

In general one can say that one will get a harmonious result when one breaks the circle into a small number of parts of say 2, 3, 4, 6 or 8 equal pieces and views together the resulting colours, we have in our 24 piece circle the distances 12, 8, 6, 4 and 3. The easiest to understand are the "opposite" colours which lie 12 units from each hue. Then there are those which are formed by taking one third of the circle with distance values of 8. These are the triads which had played a great role for the painters of the sixteenth century who had very approximately defined them and asserted that they were harmonious. However the distances of 6, 4, and 3 also yield recognisably harmonious results at least when only two or three of the colours are considered and the rest are ignored.

In general it turns out that there is no real dissonance if each colour is paired with any other from that same equal value circle. The equivalence of the white and black contents is sufficient to establish this rule. Here too artists had reached an approximation of this result because these equal value colours are what they refer to as "valour" though they were of course not in a position to properly define what was meant by this and a degree of feeling for colours was involved in finding such harmonious pairs. Once again we can see the progress that has been achieved by moving from the uncertain feelings of artists to the certain knowledge and competence of the scientist.

Colour with grey. I don't want to omit another class of harmonies which one has so far not recognised as such even if sometimes examples may be experienced as harmonious. These are the harmonies between greys and colours.

In practical colour harmonics, which plays such a central role in women's clothing, one often hears the saying "white and black go with any colour" - Let us ask ourselves what the scientific basis for this. When we do that we get a very different and very clear answer which opens up a whole new set of wonderful harmonies.

Take any colour, for example 8ie. We analyse this: 8 is the 2nd red which is roughly carmine; i is a middling content of white so that the red is halfway between pale and deep; e is a marked black content so the red is "unclear". Can one place this red together with black velvet? A lady of taste would not do so because then the red would look insipid. Could one perhaps put it with white silk? Once again the answer is no, because the red colour would look somehow dirty. However there are certain shades of grey which would fit well to it. Can one define this grey a little better?

The answer comes when one asks which grey is related to the colour 8ie. The rules say it is i and e, because i has the same amount of white as does 8ie and e the same amount of black. If one brings 8ie together with grey i or grey e or with both of them then one will have a pleasant harmonious result. No other grey will give a similarly wonderful effect.

So that is one further new discovery from the many which the rules make possible. With so much confirmation one can regard the system as proven.

What I have here very briefly presented as an outline of the extent of the discoveries in the new area of colour harmony made possible by the work on colour organisation and colour quantitation has been made available in a more extensive form in the little book I published in 1918 called "The Harmony of Colours". There was no echo to it in the art press apart from a few short low key words of rejection. Despite this the large first edition was soon sold out. In the meantime I'd undertaken a lot of experiments to determine whether the harmonies were indeed more pleasant to the eye. This was indeed the case and I was able in all the years since then to enjoy the thought that I'd been able to unveil the beauty of colour which no one before me had ever seen.

These new results together with some important conceptual advances were included in a new edition which sold around a thousand copies per year. From this I conclude that the book found quiet but dedicated readers the majority of whom were probably involved in technical matters. One doesn't get to hear from them how and to what extent they made use of the content because if the master dyer, the illustrator or the wall paper manufacturer has some success with it then he'll keep quiet so as not to give anything away to the competition.

Nevertheless I see in my annual visit to Karlsbad, both in what is offered in the shop windows and in what the ladies wear, a rapid increase in the use of my calculated harmonies. In one half hour's walk in 1925 I counted five correct triads of equal value hues as well as many more modest attempts. So I'm satisfied that the new possibilities which this work has opened up have taken root where they ought to in the design of fabrics in the widest sense and that this will help German Industry in the production of all sorts of luxury goods. Whether artists will ever accept these fundamental advances is not a question that particularly interests me, and so I leave it instead to be answered by that rather idiosyncratic bunch. Once the desert wastes of subjectivism of the last years have finally run their course, the wave law of history will ensure a sort of antipodal trend leading in design to more precise work and accentuated rhythm and to clear harmonies in the use of colour. These last can only be achieved by the application of the laws of harmony which I described. Of course it is quite possible to come up with them using a simple feeling for colour but that is a difficult and long winded approach which in any case yields results that are only approximate. However the artist who has learnt to play this "colour organ" will have freely available to him the complete range of all possible harmonies with which to bring his works to the highest level. One cannot make harmonies; one can only discover and apply them.

These remarks are not just dreams for the future. Apart from the few hundred plates which I myself prepared, some famous artists are already deliberately applying the new concepts and they tirelessly assure me how much this has improved their work. From the well known established painters I can name the Dresden master Wolfgangmüller. I won't name the younger ones since I told them that in their own interests they should better keep secret the fact that they were applying my concepts.

The problem of form. I was soon able to try out in a new area the idea that beauty derives from laws. In order to see if two or more colours worked harmoniously together I'd laid surfaces painted with the colours side by side. I tried this and that and soon realised that the achievement of harmony was strongly dependent on the shape of the surfaces used. Once again there was no other way out but to attack the problem scientifically.

It had been clear to me ever since my time as an assistant in Dorpat that this was a problem whose solution would be a joy to me. Matters of art were among the many things that Oettinger had now and then involved himself with. He held non specialist lectures about art and used for them plates in books from the university library. One of these was "The Grammar of Ornament" by Owen Jones whose numerous colour plates fascinated me. I noticed that among the ornaments of primitive peoples one which showed the outline of a human figure with legs apart with the remarkable property that the vertical and horizontal elements of the design completely filled the entire space. I told myself that this property (I later referred to it as "space coherency") must have its own geometrical causes. However since I couldn't immediately see what they might be I put the matter aside with the quiet conviction that I would return to it once the time was ripe.

Now the time was ripe, for I said to myself that if law = harmony then everything that is based on laws must be beautiful. The beauty of the ornament whose "space coherency" had so deeply impressed me must be due to these laws and if I could discover the geometric basis for this then I would have discovered the key to a whole new source of beautiful forms.

In connection with my work on colour I'd got hold of colour plates which illustrated different types of coloured surface. I learnt from coloured Japanese woodcuts how one can use the natural forms in an artistic way not merely to produce an exact image of an object but rather to alter the image so that it would fit well into the overall picture. The means of doing this must be regulated by certain laws which would have to be special cases of the general laws of harmony. I'd earlier got hold of Racinet's older works on coloured ornaments and I borrowed others from public libraries. I asked myself if all of these thousands of ornaments could really be organised under a set of rules and came to the conclusion that without exception this was indeed the case.

This discovery was in some ways quite unwelcome because at the time I was in the middle of the work on colour theory and found the new flow of ideas not just distracting but almost destructive. But I had to accept it because it would have been impossible to ignore this flow of ideas. In any case I needed the results for the work on colour.

The harmony of form. In the winter of 1921–1922 I devoted myself entirely to this new idea. I forced through the development of the basic idea by a more than average act of will because I felt it important to get the matter cleared up so that I would have secure data for my studies of the harmony of colours. The earlier work I'd done on the science of organisation were of value here and allowed me to rather quickly achieve a broad overview and even to work out a few cases in detail. The line of thought was as follows.

What is the general nature of the law? The answer is repetition. Once again I was astonished at this definitive sounding answer but I was soon able to convince myself that it was both correct and sufficient.

Every law, whether it is one proclaimed by the authorities or a law of nature, takes the form of "If A, then B" -It doesn't matter whether we are dealing with crime and punishment, income and tax, nutrition and assimilation, heat and evaporation, friction and electrical charge; it is always a case of "B" happens when "A" appears.

Thus the law and hence the harmony of form must lie in the repetition of the same or similar things. More precisely one should say more or less similar because an exact repetition never takes place since the two events are always different at least in terms of time and space.

What sorts of repetition are possible? Answer: There are three forms: spatial displacement; rotation and mirror image reflection. All individual cases can be traced back to these three processes. The application of these forms of repetition can be either stringent or relaxed. Historically it turns out that the stringent forms, whose adherence to the law is the simplest, appear early and that the more relaxed and complex forms come to the fore when the appeal of the simpler forms has worn off. Once this development has reached a certain point then there comes a time when people begin to believe that the rules can be ignored and that the artist can flourish only in the complete freedom and in the absence of any law. It is instructive and amusing to see how such outgrowths of wishing (for here we are not dealing with a question of thinking) go too far and are washed away by a new wave which sets the old rules once more prominently in place.

I summarised the concept of a law of beautiful form in a short book "The harmony of Form" (Die Harmonie der Formen) which was published in 1922. The huge amount of new results which it contained so astounded the representatives of the academic art scene that as far as I am aware not one of them dared to take this amazing structure seriously. The few comments in the journals were of the sort which Goethe characterised as:

They say: it doesn't appeal to me
And mean, they laughed it off

For myself I can say that this work has in the past 5 years been of enormous value. First of all it helped in understanding ornaments and paintings and in fact since then I find it easy having seen one example to understand its form and come up with as many variants as I want. Secondly it serves as a source of ever new pleasure in form. The comparison of an actual ornament with that which is theoretically possible and which the system of organisation permits, lets one see that only a tiny fraction of the possibilities that science has now laid open have been discovered by all the artists who have ever lived, for they merely worked on the principle of trial and error (art historians call this "inspiration").

I worked part of this area out in sketches and till now (1927) have accumulated and published 4 folders with a total of 240 sheets. Because combinations of two or more of these groups of designs and variations on them are possible, the number of demonstrated forms reaches enormous numbers and goes into the hundreds of thousands. This all goes to show how rudimentary the natural development of art is because the ornaments which are currently produced seem very primitive in the light of what is now possible and they are also not as good as those produced by the Indians, the Arabs and in particular the Moors. The modern European never has the chance to experience the joy of contemplating the complexity of a developed ornament which reminds one of the joy that we experience when listening to music.

Coloured ornaments. A new world of beauty opens up when one combines the law of colour harmony with that of the harmony of form. By producing the "colour organ" as a tool I can now just as easily and quickly pin down my ideas as the composer can translate his into musical notation. I have to relinquish the dimension of time which the composer has (in earlier years I often quite nicely improvised on the piano and the harmonium), for a picture once painted remains unchanged, but I do have the advantage that a picture once finished will continue to exert its fascination.

The lack of the dimension of time can to some extent be removed by making a sequence of pictures. This is a way that was often used in earlier centuries, particularly to depict biblical scenes and legends. It still works today. When M. von Schwind achieved an unexpected success with his sequences of pictures illustrating the fairy tales "The seven crows" and "The beautiful Melusine" he wasn't happy at all. He was trapped in the artists' belief that only with frescoes on large walls could a painter really show his worth and he considered the illustrations as inferior works. Obviously he hadn't understood the great power of sequences of pictures. The even greater popular success of Wilhelm Busch with his illustrations is due in no small part to the fact that a good three quarters of them are not individual pictures but rather sequences, some of which are quite long.

Films. If one compares the impression made by the best works in music and in painting then music turns out to have a huge advantage. It can stir people's feelings and keep them fascinated for hours on end. A picture, in contrast, can at best exercise a strong effect for a moment before the feelings it inspires die down and the initial strong interaction with the observer ends.

The reason is that there is a time frame to every emotional experience. It may begin faint or strong, develop and change and then end either quietly or with an explosion. No picture can generate this temporal succession and even a sequence of pictures can only roughly approximate it. Because of this it can't call forth a strong emotional response in the observer. Music and poetry can do this because they are art forms which have a dimension of time. This leads us then to the question, can't the art of light be given a temporal dimension?

The answer is: that is exactly what is happening at the moment. Films owe their attraction solely to their ability to offer the eye a temporal sequence. It is typical of the sterile scholasticism of current academic art that it is totally incapable of grasping this new dimension that film offers to the representational arts. Just because the Geeks and Romans didn't have cinemas these people regard an art form without historical roots as something completely inferior and this is an attitude that a scientist simply cannot accept. Since our educated classes are bound to follow the views of these tired old high priests of art they keep away from this new art from and thus it is left to the lower classes of large cities to determine which films will be successful. And even if now and then something really good is produced there still hangs unmistakeably in the air some doubts about the value of the whole field because it has no tradition.

This is where I see a new art form coming into being. The artistic deficits of current films are due largely to the fact that they depend on photography to depict realistic seeming events. However one could well imagine an art of light which is as naturalistic as music and in which a temporal use of light, form and colour are used to depict emotions and thus to simulate the viewer in much the same way as the listener is stimulated by music.

There are two principle problems which have prevented this being achieved. First of all there were no rules governing form and colour like those which have long been established to govern harmony in music. Secondly there is the technical problem of displaying light in all its forms that is to say in terms of intensity, colour and form and to manipulate these over time so that the image of an emotional experience can be produced.

Of these two problems one has now been removed for there is now a concept of the harmony of colour and form and hence one could now produce a poem of light or a symphony of light which could be written down and repeated whenever one wanted just as one can always read again a poem from a book.

As to the second problem—the optical one—one must accept that this is quite different from what one sees in current films because we would not be dealing with real objects but rather with virtual, subjective ornaments. New optical ideas will have to be developed for this new challenge but I believe that I will live to see this realised.

As one can see these are very broad goals. If one takes into account what has already been achieved then we can hope that the way towards a general concept of beauty has finally been found and that the call to implement it has already been articulated. One may then begin to understand the strange emotions I now feel. Physiology leaves me with the expectation of only a few years and the steady loss of my free energy limits what I can do. I can't even begin to imagine myself collecting in even a tiny part of the harvest which is now almost ripe.

And even that little is not certain because I can't foresee what new ideas I'll stumble across as I start to work in this vast and fruitful area. I feel like Moses who after overcoming endless problems had led his endlessly argumentative Jews to the border of the Promised Land which he could look out over but which he knew he would never enter. One must forgive me if I fail to find that tragic and think instead: "As I know my people my successor will have it no better!"

Chapter 44
The Noise of the Streets and the Peace of the Garden

The two fields of endeavour. When I look back over the many aspects of my life I can make out two major areas of work that I was involved in at various times which brought me a large part of the happiness that lit up my life. The first and oldest area is science. I didn't pursue it, as the common idiom holds, "for its own sake" (for science has no sake) but simply because nothing in the world brought me deeper, more enduring and unspoilt joy.

The second area was in influencing other people with the purpose of letting them share, at least to some extent, the flood of joy which I took from science. I can't pretend that this altruistic work sprang out of a feeling of duty in the Kantian sense. It was something purely instinctive just as the scientific work was. While my scientific work offered me complete joy, because overcoming difficulties and opposition is also enjoyable, my feelings about the organisational work are more mixed. In addition to great joy this work also brought me a lot of vexation and this led to stomach aches until I managed to forget the vexation. However, it did leave me with the feeling of unease that those who often enough acted stupidly, narrow-mindedly and badly behaved were nevertheless also human beings and thus so to speak my equals.

If you ask me which of these areas of work I consider the most important then I have to hesitate before answering. I think that the way things are at the present the organiser is more important than the discoverer. The history of culture is full of examples in which absolutely fundamental discoveries were ignored for years and had no other effect than to make the discoverer despised, hated and persecuted while the results remained totally unknown to the rest of humanity. Only after their re-discovery, sometimes after several re-discoveries did the man turn up who was able to convince not only himself but also his contemporaries of the value of the discovery and then to organise its practical application. In this sense a discovery in terms of the advancement of culture is only half the battle for there still remains the problem of hammering it into the sluggish minds of the masses. And the greater the discovery has been the harder this will be because the resistance to a discovery increases with the square of its value.

R.S. Jack and F. Scholz (eds.), *Wilhelm Ostwald*, Springer Biographies,
DOI 10.1007/978-3-319-46955-3_44

For this reason I think the work of the organiser is more important because it's more difficult than the work of the researcher. I've experienced this in my own life in which organisational work started considerably later than scientific research—and finished earlier. While I can contemplate carrying on my scientific work for as long as my brain holds out, I have drawn a firm line under organisational matters.

The organiser works on the street, the researcher looks after his garden. The organiser wants to influence and therefore must accept all the discomforts that go with dealing with large collections of people. He has to get in the way of those who have been in charge of things till now and is treated by them as an opponent and perhaps even as an enemy. And whoever works in public is always surrounded by lots of people who will constantly attack him usually for things he didn't say or do, and when he tries to explain this then he will be attacked for not having said or done them. In short he has to put up with all the noise and dust and bad smells that are part and parcel of life on the street. So long as one has some excess energy then one can cope with all this but once the energy reserves start to run low then it is more sensible to retreat into the quiet garden of pure research where one can match what one does to the energy still available.

Only a discoverer can also be an organiser of science for otherwise he has no idea of what he is trying to organise. This doesn't work the other way round, for many researchers have no organisational talent at all. This is particularly true of the majority of the "classic" type of scientist who always try to leave organisational matters in the background.

One often hears the opinion that pure science is something higher, finer and better than organisational work. What is true is that in research it easy to keep one's hands and clothes clean while the business of convincing the masses is often enough rough, dusty and dirty work, depending on the type of person who has to be convinced. However since culture is a social matter, a scientific discovery can only become a useful part of it once it has been organised into its proper place in the structure of humanity's intellectual capital.

The reason for the happiness of the researcher is not that he doesn't have to face inertia and lack of understanding because he brings enough of that himself. But for obvious reasons one tends to judge one's own stupidity a little milder than the stupidity of others. I don't know how it is with my scientific colleagues but I can say for myself that during my work I often exclaimed out loud "Oh you idiot" even when no one else was around.

The reception of the concept of colour. I'll start with a description of work on the street. As I mentioned I had arranged with the German Association of Craftsmen (Deutscher Werkbund) to try to achieve some order in the then chaotic world of colour. Over the next years I reported on my progress at their meetings and sought no compensation for the thousands of Marks which I spent on this effort. To begin with my reports were accepted with thanks even if most of the people didn't understand much. However when I presented my quantitative method which made one independent of collections of standards I was told by influential members that I should concentrate on producing immutable colour samples for example for pottery.

In 1919 there was for the first time since the war again an annual meeting—this time in Stuttgart. I'd by this time reached a point at which quantitative measurement, organisation and standardisation had all been completed and I could present the results in the form of the hue plates (Part III, Chap. 43, p. 594) which had given me the laws of the harmony of colour. Quite naturally the lecture was in two parts the larger of which was concerned with questions of the organisation of colour and a small part was devoted to harmony.

It was easy to remember the date of the lecture which was held on the 9th of September, i.e. on 9.9.1919. There was a large audience. I felt enthusiastic about my subject and was able to inspire most of the audience as judged by the applause. However a section of the audience was immediately completely against me. These were the academic art experts as well as a large fraction of the artists. The main argument—the organisation of colour—was accepted in theory and no one dared to question it because none of them understood anything about it and they knew this. No one understood anything about the other part of my talk either because it all dealt with matters that were entirely new. They however believed that they knew not just something but a very great deal about this. Since what they thought they knew was not only in stark opposition to my laws of harmony but also to the idea that any such laws could exist they immediately mounted a frontal attack in defence of the holy grail of their ignorance. I was later driven out of these circles and publically branded as the "destroyer of the innocence of colour".

The fear of a threatened invasion of science into the high holy place of their mystical beliefs was so great that after the Stuttgart meeting a sort of defensive association was formed. The danger was loudly though erroneously depicted in vivid but faulty colours and petitions were organised to warn all the German culture ministers of the impending danger. The petitions were submitted (Part III, Chap. 42, p. 586) and as far as I could see they were not without effect for a little later the Prussian minister of culture issued a proclamation phrased rather indirectly which in fact led to a prohibition of the teaching of the new theory of colour.

At the same time drawing teachers were mobilised. As is well known drawing is a Cinderella subject in secondary schools because the marks in it do not affect one's progress through the school nor do they show up in the final exams. I'd tried to argue that this was due to the completely mistaken goal of trying to reach a high level of artistic achievement which will always be reached by only a few. If drawing and painting were scientific subjects then it could be considered as important as arithmetic or geography. Now, thanks to the new concepts of form and colour there is finally the possibility to make drawing scientific and hence to give the subject the dignity it deserves.

With few exceptions, of whom R. Dorias in Chemnitz who constructively supported this line of thought should be mentioned, the rest of the art teachers were totally opposed to the idea that an outsider air his views about their affairs. And I wouldn't have done it if it had indeed really just been their affair but it is in fact a matter which concerns all of German youth and their interests seem to me to be more important than the turf wars of the organised art teachers.

I'll spare you a description of all the opposition which was mounted against the colour theory. In fact none of the discoveries I made gave rise to such massive and unanimous opposition and that seems to me to be a confirmation that in this case the discovery is of unprecedented magnitude.

The chemical industry. Finally I have to tell you about opposition from a completely different direction, opposition which was more unexpected and perhaps also more effective. It is well known that there is and always has been in this country a strong bond between science and industry particularly in chemistry. The massive growth of the German aniline dye industry was due principally to the courage which the industrialists showed in their decision to transfer the difficult chemical synthesis from a laboratory scale to an industrial one. During the time I was a professor the university laboratories were regularly visited by representatives of the major chemical companies who were looking to see what advances had been made that might be of value to industry. On the other side the professors benefited from the help that industry offered them free of charge for their research. I, for example, asked for and received all the dyes that I needed for my investigation of colour.

However once the colour theory was presented publically as a rounded off piece of work things suddenly changed. It was not only that industry did not support the new results but instead fought bitterly against them. I was shown circulars from the major factories at that time (1922) in which the directors instructed their representatives in all the various towns to tell their customers that the new concept was wrong or unimportant or otherwise suspect and that the aniline dye industry had no intention of having anything to do with it. Later their combat strategy became more indirect but even today (1927) there was not a trace of any readiness to take the greatest advance in colour theory since Newton and Lambert seriously and to make use of it. Quietly however it has smuggled its way into some places.

I am quite unable to explain this odd behaviour which is so at odds with the readiness, even eagerness, industry normally seizes on scientific advance. It can scarcely be explained by any doubts as to the scientific value of the work which after all was done by the same researcher who had given industry catalysis and the chemical fixation of nitrogen. For the moment I simply gather the necessary materials. Perhaps sometime later it will be possible to explain this mystery. That applies equally to the many other strange fates of the colour theory in the public mind of which I could tell some very odd stories.

For the moment I prefer not to burden myself and my readers with these awkward matters. From my work on chemical kinetics I know that an energetically favourable reaction cannot be prevented but at best can only be slowed up or accelerated. For that reason I can simply sit back and wait for the new concepts to permeate into science and industry and the question of when this will happen is much more important for them than for me.

In addition one should not lose sight of the fact that apart from personal animosities there are social currents at work in these times which are more than inauspicious for the acceptance of this sort of progress.

A mental epidemic. Nothing shows more clearly that war is an outmoded remnant of the barbaric past than the relapse of whole peoples to a lower mental level which, in times of peace, one had considered to have been overcome. This was the mental epidemic which hit us after the Napoleonic wars a hundred years or so ago and which in literature resulted in romantic mysticism and in science led to the empty rhetoric of natural philosophy and we needed half a century to overcome all this. The fact that the war of 1870/71[1] had no such effect was due to its brevity and to the fact that we Germans, as a result of our cultural eminence, did everything once hostilities were at an end to come to a peaceful agreement with the French. It was in fact one of the first concerns of the German Empire to pick up on the negotiations for the foundation of an International Office of Weights and Measures which had been interrupted by the war and to place the headquarters of this important body in Paris.

On the other hand, the so-called peace agreement of Versailles at the end of hostilities in 1918 had no other purpose than to continue the war against Germany by other means—and it continues to this day (1927)—and because of this the mental damage which has been done has been enormously increased. In Germany at the moment we suffer from a rampant outbreak of mysticism which once more has turned against its most dangerous enemies—science and reason.

Of course there are always many people who subconsciously feel (though they would never admit it) that they were not fairly treated when it came to the matter of parcelling out intelligence. All these people are more than happy when intelligence is held to be of little value and they will happily lend their support to this sort of mental epidemic. In addition one should not forget that there is a large number of honest people who also feel that they have missed out in the joy of understanding things and who are unable to summon up the energy do something about this by patiently studying the basic levels of a precise science. They much prefer to seize the possibility of getting some strong emotional experience through the practice of mysticism which doesn't require any great intellectual effort. In addition to the joy of leadership there is also joy to be got from a readiness to submit passively and blindly to being led. This is the form of joy which people thirst for in times like these.

There is one further influential stream of thought which should be mentioned that leads in the same direction. It comes from the fact that the humanities are most important since they directly affect the life of nations and indeed of everybody but, on the other hand, since they are placed at the pinnacle of the pyramid of sciences they are quite naturally the most underdeveloped. Where knowledge is missing,

---

[1]German French war.

belief will step in to fill the gap. While it is a fundamental feature of science to be always open to criticism and, where necessary, to improve each and every one of its propositions it is, in contrast, a fundamental feature of systems of belief that every criticism and every attempt to alter it is seen as morally outrageous and will be instantly condemned. The result is that the representatives of such under developed intellectual areas, in which science will in the future hold sway, see it as their clear duty to deny the precise sciences entry into the temples of their belief. They therefore happily welcome anything which promises to minimise the reputation of science.

Posthumous fame. All of these factors work powerfully together to oppose my efforts and they are reinforced by all the burdens of the aftermath of the war (1927). So I keep on working, well knowing that my results will for the moment have no or little effect and that the flowering of my work will happen only after my death. But all this does not really reduce my joy in my work because I know why I have to be satisfied with sowing the seed and leaving the rest to the powers of time.

The following story from the Viennese comedian Nestroy made a great impression on me. Nestroy was a wonderful character player and so one of his friends told him that he should not waste his talent on farce but turn instead to serious works. "Why", asked Nestroy. "You have to do it for posterity", the friend replied. "So, so", said Nestroy, "And what has posterity ever done for me?" He got no answer from his friend and I too couldn't give him one.

The Greeks had a good reason to worry about their posthumous fame because they believed that their existence in the underworld would be more tolerable and even enjoyable the more they and their deeds were remembered by the living. This idea formed the basis of Goethe's wonderful elegy "Euphrosyne". Today we no longer believe in that sort of thing though many of our most prominent people behave instinctively as if they did.

I've sometimes asked myself if I too am driven by such atavistic thoughts when I'm engaged in work—often enough until my aching back demands a pause—that will change the future. I really don't think that that is so. Instead I think that I have a strong need to bring all the elements of my intellectual life into some form of order. I can only compare this with the cleaning instinct of a good housewife even if I can't say from which side of the family I inherited this characteristic. Maybe it's the result of the mutation I underwent (Part II, Chap. 28, p. 349). That it is always left unsatisfied, for example by the question we are dealing with now, serves only to keep it awake and active. Wherever this drive comes from it has served to ensure that I was always interested in things and happy at each little advance that was made.

From the street into the garden. Whenever the noise on the street got too loud or continued for too long I always felt the need to flee into the peace of the garden. I've described in detail (Part II, Chap. 22, p. 248) how I could achieve the long sought garden peace by absorbing nature and playing it back in sketching and painting. How much stronger do I now feel the wish to turn my life entirely to the garden and to abandon the work on the street. Naturally this sort of change of direction doesn't happen overnight. Every holiday had prepared me for the next bout of work and I can't absolutely rule out the possibility that I may now and again venture out into the street even if I run the anger of no longer understanding the traffic rules.

Because of this it is worth describing in a little more detail how this transformation came about.

Tenerife. As a result of all the stress associated with the world language, the "Bridge", the Monist Society and so on I found in 1913 that my energy reserves were so low that I prescribed myself a period of complete rest. I wrote a stock of half a dozen Monist sermons and then set off with my eldest daughter in the late autumn to swim and paint on Tenerife. As expected from earlier trips, the sea voyages there and back were already refreshing and the range of wonderful scenery which awaited us promised a rich horde of paintings and photographs. Indeed these promises were richly fulfilled and I look back happily on the six sunny weeks we spent there.

On the voyage there we found ourselves in the company of a contingent of soldiers on their way to a West African colony. Despite my pacifism I came into social contact with the colonel who commanded them and found him to be a very pleasant fellow traveller who was completely free of the narrow mindedness that one associates with our military officers.

Another amusing incident involved a dining companion who I'd got to know a little. I came across him one afternoon fast asleep in his deckchair with a book open on his lap which threatened to fall to the deck. It looked somehow familiar and when it did fall down I picked it up. It turned out to be "Great Men" by Wilhelm Ostwald.

We landed at the expected time in Santa Cruz where we stayed for a few days in order to accustom ourselves to the new surroundings. When we'd left Hamburg we'd had tolerable sunny autumn weather which nevertheless clearly signalled the approaching winter. On the voyage the days had become ever warmer and longer and on Tenerife, despite the fresh ocean breeze, it was as hot as July at home. The shape of the rocks was different from anything I'd ever seen so far because the island was volcanic and consisted entirely of lava which had been spewed out of Mount Teide. Because of this horizontal stratification, which forms the basic structure of the landscapes I'd known up till then, is entirely absent. The rock has a rusty black colour.

The flora was just as unusual. Huge cactuses, agaves and euphorbia seem to have been the original plants and after colonisation came palms, oil trees, cypresses and other plants of the south. Large banana plantations, which were the basis of the island's economy, covered all the areas wherever earth had managed to form on the lava.

The most beautiful part of it all was the sea. The huge rolling swell, which had been able to develop across thousands of kilometres, was of a pure ice blue colour unclouded by any organic matter or inorganic colloids. It crashed with a noise like thunder on the stony beach and, wherever it was flat, covered it with foam. In these places where the waves were caught between cliffs breakers were formed which were more impressive than those I have seen anywhere else.

We soon left for Port Orotava on the other side of the island and it was there that we saw the sea in all its climactic beauty. The road there passed over a quite high ridge and on into an area of the most wonderful scenery: Alexander von Humboldt counted it as one of the most beautiful places he had seen. This impression is made by the fact that the coast is broken up by a number of bays which form a blue backdrop to the beautifully formed old lava flows. On this side too, the island is rich in many shades of green.

We took lodgings in a quite good hotel which once had been a rich man's mansion. We'd arrived in the evening and after dinner we set off for the sea which one could hear crashing unceasingly on the beach. As always in the tropics it was suddenly dark but an almost full moon rose behind the clouds and we could see an almost ghostly scene as the breakers shone silver-topped between the high black rocks. It was an unforgettable scene which I tried several times without success to capture in a painting.

Thus began very comfortable weeks of holiday in which I didn't miss the usual round of work at all. Every hike offered ever new motifs for painting and I found it a real challenge to try to deal with the grotesque forms of the lava, for normally the outline of a painting is defined by the stratification of the rock even when this is disrupted by glaciers, waterfalls and weathering. Here, however, there was just lava which does not settle down once its surface has cooled but rather is carried along and broken into strangely shaped blocks by the force of the slowing flow until the temperature has dropped sufficiently to prevent any further movement. In the meantime the rain which suddenly pours down in buckets has gouged out deep gorges which are usually dry but after new rain carry a wild mixture of water and rocks down to the sea. Perhaps it was these experiences which gelled the slowly forming view that beauty can only be found where laws are obeyed, because the haphazardness of the forms of the larva was a real hindrance in converting them into art and forced me to choose to paint broader landscapes where these chaotic details disappeared and only the overall forms remained.

It was a very pleasant surprise when the excellent Munich biologist Richard Hertwig came with his wife and small daughter to stay a few days. He'd had a particularly strenuous time as rector and was wise enough to fend off an imminent breakdown by taking a pause. We hadn't met before though I knew his name from the excellent work which he'd done in collaboration with his brother Oskar (Part II, Chap. 29, p. 366). There soon developed a close relationship also between the ladies so that also from this side the journey was considered a real success.

We several times visited the botanic garden near Orotava which I would appreciate much more today now that I have developed a closer affinity for plants. Just by chance the bird of paradise flower, which has one of the strangest forms I have ever seen, was in bloom. My attempts to paint it showed me just how much I still had to learn. Only in the last couple of years since 1926 have I really applied myself to this.

We'd come on a steamer of the Wöhrmann line where the head steward had regarded us as being not particularly important people—though neither of us had had anything against that. For the voyage back we caught a ship of the Hamburg-America line where we were treated very differently. The captain sat us at his side for meals and made every effort to make the voyage as pleasant as possible for us. In this he was successful because he was an engaging and thoughtful person with whom I had several fascinating conversations. I met him once later during the war in Hamburg or maybe Kiel. He was then commander of a minesweeper and told me what a strange feeling it was to be constantly aware that one could be blown up at any moment. We also met the friendly colonel who had delivered his detachment of troops and was now returning to service in Germany.

We did, however suffer from the change in the weather from the sunny south to the dark, misty December of home and with every day's sailing it got colder and darker. We left the ship in Antwerp since from there we could get a rail connection which would bring us home two days earlier than if we had continued to Hamburg. There was just time to visit the museum to have a look at the famous Rubens pictures. They seemed to me to be rather theatrical. One technical point I saw was the speck of pure vermillion that he had placed in the deepest shadow of his subject's skin which had a very good effect. It was only later that my research on colours showed me that this was well considered.

Salzburg. I also have to count as part of my garden-island the 2 weeks I spent at the summer university in Salzburg. These were informal gatherings to which professors with something new to say were invited. Seminars from each speaker were spread over no more than 1 week and the whole event lasted two, so that after the first week there was a complete change of speakers. The outer form was a humorous take off of academic traditions so that for example a rector had to be elected for each week. This honour fell on me—it was the only time that I had the office of rector.

Since for these meetings the people invited mostly had lively minds one had lots of opportunity to meet interesting people especially as the range of topics to be discussed was very wide. The audience too was in a holiday mood and thus particularly receptive. Although my attendance at these meetings involved quite a bit of work they nevertheless remain in my memory as special events and it would be a good thing to mount something similar again.

Karlsbad. My annual pilgrimages to Karlsbad were a whole set of garden islands, like a string of pearls, for there I always meet the closest friend of my old age whose warm company contributes so much to the recreation I bring back from there.

In January of 1911 Wilhelm Exner, the talented organiser of the commercial and technical education system in Austria, had accepted an invitation to hold a lecture at the museum in Munich. I'd already seen him several times in Vienna and at the museum meeting in Munich but didn't know him well. I'd heard about his creative organisational work and so I thought I'd take the opportunity to get to know him better in Munich. By chance we were both staying at the Hotel Rheinischer Hof and just as I was going in he came out holding a booklet in his hand. He recognised me and showing me the booklet said, “Your name is in this”. I looked puzzled and he explained: my lecture is concerned with international organisations and naturally I have to take your work in this area into account.

This was the start of a friendship which has bound me to this good man ever since. Though he is 14 years older than me and now in the spring of 1927 is about to reach the end of his 87th year he still carries out his demanding duties as director and organiser of the Austrian Technical and Experimental Research Centres and every year manages to add and breathe life into a new link to this rich complex.

Vienna and Großbothen are so far apart that this relationship though it was important to me, would have suffered from lack of interaction and would have quietly faded away into the mists of memory had it not been for our visits to Karlsbad which at the beginning were once a year and later became twice a year.

Around my 60th birthday I suffered from bilious attacks which at the beginning passed off after a few hours but later led to painful attacks. In a chance encounter Exner had earnestly suggested that I should take the waters at Karlsbad. He himself ascribed his freshness and energy, which indeed astonished me and everyone else, to the fact that he had been going to Karlsbad every year for the last 30 years and intended to keep on going there. It didn’t take a lot to persuade me to join him the next autumn in Karslbad. I got there in good time and that was the first of an unbroken series of meetings in the lovely spa that I found to be a real refreshment for both soul and body. To be honest at the end of the first occasion I suffered a pretty painful attack but I learned that this is quite a common occurrence and is caused by the outflow of the gall stones. Since then the problem has diminished though with one or two relapses and it now seems to have entirely disappeared.

In the first years in Exner’s company I met the architect Wilhelm Doderer who was one of the builders of the Kiel Canal. We three Wilhelms spent the days drinking the water, going on hikes and eating together and in all of this Exner was unquestionably the leader. He made us take a daily excursion of at least 20,000 steps which he checked minutely on a pedometer which was one of the many gadgets he always carried about with him and he introduced such a strict rhythm into our days that we felt no wish to waste our time with any of the usual distractions.

Exner’s had a huge circle of friends and acquaintances so that I met through him a large number of first rate people most of whom were quite naturally Austrian. I won’t make a list of them because no long term relationships developed from them.

An experiment in the teaching of art. The 2–3 weeks that I usually spent in Karlsbad were not completely filled with the idleness one associates with spa life. Chance often saw to it that I received an invitation to write an essay for some newspaper or journal and I almost always accepted. For this sort of work its good to be away for a while where new images and points of view come easier to mind than when sitting at one's own desk at home. I often used the time during hour long walks to put together arguments which till then had been only vague ideas.

On one such occasion I tried to work out how I could get people quickly to accept my colour and harmony ideas and it struck me that poetry always makes a strong and lasting impression. Rhyme and rhythm, for example, play important roles in advertising.

During my youth I had now and then written doggerel, but I'd never tried my hand at real lyrical poem. In later years the only verse I ever composed was the energetic imperative—"don't waste your energy, ennoble it and use it". Which is usually expressed in a shorter form and, in any case, the opponents of Energetics claim that it's a bad attempt at poetry, though I, of course, do not agree.

Since during the course of all my writings I'd developed my capacity for linguistic expression I saw no reason why I shouldn't be able to write more than just couplets. I learned from my reading of Goethe a great deal about the technical aspects and having learned from my research into colour and form much about the nature of works of art I reasoned that I now fulfilled all the prerequisites necessary for the generation of poetry.

To collect a bit of experience I began to write a teaching poem about how the scientific findings about colour could be translated into practice. I soon developed a technique which is quite similar to that used to design a picture. Once one has the content clear, one makes a sketch in which the verse form and the most important rhymes are initially roughly set out irrespective of the detailed structure which comes only later. This sketch is then filled out line by line whereby one mustn't shy away from throwing away some well turned lines if they don't fit to the others. The compulsion imposed by the rhymes often leads to new ideas and images, though one has to watch out that this doesn't end giving the impression that things were done just to force the rhyme. The organic relationship of form and content is the main point. The better this works, the better will be the work of art.

Once I was clear about all this, I realised that the great periods of art are characterised by an emphasis either on form or on content. I tried to express this thought in poetic form and came up with:

Content and Form start to fight.
Form claims victory (bang, bang you're done)
But Content gives no respite
And boasts that it has won
War ends. Peace starts.
But love only wins in a meeting of hearts.
Form and Content fuse
And only then is born a work of art

Anyone who understands something of how poetry is written will see that I have used here a whole set of technical aids beyond just rhythm and rhyme to heighten the effect. In this way I had made a well sounding and impressive poem.

To eliminate the possibility that this had happened just by chance I then set myself the following challenge. I'd read an essay by A. Holz in which he blamed the use of rhyme for limiting the possibilities open to the poet. I'd found in my experiments that the need to form a rhyme had often been helpful to make the content livelier and that it usually wasn't that difficult to find a good strong rhyming word which could be used. Now I noticed that not so many words rhyme with rhyme. I only found a few like slime and prime, and so I set myself the task of using these rhymes to construct a poem. For this it was clearly necessary to find an interesting line of thought which would accommodate these rather diverse terms. This looked to be difficult as at 11 o'clock I set out on a long walk to think. At first I didn't get anywhere beyond the line "Rhyme is fate". However by the time I got back for lunch at 1 pm I'd only solved around two thirds of the problem. However, I didn't give up, as even Goethe had succeeded in his first Sonnet to find a rhyme for rhyme.

After a short siesta I got back to work with renewed energy at and in two hours I was happily finished. Here is the result:Rhyme is fate.

They say it restricts.
But rhyme is fine to open the gate
To the mind's pulsing thickets.
Rhyme is a straight line to the mind.
And it gives rhythm melody.
Rhyme is the prime
Component of beautiful poetic elogy.
Rhyme is like the glue of pine.
And it provides cohesion.
But rhyme is also slime, and the unwary
Will harvest derision!

I carried out this and several other such experiments in the autumn of 1925 once I'd finished the first volume of these memoirs. They were enough to convince me that my views of this art were correct, but still I decided not to try to win my spurs as a poet and I haven't written any poems since. Instead I turned refreshed to attack further problems in the newly opened field of light and colour.

The new way. If you asked me how I intend to steer the small boat of my remaining life then I'd say that its time to leave the stormy sea of activities which involve convincing other people and instead to chose a way of life which seeks only the joy of the hearth and which allows me to work free of resistance and conserve my last reserves of energy. To begin with I have made the process of aging a subject for scientific investigation and this is an area which, not surprisingly, is both of constant interest and great service to me. For most people these things are a source of worry. They make things worse by trying to ignore them and they feel they have to make an extra effort instead of just sitting back. The scientific approach is completely different. One doesn't get upset at the workings of natural relationships, particularly if one has managed to delve deeply into a scientific appreciation of the laws governing the rhythm of the world and the laws of nature. Then one can feel that one is privileged to be part of this universal rhythm and is happy to be swept along by this great wave below the surface of consciousness.

But this sort of thought is not enough to fill the day and my life-long habits demand a practical occupation that will lead to good and happy results. In this respect I find that the many strands which went to make up my life now come together and form a ribbon, as if this was where things had been going right from the start.

Even today I have difficulty not to regard the more or less trivial occupations of the day as challenges and I feel a sense of guilt if I neglect them. In other words I have always regarded life as a duty and meeting these obligations was always a source of joy to me. This joy was more than just a transient feeling of success at having achieved something but rather the joy that comes from fulfilling one's reason for living. Practically, of course, it didn't matter what my attitude was because the feeling of joy was for me a sure sign that my work was done for the best and hence satisfied the energetic imperative. Duty and joy were to be found in the same activity.

Now, however, as the daily ration of well used energy gets ever smaller, I decided that joy must become the central issue in life and that duty has to serve merely as a way to joy. In the same way that a very general mental process connects means to ends, so real happiness is the result of linking means to joy. This is particularly true of work which throughout my life always brought me luck. This now has to fade into the background and I face the question as to which sources of joy are still open to me.

The best known and also perhaps the best way would be to work to make as many people in my surroundings as possible happy, or at least to moderate their unhappiness. But this option is no longer open to me for I lack the ability and don't believe that I will ever acquire it. People have always only been of interest to me in the plural and I regard each individual merely as an example of his species.

But there is another way and that is what I have taken for myself—art. Art is the means by which one reaches feelings of joy in purely synthetic ways. The discovery of the laws of harmony of form and colour which has opened up such broad new approaches to art was for me a field, or perhaps better, a garden, in which I could find beauty and by doing so could breed joy as a gardener breeds flowers. The practical application of these laws in the generation of beautiful things was completely unknown. It first had to be methodically worked out under the rules of organisation and each individual application provided not only scientific clarification but also immediate pleasure. It was not only me who felt this pleasure because I was able to make my pictures available to the public and in doing so I didn't need to pay any regard to praise or blame since in this I was not acting as an artist. Harmony was not some personal achievement I'd invented but rather an explanation of natural law which I had carried out in the service of science. Praise and blame should be reserved for the extent of understanding that these laws achieve with the observer.

Artist and researcher. This is something one has to think about. Anybody who knows real artists personally knows that they are almost always complaining and that their complaints are only now and then interrupted by short periods of soaring achievement. These brief moments do however make the artist so happy that he is prepared to stick to his profession.

The reason for this is that the artist works entirely unconsciously and he feels a real urge to work. Once the way of doing what he does is explained then we talk no more of art but of science. The numerous internal (and external) prerequisites for creativity, including his subconscious abilities, must be accessed by the artist before he can achieve anything. Between the wonderful moments when this all happens lie long periods of reduced energy, which have to be endured until things build up again to another happy moment. This is what happens with the young artist. The old one, if he's really good has managed to make a lot of the prerequisites habitual even if he is not consciously aware of them so that he can achieve a lot and is not so affected by the periods of reduced energy.

Once the work of art has been completed the artist (here I am speaking mainly about painters) has to accept that it is far removed from the ideal form he had in his mind's eye. He wants to improve it but bitter experience has taught him that that frequently ends up just making things worse and that trying to reverse the "improvement" will turn out at best to be very difficult and will often be simply impossible. So he leaves it as it was and will be left with an abiding feeling of dissatisfaction. The poet and the composer are here in a better situation because they can alter their work and then choose the best solution without losing the original version.

All these difficulties disappear when one scientifically constructs a beautiful harmonious piece of work. The production itself is a source of joy as one sees the beauty emerge ever more clearly from the raw outline and one looks forward ever more eagerly to the final finished work. Its beauty usually surpasses all that one had expected. The fantasy of the artist is restricted to what he has experienced. The scientist, in contrast, can open up completely new and unexpected possibilities simply by applying the known laws and what he produces is perfect in its way and cannot be bettered, just as a scientifically precisely defined musical triad cannot be beaten.

Often enough I've played on my "colour organ" in the evening to generate harmonies by lamplight and found that they were not in fact particularly beautiful because they were thrown out of kilt by the dominating red and yellow of the artificial light. When, however, I looked at them in the morning I was astonished and captivated at their beauty. Once one has had some experience in generating harmonies then one can make them so that they appear beautiful even by lamplight just as triads on a piano can sound harmonious even if the instrument is out of tune.

A fundamental cultural advance. If one looks at the situation I have described from the point of view of general cultural science then one will arrive at the following conclusion.

All sciences started off as art forms, that is to say as the expression of skills which were not consciously understood. There had been an art of metal working, an art of building mills, an art of writing, an art of counting and many others which in the meantime have developed into sciences. This is true particularly for the lower levels of the pyramid of sciences. The higher one comes up the pyramid the lower is the degree of scientific development and the higher is the input of the artistic elements that one finds in its foremost practitioners. For example in medicine, despite the established scientific basis, the low degree of scientific development of psychology results in a doctor's work being to a very substantial extent an art. If you watch a famous surgeon operating then one will be immediately convinced that he is practising an art.

Now what one generally understands by the word "art" is the ability to generate welcome feelings in an observer and this is just as much an applied science as is medicine. The only difference is that it is not something physiological like medicine but rather it is a psychological technique. Because of this its scientific component is at a much lower level than in medicine and is indeed so low that one can fairly talk of an "art" that is to say of an ability which is not consciously understood.

There can, however, be no doubt that sooner or later there will be an advance from art to science so that a beautiful picture will be just as scientifically created as a diphtheria patient is treated with a scientifically based antiserum. The process towards this state will be slow so that the artistic component in the generation of beautiful pictures will remain important for a long time to come. However, it is in the nature of things that the scientific component will gradually increase.

It was my great good luck to have been able as discoverer and researcher to have initiated the development of this new cultural epoch and I owe it to my life long love of science. My happiness at this is not troubled by the unified rejection by those involved in art at all levels from journalists to professors—quite the contrary. The unanimity of the criticism merely goes to show how enormous and paradigm changing the new step forward is in comparison to everything that has heretofore been produced by artists at either the practical or theoretical levels. And since no one can stop me from enjoying producing my harmonious works of art their criticism has had no other effect than to deny so many of their contemporaries, who never did them any harm, access to an inexhaustible source of pure joy. In Lucas 11, 52 is stated "Woe unto you, lawyers! For ye have taken away the key of knowledge: ye entered not in yourselves, and them that were entering in ye hindered."

The next challenges. To begin with I studied the colour harmonies by exploiting my knowledge of the harmony of forms and colouring in simple geometric patterns. This soon got boring even when I used really complicated patterns. I learnt from the Japanese (who in turn learnt it from the Chinese) how to produce a naturalistic effect by the use of shapes put together by following rules. This requires the participation of the observer who remembers the natural form and appreciates the way it has been transformed by the rules.

The most obvious and profuse motifs for this are flowers, plants in general, and lower animals. I was able to make a huge advance over the Japanese by choosing the colours strictly according to the rules of harmony. The Japanese too have managed to achieve by trial and error some good uses of colour in their woodcuts and these effects were then preserved by tradition. But they also make some awful mistakes especially since the new dyes became available for which no traditional rules of thumb are available.

I therefore have painted a large number of pictures of flowers in which I used the flower itself only as an inspiration and then produced both the form and the colours strictly according to the laws of harmony. I, and lots of other people who have seen them, liked the results. Nobody thought them ugly and nobody was left cold by them. This means that my opponents, of whom there are plenty, will have to find some other point of attack. I have heard the comment, "He tells us he's going to revolutionise art, and all he does is paint little flowers!"

A few years ago I saw a modern painting. It was about one meter square with a dark grey background on which had been splashed in horrible garish colours cloudy, ray like and round blobs which didn't make any sense. The catalogue gave its title as "Cosmic Events". It was of course incomparably bigger than my little flowers but I don't believe that anybody really enjoyed it—not even the artist who painted it.

Apart from these simple products of nature I have also started to work on larger pictures which are idiosyncratic landscapes produced using the laws of form and colour which, even if incomplete, are the reason why we like them. I think that this challenge is rather like that of writing a poem (Part III, Chap. 44, p. 617) and see possibilities for work for me here which will more than fill the remaining time of my life.

But even my little flower pictures had further happy consequences. To begin with I used whatever the garden and meadow offered but after a while that source was exhausted. During the war years I'd let the garden of "Energy" run wild and afterwards I didn't have the money to put it back into shape. Occasional visits to other gardens and numerous visits to the exhibition of gardening in Dresden (1926) showed me the rich variety of forms and colours which our highly developed gardeners can and do produce. As a result I began to develop an ever stronger interest in developing the extensive grounds which belong to "Energy". From viewing the garden chiefly as a source of motifs for my paintings I now more and more come to regard it as a work of art in its own right and to find a real life's goal in looking after my flowers.

In this way my work in and out of the house becomes one and brings me doubled joy. I don't need to regret the transience of a flower's beauty because I can transform a large part of that to pure harmony in my pictures. And since I've developed the habit of writing down the codes of the colours I use I always have a sort of score of all of my pictures. This will make it possible, even if time alters the pigments, to call back the original harmonies and thus secure for these harmless but pretty first members of a new artistic epoch an eternal existence for they are now as secure and unchangeable as the great compositions in music or poetry, while the great paintings of times past are inevitably doomed to destruction.

And so, in these last days of my life, I feel myself blessed by the touch of the wings of eternity.

Appendix
Pedigree of Ostwald Family

R.S. Jack and F. Scholz (eds.), *Wilhelm Ostwald*, Springer Biographies,
DOI 10.1007/978-3-319-46955-3

Ostwald, Wilhelm Gottfried (1824 – 1903). Cooper.
Father: GottfriedOstwald (1785 – 1860). Cooper in Moscow.

Ostwald, Elisabeth (neé Leukel) (1832 – 1920)
Mother: Johanna Christine Leukel (neé Braun) (1808 – 1869)
Father: Johann Heinrich Leukel (1804 – 1862). Master baker.

Ostwald, Eugen
Riga 23 October 1851 – Riga 12 February 1932. Professor of Forestry in Riga.
Son: Heinrich Ostwald (Riga 27 July 1877 – Eberswalde 23 March 1950). Professor of Forestry in Eberswalde and Berlin

Ostwald, Gottfried
1855 – 1918. Entrepreneur possessing an iron foundry and machine factory

Ostwald, Friedrich Wilhelm

Riga 2 September 1853 – Leipzig 4 April 1932.

Ostwald, Flora Helene Mathilde, neé von Reyher

Riga 14 January 1854 – Großbothen 2 April 1946. Teacher
Father: Carl Christoph von Reyher (Riga 22 June 1817 – Riga 1 January 1890). Courtyard councillor
Mother: Maria Helena Mathilde von Reyher (neé Ulmann) (25 November 1822 – 6 June 1904). Teacher
Brother: Carl Dietrich Christoph von Reyher (Riga, October 23 1846 – St. Petersburg 11 January 1891). Surgeon
Brother: Carl Gustav Paul von Reyher (Riga 30 May 1848 – ?)
Brother: Carl Max Christopher von Reyher (Riga 15 March 1850 – ?)
Brother: Carl Wilhelm Emil von Reyher (Riga 22 January 1852 – ?)
Brother: Carl Friedrich Otto von Reyher (Riga 2 October 1855 – ?)
Brother: Carl Reinhold Hans von Reyher (20 June 1859 – 25 November 1932). Physician
Brother: Piet von Reyher (Riga 2 May 1862 – Berlin (Weißensee) 5 January 1927). Writer
and other brothers

Ostwald, Grete (Margarete) (Riga 13 February 1882 – Großbothen 1 August 1960). She studied from 1905-1907 at „Großherzoglich-Sächsischen Kunstschule" (Art Academy) in Weimar. Since 1918 she suffered from severe arthritis. Since 1932 she managed the Ostwald property "Haus Energie" in Großbothen, where she founded the Wilhelm Ostwald Archive.

Ostwald, Carl Wilhelm Wolfgang (Riga 27 May 1883 – Dresden 22 November 1943). Studied natural sciences, esp. biology in Leipzig. Student of Jaques Loeb (1904–1906). 1923 Professor of Colloid Chemistry in Leipzig

Ostwald, Elisabeth (Riga 19 June 1884 – Großbothen 10 November 1968). Married to Dr. Eberhard Brauer (1875 – Großbothen 1 May 1958)
Son: Georg Brauer (Bochum 11 April 1908 – 26 February 2001). Chemist. Professor in Freiburg
Son: Peter Brauer (Leipzig 1911 – 9 January 1995). Physicist. Professor in Freiburg
Son: Hellmut Brauer (1913 – 1940). Engineer, PhD. Died in WW II
Daughter: Margarete Brauer (Miltitz 23 December 1918 – 6 September 2008). Kindergarten worker, X-ray assistant, study of chemical engineering, finally archivist of Haus Energie in Großbothen

Ostwald, Walter (Riga 20 May 1886 – Freiburg, Breisgau 12 July 1958). Studied chemistry in Leipzig. Student of William Ramsay (1907). He worked in industry and as a science journalist.

Ostwald, Carl Otto (Leipzig 23 June 1890 – Leipzig 5 August 1958). Engineer.

Descriptive Name Index

Fritz Scholz

Wilhelm Ostwald mentioned numerous individuals in his autobiography, ranging from the famous, like Newton and Beethoven, to the less well known people whom he met in his life. Since we did not wish to make an arbitrary selection, *all* mentioned persons for whom we could retrieve some relevant information are listed in this index (630 persons). This autobiography should be understandable to people who have no scientific background, especially not in chemistry. It should also be understandable to people who do not have a cultural background in Europe or the US and who might not know the respective politicians and artists from these areas. Where ever possible, the places and dates of birth and death are given, because they allow one to relate the person to Ostwald's life. In the case of persons who are well known world wide (like emperors, and famous artists), no further details are given, because that information is readily available from many sources, especially the Internet. In case of less well known people, more details, often collected from rather remote sources, are given. Because of this the entries about famous people are much shorter than the entries of about the less well known. This was deliberately done to make this book a productive source for the historian and interested layman.

When people with more than one christian name are commonly known and cited under a second christian name, that has been underlined.

A

Abel, Sir Frederick Augustus: Woolwich, London 17 July 1827–Westminster, London 6 September 1902. British chemist. Studied at the Royal College of Chemistry (Imperial College London) with August Wilhelm von Hofmann. From 1854 to 1888 Abel was an ordnance chemist at the Chemical Establishment of the Royal Arsenal at Woolwich. After this he was a chemist at the War Department. His special field was the chemistry of explosives.

Abbe, Ernst Karl: Eisenach 23 January 1840–Jena 14 January 1905. German physicist, developer of optical instruments and entrepreneur.

Adler. Friedrich Wolfgang: Vienna 9 July 1879–Zurich 2 January 1960. Austrian social democrat and scientist. Studied sciences in Zurich.

R.S. Jack and F. Scholz (eds.), *Wilhelm Ostwald*, Springer Biographies,
DOI 10.1007/978-3-319-46955-3

Albert von Sachsen: Dresden 23 April 1828–Sibyllenort 19 June 1902. King of Saxony 1873 to 1902.

Alesch, Gustav Johann von: Graz 25 October 1882–Göttingen 11 June 1967. German psychologist. Studied humanities in Graz, Munich and Berlin. 1931 Professor in Greifswald, 1938 in Halle, 1941 in Göttingen.

Alexander II (Alexander II Nikolayevich; Александр II Николаевич): Moscow 29 April 1818–St. Petersburg 13 March 1881 (assassinated). Russian Tsar who introduced important reforms, such as the abolishment of serfdom in 1861.

Althoff, Friedrich Theodor: Dinslaken 19 February 1839–Steglitz 20 October 1908. German (Prussian) politician who was most influential in University politics, both in the establishment of Universities and in the appointment of professors.

Aristotle: Stageira 384 BC–Chalkis 322 BC. Greek philosopher and scientist.

Armstrong, Henry Edward: London 6 May 1848–London 13 July 1937. British chemist. Studied at the Royal College of Chemistry (Imperial College London) with August Wilhelm von Hofmann and with Adolph Wilhelm Hermann Kolbe in Leipzig. 1871 Professor of Chemistry in London.

Arrhenius, Svante: Wik/Uppsala 19 February 1859–Stockholm 2 October 1927. Swedish chemist. Studied mathematics and natural sciences in Uppsala and Stockholm. In his Ph.D. thesis he demonstrated the existence of ions as a result of the dissolution of salts. He worked with Friedrich Wilhelm Ostwald in Riga, with Friedrich Wilhelm Georg Kohlrausch in Würzburg, with Ludwig Boltzmann in Graz, with Jacobus Henricus van't Hoff in Amsterdam and then again with Friedrich Wilhelm Ostwald in Leipzig. 1895 Professor in Stockholm. 1905 Nobel Prize in Chemistry.

Wilbur Olin Atwater: Johnsburg 3 May 1844–Middletown 22 September 1907. US chemist. Studied at Wesleyan University and Yale University. 1873 Professor of Chemistry at Wesleyan.

B

Backlund, Johan (Jons) Oskar: Länghem (Sweden) 28 April 1846–Pulkovo (near Petrograd) 29 August 1916. Swedish-Russian astronomer. Studied mathematics in Uppsala, settled in Russia in 1876, where he first worked at the observatory in Dorpat, and, since 1879, at that of Pulkovo.

Bacon, Francis (1st Viscount St. Albans, 1st Baron Verulam: London 22 January 1561–Highgate 9 April 1626. English philosopher.

Baeyer, Johann Friedrich Wilhelm Adolf von: Berlin 31 October 1835–Starnberg 20 August 1917. German chemist. Student of Robert Wilhelm Bunsen and Friedrich August Kekulé. Professorships: 1866 in Berlin, 1872 in Straßburg, 1875 Munich (following Justus Liebig). 1905 Nobel Prize in chemistry.

Bajer, Fredrik: Vesteregede 21 April 1837–Copenhagen 22 January 1922. Danish politician. Nobel Peace Prize in 1908.

Bamberger, Eugen: Berlin 19 July 1857–Ponte Tresa 10 December 1932. German chemist. Assistant to the mineralogist Karl Friedrich Rammelsberg in Berlin and to Adolf von Baeyer in Munich. 1891 Professor in Munich, 1893 Professor at the Polytechnic in Zurich.

Barth zu Barthenau, Ludwig: Rovereto 17 January 1839–Vienna 3 August 1890. Austrian organic chemist. Worked with Justus Liebig and Max Pettenkofer in Munich. 1867 Professor in Innsbruck, 1876 Professor in Vienna.

Baudouin de Courtenay, Jan Ignacy Niecisław: Radzymin 13 March 1845–Warsaw 3 November 19929. Polish linguist and Slavicist (with French roots). Studied in Warsaw, Prague, Jena and Berlin. Professor in Kazan, Dorpat, Cracow (1893–1900), St. Petersburg (1900–1918) and finally in Warsaw.

Bauer, Stephan: Vienna 20 May 1865–Basel 15 November 1934. Swiss national economist. 1900 to 1925 Secretary General of the International Association for Labour Legislation.

Beaufront, Louis de (real name: Louis Chevreux): 3 October 1855–8 January 1935. French amateur linguist and writer. One of the first French propagandists of the synthetic language Esperanto. Later advocated Ido. His biography is still incompletely researched.

Beck, Hermann: Mülheim an der Ruhr 25 August 1879–around 1920. Studied mechanical engineering and social sciences in Dresden, Berlin, and Heidelberg. In 1905 he established the *Internationales Institut für Sozial-Bibliographie*, and in 1908 the *Institut für Techno-Bibliographie*.

Beck (Ritter) von Mannagetta und Lerchenau, Günther: Pressburg 25 August 1856–Prague 23 June 1931. Austrian botanist.

Beckurts, Heinrich: Braunschweig 23 August 1855–Bargteheide 15 September 1929. German pharmacist and chemist. Appointed Professor in Braunschweig in 1886.

Beckmann, Ernst Otto: Solingen 4 July 1853–Berlin 13 July 1923. German Chemist. An initial apprenticeship as an apothecary was followed by studies of pharmacy and chemistry in Leipzig. Student of Carl Remigius Fresenius (Wiesbaden), Robert Otto (Braunschweig), Hermann Kolbe (Leipzig). Professorships: 1892 Erlangen, 1897 Leipzig. 1912 Director at the Kaiser-Wilhelm Institute of Chemistry, Berlin Dahlem.

Beernaert, Auguste Marie François: Ostende 26 July 1829–Luzern 6 October 1912. Belgian jurist and politician. Nobel Peace Prize in 1909.

Beethoven, Ludwig van: Baptised in Bonn 17 December 1770–Vienna 26 March 1827. German-Austrian composer (with Dutch ancestors).

Béhal, Auguste: Lens (Pas-de-Calais) 29 March 1859–Mennecy 1 February 1941. French chemist and pharmacologist. Studied in Paris with Charles Adolphe Wurtz. 1901 Professor of Toxicology at *École Supérieure de Pharmacie* in Paris.

Behrens, Peter: Hamburg 14 April 1868–Berlin 27 February 1940. German architect and designer.

Behring, Emil Adolf: Hansdorf 15 March 1854–Marburg 31 March 1917. German physiologist. Studied medicine in Berlin. 1894 Professor in Halle, 1895 in Marburg. Nobel Prize for Medicine in 1909.

Bell, Alexander Graham: Edinburgh 3 March 1847–Baddeck (Canada) 2 August 1922. Scientist, inventor, engineer and innovator of the telephone.

Bergman, Torbern Olof: Låstads 20 March 1735–Medevi 8 July 1784. Swedish chemist and mineralogist. He is the author of "A Dissertation on Elective Attraction", London, Murray and Elliot, 1785.

Bergmann, Ernst Gustav Benjamin von: Riga 16 December 1836–Wiesbaden 25 March 1907. German surgeon and advocate of aseptic treatment of wounds. Professorships: 1871 to 1877 in Dorpat, 1878 to 1882 in Würzburg, 1882 in Berlin.

Berthelot, Marcelin (also: Marcellin) Pierre Eugène: Paris 25 October 1827–Paris 18 March 1907. French chemist. Studied natural sciences in Paris. 1859 Professor at *École Supérieure de Pharmacie* and in 1865 Professor at *Collège de France*. He assumed that chemical affinity is defined by the reaction heat.

Berthollet, Claude Louis, comte: Talloires 9 December 1748–Arcueil 6 November 1822. French chemist and physician. Together with Louis Bernard Guyton de Morveau, Antoine Laurent de Lavoisier and Antoine François de Fourcroy he proposed a new nomenclature for chemistry (*Méthode de nomenclature chimique*) in 1787, which had a far-reaching effect on science.

Bertrand, Gabriel Emile: Paris 17 May 1867–Paris 20 June 1962. French pharmacologist and biochemist.

Berzelius, Jöns Jacob: Väversunda 20 August 1779–Stockholm 7 August 1848. Swedish chemist. He studied initially medicine and moved later to chemical studies. 1807 Professor of Chemistry and Pharmacy at the *Karolinska Institut* in Stockholm.

Bessel, Friedrich Wilhelm: Minden 22 July 1784–Königsberg 17 March 1846. German mathematician, astronomer, physicist and geodesist. He was trained as a merchant, became interested in astronomy and was self-educated in mathematics. 1810 Professor of Astronomy in Königsberg.

Bidder, Georg Friedrich Karl Heinrich: Manor Laudohn / Livonia 9 November 1810–Dorpat 27 August 1894. German physiologist and pathologist.

Bigelow, Samuel Lawrence: Boston, Massachusetts, 23 February 1870–West Hartford 3 December 1947. US chemist. Studied at Harvard University and Massachusetts Institute of Technology. Ph.D. with Wilhelm Ostwald in Leipzig in 1898. He joined the University of Michigan as an Instructor of General Chemistry in 1898. 1901 Assistant Professor and Acting Director of the Laboratory of General Chemistry. Full Professor in 1907.

Bischof, Karl Gustav: Wöhrd 18 January 1792–Bonn 30 November 1870. German geologist and chemist. Studied mathematics, astronomy, chemistry and physics in Erlangen, where he became Professor of Technical Chemistry in 1819, and Professor of Chemistry in 1822.

Bischoff, Carl: Würzburg 8 April 1855–Munich 18 October 1908. German chemist. Studied chemistry in Würzburg and Wiesbaden (with Carl Remigius Fresenius). Assistant to Johannes Adolf Wislicenus in Leipzig. In 1887 he followed Friedrich Wilhelm Ostwald as Professor in Riga.

Bismarck, Otto Eduard Leopold von: Schönhausen 1 April 1815–Friedrichsruh 30 July 1898. German politician. From 1871 to 1890 he was the first Chancellor of the German Reich.

Bloh, Friedrich: Wardenburg 23 October 1854–Hamburg 19 June 1941. German school director, publicist and freethinker.

Blomeyer, Adolf (Adolph): Frankenhausen 24 February 1830–Leipzig 18 December 1889. German agronomist. Studied law in Heidelberg and Marburg. After some years of apprenticeship and administration in agricultural enterprises, he was appointed Professor at the Agricultural Academy in Proskau in 1865, and in 1869 at the University of Leipzig.

Böcklin, Arnold: Basel 16 October 1827–San Domenico 16 January 1901. Swiss painter.

Bodenstein, Max Ernst August: Magdeburg 15 July 1871–Berlin 3 September 1942. German chemist. Studied in Wiesbaden with Carl Remigius Fresenius and in Heidelberg with Victor Meyer. 1904 Professor in Leipzig, 1906 in Berlin, 1908 in Hannover 1923 again in Berlin.

Bodländer, Guido: Breslau 31 July 1855–Braunschweig 25 December 1904. German chemist. Studied in Breslau. 1899 Professor in Braunschweig.

Boirac, Émile: Guelma (Algeria) 26 August 1851–Dijon 20 September 1917. French philosopher and parapsychologist.

Bollack, Léon: 4 May 1859–1925. French trader and inventor of the synthetic language Bolak.

Boltzmann, Ludwig Eduard: Vienna 20 February 1844–Duino 5 September 1906. Austrian physicist. Studied mathematics and physics in Vienna. Professorships: 1869–1873 in Graz, 1873–1876 in Vienna, 1876–1890 again in Graz, 1890–1894 In Munich, 1894–1900 again in Vienna, 1900–1902 in Leipzig, 1902–1906 again in Vienna.

Boor, Julie de: Hamburg 21 July 1848–Hamburg 4 June 1932. German portraitist.

Bosse, Julius Robert: Quedlinburg 12 July 1832–Berlin 31 July 1901. German politician. Studied law in Heidelberg, Halle and Berlin. 1892 to 1899 Minister of Education of Prussia.

Böttinger, Henry Theodore: Burton-upon-Trent 10 July 1848–Charlottenburg (Berlin) 9 June 1920. German Industrialist. Studied sciences in Freiburg and Würzburg. In 1882 he joined the Bayer AG in Elberfeld.

Bozi, Alfred: Bielefeld 26 December 1857–Bielefeld 5 May 1938. Studied law in Freiburg and Berlin. He was a judge at several courts

Brahms, Johannes: Hamburg 7 May 1833–Vienna 3 April 1897. German composer.

Brauer, Otto Eberhard Herrmann: 1875–Großbothen 1 May 1958. Private assistant of Friedrich Wilhelm Ostwald and husband of Ostwald's daughter Elisabeth.

Braun, Karl Ferdinand: Fulda 6 June 1850–New York 20 April 1918. German physicist. Student of Heinrich Gustav Magnus. Professorships: 1877 in Marburg, 1880 in Straßburg, 1883 in Karlsruhe, 1887 in Tübingen, 1895 in Straßburg. 1909 Nobel Prize in Physics for wireless telegraphy (together with Guglielmo Marconi).

Bredig, Georg: Glogau 1 October 1868–24 April 1944. German chemist. Studied chemistry in Freiburg and Berlin. Ph.D. in Wilhelm Ostwald's Institute. 1895 private assistant of Wilhelm Ostwald in Leipzig. 1901 Professor in Heidelberg, 1910 in Zurich, 1911 in Karlsruhe. In 1933 he was forced to retire because of his Jewish ancestry. He emigrated via the Netherlands to the USA. Bredig prepared aqueous colloidal dispersions of metals by using an electric arc in water. He studied the catalytic properties of the metal colloids and compared their action with that of ferments (enzymes). His work was fundamental to catalysis.

Brinckmann, Justus: Hamburg 23 May 1843–Hamburg 8 February 1915. German Fine Arts scientist. Studied sciences, law and economics in Leipzig and Vienna. 1877 to 1915 Director of the Museum of Arts and Crafts in Hamburg.

Brown, Alexander Crum: Edinburgh 26 March 1838–Edinburgh 28 October 1922. Scottish chemist. Studied in London, in Heidelberg with Robert Wilhelm Bunsen and in Marburg with Adolph Wilhelm Hermann Kolbe. 1869 to 1908 Professor of Chemistry in Edinburgh.

Brunck, Heinrich von: Winterborn 26 March 1847–Ludwigshafen 4 December 1911. German chemist. Studied in Zurich and Gent. He was a leading technical director of BASF, where he was much engaged in the welfare of workers.

Brunialti, Attilio: Vicenza 2 April 1849–Rome 2 December 1920. Italian jurist and politician. Studied law at Padua. 1879 to 1881 Professor of Law in Pavia and then in Torino.

Brüning, Gustav von: Höchst 5 August 1864–Höchst 8 February 1913. German chemist. Studied in Würzburg and Munich. 1910 Director of *Farbwerke Höchst*.

Bruno, Giordano: Nola January 1548–Rome 17 February 1600. Italian priest and astronomer

Bruns, Ernst Heinrich: Berlin 4 September 1848–Leipzig 23 September 1919. German mathematician and astronomer. Studied mathematics, astronomy and physics in Berlin. 1872/73 calculator at the Pulkovo observatory, 1873 to 1876 at the observatory in Dorpat, 1876 Professor in Berlin, and 1882 Professor in Leipzig.

Brühl, Julius Wilhelm: Warsaw 13 February 1850–Heidelberg 5 February 1911. Studied chemistry in Zurich and Berlin. 1888 Professor in Heidelberg. His research concerned the relationships between the structure of organic compounds and refraction and dispersion.

Bryce, James (1st Viscount Bryce): Belfast 10 May 1838–Sidmouth 22 January 1922. British jurist, historian and politician. Studied at Belfast Academy, Glasgow High School, the University of Glasgow, the University of Heidelberg and Trinity College. 1870–1893 Regius Professor of Civil Law at Oxford University. Member of British Parliament, and UK Ambassador to the USA from 1907 to 1913.

Buber, Martin Mordechai: Vienna 8 February 1878–Jerusalem 8 February 1878. Austrian-born Israeli Jewish philosopher.

Bücher, Karl (Carl) Wilhelm: Kirberg 16 February 1847–Leipzig 12 November 1930. German economist. Studied in Bonn and Göttingen. He held professorships in Dorpat, Basel, and Karlsruhe and from 1892 in Leipzig. He founded journalism as an academic discipline.

Budde, Emil Arnold: Geldern 28 July 1842–Feldafing 15 August 1921. German physicist. Studied physics and mathematics in Bonn. From 1872 to 1887 he worked as correspondent for a German newspaper in Paris, Rom and Constantinople. From 1890 to 1893 he was editor of *Fortschritte der Physik*, a review journal of advances in physics. He was director of the Charlottenburg factory of Siemens & Halske, and President of the International Electrotechnical Commission from 1911 to 1913.

Buff, Johann Heinrich: Rödelsheim 23 May 1805–Gießen 24 December 1878. German physicist and chemist. Together with Friedrich Zamminer and Hermann Kopp he published the first book about what is now called Physical Chemistry (*Lehrbuch der physikalischen und theoretischen Chemie*) in 1857.

Bührer, Karl Wilhelm: Bibern 1 June 1861–Berlin 1917. Swiss writer and entrepreneur. Advocate of standardization of printing formats. In 1905 he founded the *Internationale Mono-Gesellschaft*.

Bunge, Gustav Piers Alexander von: Dorpat 19 January 1844–Basel 5 November 1920. German physiologist and biochemist. Studies in Dorpat, Dr. med. (MD) in Leipzig (1882), 1886 to 1920 Professor of Physiological Chemistry in Basel.

Bunsen, Robert Wilhelm: Göttingen 30 (or 31) March 1811–Heidelberg 16 August 1899. German chemist. Studied sciences and mathematics in Göttingen. Positions: 1833 lecturer in Göttingen, 1836 he followed Friedrich Wöhler at the Polytechnic in Kassel, 1837 Professor in Marburg, 1851 Professor in Breslau, 1852 Professor in Heidelberg, where he followed Leopold Gmelin. Together with Gustav Robert Kirchhoff he established spectroscopic analysis.

Bunte, Hans Hugo Christian: Wunsiedel 25 December 1848–Karlsruhe 17 August 1925. German technical chemist. He worked on gas technologies for illumination.

Busch, Heinrich Christian Wilhelm: Wiedensahl 15 April 1832–Mechtshausen 9 January 1908. German humorous poet, illustrator and painter.

C

Campbell, William Wallace: Hancock County 11 April 1862–San Francisco 14 June 1938. US astronomer. Studied engineering and astronomy at the University of Michigan. 1923 to 1930 President of the University of California at Berkeley.

Candolle, Alphonse Louis Pierre Pyrame de: Paris 28 October 1806–Geneva 4 April 1893. Swiss botanist. Studied in Geneva. 1831 Professor in Geneva. Son of Augustin Pyrame de Candolle, and father of Casimir Pyramus de Candolle both of whom were also botanists.

Candolle, Augustin Pyramus de (Augustin Pyrame de Candolle): Geneva 4 February 1778–Geneva 9 September 1841. Swiss botanist. Father of Alphonse Louis Pierre Pyrame de Candolle.

Candolle, Anne Casimir Pyrame de: Geneva 20 February 1836–Chêne-Bougeries 3 October 1918. Swiss botanist. Son of Alphonse Louis Pierre Pyrame de Candolle. Studied chemistry, physics and mathematics in Paris.

Cannizzaro, Stanislao: Palermo 13 July 1826–Rome 10 May 1910. Italian chemist. Studied medicine in Palermo. 1851 Professor of Physical Chemistry in Alessandria, 1855 in Genoa, 1861 in Palermo, 1871 in Rome.

Carnelley Thomas: Manchester 22 October 1852–Aberdeen 27 August 1890. British chemist. He published tables of melting and boiling points.

Carnot, Nicolas Léonard Sadi: Paris 1 June 1796–Paris 24 August 1832. French physicist and engineer. Studied at the *École Polytechnique* in Paris. His publication "*Réflexions sur la puissance motrice du feu et sur les machines propres à développer cette puissance* » is part of the foundation of thermodynamics.

Carstanjen, Ernst: Duisburg 2 July 1826–Leipzig 13 July 1884. German chemist. Studied montane sciences and chemistry in Bonn, Essen, Freiberg, and Berlin. 1873 Professor in Leipzig.

Carus, Paul: (Ilsenburg 18 July 1852–La Salle (Ill.) 11 February 1919. German-American author and philosopher.

Cattell, James McKeen: Easton 25 May 1860–Lancaster 20 January 1944. US psychologist and publisher of scientific literature. Studied at Lafayette College and in Göttingen and Leipzig (with Wilhelm Wundt), and on his return to the US at Johns Hopkins University. 1886 doctorate with Wilhelm Wundt in Leipzig. 1888 Professor of Psychology at the University of Pennsylvania (first Professor of Psychology in the US). 1891 Professor of Philosophy at Columbia University.

Chandler, Charles Frederick: Lancaster 6 December 1836–East Hartford 25 August 1925. US chemist. Studied at Lawrence Scientific School, Harvard University, and in Göttingen with Friedrich Wöhler and Heinrich Rose. 1857 Associate Professor, 1859 full Professor at Union College in Schenectady. 1864 Professor at the Columbia School of Mines in New York.

Chatelier, Henry Louis Le: Paris 8 October 1850–Miribel-les-Èchelles 17 June 1936. French chemist. Studied sciences and mathematics in Paris. 1897 Professor of Industrial Chemistry at *École nationale supérieure des mines de Paris*, and he was also Professor of Chemistry at the *École Polytechnique* from 1884 until 1897. Later he was Professor of Chemistry at *Collège de France* and at *Faculté des sciences de Paris*. Discoverer of the thermodynamic principle now known as Le Chatelier's principle.

Chevreul, Michel Eugène: Angers 31 August 1786–Paris 9 April 1889. French chemist. Studied at *École Centrale de Angers*, and in Paris with Antoine François de Fourcroy und Louis-Nicolas Vauquelin. Founder of fat chemistry and contributor to colour theory.

Christy, Samuel Benedict: San Francisco 8 August 1853–Berkeley 30 November 1914. US chemist. Studied at the University of California. 1885 Professor of Mining and Metallurgy at the University of California at Berkeley.

Clarke, Frank Wigglesworth: Boston 19 March 1847–23 May 1931. US geologist and chemist. Studied at Harvard College's Lawrence Scientific School. 1873 Professor at Howard University and 1874 at University of Cincinnati. Chief Chemist of the U.S. Geological Survey from 1883 to 1925.

Classen, Alexander: Aachen 13 April 1843–Aachen 28 January 1934. German analytical chemist. 1878: Professor at *Technische Hochschule Aachen*. Made many contributions to the development of electrogravimetry.

Claude, Georges: Paris 24 September 1870–Saint-Claude 23 May 1960. French physicist. Studied in Paris at *École supérieure de physique et de chimie industrielles de la ville de Paris*. Inventor of a device to liquify air.

Clausius, Rudolf Julius Emanuel: Köslin 2 January 1822–Bonn 24 August 1888. German physicist. Studied mathematics and physics in Berlin. Student of Heinrich Gustav Magnus. Professorships in Berlin, Zurich (1855), Würzburg (1867) and finally Bonn (1869). Discoverer of the Second Law of Thermodynamics.

Cleve, Per Teodor: Stockholm 10 February 1840–Uppsala 18 June 1905. Swedish natural scientist. Discoverer of the elements holmium and thulium. 1860 Professor in Uppsala.

Cohn, Emil (Georg): Neustrelitz 28 September 1854–Ringgenberg 28 January 1944. German physicist. Studied chemistry in Leipzig, Heidelberg and Straßburg, and was assistant to August Kundt. 1884 Professor in Straßburg. Driven out of Straßburg, which became French after WW I he was appointed Professor in Rostock and in 1920 Professor in Freiburg.

Comte, Isidore Marie Auguste François Xavier: Montpellier 19 January 1798–Paris 5 September 1857. French mathematician and philosopher. Studied at *École polytechnique* in Paris and at the University of Montpellier. Founder of sociology as a discipline and of the doctrine of positivism.

Coudres, Theodor des: Veckerhagen 13 March 1862–Leipzig 8 October 1926. German physicist. Studied in Geneva, Leipzig, Munich and Berlin. 1895 Professor in Göttingen, 1901 in Würzburg, 1903 in Leipzig.

Couturat, Louis: Ris-Orangis 17 January 1868–Melun 3 August 1914. French logician, mathematician, philosopher and linguist. Studied philosophy and mathematics in Paris.

Credner, Carl Hermann: Gotha 1 October 1841–Leipzig 21 July 1913. German scientist. Studied in Clausthal, Breslau and Göttingen. 1870 Professor of Historic Geology and Palaeontology in Leipzig, 1872 Director of the Geological Survey of Saxony.

Czapski, Siegfried: Domain Obra (near Koschmin) 28 May 1861–Weimar 29 June 1907. German physicist. Studied in Göttingen, Breslau and Berlin. Following his Ph.D. he worked as assistant of Ernst Abbe in the Zeiss Company in Jena, where he later held high administrative positions.

D

D'Annunzio, Gabriele: Pescara 12 March 1863–Gardone 1 March 1938. Italian writer.

Daller, Balthasar: Niklasreuth 22 January 1835–Freising 3 March 1911. Catholic cleric and politician. Studied philosophy and theology in Freising and Munich.

Dalton, John: Eaglesfield 6 September 1766–Manchester 27 July 1844. English scientist. Autodidact who started with meteorological studies. In chemistry he is famed as the founder of the atomic theory.

Damaschke, Adolf Wilhelm Ferdinand: Berlin 24 November 1865–Berlin 30 July 1935. German pedagogue and economist.

Darboux, Jean Gaston: Nîmes 14 August 1842–Paris 23 February 1917. French mathematician. Studied at *École polytechnique* and *École normale supérieure* in Paris. 1873 Professor at Sorbonne.

Darwin, Charles Robert: Shrewsbury 12 February 1809–Downe 19 April 1882. British natural scientist.

Davy, Humphry: Penzance 17 December 1778–Geneva 29 May 1829. English chemist. Apprentice to a surgeon and apothecary, he studied chemistry as an autodidact. Discoverer of several elements and one of the founders of modern chemistry.

Day, Arthur Louis: Brookfield 30 October 1869–Wahington DC 2 March 1960. US geophysicist. Studied at Yale University, where he taught until 1897. Ph.D. University of Groningen 1914. 1894 and 1895 he worked with German physicist Friedrich Kohlrausch studying the conductive properties of electrolytes. From 1897 to 1900 he worked at *Physikalisch-Technische Bundesanstalt* in Berlin. From 1900 to 1907 at U.S. Geological Survey, and 1907 to 1936 director of the Geophysical Laboratory of the Carnegie Institution for Science.

Defregger, Franz: Ederhof near Stronach 30 April 1835–Munich 2 January 1921. Austrian and Bavarian painter.

Delbrück, Berthold Gustav Gottlieb: Putbus 26 July 1842–Jena 3 January 1922. German philologist. Studied in Halle and Berlin. 1870 Professor in Jena.

Despretz, César-Mansuète: Lessines 4 May 1791–Paris 15 March 1863. Belgian (later French citizen) chemist and physicist. Studied chemistry and physics in Paris. Worked with Louis Joseph Gay-Lussac. 1824 Professor of Physics in Paris. His research concerned heat, latent heat, heat conductivity, animal heat, etc., and also the voltaic cell and electric arc.

Deville, Henri Étienne Sainte-Claire: Saint Thomas/Caribbean 11 March 1818–Boulogne-sur-Seine 1 July 1881. French chemist. Student of Louis Jacques Thénard. 1844 doctor of medicine. 1845 to 1851 Professor in Besançon, 1852 Professor in Paris. Best known for the method developed for the technical production of aluminium.

Dewar, James: Kincardine 20 September 1842–London 27 March 1923. Scottish physical chemist. Studied at the Universities of Edinburgh and Gent. 1875 Professor at the University of Cambridge. He invented the double-wall vacuum flask, which bear his name.

Dewey, Melvil Louis Kossuth: Adams Center 10 December 1851–Lake Placid 26 December 1931. US librarian and educator. Inventor of the Dewey Decimal system of library classification.

Diesel, Rudolf Christian Karl: Paris 18 March 1858–disappeared from a steamer on the way from Antwerp to London 29 September 1913. German engineer and inventor of an engine now bearing his name.

Dittrich, Rudolf Bernhard August: Bärenwalde 2 January 1855–Berlin 15 February 1929. German politician. 1908 to 1917 mayor of Leipzig.

Dixon, Harold Baily: Marylebone, London 11 August 1852–Lytham 18 September 1930. British chemist. Studied in Oxford, 1887 Professor at Owen's College, Manchester. He founded a school of chemists studying gas explosions.

Doderer, Wilhelm Carl Gustav Ritter von: Klosterbruck 16 August 1854–Vienna 1 November 1932. Austrian architect, engineer, and entrepreneur.

Dom Pedro: see **Pedro II of Brazil**

Donnan, Frederick George: Colombo (Ceylon) 5 September 1870–Canterbury 16 December 1956. British chemist. Studied at Queen's College, Belfast, Ph.D. with Wilhelm Ostwald in Leipzig, then worked with Jacobus Henricus van't Hoff in Berlin, and with William Ramsay in London. 1904 Professor of Physical Chemistry in Liverpool, 1913 University College, London.

Dosenheimer, Emil: (Ungstein 11 February 1870–Heidelberg 16 February 1936. German jurist and director of the provincial court at Landau. Strong engagement for national education and peace. Because of his Jewish descent he lost his position in 1933.

Du Bois, Henri Éduard Johan Godfried: Velp 24 June 1863–Utrecht 21 October 1918. Dutch physicist.

du Bois Reymont, Emil Heinrich: Berlin 7 November 1818–Berlin 26 December 1896. German physiologist and physician (from a Huguenot family). Studied theology, philosophy, mathematics, geology and finally medicine in Berlin. 1855 Professor of Physiology in Berlin. Founder of experimental electrophysiology. He had a materialistic world view and was a Darwinist.

Dühring, Eugen Karl: Berlin 12 January 1833–Nowawes (Potsdam-Babelsberg) 21 September 1921. Philosopher and economist. From 1864 to 1877 he taught at the University of Berlin, but then lost his position because he affronted the University. He then lived as a *Privatgelehrter* (private scholar). Infamous for his antisemitism.

Duisberg, Carl: Barmen 29 September 1861–Leverkusen 19 March 1935. German chemist and industrialist. Studied in Göttingen and Jena. He started his career in *Farbenfabriken vorm. Friedr. Bayer & Co AG* in Wuppertal-Elberfeld. He was involved in the development and use of chemical weapons in WW I.

Duttenhofer, Max Wilhelm Heinrich: Rottweil 20 May 1843–Tübingen 14 August 1903. German entrepreneur and industrialist. Inventor of smokeless gunpowder.

Dyck, Walther Franz Anton von: Munich 6 December 1856–Munich 5 November 1934. German Mathematician. Studied in Munich, Berlin, and Leipzig. 1884 Professor in Munich.

E

Ebner von Eschenbach, Marie Freifrau: Castle Zdislawitz 13 September 1830 – Vienna 12 March 1916. Austrian author.

Edison, Thomas Alva: Milan (Ohio) 11 February 1847–East Orange (New Jersey) 18 October 1931. American inventor and entrepreneur. Without much formal education, he became famous for various inventions, e.g., the phonograph and the incandescent lamp.

Edlund, Erik Edsberg 14 March 1819–Stockholm 19 August 1888. Swedish physicist and meteorologist. 1846 Professor in Uppsala, later member of the Royal Academy. Edlund initiated meteorological stations in Sweden. He was the formal advisor of the Ph.D. thesis of Svante Arrhenius.

Edward VII (Albert Edward): London 9 November 1841–London 6 May 1910. King of the UK from 1901 to 1910.

Eliot, Charles William: Boston 20 March 1834–Notheast Harbor 22 August 1926. US chemist. Graduated from Boston Latin School in 1849 and from Harvard University in 1853. 1869 to 1909 President of Harvard University.

Emerson, Ralph Waldo: Boston 25 May 1803–Concord 27 April 1882. US author and philosopher.

Engelmann, Friedrich Wilhelm Rudolf: Leipzig 1 June 1841–Leipzig 28 March 1888. German astronomer and publisher. Son of Wilhelm Engelmann.

Engelmann, Wilhelm: Lemgo 1 August 1808–Leipzig 23 December 1878. German publisher. Father of Friedrich Wilhelm Rudolf Engelmann

Engler, Carl Oswald Viktor: Weisweil 5 January 1842–Karlsruhe 7 February 1925. German chemist. 1872 Professor in Halle. His main research topic was the chemistry of natural oil. Inventor of the so-called Engler viscometer.

Enneking, John Joseph: Minster (Ohio) 4 October–Hyde Park (Mass.) 16 November 1916. American painter.

Erdmann, Benno: Guhrau 30 May 1851–Berlin 7 January 1921. German philosopher, logician and psychologist. Studied in Berlin and Heidelberg. 1878 Professor in Kiel, 1884 in Breslau, 1890 in Halle, 1898 in Bonn, and in 1909 in Berlin.

Erdmann, Otto Linné: Dresden 11 April 1804–Leipzig 9 October 1869. German chemist. Studied medicine and chemistry. 1827 Professor of Technical Chemistry in Leipzig. Research on nickel and cobalt (analysis and separation), organic dyestuffs, determination of atomic masses. He developed a method for desulfurization of Saxonian coke. He founded the „*Journal für Praktische Chemie*" (in 1928 named „*Journal für technische und ökonomische Chemie*"), which is now published as "Advanced Synthesis & Catalysis".

Erlenmeyer, Richard August Carl Emil: Wiesbaden 28 June 1825–Aschaffenburg 22 January 1909. German pharmacist and chemist. Student of Justus Liebig in Gießen and Leopold Gmelin in Heidelberg, where he later worked with Robert Wilhelm Bunsen. 1863 Professor in Heidelberg and in 1877 in Munich.

Escherich, Theodor: Ansbach 29 November 1857–Vienna 15 February 1911. German-Austrian paediatrician. Studied medicine in Würzburg, Kiel and Berlin. 1890 Professor in Graz. Discoverer of the bacterium known as *Escherichia coli.*

Eschke, Hermann Wilhelm Benjamin: Berlin 6 May 1823–Berlin 15 January 1900. German painter.

Ettingshausen, Albert von: Vienna 30 March 1850–Granz 9 June 1932. Austrian physicist. Assistant to Ludwig Boltzmann in Graz. 1886 Professor in Graz. With Walther Hermann Nernst he discovered the Ettinghausen-Nernst Effect.

Euclid of Alexandria: Greek mathematician, who presumably lived in the 3rd century BC.

Exner, Franz Serafin: Vienna 24 March 1849–Vienna 15 November 1926. Austrian physicist. Student of August Kundt and Conrad Röntgen (Zurich). 1879 Professor of Physics in Vienna. He worked on spectroscopy, electrochemistry, radioactivity, and electricity of the atmosphere.

Exner, Wilhelm Franz: Gänserndorf 9 April 1840–Vienna 25 May 1931. Austrian technician and forestry scientist.

Eykman (Eijkman), Johan (Johann) Fredrik: Nijkerk 19 January 1851–Groningen 7 January 1915. Dutch chemist. First apprenticed to an apothecary, later studied pharmacy in Amsterdam. 1876 Director of the Hygiene and Pharmacy Laboratory in Nagasaki, 1878 a similar position in Tokyo. he was in Java from 1885 to 1886 and later returned to Amsterdam.

F

Falkenstein, Johann Paul Freiherr von: Pegau 15 June 1801–Dresden 14 January 1882. High state official in Saxony contributed to the development of the University of Leipzig. 1853 Minister of Education of Saxony.

Faraday, Michael: Newington 22 September 1791–Hampton Court Green 25 August 1867. English scientist. He started his scientific life as an apprentice to a bookbinder and bookseller, where he read science books. Later he attended lectures of Humphry Davy who employed him as an apprentice. Faraday made fundamental discoveries in science, especially chemistry and electricity.

Fechner, Gustav Theodor: Groß Särchen/Muskau 19 April 1801–Leipzig 18 November 1887. German physicist, psychologist and philosopher. Studied medicine in Leipzig. 1828 to 1839 Professor of Physics in Leipzig. After an interval caused by weak health, he was appointed Professor of *Naturphilosophie* (natural philosophy) and Anthropology in Leipzig. He is well known for his research in galvanism (electrochemistry), physiological optics and is regarded as the founder of psychophysics.

Fischer, Hermann Emil: Euskirchen 9 October 1852–Berlin 15 July 1919. German chemist. Studied chemistry in Bonn and Straßburg. Student of Johann Friedrich Wilhelm Adolf von Baeyer. Professorships: 1879 Munich, 1882 Erlangen, 1885 Würzburg, 1892 Berlin (following August Wilhelm von Hofmann). Nobel Prize for Chemistry in 1906. Cousin of Otto Philipp Fischer.

Fischer, Otto Philipp: Euskirchen 28 November 1852–Erlangen 4 April 1932. German chemist. Cousin of Hermann Emil Fischer. The Fischer-Hepp rearrangement in organic chemistry is named in his honour.

Fittig, Wilhelm Rudolph: Hamburg 6 December 1835–Straßburg 19 November 1910. German chemist. Student of Friedrich Wöhler. Professorships: 1870 in Tübingen, 1876 in Straßburg.

Fischer, Ernst Kuno Berthold: Sandewalde 23 July 1824–Heidelberg 5 July 1907. German philosopher. Studied in Leipzig and Halle. 1850 *Dozent* in Heidelberg. In 1853 he lost the right to teach after being accused of pantheism. 1856 Professor in Jena.

FitzGerald, George Francis: Dublin 3 August 1851–Dublin 22 February 1901. Irish physicist. Studied at Trinity College Dublin. In 1877 he became a Fellow of this college. He worked on electromagnetic fields and waves.

Flechsig, Paul Emil: Zwickau 29 June 1847–Leipzig 23 July 1929. German neuroanatomist, neuropathologist and psychiatrist. 1877 Professor in Leipzig. Mainly known for his research on myelinogenesis.

Fleischl Edler von Marxow, Ernst: Vienna 5 August 1846–Vienna 22 October 1891. Austrian physiologist. Studied mathematics, physics, chemistry and medicine in Vienna and Leipzig, and worked with Ernst Wilhelm von Brücke and Carl von Rokitansky in Vienna and with Carl Ludwig in Leipzig. 1880 Professor in Vienna. Research on neurophysiology and bioelectricity.

Foerster, Wilhelm Julius: Grünberg 16 December 1832–Bornim 18 January 1921. German astronomer. 1863 Professor in Berlin. 1865 Director of the Observatory in Berlin.

Francke, Ernst: Coburg 10 November 1852–Freiburg im Breisgau 23 December 1921. German journalist and politician. Studied philosophy, sciences, and economics.

Francke, Kuno: Kiel 27 September 1855–Cambridge (Mass.) 25 June 1930. German-born US educator, historian and poet (in German language). Studied in Berlin, Köln, Jena and Munich. 1884 instructor in German at Harvard. Professor of History and German culture in 1896 (at Harvard). In 1902 he proposed to Charles W. Eliot that an exchange program for professors be established with Germany.

Frankland, Sir Edward: Churchtown (near Lancaster) 18 January 1825–Golaa, Gudbrandsdalen 9 August 1899. British chemist. First apprenticed to an apothecary, then studied geology in London. Later he studied with Robert Wilhelm Bunsen in

Marburg. 1851 Professor at Owen's College, Manchester, in 1863 he succeeded Michael Faraday at the Royal Institution of Great Britain in London, and in 1865 he succeeded August Wilhelm von Hoffman at the Royal College of Chemistry.

Fraunhofer, Joseph: Straubing 6 March 1787–Munich 7 June 1826. German optician and physicist. Discoverer of the dark lines in the solar spectrum (Fraunhofer lines) and inventor of the first modern spectroscope.

Frick, Joseph: Staufen 16 June 1806–Karlsruhe 11 October 1875. Studied medicine, physics, and mathematics, and held a doctorate in medicine. He worked as physician and teacher.

Friedrich Wilhelm IV: Berlin 15 October 1795–Potsdam 2 January 1861. King of Prussia from 1840 to 1861.

Fröbel, Friedrich Wilhelm August: Oberweißbach 21 April 1782–Marienthal 21 June 1852. German progressive pedagogue. Founder of the first kindergarten.

Fromm, Johann (Hans): 1812–1904. Teacher in Riga. Because of the strong pressure of Russification in Riga, he moved at the end of his life with his two daughters to Merano (South Tyrol).

G

Galenus, Aelius (or **Claudius Galenus**): Pergamum September 129 AD–c. 200 or 216 AD. Greek physician and philosopher.

Gärtner, August Anton Hieronymus: Ochtrup 18 April 1848–Jena 21 December 1934. German physician and microbiologist. Studied medicine in Berlin. 1886 Professor in Jena. He focussed on hygiene, especially water quality.

Gauß, Johann Carl Friedrich: Braunschweig 30 April 1777–Göttingen 23 February 1855. German mathematician, astronomer and physicist. Studied at Collegium Carolinum in Braunschweig and in Göttingen. 1807 Professor in Göttingen.

Gautier, Armand: Narbonne 23 September 1837–Cannes 27 July 1920. French chemist. Studied with Charles Adolphe Wurtz in Paris. 1876, 1891 and 1906 President of *Société chimique de Paris*.

Gay-Lussac, Joseph Louis: Saint-Léonard-de-Noblat 6 December 1778–Paris 9 May 1850. French chemist and physicist. Studied in Paris. 1809 Professor of Chemistry at the Sorbonne, Paris.

Gerber, Karl Friedrich Wilhelm: Ebeleben 11 April 1823–Dresden 23 December 1891. German jurist and politician in Saxony. Professorships: 1847 in Erlangen, 1851 in Tübingen, 1862 Professor in Jena, 1863 Professor in Leipzig. He was Royal Saxon State Minister and Minister of Cultural Affairs.

Gerdes, Heinrich Bernhard: Accum 16 September 1856–Berlin 3 September 1932. Studied at the technical college Mittweida. As a specialist in lighting engineering he advanced to the position of Technical Director of the company *Julius Pintsch* AG, Berlin. Dr. honoris causa in 1928 (Berlin).

Gerhardt, Charles Frédéric: Straßburg 21 August 1816–Straßburg 19 August 1856. Chemist from Alsace. Studied chemistry at the Polytechnic in Karlsruhe, in Leipzig, Gießen, Dresden and Paris. In 1848 he founded the *École de chimie pratique* in Paris. 1855 Professor of Chemistry in Straßburg.

Gibbs, Josiah Willard: New Haven 11 February 1839–New Haven 28 April 1903. US-physicist. Studied mathematics and natural sciences at the University of New Haven. 1871 Professor at Yale University.

Glöckel, Otto: Pottendorf 8 February 1874–Vienna 23 July 1935. Austrian politician and school reformer.

Gmelin, Leopold: Göttingen 2 August 1788–Heidelberg 13 April 1853. German chemist. Studied in Tübingen and Vienna. In 1817 he published the first volume of his *Handbook of Chemistry*, which remained in print until 1997 in 760 volumes + 35 registry volumes. This inorganic chemistry data collection is now incorporated in the electronic data bank Reaxys.

Gobat, Charles Albert: Tramelan 21 May 1843–Bern 16 March 1914. Swiss politician. Nobel Peace Prize in 1910.

Goodale, George Lincoln: Saco 3 August 1839–Cambridge (Mass.) 12 April 1923. US botanist. Graduated from Amherst College. 1868 Professor at Bowdoin College, 1873 Professor at Harvard.1879 to 1909 Director of the Botanic Garden of Harvard University.

Goodwin, Harry Manley: Boston 18 April 1870–Squam Lake 26 June 1949. US physicist. Studied at Massachusetts Institute of Technology and Harvard. Ph.D. with Wilhelm Ostwald in Leipzig in 1894. 1906 Professorship at the Massachusetts Institute of Technology.

Goodwin, William Watson: Concord 9 May 1831 –Cambridge (Mass.) 15 June 1912. US philologist. Arts Bachelor from Harvard, then studies in Göttingen, Bonn and Berlin. 1860 Professor at Harvard.

Goethe, Johann Wolfgang von: Frankfurt am Main 28 Aug. 1749–Weimar 22 March 1832. German writer and poet.

Gottschall, Rudolf Karl von: Breslau 30 September 1823–Leipzig 21 March 1909. German writer.

Graham, Thomas: Glasgow 21 December 1805–London 16 September 1869. British chemist. 1830 Professor of Chemistry at Andersons College, Glasgow, 1837 University College, London. First President of the Chemical Society (then the Chemical Society of London).

Graebe (Gräbe), Carl James Peter: Frankfurt am Main 24 February 1841–Frankfurt am Main 19 January 1927. German chemist. Studied with Robert Wilhelm Bunsen in Heidelberg. 1870 Professor in Königsberg, 1878 in Geneva.

Grönberg, Theodor: Mižirica (Gouv. Kiev) 9 August 1845–Freiburg/Breisgau 25 July 1910. German physicist. Studied physics in Dorpat and Heidelberg. 1875 Professor in Riga.

Groth, Paul Heinrich Ritter von: Magdeburg 23 June 1843–Munich 2 December 1927. German mineralogist and crystallographer. Professorships: 1872 Straßburg, 1883 Munich.

Goldscheid, Rudolf: (pseudonym *Rudolf Golm*) Vienna 12 August 1870–Vienna 6 October 1931. Austrian sociologist. 1912 to 1917 President of the Austrian *Monistenbund* (League of Monists).

Goldschmiedt, Guido: Triest 29 May 1850–Gainfarn 6 August 1915. Austrian chemist. Studied chemstry in Frankfurt am Main, Vienna and Heidelberg. Ph.D. student of Robert Wilhelm Bunsen. Worked with Johann Friedrich Wilhelm Adolf von Baeyer in Straßburg. 1890 Professor at *Hochschule für Bodenkultur* (agricultural university) Vienna, 1892 Professor in Prague, 1911 Professor in Vienna.

Goldschmidt, Hans (Johannes) Wilhelm: Berlin 18 January 1861–Baden-Baden 21 May 1923. German chemist. Studied chemistry in Heidelberg with Robert Wilhelm Bunsen. Later he worked in the family factory „*Chemische Fabrik Th. Goldschmidt*“. Inventor of the thermite process for producing (carbon-free) metals by the exothermic reaction of the oxides with metallic aluminium (aluminothermic reaction). The process is applied for welding (exothermid welding, or thermite welding).

Gudden, Johann Bernhard Aloys von: Kleve 7 June 1824–shore of Lake Starnberg 13 June 1886. German neuroanatomist and psychiatrist. Studied in Halle. 1869 Professor of Psychiatry in Zurich, later in Munich. Gudden was personal physician to King Ludwig II of Bavaria. Both were found dead at the shore of Lake Starnberg on June 13, 1886.

Guericke, Otto von: Magdeburg 20 November 1602–Hamburg 21 May 1686. German scientist, inventor, and politician. Known for his experiments with a vacuum sphere

Guldberg; Cato Maximilian: Christiana 11 August 1836–Kristiana 14 January 1902. Norwegian mathematician and chemist. Together with his brother-in-law Peter Waage he has formulated the law of mass action.

Guye, Philippe Auguste: Saint-Christophe 12 June 1862–Geneva 27 March 1922. Studied in Geneva and Paris. 1895 Professor in Geneva.

H

Haber, Fritz: Breslau 9 December 1868–Basel 29 January 1934. German chemist. Studied in Heidelberg (with Robert Wilhelm Bunsen) and in Berlin (with August Wilhelm von Hofmann). 1906 Professor in Karlsruhe. 1911 foundation director of the *Kaiser-Wilhelm-Institut für physikalische Chemie und Elektroche*mie in Berlin-Dahlem. Nobel Prize in Chemistry for 1918.

Haeckel, Ernst Heinrich Philipp August: Potsdam 16 February 1834–Jena 9 August 1919. German zoologist and philosopher. Studied medicine in Berlin and Würzburg. 1865 Professor in Jena. Strong supporter of the ideas of Charles Darwin, and father of the theory of recapitulation (biogenetic law).

Haffner, Johann Samuel Eduard von: Riga 15 August 1804–Riga 13 January 1889. Doctor of Philosophy. Haffner worked as a teacher at different places; became in 1850 rector of the Dorpat University from which post he was released at his own request in 1857. It is said that he had problems as rector because he was not a professor. From 1860 to 1877 he was rector of the *Realgymnasium*. He was also chairman of the censor committee, and was highly decorated by the Russian authorities.

Hale, George Ellery: Chicago 29 June 1868–Pasadena 21 February 1938. US astronomer. Studied at Massachusetts Institute of Technology. 1892 Professor at the University of Chicago.

Hall, Edwin Herbert: Great Falls 7 November 1855–Cambridge (Mass.) 20 November 1938. US physicist. Studied at Bowdoin College. 1895 Professor at Harvard University. Discoverer of the Hall Effect in 1879.

Haller, Albin: Felleringen 7 March 1849–Paris 29 April 1925. French chemist. First apprenticed to an apothecary, then studied pharmacy in Nancy. 1885 Professor in Nancy, 1889 in Paris (Sorbonne). 1905 Director of *École supérieure de physique et de chimie industrielles.*

Hankel, Wilhelm Gottlieb: Ermsleben 17 May 1814–Leipzig 17 February 1899. German physicist. 1847 Professor in Halle. Worked on thermoelectricity and photoelectricity.

Hanriot, Armand Maurice: Conflans-Sainte-Honorine 29 March 1854–Lisses 31 August 1933. French chemist. 1880 Professor at *École Municipale de Physique et de Chimie de Paris.*

Hantzsch, Arthur Rudolf: Dresden 7 March 1857–Dresden 14 March 1935. German chemist. Studied chemistry in Dresden and Würzburg. Professorships: 1885–1893 Zurich, 1893–1903 Würzburg, 1903–1927 Leipzig.

Harnack, Karl Gustav Adolf von: Dorpat 25 April 1851–Heidelberg 10 June 1930. German theologian (Lutheran).

Hartknoch, Johann Friedrich: Goldap 28 September 1740 (?)–Riga 1 April 1789. Played an important role as publisher of books and sheet music in the Baltic countries and Germany.

Hayden, Franz Joseph: Rohrau 31 March (or 1 April) 1732–Vienna 31 May 1809. Austrian Composer.

Hearst, Phoebe Elizabeth Appserson: (Franklin County 3 December 1842 Pleasanton 14 April 1919. US philantropist. Mother of William Randolph Hearst.

Hearst, William Randolph: San Francisco 29 April 1863–Beverly Hills 14 August 1951. Son of Phoebe Elizabeth Appserson Hearst. US newspaper publisher.

Heimrod, George Willram: 1876–1917. US chemist. Studied from 1901 to 1903 with Wilhelm Ostwald in Leipzig. Worked with Theodore William Richards.

Heine, Christian Johann Heinrich: Düsseldorf 13 December 1797–Paris 17 February 1856. German poet.

Helm, Georg Ferdinand: Dresden 15 March 1851–Dresden 13 September 1923. German mathematician and physicist. Studied mathematics and natural sciences in Dresden, Leipzig and Berlin. 1888 Professor of Mathematics and Theoretical Physics in Dresden. Helms strongly advocated the concept of Energetics and was a supporter of Friedrich Wilhelm Ostwald.

Helmholtz, Hermann Ludwig Ferdinand von: Potsdam 31 August 1821–Charlottenburg 8 September 1894. German physicist and physiologist. Studied medicine in Berlin. Professorships: 1848 (Physiology) Berlin, 1870 (Physics) Berlin.

Helmling, Peter: Erbach near Heppenheim 9 September 1817–Reval 11 April 1901. German mathematician. Studied mathematics in Heidelberg. 1859 Professor in Dorpat.

Hempel, Walther Mathias: Pulsnitz 5 May 1851–Dresden 1 December 1916. German chemist. Studied chemistry in Berlin (with August Wilhelm von Hofmann, Johann Friedrich Wilhelm Adolf von Baeyer), and in Heidelberg (with Robert Wilhelm Bunsen). Best known for his contributions to gas analysis.

Hentschel, Willibald: Łódź 7 November 1858–Leoni 2 February 1947. German scientist. Studied physics, chemistry and biology (in Dresden and Jena). 1888/89 he worked with Johannes Adolf Wislicenus in Leipzig. Later he was much engaged in eugenics and in propagating the supposed superiority of the Aryan race.

Herder, Johann Gottfried: Mohrungen 25 August 1744–Weimar 18 December 1803. German theologian, philosopher, and poet.

Hering, Karl Ewald Konstantin: Altgersdorf 5 August 1834–Leipzig 26 January 1918. German physiologist. Studied in Leipzig. 1870 Professor in Prague.

Herter, Christian Archibald: Glenville 3 September 1865–5 December 1922. US physician and pathologist. Studied at Columbia and Johns Hopkins University and in Zurich. 1897 Professor at University and Bellevue Hospital Medical College, 1903 at Columbia University.

Hertwig, Oskar Wilhelm Ausgust: Friedberg 21 April 1849–Berlin 25 October 1922. German zoologist. Studied medicine in Jena and Bonn. 1881 Professor in Jena, 1888 in Berlin.

Hertz, Heinrich Rudolf: Hamburg 22 February 1857–Bonn 1 January 1894. German physicist. Studied mathematics and physics in Dresden, Munich and Berlin. Student of Hermann Ludwig Ferdinand von Helmholtz. 1885 Professor in Karlsruhe, 1889 Professor in Bonn. Hertz confirmed Maxwell's theory of electromagnetism by producing for the first time radio waves at very high frequencies.

Herzig, Josef: Sanok 25 September 1853–Vienna 4 July 1924. Austrian chemist. Student of August Wilhelm von Hofmann (Berlin) and Robert Wilhelm Bunsen (Heidelberg). 1897 Professor in Vienna.

Heymans, Gerardus: Ferwert 17 April 1857–Groningen 18 February 1930. Dutch philosopher and psychologist. Studied in Leiden and Freiburg. 1890 Professor in Groningen.

Hippocrates of Kos: Kos c. 460–Larissa c. 370 BC. Greek physician.

Hirth, Georg: Gräfentonna 13 July 1841–Tegernsee 28 March 1916. German writer journalist and publisher.

Hirschfeld, Magnus: Kolberg 14 May 1868–Nizza 14 May 1935. German physician and sexologist.

Hittorf, Johann Wilhelm: Bonn 27 March 1824–Münster 28 November 1914. Studied natural sciences and mathematics in Bonn. 1847 Professor in Münster. He is best known for his work on the mobility of ions (Hittorf transference numbers).

Hobbema, Meindert: Amsterdam 31 October 1638–Amsterdam 7 December 1709. Dutch painter.

Hoff, Jacobus Henricus van't: Rotterdam 30 August 1852–Steglitz 1 March 1911. Dutch physical and organic chemist. Student of Friedrich August Kekulé in Bonn, and Charles Adolphe Wurtz (Paris). 1878 Professor in Amsterdam, 1896 in Berlin. His most fundamental contributions to chemistry where the discovery of chirality of carbon in organic compounds, development of thermodynamics of solutions, and reaction kinetics. 1n 1887 he founded together with Friedrich Wilhelm Ostwald the *Zeitschrift für physikalische Chemie (Z Phys Chem (Leipzig))*. 1901 Nobel Prize for Chemistry.

Hoffman, Friedrich Albin: Ruhrort 13 November 1843–Leipzig 13 November 1924. German anatomist and internist. Studied medicine in Berlin, Tübingen and Würzburg. 1874 Professor in Dorpat, 1886 Professor in Leipzig.

Hoffmann, Ernst Theodor Amadeus (real names Ernst Theodor Wilhelm Hoffmann): Königsberg 24 January 1776–Berlin 25 June 1822. German romantic author.

Hofmann, August Wilhelm von: Gießen 8 April 1818–Berlin 5 May 1892. German chemist. Studied chemistry in Gießen. Student of Justus Liebig. 1845 director of the newly established Royal College of Chemistry in London (and parallel professorship in Bonn). 1865 Professor in Berlin.

Höft, Gustav: 1864–1935. German pedagogue.

Holls, Frederick William: 1857–1903. US jurist and publicist. Author of a part of the The Hague Conference on Private International Law.

Holtz, Julius Friedrich: Prenzlau 2 September 1836–Berlin 11 June 1911. German pharmacist, industrialist and administrator of the German Chemical Society.

Holz, Arno: Rastenburg 26 April 1863–Berlin 26 October 1929. German poet and dramatist.

Horneffer, Ernst: Stettin 7 September 1871–Iserlohn 5 September 1954. German philologist and philosopher.

Horstmann, August Friedrich: Mannheim 20 November 1842–Heidelberg 8 October 1929. German chemist who studied in Heidelberg (with Emil Erlenmeyer) and in in Zurich (with Johannes Wislicenus and Rudolf Clausius). He applied the Second Law of Thermodynamics to chemical problems and is regarded as one of the founders of chemical thermodynamics.

Hozumi, Nobushige: Uwajima-Nakanomachi 23 August 1855–Shinjuku,Tokyo (then Ushigome, Tokyo City) 7 April 1926. Japanese jurist and politician. Studied languages in Tokyo and law in London and Berlin. 1881 Professor at the Imperial University of Tokyo. He drafted the Japanese Civil Code.

Hübschmann, Hermann Max Johannes: Löbtau 25 February 1867–Chemnitz 22 November 1930. German jurist and politician.

Humboldt, Friedrich Wilhelm Christian Carl Ferdinand von: Potsdam 22 June 1767–Tegel 8 April 1835. German diplomat, state philosopher and reformer of the education system in Germany. He introduced numerous progressive and very successful reforms e.g., the introduction of the *neuhumanistisches Gymnasium* (new-humanistic high school). In these schools, Greek and Latin played a central role, and the *Abitur* (school leaving examination) of these schools was the entrance ticket for the Universities. For a long time, pupils of other schools could not enter a university, or only after taking additional exams. This situation was strongly criticized by Wilhelm Ostwald. Friedrich Wilhelm von Humboldt was the brother of Friedrich Wilhelm Heinrich *Alexander* von Humboldt (Berlin 14 September 1769–Berlin 6 May 1859), the geographer and natural scientist.

Huntington, Edward Vermilye: Clinton 26 April 1874–Cambridge 25 November 1952. US mathematician and physicist. Studied at Harvard University, Ph.D. in Straßburg, 1919 Professor at Harvard.

Hüfner Carl Gustav von: Köstritz 13 May 1840 –Tübingen 14 March 1908. German physiological chemist. 1872 Professor in Tübingen. Research on haemoglobin. In 1894 Hüfner determined that a gram of hemoglobin could maximally bind 0.0598 milimoles of oxygen gas.

J

Jäckh, Ernst Friedrich Wilhelm: Urach 22 February 1875–New York 17 August 1959. German journalist.

Jacobson, Paul Heinrich: Königsberg 5 October 1859–Berlin 25 January 1923. German chemist. Studied in Göttingen with Victor Meyer. 1891 Professor of Pharmaceutical Chemistry in Heidelberg. 1896 Secretary General of the German Chemical Society.

James, William: New York 11 January 1842–Chocorua 26 August 1910. US psychologist and philosopher. Studied various subjects in the USA, including painting, chemistry, medicine, and in Berlin psychology and philosophy. After returning to USA he finished his medical studies with a MD in 1869. From 1872 to 1907 he taught at Harvard University.

Jespersen, Jens Otto Harry: Randers 16 July 1860–Roskilde 30 April 1943. Danish linguist. Studied in Copenhagen and Oxford. 1893 Professor of English in Copenhagen.

Jodl, Friedrich: Munich 23 August 1849–Vienna 26 January 1914. German philosopher and psychologist. Studied in Munich. 1885 Professor at the German University in Prague, 1896 in Vienna.

Johann von Sachsen: Dresden 12 December 1801–Pillnitz 29 October 1873. 1854 to 1873 King of Saxony.

Jones, Harry Clary: New London 11 November 1865–Baltimore 9 April 1916. US chemist. Graduated from Johns Hopkins University. Ph.D. in 1892. Postdoc with Wilhelm Ostwald in Leipzig. He also worked with Svante Arrhenius and Jacobus Henricus van't Hoff. 1903 Professor of Physical Chemistry at Johns Hopkins University.

Jones, Owen: London 15 February 1809–London 19 April 1874. British architect and designer.

Jordan, David Starr: Gainesville, N.Y. 19 January 1851–Stanford 19 September 1931. US zoologist and peace activist. Studied at Cornell and Butler University. 1891 first President of Stanford University.

Juliusburger, Otto: Breslau 26 September 1867–New York 7 June 1952. German-US psychiatrist. Emmigrated in 1941 to USA.

K

Kammerer, Paul: Vienna 17 August 1880–Puchberg 23 September 1926. Austrian biologist. Studied zoology in Vienna, and at the same time Counterpoint at the Vienna Academy.

Kämmerer, Hermann: Mutterstadt 7 April 1840–Nürnberg 10 April 1898. German chemist. Studied in Nürnberg, Heidelberg (with Robert Wilhelm Bunsen) and Leipzig. In Munich he was assistant to Justus Liebig. Later director of the Industrial School in Nürnberg.

Kant, Immanuel: Königsberg 22 April 1724–Königsberg 12 February 1804. German philosopher.

Kappeler, Johann Karl: Frauenfeld 23 March 1816–Zurich 20 October 1888. Swiss politician contributed to the development of the Zurich Polytechnic to one of the leading universities world wide.

Keith, William: Oldmeldrum 18 November 1838–Berkeley 13 April 1911. US painter.

Kekulé, Friedrich August (later Friedrich August Kekulé von Stradonitz): Darmstadt 7 September 1829–Bonn 13 July 1896. German chemist. Studied chemistry in Gießen. Student of Justus Liebig and Jean Baptiste Dumas (Paris).

Kelvin: see: **Thomson, William**

Kenrick, Frank Boteler: England 1874–1944. English-born Canadian chemist. Studied at Upper Canada College Toronto and the University of Toronto. 1894 to 1896 he studied with Wilhelm Ostwald in Leipzig. Around 1930 he became Professor of Chemistry at the University of Toronto.

Kerschensteiner, Georg Michael Anton: Munich 29 July 1854–Munich 15 January 1932. German reform pedagogue.

Kienzl, Wilhelm: Waizenkirchen 17 January 1857–Vienna 3 October 1941. Austrian composer.

Kieseritzky, Johann Georg Gustav: Wenden 28 February 1830–Sassenhof/Riga 31 August 1896. German mathematician. Studied mathematics in Dorpat. 1864 Professor in Riga.

Kirchhoff, Gustav Robert: Königsberg 12 March 1824–Berlin 17 October 1887. German physicist. Studied mathematics and physics in Königsberg. Professorships: 1854 Heidelberg, 1875 Berlin. Research on electricity, spectroscopy and radiation. Together with Robert Wilhelm Bunsen he is the father of spectroscopic analysis.

Kirchhoff, Gottlieb Sigismund Constantin: Teterow 15 February 1764–St. Petersburg 4 February 1833. German pharmacist and chemist at the Imperial Court of Russia. He discovered the acid hydrolysis of starch to glucose.

Kirchner, Wilhelm: Göttingen 9 July 1848–Leipzig 22 August 1921. German agronomist. Studied in Göttingen and Halle. 1890 Professor in Leipzig. 1899 to 1900 he was rector of the University of Leipzig.

Klinger, Max: Leipzig 18 February 1857–4 July 1920. German painter, sculptor and writer.

Knietsch, Rudolf Theophil Josef: Oppeln 13 December 1854–Ludwigshafen 28 May 1906. German chemist. Studied chemistry in Berlin. In 1884 he joined BASF. He developed the technology of contact syntheses of sulfuric acid, and production of synthetic indigo.

Knop, Wilhelm: Altenau 28 June 1817–Leipzig 28 January 1891. German agrochemist. Student of Friedrich Wühler (Göttingen), Leopold Gmelin (Heidelberg), Otto Linné Erdmann (Leipzig). 1861 Professor in Leipzig. Research on plant physiology and fertilizers. One of the founders of hydroculture.

Knorr, Ludwig: Munich 2 December 1859–Jena 4 June 1921. German chemist. Student of Robert Wilhelm Bunsen (Heidelberg) and Emil Fischer (Erlangen). 1889 Professor in Jena. The Paal-Knorr synthesis of heterocycles bears his name.

Knorre, Georg Karl von: Mykolaiv (Nikolaev), Ukraine 18 March 1859–Berlin 29 December 1910. German chemist. Studied chemistry at *Technische Hochschule Berlin* (Now Technical University), where he later became Professor and head of the electrochemical laboratory.

Kocher, Emil Theodor: Bern 25 August 1841–Bern 27 July 1917. Swiss physician. Studied in Bern and Zurich. Nobel Prize in Medicine of 1909.

Kohler, Josef: Offenburg 9 March 1849–Charlottenburg 3 August 1919. German jurist.

Kohlrausch, Friedrich Wilhelm Georg: Rinteln 14 October 1840–Marburg 17 January 1910. German physicist. 1867 Professor in Göttingen, 1870 at the Polytechnic in Zurich, 1871 in Darmstadt, 1875 in Würzburg, 1888 Straßburg. Known for ground-breaking work on electrolyte conductivity (Kohlrausch square root law).

Kolbe, Adolph Wilhelm Hermann: Elliehausen/Göttingen 27 September 1818–Leipzig 25 November 1884. German chemist. Studied chemistry in Göttingen, Student of Friedrich Wöhler, Assistant to Robert Wilhelm Bunsen (Marburg) and Lyon Playfair (London). Professorships: 1851 Marburg, 1865 Leipzig. Kolbe contributed to establishing organic electrochemistry (Kolbe synthesis), but is also remembered as a furious opponent of modern structural chemistry, as fostered by Kekulé and van't Hoff.

Kopp, Hermann Franz Moritz: Hanau 30 October 1817–Heidelberg 20 February 1892. German chemist and historian of chemistry (author of the 4-volume treatise *Geschichte der Chemie* (History of Chemistry), Vieweg, Braunschweig 1843–1847).

Kortum, Carl Arnold: Mühlheim an der Ruhr 5 July 1745–Bochum15 August 1824. German physician and writer.

Krause, Max: Breslau 23 May 1853–Berlin 11 July 1918. German engineer. Studied at *Königliche Gewerbeakademie Berlin* (later this became the Technical University). Later Director of the *A. Borsig Berg- und Hüttenverwaltung.*

Kraut, Karl Johann: Lüneburg 29 September 1829–Hannover 13 January 1912. German chemist. 1868: Professor in Hannover. He became known as an editor of *Gmelins Handbuch der anorganischen Chemie* (Gmelin Handbook of Inorganic Chemistry), which is now the Gmelin Database.

Kronecker, Leopold: Liegnitz 7 December 1823–Berlin 29 December 1891. German mathematician. Studied philosophy, mathematics and sciences in Berlin, Bonn and Breslau. 1883 Professor in Berlin.

Kundt, August: Schwerin 18 November 1839–Israelsdorf 21 May 1894. German physicist. Studied mathematics and physics in Leipzig and Berlin. Student of Heinrich Gustav Magnus. Professorships: 1868 Zurich, 1870 Würzburg, 1872 Straßburg, 1888 Berlin.

Külpe, Oswald: Kandau (Courland) 3 August 1862–Munich 30 December 1915. German phsychologist and philosopher. Ph.D. with Wilhelm Wundt (Leipzig). He became Professor in 1894 in Würzburg, in 1909 in Bonn, and in 1912 in Munich.

L

Ladenburg, Albert: Mannheim 2 July 1842–Breslau 15 August 1911. German chemist.

Lagerlöf, Selma Ottilia Lovisa: Mårbacka (Sunne) 20 November 1858–Mårbacka (Sunne) 16 March 1940. Swedish author. Nobel Prize for Literature in 1909.

Lagorio, Alexander Yevgenyevich (von), (Александр Евгеньевич Лагорио): Feodossiya/Crimean 27 August 1852–Berlin 1 August 1922. petrographer, crystallographer and mineralogist, was later Professor at the Imperial University of Warsaw and director of the Warsaw Polytechnic Institute. His uncle Feliks Lagorio, was a General in Napoleon's army and later vice-consul of the Kingdom of the Two Sicilies in Russia. His son Lev Felixovich Lagorio (Лев Феликсович Лагорио) (1828–1905) was a famous Russian painter. The Lagorio's descend from a Genoese noble family.

Lambert, Johann Heinrich: Mülhausen 26 August 1728–Berlin 25 September 1777. Swiss polymath.

Lamprecht, Karl Gotthard: Jessen 25 February 1856–Leipzig 10 May 1915. German historian. Studied in Göttingen, Leipzig and Munich. 1890 Professor in Marburg, 1891, Professor in Leipzig.

Landolt, Hans Heinrich: Zurich 5 December 1831–Deutsch-Wilmersdorf / Berlin 15 March 1910. Swiss chemist. Studied chemistry in Zurich and Breslau. Student of Robert Wilhelm Bunsen. Professorships: 1858 Bonn, 1870 Aachen,

1881 Berlin. Research on physical chemistry. Known as the founder of the *Landolt-Börnstein*, the data basis of physical chemistry.

Landmann, Robert August von: Großweingarten 12 January 1845–Munich 11 March 1926. Jurist and politician in Bavaria. 1895 to 1902 he was Minister of Internal, Church and Education Affairs of Bavaria.

Lang, Viktor von: Vienna Neustadt 2 March 1838–Vienna 3 July 1921. Austrian physicist. Worked with Gustav Robert Kirchhoff and Robert Wilhelm Bunsen in Heidelberg, and with Henri Victor Regnault in Paris.

Langhans, Paul: Hamburg 1 April 1867–Gotha 17 January 1952. German cartographer and geographer.

Laplace, Pierre-Simon (Marquis de): Beaumont-en-Auge 28 March 1749–Paris 5 March 1827. French mathematician physicist and astronomer.

Larmor, Sir Joseph: Magheragall, Northern Ireland 11 July 1857–Holywood, County Down, Northern Ireland 19 May 1942. Irish physicist and mathematician. Studied in Belfast (Queens College) and Cambridge (St. Johns College). 1903 to 1932 Lucasian Professor of Mathematics at Cambridge. He was the first to publish the Lorentz transformation and calculated the energy radiation from an accelerated electron. He believed in an ether as a homogeneous, incompressible and elastic fluid.

Laubenheimer, August: Gießen 9 August 1848–Gießen 22 July 1904. German chemist and entrepreneur. Studied in Gießen. Director of *Farbwerke Höchst*.

Leau, Léopold: 6 April 1868–28 December 1943. French mathematician and linguist. Professor in Nancy.

Lellmann, Eugen: New York 15 October 1856–Gießen 10 December 1893. German chemist. Chemistry studies in Göttingen. 1890 Professor in Tübingen, 1892 in Gießen.

Lehmann, Otto: Konstanz 13 January 1855–Karlsruhe 17 June 1922. German physicist. Studied in Straßburg. 1888 Professor in Dresden, same year move to Karlsruhe.

Lehne, Adolf: Winkel 6 May 1856–Munich 8 February 1930. German textile chemist. Studied chemistry in Gießen, Heidelberg und Freiburg. From1919 to 1925 he was Professor and head of the Department of Textile Chemistry at *Technische Hochschule Karlsruhe*. Founder of *Färber-Zeitung* (Journal of Dying).

Leibniz, Gottfried Wilhelm: Leipzig 1 July 1646–Hannover 14 November 1716. German polymath and philosopher.

Lemberg, Johann Theodor: Reval 25 August 1842–Dorpat 7 November 1902. Chemist, mineralogist and geologist.

Lenbach, Franz Seraph: Schrobenhausen 13 December 1836–Munich 6 May 1904. German painter.

Lessing, Gotthold Ephraim: Kamenz 22 January 1729–Braunschweig 15 February 1781. German writer, philosopher and dramatist.

Leuckart, Karl Georg Friedrich Rudolf: Helmstedt 7 October 1822–Leipzig 6 February 1898. Father of Leuckart, Carl Louis Rudolf Alexander. German zoologist. Studied medicine and natural sciences in Göttingen. 1850 Professor in Gießen, 1869 Professor in Leipzig.

Leuckart, Carl Louis Rudolf Alexander: Gießen 23 June 1854–Leipzig 24 July 1889. Son of Leuckart, Karl Georg Friedrich Rudolf. German chemist. Studied chemistry, physics and mineralogy in Heidelberg with Robert Wilhelm Bunsen and Gustav Robert Kirchhoff, and with Hermann Kolbe in Leipzig. Following work with Adolf von Baeyer in Munich, he was appointed Professor in Göttingen.

Leukel; Johanna Christine, neé Braun: 1809–1869. Maternal grandmother of Friedrich Wilhelm Ostwald.

Leukel, Johann Heinrich: 1804–1862. Maternal grandfather of Friedrich Wilhelm Ostwald.

Lewald, Theodor: Berlin 18 August 1860–Berlin 15 April 1947. German top civilian administrator. Studied law in Heidelberg, Leipzig, and Berlin.

Lichtenberg, Georg Christoph: Ober-Ramstadt 1 July 1742–Göttingen 24 February 1799. German mathematician and physicist.

Lick, James: Stumpstown 25 August 1796–San Francisco 1 October 1876. US carpenter, piano builder, land owner and sponsor of sciences.

Lie, Marius Sophus: Nordfjordeid 17 December 1842–Kristiania 18 February 1899. Norwegian mathematician. Studied in Christiania (later Kristiania, now Oslo) mathematics. 1872 Professor in Christiania, 1886 Professor in Leipzig. 1894 Professor in Christiania.

Liebig, Justus: Darmstadt 12 May 1803–Munich 18 April 1873. German chemist. Initially apprenticed to an apothecary. Later studied chemistry in Bonn and Erlangen. Student of Karl Wilhelm Gottlob Kastner, Joseph Louis Gay-Lussac, Louis Jacques Thénard and Louis-Nicolas Vauquelin (Paris). Professorships: 1824 Gießen, 1852 Munich.

Liesegang, Raphael Eduard: Elberfeld 1 November 1869–Bad Homburg 13 November 1947. German chemist and writer. He did not have a university degree. After a course in analytical chemistry with Carl Remigius Fresenius (Wiesbaden) and studies in Freiburg, he worked in the family photographic factory. He discovered periodic precipitation rings in gels (Liesegang rings), advanced the photographic development process and contributed to colloid science.

Lieventhal, Karl August: Groß-Buschhof (Courland) 23 November 1844–Riga 18 May 1900. German economist.

Linde, Carl Paul Gottfried: Berndorf 11 June 1842–Munich 16 November 1934. German engineer. Studied in Zurich with Rudolf Julius Emanuel Clausius. Inventor of the first industrial-scale air separation and gas liquefaction processes.

Lindet, Gaston Aimé Léon: Paris 10 April 1857–Gaillon 15 June 1929. French chemist.

Lindstedt, Anders: Församling 27 June 1854–Stockholm 16 May 1939. Swedish mathematician and astronomer. From 1879 he became a *Dozent* at the University of Dorpat, in charge of the astronomical observatory. In 1887 he became Professor of Mathematics. In 1886 he returned to Stockholm where he became Professor of Mathematics and Technical Mechanics at the Royal Institute of Technology (*Kungliga Tekniska högskolan*). Later he turned completely to Actuarial Science and in 1909 he joint an insurance company.

Lippmann, Jonas Ferdinand Gabriel: Bonneweg 16 August 1845–(on a ship while crossing the Atlantic) 13 July 1921. French physicist. Appointed Professor of Mathematical Physics in 1878 at the Faculty of Science in Paris. Later Professor of Experimental Physics. He worked with Kühne and Kirchhoff in Heidelberg and with Hermann Ludwig Ferdinand von Helmholtz in Berlin. In Heidelberg he studied electrocapillary phenomena and invented the first version of what later became known as the Lippmann capillary electrometer. His research on electrocapillaries led to the Lippmann equation, and were fundamental for the entire development of electrochemistry, especially the structure and properties of electrode interfaces. Nobel Prize for Physics in 1908 for the development of interference-based colour photography.

Lipsius, Justus Hermann: Leipzig 9 May 1834–Leipzig 5 September 1920. German philologist. Studied philology and theology in Leipzig. 1869 Professor of Classic Philology in Leipzig.

Lobe, Johann Christian: Weimar 30 May 1797–Leipzig 27 July 1881. German composer and music theoretician.

Lodge, Sir Oliver Joseph: Penkhull 12 June 1851–22 August 1940. British physicist. Studied science in London. 1881 Professor in Liverpool, 1900 Professor in Birmingham. His research concerned among other things electromagnetism, electrolysis, psychical phenomena and spiritualism.

Loeb, Jacques: **Mayen 7 April 1859**–Hamilton (Bermuda Islands) 11 February 1924. German-born American biologist and physiologist. Studied philosophy and medicine in Berlin, Munich, and Straßburg. 1892 Professor of Physiology at the University of Chicago, 1902 Professor at the University of California at Berkeley, and from 1910 Professor at the Rockefeller Institute for Medical Research.

Loeschcke, Georg: Penig 28 June 1852–Baden-Baden 26 November 1915. German archaeologist. Studied philology, history and archaeology in Leipzig and Bonn. 1879 Professor in Dorpat, 1889 Professor in Bonn, 1912 Professor in Berlin.

Lommel, Eugen von: Edenkoben 19 March 1837–Munich 19 June 1899. German physicist. Studied mathematics and physics in Munich. 1867 Professor in Hohenheim, 1868 in Erlangen, and 1886 in Munich.

Loschmidt, Johann Joseph: Putschirn 15 March 1821–Vienna 8 July 1895. Austrian physicist and chemist. He studied physics and chemistry. He first worked in a steel factory and then founded a chemical factory which went bankrupt. he was appointed Professor in Vienna only in 1868. He is remembered to be the first to calculate the number of particles (molecules or atoms) in a given volume of an ideal gas. This number now bears the name "Loschmidt constant".

Ludwig, Carl Friedrich Wilhelm: Witzenhausen 29 December 1816–Leipzig 23 April 1895. German physician and physiologist. Studied medicine in Marburg. Professorships: 1846 in Marburg, 1849 in Zurich, 1855 Vienna, 1865 Leipzig. Very wide research interests including blood pressure, secretory nerves, urinary excretion and anaesthesia.

Luitpold Karl Joseph Wilhelm von Bayern: Würzburg 12 March 1821–Munich 12 December 1912. He was the *de facto* ruler of Bavaria from 1886 to 1912.

Luthardt, Christoph Ernst: Maroldweisach 22 March 1823–Leipzig 21 September 1902. German Lutheran theologian. 1854 Professor in Marburg, 1856 Professor in Leipzig.

M

Macallum, Archibald Byron: Belmont (Ontario) 7 April 1858–London (Ontario) 5 April 1934. Canadian biochemist. Studied at the University of Toronto and Johns Hopkins University. 1890 Professor in Toronto, 1920 Professor at McGill University, Montreal.

Mach, Ernst Waldfried Josef Wenzel: Chirlitz 18 February 1838–Vaterstetten 19 February 1916). Austrian physicist and philosopher. Studied natural sciences and mathematics in Vienna. Professorships: 1864 Professor of Mathematics in Graz, 1866 Professor of Physics, 1867 Professor in Prague, 1895 he had the chair of "history and philosophy of the inductive sciences" in Vienna.

Magnus, Heinrich Gustav: Berlin 2 May 1802–Berlin 4 April 1870. German physicist. Studied chemistry physics and technology in Berlin. Student of Eilhard Mitscherlich (Berlin), Jöns Jacob Berzelius (Stockholm), Joseph Louis Gay-Lussac, and Louis Jacques Thénard (Paris). 1834 appointed Professor in Berlin.

Mallinckrodt, Edward: St. Louis 21 January 1845–St. Louis 1 February 1928. US chemical industrialist. Together with his brother Otto he studied chemistry in Wiesbaden, Germany with Carl Remigius Fresenius. After this he had a traineeship in *Eugen de Haen Chemische Fabrik List GmbH*. In 1867, the three Mallinckrodt brothers (Edward, Otto and Gustav) formed the company G. Mallinckrodt & Co.

Manet, Édouard: Paris 23 January 1832–Paris 30 April 1883. French painter.

Maquenne, Léon-Gervais-Marie: Paris 2 December 1853–Paris 10 January 1925. French chemist and plant physiologist.

Marconi, Guglielmo: Bologna 25 April 1874–Rom 20 July 1937. Italian electrical engineer and entrepreneur. Inventor of long-distance radio transmission. Nobel Prize in Physics in 1909.

Maurenbrecher, Karl Peter Wilhelm: Bonn 21 December 1838–Leipzig 6 November 1892. German historian. Studied in Bonn, Munich and Berlin, where he was a student of Leopold von Ranke. 1867 Professor in Dorpat, 1869 Professor in Königsberg, 1877 in Bonn, 1884 in Leipzig.

Maxwell, James Clerk: Edinburgh 13 June 1831–Cambridge 5 November 1879. Scottish physicist. Started his studies in 1847 in Edinburgh, continued in 1850 in Cambridge. 1856 Professor in Aberdeen, 1860 Professor at King's College, London.

Mayer, Christian Gustav Adolf (Adolph): Leipzig 15 February 1839–Gries 11 April 1908. German mathematician. Studied mathematics together with chemistry mineralogy in Heidelberg, Göttingen, Königsberg and Leipzig. 1890 to 1900 Professor in Leipzig.

Mayer, Julius Robert von: Heilbronn 25 November 1814–Heilbronn 20 March 1878. German physician and physicist. One of the fathers of the first Law of Thermodynamics stating the conservation of energy as early as 1841.

Mayer, Tobias: Marbach 17 February 1723–Göttingen 20 February 1762. German astronomer, geographer, physicist and mathematician.

Meldola, Raphael: Islington 19 July 1849–London 16 November 1915. British chemist. 1885 Professor in London.

Mendeleev, Dmitri Ivanovich: (Николай Александрович Меншуткин). Tobolsk 8 February 1834–St. Petersburg 2 February 1907. Russian Chemist. He discovered the periodic system of elements (parallel to Lothar Meyer).

Mendelssohn Bartholdy, Jakob Ludwig Felix: Hamburg 3 February 1809–Leipzig 4 November 1847. German composer.

Menzel, Adolph Friedrich Erdmann von: Breslau 8 December 1815–Berlin 9 February 1905. German painter.

Menshutkin, Nikolai Aleksandrovich: (Николай Александрович Меншуткин). St. Petersburg 24 October 1842–St. Petersburg 5 February 1907. Studied chemistry in St. Petersburg. Professor of Analytical Chemistry 1869, and Professor of Organic Chemistry in 1886.

Metschnikoff (Metchnikov), Ilja (Elie) Iljitsch (Илья Ильич Мечников): Ivanovka 15 March 1845–Paris 15 July 1916. Russian zoologist and immunologist. Nobel Prize for Medicine in 1908.

Meyer, Ernst Sigismund Christian von: Kassel 25 August 1847–Dresden 11 April 1916. German chemist and chemistry historian. He studied in Heidelberg with Robert Wilhelm Bunsen and attended lectures of Hermann Kopp. From 1878 to 1885 he was Professor in Leipzig and after 1895 Professor in Dresden.

Meyer, Julius Lothar (since 1892: von): Varel 19 August 1830–Tübingen 11 April 1895. German physician and chemist. 1868 Professor at the Polytechnic Karlsruhe, 1877 Professor in Tübingen. He is one of the discoverers of the Periodic System of Elements (the other is Dmitri Mendeleev).

Meyer, Victor: Pankow (now part of Berlin) 8 September 1848–Heidelberg 8 August 1897. German chemist. 1871 Professor in Stuttgart, 1872 Professor in Zurich, 1885 Professor in Göttingen, 1889 Professor in Heidelberg.

Michelangelo di Lodovico Buonarroti Simoni: Caprese 6 March 1475–Rome 18 February 1564. Italian sculptor, painter, architect and poet.

Michelson, Albert Abraham: Strelno (then Germany) 19 December 1852–Pasadena 9 May 1931. German-born US physicist. Studied at the United States Naval Academy. 1889 Professor at the Clark University in Worcester, 1892 Professor at the University of Chicago. Nobel Prize for Physics in 1907 for the development of an interferometer and the precise measurements performed with this instrument (now called Michelson interferometer).

Miller Oskar von: Munich 7 May 1855–Munich 9 April 1934. Brother of Wilhelm von Miller. German engineer, entrepreneur and founder of the *Deutsches Museum* in Munich. Studied construction engineering in Munich and later was employed by the state administration of Bavaria.

Miller, Wilhelm von: Munich 9 December 1848–Munich 1 March 1899. Brother of Oskar von Miller. German chemist. Studied in Munich with Richard August Carl Emil Erlenmeyer. 1883 Professor in Munich, where he established the first electrochemical laboratory at a German university.

Miller, William Lash: Galt 10 September 1866–Totonto 1 September 1940. Canadian chemist. Studied at the University of Toronto and in Berlin, Göttingen, Munich and Leipzig (with Wilhelm Ostwald). 1900 Associate Professor, 1908 full Professor of Physical Chemistry in Toronto

Minding, Ernst Ferdinand Adolph: Kalisch 11 January 1806–Dorpat 13 May 1885. Studied in Halle and Berlin philology, philosophy and physics, and taught himself mathematics. 1843 Professor of Mathematics in Dorpat.

Mitscherlich, Eilhard: Neuende/Wilhelmshaven 7 January 1794–Berlin 28 August 1863. German chemist and mineralogist. Studied philology (Persian Language) in Heidelberg, then medicine in Göttingen, where he received a doctorate for Persian language for his previous studies. After working with Jöns Jacob Berzelius in Stockholm, he became Professor of Chemistry in Berlin in 1822.

Mittasch, Paul Alwin: Großdehsa 27 December 1869–Heidelberg 4 June 1953. German chemist. He first became a teacher, later studied chemistry in Leipzig. Ph.D. with Max Bodenstein in Ostwald's institute. Since he did not have the *Abitur* (German school leaving examination to qualify for University studies), he could not become a University professor and so he took a position in industry.

Moch, Gaston: Saint-Cyr-l'École 6 March 1859–1935. French-Jewish pacifist and Esperantist.

Moissan, Ferdinand Frederic Henri: Paris 28 September 1852–Paris 20 February 1907. French chemist. 1906 Nobel Prize for Chemistry in recognition of the preparation of elemental fluorine.

Monckhoven, Désiré Charles Emmanuel van: Gent 25 September 1834–Gent 25 September 1882. Belgian chemist and photographer. Author of several books on photography.

Mond, Ludwig: Kassel 7 March 1839–London 11 December 1909. German-British chemist and industrial entrepreneur. Studied in Marburg and Heidelberg with Hermann Kolbe and Robert Wilhelm Bunsen. Inventor of nickel refining via nickel carbonyl (Mond process).

Moltke, Helmuth Karl Bernhard von: Parchim 26 October 1800–Berlin 24 April 1891. German Field Marshal.

Monet Claude: Paris 14 November 1840–Giverny 5 December 1926. French painter.

Monge, Gaspard, Comte de Péluse: Beaune 9 May 1746–Paris 28 July 1818. French mathematician, physicist and chemist. Friedrich Wilhelm Ostwald published Monge's *Darstellende Geometrie* (Géométrie descriptive) in the series *Ostwalds Klassiker der exakten Wissenschaften* as volume 117 in 1900.

Montgelas, Maximilian Carl Joseph Franz de Paula Hieronymus Graf von: Munich 12 September 1759–Munich 14 June 1838. Bavarian politician, jurist, and historian. From 1799 to 1817 Minister in Bavaria. Responsible for reforms including those aimed at secularisation.

Morelli, Giovanni (pseudonyms: Iwan (or Ivan) Lermolieff, Nicholas (or Nicolaus) Schäffer, Johannes Schwarze): Verona 25 February 1816–Milan 28 February 1891. Italian politician, physician and art critic.

Morgan, John Livingston Rutgers: New Brunswick 27 June 1872–New York 13 April 1935. US chemist. Studied at Rutgers University and in Leipzig with Wilhelm Ostwald. 1897 appointed Professor at Columbia University.

Morse, Harry Wheeler: San Diego 25 February 1873–Stanford 12 March 1936. Graduated from Leland Stanford Jr. University. 1899 to 1902 he worked for a Ph.D. (1901) with Wilhelm Ostwald and Robert Luther in Leipzig.

Moses, Bernard: Burlington, 28 August 1846–3 March 1930. US historian. Studied at the University of Michigan and Heidelberg (Germany). 1876 appointed Professor of History and Political Science at the University of California.

Mozart, Wolfgang Amadeus: Salzburg 27 January 1756–Vienna 5 December 1791. Austrian composer.

Muir, Matthew Moncrieff Pattison: Glasgow 1 November 1848–Epsom 2 Sept 1931. Scottish chemist. Studied in Glasgow and Tübingen. Was Demonstrator in Chemistry at Anderson's College, Glasgow, 1871–3, Assistant Lecturer in Chemistry at Owens College, Manchester, 1873–7, and in charge of Caius College Chemical Laboratory until his retirement in 1908.

Müller, Johannes: Koblenz 14 July 1801–Berlin 28 April 1858. German physiologist, comparative anatomist, ichthyologist, and herpetologist. Studied medicine in Bonn and Berlin. 1826 Professor in Berlin. Müller strongly emphasized the physical and chemical basis of physiology.

Müller, Wolfgang (also known as Wolfgangmüller): Dresden 1877–Dresden 1949. German painter

Münsterberg, Hugo: Danzig 1 June 1863–Cambridge, Massachusetts 16 December 1916. German-US psychologist and philosopher. Studied physchology, philosophy and medicine in Leipzig, where he did his Ph.D. with Wilhelm Wundt. 1882 to 1892 *Dozent* in Heidelberg. 1892 to 1894 guest Professor at Harvard University, 1895 to 1897 Freiburg, Germany, and then Professor at Harvard University.

Munsell, Albert Henry: Boston 6 January 1858–Brookline 28 June 1918. US painter and inventor of a numerical colour system.

Muspratt, Edmund Knowles: Liverpool 6 November 1833–Seaforth 1 September 1923. British industrialist. Son of James Muspratt, and brother of James Sheridan Muspratt. Studied at Owens College, Manchester, and in Gießen and Munich with Justus Liebig.

Muspratt, James: Dublin 12 August 1793–Liverpool 4 May 1886. British chemist and industrialist. Father of Edmund Knowles and James Sheridan Muspratt.

Muspratt, James Sheridan: Dublin 8 March 1821–Liverpool 3 February 1871. British chemist. Brother of Edmund Knowles Muspratt and son of James Muspratt.

Studied in Glasgow, London and Gießen (with Justus Liebig). Famous for his handbook on industrial chemistry. In 1848 he founded the Liverpool College of Chemistry.

Muthesius, Adam Gottlieb Hermann: Großneuhausen 20 April 1861–Berlin 26 October 1927. German architect, writer and diplomat.

Muther, Albert Carl Richard: Ohrdruf 25 February 1860–Wölfelsgrund 28 June 1909. German art historian. Studied in Heidelberg and Leipzig. 1895 Professor in Breslau.

N

Naumann, Alexander: Eudorf 31 July 1837–Gießen 16 March 1922. German chemist. Studied chemistry in Gießen. 1882 Appointed Professor there as the successor of Justus Liebig.

Natanson, Władysław (Ladislaus): Warsaw 18 June 1864–Kraków 26 February 1937. Studied chemistry in St. Petersburg (with Dmitri Mendeleev) and mathematics. He continued his studies at Cambridge (with Joseph John Thomson) and in Dorpat (with Arthur Joachim von Oettingen). 1894 Professor in Kraków.

Natorp, Paul Gerhard: Düsseldorf 24 January 1854–Marburg 17 August 1924. German philosopher. Studied music, history, philology and philosophy in Berlin, Bonn, and Straßburg. 1893 Professor in Marburg.

Nernst, Walther Hermann: Briesen 25 June 1864–Zibelle 18 November 1941. German physicist and chemist. Student of Heinrich Friedrich Weber, Arnold Meyer, Viktor Merz, Richard Börnstein, Georg Hettner, Hans Heinrich Landolt, Ludwig Boltzmann. Habilitation with Friedrich Wilhelm Ostwald (derivation of the so-called Nernst Equation in electrochemistry). 1894 Professor of Physical Chemistry in Göttingen, 1905–1924 Professor of Chemistry at the University of Berlin. In 1905 he discovered the third law of thermodynamics, 1920 Nobel Prize in Chemistry, 1924–1933 Professor of Physics at the University of Berlin.

Nestroy, Johann Nepomuk Eduard Ambrosius: Vienna 7 December 1801–Graz 25 May 1862. Austrian singer, actor and playwright.

Newcomb, Simon: Wallace (Canada) 12 March 1835–Washington D.C. 11 June 1909. US astronomer and mathematician. Autodidact. Graduated from Lawrence Scientific School at Harvard University. 1861 Professor of Mathematics and astronomer at the United States Naval Observatory, Washington D.C. 1884 Professor of Mathematics and Astronomy at Johns Hopkins University.

Newton, Isaak: Woolsthorpe-by-Colsterworth 4 January 1643–31 March 1727. English physicist and mathematician.

Nichols, William Henry: New York 1852–1930. US chemist and businessman. Studied at New York University.

Nietzki, Rudolf Hugo: Heilsberg 9 March 1847–Neckargemünd 28 September 1917. German pharmacist and chemist. He was Assistant toAugust Wilhelm von Hofmann in Berlin. 1887 Professor in Basel.

Nicholas I Pavlovich (Николай I Павлович): Tsarskoye Zelo 6 July 1796–St. Petersburg 2 March 1855. Russian Tsar.

Nilson, Lars Fredrik: Söderköping 27 May 1840–14 May 1899. Swedish chemist. 1878 Professor in Uppsala. Discoverer of scandium, the element, which Dmitri Mendeleev had predicted as 'eka-boron'.

Nobel, Alfred Bernhard: Stockholm 21 October 1833–Sanremo 10 December 1896. Swedish chemist and inventor, among others of dynamite. Established the Nobel Prize.

Norton, Charles Eliot: Cambridge (Mass.) 16 November 1827–Cambridge (Mass.) October 21, 1908. US author, social critic, and Professor of Art. Graduated from Harvard University. 1875 Professor at Harvard.

Noyes, Arthur Amos: Newburyport 13 September 1866–California, 3 June 1936. US chemist. Studied at **the** Massachusetts Institute of Technology. Ph.D. under the supervision of Friedrich Wilhelm Ostwald in Leipzig. 1894 Professor at the Massachusetts Institute of Technology. In 1920 he moved to Caltech.

O

Öchelhäuser (Oechelhaeuser) Justus Wilhelm jun.: Frankfurt 4 January 1850–Dessau 31 May 1923. German engineer and entrepreneur. Developed a technique to couple gas incineration with electricity generation.

Oettingen (Öttingen), Arthur Joachim von: Manor Ludenhof near Dorpat 28 March 1836–Bensheim 5 September 1920. German physicist and music theoretician. Studied physics in Dorpat, Paris (student of Antoine César Becquerel und Henri Victor Regnault) and Berlin (student of Heinrich Gustav Magnus, Johann Christian Poggendorff, Heinrich Wilhelm Dove and Karl Adolph Paalzow). 1864 *Dozent* at the Imperial Observatory in St. Petersburg. 1867 Professor in Dorpat, 1893 Leipzig. Research in acoustics. Brother of Georg Philipp von Oettingen.

Oettingen (Öttingen), August Georg Friedrich von: Wissust 5 July 1823–Dorpat 7 April 1908. German-Baltic judge and politician. Studied law in Dorpat and Heidelberg. Brother of Arthur Joachim von Oettingen, Georg Philipp von Oettingen and Alexander Konstantin von Oettingen (theologian).

Oettingen (Öttingen), Georg Philipp von: Manor Wissust near Dorpat 22 November 1824–Dorpat 16 February 1916. German physician, surgeon, ophthalmologist. Studied in Dorpat first law, but soon moved to medicine. He Studied and worked in Vienna, Paris, London, Edinburgh, Prague, Berlin and St Petersburg. 1857 Professor of Surgery in Dorpat, 1871 Professor of Ophthalmology. Brother of Arthur Joachim von Oettingen.

Oettingen (Öttingen), Helmut Alexander Woldemar von: 1871–1921. German chemist. Son of Arthur Joachim von Oettingen (Öttingen). 1893 to 1899 at the institute of Friedrich Wilhelm Ostwald in Leipzig.

Oppenheimer, Carl Nathan: Berlin 21 February 1874–Zeist 24 December 1941. German biochemist.

Oppenheimer, Franz: Berlin 30 March 1864–Los Angeles 30 September 1943. German physician, sociologist and economist. Studied in Freiburg and Berlin. 1919 Professor of Sociology in Frankfurt am Main, 1934/35 in Palestine, then in Japan, and finally in the USA.

Ostwald, Carl Wilhelm Wolfgang: Riga 27 May 1883–Dresden 22 November 1943. Son of Friedrich Wilhelm Ostwald (author of this autobiography). Studied natural sciences, esp. biology in Leipzig. Student of Jaques Loeb (1904–1906). 1923 Professor of Colloid Chemistry in Leipzig.

Ostwald, Elisabeth (neé Leukel): 1832–1920. Daughter of Johanna Christine Leukel and Johann Heinrich Leukel. Wife of Wilhelm Gottfried Ostwald. Mother of Friedrich Wilhelm Ostwald (author of this autobiography).

Ostwald, Elisabeth: Riga 19 June 1884–Großbothen 10 November 1968. Third child of Friedrich Wilhelm Ostwald (author of this autobiography). She was the husband of Eberhard Brauer (1875–1958), who was a co-worker of Friedrich Wilhelm Ostwald.

Ostwald, Eugen: Riga 23 October 1851–Riga 12 February 1932. Elder brother of Friedrich Wilhelm Ostwald (author of this autobiography). Professor of Forestry in Riga. His son Heinrich Ostwald (1877–1950) was Professor of Forestry at the Humboldt University Berlin

Ostwald, Friedrich Wilhelm: Riga 2 September 1853–Leipzig 4 April 1832. Son of Wilhelm Gottfried Ostwald. Author of this autobiography.

Ostwald, Gottfried: 1855–1918. Younger brother of Friedrich Wilhelm Ostwald (author of this autobiography). He was an entrepreneur with an iron foundry and machine factory.

Ostwald Gottfried: 1785–1860. Cooper in Moscow. Grandfather of Friedrich Wilhelm Ostwald (author of this autobiography).

Ostwald, Grete (Margarete): Riga 13 February 1882–Großbothen, 1 August 1960. Daughter of Friedrich Wilhelm Ostwald (author of this autobiography). She studied from 1905 to 1907 at *Großherzoglich-Sächsischen Kunstschule* (Art Academy) in Weimar. Since 1918 she suffered from severe arthritis. Since 1932 she managed the Ostwald property *Haus Energie* in Großbothen, where she founded the Wilhelm Ostwald Archive.

Ostwald, Flora Helene Mathilde (neè von Reyher): Riga 14 January 1854–Großbothen 2 April 1946. Spouse of Friedrich Wilhelm Ostwald (author of this autobiography).

Ostwald, Walter: Riga 20 May 1886–Freiburg, Breisgau 12 July 1958. Son of Friedrich Wilhelm Ostwald (author of this autobiography). Studied chemistry in Leipzig. Student of William Ramsay (1907). He worked in industry and as a science journalist.

Ostwald, Wilhelm Gottfried: 1824–1903. Cooper. Father of Friedrich Wilhelm Ostwald (author of this autobiography).

Otlet, Paul: Brussels 23 August 1868–Brussels 10 December 1944. Belgian author, entrepreneur and peace activist. Pioneer of information science. Inventor of the Universal Decimal Classification for bibliographic and library classification.

Otto, Friedrich Wilhelm Robert: Braunschweig 18 August 1837–Braunschweig 14 February 1907. German chemist. 1870: Professor of Pharmacy and Applied Chemistry in Braunschweig, where he succeeded his father Julius Otto (1809–1870). He is e known for the translation of Thomas Graham's textbook on chemistry, which was published in several editions as *Graham-Ottos Lehrbuch der Chemie*.

Otto, Berthold: Bienowitz 6 August 1859–Berlin 29 June 1933. German reform-pedagogue.

P

Parseval, August von: Frankenthal 5 February 1861–Berlin 22 February 1942. German airship designer.

Paschen, Louis Carl Heinrich Friedrich: Schwerin 22 January 1865–Potsdam 25 February 1947. German Physicist. Studied in Berlin and Straßburg. Student of August Kundt and Johann Wilhelm Hittorf. 1901 Professor in Tübingen, 1924 to 1933 President of *Physikalisch-technische Reichsanstalt* in Berlin. Research on spectroscopy.

Paul, Theodor: Lorenzkirch 13 February 1862–Munich 30 September 1928. First apprenticed to an apothecary, later studied pharmacy and chemistry in Leipzig. Ph.D. with Ernst Otto Beckman. 1894 habilitation with Friedrich Wilhelm Ostwald. 1895 approbation as food chemist. 1898 Professor in Tübingen, where he obtained an MD degree. 1902 director of the *Kaiserliches Gesundheitsamt* (Imperial Health Authority) in Berlin. 1905 Professor in Munich and Director of the *Kgl. Untersuchungsanstalt für Nahrungs- und Genußmittel* (Royal Institution of Food Survey).

Peabody, Francis Greenwood: Boston 4 December 1847–Cambridge (Mass.) 28 December 1936. US theologian. Graduated from Harvard College. 1872–73 at the University of Halle, Germany. Parkman Professor of Theology at Harvard (1881–1886), University preacher (1886–1906), Plummer Professor of Christian Morals (1886–1912), and Dean of the Divinity School (1901–1906).

Peano, Giuseppe: Spinetta 27 August 1858–Turin 20 April 1932. Italian mathematician. Studied in Turin. 1890 Professor in Turin.

Pechmann, Hans Freiherr von: Nürnberg 1 April 1850–Tübingen 19 April 1902. German chemist. Professor in Munich and Tübingen.

Pedro II of Brazil: Rio de Janeiro 2 December 1825–Paris 5 December 1891. From 1831 to 1889 Emperor of Brazil. He was educated in ancient and modern languages, as well as in sciences. He was a co-founder of the Institute Pasteur in Paris.

Pebal, Leopold von: Seckau 29 December 1826–Graz 17 February 1887. Austrian chemist. Student of Robert Wilhelm Bunsen (Heidelberg). 1857 to 1865 Professor in Lemberg, after that in Graz. He was murdered by a laboratory technician.

Peirce, Benjamin Osgood: Beverly (Mass.) 11 February 1854–Cambridge (Mass.) 14 January 1914. US mathematician. Graduated from Harvard College. He worked with Eilhard Ernst Gustav Wiedemann and Wilhelm Gottlieb Hankel in Leipzig, and with Hermann Ludwig Ferdinand von Helmholtz in Berlin. 188 Hollis Professor of Mathematics and Natural Philosophy at Harvard.

Penzig, Rudolf: Samnitz 30 January 1855–Berlin 20 April 1931. German author, politician and reform pedagogue.

Peter the Great (Pyotr Alekseyevich Romanov, Пётр I): Алексеевич Moscow 9 June 1672–St. Petersburg 8 February 1725. Russian Tsar who attempted to modernize Russia by introducing a number of reforms and opened it to the West.

Pettenkofer, Max Josef: Lichtenheim 3 December 1818–Munich 10 February 1901. German chemist and hygienist. Studied sciences, pharmacy and medicine. 1847 Professor in Munich.

Pettersson, Sven Otto: Göteborg 12 February 1848–Göteborg 16 January 1941. Swedish chemist. 1872 Professor in Uppsala, later in Stockholm. Research in physical inorganic and organic chemistry. He also developed oceanographic instruments.

Pëus, Wilhelm Heinrich: Elberfeld 24 July 1862–Dessau 10 April 1937. German politician.

Pfaundler von Hadermur, Leopold: Innsbruck 14 February 1839–Graz 6 March 1920. Austrian physicist (and alpinist). 1867 Professor of Physics in Innsbruck, 1891 in Graz.

Pfeffer, Wilhelm Friedrich Philipp: Grebenstein 9 March 1845–Leipzig 31 January 1920. German botanist and plant physiologist. He studied botany, physics and pharmacology in Göttingen, Marburg and Berlin. Student of Friedrich Wöhler and Wilhelm Rudolph Fittig. He developrd the *Pfeffer cell* to measure osmotic pressure.

Pickering, Percival Spencer Umfreville: 6 March 1858–Harpenden 5 December 1920. British chemist and horticulturist. Studied at Balliol College Oxford. Lecturer in Chemistry at Bedford College, 1881–1888. Director of the Woburn Experimental Fruit Farm

Planck, Max Karl Ernst Ludwig: Kiel 23 April 1858–Göttingen 4 October 1947. Studied mathematics and natural sciences in Munich and Berlin. Student of Gustav Kirchhoff and Hermann Ludwig Ferdinand von Helmholtz. Professorships: 1885 Kiel, 1889 Berlin. Nobel Prize for Physics 1918.

Poelchau, Harald Oskar Georg: Riga 20 February 1835–Riga 9 May 1882. German painter and teacher at the Polytechnic and at the *Realschule* in Riga. Since 1872 Director of the *Gewerbeschule* (vocational school) in Riga.

Poggendorff, Johann Christian: Hamburg 29 December 1796–Berlin 24 January 1877. German physicist. First did an apprenticeship to an apothecary in Hamburg, later moved to Berlin to study at the university. Research mainly in electricity. Served for a long time as editor of the journal *Annalen der Physik.*

Poincaré, Jules Henri: Nancy 29 April 1854–Paris 17 July 1912. French mathematician, physicist and philosopher. Studied in Nancy and Paris. 1881 Professor at the Sorbonne, Paris.

Polako, Isaac: Izmir 15 March 1864–?. Was owner of the British carpet trading company Habif & Polako in Izmir (Smyrne) and a member of *Société Positiviste Internationale* and the *Société de la morale de la nature*. Since 1913 citizen of France.

Poynting, John Henry: Monton. 9 September 1852–Birmingham 30 March 1914. British physicist. Studied at Owen's College, Manchester and at Trinity College, Cambridge. 1880 to 1914 Professor of Physics in Birmingham. Student of James Clerk Maxwell.

Prescott, Albert Benjamin: Hastings 12 December 1832–Ann Arbor 25 February 1905. US chemist. Studied medicine at the University of Michigan. 1865 Assistant Professor of Chemistry at the University of Michigan, 1870 dean of the school of pharmacy, and 1884 director of the chemical laboratory.

Pythagoras of Samos: Samos c. 570–Metapontum c. 495 BC. Greek philosopher and mathematician.

R

Racinet, Albert Charles Auguste: Paris 20 July 1825–Montfort-l'Amaury 29 October 1893. French costume designer, illustrator and painter.

Raffaello Sanzio da Urbino (known as **Raphael)** Urbino 6 April (or March 28) 1483–Rome 6 April 1520. Italian painter and architect.

Ramsay William: Glasgow 2 October 1852–High Wycombe 23 July 1916. Scottish chemist. Studied in Glasgow, Heidelberg (with Robert Wilhelm Bunsen) and Tübingen (with Wilhelm Rudolf Fittig). 1880 Professor in Bristol, 1887 Professor at the University College London. 1904 Nobel Prize for chemistry.

Ranke, Franz Leopold: Wiehe 21 December 1795–Berlin 23 May 1886. German historian. Studied theology and philology in Leipzig. 1824 Professor in Berlin.

Raoult, François-Marie: Fournes-en-Weppes 10 May 1830–Grenoble 1 April 1901. 1862 Professor at the Lycée de Sens (Yonne). Known for the discovery of the freezing point and vapour pressure depression of solvents when substances are dissolved in them.

Raphael: **Raffaello Sanzio da Urbino**: Urbino April 6 (or March 28) 1483–Rom 6 April 1520. Italian painter and architect.

Rath, Carl Rudolf Walther vom: Amsterdam 11 Sepember 1857–Kronberg 2 February 1940. German scientist, jurist and entrepreneur.

Rathenau, Walther: Berlin 29 September 1867–Berlin-Grunewald 24 June 1922 (murdered). German industrialist, writer and politician. Studied physics, philosophy and chemistry in Straßburg. He had leading positions in the AEG, and held minister positions in Germany during the 20ies.

Ratzel, Friedrich: Karlsruhe 30 August 1844–Ammerland am Starnberger See 9 August 1904. German zoologist and geographer. Studied geology and zoology in Heidelberg, Jena and Berlin. Voyages to Italy, Cuba, Mexico, and the USA.1876 Professor in Munich, 1886 Professor in Leipzig.

Ratzenhofer, Gustav: Vienna 4 July 1842–on the Atlantic, 8 October 1904. Austrian officer, philosopher and sociologist.

Rayleigh, i.e. John William Strutt, 3rd Baron Rayleigh: Langford Grove 12 November 1842–Termins Place 30 June 1919. Studied mathematics at Trinity College, Cambridge, and after some interruption he studied science. In 1879 he followed James Clerk Maxwell as Professor of Experimental Physics and was head of the Cavendish Laboratory in Cambridge. 1887 to 1905 Professor at the *Royal Institution of Great Britain*. 1904 Noble Prize in Physics.

Rée, Anton: (Hamburg 9 November–Hamburg 13 January 1891. Reform pedagogue and politician in Hamburg.

Regnault, Henri Victor: Aachen 21 July 1810–Paris 19 January 1878. French chemist and physicist. Studied in Paris. Later student of Justus Liebig (Gießen). 1840 Professor at *École Polytechnique* in Paris. Robert Wilhelm Bunsen was his student.

Reicher, Lodewijk Theodorus: Kampen 16 April 1857–Sobibor (Nazi German extermination camp), 30 April 1943. Dutch-Jewish chemist. Studied in Amsterdam. 1883 Ph.D. for the thesis "Temperatuur der allotropische verandering van de zwavel en haar afhankelijkheid van den druk" (temperature of the allotropic changes of sulphur and its dependence on pressure). 1883 to1893 private assistant of Jacobus Henricus van't Hoff.

Remsen, Ira: New York 10 February 1846–Carmel 4 March 1927. US chemist. Studied medicine at Columbia University (1867 MD), and chemistry in Germany (Göttingen and Tübingen). 1875 Professor at Williams College, Massachusetts. He founded the Chemistry Department at Johns Hopkins University. Together with Constantin Fahlberg he discovered the sweetener saccharin.

Reyher, Carl Dietrich Christoph von: Riga, October 23 1846–11 January 1891 St. Petersburg. Surgeon in the Russian army who contributed to the development of antiseptic surgery. Cousin of Friedrich Wilhelm Ostwald.

Reyher, Helene von: see: **Ostwald, Flora Helene Mathilde**

Ribbeck, Johann Carl Otto: Erfurt 23 July 1827–Leipzig 18 July 1898. German philologist. Professorships: 1856 Bern, 1861 Basel, 1862 Kiel, 1872 Heidelberg, 1877 Leipzig.

Richards, Theodore William: Germantown 31 January 1868–Cambridge (Mass.) 2 April 1928. US chemist. Studied at Haverford College and Harvard University. Postdoc in Germany with Viktor Meyer. 1901 Professor at Harvard. 1914 Nobel Prize in Chemistry for exact determinations of atomic masses of elements.

Richter, Jeremias Benjamin: Hirschberg 10 March 1762–Berlin 14 April 1807. German chemist. Self-educated in science and chemistry he studied philosophy with Immanuel Kant. In 1789 doctor of mathematics and chemistry. His most important contribution to chemistry was a treatise on stoichiometry.

Ridder, August Cornelius de: Antwerpen 4 May 1837–Paris 13 May 1911. Belgian-German merchant and art collector. Board member of *Farbwerke Höchst*.

Riehl Alois (Aloys) Adolf: Rielhof (near Bozen) 27 April 1844–Neubabelsberg 21 November 1924. Austrian philosopher. Studied philosophy, geography and history in Vienna, Munich, Innsbruck and Graz. 1878 Professor in Graz, 1882 in Freiburg, 1896 in Kiel, 1898 in Halle, and 1905 in Berlin.

Rieß (Riess), Carl: 1875–1929. German merchant. Headed the Hamburg branch of the Monist Society from 1923 to 29.

Ristenpart, Eugen Karl (Carl) Emil: Frankfurt am Main 22 November 1873–Wiesbaden 2 May 1953. German chemist. Studied sciences in Freiburg and Berlin. Ph.D. in Berlin. From 1897 to 1899 he was assistant lecturer at *Königliche Färberei- und Agenturschule* (Royal Dying College) in Krefeld, 1899 to 1901 chemist at Knipscher & Maas Silk Company, Paterson, New Jersey. Thereafter until 1908 chemist at the silk dying factory in Krefeld. 1912 Professor with teaching duties at several colleges. 1920 foundation and directorship of the Chemnitz dependence of Friedrich Wilhelm Ostwald's dye factory.

Ritter, Johann Wilhelm: Samitz 16 December 1776–Munich 23 January 1810. German physicist and philosopher. First apprenticed to an apothecary, then private-studies in Jena. Only in 1804 did he get a permanent position in the Academy of Science of Bavaria in Munich.

Rood, Ogden Nicholas: Danbury 3 February 1831–Manhattan 12 November 1902. US physicist. Studied at Yale and Princeton, and in Berlin and Munich (with Justus Liebig). 1861 Professor at Columbia University.

Roscher, Georg Friedrich Wilhelm: Hannover 21 October 1817–Leipzig 4 June 1894. German historian and economist. Studied in Göttingen, after the Ph.D. also in Berlin where he was a pupil of Leopold von Ranke. 1844 Professor in Göttingen, 1848 Professor in Leipzig.

Roscoe, Henry Enfield: London 7 January 1833–West Horsley 18 December 1915. English chemist. Studied at University College, London and in Heidelberg with Robert Wilhelm Bunsen. 1857 Professor at Owens College, Manchester.

Rose, Heinrich: Berlin 6 August 1795–Berlin 27 January 1864. German analytical chemist and mineralogist. First apprenticed to the apothecary of Martin Heinrich Klaproth; later student of Jöns Jacob Berzelius (Stockholm). 1823 Professor in Berlin.

Ross, Ronald: Almora (British India) 13 May 1857–London 16 September 1932. British physician. Studied medicine at St Bartholomew's Hospital in London. Following various posts in the medical service in India, he was appointed Professor with the Chair of Tropical Medicine of the Liverpool School of Tropical Medicine from 1902 to 1912. Discoverer of the transmission route of malaria. Nobel Prize for Medicine in 1902.

Rood, Ogden Nicholas: Dunbury 3 February 1831–Manhattan 12 November 1902. US physicist. Studied at Yale and Princeton University, then in Berlin and in Munich (with Justus Liebig). 1863 Professor of Physics at Columbia University.

Roozeboom, Hendrik Willem Bakhuis: Alkmaar 24 October 1854–Amsterdam 8 February 1907. Dutch chemist. Before studying chemistry in Leiden he worked in a chemical factory. 1896 Professor in Amsterdam.

Roosevelt, Theodore: New York 27 October 1858–Oyster Bay 6 January 1919. 26th President of the USA.

Röntgen, Wilhelm Conrad: Lennep 27 March 1845–Munich 10 February 1923. German physicist. Studied in Zurich (Switzerland) and worked as assistant to August Kundt in Würzburg. Professor of Physics in Hohenheim, Straßburg, Gießen, Würzburg and Munich. Discoverer of X-rays. 1901 Nobel Prize for Physics.

Röthlisberger, Ernst: Burgdorf 20 November 1858–Bern 29 January 1926. Swiss jurist. Studied theology, philosophy and history of law in Bern. 1881–85 Professor in Bogota (Columbia). 1888 to 1921 Secretary and vice-president of the International Bureau of Intellectual Property. 1912 to 1926 Professor in Bern.

Raoult, François Marie: Fournes-en-Weppes 10 May 1830–Grenoble 1 April 1901. French physicist and chemist. Studied at Lycée of Reims. In 1870 Professor in Grenoble.

Royce, Josiah: Grass Valley 20 November 1855–Sambridge (Mass.) 14 September 1916. US philosopher. Studied at University of California, Berkeley, than in Germany (Heidelberg, Leipzig, Göttingen). 1892 Professor at Harvard University.

Ruhemann Siegfried: Johannesburg (Eastern Prussia) 4 January 1859–London, 22 August 1943. German-English chemist. Studied in Berlin Charlottenburg at the forerunner of the Technical University. On suggestion of August Wilhelm von Hoffman he became assistant to James Dewar at the University of Cambridge.

Runge, Philipp Otto: Wolgast 23 July 1777 Hamburg 2 December 1810. German painter.

Ruskin, John: London 8 February 1819–Brantwood 20 January 1900. British art critic, social thinker and philantropist.

Russell, Bertrand Arthur William, 3rd Earl Russell: Trellech 18 May 1872–Penrhyndeudraeth 2 February 1970. British philosopher. Studied mathematics at the University of Cambridge. He was Professor at the University of Cambridge (with a long interruption). 1950 Nobel Prize for Literature.

S

Saager, Adolf: Stuttgart 20 April 1879–Massagno 31 August 1949. Swiss writer and journalist. Studied sciences in Munich, Geneva and Heidelberg (Ph.D.).

Sachs, Julius: Breslau 2 October 1832–Würzburg 29 May 1897. German botanist and plant physiologist. Studied science in Prague. 1861 Professor of Botany in Poppelsdorf (near Bonn), 1867 Professor in Freiburg, 1868 in Würzburg.

Schacht, Horace Greeley Hjalmar: Tingleff 22 January 1877–Munich 3 June 1970. German politician and banker.

Scheffel, Joseph Victor: Karlsruhe 16 February 1826–Karlsruhe 9 April 1886. German writer.

Scheye, Anton: Berlin, 1873–?. Studied physics and mathematics in Berlin, where he listened to the lectures of Hermann Ludwig Ferdinand von Helmholtz. His lecture notes were the basis of the book *Vorlesungen über die mathematischen Prinzipien der Akustik von H. von Helmholtz*, edited by A. König and C. Runge, Leipzig, Johann Ambrosius Barth, 1898. Scheye's Ph.D. thesis was entitled *Über die Vorgänge in Elektrolyten welche vom galvanischen Strome durchflossen werden und von unpolarisierbaren Elektroden begrenzt sind*, published: Berlin, Friedländer, 1895.

Schiff, Robert: Frankfurt am Main 25 July 1854–Massa 1940. German/Italian chemist. Studied in Heidelberg and Zurich. 1879 Professor in Modena. Hugo Schiff, the discoverer of Schiff's bases was his uncle.

Schiller, Johann Christoph Friedrich: Marbach 10 November 1759–Weimar 9 May 1805. German author, poet, historian and philosopher.

Schilling, Johannes: Mittweida 23 June 1828–Klotzsche 21 March 1910. German sculptor.

Schleyer, Johann Martin: Oberlauda 18 July 1831–Konstanz 16 August 1912. German Catholic priest and inventor of the synthetic language Volapük. Studied Catholic theology, history, philosophy and some medicine in Freiburg.

Schlichtegroll, Adolf Heinrich Friedrich von: Waltershausen 8 December 1765–Munich 4 December 1822. German philologist and archaeologist.

Schlomann, Alfred: Malchow 10 August 1878–New York 31 March 1952. German engineer. Because of his Jewish descent he emmigrated to the US in 1937. He published several illustrated multilingual technical dictionaries.

Schmidt, Carl Ernst Heinrich: Mitau 13 June 1822–Dorpat 27 February 1894. German chemist and physician. First apprenticed to an apothecary in Berlin, then studied chemistry and medicine. Student of Justus Liebig in Gießen (Ph.D. in 1844). In 1845 doctor of medicine in Göttingen. Thereafter he worked in St. Petersburg at the Military Medical Academy and acquired Russian academic degrees.1850 Professor of Pharmacy (Dorpat). In 1878 Friedrich Wilhelm Ostwald did his Ph.D. in Schmidt's group.

Schmidt, Heinrich: Heubach 18 December 1874 – Jena 2 May 1935. German archivist and philosopher. In 1900 he became private secretary of Ernst Haeckel.

Schmidt, Hermann Adolf Alexander: Liiva 27 May 1831–Dorpat 22 April 1894. German physiologist. Studied history, later medicine in Dorpat, and thereafter at various places in Europe. Student of Ernst Felix Immanuel Hoppe-Seyler. 1869 Professor of Physiology in Dorpat.

Schmidt (since 1938 Schmidt-Hellerau), Karl Camillo: Zschopau 1 February 1873–Hellerau 6 November 1948. German carpenter, furniture manufacturer, and social reformer. Founded the first German Garden City Hellerau.

Schmitt, Rudolf Wilhelm: Wippershain 5 August 1830–Radebeul 18 February 1898. German chemist. In 1870 he was appointed Professor of Organic Chemistry at the *Königlich-Sächsische Polytechnikum* in Dresden (a forerunner of the present Technical University). After the death of Hermann Kolbe he became the technical supervisor of the *Salicylsäure-Fabrik Dr. F. v. Heyden* in Radebeul. This factory was the first factory worldwide to produce pharmaceuticals on a large scale.

Schneeberger, Friedrich: Biel 10 October 1875–Bern 18 March 1926. Swiss priest.

Schoop, Max Ulrich: Frauenfeld 10 April 1879–Zurich 29 February 1956. Swiss inventor and entrepreneur. He invented the metal spraying technique.

Schotten, Carl: Marburg 12 July 1853–(Berlin?) 9 January 1910. German chemist, known for having discovered, together with Eugen Baumann, the Schotten-Baumann synthesis of amides.

Schönbein, Christian Friedrich: Metzingen 18 October 1799–Baden-Baden 29 August 1868. German-Swiss chemist and physicist. Studied chemistry in Erlangen. He discovered the electrochemical fuel cell (shortly before William Robert Grove), ozone and guncotton. He coined the term *passivity* for metals, which do not dissolve in acids.

Schopenhauer, Arthur: Danzig 22 February 1788–Frankfurt am Main 21 September 1860. German philosopher.

Schreber, Daniel Gottlob Moritz: Leipzig 15 October 1808–Leipzig 10 November 1861. German physician. Studied in Leipzig. 1844 director of the Leipzig *Heilanstalten* (sanatorium).

Schröder, Paul Woldemar Viktor von: Dorpat 15 August 1850–Heidelberg 28 January 1898. German pharmacologist. Studied in Dorpat, Riga, Berlin and Leipzig. 1890 Professor of Pharmacology in Heidelberg.

Schuster, Franz Arthur Friedrich: Frankfurt am Main 12 September 1851–Yeldall 17 October 1934. German-born British physicist. Studied in Geneva, Manchester (with Henry Enfield Roscoe), and Heidelberg (with Gustav Robert Kirchhoff), in Göttingen (with Wilhelm Eduard Weber), in Berlin with Hermann Ludwig Ferdinand von Helmholtz. In Cambridge he worked with James Clerk Maxwell and later with John William Strutt (Baron Rayleigh).

Schweder, Gotthard: (Loddiger / Livonia 30 December 1831–Riga 3 January 1913) was teacher at the 'Realgymnasium' and later *Dozent* of physics at the Dorpat Polytechnic.

Schwind, Moritz Ludwig von: Vienna 21 January 1804–Niederpöcking 8 February 1871. Austrian painter.

Scott, Walter: Edinburgh 15 August 1771–Abbotsford 21 September 1832. Scottish novelist.

Seeck, Fritz: 1851–1875. Brother of Otto Karl Seeck. In 1872/73 student of Robert Bunsen. He was a school and University friend of Friedrich Wilhelm Ostwald.

Seeck, Otto Karl: Riga 2 February 1850–Münster 29 June 1921. Studied chemistry in Dorpat from 1867 to 1868, and then history in Berlin till 1872. From 1881 to 1908 Professor of History in Greifswald, since 1908 in Münster.

Seffner, Carl Ludwig: Leipzig 19 June 1861–Leipzig 2 October 1932. German sculptor.

Seidl, Gabriel: Munich 9 December 1848–Munich 27 April 1913. German architect. Studied at the Academy of Fine Arts in Munich.

Semon, Felix: Danzig 8 December 1849–Rignalls (UK) 1 March 1921. German-British physician. Studied in Heidelberg and Berlin. Since his Jewish descent prevented a University carrier in Germany, he settled in London in 1874, where he became private physician to King Edward VII.

Senhofer, Carl Griesbruck, Tirol 29 September 1841–Innsbruck 17 October 1904. Austrian pharmacist and chemist. Studied pharmacy in Innsbruck, Brixen, and Klausen. 1874 Professor in Innsbruck.

Seydewitz, Kurt Damm Paul von: Lauterbach 3 May 1843–Dresden 17 December 1910. Studied Law in Leipzig. In 1892 he became minister of education (*Kultusminister*) of Saxony.

Shaler, Nathaniel Southgate: Newport 20 February 1841–Cambridge (Mass.) 10 April 1906. US palaeontologist and geologist. Graduated from Harvard where he was appointed Professor of Palaeontology in 1869 and Professor of Geology in 1888.

Siedentopf, Henry Friedrich Wilhelm: Bremen 22 September 1872–Jena 8 May 1940. German physicist. From 1899 to 1938 he worked for the optical company Carl Zeiss in Jena. Since 1919 he was also Professor at the University of Jena.

Siemens, Ernst Werner: Lenthe 13 December 1816–Berlin 6 December 1892. German engineer, inventor and entrepreneur. Student of Heinrich Gustav Magnus at the military academy in Berlin, where he studied mathematics and sciences.

Sievers, Eduard: Lippoldsberg 25 November 1850–Leipzig 30 March 1932. Professor in Jena, Tübingen, Halle and Leipzig. The Sievers' law in Indo-European linguistics bears his name.

Sigismund, Karl: 1861–1932. German publisher and first chairman of *Börsenverein der Deutschen Buchhändler* (Association of Capital Market Groups of German Book Sellers).

Simroth, Heinrich Rudolf: Riestedt 10 May 1851–Gautzsch 31 August 1917. German zoologist. Studied in Straßburg. From 1895 Professor in Leipzig.

Skiff, Frederick James Volney: Chicopee 5 November 1851–Chicago 24 February 1921. He worked for newspapers and as a local politician. He was commissioner general of the Columbian World's Fair in 1892, chief of staff commissioner general of the US to the Paris exhibition in 1900, and director general of the St. Louis exposition in 1901. Skiff was the first director of the Field Museum of Natural History in Chicago.

Skraup, Zdenko Hans: Prague 3 March 1850–Vienna 10 September 1910. Czech-Austrian chemist. Chemistry studies in Prague, Ph.D. in Gießen. 1881 Professor at the Trade Academy Vienna, 1886 at the University of Graz and since 1906 University of Vienna.

Small, Albion Woodbury: Buckfield 11 May 1854–Chicago 24 March 1926. US sociologist. Studied theology at Andover Newton Theological School, and history social economics and politics in Leipzig and Berlin, followed by studies at Johns Hopkins University, Baltimore. In 1892 he founded the first Department of Sociology at the University of Chicago.

Smith Alexander: Edinburgh 11 September 1865–Edinburgh 8 September 1922. US chemist of Scottish birth. Studied in Edinburgh and Munich. 1903–11, Professor of Chemistry at Chicago University; 1911–21 Professor of Chemistry at Columbia University, New York.

Smithells, Arthur: Bury 24 May 1860–Highgate 24 February 1939. British chemist. Studied in Glasgow and Manchester (with Henry Enfield Roscoe and Carl Schorlemmer), in London and Munich, and in Heidelberg with Robert Wilhelm Bunsen. 1885 Professor in Leeds.

Snell, Karl: Dachsenhausen 19 January 1806–Jena 12 August 1886. German mathematician physicist and natural philosopher. Studied philosophy and mathematics in Gießen, Halle, Göttingen and Berlin. 1844 Professor of Mathematics and Physics in Jena.

Sohncke, Leonhard: Halle 22 February 1842–Munich 1 November 1897. German mathematician and physicist. Professorships: 1871 Karlsruhe, 1883 Jena, 1886 Munich.

Solvay, Ernest Gaston Joseph: Rebecq-Rognon 16 April 1838–Ixelles 26 May 1922. Belgian chemist. Together with his brother Alfred he founded the Solvay Company. Inventor of the Solvay process of soda production.

Sömmering, Samuel Thomas: Thorn 28 January 1755–Frankfurt am Main 2 March 1830. German physician, anatomist, anthropologist, paleontologist and inventor. Studied in Göttingen. Later Professor in Kassel and Mainz. 1804 counsellor to the court of Bavaria.

Sorley, William Ritchie: Selkirk 4 November 1855–Cambridge 28 July 1935. Scottish philosopher. Studied in Edinburgh and Cambridge. 1909 to 1933 he was Knightsbridge Professor of Philosophy at the University of Cambridge.

Speck von Sternburg, Hermann Freiherr: Leeds 21 August 1852–Heidelberg 23 August 1908. German diplomat. 1898 First Secretary at the German Embassy in Washington.

Spielhagen, Friedrich: Magdeburg 24 February 1829–Berlin 25 February 1911. Anti-feudal, radical democratic and liberal writer.

Spring, Walter: 1848–1911. Chemist in Liège.

Spitzweg, Franz Carl: Unterpfaffenhofen 23 September 1808–Munich 23 September 1885. German painter.

Staedel, Wilhelm: Darmstadt 18 March 1843–Darmstadt 14 May 1919. German chemist. Student of Adolph Strecker (Tübingen), Carl Remigius Fresenius (Wiesbaden) 1881. Professor in Darmstadt. Research on physical and theoretical organic chemistry.

Stanford, Amasa Leland: Watervliet 9 March 1824–Palo Alto 21 June 1893. US industrialist and politician. Founder (together with his wife) of Stanford University.

Stas, Jean Servais: Louvain 21 August 1813–Brussels 13 December 1891. Belgian chemist. Studied at *École polytechnique* in Paris. 1869 Professor at *École Royal Militaire* in Brussels. Famed for precise determinations of atomic masses of elements.

Staudinger, Franz: Groß-Gerau 15 February 1849–Darmstadt 18 November 1921. German teacher, philosopher and activist of consumer cooperatives.

Stefan, Josef: (Slovene version: **Jožef Štefan**): St. Per (near Ebenthal) 24 March 1835–Vienna 7 January 1893. Carinthian Slovene/Austrian mathematician and physicist. He studied in Vienna and in 1863 became Professor of Physics there. He suggested a dependence of black body radiation on temperature, which is now referred to as Stefan's Law.

Stein, Ludwig: Erdö-Bénye 12 November 1859–Salzburg 13 July 1930. Hungarian-Swiss philosopher, sociologist, rabbi, and peace activist.

Steiner, Jakob: Utzenstorf 18 March 1796–Bern 1 April 1863. Swiss mathematician. Studied mathematics in Heidelberg, from 1820 he lived in Berlin as a private teacher and later was appointed Professor at the University. He contributed mainly to geometry.

Stohmann, Friedrich Carl Adolf: Bremen 25 April 1832–Leipzig 1 November 1897. German agricultural chemist. Student of Friedrich Wöhler (Göttingen), 1853–55 Assistant of Thomas Graham (London), 1862 Professor in Braunschweig, 1871 Professor in Leipzig, where he founded a *landwirtschaftlich-physiologisches Institut* (agricultural and physiological institute).

Stokes, George Gabriel: Skreen 13 August 1819–Cambridge 1 February 1903. Irish/British physicist. Studied at Pembroke College, Cambridge. 1849 Professor of Mathematics, Cambridge.

Stöcker, Helene: (Elberfeld 13 November 1869–New York 24 February 1943. German suffragette and founder of the German Society for Maternity Protection.

Stöckhardt, Julius Adolph: Röhrsdorf 4 January 1809–Tharandt 1 June 1886. German agricultural chemist. First apprenticed to an apothecary, then studied sciences in Berlin. 1847 Professor at *Akademie für Forst- und Landwirte* (Royal Saxon Academy of Forestry) in Tharandt. He is best known for his book *Schule der Chemie*, which had a tremendous influence in Germany and abroad.

Strecker, Adolph: Darmstadt 21 October 1822–Würzburg 7 November 1871. German Chemist famed for a synthesis of amino acids, which now bears his name. He translated the textbook *Eléments de Chimie* of Henri Victor Regnault and published it as an extended book with the title *Kurzes Lehrbuch der Chemie*, Vieweg, Braunschweig, 1851 (there were several later editions).

Strutt, John William, 3rd Baron Rayleigh: see **Rayleigh**

Stürgkh, Karl Graf: Graz 30 October 1859–Vienna 21 October 1916. Austrian politician.

Suttner, Bertha Sophia Felicita Freifrau von: Prag 9 June 1843–Vienna 21 June 1914. Austrian pacifist and writer. Nobel Peace Prize in 1905.

Swan, Joseph Wilson: Sunderland 31 October 1828–Warlingham 27 May 1914. British physicist, chemist and inventor. He invented the first incandescent lamp.

T

Tait, Peter Guthrie: Dalkeith (Midlothian) Scotland 28 April 1831–Edinburgh 4 July 1901. Scottish physicist. Studied in Edinburgh and Cambridge. 1854 Professor of Mathematics at Queens College, Belfast, 1860 Professor of Natural Philosophy in Edinburgh.

Thalén, Tobias Robert: Köping 28 December 1827–Uppsala 27 July 1905. Swedish physicist. 1856 Professor in Uppsala, 1874 Professor in Stockholm.

Than, Carl von (Hungarian version Károly Than): Altbetsche (Hungarian: Óbecse, Serbian: Bečej) 20 December 1834–Budapest 5 July 1908. Hungarian chemist. Studied chemistry in Vienna. Student of Robert Wilhelm Bunsen (Heidelberg) and Charles Adolphe Wurtz (Paris). 1860 Professor of Chemistry at the Technical University, Budapest. His main field of research was analytical chemistry, He is the founder of the first Hungarian journal of chemistry (Magyar Chémiai Folyóirat).

Therese Charlotte Marianne Auguste, Prinzessin von Bayern: Munich 12 November 1850–Lindau 19 September 1925. German ethnologist, zoologist, botanist and writer. Daughter of prince regent Luitbold von Bayern. She was self-educated, as neither higher schools (gymnasium), nor universities had at that time yet opened their doors for women.

Thompson, Benjamin (Count Rumford): Wobum (Mass.) 26 March 1753–Paris 21 August 1814. American born British physicist and inventor.

Thoms, Georg: Riga 12 February 1843 –Riga 2 November 1902. German agricultural chemist. Thoms studied in Heidelberg (with R. Bunsen), worked in Berlin and Bonn (with A. Kekulé) and from 1868 to 71 he was in the US, where he was the co-founder of a factory for meat extract in Western Texas. Later pharmacist in Victoria and Indianola. Back to Riga in 1872 he held high academic and administrative positions.

Thomsen, Hans Peter Jørgen Julius: Copenhagen 16 February 1826–Frederiksberg 13 February 1909. Danish chemist. As a child he was a pupil of the Petri School in St. Petersburg. Studied chemistry in Copenhagen. 1865 appointed Professor of Chemistry in Copenhagen.

Thomson, Joseph John: Cheetham Hill 18 December 1856–Cambridge 30 August 1940. British physicist. Studied mathematics and physics at Trinity College, Cambridge. 1894 Professor in Cambridge. Nobel Prize for Physics in 1906.

Thomson, William (1st Baron Kelvin): Belfast 26 June 1824–Netherhall (near Largs) 17 December 1907. Irish/Uk physicist and engineer. Studied at Cambridge. 1846 Chair of Natural Philosophy in Glasgow.

Thorpe, Sir Thomas Edward: Harpurhey 8 December 1845–Salcombe 23 February 1925. British chemist. Studied in Manchester and Heidelberg (with Friedrich August Kekulé). 1870 Professor in Glasgow, then Leeds, 1885 in South Kensington (at what was later Imperial College London).

Tiemann: **Johann Karl Wilhelm Ferdinand**: Rübeland 10 June 1848–Meran 14 November 1899. German chemist, who specialised in chemistry of aroma compounds. In 1882 he was appointed Professor in Berlin.

Toepler, August Joseph Ignaz: Brühl 7 September 1836–Dresden 6 March 1912. German physicist. Studied physics, mathematics and chemistry in Berlin. 1865 Professor of Chemistry and Chemical Technology in Riga, 1869 Professor in Graz.

Tolstoy, Lev Nikolayevich (Лев Николаевич Толстой): Yasnaya Polyana 9 September 1828–Astapovo 20 November 1910. Russian novelist and pacifist.

Tönnies, Ferdinand: Oldenswort 26 July 1855–Kiel 9 April 1936. German sociologist, economist and philosopher. Studied in Jena, Leipzig (with Wilhelm Wundt), Bonn, Berlin and Tübingen. 1909 apl-Professor, 1913 full Professor in Kiel. Because of his criticism of the Nazi regime, he lost his teaching approbation in 1933.

Trey, Heinrich Peter Friedrich: Riga 8 October 1851–Dorpat 29 December 1916. German Chemist. Studied in Dorpat and Munich. 1903 Professor in Riga. Student of Friedrich Wilhelm Ostwald.

Tröndlin, Carl Bruno: Leipzig 26 May 1835–Dresden 27 May 1908. German jurist and politician. 1877 to 1908 mayor of Leipzig.

Türin, Vladislav Aleksandrovich von (Тюрин, Владислав Александрович): S. Petersburg 19 April 1862^{jul} (1 May 1862^{greg}) – S. Ptersburg 28 January 1907^{jul} (10 February 1907). Taught thermodynamics at the Mining Institute in St. Petersburg. Worked on airship aviation.

Tyndall, John: Leighlin Bridge 2 August 1820–Hindhead 4 December 1893. British physicist. Studied with Robert Wilhelm Bunsen.

U

Unna, Paul Gerson: Hamburg 8 September 1850–Hamburg 29 January 1929. German dermatologist.

Unold, Johannes: Memmingen 1860–? German philosopher.

Urbain, Georges: Paris 12 April 1872–Paris 5 November 1938. French chemist. Studied in Paris. From 1899 Professor in Paris.

Urech, Friedrich Wilhelm Karl: 1844–1904. German teacher at a technical college in Munich and publicist of sociology books.

Utzschneider, Joseph von: castle Rieden 2 March 1763–Munich 31 January 1840. Bavarian entrepreneur, technician, and politician. Co-founder of the *Mathematisch-Feinmechanische Institut* in Munich, joined by Joseph Fraunhofer in 1812.

V

Vaihinger, Hans: Tübingen 25 September 1852–Halle/Saale 18 December 1933. German philosopher. Known for his "als ob" (as if) philosophy. 1892 Professor in Halle.

Vinci, Leonardo da: Anchiano 15 April 1452–Castle Vlos Lucé 2 May 1519. Italian polymath.

Vladimir Sviatoslavich the Great (Владимир I Святославич): c. 958–15 July 1015, Berestove. Prince of Novgorod, Grand Prince of Kiev, ruler of the Kievan Rus from 980 to 1015. He was baptised in 987 and he is venerated as a Saint (of the Eastern Orthodox Churches) for the Christianisation of the Slavonic population of the Rus in 988.

Volhard, Jacob: Darmstadt 4 June 1834–Halle/Saale 14 January 1910. German chemist. Nephew of Justus Liebig. Student of Robert Wilhelm Bunsen. 1856: assistant to Justus Liebig, 1858 work with August Wilhelm von Hoffmann in London, 1881 Professor in Halle.

Volkmann, Paul: Bladiau 12 January 1856–Königsberg 15 April 1938. Studied mathematics and physics in Königsberg, where he became Professor in 1886.

Vries, Hugo Marie de: Haarlem 16 February 1848–Lunteren 21 May 1935. Dutch biologist. Worked on mutagenesis.

W

Waage, Peter: Flekkefjord 29 January 1833–Kristiania 13 January 1900. Norwegian chemist. 1862 Professor in Christiania. Together with his brother-in-law Cato Maximilian Guldberg he formulated the law of mass action.

Wach, Adolf Eduard Ludwig Gustav: Culm 11 September 1843–Leipzig 4 April 1926. German jurist. Studied in Berlin, Heidelberg, Königsberg, and Göttingen. 1869 Professor in Rostock, 1871 in Tübingen, 1872 in Bonn und 1875 in Leipzig.

Wachsmuth, Kurt (Curt): Naumburg (Saale) 27 April 1837–Leipzig 8 June 1905. Studied in Jena and Bonn. Professorships: 1864 in Marburg, 1868 in Göttingen, 1877 in Heidelberg, and 1886 in Leipzig.

Waentig (Wäntig) Karl Heinrich Moritz: Leipzig 13 March 1843–Radebeul 19 April 1917. Ministerial official in Saxony (Germany). Studied law in Heidelberg and Leipzig.

Wagner, Julius: Hanau 3 July 1857–Leipzig 17 July 1924. German chemist. Studied natural sciences in Straßburg, Gießen and Leipzig. 1904 Professor in Leipzig. In Germany, he was the first Professor of Didactics of Chemistry.

Wagner, Wilhelm Richard: Leipzig 22 May 1813–Venice 13 February 1883. German composer.

Wahrmund, Ludwig: Vienna 21 August 1860–Prague 10 September 1932. Austrian jurist.

Wald, František (Franz): Brandýsek 9 January 1861–Moravská Ostrava 19 October 1930. Czech chemist. Studied technical chemistry at the German Technical University of Prague. 1908 Professor at the Czech Technical University of Prague.

Walden, Paul: Latvian version Pauls Valdens, (Rosenbeck 26 July 1863–Gammertingen 22 January 1957. Latvian/German/Russian chemist. He studied chemistry in Riga, Leipzig and Munich. Ph.D. with Friedrich Wilhelm Ostwald in Leipzig 1891. 1893 Professor at the Polytechnic in Riga, 1908 Professor in St. Petersburg (he succeeded Dmitri Mendeleev). In 1910 he succeeded Friedrich Konrad Beilstein at the Academy of Science in St. Petersburg. After the October Revolution he left Russia and was appointed Professor in Rostock (1919–1934). 1927/28 he was Guest Professor at the University Ithaca, USA. In 1942 his house in Rostock was completely destroyed by British bombing and he lost all his books and papers. Walden made numerous contributions to chemistry: he discovered the inversion of chiral centers in chemical reactions (Walden inversion), studied non-aqueous electrolytes (coined the term solvation), he synthesized the first room-temperature ionic liquid, and he discovered the empirical rule that the product of the equivalent *conductivity* and the *viscosity* of the solvent is a constant for a particular electrolyte at a given temperature (Walden's rule). He is also known for his writings on the history of chemistry, among which is a biography of Friedrich Wilhelm Ostwald.

Waldeyer, Heinrich Wilhelm: Hehlen 6 October 1836–Berlin 23 January 1921. German anatomist. Studied mathematics, sciences and then medicine in Göttingen, and continued this at Greifswald and Berlin. 1865 Professor in Breslau, 1872 Professor in Strassburg, 1883 in Berlin. He coined the terms neurone and chromosome.

Walker, Sir James: Dundee 6 April 1863–Edinburgh 6 May 1935. Scottish chemist. Studied physical sciences at the University of Edinburgh (D.Sc. in 1886). He worked in Germany with Johann Friedrich Wilhelm Adolf von Baeyer and Ludwig Rainer Claisen, later with Friedrich Wilhelm Ostwald in Leipzig, where he obtained a Ph.D. in 1889. Since 1894 he held the Chair of Chemistry at University College, Dundee.

Wallach, Otto: Königsberg 27 March 1847–Göttingen 26 February 1931. German chemist. Student of Friedrich Wöhler (Göttingen), August Wilhelm von Hofmann (Berlin), and co-worker of Friedrich August Kekulé (Bonn). 1876: Professor in Bonn, 1889: Professor in Göttingen. 1910 Nobel Prize for Chemistry in recognition of his work on alicyclic compounds.

Weber, Ernst Heinrich: Wittenberg 24 June 1795–Leipzig 26 January 1878. German physiologist and anatomist. Studied in Wittenberg and Leipzig. 1821 Professor in Leipzig.

Weber, Heinrich Franz: Rettershain (Nassau) 24 July 1834–Riga 27 October 1881. German chemist. He was first a teacher. In 1860 he started studies of natural sciences in Bonn and after 2 years continued in Berlin, where Heinrich Rose was one of his professors. In 1865 he started as chemist at the agricultural-chemical experimental station (*Landwirtschaftlich-chemische Versuchsstation*) of the Riga Polytechnic. 1869–81 Professor at the Riga Polytechnic. There he was succeeded by Friedrich Wilhelm Ostwald.

Weber, Heinrich Friedrich: Magdala, near Weimar, 7 November 1843–Zurich 24 May 1912. German/Swiss physicist. In 1870 he was Assistant to Gustav Heinrich Wiedemann in Karlsruhe, in 1871–74 Assistant to Hermann Ludwig Ferdinand von Helmholtz in Berlin. Later he was Professor in Zurich.

Weber, Wilhelm Eduard: Wittenberg 24 October 1804–Göttingen 23 June 1891. German physicist. Studied in Halle. 1831 Professor in Göttingen, where he cooperated with Gauß.

Websky, Christian Friedrich Martin: Nieder-Wüstegiersdorf / Silesia17 July 1824–Berlin 27 Nov. 1886. German mineralogist. Worked as mining officer in Silesia. 1868 Professor in Breslau, 1874 Professor in Berlin.

Wegscheider, Rudolf Franz Johann von: Großbetschkerek 8 October 1859–Vienna 8 January 1935. Austrian chemist. 1902 Professor in Vienna. Research on energetics and kinetics of chemical reactions.

Wehrenpfennig, Wilhelm: Blankenburg 25 March 1829–Berlin 25 July 1900. German political official. Studied theology in Jena and Berlin. Since 1879 he worked in the education ministry, responsible for technical universities.

Wenzel, Carl (Karl) Friedrich: Dresden 1740–Freiberg 26 February 1793. German chemist. Studied in Leipzig. 1785 mining officer in Freiberg, and 1786 chemist (*Arkanist* = secret chemist, i.e., chemist who knows the secrets) in the Meissen porcelain factory.

Weidel, Hugo: (Vienna 13 November 1849–Vienna 7 June 1899) Austrian chemist. Studied chemistry in Vienna and Heidelberg. 1874 Professor in Vienna. Worked on the chemistry of natural products.

Weigt, Karl: 15 November 1862–Hannover 29 August 1932. German protestant pastor and later prominent freethinker.

Weisman, Friedrich Leopold August: Frankfurt am Main 17 January 1834–Freiburg im Breisgau 5 November 1914. German Biologist. Neodarwinist.

Wheeler, Benjamin Ide: Randolph 15 July 1854–Vienna 3 May 1927. US philologist. Studied at Brown University; Ph.D. Heidelberg (Germany) 1885. 1887 Professor at Cornell University. 1899 to 1919 President of the University of California at Berkeley.

Wichelhaus, Karl Hermann: Elberfeld 8 January 1842–Heidelberg 28 February 1927. German chemist. Studied in Bonn, Göttingen, Gent London and Heidelberg. Later he had a private laboratory in Berlin and was one of the co-founders of the German Chemical Society.

Wiedemann, Alfred: Berlin 18 July 1856–Bad Godesberg 7 December 1936. German Egyptologist. Son of Gustav Heinrich Wiedemann and brother of Eilhard Ernst Gustav Wiedemann. From 1891 to 1820 Professor at the University of Bonn.

Wiedemann, Gustav Heinrich: Berlin 2 October 1826–Leipzig 23 March 1899. German physicist. W. had the first chair in Physical Chemistry in Leizig 1871. In 1887 he moved to the chair of Physics, and Friedrich Wilhelm Ostwald followed him as head of the Institute of Physical Chemistry. He followed Johann Christian

Poggendorff as editor of the journal "*Annalen der Physik*". Father of Alfred Wiedemann and of Eilhard Ernst Gustav Wiedemann.

Wiedemann, Eilhard Ernst Gustav: Berlin 1 August 1852–Erlangen 7 January 1928. German physicist. Son of Gustav Heinrich Wiedemann, brother of Alfred Wiedemann. 1878 Professor of Physics in Leipzig, 1886 in Erlangen, 1886 in Darmstadt.

Wieland, Christoph Martin: Oberholzheim 5 September 1733–Weimar 20 January 1813. German writer.

Viennaer, Otto Heinrich: Karlsruhe 15 June 1862–Leipzig 18 January 1927. German physicist. Studied in Straßburg. 1895 Professor in Gießen, 1890 in Leipzig.

Wiesner, Julius von: Tschechen 20 January 1838–Vienna 9 October 1916. Austrian botanist and plant physiologist. 1868 Professor in Vienna. Research on phototropism and chlorophyll.

Wilhelm II (Friedrich Wilhelm Viktor Albert von Preußen): Berlin 27 January 1859–Doorn 4 June 1941. Last Emperor of Germany.

Wilhelmy, Ludwig Ferdinand: Stargard/Pommern 25 December 1812–Berlin 18 February 1864. German pharmacist and chemist. W. He was an academic lecturer in Heidelberg from 1849 to 1854. Thereafter he lived as private scholar in Berlin. He was the first to formulate a chemical rate law as a differential equation, which he integrated, and then compared the results with experimental data. The first reaction studied in this way was the inversion of sucrose by acid. In 1863 he published his well known paper on the Wilhelmy plate method for measuring the interfacial tension of liquids.

Wilke, Arthur: 1853–1913. German electro-technical engineer. Author of several books to popularize knowledge about electricity and its technical use.

Will; Karl Wilhelm: Gießen 12 April 1854–Berlin 30 December 1919. German chemist. His research was focussed on explosives.

Wilson, Thomas Woodrow: Staunton 28 December 1956–Washington DC 3 February 1924. US politician and 28th President of the USA. 1919 Nobel Peace Prize.

Windelband, Wilhelm: Potsdam 11 May 1848–Heidelberg 22 October 1915. German philosopher. Studied medicine and natural sciences in Jena, Berlin, and Göttingen, later history and philosophy. 1876 Professor in Zurich, 1877 in Freiburg, 1882 in Straßburg, 1903 in Heidelberg.

Windscheid, Bernhard Joseph Hubert: Düsseldorf 26 June 1817–Leipzig 26 October 1892. German jurist. Studied in Berlin and Bonn. 1847 Professor in Basel, 1852 in Greifswald, 1857 in Munich, 1871 in Heidelberg and 1874 in Leipzig.

Winkler, Clemens Alexander: Freiberg 26 December 1838–Dresden 8 October 1904. German chemist. 1873 Professor at Bergakademie (mining academy) Freiberg. He discovered germanium, the element, whose existence had been proposed by Dmitri Mendeleev as eka-silicon.

Wislicenus, Johannes Adolf: Kleineichstädt 24 June 1835–Leipzig 5 December 1902. German chemist. 1864 to 1870 Professor at the University of Zurich, 1870 to 1872 at the Polytechnic in Zurich, 1885 Professor in Leipzig (successor of Hermann Kolbe). Research on organic chemistry and stereochemistry.

Wolff, Jacob: 1868–4 December 1926. German entrepreneur.

Woodward, Robert Simpson: Rochester 21 July 1849–Washington DC 29 June 1924. US physicist and astronomer. Studied at the University of Michigan. 1884 to 1890 astronomer to the United States Geological Survey, 1890 assistant in the United States Coast and Geodetic Survey, 1893 Professor at Columbia University, 1905 president of the Carnegie Institution, Washington.

Wöhler, Friedrich: Eschersheim 31 July 1800–Göttingen 23 September 1882. German chemist. He studied medicine and chemistry and was a student of Leopold Gmelin (Heidelberg) and Jöns Jacob Berzelius (Stockholm). 1836 Professor of Chemistry and Pharmacy in Göttingen. In 1828 he synthesized oxalic acid by hydrolysis of cyanogen, and urea from ammonium cyanate and thus showed that inorganic compounds can be converted to organic compounds (i.e., without involvement of a special *agens vitalis*). Wöhler translated Berzelius' famous multi-volume textbook of chemistry to German.

Wright, John Henry: Urumiah (Iran) 4 February 1852–Cambridge (Mass.) 25 November 1908. US classical scholar. Studied at Dartmouth College. 1886 Professor at Johns Hopkins, and 1887 at Harvard University.

Wundt, Wilhelm Maximilian: Neckarau 16 August 1832–Großbothen 31 August 1920. German physician, psychologist and philosopher who became a close friend of Friedrich Wilhelm Ostwald when they both were in Leipzig. He is one of the founders of experimental psychology.

Wüllner, Adolf: Düsseldorf 13 June 1835–Aaachen 6 October 1908. German experimental physicist. 1870 appointed Professor in Aachen

Y

Young, Stewart Woodford: Orient (New York) 14 March 1869–Stanford 9 April 1931. American chemist. Studied at Cornell, and from 1899 to 1901 at the Institute of Wilhelm Ostwald in Leipzig. Assistant Professor 1894; Associate Professor 1901; Professor of Physical Chemistry at Stanford University from 1906 to 1931.

Z

Zamenhof, Ludwik Lejzer (born as Eliezer Levi Samenhof) Białystok 15 December 1859–Warsaw 14 April 1917. Polish-Jewish ophthalmologist and philologer, inventor of the synthetic language Esperanto. Studied medicine in Moscow. Warsaw and Vienna.

Zamminer, Friedrich: Darmstadt 26 October 1817–Gießen 15 August 1858. Physicist and chemist. Together with Johann Heinrich Buff and Hermann Kopp he published the first book about what is now called Physical Chemistry (*Lehrbuch der physikalischen und theoretischen Chemie*) in 1857.

Zeisel, Simon: Lomnitz 10 April 1854–Vienna 10 January 1933. Czech/Austrian chemist. 1891 Professor at *Hochschule für Bodenkultur* (agricultural university) of Vienna.

Zeppelin, Ferdinand Adolf Heinrich August Graf von: Konstanz 8 July 1838–Berlin 8 March 1917. German general and aircraft manufacturer (founder of the Zeppelin airship company).

Zirkel, Ferdinand: Bonn 20 May 1838–Bonn 11 June 1912. German geologist. 1863: Professor of Geology in Lemberg (now Lviv, Ukraine), 1868: Professor in Kiel, 1870: Professor of Mineralogy and Geology in Leipzig.

Zöllner, Johann Karl Friedrich: Berlin 8 November 1834–Leipzig 25 April 1882. German physicist and astronomer. Studied in Berlin and Basel. 1872 Professor in Leipzig.

Zschokke, Johann Heinrich Daniel: Magdeburg 22 March 1771–Aarau 27 June 1848. Pedagogue and writer.

Zsigmondy, Richard Adolf: Vienna 1 April 1865–Göttingen 23 September 1929. Austro-Hungarian chemist. Studied in Vienna, Munich and Berlin. 1908 Professor in Göttingen. Known for his study of colloids, for which he received the Nobel Prize in Chemistry in 1925.

Name Index

R.S. Jack and F. Scholz (eds.), *Wilhelm Ostwald*, Springer Biographies,
DOI 10.1007/978-3-319-46955-3

CPSIA information can be obtained
at www.ICGtesting.com
Printed in the USA
LVHW050049280119
605463LV00004B/220/P

9 783319 469539